Introduction to Petroleum Biotechnology

Introduction to Petroleum Biotechnology

James G. Speight

Laramie, Wyoming, United States

Nour Shafik El-Gendy

Egyptian Petroleum Research Institute, Nasr City, Cairo, Egypt

Gulf Professional Publishing
An imprint of Elsevier

Gulf Professional Publishing is an imprint of Elsevier
50 Hampshire Street, 5th Floor, Cambridge, MA 02139, United States
The Boulevard, Langford Lane, Kidlington, Oxford, OX5 1GB, United Kingdom

Notices
Knowledge and best practice in this field are constantly changing. As new research and experience broaden our
understanding, changes in research methods, professional practices, or medical treatment may become necessary.

Practitioners and researchers must always rely on their own experience and knowledge in evaluating and using
any information, methods, compounds, or experiments described herein. In using such information or methods
they should be mindful of their own safety and the safety of others, including parties for whom they have a
professional responsibility.

To the fullest extent of the law, neither the Publisher nor the authors, contributors, or editors, assume any liability
for any injury and/or damage to persons or property as a matter of products liability, negligence or otherwise, or
from any use or operation of any methods, products, instructions, or ideas contained in the material herein.

Library of Congress Cataloging-in-Publication Data
A catalog record for this book is available from the Library of Congress

British Library Cataloguing-in-Publication Data
A catalogue record for this book is available from the British Library

ISBN: 978-0-12-805151-1

For information on all Gulf Professional Publishing publications visit our website at
https://www.elsevier.com/books-and-journals

Working together
to grow libraries in
developing countries

www.elsevier.com • www.bookaid.org

Publishing Director: Joe Hayton
Senior Acquisition Editor: Katie Hammon
Senior Editorial Project Manager: Kattie Washington
Production Project Manager: Anitha Sivaraj
Designer: Mark Rogers

Typeset by TNQ Books and Journals

Contents

Biography by Dr. James G. Speight

Dr. James G. Speight has doctorate degrees in Chemistry, Geological Sciences, and Petroleum Engineering and is the author of more than 60 books in petroleum science, petroleum engineering, and environmental sciences.

Dr. Speight has 50 years of experience in areas associated with (1) the properties, recovery, and refining of reservoir fluids, conventional petroleum, heavy oil, and tar sand bitumen, (2) the properties and refining of natural gas, gaseous fuels, and (3) the properties and refining of biomass, biofuels, biogas, and the generation of bioenergy. His work has also focused on safety issues, environmental effects, remediation, and safety issues associated with the production and use of fuels and biofuels.

Although he has always worked in the private industry, which focused on contract-based work, he has served as Adjunct Professor in the Department of Chemical and Fuels Engineering at the University of Utah and in the Departments of Chemistry and Chemical and Petroleum Engineering at the University of Wyoming. In addition, he was a Visiting Professor in the College of Science, University of Mosul, Iraq, and has also been a Visiting Professor in Chemical Engineering at the following universities: University of Missouri–Columbia, the Technical University of Denmark, and the University of Trinidad and Tobago.

Dr. Speight was elected to the Russian Academy of Sciences in 1996 and awarded the Gold Medal of Honor that same year for outstanding contributions to the field of petroleum sciences.

He has also received the Scientists without Borders Medal of Honor of the Russian Academy of Sciences. In 2001, the Academy also awarded Dr. Speight the Einstein Medal for outstanding contributions and service in the field of Geological Sciences.

Biography by Dr. Nour Shafik El-Gendy

Dr. Nour Shafik El-Gendy is Professor in the field of Petroleum and Environmental Biotechnology, She is the Vice Head of Process Design & Development Department and former Head Manager of Petroleum Biotechnology Lab, Egyptian Petroleum Research Institute, Cairo, Egypt. She is an author of two chapters and two books in the fields of biofuels and petroleum biotechnology and more than 95 research papers in the fields of oil pollution, bioremediation, biosorption, biofuels, macro- and micro-corrosion, green chemistry, wastewater treatment, biodesulfurization, biodenitrogenation and nano-biotechnology, and its applications in petroleum industry and biofuels. Dr. El-Gendy is also an
editor of 27 international journals and reviewer for 51 international journals, in the field of environmental biotechnology, bioenergy, and petroleum microbiology and biotechnology and has supervised 26 MSc and PhD theses in the field of biofuels, micro-macro-fouling, bioremediation, nano-biotechnology, wastewater treatment, biodenitrogenation, and biodesulfurization. Dr. El-Gendy participated in 36 international workshops and training courses, and 52 international conferences. Dr. El-Gendy is a member in many international associations concerned with petroleum industry, environmental health, and sciences. She also has teaching experience as lecturer and supervisor for undergraduates' research projects at Faculty of Biotechnology in Modern Sciences and Arts, MSA University, Egypt (2007–08), and Chemical Engineering Department, Faculty of Engineering in The British University in Egypt, BUE, and Cairo University, Egypt (2011–13) and also has taught Environmental Biotechnology course for post-graduates at Faculty of Science, Monufia University, Egypt (2014–15). Her biography is recorded in *Who's Who in Science and Engineering,* ninth edition, 2006–07. She is also the Vice Coordinator of the Scientific Research Committee, National Council for Women in Egypt.

Preface

The term *petroleum biotechnology* in the context of this book refers to the implementation of biological processes in the oil industry to explore, produce, and transform petroleum into valuable derivatives. Environmental biotechnology utilizes the biochemical potential of microorganisms and plants for the preservation and restoration of the environment. Biotechnology has played a significant role in the transformation (biotransformation) of petroleum constituents and petroleum products after spills. The technology also moved into the realm of petroleum oil recovery from depleted reservoirs as a means of solving the problem of stagnant petroleum production, after a three-stage recovery process employing mechanical, physical, and chemical methods of recovery. Biotechnologically enhanced oil recovery process, known as microbial enhanced oil recovery(MEOR), involves stimulating indigenous reservoir microbes or injecting specially selected consortia of natural bacteria into the reservoir to produce specific metabolic events (biotransformations) that lead to improved oil recovery.

Petroleum (crude oil) is a complex mixture of thousands of different chemical compounds. In addition, the composition of each accumulation of oil is unique, varying in different producing regions and even in different unconnected zones of the same formation. The composition of petroleum also varies with the amount of refining. Significantly, the many constituents of petroleum differ markedly in volatility, solubility, and susceptibility to biotransformation—some constituents are susceptible to microbial biotransformation while others cannot be transformed by biotic means. Furthermore, the biotransformation of different petroleum constituents can occur simultaneously but at very different rates. This leads to the sequential transformation of individual constituents of petroleum and, because different microbial species preferentially attack different compounds, to successional changes in the degrading microbial community. Thus, to evaluate the effectiveness of biotransformation through the application of microbial technologies it is necessary to know the molecular composition of petroleum and the effects of the bioprocesses.

Furthermore, with the depletion of light and sweet crude oils, reliance and dependence on heavier crude oils will increase. However, refining of heavier crudes generates not only larger amounts of high-molecular weight products, but is also costly and involves problems of corrosion of equipment and poisoning of costly catalysts. The presence of asphaltene constituents, as well as organic heterocyclic derivatives of sulfur and nitrogen and organometallic derivatives in crude oils, results in poisoning of costly catalysts used for fluid catalytic cracking and hydrocracking. Thus, there is a need to remove sulfur, nitrogen, and toxic metals from refinery feedstocks and there is also the necessity of degrading larger higher molecular weight-bearing complex molecules to lower molecular weight-bearing lighter crudes. Conventional physical and chemical refining techniques may not be easily applied to heavier crudes. Thus, there is a need to develop simpler biorefining techniques to not only biotransform the constituents of the resin and asphaltene fractions, but to remove sulfur and nitrogen derivatives and toxic metal constituents from crude oil as well.

This book focuses on biotechnology as applied to petroleum and brings together the various biotechnology methods (biotransformations) applied to recovery, refining, and remediation of petroleum and petroleum products.

In summary, the book will present to the reader an overview (with some degree of detail) of the area of petroleum biotechnology.

Dr. Nour Sh. El-Gendy
Cairo, Egypt

Dr. James G. Speight
Laramie, Wyoming, United States

July 2017

PETROLEUM COMPOSITION AND PROPERTIES

1.0 INTRODUCTION

Historically, petroleum and its derivatives have been known and used for millennia when ancient technologists recognized that certain derivatives of petroleum, such as the nonvolatile residuum or asphalt, could be used for civic and decorative purposes. It was also recognized that other derivatives (especially the volatile naphtha fraction was used in what came to be called *Greek fire* but originally was invented in Arabian countries) could provide certain advantages in warfare (Abraham, 1945; Pfeiffer, 1950; Van Nes and van Westen, 1951; Forbes, 1958a, 1958b, 1959, 1964; Hoiberg, 1964; Berger and Anderson, 1978; Speight, 1980, 2014a; Cobb and Goldwhite, 1995).

Scientifically, petroleum is a carbon-based resource that varies in viscosity and which is an extremely complex mixture of hydrocarbon compounds, usually with minor amounts of nitrogen-containing compounds and sulfur-containing compounds, as well as smaller amounts of oxygen-containing compounds (typically naphthenic acid derivatives and phenol derivatives) and trace amounts of metal-containing constituents (Speight, 2014a,b). Heavy oil is a subcategory of petroleum that contains a greater proportion of the higher-boiling constituents and heteroatom compounds. On the other hand, tar sand bitumen is different to petroleum and heavy oil insofar as it cannot be recovered by any of the conventional methods, including enhanced recovery methods (Speight, 2014a, 2016). For the purposes of this book and for convenience, residua (also called resids, which are the nonvolatile distillation products obtained from crude oil), heavy oil, extra heavy oil, and tar sand bitumen are (for convenience) included in the term *heavy feedstocks*.

In the crude state, petroleum, heavy oil, extra heavy oil, and tar sand bitumen have minimal value, but when refined they provide high-value products such as gaseous fuels, liquid fuels, solvents, lubricants, as well as various asphalt products. The fuels derived from petroleum contribute approximately one-third to one-half of the total world energy supply and are used not only for transportation fuels (i.e., gasoline, diesel fuel, and aviation fuel, among others) but also to heat buildings. Petroleum products have a wide variety of uses that vary from gaseous and liquid fuels to near-solid lubricants for industrial machinery. In addition, asphalt (a once-maligned by-product and the residue of many refinery processes) is now a premium value product for highway surfaces, roofing materials, and miscellaneous waterproofing uses (Speight, 2014a, 2016).

Petroleum occurs underground, at various pressures depending on the depth. Because of the pressure, it contains considerable natural gas in solution. The oil underground is much more fluid than it is on the surface and is generally mobile under reservoir conditions because of the elevated temperatures in subterranean formations. Generally, the subterranean temperature rises 1°C for every 100 feet of depth (1.8°F for every 33 m of depth) resulting in a decrease in the viscosity of the oil.

In terms of petroleum origin, petroleum is derived from the remains of aquatic plants and animals that lived and died hundreds of millions of years ago. The floral and faunal remains mixed with mud

Introduction to Petroleum Biotechnology. https://doi.org/10.1016/B978-0-12-805151-1.00001-1

and sand in layered deposits that, over the millennia, were geologically transformed into sedimentary rock. At the same time, the organic matter decomposed and eventually formed petroleum (or a related precursor often referred to as *protopetroleum*), which migrated from the original source beds (the original deposits of the decomposed floral and faunal remains) to more porous and permeable rocks, such as *sandstone* where it finally became trapped as an accumulation in a *reservoir*. Generally, the crude oil and natural gas migrated upward through the porous strata where it was trapped by the sealing cap rock and the shape of the structure. A series of reservoirs within a common rock structure or a series of reservoirs in separate but neighboring formations is commonly referred to as an *oil field*. In addition, a group of fields is often found in a single geologic environment often referred to as *sedimentary basin* or *province*.

Petroleum reservoirs are generally classified according to their geologic structure and the mechanism by which petroleum (the drive mechanism) is produced from the reservoir. However, it is essential to recognize that petroleum reservoirs exist in many different sizes and forms of geologic structures (Speight, 2014a). Typically, a reservoir is named according to the conditions of the formation of the reservoir. For example, *dome-shaped* and *anticline reservoirs* are formed by the folding of the rock. The dome-shaped reservoir is typically circular in outline while the anticline reservoir is long and narrow. On the other hand, *fault reservoirs* are formed by shearing and offsetting of the geological strata (faulting). The movement of the nonporous rock adjacent to or opposite to the porous formation containing the oil/gas creates the sealing. *Salt-dome reservoirs* take the shape of a dome which was formed due to the upward movement of an impermeable salt dome that deformed and lifted the overlying layers of rock. Other reservoir types include (1) unconformity reservoir, (2) lens reservoir, and (3) combination reservoir.

The *unconformity* type of reservoir is formed because of an unconformity where the impermeable cap rock was laid down across the cutoff surfaces of the lower beds. In the *lens-type reservoir*, the petroleum-bearing porous formation is sealed by the surrounding, nonporous formation. Irregular deposition of sediments at the time the formation was laid down is the probable cause for the change in the porosity of the formation. Finally, a *combination reservoir* is, as the name implies, a combination of folding, faulting, abrupt changes in porosity, or other conditions which create the trap that exists as this type of petroleum reservoir. In recent years, the shale-type of reservoir has become important because of the potentially large amount of reserves of crude oil and natural gas that exist in such reservoirs.

Natural gas and crude oil from shale formations (as well as other tight formations—impermeable formations) offer additional energy-producing resources. These formations typically function as both the reservoir rock and the source rock. In terms of chemical makeup, shale gas is typically a dry gas composed primarily of methane but some formations do produce wet gas; crude oil from tight formations is typically more volatile than many crude oils from conventional reservoirs. The shale formations that yield gas and oil are organic-rich shale formations that were previously regarded only as source rocks and seals for gas accumulating in the strata near sandstone and carbonate reservoirs of traditional onshore gas development. On the other hand, the crude oil from shale formations is a light highly volatile crude oil that contains more of the volatile hydrocarbons (C_1 to C_4 hydrocarbons, i.e., methane to butane hydrocarbons) than crude oil from the more conventional reservoirs. In the case of shale formations, the natural gas and crude oil present in the matrix system of pores—similar to that found in conventional reservoir rocks—is also accompanied by gas or oil that is bound to, or adsorbed on, the surface of inorganic minerals in the shale. The relative contributions and combinations of free gas and oil from matrix porosity and from desorption of adsorbed constituents is a key determinant of the production profile of the well.

Reservoir nomenclature aside, it is pertinent to note that throughout the millennia in which petroleum has been known and used, it is only in the last four decades that some attempts have been made to standardize petroleum nomenclature and terminology. But confusion may still exist not only for the neophyte researcher but also for the seasoned researcher. Therefore, it is the purpose of this chapter to provide some semblance of order into the disordered state that exists in the segment of petroleum technology that is known as *terminology*. There is no effort here to define the individual processes since they will be defined in the relevant chapters.

For the purposes of definitions and terminology, it is preferable to subdivide petroleum and related materials into three major classes: (1) materials that are of natural origin, (2) materials that are manufactured, and (3) materials that are integral fractions derived from the natural or manufactured products. This will guide the reader to a better understanding of the nature of the various petroleum-related and non-related materials so that a better understanding of petroleum biotechnology can be realized.

2.0 **CRUDE OIL**

The *definition* of crude oil has been varied, unsystematic, and diverse with many of the names coming from ancient time and then from the beginnings of the petroleum industry in the 19th century. Thus, the terminology of petroleum is a product of many decades of evolution and the long-established use of an expression.

If there is to be a thorough understanding of petroleum and the related biotechnologies, it is essential that the definitions and the terminology of petroleum science and technology be given prime consideration as they relate to composition and properties (Meyer and DeWitt, 1990; Speight, 2014a). This will aid in a better understanding of petroleum, its constituents, and its various fractions. However, particularly troublesome, and more confusing, are those terms that are applied to the more viscous materials such as *bitumen* and *asphalt*. This part of the text attempts to alleviate much of the confusion that exists, but it must be remembered that the terminology of petroleum is still open to personal choice and historical usage (van der Have and Verver, 1957; Clark and Brown, 1977; Berger and Anderson, 1978; Speight, 1984; Reynolds, 1998).

Petroleum and the equivalent term *crude oil*, cover a wide assortment of materials consisting of mixtures of hydrocarbons and other compounds containing variable amounts of sulfur, nitrogen, and oxygen, which may vary widely in API (American Petroleum Institute) gravity and sulfur content (Table 1.1), as well as viscosity and the amount of residuum (the portion of crude oil boiling above 510°C, 950°F). For simplicity and convenience, the data in this table are restricted to API gravity and sulfur content which indicate the variations in the crude oil feedstocks (from one country or region to another) and therefore the differences that can be anticipated when biotechnology is applied to these crude oils.

By general definition, *crude oil* is a mixture of gaseous, liquid, and solid hydrocarbon compounds that occur in sedimentary rock deposits throughout the world and contains small quantities of nitrogen-, oxygen-, and sulfur-containing compounds, as well as trace amounts of metallic constituents (Speight, 1984, 2012, 2014a). These constituents boil at different temperatures that can be separated into a variety of different generic fractions by distillation. And the terminology of these fractions has been bound by utility and often bears little relationship to composition. Furthermore, there is a wide variation in the

Table 1.1 API Gravity and Sulfur Content of Selected Crude Oils

Country	Crude Oil	API	Sulfur % w/w
Abu Dhabi (UAE)	Abu Al Bu Khoosh	31.6	2.00
Abu Dhabi (UAE)	Murban	40.5	0.78
Angola	Cabinda	31.7	0.17
Angola	Takula	32.4	0.09
Brunei	Champion Export	23.9	0.12
Brunei	Seria	40.5	0.06
Canada (Alberta)	Pembina	38.8	0.20
Canada (Alberta)	Wainwright-Kinsella	23.1	2.58
China	Nanhai Light	40.6	0.06
China	Shengli	24.2	1.00
Dubai (UAE)	Fateh	31.1	2.00
Dubai (UAE)	Margham Light	50.3	0.04
Egypt	Gulf of Suez	31.9	1.52
Egypt	Ras Gharib	21.5	3.64
Indonesia	Badak	49.5	0.03
Indonesia	Bima	21.1	0.25
Iran	Foroozan (Fereidoon)	31.3	2.50
Iran	Iranian Light	33.8	1.35
Iraq	Basrah Heavy	24.7	3.50
Iraq	Basrah Light	33.7	1.95
Libya	Bu Attifel	43.3	0.04
Libya	Buri	26.2	1.76
Malaysia	Dulang	39.0	0.12
Malaysia	Miri Light	32.6	0.04
Mexico	Maya	22.2	3.30
Mexico	Olmeca	39.8	0.80
Nigeria	Antan	32.1	0.32
Nigeria	Brass River	42.8	0.06
North Sea (UK)	Alba	20.0	1.33
North Sea (UK)	Innes	45.7	0.13
Saudi Arabia	Arab Extra Light (Berri)	37.2	1.15
Saudi Arabia	Arab Heavy (Safaniya)	27.4	2.80
USA (Alaska)	Drift River	35.3	0.09
USA (Florida)	Sunniland	24.9	3.25
USA (Texas)	West Texas Semi-Sweet	39.0	0.27
Venezuela	Temblador	21.0	0.83
Venezuela	Tia Juana	25.8	1.63

properties of crude petroleum because the proportions in which the different constituents occur vary with origin. Thus, some crude oils have higher proportions of the lower boiling components and others (such as heavy oil and bitumen) have higher proportions of higher boiling components (asphaltic components and residuum).

The molecular boundaries of petroleum cover a wide range of boiling points and carbon numbers of hydrocarbon compounds and other compounds containing nitrogen, oxygen, and sulfur, as well as metallic (porphyrin) constituents which dictate the options that can be used not only in a refinery (Long and Speight, 1998; Parkash, 2003; Hsu and Robinson, 2006; Gary et al., 2007; Speight, 2014a, 2017) but also to guide any applications of biotechnology. However, the actual boundaries of such a *petroleum map* can only be arbitrarily defined in terms of boiling point and carbon number. In fact, petroleum is so diverse that materials from different sources exhibit different boundary limits, and for this reason alone it is not surprising that petroleum has been difficult to *map* in a precise manner.

The major components of petroleum are hydrocarbons, compounds of hydrogen and carbon that display great variation in their molecular structure. The simplest hydrocarbons are a large group of chain-shaped molecules known as the paraffins. This broad series extends from methane, which forms natural gas, through liquids that are refined into gasoline, to crystalline waxes. A series of saturated hydrocarbons containing a (usually six-membered) ring, known as the naphthenes, ranges from volatile liquids such as naphtha to high molecular weight substances condensed polycyclic species (Speight, 2014a). The alkanes, or aliphatic hydrocarbons consist of fully saturated normal alkanes and branched alkanes of the general formula ($C_nH_{2n}+2$), with n ranging from 1 to usually around 40, although compounds with 60 carbons have been reported. Many of the cycloalkanes or saturated ring structures, also called cycloparaffins or naphthenes, are important constituents of the isoprenoids. They have specific animal or plant precursors (e.g., sterane derivatives, diterpane derivatives, and triterpane derivatives) that serve as important molecular markers in oil spill and geochemical studies (Albaiges and Albrecht, 1979).

Another group of hydrocarbons containing a single or condensed aromatic ring system is known as the aromatics; the chief compound in this series is benzene, a popular raw material for making petrochemicals. Aromatics are a class of hydrocarbons characterized by rings with six carbon atoms. They are usually less abundant than the saturated hydrocarbons, contain one or more aromatic (benzene) rings connected as fused rings (e.g., naphthalene) or lined rings (e.g., biphenyl). Aromatic derivatives are the most acutely toxic components of crude oil, and are also associated with chronic and carcinogenic effects. Many low-weight aromatics are soluble in water, increasing the potential for exposure to aquatic resources. Aromatics with two or more rings are referred to as polycyclic aromatic hydrocarbons. Petroleum contains many homologous series of aromatic hydrocarbons. Polycyclic aromatic hydrocarbons (PAHs) are a group of lipophilic anthropogenic compounds that are ubiquitously distributed in the environment. PAHs consist of two or more fused benzene rings in various arrangements. They form an important class of environmental contaminants, because some exhibit carcinogenic or mutagenic potential. There are several reports of increased incidence of cancer in marine animals from the vicinity of oil spills (Colombo et al., 2005; Tfouni et al., 2007). Concern about PAHs in the environment arises also from the fact that many of them are persistent and are not resistant to any form of biodegradation (Samanta et al., 2002; Maskaoui et al., 2002; Boonyatumanond et al., 2006; Qiao et al., 2006).

Non-hydrocarbon constituents of petroleum include organic derivatives of nitrogen, oxygen, sulfur, and the metals nickel and vanadium. Most of these impurities are removed during refining. They are constituents of petroleum which can be grouped into six classes; sulfur compounds, nitrogen compounds, oxygen compounds, porphyrins, asphaltene constituents, and trace metals. The formula for the asphaltene fraction has been left blank (Fig. 1.1) because the structures of the constituents are unknown and subject to speculation. In addition, an average structural formula for the fraction has little meaning and bears no relationship to the reality of the chemistry and physics of the reactions of these constituents (Speight, 2014a).

Nitrogen is present in all crude oils in compounds as pyridines, quinolines, benzoquinoline derivatives, acridine derivatives, pyrrole derivatives, indole derivatives, carbazole derivatives, and benzocarbazole derivatives. Sulfur compounds comprise the most important group of non-hydrocarbon constituents.

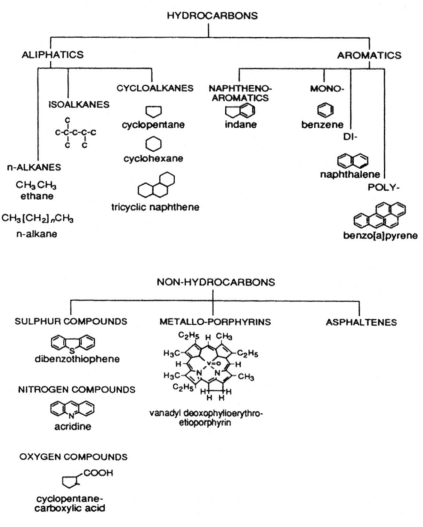

FIGURE 1.1

Examples of the chemical structure of some common components of crude oil.

In many cases, they have harmful effects and must be removed or converted to less harmful compounds during refining process. The asphaltene fraction is made up of the nonvolatile, high molecular weight constituents of petroleum that vary in molecular weight up to several thousand (although some of the higher molecular weight species may be associated lower molecular weight species). The asphaltene constituents are insoluble in low-boiling liquid *n*-alkanes, such as *n*-heptane and remain dispersed in the crude oil though integration with the resin constituents (Shirokoff et al., 1997; Scholz et al., 1999; Speight, 2014a).

The asphaltene fraction (Fig. 1.2), which appears as dark brown to black, friable solid, is largely responsible for adverse properties of crude oil, such as catalyst-poisoning and high yields of thermal coke (Jacobs and Filby, 1983; El-Gendy, 2004; Speight, 2014a). The structures of the asphaltene constituents are unknown but do contain nitrogen, oxygen, and sulfur atoms (all predominantly in ring systems), as well as vanadium and nickel complexes (predominantly as porphyrin derivatives) (Murgich et al., 1999; Strausz et al., 1999; Speight, 2014a). Also, many of the problems associated with recovery, separation, or processing of crude oil are related to the presence of high concentration of asphaltene constituents.

One of the problems related to the presence of asphaltene constituents, as well as nitrogen, oxygen, and sulfur ring compounds in the environment, resides in their resistance to biodegradation by microbial metabolic activity. Thus, it is not surprising that metabolic routes involved in this process are the less well-known, although, there is evidence suggesting that some microorganisms have the potential capability of transforming asphaltenes, and in the best case, eliminating them (Atlas, 1981; Speight and Arjoon, 2012).

These variations in properties, which are dependent upon composition, affect the way crude oil will interact with biotechnology agents (such as biodegradation techniques and bio-refining) and because of the variation in properties, the manner of the interaction, and the extent of the interaction can be

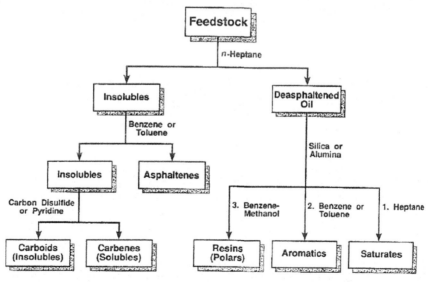

FIGURE 1.2

Separation scheme for various feedstocks.

expected to vary. In addition, there are other members of the crude oil family that are worthy of mention here and they include: (1) crude oil from tight formations, (2) opportunity crudes, (3) high acid crude oils, and (4) foamy oil.

2.1 CRUDE OIL FROM TIGHT FORMATIONS

Generally, unconventional tight oil resources are found at considerable depths in sedimentary rock formations that are characterized by very low permeability. While some of the tight oil plays produce oil directly from shales, tight oil resources are also produced from low-permeability siltstone formations, sandstone formations, and carbonate formations that occur in close association with a shale source rock. Tight formations (low permeability formations) that are scattered throughout North America have the potential to produce crude oil (*tight oil*) (US EIA, 2011; Mayes, 2015). Such formations might be composted of shale sediments or sandstone sediments. In a conventional sandstone reservoir, the pores are interconnected so gas and oil can flow easily from the rock to a wellbore. In tight sandstones, the pores are smaller and are poorly connected by very narrow capillaries which results in low permeability. Tight oil occurs in sandstone sediments that have an effective permeability of less than 1 milliDarcy (<1 mD).

By definition, a shale play is a geographic area containing an organic-rich, fine-grained sedimentary rock that has undergone physical and chemical compaction during diagenesis to produce the following characteristics: (1) the presence of clay to silt-sized particles, (2) a high proportion of silica and, sometimes, carbonate minerals, (3) it is thermally mature, (4) hydrocarbon-filled porosity—on the order of 6%–14% by volume, (5) low permeability—on the order of <0.1 mD, and (6) large areal distribution. Because of these properties, such formations require fracture stimulation for the economic production of natural gas and/or crude oil.

The most notable tight oil plays in North America include the Bakken shale, the Niobrara formation, the Barnett shale, the Eagle Ford shale, the Miocene Monterey play of California's San Joaquin Basin and the Cardium play (Alberta, Canada). Other known tight formations (on a worldwide basis) include the R'Mah formation (Syria), the Sargelu formation (northern Persian Gulf region), the Athel formation (Oman), the Bazhenov formation, Achimov formation (West Siberia, Russia), the Coober Pedy formation (Australia), the Chicontepex formation (Mexico), and the Vaca Muerta field (Argentina) (US EIA, 2011, 2013). In many of these tight formations, the existence of large quantities of crude oil has been known for decades and efforts to commercially produce those resources have occurred sporadically with typically disappointing results. However, starting in the mid-2000s, advancements in well drilling and stimulation technologies combined with high oil prices have turned tight oil resources into one of the most actively explored and produced targets in North America.

Typical of the crude oil from tight formations (*tight oil, tight light oil, and tight shale oil* have been suggested as alternate terms) is the Bakken crude oil which is a light, highly volatile crude oil. Briefly, Bakken crude oil is a light, sweet (low-sulfur) crude oil that has a relatively high proportion of volatile constituents. The production of the oil also yields a significant amount of volatile gases (including propane and butane) and low-boiling liquids (such as pentane and natural gasoline), which are often referred to collectively as light naphtha (low-boiling naphtha). Because of the presence of low-boiling hydrocarbons, low-boiling naphtha (and crude oil containing such constituents) can become extremely explosive, even at relatively low ambient temperatures. Some of these gases may be burned off (flared) at the field wellhead, but others remain in the liquid products extracted from the well (Speight, 2014a).

Oil from tight shale formations is characterized by low-asphaltene content and a high API gravity, low sulfur content, and a significant molecular weight distribution of the paraffinic wax content (Table 1.2) (Speight, 2014a, 2015b, 2016). Paraffin carbon chains of C_{10} to C_{60} have been found, with some tight oils containing carbon chains up to C_{72} that can result in deposition and plugging in formations (Speight, 2014a). However, the properties of crude oils from tight formations are highly variable. Density and other properties can show wide variation, even within the same field. The Bakken crude has an API of 42 degrees and a sulfur content of 0.19% w/w. Similarly, Eagle Ford crude oil has a sulfur content on the order of 0.1% w/w and an API gravity falling in the range 40–62 degrees.

Table 1.2 Simplified Differentiation Between Conventional Crude Oil, Tight Oil, Heavy Crude Oil, Extra Heavy Crude Oil, and Tar Sand Bitumen[a]

Conventional Crude Oil

Mobile in the reservoir; API gravity: >25°
High-permeability reservoir
Produced by primary and secondary recovery methods

Tight Oil

Similar properties to the properties of conventional crude oil; API gravity: >25°
Immobile in the reservoir
Low-permeability reservoir
Produced by horizontal drilling into reservoir and fracturing to release fluids/gases

Heavy Crude Oil

More viscous than conventional crude oil; API gravity: 10°–20°
Mobile in the reservoir
High-permeability reservoir
Produced by secondary and tertiary recovery methods (enhanced oil recovery)

Extra Heavy Crude Oil;

Similar properties to the properties of tar sand bitumen; API gravity: <10°
Mobile in the reservoir
High-permeability reservoir
Produced by secondary and tertiary recovery methods (enhanced oil recovery)

Tar Sand Bitumen

Immobile in the deposit; API gravity: <10°
High-permeability reservoir
Mining (often preceded by explosive fracturing)
Steam-assisted gravity draining (SAGD); solvent methods (VAPEX)
Innovative methods[b]

[a]*This list is not intended for use as a means of classification.*
[b]*The term* innovative methods *does not include tertiary (enhanced) recovery methods but does include methods such as SAGD, vapor-assisted extraction (VAPEX), and variants or hybrids thereof, as well as extreme heating methods such as in-situ combustion.*

2.2 OPPORTUNITY CRUDES

Opportunity crude oils are either relatively new-to-the-market crude oils with unknown or poorly understood properties relating to processing issues or are existing on-the-market crude oils with well-known properties that cause processing problems (Speight, 2014a,b; Ohmes, 2014; Yeung, 2014). Opportunity crude oils are often, but not always, heavy crude oils but in either case are more difficult to process due to high levels of solids (and other contaminants) produced with the oil, high levels of acidity, and high viscosity.

2.3 HIGH ACID CRUDES

High acid crude oils contain considerable proportions of naphthenic acids which, as commonly used in the petroleum industry, refers collectively to all of the organic acids present in the crude oil (Shalaby, 2005; Speight, 2014b). In many instances, high-acid crude oils are the heavier (more viscous) crude oils (Speight, 2014a,b). The total acid matrix is therefore complex, and it is unlikely that a simple titration, such as the traditional methods for measurement of the total acid number (TAN), can give meaningful results to use in predictions of problems. An alternative way of defining the relative organic acid fraction of crude oils is therefore a real need in the oil industry, both upstream and downstream.

By the original definition, a naphthenic acid is a monobasic carboxyl group attached to a saturated cycloaliphatic structure. However, it has been a convention accepted in the oil industry that all organic acids in crude oil are called naphthenic acids. Naphthenic acids in crude oils are now known to be mixtures of low- to high molecular weight acids and the naphthenic acid fraction also contains other acidic species. Naphthenic acids, which are not *user friendly* in terms of behavior and refining (Kane and Cayard, 2002; Ghoshal and Sainik, 2013), can be either (or both) water-soluble to oil-soluble depending on their molecular weight, process temperatures, salinity of waters, and fluid pressures. In the water phase, naphthenic acids can cause stable reverse emulsions (oil droplets in a continuous water phase). In the oil phase with residual water, these acids have the potential to react with a host of minerals, which can neutralize the acids. The main reaction product found in practice is the calcium naphthenate (the calcium salt of naphthenic acids). The total acid matrix is therefore complex and it is unlikely that a simple titration, such as the traditional methods for measurement of the TAN, can give meaningful results to use in predictions of problems.

It is commonly assumed that acidity in crude oils is related to carboxylic acid species, i.e., components containing a –COOH functional group. While carboxylic acid functionality is an important feature (60% of the ions have two or more oxygen atoms), a major portion (40%) of the acid types are not carboxylic acids. In fact, naphthenic acids are a mixture of different compounds, including phenol derivatives, some of which may be polycyclic (Kane and Cayard, 2002; Ghoshal and Sainik, 2013; Speight, 2014b).

2.4 FOAMY OIL

Foamy oil is oil-continuous foam that contains dispersed gas bubbles produced at the wellhead from heavy oil reservoirs under solution gas drive. The nature of the gas dispersions in oil distinguishes foamy oil behavior from conventional heavy oil. The gas that comes out of solution in the reservoir does not coalesce into large gas bubbles or into a continuous flowing gas phase. Instead it remains as small bubbles entrained in the crude oil, keeping the effective oil viscosity low while providing expansive energy that helps drive the oil toward the production well. Foamy oil accounts for unusually high production in heavy oil reservoirs under solution-gas drive (Chugh et al., 2000).

However, the actual structure of foamy oil flow and its mathematical description are still not well understood. Much of the earlier discussion of such flows was based on the concept of microbubbles (i.e., bubbles that are much smaller than the average pore-throat size and are thus free to move with the oil during flow (Sheng et al., 1999). Dispersion of this type can be produced only by nucleation of a very large number of bubbles (explosive nucleation) and by the availability of a mechanism that prevents these bubbles from growing into larger bubbles with decline in pressure. Another hypothesis for the structure of foamy oil flow is that much larger bubbles migrate with the oil, and the dispersion is created by breakup of bubbles during migration. The major difference between conventional solution gas drive and foamy solution gas drive is that the pressure gradient in the latter is strong enough to mobilize gas clusters once they have grown to a certain size (Poon and Kisman, 1992; Maini, 1999, 2001).

3.0 HEAVY OIL AND EXTRA HEAVY OIL

There are also other *types* of petroleum that are different from the conventional petroleum insofar as they are much more difficult to recover from the subsurface reservoir. These materials have a higher viscosity (and lower API gravity) than conventional petroleum, and recovery of these petroleum types usually requires thermal stimulation of the reservoir leading to application of various thermal methods (such as coking processes) for conversion of crude oil to low-boiling distillates.

For example, petroleum and heavy oil have been arbitrarily defined in terms of physical properties. Thus, heavy oil has been considered to be the type of crude oil that had an API gravity less than 20 degrees; an API gravity equal to 12 degrees signifies a heavy oil, while extra heavy oil and tar sand bitumen, usually have an API gravity less than 10 degrees (e.g., Athabasca tar sand bitumen = 8 degrees API). Residua would vary depending upon the temperature at which distillation was terminated but usually vacuum residua are in the range 2–8 degrees API. The term *heavy oil* has also been used collectively to describe both the heavy oils that require thermal stimulation of recovery from the reservoir and the bitumen in bituminous sand formations from which the viscous bituminous material is recovered by a mining operation (Speight, 2013b, 2013c, 2014a). Convenient as this may be, it is scientifically and technically incorrect.

3.1 HEAVY OIL

Heavy oil is a *type* of petroleum that is different from conventional petroleum insofar as it is very viscous and does not flow easily. The common characteristic properties (relative to conventional crude oil) are high specific gravity, low hydrogen to carbon ratio, high carbon residue, and high content of asphaltene constituents, sulfur, nitrogen, and metals. The term *heavy oil* has also been arbitrarily (incorrectly) used to describe both the heavy oils that require thermal stimulation of recovery from the reservoir and the bitumen in bituminous sand (tar sand) formations from which the heavy bituminous material is recovered by a mining operation.

Heavy oil is more difficult to recover from the subsurface reservoir than conventional or light oil. A very general definition of heavy oils has been and remains based on API gravity or viscosity. For example, heavy oil was considered to be crude oil with an API gravity somewhat less than 20 degrees. For example, Cold Lake heavy crude oil has an API gravity equal to 12 degrees and tar sand bitumen usually have an API gravity in the range 5–10 degrees (Athabasca bitumen = 8 degrees API) (Speight, 2000, 2014a; Parkash, 2003; Hsu and Robinson, 2006; Gary et al., 2007).

3.2 EXTRA HEAVY OIL

The term *extra heavy oil* is a recently evolved term (related to viscosity) of little scientific meaning and generally refers to a material that occurs in the solid or near-solid state and but has mobility under reservoir conditions. While this type of oil may resemble tar sand bitumen and does not flow easily, extra heavy oil is generally recognized as having mobility in the reservoir compared to tar sand bitumen, which is typically incapable of mobility (free flow) under reservoir conditions. For example, the tar sand bitumen located in Alberta, Canada, is not mobile in the deposit and requires extreme methods of recovery to recover the bitumen. The general difference between extra heavy oil and tar sand bitumen is that extra heavy oil, which may have properties similar to tar sand bitumen in the laboratory but, unlike tar sand bitumen in the deposit, has some degree of mobility in the reservoir or deposit (Table 1.2) (Delbianco and Montanari, 2009; Speight, 2014a).

In the context of this book, the methods outlined in this book for heavy oil have an API gravity of less than 20 with a variable sulfur content (Table 1.3) (Speight, 2000, 2014a). Again, for simplicity and convenience, the data in this table are restricted to API gravity and sulfur content which indicate

Table 1.3 API Gravity and Sulfur Content of Selected Heavy Oils and Tar Sand Bitumen

Country	Crude Oil	API	Sulfur% w/w
Brazil	Albacor Leste	18.9	0.66
Canada (Alberta)	Athabasca	8.0	4.8
Canada (Alberta)	Cold Lake	13.2	4.11
Canada (Alberta)	Lloydminster	16.0	2.60
Canada (Alberta)	Wabasca	19.6	3.90
Chad	Bolobo	16.8	0.14
Chad	Kome	18.5	0.20
China	Bozhong	16.7	0.30
China	Zhao Dong	18.4	0.25
Colombia	Castilla	13.3	0.22
Colombia	Chichimene	19.8	1.12
Ecuador	Ecuador Heavy	18.2	2.23
Ecuador	Napo	19.2	1.98
USA (California)	Hondo Monterey	17.2	4.70
USA (California)	Huntington Beach	14.4	0.90
Venezuela	Boscan	10.1	5.50
Venezuela	Pilon	14.1	1.91
Venezuela	Tremblador	19.0	0.80

the variations in the crude oil feedstocks (from one country or region to another) and therefore the differences that can be anticipated when biotechnology is applied to these crude oils. As with the lighter crude oils, the behavior of heavy oil to the application of biotechnology depends on their characteristics (properties) since these heavy oils fall into a range of high viscosity, which is subject to temperature effects (Speight, 2014a) but is a parameter that can dictate the application of biotechnology to the oil.

4.0 TAR SAND BITUMEN

In addition to conventional petroleum and heavy crude oil, there remains an even more viscous material that offers some relief to the potential shortfalls in supply (Meyer and De Witt, 1990; BP, 2016). This is the *bitumen* found in *tar sand* (*oil sand*) deposits. However, many of these reserves are only available with some difficulty and optional biotechnology scenarios will be necessary for bioconversion (bioremediation, biodegradation) of these materials because of the substantial differences in character between conventional petroleum and tar sand bitumen (Speight and Arjoon, 2012; El-Gendy and Speight, 2015). *Tar sand*, also variously called *oil sand* or *bituminous sand* is a loose-to-consolidated sandstone or a porous carbonate rock, impregnated with bitumen, a heavy asphaltic crude oil with an extremely high viscosity under reservoir conditions.

The term *tar sand bitumen* (also, on occasion referred to as *extra heavy oil* and *native asphalt*, although the latter term is incorrect) includes a wide variety of reddish brown to black materials of near solid to solid character that exist in nature either with no mineral impurity or with mineral matter content that exceeds 50% by weight. Bitumen is frequently found filling pores and crevices of sandstone, limestone, or argillaceous sediments, in which case the organic and associated mineral matrix is known as *rock asphalt*.

Bitumen is also a naturally occurring material that is found in deposits that are incorrectly referred to as *tar sand* since tar is a product of the thermal processing of coal (Speight, 2013a). The permeability of a tar sand deposit is low and passage of fluids through the deposit can only be achieved by prior application of fracturing techniques. Alternatively, bitumen recovery can be achieved by conversion of the bitumen to a product in situ (in situ upgrading) followed by product recovery from the deposit (Speight, 2013b,c, 2014a, 2016). Tar sand bitumen is a high-boiling material with little, if any, material boiling below 350°C (660°F) and the boiling range approximates the boiling range of an atmospheric residuum.

There have been many attempts to define tar sand deposits and the bitumen contained therein. To define conventional petroleum, heavy oil, and bitumen, the use of a single physical parameter such as viscosity is not sufficient. Other properties such as API gravity, elemental analysis, composition, and, most of all, the properties of the bulk deposit must also be included in any definition of these materials. In fact, the most appropriate definition of *tar sands* is found in the writings of the United States government, viz.:

> Tar sands are the several rock types that contain an extremely viscous hydrocarbon which is not recoverable in its natural state by conventional oil well production methods including currently used enhanced recovery techniques. The hydrocarbon-bearing rocks are variously known as bitumen-rocks oil, impregnated rocks, oil sands, and rock asphalt.
>
> **US Congress. 1976. Public Law FEA-76-4. United States Library of Congress, Washington, DC.**

This definition speaks to the character of the bitumen through the method of recovery. Thus, the bitumen found in tar sand deposits is an extremely viscous material that is *immobile under reservoir conditions* and cannot be recovered through a well by the application of secondary or enhanced recovery techniques. Mining methods match the requirements of this definition (since mining is not one of the specified recovery methods) and the bitumen can be recovered by alteration of its natural state such as thermal conversion to a product that is then recovered. By inference and by omission, conventional petroleum and heavy oil are also included in this definition. Petroleum is the material that can be recovered by conventional oil well production methods, whereas heavy oil is the material that can be recovered by enhanced recovery methods.

The application of biotechnologies to tar sand bitumen depends to a large degree on the composition of the bitumen, such as the sulfur content and API gravity and the structure of the bitumen (Speight, 2014a). On the other hand, extra heavy oil, which is often likened to tar sand bitumen because of similarities in the properties of the two, has a degree of mobility under reservoir or deposit conditions but typically suffers from the same drawbacks as the immobile bitumen in biotechnology applications.

5.0 COMPOSITION

The analysis of petroleum, heavy oil, extra heavy oil, and tar sand bitumen for the percentages by weight of carbon, hydrogen, nitrogen, oxygen, and sulfur (elemental composition, ultimate composition) is perhaps the first method used to examine the general nature, and perform an evaluation, of a feedstock. The atomic ratios of the various elements to carbon (i.e., H/C, N/C, O/C, and S/C) are frequently used for indications of the overall character of the feedstock. It is also of value to determine the amounts of trace elements, such as vanadium and nickel, in a feedstock since these materials can have serious deleterious effects on catalyst performance during refining by catalytic processes.

5.1 ELEMENTAL COMPOSITION

Carbon content can be determined by the method designated for coal and coke (ASTM D3178) or by the method designated for municipal solid waste (ASTM E777). There are also methods designated for: (1) *hydrogen content*—ASTM D1018, ASTM D3178, ASTM D3343, ASTM D3701, and ASTM E777, (2) *nitrogen content*—ASTM D3179, ASTM D3228, ASTM E258, and ASTM E778 (3) *oxygen content*—ASTM E385), (4) *sulfur content*—ASTM D1266, ASTM D1552, ASTM D1757, ASTM D2622, ASTM D3177, ASTM D4045 and ASTM D4294, and (5) metals—ASTM D1318, ASTM D3340, ASTM D3341, ASTM D3605. For all oils, the higher the atomic hydrogen-carbon ratio, the lower the sulfur and nitrogen content, and the lower the metals, the more suitable (or susceptible) the oil will be for biotechnology applications (El-Gendy, 2004; El-Gendy and Speight, 2015).

The heteroatoms (nitrogen, oxygen, sulfur, and metals) in oil affect every aspect of biotechnology. During the fractionation of petroleum, the heteroatoms are concentrated in the asphaltene fraction. The de-asphaltened oil contains smaller concentrations of porphyrins than the parent materials and usually very small concentrations of non-porphyrin metals.

5.2 CHEMICAL COMPOSITION

In very general terms (and as observed from elemental analyses), petroleum, heavy oil, bitumen, and residua are a complex composition of: (1) hydrocarbons; (2) nitrogen compounds; (3) oxygen compounds; (4) sulfur compounds; and (5) metallic constituents (Table 1.4). However, this general definition is not adequate to describe the composition of petroleum et al. as it relates to the behavior of these feedstocks. Indeed, the consideration of hydrogen-to-carbon atomic ratio, sulfur content, and API gravity are no longer adequate to the task of determining refining behavior. Nevertheless, some of these constituents—especially the saturated hydrocarbon constituents—are more amenable to bioreactions than the aromatic constituents (Speight and Arjoon, 2012; El-Gendy and Speight, 2015).

Furthermore, the molecular composition of petroleum can be described in terms of three classes of compounds: saturates, aromatics, and compounds bearing heteroatoms (sulfur or nitrogen) (Tables 1.5 and 1.6). Within each class, there are several families of related compounds: (1) saturated constituents include normal alkanes, branched alkanes, and cycloalkanes (paraffins, iso-paraffins, and naphthenes, in petroleum terms), (2) alkene constituents (olefins) are rare to the extent of being considered an oddity, (3) monoaromatic constituents range from benzene to multiple fused ring analogs (naphthalene, phenanthrene, etc.), (4) thiol constituents—mercaptan—contain sulfur as do thioether derivatives and

Table 1.4 Hydrocarbon and Heteroatom Types in Crude Oil, Heavy Oil, and Tar Sand Bitumen

Class	Compound Types
Saturated hydrocarbons	n-Paraffins
	iso-Paraffins and other branched paraffins
	Cycloparaffins (naphthenes)
	Condensed cycloparaffins (including steranes, hopanes)
	Alkyl side chains on ring systems
Unsaturated hydrocarbons	Olefins: non-indigenous;
	present in products of thermal reactions
Aromatic hydrocarbons	Benzene systems
	Condensed aromatic systems
	Condensed naphthene-aromatic systems
	Alkyl side chains on ring systems
Saturated heteroatomic systems	Alkyl sulfides
	Cycloalkyl sulfides
Sulfides	Alkyl side chains on ring systems
Aromatic heteroatomic systems	Furans (single-ring and multi-ring systems)
	Thiophenes (single-ring and multi-ring systems)
	Pyrroles (single-ring and multi-ring systems)
	Pyridines (single-ring and multi-ring systems)
	Mixed heteroatomic systems
	Amphoteric (acid-base systems)
	Alkyl side chains on ring systems
	Porphyrins
	Organometallic systems and metal salts

Table 1.5 Nomenclature and Types of Organic Nitrogen Compounds

Nonbasic

Pyrrole	C_4H_5N	
Indole	C_8H_7N	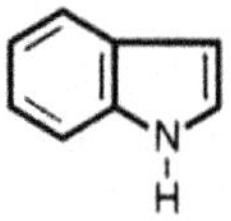
Carbazole	$C_{12}H_9N$	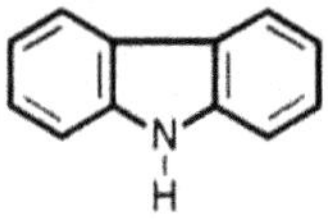
Benzo(a)carbazole	$C_{16}H_{11}N$	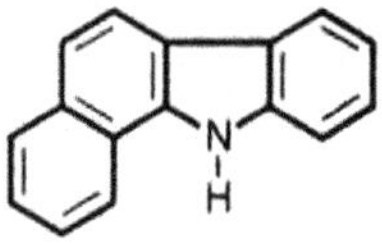 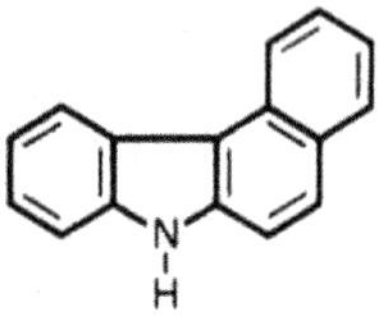

Basic

Pyridine	C_5H_5N	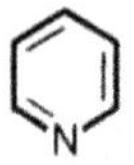
Quinoline	C_9H_7N	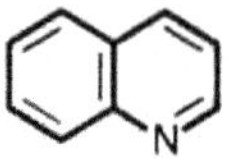
Indoline	C_8H_9N	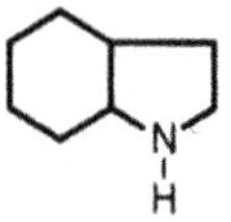
Benzo(f)quinoline	$C_{13}H_9N$	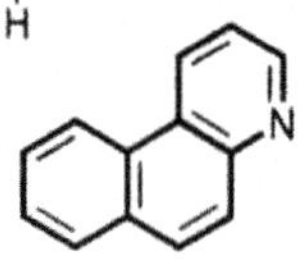

thiophene derivatives, (5) nitrogen-containing and oxygen-containing constituents are more likely to be found in polar forms (pyridines, pyrroles, phenols, carboxylic acids, amides, etc.) than in nonpolar forms (such as ethers).

The occurrence of amphoteric species (i.e., compounds having a mixed acid/base nature) is rarely addressed nor is the phenomenon of molecular size or the occurrence of specific functional types (Fig. 1.3) which can play a major role in the properties of the constituents, properties of the oil through the interactions between the constituents of the oil, and the success or failure of the application

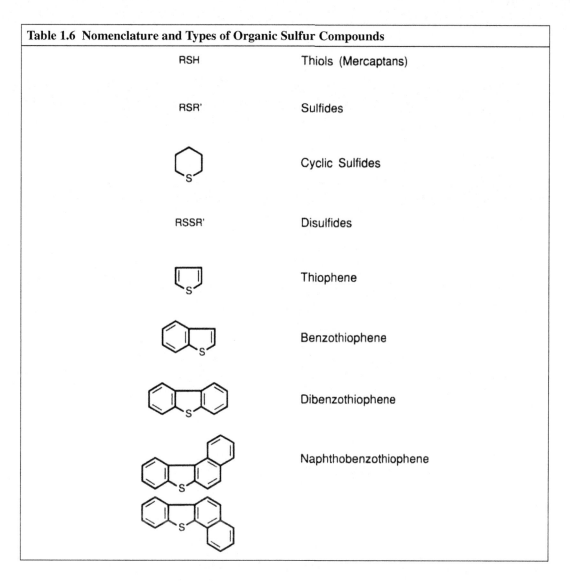

Table 1.6 Nomenclature and Types of Organic Sulfur Compounds

RSH	Thiols (Mercaptans)
RSR'	Sulfides
	Cyclic Sulfides
RSSR'	Disulfides
	Thiophene
	Benzothiophene
	Dibenzothiophene
	Naphthobenzothiophene

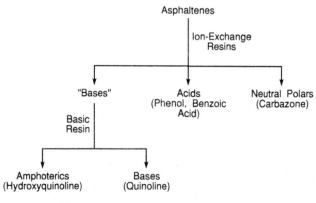

FIGURE 1.3

Separation of asphaltene constituents based on functionality (polarity).

of biotechnology. In addition, as the feedstock series progresses to higher molecular weight feedstocks from crude oil to heavy crude oil to tar sand bitumen, not only does the number of the constituents increase but the molecular complexity of the constituents also increases (Speight, 2014a).

5.3 COMPOSITION BY VOLATILITY

Volatility plays a major role when biotechnology is applied to crude oil. By the nature of constituent volatility, petroleum can be separated into a variety of fractions based on the boiling points of the petroleum constituents. Such fractions are primarily identified by their respective boiling ranges and, to a lesser extent, by chemical composition. However, it is often obvious that as the boiling ranges increase, the nature of the constituents remains closely similar and it is the number of the substituents that caused the increase in boiling point. It is through the recognition of such phenomena that molecular design of the higher boiling constituents can be achieved (Speight, 2014a).

Invoking the existence of structurally different constituents in the nonvolatile fractions from those identifiable constituents in the lower boiling fractions is unnecessary and (considering the nature of the precursors and maturation paths) irrational (Speight, 1994, 2014a). For example, the predominant types of condensed aromatic systems in petroleum are derivatives of phenanthrene and there it is to be anticipated that the higher peri-condensed homologs (Fig. 1.4) shall be present in resin constituents and asphaltene constituents rather than the derivatives of kata-condensed polynuclear aromatic system (Fig. 1.5). Either of these condensed polynuclear aromatic derivatives can cause problems when biotechnological methods are applied to crude oil and crude oil products (Speight and Arjoon, 2012; El-Gendy and Speight, 2015).

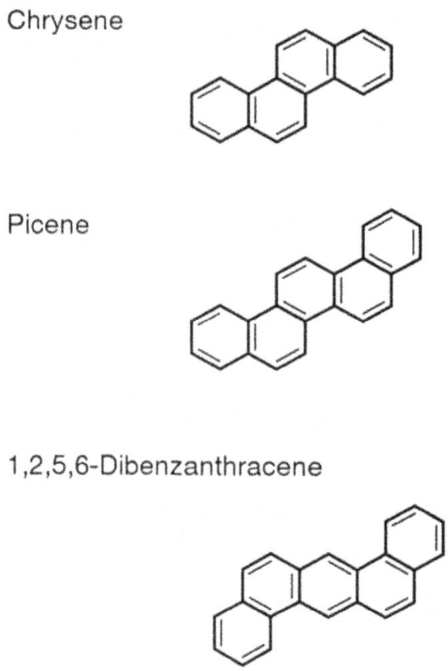

Chrysene

Picene

1,2,5,6-Dibenzanthracene

FIGURE 1.4

Examples of peri-condensed aromatic systems.

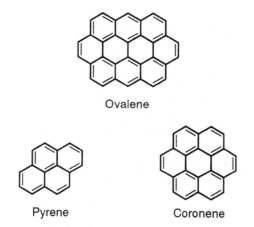

Ovalene

Pyrene   Coronene

FIGURE 1.5

Examples of kata-condensed (cata-condensed) aromatic systems.

5.4 COMPOSITION BY FRACTIONATION

Petroleum can be defined (on a *relative* or *standard* basis) in terms of three or four general fractions: asphaltenes, resins, saturates, and aromatics (Fig. 1.2). Thus, it is possible to compare inter-laboratory investigations and thence to apply the concept of predictability to refining sequences and potential products (Speight, 1994; Long and Speight, 1998).

The fractionation methods available for application to petroleum allow a reasonably effective degree of separation of hydrocarbon mixtures (Speight, 2014a, 2015b). However, the problems are separating the petroleum constituents without alteration of their molecular structure and obtaining these constituents in a substantially pure state. Thus, the general procedure is to employ techniques that segregate the constituents according to molecular size and molecular type.

It is more generally true, however, that the success of any attempted fractionation procedure involves not only the application of one technique but also the utilization of several integrated techniques, especially those techniques involving the use of chemical and physical properties to differentiate among the various constituents. For example, the standard processes of physical fractionation used in the petroleum industry are those of distillation and solvent treatment, as well as adsorption by surface-active materials. Chemical procedures depend on specific reactions, such as the interaction of olefins with sulfuric acid or the various classes of adduct formation. Chemical fractionation is often but not always successful because of the complex nature of crude oil. This may result in unprovoked chemical reactions that have an adverse effect on the fractionation and the resulting data. Indeed, caution is advised when using methods that involve chemical separation of the constituents.

5.5 COMPOSITION BY SPECTROSCOPY

Spectroscopic studies have played an important role in the evaluation of petroleum and of petroleum products for the last three decades and many of the methods are now used as standard methods of analysis for crude oil and crude oil products. The methods include the use of *mass spectrometry* to determine the (1) hydrocarbon types in middle distillates (ASTM D2425); (2) hydrocarbon types of gas oil saturate fractions (ASTM D2786); (3) hydrocarbon types in low-olefin gasoline; and (4) aromatic types of gas oil aromatic fractions (ASTM D3239). *Nuclear magnetic resonance spectroscopy* has been developed as a standard method for the determination of hydrogen types in aviation turbine fuels (ASTM D3701). *X-ray fluorescence spectrometry* has been applied to the determination of sulfur in various petroleum products (ASTM D2622; ASTM D4294).

Infrared spectroscopy is used for the determination of benzene in motor and/or aviation gasoline (ASTM D4053), whilst ultraviolet spectroscopy is employed for the evaluation of mineral oils (ASTM D2269) and for determining the naphthalene content of aviation turbine fuels (ASTM D1840). Other techniques include the use of *flame emission spectroscopy* for determining trace metals in gas turbine fuels (ASTM D3605) and the use of *absorption spectrophotometry* for the determination of the alkyl nitrate content of diesel fuel (ASTM D4046). *Atomic absorption* has been employed as a means of measuring the lead content of gasoline (ASTM D3237) and for the manganese content of gasoline (ASTM D3831) as well as for determining the barium, calcium, magnesium, and zinc contents of lubricating oils (ASTM D4628). *Flame photometry* has been employed as a means of measuring the lithium/sodium content of lubricating greases (ASTM D3340) and the sodium content of residual fuel oil (ASTM D1318).

5.6 COMPOSITION BY CHROMATOGRAPHY

Chromatography is the collective term for a collection of laboratory techniques for the separation of mixtures. Typically, the mixture is dissolved in a fluid (*mobile phase*) which carries it through a structure holding another material (*stationary phase*). The various constituents of the mixture travel at different speeds, causing them to separate. The separation is based on differential partitioning between the mobile and stationary phases. Subtle differences in the partition coefficient of different compounds result in differential retention on the stationary phase and thus changing the separation.

A chromatographic technique may be preparative or analytical. The purpose of preparative chromatography is to separate the components of a mixture for more advanced use (and is thus a form of purification). Analytical chromatography generally requires smaller amounts of material and is for measuring the relative proportions of analytes in a mixture. The two are not mutually exclusive.

5.6.1 Adsorption Chromatography

Adsorption chromatography has helped to characterize the group composition of crude oils and hydrocarbon products since the beginning of the 20th century. The type and relative amount of certain hydrocarbon classes in the matrix can have a profound effect on the quality and performance of the hydrocarbon product and two standard test methods have been used predominantly over the years (ASTM D2007; ASTM D4124). The fluorescent indicator adsorption (FIA) method (ASTM D1319) has served for over 30 years as the official method of the petroleum industry for measuring the paraffinic, olefinic, and aromatic content of gasoline and jet fuel. The technique consists of displacing a sample under *iso*-propanol pressure through a column packed with silica gel in the presence of fluorescent indicators specific to each hydrocarbon family. Despite its widespread use, fluorescent indicator adsorption has numerous limitations (Suatoni and Garber, 1975; Miller et al., 1983; Norris and Rawdon, 1984).

The segregation of individual components from a mixture can be achieved by application of adsorption chromatography in which the adsorbent is either packed in an open tube (column chromatography) or shaped in the form of a sheet (thin-layer chromatography, TLC). A suitable solvent is used to elute from the bed of the adsorbent. Chromatographic separations are usually performed for determining the composition of a sample. Even with such complex samples as petroleum, some information about the chemical structure of a fraction can be gained from the separation data.

5.6.2 Gas Chromatography

Gas-liquid chromatography (*GLC*) is a method for separating the volatile components of various mixtures. It is, in fact, a highly efficient fractionating technique, and it is ideally suited to the quantitative analysis of mixtures when the possible components are known and the interest lies only in determining the amount of each component present. In this type of application, gas chromatography has taken over much of the work previously done by the other techniques; it is now the preferred technique for the analysis of hydrocarbon gases.

Thus, it is not surprising that gas chromatography has been used extensively for individual component identification, as well as percentage composition, in the gaseous boiling ranges (ASTM D2163; ASTM D2504; ASTM D2505; ASTM D2593; ASTM D2597; ASTM D2712; ASTM D4424; ASTM D4864; ASTM D5303; ASTM D6159), in the gasoline boiling range (ASTM D2427; ASTM D3525; ASTM D3606; ASTM D3710; ASTM D4815; ASTM D5134; ASTM D5441; ASTM D5443; ASTM D5501; ASTM D5580; ASTM D5599; ASTM D5623; ASTM D5845; ASTM D5986), in higher boiling ranges such as diesel fuel, aviation gasoline (ASTM D3606), engine oil, motor oil, and wax (ASTM D5442), as well as for the boiling range distribution of petroleum fractions (ASTM D2887; ASTM D5307) or the purity of solvents using capillary gas chromatography (ASTM D2268).

The technique has proved to be an exceptional and versatile instrumental tool for analyzing compounds that are of low molecular weight and that can be volatilized without decomposition. However, these constraints limit the principal applicability in petroleum science to feedstock identification when the composition is known to be in the low to medium boiling range. The use of this technique for direct component analysis in the heavy fractions of petroleum is subject to many limitations (Speight, 2001).

A more recent, very important development in gas chromatography is its combination with a mass spectrometer as the detector. The technique in which gas chromatography is combined with spectrometry (GC/MS) has proved to be a powerful tool for identifying many compounds at very low levels in a wide range of boiling matrix. By the combination of the two techniques in one instrument, the onerous trapping of fractions from the gas chromatographic column is avoided and higher sensitivities can be attained. In passing through the gas chromatographic column, the sample is separated according to the boiling points of the individual constituents.

In view of the molecular characterizing nature of spectrometric techniques, it is not surprising that considerable attention has been given to the combined use of GLC and these techniques. In recent years, the use of the *mass spectrometer* to monitor continuously the effluents of a chromatographic column has been reported, and considerable progress has been made in the development of rapid scan infrared spectrometers for this purpose. The mass spectrometer, however, has the advantage that the quantity of material required to produce a spectrum is considerably less than that necessary to produce an infrared spectrum.

Pyrolysis gas chromatography can be used for information on the gross composition of heavy petroleum fractions. In this technique, the sample under investigation is pyrolyzed and the products are introduced into a gas chromatography system for analysis. There has also been extensive use of pyrolysis gas chromatography by geochemists to correlate crude oil with source rock and to derive geochemical characterization parameters from oil-bearing strata.

In the technique of inverse GLC, the sample under study is used as the stationary phase and several volatile test compounds are chromatographed on this column. The interaction coefficient determined for these compounds is a measure of certain qualities of the liquid phase. The coefficient is therefore indicative of the chemical interaction of the solute with the stationary phase. The technique has been used largely for studies of asphalt.

5.6.3 Gel Permeation Chromatography

There are two additional techniques that have evolved from the more recent development of chromatographic methods.

The first technique, *gel filtration chromatography (GFC)*, has been successfully employed for application to aqueous systems by biochemists for more than five decades. The technique was developed using soft, cross-linked dextran beads. The second technique, *gel permeation chromatography (GPC)*, employs semi-rigid, cross-linked polystyrene beads. In either technique, the packing particles swell in the chromatographic solvent and form a porous gel structure. The distinction between the methods is based on the degree of swelling of the packing; the dextran swells to a much greater extent than the polystyrene. Subsequent developments of rigid porous packings of glass, silica, and silica gel have led to their use and classification as packings for GPC.

GPC, also called *size exclusion chromatography (SEC)*, in its simplest representation consists of employing column(s) packed with gels of varying pore sizes in a liquid chromatograph (Carbognani, 1997). Under conditions of constant flow, the solutes are injected onto the top of the column, whereupon they appear at the detector in order of decreasing molecular weight. The separation is based on the fact that the larger solute molecules cannot be accommodated within the pore systems of the gel beads and thus are eluted first. On the other hand, the smaller solute molecules have increasing volume within the beads, depending upon their relative size, and require more time to elute.

In theory, GPC is an attractive technique for the determination of the number of average molecular weight (M_n) distribution of petroleum fractions. However, it is imperative to recognize that petroleum contains constituents of widely differing polarity, including nonpolar paraffins and naphthenes (alicyclic compounds), moderately polar aromatics (mononuclear and condensed), and polar nitrogen, oxygen, and sulfur species. Each compound type interacts with the gel surface to a different degree. The strength of the interaction increases with increasing polarity of the constituents and with decreasing polarity of the solvent. Therefore, the ideal linear relationship of log M_n against elution volume V_e that may be operative for nonpolar hydrocarbon species cannot be expected to remain in operation. It must also be recognized that the lack of realistic standards of known number average molecular weight distribution and of chemical nature similar to that of the constituents of petroleum for calibration purposes may also be an issue. However, GPC has been employed in the study of petroleum constituents, especially the heavier constituents, and has yielded valuable data (Baltus and Anderson, 1984; Reynolds and Biggs, 1988; Speight, 2001, 2015b).

SEC is usually practiced with refractive index (RI) detection and yields a mass profile (concentration vs. time or elution volume) that can be converted to a mass versus molecular weight plot by means of a calibration curve. The combination of SEC with element-specific detection has widened this concept to provide the distribution of heterocompounds in the sample as a function of elution volume and molecular weight.

The combination of GPC with another separation technique also allows the fractionation of a sample separately by molecular weight and by chemical structure. This is particularly advantageous for the characterization of the heavier fractions of petroleum materials because there are limitations to the use of other methods. Thus, it is possible to obtain a matrix of fractions differing in molecular weight and in chemical structure.

In summary, the gel permeation chromatographic technique concentrates all of a specific functional type into one fraction, recognizing that there will be a wide range of molecular weight species in that fraction. This is especially true when the chromatographic feedstock is a whole feed rather than a distillate fraction.

5.6.4 High-Performance Liquid Chromatography

High-performance liquid chromatography (HPLC), particularly in the normal phase mode, has found great utility in separating different hydrocarbon group types and identifying specific constituent types (Colin and Vion, 1983; Miller et al., 1983). The general advantages of HPLC method are (1) each sample is analyzed *as received*; (2) the boiling range of the sample is generally immaterial; (3) the total time per analysis is usually of the order of minutes; and (4) the method can be adapted for on-stream analysis.

However, a severe shortcoming of most HPLC approaches to a hydrocarbon group type of analysis is the difficulty in obtaining accurate response factors applicable to different distillate products. Unfortunately, accuracy can be compromised when these response factors are used to analyze hydrotreated and hydrocracked materials having the same boiling range. In fact, significant changes in the hydrocarbon distribution within a certain group type cause the analytic results to be misleading for such samples because of the variation in response with carbon number exhibited by most routinely used HPLC detectors. Of particular interest is the application of the HPLC technique to the identification of the molecular types in nonvolatile feedstocks such as residua. The molecular species in the asphaltene fraction have been of interest (Colin and Vion, 1983; Felix et al., 1985) leading to identification of the size of polynuclear aromatic systems in the asphaltene constituents (Speight, 1986).

Several recent HPLC separation schemes are particularly interesting since they also incorporate detectors not usually associated with conventional hydrocarbon group types of analyses (Miller et al., 1983; Norris and Rawdon, 1984; Rawdon, 1984; Schwartz and Brownlee, 1986). The ideal detector for a truly versatile and accurate hydrocarbon group type of analysis is one that is sensitive to hydrocarbons but demonstrates a response independent of carbon number.

In general, the amount of information that can be derived from any chromatographic separation, however effective, depends on the detectors. As the field of application for HPLC has increased, the limitations of commercially available conventional detectors, such as ultraviolet/visible absorption (UV/VIS) and RI have become increasingly restrictive to the growth of the technique. This has led a search for detectors capable of producing even more information. The so-called hyphenated techniques are the outcome of this search.

5.6.5 Ion-Exchange Chromatography

Ion-exchange chromatography is widely used in the analyses of petroleum fractions for the isolation and preliminary separation of acid and basic components (Speight, 2001). This technique has the advantage of greatly improving the quality of a complex operation, but it can be a very time-consuming separation.

Ion-exchange resin constituents are prepared from aluminum silicates, synthetic resin constituents, and polysaccharides. The most widely used resin constituents have a skeletal structure of polystyrene cross-linked with varying amounts of divinylbenzene derivatives. They have a loose gel structure of cross-linked polymer chains through which the sample ions must diffuse to reach most of the exchange sites. Since ion-exchange resin constituents are usually prepared as beads that are several hundred micrometers in diameter, most of the exchange sites are located at points quite distant from the surface. Because of the polyelectrolyte nature of these organic resin constituents, they can absorb large amounts of water or solvents and swell to volumes considerably larger than the dried gel. The size of the species that can diffuse through the particle is determined by the intermolecular spacing between the polymeric chains of the three-dimensional polyelectrolyte resin.

Cation-exchange chromatography is now used primarily to isolate the nitrogen constituents in a petroleum fraction. The relative importance of these compounds in petroleum has arisen because of their deleterious effects in many petroleum refining processes. They reduce the activity of cracking and hydrocracking catalysts and contribute to gum formation, color, odor, and poor storage properties of the fuel. However, not all basic compounds isolated by cation-exchange chromatography are nitrogen compounds. Anion-exchange chromatography is used to isolate the acid components (such as carboxylic acids and phenols) from petroleum fractions.

6.0 PROPERTIES

For the purposes of this text, a *physical property* is any property that is measurable and the value of which describes the physical state of petroleum that does not change the chemical nature of petroleum. The changes in the physical properties of a system can be used to describe its transformations (or evolutions between its momentary states). Physical properties are contrasted with chemical properties, which determine the way a material behaves in a chemical reaction, especially in terms of the application of biotechnology to crude oil.

6.1 ACID NUMBER

Another characteristic of crude oil is the *total acid number* (TAN), which represents a composite of acids present in the crude oil (ASTM D664) and is often also expressed as the *neutralization number*. Crude oils having a high acid number (high TAN number) account for an increasing percentage of the global crude oil market, and, in fact, the increase in world production of heavy, sour, and high TAN crude oils will impact many world (especially US) refineries (Shafizadeh et al., 2003; Sheridan, 2006).

High acid crude oils are those with an acid content >1.0 mg KOH/g sample (many refiners consider TAN number greater than 0.5 mg if KOH/g is found to be high) and refiners looking for discounted crude supplies will import and use greater volumes of high TAN crude oils.

In the *potentiometric titration* method (ASTM D664), the sample is normally dissolved in toluene and propanol with a little water and titrated with alcoholic KOH (if sample is acidic). A glass electrode and reference electrode is immersed in the sample and connected to a voltmeter/potentiometer. The meter reading (in millivolts) is plotted against the volume of titrant. The end point is taken at the distinct inflection of the resulting titration curve corresponding to the basic buffer solution. In the *color indicating titration* method, an appropriate pH color indicator (such as phenolphthalein) is used. The titrant is added to the sample by means of a burette and the volume of titrant used to cause a permanent color change in the sample is recorded from which the TAN is calculated. It can be difficult to observe color changes in crude oil solutions. It is also possible that the results from the color indicator method may or may not be the same as the potentiometric results.

Test method ASTM D1534 is similar to ASTM D974 in that they both use a color change to indicate the end point. ASTM D1534 is designed for electric insulating oils (transformer oils), where the viscosity will not exceed 24 cSt at 40°C. The standard range of applications is for oils with an acid number between 0.05 mg KOH/g and 0.50 mg KOH/g, which is applicable to the transformer oils. Test method ASTM D3339 is also similar to ASTM D974, but is designed for use on smaller oil samples. ASTM D974 and D664 roughly use a 20-g sample; ASTM D3339 uses a 2.0-g sample.

6.2 DENSITY AND SPECIFIC GRAVITY

The *density* and *specific gravity* of crude oil (ASTM D70; ASTM D71; ASTM D287; ASTM D1217; ASTM D1298; ASTM D1480; ASTM D1481; ASTM D1555; ASTM D1657; ASTM D4052, IP 235; IP 160; IP 249; IP 365) are two properties that have found wide use in the industry for preliminary assessment of the character and quality of crude oil.

Density is the mass of a unit volume of material at a specified temperature and has the dimensions of grams per cubic centimeter (a close approximation to grams per milliliter). *Specific gravity* is the ratio of the mass of a volume of the substance to the mass of the same volume of water and is dependent on two temperatures, those at which the masses of the sample and the water are measured. When the water temperature is 4°C (39°F), the specific gravity is equal to the density in the centimeter-gram-second (cgs) system, since the volume of 1 g of water at that temperature is, by definition, 1 mL. Thus, the density of water, for example, varies with temperature, and its specific gravity at equal temperatures is always unity. The standard temperatures for a specific gravity in the petroleum industry in North America are 60/60°F (15.6/15.6°C).

In the early years of the petroleum industry, density was the principal specification for crude oil and crude oil products; it was used to give an estimation of the gasoline and, more particularly, the kerosene present in the crude oil. However, the derived relationships between the density of petroleum and its fractional composition were valid only if they were applied to a certain type of petroleum and lost some of their significance when applied to different types of petroleum. Nevertheless, density is still used to give a rough estimation of the nature of petroleum and petroleum products. Although density and specific gravity are used extensively, the API gravity is the preferred property:

Degrees API = (141.5/sp gr @ 60/60°F) − 131.5.

The density or specific gravity of petroleum, petroleum products, heavy oil, and bitumen may be measured by means of a hydrometer (ASTM D287; ASTM D1298, ASTM D1657; IP 160), a pycnometer (ASTM D70; ASTM D1217; ASTM D1480, and ASTM D1481), or by means of a digital density meter (ASTM D4052) and a digital density analyzer (ASTM D5002). Not all of these methods are suitable for measuring the density (or specific gravity of heavy oil and bitumen) although some methods lend themselves to adaptation. The API gravity of a feedstock (ASTM D287) is calculated directly from the specific gravity.

The density of petroleum usually ranges from approximately 0.8 (45.3 API) for the lighter crude oils to over 1.0 (less than 10 API) for heavy crude oil and bitumen. The variation of density with temperature, effectively the coefficient of expansion, is a property of great technical importance, since most petroleum products are sold by volume and specific gravity is usually determined at the prevailing temperature (21°C, 70°F) rather than at the standard temperature (60°F, 15.6°C). The tables of gravity corrections (ASTM D1555) are based on an assumption that the coefficient of expansion of all petroleum products is a function (at fixed temperatures) of density only.

The API gravity of crude oil is calculated directly from the specific gravity (ASTM D287). The specific gravity of bitumen shows a fairly wide range of variation. The largest degree of variation is usually due to local conditions that affect material close to the faces, or exposures, occurring in surface oil sand beds. There are also variations in the specific gravity of the bitumen found in beds that have not been exposed to weathering or other external factors. The range of specific gravity usually varies over the range of the order of 0.995–1.04.

6.3 SURFACE AND INTERFACIAL TENSION

Surface tension is a measure of the force acting at a boundary between two phases. If the boundary is between a liquid and a solid or between a liquid and a gas (air) the attractive forces are referred to as surface tension, but the attractive forces between two immiscible liquids are referred to as *interfacial tension*.

Temperature and molecular weight have a significant effect on surface tension of crude oil and crude oil products (Speight, 2014a, 2015b). For example, in the normal hydrocarbon series, a rise in temperature leads to a decrease in the surface tension, but an increase in molecular weight increases the surface tension. A similar trend, that is, an increase in molecular weight causing an increase in surface tension, also occurs in the acrylic series and, to a lesser extent, in the alkyl benzene series.

The narrow range of values (approximately 24–38 dyne/cm) for such widely diverse materials as gasoline (26 dyn/cm), kerosene (30 dyn/cm), and the lubricating fractions (34 dyn/cm) has rendered the surface tension of little value for any attempted characterization. However, it is generally acknowledged that non-hydrocarbon materials dissolved in an oil reduce the surface tension: polar compounds, such as soaps and fatty acids, are particularly active. The effect is marked at low concentrations up to a critical value, beyond which further additions cause little change; the critical value corresponds closely with that required for a monomolecular layer on the exposed surface, where it is adsorbed and accounts for the lowering.

On the other hand, although petroleum products show little variation in surface tension, within a narrow range the *interfacial tension* of petroleum, especially of petroleum products, against aqueous solutions provides valuable information (ASTM D971). Thus, the interfacial tension of petroleum is subject to the same constraints as surface tension, that is, differences in composition, molecular weight, and so on. When oil-water systems are involved, the pH of the aqueous phase influences the tension at the interface; the change is small for highly refined oils, but increasing pH causes a rapid decrease for poorly refined, contaminated, or slightly oxidized oils.

6.4 VISCOSITY

Viscosity is the force in dynes required to move a plane of $1\,cm^2$ area at a distance of 1 cm from another plane of $1\,cm^2$ area through a distance of 1 cm in 1 s. In the centimeter-gram-second (cgs) system, the unit of viscosity is the poise or centipoise (0.01 P). Two other terms in common use are *kinematic viscosity* and *fluidity*. The kinematic viscosity is the viscosity in centipoises divided by the specific gravity, and the unit is the stoke (cm^2/s), although centistokes (0.01 St) is in more common usage; fluidity is simply the reciprocal of viscosity. The viscosity (ASTM D445; ASTM D88; ASTM D2161; ASTM D341; ASTM D2270) of crude oils varies markedly over a very wide range. Values vary from less than 10 cP at room temperature to many thousands of centipoises at the same temperature.

Many types of instruments have been proposed for the determination of viscosity. The simplest and most widely used are capillary types (ASTM D445), and the viscosity is derived from the equation:

$$\mu = \pi r4P/8nl$$

where r is the tube radius, l the tube length, P the pressure difference between the ends of a capillary, and n the *coefficient of viscosity*. Not only are such capillary instruments the most simple, but when designed in accordance with known principles and used with known necessary correction factors, they are probably the most accurate viscometers available. It is usually more convenient, however, to use relative measurements, and for this purpose the instrument is calibrated with an appropriate standard liquid of known viscosity.

The Saybolt universal viscosity (SUS) (ASTM D88) is the time in seconds required for the flow of 60 mL of petroleum from a container, at constant temperature, through a calibrated orifice. The Saybolt furol viscosity (ASTM D88) is determined in a similar manner except that a larger orifice is employed.

As a result of the various methods for viscosity determination, it is not surprising that much effort has been spent on interconversion of the several scales, especially converting Saybolt to kinematic viscosity (ASTM D2161),

$$\text{Kinematic viscosity} = a \times \text{Saybolt s} + b/\text{Saybolt s}$$

In this equation, a and b are constants.

The SUS equivalent to a given kinematic viscosity varies slightly with the temperature at which the determination is made because the temperature of the calibrated receiving flask used in the Saybolt method is not the same as that of the oil. Conversion factors are used to convert kinematic viscosity from 2 to 70 cSt at 38°C (100°F) and 99°C (210°F) to equivalent SUS in seconds. Appropriate multipliers are listed to convert kinematic viscosity over 70 cSt. For a kinematic viscosity determined at any other temperature the equivalent Saybolt universal value is calculated by use of the Saybolt equivalent at 38°C (100°F) and a multiplier that varies with the temperature:

$$\text{Saybolt s at } 100°F\ (38°C) = \text{cSt} \times 4.635$$

$$\text{Saybolt s at } 210°F\ (99°C) = \text{cSt} \times 4.667$$

Various studies have also been made on the effect of temperature on viscosity since the viscosity of petroleum, or a petroleum product, decreases as the temperature increases. The rate of change appears to depend primarily on the nature or composition of the petroleum, but other factors, such as volatility, may also have a minor effect. The effect of temperature on viscosity is generally represented by the equation

$$\log \log (n+c) = A + B \log T$$

where n is absolute viscosity, T is temperature, and A and B are constants. This equation has been sufficient for most purposes and has come into very general use. The constants A and B vary widely with different oils, but c remains fixed at 0.6 for all oils having a viscosity over 1.5 cSt; it increases only slightly at lower viscosity (0.75 at 0.5 cSt). The viscosity-temperature characteristics of any oil, so plotted, thus create a straight line, and the parameters A and B are equivalent to the intercept and slope of the line. To express the viscosity and viscosity-temperature characteristics of an oil, the slope and the viscosity at one temperature must be known; the usual practice is to select 38°C (100°F) and 99°C (210°F) as the observation temperatures.

Suitable conversion tables are available (ASTM D341), and each table or chart is constructed in such a way that for any given petroleum or petroleum product the viscosity-temperature points result in a straight line over the applicable temperature range.

Since the viscosity-temperature coefficient of lubricating oil is an important expression of its suitability, a convenient number to express this property is very useful, and hence, a viscosity index (ASTM D2270) was derived. It is established that naphthenic oils have higher viscosity-temperature coefficients than do paraffinic oils at equal viscosity and temperatures. The Dean and Davis scale was based on the assignment of a zero value to a typical naphthenic crude oil and that of 100 to a typical paraffinic crude oil; intermediate oils were rated by the formula:

$$\text{Viscosity index} = (L - U) / (L - H \times 100)$$

where L and H are the viscosities of the zero and 100 index reference oils, both having the same viscosity at 99°C (210°F), and U is that of the unknown, all at 38°C (100°F). Originally the viscosity index was calculated from Saybolt viscosity data, but subsequently data were provided for kinematic viscosity.

6.5 VOLATILITY

The volatility of a liquid or liquefied gas may be defined as its tendency to vaporize, that is, to change from the liquid to the vapor or gaseous state. Because one of the three essentials for combustion in a flame is that the fuel be in the gaseous state, volatility is a primary characteristic of liquid fuels.

Before any volatility tests are carried out it must be recognized that the presence of more than 0.5% water in test samples of crude oil can cause several problems during various test procedures and produce erroneous results. For example, during various thermal tests, water (which has a high heat of vaporization) requires the application of additional thermal energy to the distillation flask. In addition, water is relatively easily superheated and therefore excessive *bumping* can occur, leading to erroneous readings and the potential for destruction of the glass equipment is real. Steam formed during distillation can act as a carrier gas and high boiling point components may end up in the distillate (often referred to as *steam distillation*).

Removal of water (and sediment) can be achieved by centrifugation if the sample is not a tight emulsion. Other methods that are used to remove water include: (1) heating in a pressure vessel to control loss of light ends, (2) addition of calcium chloride as recommended in ASTM D1160, (3) addition of an azeotroping agent such as *iso*-propanol or *n*-butanol, (4) removal of water in a preliminary low-efficiency or flash distillation followed by re-blending the hydrocarbon which co-distills with the water into the sample (see also IP 74), and (5) separation of the water from the hydrocarbon distillate by freezing.

The vaporizing tendencies of petroleum and petroleum products are the basis for the general characterization of liquid petroleum fuels, such as liquefied petroleum gas, natural gasoline, motor and aviation gasoline, naphtha, kerosene, gas oil, diesel fuel, and fuel oil (ASTM D2715). A test (ASTM D6) also exists for determining the loss of material when crude oil and asphaltic compounds are heated. Another test (ASTM D20) is a method for the distillation of road tars that might also be applied to estimating the volatility of high molecular weight residues.

The *flash point* of petroleum or a petroleum product is the temperature to which the product must be heated under specified conditions to give of sufficient vapor to form a mixture with air that can be ignited momentarily by a specified flame (ASTM D56; ASTM D92; ASTM D93). The *fire point* is the temperature to which the product must be heated under the prescribed conditions of the method to burn continuously when the mixture of vapor and air is ignited by a specified flame (ASTM D92).

A further aspect of volatility that receives considerable attention is the vapor pressure of petroleum and its constituent fractions. The *vapor pressure* is the force exerted on the walls of a closed container by the vaporized portion of a liquid. Conversely, it is the force that must be exerted on the liquid to prevent it from vaporizing further (ASTM D323). The vapor pressure increases with temperature for any given gasoline, liquefied petroleum gas, or other product. The temperature at which the vapor pressure of a liquid, either a pure compound or a mixture of many compounds, equals 1 atm (14.7 psi, absolute) is designated as the boiling point of the liquid.

In each homologous series of hydrocarbons, the boiling points increase with molecular weight, and structure also has a marked influence since it is a general rule that branched paraffin isomers have lower boiling points than the corresponding *n*-alkane. In any given series, steric effects notwithstanding, there is an increase in boiling point with an increase in carbon number of the alkyl side chain. This particularly applies to alkyl aromatic compounds where alkyl-substituted aromatic compounds can have higher boiling points than polycondensed aromatic systems. And this fact is very meaningful when attempts are made to develop hypothetical structures for asphaltene constituents (Speight, 1994, 2014a).

As an early part of characterization studies, a correlation was observed between the quality of petroleum products and the hydrogen content since gasoline, kerosene, diesel fuel, and lubricating oil are made up of hydrocarbon constituents containing high proportions of hydrogen. Thus, it is not surprising that test to determine the volatility of petroleum and petroleum products was among the first to be defined. Indeed, volatility is one of the major tests for petroleum products, and it is inevitable that all products will, at some stage of their history, be tested for volatility characteristics.

Distillation involves the general procedure of vaporizing the petroleum liquid in a suitable flask either at *atmospheric pressure* (ASTM D86; ASTM D2892) or at *reduced pressure* (ASTM D1160), and the data are reported in terms of one or more of the following seven items:

1. *Initial boiling point* is the thermometer reading in the neck of the distillation flask when the first drop of distillate leaves the tip of the condenser tube. This reading is materially affected by a number of test conditions, namely, room temperature, rate of heating, and condenser temperature.
2. *Distillation temperatures* are usually observed when the level of the distillate reaches each 10% mark on the graduated receiver, with the temperatures for the 5% and 95% marks often included. Conversely, the volume of the distillate in the receiver, that is, the percentage recovered, is often observed at specified thermometer readings.
3. *End-point* or *maximum temperature* is the highest thermometer reading observed during distillation. In most cases it is reached when the entire sample has been vaporized. If a liquid residue remains in the flask after the maximum permissible adjustments are made in heating rate, this is recorded as indicative of the presence of very high boiling compounds.
4. *Dry point* is the thermometer reading at the instant the flask becomes dry and is used for special purposes, such as for solvents and for relatively pure hydrocarbons. For these purposes dry point is considered more indicative of the final boiling point than end point or maximum temperature.

5. *Recovery* is the total volume of distillate recovered in the graduated receiver and *residue* is the liquid material, mostly condensed vapors, left in the flask after it has been allowed to cool at the end of distillation. The residue is measured by transferring it to an appropriate small graduated cylinder. Low or abnormally high residues indicate the absence or presence, respectively, of high-boiling components.

6. *Total recovery* is the sum of the liquid recovery and residue; *distillation loss* is determined by subtracting the total recovery from 100%. It is, of course, the measure of the portion of the vaporized sample that does not condense under the conditions of the test. Like the initial boiling point, distillation loss is affected materially by a number of test conditions, namely, condenser temperature, sampling and receiving temperatures, barometric pressure, heating rate in the early part of the distillation, and others. Provisions are made for correcting high distillation losses for the effect of low barometric pressure because of the practice of including distillation loss as one of the items in some specifications for motor gasoline.

7. *Percentage evaporated* is, the percentage recovered at a specific thermometer reading or other distillation temperatures, or the converse. The amounts that have been evaporated are usually obtained by plotting observed thermometer readings against the corresponding observed recoveries plus, in each case, the distillation loss. The initial boiling point is plotted with the distillation loss as the percentage evaporated. Distillation data are considerably reproducible, particularly for the more volatile products.

There is also another method that is increasing in popularity for application to a variety of feedstocks and that is the method commonly known as *simulated distillation* (ASTM D2887) (Carbognani et al., 2012). The method has been well researched in terms of method development and application (Romanowski and Thomas, 1985; MacAllister and DeRuiter, 1985; Schwartz et al., 1987; Neer and Deo, 1995). The benefits of the technique include good comparisons with other ASTM distillation data, as well as the application to higher boiling fractions of petroleum. In fact, data output include the provision of the corresponding Engler profile (ASTM D86), as well as the prediction of other properties such as vapor pressure and flash point. When it is necessary to monitor product properties, as is often the case during refining operations, such data provide a valuable aid to process control and on-line product testing.

For a more detailed distillation analysis of feedstocks and products, a low-resolution, temperature-programmed gas chromatographic analysis has been developed to simulate the time-consuming true boiling point distillation. The method relies on the general observation that hydrocarbons are eluted from a nonpolar adsorbent in the order of their boiling points. The regularity of the elution order of the hydrocarbon components allows the retention times to be equated to distillation temperatures and the term *simulated distillation by gas chromatography* (or *simdis*) is used throughout the industry to refer to this technique.

Simulated distillation by gas chromatography is often applied in the petroleum industry to obtain true boiling point data for distillates and crude oils (Speight, 2001). Two standardized methods (ASTM D2887; ASTM D3710) are available for the boiling point determination of petroleum fractions and gasoline, respectively. The ASTM D2887 method utilizes nonpolar, packed gas chromatographic columns in conjunction with flame ionization detection. The upper limit of the boiling range covered by this method is approximately 540°C (1000°F) atmospheric equivalent boiling point. Recent efforts in which high temperature gas chromatography were used have focused on extending the scope of the ASTM D2887 method for higher boiling petroleum materials to 800°C (1470°F) atmospheric equivalent boiling point.

REFERENCES

Abraham, H., 1945. Asphalts and Allied Substances. Van Nostrand Scientific Publishers, New York.

Albaiges, J., Albrecht, P., 1979. Fingerprinting marine pollutant hydrocarbons by computerized gas chromatography-mass spectrometry. International Journal of Environmental Analytical Chemistry 6 (2), 171–190.

ASTM D1018, 2017. Standard Test Method for Hydrogen in Petroleum Fractions. Annual Book of Standards. ASTM International, West Conshohocken, Pennsylvania.

ASTM D1160, 2017. Standard Test Method for Distillation of Petroleum Products at Reduced Pressure. Annual Book of Standards. ASTM International, West Conshohocken, Pennsylvania.

ASTM D1217, 2017. Standard Test Method for Density and Relative Density (Specific Gravity) of Liquids by Bingham Pycnometer. Annual Book of Standards. ASTM International, West Conshohocken, Pennsylvania.

ASTM D1266, 2017. Standard Test Method for Sulfur in Petroleum Products (Lamp Method). Annual Book of Standards. ASTM International, West Conshohocken, Pennsylvania.

ASTM D1298, 2017. Standard Test Method for Density, Relative Density, or API Gravity of Crude Petroleum and Liquid Petroleum Products by Hydrometer Method. Annual Book of Standards. ASTM International, West Conshohocken, Pennsylvania.

ASTM D1318, 2017. Standard Test Method for Sodium in Residual Fuel Oil (Flame Photometric Method). Annual Book of Standards. ASTM International, West Conshohocken, Pennsylvania.

ASTM D1319, 2017. Standard Test Method for Hydrocarbon Types in Liquid Petroleum Products by Fluorescent Indicator Adsorption. Annual Book of Standards. ASTM International, West Conshohocken, Pennsylvania.

ASTM D1480, 2017. Standard Test Method for Density and Relative Density (Specific Gravity) of Viscous Materials by Bingham Pycnometer. Annual Book of Standards. ASTM International, West Conshohocken, Pennsylvania.

ASTM D1481, 2017. Standard Test Method for Density and Relative Density (Specific Gravity) of Viscous Materials by Lipkin Bicapillary Pycnometer. Annual Book of Standards. ASTM International, West Conshohocken, Pennsylvania.

ASTM D1534, 2017. Standard Test Method for Approximate Acidity in Electrical Insulating Oils by Color-Indicator Titration. Annual Book of Standards. ASTM International, West Conshohocken, Pennsylvania.

ASTM D1552, 2017. Standard Test Method for Sulfur in Petroleum Products (High-Temperature Method). Annual Book of Standards. ASTM International, West Conshohocken, Pennsylvania.

ASTM D1555, 2017. Standard Test Method for Calculation of Volume and Weight of Industrial Aromatic Hydrocarbons and Cyclohexane. Annual Book of Standards. ASTM International, West Conshohocken, Pennsylvania.

ASTM D1657, 2017. Standard Test Method for Density or Relative Density of Light Hydrocarbons by Pressure Hydrometer. Annual Book of Standards. ASTM International, West Conshohocken, Pennsylvania.

ASTM D1757, 2017. Standard Test Method for Sulfur in Ash from Coal and Coke. Annual Book of Standards. ASTM International, West Conshohocken, Pennsylvania.

ASTM D1840, 2017. Standard Test Method for Naphthalene Hydrocarbons in Aviation Turbine Fuels by Ultraviolet Spectrophotometry. Annual Book of Standards. ASTM International, West Conshohocken, Pennsylvania.

ASTM D20, 2017. Standard Test Method for Distillation of Road Tars. Annual Book of Standards. ASTM International, West Conshohocken, Pennsylvania.

ASTM D2007, 2017. Standard Test Method for Characteristic Groups in Rubber Extender and Processing Oils and Other Petroleum-Derived Oils by the Clay-Gel Absorption Chromatographic Method. Annual Book of Standards. ASTM International, West Conshohocken, Pennsylvania.

ASTM D2161, 2017. Standard Practice for Conversion of Kinematic Viscosity to Saybolt Universal Viscosity or to Saybolt Furol Viscosity. Annual Book of Standards. ASTM International, West Conshohocken, Pennsylvania.

ASTM D2163, 2017. Standard Test Method for Determination of Hydrocarbons in Liquefied Petroleum (LP) Gases and Propane/Propene Mixtures by Gas Chromatography. Annual Book of Standards. ASTM International, West Conshohocken, Pennsylvania.

ASTM D2268, 2017. Standard Test Method for Analysis of High-purity N-Heptane and Isooctane by Capillary Gas Chromatography. Annual Book of Standards. ASTM International, West Conshohocken, Pennsylvania.

ASTM D2269, 2015. Standard Test Method for Evaluation of White Mineral Oils by Ultraviolet Absorption. Annual Book of Standards. ASTM International, West Conshohocken, Pennsylvania.

ASTM D2270, 2017. Standard Practice for Calculating Viscosity Index from Kinematic Viscosity at 40 and 100°C. Annual Book of Standards. ASTM International, West Conshohocken, Pennsylvania.

ASTM D2425, 2017. Standard Test Method for Hydrocarbon Types in Middle Distillates by Mass Spectrometry. Annual Book of Standards. ASTM International, West Conshohocken, Pennsylvania.

ASTM D2427, 2017. Standard Test Method for Determination of C2 through C5 Hydrocarbons in Gasolines by Gas Chromatography. Annual Book of Standards. ASTM International, West Conshohocken, Pennsylvania.

ASTM D2504, 2017. Standard Test Method for Non-condensable Gases in C2 and Lighter Hydrocarbon Products by Gas Chromatography. Annual Book of Standards. ASTM International, West Conshohocken, Pennsylvania.

ASTM D2505, 2017. Standard Test Method for Ethylene, Other Hydrocarbons, and Carbon Dioxide in High-Purity Ethylene by Gas Chromatography. Annual Book of Standards. ASTM International, West Conshohocken, Pennsylvania.

ASTM D2593, 2017. Standard Test Method for Butadiene Purity and Hydrocarbon Impurities by Gas Chromatography. Annual Book of Standards. ASTM International, West Conshohocken, Pennsylvania.

ASTM D2597, 2017. Standard Test Method for Analysis of Demethanized Hydrocarbon Liquid Mixtures Containing Nitrogen and Carbon Dioxide by Gas Chromatography. Annual Book of Standards. ASTM International, West Conshohocken, Pennsylvania.

ASTM D2622, 2017. Standard Test Method for Sulfur in Petroleum Products by Wavelength Dispersive X-ray Fluorescence Spectrometry. Annual Book of Standards. ASTM International, West Conshohocken, Pennsylvania.

ASTM D2712, 2012. Standard Test Method for Hydrocarbon Traces in Propylene Concentrates by Gas Chromatography. Annual Book of Standards. ASTM International, West Conshohocken, Pennsylvania.

ASTM D2715, 2017. Standard Test Method for Volatilization Rates of Lubricants in Vacuum. Annual Book of Standards. ASTM International, West Conshohocken, Pennsylvania.

ASTM D2786, 2017. Standard Test Method for Hydrocarbon Types Analysis of Gas-Oil Saturates Fractions by High Ionizing Voltage Mass Spectrometry. Annual Book of Standards. ASTM International, West Conshohocken, Pennsylvania.

ASTM D287, 2017. Standard Test Method for API Gravity of Crude Petroleum and Petroleum Products (Hydrometer Method). Annual Book of Standards. ASTM International, West Conshohocken, Pennsylvania.

ASTM D2887, 2017. Standard Test Method for Boiling Range Distribution of Petroleum Fractions by Gas Chromatography. Annual Book of Standards. ASTM International, West Conshohocken, Pennsylvania.

ASTM D2892, 2017. Standard Test Method for Distillation of Crude Petroleum (15-Theoretical Plate Column). Annual Book of Standards. ASTM International, West Conshohocken, Pennsylvania.

ASTM D3177, 2017. Standard Test Methods for Total Sulfur in the Analysis Sample of Coal and Coke. Annual Book of Standards. ASTM International, West Conshohocken, Pennsylvania.

ASTM D3178, 2017. Standard Test Methods for Carbon and Hydrogen in the Analysis Sample of Coal and Coke. Annual Book of Standards. ASTM International, West Conshohocken, Pennsylvania.

ASTM D3179, 2017. Standard Test Methods for Nitrogen in the Analysis Sample of Coal and Coke. Annual Book of Standards. ASTM International, West Conshohocken, Pennsylvania.

ASTM D3228, 2017. Standard Test Method for Total Nitrogen in Lubricating Oils and Fuel Oils by Modified Kjeldahl Method. Annual Book of Standards. ASTM International, West Conshohocken, Pennsylvania.

ASTM D323, 2017. Standard Test Method for Vapor Pressure of Petroleum Products (Reid Method). Annual Book of Standards. ASTM International, West Conshohocken, Pennsylvania.

ASTM D3237, 2017. Standard Test Method for Lead in Gasoline by Atomic Absorption Spectroscopy. Annual Book of Standards. ASTM International, West Conshohocken, Pennsylvania.

ASTM D3239, 2017. Standard Test Method for Aromatic Types Analysis of Gas-Oil Aromatic Fractions by High Ionizing Voltage Mass Spectrometry. Annual Book of Standards. ASTM International, West Conshohocken, Pennsylvania.

ASTM D3339, 2017. Standard Test Method for Acid Number of Petroleum Products by Semi-micro Color Indicator Titration. Annual Book of Standards. ASTM International, West Conshohocken, Pennsylvania.

ASTM D3340, 2017. Standard Test Method for Lithium and Sodium in Lubricating Greases by Flame Photometer. Annual Book of Standards. ASTM International, West Conshohocken, Pennsylvania.

ASTM D3341, 2017. Standard Test Method for Lead in Gasoline – Iodine Monochloride Method. Annual Book of Standards. ASTM International, West Conshohocken, Pennsylvania.

ASTM D3343, 2017. Standard Test Method for Estimation of Hydrogen Content of Aviation Fuels. Annual Book of Standards. ASTM International, West Conshohocken, Pennsylvania.

ASTM D341, 2017. Standard Practice for Viscosity-temperature Charts for Liquid Petroleum Products. Annual Book of Standards. ASTM International, West Conshohocken, Pennsylvania.

ASTM D3525, 2017. Standard Test Method for Gasoline Diluent in Used Gasoline Engine Oils by Gas Chromatography. Annual Book of Standards. ASTM International, West Conshohocken, Pennsylvania.

ASTM D3605, 2017. Standard Test Method for Trace Metals in Gas Turbine Fuels by Atomic Absorption and Flame Emission Spectroscopy. Annual Book of Standards. ASTM International, West Conshohocken, Pennsylvania.

ASTM D3606, 2017. Standard Test Method for Determination of Benzene and Toluene in Finished Motor and Aviation Gasoline by Gas Chromatography. Annual Book of Standards. ASTM International, West Conshohocken, Pennsylvania.

ASTM D3701, 2017. Standard Test Method for Hydrogen Content of Aviation Turbine Fuels by Low Resolution Nuclear Magnetic Resonance Spectrometry. Annual Book of Standards. ASTM International, West Conshohocken, Pennsylvania.

ASTM D3710, 2017. Standard Test Method for Boiling Range Distribution of Gasoline and Gasoline Fractions by Gas Chromatography. Annual Book of Standards. ASTM International, West Conshohocken, Pennsylvania.

ASTM D3831, 2017. Standard Test Method for Manganese in Gasoline by Atomic Absorption Spectroscopy. Annual Book of Standards. ASTM International, West Conshohocken, Pennsylvania.

ASTM D4045, 2017. Standard Test Method for Sulfur in Petroleum Products by Hydrogenolysis and Rateometric Colorimetry. Annual Book of Standards. ASTM International, West Conshohocken, Pennsylvania.

ASTM D4046, 2017. Standard Test Method for Alkyl Nitrate in Diesel Fuels by Spectrophotometry. Annual Book of Standards. ASTM International, West Conshohocken, Pennsylvania.

ASTM D4052, 2017. Standard Test Method for Density, Relative Density, and API Gravity of Liquids by Digital Density Meter. Annual Book of Standards. ASTM International, West Conshohocken, Pennsylvania.

ASTM D4053, 2017. Standard Test Method for Benzene in Motor and Aviation Gasoline by Infrared Spectroscopy. Annual Book of Standards. ASTM International, West Conshohocken, Pennsylvania.

ASTM D4124, 2017. Standard Test Method for Separation of Asphalt into Four Fractions. Annual Book of Standards. ASTM International, West Conshohocken, Pennsylvania.

ASTM D4294, 2017. Standard Test Method for Sulfur in Petroleum and Petroleum Products by Energy Dispersive X-ray Fluorescence Spectrometry. Annual Book of Standards. ASTM International, West Conshohocken, Pennsylvania.

ASTM D4424, 2017. Standard Test Method for Butylene Analysis by Gas Chromatography. Annual Book of Standards. ASTM International, West Conshohocken, Pennsylvania.

ASTM D445, 2017. Standard Test Method for Kinematic Viscosity of Transparent and Opaque Liquids (And Calculation of Dynamic Viscosity). Annual Book of Standards. ASTM International, West Conshohocken, Pennsylvania.

ASTM D4628, 2017. Standard Test Method for Analysis of Barium, Calcium, Magnesium, and Zinc in Unused Lubricating Oils by Atomic Absorption Spectrometry. Annual Book of Standards. ASTM International, West Conshohocken, Pennsylvania.

ASTM D4815, 2017. Standard Test Method for Determination of MTBE, ETBE, TAME, DIPE, Tertiary-Amyl Alcohol and C1 to C4 Alcohols in Gasoline by Gas Chromatography. Annual Book of Standards. ASTM International, West Conshohocken, Pennsylvania.

ASTM D4864, 2017. Standard Test Method for Determination of Traces of Methanol in Propylene Concentrates by Gas Chromatography. Annual Book of Standards. ASTM International, West Conshohocken, Pennsylvania.

ASTM D5002, 2017. Standard Test Method for Density and Relative Density of Crude Oils by Digital Density Analyzer. Annual Book of Standards. ASTM International, West Conshohocken, Pennsylvania.

ASTM D5134, 2017. Standard Test Method for Detailed Analysis of Petroleum Naphtha through N-Nonane by Capillary Gas Chromatography. Annual Book of Standards. ASTM International, West Conshohocken, Pennsylvania.

ASTM D5303, 2017. Standard Test Method for Trace Carbonyl Sulfide in Propylene by Gas Chromatography. Annual Book of Standards. ASTM International, West Conshohocken, Pennsylvania.

ASTM D5307, 2017. Standard Test Method for Determination of Boiling Range Distribution of Crude Petroleum by Gas Chromatography. Annual Book of Standards. ASTM International, West Conshohocken, Pennsylvania.

ASTM D5441, 2017. Standard Test Method for Analysis of Methyl Tert-butyl Ether (MTBE) by Gas Chromatography. Annual Book of Standards. ASTM International, West Conshohocken, Pennsylvania.

ASTM D5442, 2017. Standard Test Method for Analysis of Petroleum Waxes by Gas Chromatography. Annual Book of Standards. ASTM International, West Conshohocken, Pennsylvania.

ASTM D5443, 2017. Standard Test Method for Paraffin, Naphthene, and Aromatic Hydrocarbon Type Analysis in Petroleum Distillates through 200°C by Multi-dimensional Gas Chromatography. Annual Book of Standards. ASTM International, West Conshohocken, Pennsylvania.

ASTM D5501, 2017. Standard Test Method for Determination of Ethanol and Methanol Content in Fuels Containing Greater than 20% Ethanol by Gas Chromatography. Annual Book of Standards. ASTM International, West Conshohocken, Pennsylvania.

ASTM D5580, 2017. Standard Test Method for Determination of Benzene, Toluene, Ethylbenzene, P/m-Xylene, O-Xylene, C9 and Heavier Aromatics, and Total Aromatics in Finished Gasoline by Gas Chromatography. Annual Book of Standards. ASTM International, West Conshohocken, Pennsylvania.

ASTM D5599, 2017. Standard Test Method for Determination of Oxygenates in Gasoline by Gas Chromatography and Oxygen Selective Flame Ionization Detection. Annual Book of Standards. ASTM International, West Conshohocken, Pennsylvania.

ASTM D56, 2017. Standard Test Method for Flash Point by Tag Closed Cup Tester. Annual Book of Standards. ASTM International, West Conshohocken, Pennsylvania.

ASTM D5623, 2017. Standard Test Method for Sulfur Compounds in Light Petroleum Liquids by Gas Chromatography and Sulfur Selective Detection. Annual Book of Standards. ASTM International, West Conshohocken, Pennsylvania.

ASTM D5845, 2017. Standard Test Method for Determination of MTBE, ETBE, TAME, DIPE, Methanol, Ethanol and T-Butanol in Gasoline by Infrared Spectroscopy. Annual Book of Standards. ASTM International, West Conshohocken, Pennsylvania.

ASTM D5986, 2017. Standard Test Method for Determination of Oxygenates, Benzene, Toluene, C8–C12 Aromatics and Total Aromatics in Finished Gasoline by Gas Chromatography/Fourier Transform Infrared Spectroscopy. Annual Book of Standards. ASTM International, West Conshohocken, Pennsylvania.

ASTM D6, 2017. Standard Test Method for Loss on Heating of Oil and Asphaltic Compounds. Annual Book of Standards. ASTM International, West Conshohocken, Pennsylvania.

ASTM D6159, 2017. Standard Test Method for Determination of Hydrocarbon Impurities in Ethylene by Gas Chromatography. Annual Book of Standards. ASTM International, West Conshohocken, Pennsylvania.

ASTM D664, 2017. Standard Test Method for Acid Number of Petroleum Products by Potentiometric Titration. Annual Book of Standards. ASTM International, West Conshohocken, Pennsylvania.

ASTM D70, 2017. Standard Test Method for Density of Semi-solid Bituminous Materials (Pycnometer Method). Annual Book of Standards. ASTM International, West Conshohocken, Pennsylvania.

ASTM D71, 2017. Standard Test Method for Relative Density of Solid Pitch and Asphalt (Displacement Method). Annual Book of Standards. ASTM International, West Conshohocken, Pennsylvania.

ASTM D86, 2017. Standard Test Method for Distillation of Petroleum Products at Atmospheric Pressure. Annual Book of Standards. ASTM International, West Conshohocken, Pennsylvania.

ASTM D88, 2017. Standard Test Method for Saybolt Viscosity. Annual Book of Standards. ASTM International, West Conshohocken, Pennsylvania.

ASTM D92, 2015. Standard Test Method for Flash and Fire Points by Cleveland Open Cup Tester. Annual Book of Standards. ASTM International, West Conshohocken, Pennsylvania.

ASTM D93, 2017. Standard Test Methods for Flash Point by Pensky-Martens Closed Cup Tester. Annual Book of Standards. ASTM International, West Conshohocken, Pennsylvania.

ASTM D971, 2017. Standard Test Method for Interfacial Tension of Oil Against Water by the Ring Method. Annual Book of Standards. ASTM International, West Conshohocken, Pennsylvania.

ASTM D974, 2017. Standard Test Method for Acid and Base Number by Color-indicator Titration. Annual Book of Standards. ASTM International, West Conshohocken, Pennsylvania.

ASTM E258, 2017. Standard Test Method for Total Nitrogen in Organic Materials by Modified Kjeldahl Method. Annual Book of Standards. ASTM International, West Conshohocken, Pennsylvania.

ASTM E385, 2017. Standard Test Method for Oxygen Content Using a 14-MeV Neutron Activation and Direct-counting Technique. Annual Book of Standards. ASTM International, West Conshohocken, Pennsylvania.

ASTM E777, 2017. Standard Test Method for Carbon and Hydrogen in the Analysis Sample of Refuse-derived Fuel. Annual Book of Standards. ASTM International, West Conshohocken, Pennsylvania.

ASTM E778, 2017. Standard Test Methods for Nitrogen in Refuse-derived Fuel Analysis Samples. Annual Book of Standards. ASTM International, West Conshohocken, Pennsylvania.

Atlas, R.M., 1981. Microbial degradation of petroleum hydrocarbons: an environmental perspective. Microbiological Reviews 45, 180–209.

Baltus, R.E., Anderson, J.L., 1984. Comparison of GPC elution and diffusion coefficients of asphaltenes. Fuel 63, 530.

Berger, B.D., Anderson, K.E., 1978. Modern Petroleum, a Basic Primer of the Industry. PennWell Books, Tulsa, OK, USA.

Boonyatumanond, R., Wattayakorn, G., Togo, A., Takada, H., 2006. Distribution and origins of polycyclic aromatic hydrocarbons (PAHs) in riverine, estuarine, and marine sediments in Thailand. Marine Pollution Bulletin 52 (8), 942–956.

BP, June 2016. BP Statistical Review of World Energy. British Petroleum Company, London, United Kingdom.

Carbognani, L., 1997. Fast monitoring of C_{20}-C_{160} crude oil alkanes by size-exclusion chromatography-evaporative light scattering detection performed with silica columns. Journal of Chromatography A 788, 63–73.

Carbognani, L., Díaz-Gómez, L., Oldenburg, T.B.P., Pereira-Almao, P., 2012. Determination of molecular masses for petroleum distillates by simulated distillation. CT&F – Ciencia, Tecnología y Futuro 4 (5), 43–55.

Chugh, S., Baker, R., Telesford, A., Zhang, E., 2000. Mainstream options for heavy oil: Part I – Cold production. Journal of Canadian Petroleum Technology 39 (4), 31–39.

Clark, R.C., Brown, D.W., 1977. Petroleum properties and analysis in biotic and abiotic systems. In: Malins, D.C. (Ed.), Effects of Petroleum on Arctic and Subarctic, Marine Environments and Organisms. Nature and Fate of Petroleum, vol. I. Academic Press, New York, pp. 1–89.

Cobb, C., Goldwhite, H., 1995. Creations of Fire: Chemistry's Lively History from Alchemy to the Atomic Age. Plenum Press, New York.

Colin, J.M., Vion, G., 1983. Routine hydrocarbon group-type analysis in refinery laboratories by high-performance liquid chromatography. Journal of Chromatography 280, 152–158.

Colombo, J.C., Barreda, A., Bilos, C., Cappelletti, N., Demichelis, S., Lombardi, P., Migoya, M.C., Skorupka, C., Suárez, G., 2005. Oil spill in the Rio de la Plata estuary, Argentina: 1. Biogeochemical assessment of waters, sediments, soils and biota. Environmental Pollution 134 (2), 277–289.

Delbianco, A., Montanari, R., 2009. Encyclopedia of hydrocarbons. In: New Developments: Energy, Transport, Sustainability. vol. III. Eni S.P.A., Rome, Italy.

El-Gendy, N.Sh., 2004. Biodesulfurization Potentials of Crude Oil by Bacteria Isolated from Hydrocarbon Polluted Environments in Egypt. A (Ph.D. thesis). Cairo University, Egypt.

El-Gendy, N.Sh., Speight, J.G., 2015. Handbook of Refinery Desulfurization. CRC Press, Taylor & Francis Group, Boca Raton, Florida.

Felix, G., Bertrand, C., Van Gastel, F., 1985. Hydroprocessing of heavy oils and residua. Chromatographia 20, 155–160.

Forbes, R.J., 1958a. A History of Technology. Oxford Univ. Press, Oxford, United Kingdom.

Forbes, R.J., 1958b. Studies in Early Petroleum Chemistry. E.J. Brill Publishers, Leiden, Netherlands.

Forbes, R.J., 1959. More Studies in Early Petroleum Chemistry. E.J. Brill Publishers, Leiden, Netherlands.

Forbes, R.J., 1964. Studies in Ancient Technology. E.J. Brill Publishers, Leiden, Netherlands.

Gary, J.G., Handwerk, G.E., Kaiser, M.J., 2007. Petroleum Refining: Technology and Economics, fifth ed. CRC Press, Taylor & Francis Group, Boca Raton, Florida.

Ghoshal, S., Sainik, V., 2013. Monitor and minimize corrosion in high-TAN crude processing. Hydrocarbon Processing 92 (3), 35–38.

Hoiberg, A.J., 1964. Bituminous Materials: Asphalts, Tars, and Pitches. John Wiley & Sons Inc., New York.

Hsu, C.S., Robinson, P.R., 2006. Practical Advances in Petroleum Processing, vols. 1 and 2. Springer, New York.

Jacobs, F.-S., Filby, R.H., 1983. Liquid chromatographic fractionation of oil sand and crude oil asphaltenes. Fuel 62, 1186–1192.

Kane, R.D., Cayard, M.S., 2002. A Comprehensive Study on Naphthenic Acid Corrosion. Corrosion 2002. NACE International, Houston, Texas.

Long, R.B., Speight, J.G., 1998. The composition of petroleum. In: Speight, J.G. (Ed.), Petroleum Chemistry and Refining. Taylor & Francis, Washington, DC (Chapter 1).

MacAllister, D.J., DeRuiter, R.A., 1985. Further development and application of simulated distillation for enhanced oil recovery. Paper No. SPE 14335. In: 60th Annual Technical Conference. Society of Petroleum Engineers. Las Vegas. September 22–25, 1985.

Maini, B.B., 1999. Foamy oil flow in primary production of heavy oil under solution gas drive. Paper No. SPE 56541. In: Proceedings. Annual Technical Conference and Exhibition, Houston, Texas USA, October 3–6, 1999. Society of Petroleum Engineers, Richardson, Texas.

Maini, B.B., October 2001. Foamy Oil Flow. Paper No. SPE 68885. SPE J. Pet. Tech., Distinguished Authors Series. Society of Petroleum Engineers, Richardson, Texas, pp. 54–64.

Maskaoui, K., Zhou, J.L., Hong, H.S., Zhang, Z.L., 2002. Contamination by polycyclic aromatic hydrocarbons in the Jiulong river estuary and Western Xiamen Sea, China. Environmental Pollution 118, 109–122.

Mayes, J.M., 2015. What are the possible impacts on US refineries processing shale oils? Hydrocarbon Processing 94 (2), 67–70.

Meyer, R.F., De Witt Jr., W., 1990. Definition and World Resources of Natural Bitumens. Bulletin No. 1944. US Geological Survey, Reston, Virginia.

Miller, R.L., Ettre, L.S., Johansen, N.G., 1983. Quantitative analysis of hydrocarbons by structural group type in gasoline and distillates. Part II. Journal of Chromatography 259, 393.

Murgich, J., Abanero, A.J., Strausz, P.O., 1999. Molecular recognition in aggregates formed by asphaltene and resin molecules from the Athabasca oil sand. Energy Fuel 13 (2), 278–286.

Neer, L.A., Deo, M.D., 1995. Simulated distillation of oils with a wide carbon number distribution. Journal of Chromatographic Science 33, 133–138.

Norris, T.A., Rawdon, M.G., 1984. Determination of hydrocarbon types in petroleum liquids by supercritical fluid chromatography with flame ionization detection. Analytical Chemistry 56, 1767–1769.

Ohmes, R., 2014. Characterizing and Tracking Contaminants in Opportunity Crudes. Digital Refining.

Parkash, S., 2003. Refining Processes Handbook. Gulf Professional Publishing, Elsevier, Amsterdam, Netherlands.

Pfeiffer, J.H., 1950. The Properties of Asphaltic Bitumen. Elsevier, Amsterdam, Netherlands.

Poon, D., Kisman, K., 1992. Non-newtonian effects on the primary production of heavy oil reservoirs. Journal of Canadian Petroleum Technology 31 (7), 1–6.

Qiao, M., Wang, C., Huang, S., Wang, D., Wang, Z., 2006. Composition, sources, and potential toxicological significance of PAHs in the surface sediments of the Meiliang Bay, Taihu lake, China. Environment International 32, 28–33.

Rawdon, M., 1984. Modified flame ionization detector for supercritical fluid chromatography. Analytical Chemistry 56, 831–832.

Reynolds, J.G., 1998. Metals and heteroatoms in heavy crude oils. In: Speight, J.G. (Ed.), Petroleum Chemistry and Refining. Taylor & Francis Publishers, Washington, DC (Chapter 3).

Reynolds, J.G., Biggs, W.R., 1988. Analysis of residuum demetallation by size exclusion chromatography with element specific detection. Fuel Science and Technology International 6, 329.

Romanowski, L.J., Thomas, K.P., 1985. Steamflooding of Preheated Tar Sand Report No. DOE/FE/60177–62326. United States Department of Energy, Washington, DC.

Samanta, S., Singh, O.V., Jain, R.K., 2002. Polycyclic aromatic hydrocarbons: environmental pollution and bioremediation. Trends in Biotechnology 20 (6), 243–248.

Scholz, D.K., Kucklick, J.H., Pond, R., Walker, A.H., Bostrom, A., Fischbeck, P., 1999. Fate of Spilled Oil in Marine Waters: Where Does it Go? What Does it Do? How Do Dispersants Affect it? An Information Booklet for Decision-Makers. American Petroleum Institute, Washington, DC.

Schwartz, H.E., Brownlee, R.G., 1986. Use of reversed-phase chromatography in carbohydrate analysis. Journal of Chromatography 353, 77.

Schwartz, H.E., Brownlee, R.G., Boduszynski, M.M., Su, F., 1987. Simulated distillation of high-boiling petroleum fractions by capillary supercritical chromatography and vacuum thermal gravimetric analysis. Analytical Chemistry 59, 1393–1401.

Shafizadeh, A., McAteer, G., Sigmon, J., January 30, 2003. High-acid crudes. In: Proceedings. Crude Oil Quality Group Meeting. Louisiana, New Orleans.

Shalaby, H.M., 2005. Refining of Kuwait's heavy crude oil: materials challenges. In: Proceedings. Workshop on Corrosion and Protection of Metals. Arab School for Science and Technology. December 3–7, 2005, Kuwait.

Sheng, J.J., Maini, B.B., Hayes, R.E., Tortike, W.S., 1999. Critical review of foamy oil flow. Transport in Porous Media 35, 157–187.

Sheridan, M., April 2006. California Crude Oil Production and Imports Staff Paper, Report No. CERC-600-2006-006. Fossil Fuels Office, Fuels and Transportation Division, California Energy Commission, Sacramento, California.

Shirokoff, W.J., Siddiqui, N.M., Ali, F.M., 1997. Characterization of the structure of Saudi crude asphaltenes by x-ray diffraction. Energy Fuel 11 (3), 561–565.

Speight, J.G., 1984. In: Kaliaguine, S., Mahay, A. (Eds.), Characterization of Heavy Crude Oils and Petroleum Residues. Elsevier, Amsterdam, Netherlands, p. 515.

Speight, J.G., 2013a. The Chemistry and Technology of Coal 3rd Edition. CRC-Taylor & Francis Group, Boca Raton, Florida.

Speight, J.G., 2017. Handbook of Petroleum Refining. CRC Press, Taylor & Francis Group, Boca Raton, Florida.

Speight, J.G., 1986. Polynuclear aromatic systems in petroleum. Preprints Division of Petroleum Chemistry American Chemical Society 31 (4), 818.

Speight, J.G., 1994. Chemical and physical studies of petroleum asphaltenes. In: Yen, T.F., Chilingarian, G.V. (Eds.), Asphaltenes and Asphalts. I. Developments in Petroleum Science, vol. 40. Elsevier, Amsterdam, Netherlands (Chapter 2).

Speight, J.G., 2001. Handbook of Petroleum Analysis. John Wiley & Sons Inc., Hoboken, New Jersey.

Speight, J.G., 2012. Crude Oil Assay Database. Knovel. Elsevier, New York. Online version available at: http://www.knovel.com/web/portal/browse/display?_EXT_KNOVEL_DISPLAY_bookid=5485&VerticalID=0.

Speight, J.G., Arjoon, K.K., 2012. Bioremediation of Petroleum and Petroleum Products. Scrivener Publishing, Beverly, Massachusetts.

Speight, J.G., 2013b. Heavy Oil Production Processes. Gulf Professional Publishing, Elsevier, Oxford, United Kingdom.

Speight, J.G., 2013c. Oil Sand Production Processes. Gulf Professional Publishing, Elsevier, Oxford, United Kingdom.

Speight, J.G., 2014a. The Chemistry and Technology of Petroleum, fifth ed. CRC-Taylor and Francis Group, Boca Raton, Florida.

Speight, J.G., 2014b. High Acid Crudes. Gulf Professional Publishing, Elsevier, Oxford, United Kingdom.

Speight, J.G., 2015b. Handbook of Petroleum Product Analysis, second ed. John Wiley & Sons Inc., Hoboken, New Jersey.

Speight, J.G., 2016. Introduction to Enhanced Recovery Methods for Heavy Oil and Tar Sands, second ed. . Gulf Professional Publishing, Elsevier, Oxford, United Kingdom.

Speight, J.G., 2000. The Desulfurization of Heavy Oils and Residua, second ed. Marcel Dekker Inc., New York.

Strausz, P.O., Mojelsky, W.T., Faraji, F., Lown, M.E., Peng, P., 1999. Additional structural details on Athabasca asphaltene and their ramifications. Energy Fuels 13, 207–227.

Suatoni, J.C., Garber, H.R., 1975. HPLC preparative group-type separation of olefins from Synfuels. Journal of Chromatographic Science 13, 367.

Tfouni, S.A.V., Machado, R.M.D., Camargo, M.C.R., Vitorino, S.H.P., Eduardo, V., Toledo, M.C.F., 2007. Determination of polycyclic aromatic hydrocarbons in cachaça by HPLC with fluorescence detection. Food Chemistry 101 (1), 334–338.

US EIA, 2011. Review of Emerging Resources. US Shale Gas and Shale Oil Plays. Energy Information Administration, United States Department of Energy, Washington, DC.

van der Have, J.H., Verver, C.G., 1957. Petroleum and its Products, first ed. Pitman, London, UK.

Van Nes, K., van Westen, H.A., 1951. Aspects of the Constitution of Mineral Oils. Elsevier, Amsterdam, Netherlands.

Yeung, T.W., 2014. Evaluating Opportunity Crude Processing. Digital Refining).

FURTHER READING

Altgelt, K.H., Gouw, T.H., 1979. Chromatography in Petroleum Analysis. Marcel Dekker Inc., New York.

ASTM D4, 2017. Test Method for Bitumen Content. Annual Book of Standards. ASTM International, West Conshohocken, Pennsylvania.

ASTM D87, 2017. Standard Test Method for Melting Point of Petroleum Wax (Cooling Curve). Annual Book of Standards. ASTM International, West Conshohocken, Pennsylvania.

Baker, E.W., 1969. In: Eglinton, G., Murphy, M.T.J. (Eds.), Organic Geochemistry. Springer-Verlag, New York.

Baker, E.W., Louda, J.W., 1986. In: Johns, R.B. (Ed.), Biological Markers in the Sedimentary Record. Elsevier, Amsterdam.

Baker, E.W., Palmer, S.E., 1978. In: Dolphin, D. (Ed.), The Porphyrins. Volume I. Structure and Synthesis. Part a. Academic Press, New York.

Bonnett, R., 1978. In: Dolphin, D. (Ed.), The Porphyrins. Structure and Synthesis. Part A, vol. I. Academic Press, New York.

Filby, R.H., Van Berkel, G.J., 1987. Metal complexes in fossil fuels. In: Filby, R.H., Branthaver, J.F. (Eds.), Symposium Series No. 344. American Chemical Society, Washington, DC, p. 2.

Hodgson, G.W., Baker, B.L., Peake, E., 1967. In: Nagy, B., Columbo, U. (Eds.), Fundamental Aspects of Petroleum Geochemistry. Elsevier, Amsterdam (Chapter 5).

http://www.digitalrefining.com/article/1000644.

http://www.digitalrefining.com/article/1000893.

http://www.knovel.com/web/portal/browse/display?_EXT_KNOVEL_DISPLAY_bookid=5485&VerticalID=0.

Lewis, A., Aurand, D., 1997. Putting Dispersants to Work: Overcoming Obstacles. International Oil Spill Conference Issue Paper Technical Report No. IOSC-004.23. American Petroleum Institute, Washington, DC.

Matsushita, S., Tada, Y., Ikushige, T., 1981. Rapid hydrocarbon group analysis of gasoline by high-performance liquid chromatography. Journal of Chromatography 208, 429–432.

Meyer, R.F., Dietzman, W.D., 1981. World geography of heavy crude oils. In: Meyer, R.F., Steele, C.T. (Eds.), The Future of Heat Crude and Tar Sands. McGraw-Hill, New York, p. 16.

Quirke, J.M.E., 1987. Metal complexes in fossil fuels. In: Filby, R.H., Branthaver, J.F. (Eds.), Symposium Series No. 344. American Chemical Society, Washington, DC, p. 74.

Reynolds, J.G., Biggs, W.E., Bezman, S.A., 1987. Metal complexes in fossil fuels. In: Filby, R.H., Branthaver, J.F. (Eds.), Symposium Series No. 344. American Chemical Society, Washington, DC, p. 205.

Speight, J.G., 1980. The Chemistry and Technology of Petroleum 1st Edition. Marcel Dekker Inc., New York.

Speight, J.G., 1990. Tar sands. In: Fuel Science and Technology Handbook. Marcel Dekker Inc., New York (Chapter 12).

Speight, J.G., 2005. Upgrading and refining of natural bitumen and heavy oil. In: Coal, Oil Shale, Natural Bitumen, Heavy Oil and Peat. Encyclopedia of Life Support Systems (EOLSS). Developed under the Auspices of the UNESCO. EOLSS Publishers, Oxford, UK.

Speight, J.G., 2007. Natural Gas: A Basic Handbook. GPC Books, Gulf Publishing Company, Houston, Texas.

Speight, J.G., 2011. The Refinery of the Future. Gulf Professional Publishing, Elsevier, Oxford, United Kingdom.

Speight, J.G., 2015a. Handbook of Hydraulic Fracturing. John Wiley & Sons Inc., Hoboken, New Jersey.

Speight, J.G., Francisco, M.A., 1990. Studies in petroleum composition IV: changes in the nature of the chemical constituents during crude oil distillation. Revue de l'Institut Français du Pétrole 45, 733.

Speight, J.G., Wernick, D.L., Gould, K.A., Overfield, R.E., Rao, B.M.L., Savage, D.W., 1985. Molecular weights and association of asphaltenes: a critical review. Revue de l'Institut Français du Pétrole 40, 27.

Speight, J.G., 2014. The Chemistry and Technology of Petroleum, fifth Edition. CRC Press, Taylor & Francis Group, Boca Raton, Florida.

Thompson, C.J., Ward, C.C., Ball, J.S., 1976. Characteristics of World's Crude Oils and Results of API. Research Project 60. Report BERC/RI-76/8. Bartlesville Energy Technology Center, Bartlesville, Oklahoma.

Tissot, B.P., 1984. Characterization of Heavy Crude Oils and Petroleum Residues. Editions Technip, Paris.

Tissot, B.P., Welte, D.H., 1978. Petroleum Formation and Occurrence. Springer-Verlag, New York.

Vogh, J.W., Reynolds, J.W., 1988. Report No. DE88 001242. Contract FC22–83FE60149. United States Department of Energy, Washington, DC.

Weiss, V., Edwards, J.M., 1980. The Biosynthesis of Aromatic Compounds. John Wiley & Sons, Inc., New York.

Yan, J., Plancher, H., Morrow, N.R., 1997. Wettability changes induced by adsorption of asphaltenes. Paper No. Spe 37232. In: Proceedings. SPE International Symposium on Oilfield Chemistry. Houston, Texas. Society of Petroleum Engineers, Richardson, Texas.

REFINERY PRODUCTS AND BY-PRODUCTS

1.0 INTRODUCTION

Petroleum, in the unrefined or crude form, like many industrial feedstocks has little or no direct use and its value as an industrial commodity is only realized after the production of saleable products by a series of processing steps as performed in a refinery (Fig. 2.1). Each processing step is, in fact, a separate process and thus a refinery is a series of integrated processes that generate the desired products according to the market demand. Therefore, the value of petroleum is directly related to the yield of saleable products and is subject to the market pull. In general, crude oil, once refined, yields four basic groupings of distillation products that are produced when it is broken down into fractions: naphtha, middle distillates (kerosene and light gas oil), heavy gas oil/vacuum gas oil, and the residuum (Table 2.1). It is from these unrefined fractions that saleable petroleum products are produced by a series of refinery processes (Hsu and Robinson, 2006; Gary et al., 2007; Speight, 2011, 2017).

Petroleum products, in contrast to *petrochemical products* (Speight, 2014, 2017), are those bulk fractions that are derived from petroleum and have commercial value as a bulk product. In the strictest sense, petrochemical products are also petroleum products but they are individual well-defined chemicals (rather than complex mixtures) that are used as the basic building blocks of the chemical industry. They are chemicals (as distinct from fuels and petroleum products) manufactured from petroleum (and natural gas) and used for a variety of commercial purposes. The definition, however, has been broadened to include the whole range of aliphatic, aromatic, and naphthenic organic chemicals, as well as carbon black and such inorganic materials as sulfur and ammonia. An aromatic petrochemical is also an organic chemical compound but one that contains, or is derived from, the basic benzene ring system (Hsu and Robinson, 2006; Gary et al., 2007; Speight, 2011, 2017).

Product complexity, and the means by which the product is evaluated, has made the petroleum industry unique among industries. But product complexity has also brought to the fore issues such as instability and incompatibility. To understand the evolution of the products, it is essential to understand the composition of the various products. Product complexity becomes even more meaningful when various fractions from different types of crude oil, as well as fractions from synthetic crude oil are blended with the corresponding petroleum feedstock.

There is a myriad of products that have evolved through the short life of the petroleum industry and the complexities of product composition have matched the evolution of the products (Speight, 2014, 2017). In fact, it is the complexity of product composition that has served the industry well and, at the same time, has had an adverse effect on product use. Product complexity has made the industry unique among industries. Indeed, current analytical techniques that are accepted as standard methods for, as an example, the aromatics content of fuels (ASTM D1319; ASTM D2425; ASTM D2549; ASTM D2789), as well as proton and carbon nuclear magnetic resonance methods, yield different information.

Introduction to Petroleum Biotechnology. https://doi.org/10.1016/B978-0-12-805151-1.00002-3

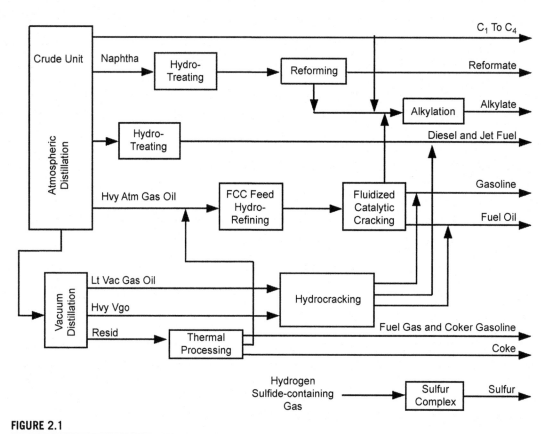

FIGURE 2.1

Refinery schematic.

The customary processing of petroleum (Fig. 2.1) does not usually involve the separation and handling of pure hydrocarbons. Indeed, petroleum-derived products are always mixtures: occasionally simple but more often very complex. Thus, for the purposes of this chapter, such materials as the gross fractions of petroleum (e.g., gasoline, naphtha, kerosene, and the like) which are usually obtained by distillation and/or refining are classified as *petroleum products*; asphalt and other solid products (e.g., wax) are also included in this division. This type of simplified classification separates this group of products from those obtained as petroleum chemicals (petrochemical products), for which the emphasis is on separation and purification of single chemical compounds and which are starting materials for a long list of other chemical products.

It is the purpose of this chapter to present a description of the various types of petroleum products (as well as petroleum wastes) and the properties and behavior of these products and wastes. This will guide the reader to a better understanding of the nature of the various petroleum-based products and petroleum-based wastes so that a better understanding of petroleum product biotechnology can be realized.

Table 2.1 Petroleum Products

Product	Lower Carbon Limit	Upper Carbon Limit	Lower Boiling Point (°C)	Upper Boiling Point (°C)	Lower Boiling Point (°F)	Upper Boiling Point (°F)
Refinery gas	C_1	C_4	−161	−1	−259	31
Liquefied petroleum gas	C_3	C_4	−42	−1	−44	31
Naphtha	C_5	C_{17}	36	302	97	575
Gasoline	C_4	C_{12}	−1	216	31	421
Kerosene/diesel fuel	C_8	C_{18}	126	258	302	575
Aviation turbine fuel	C_8	C_{16}	126	287	302	548
Fuel oil	C_{12}	$>C_{20}$	216	421	>343	>649
Lubricating oil	$>C_{20}$		>343		>649	
Wax	C_{17}	$>C_{20}$	302	>343	575	>649
Asphalt	$>C_{20}$		>343		>649	
Coke	$>C_{50}$[a]		>1000[a]		>1832[a]	

[a]Carbon number and boiling point difficult to assess; inserted for illustrative purposes only.

2.0 GASEOUS PRODUCTS

Petroleum product gases and refinery gases (process gases), produced from petroleum refining, upgrading, or natural gas processing facilities, are a category of saturated and unsaturated light hydrocarbons, predominantly C_1 to C_6. Some gases may also contain inorganic compounds, such as hydrogen, nitrogen (N_2), hydrogen sulfide (H_2S), carbon monoxide, and carbon dioxide (CO_2). As such, petroleum and refinery gases (unless produced as a saleable product that must meet specifications prior to sale) are considered to be of unknown or variable composition. The site-restricted petroleum and refinery gases (i.e., those not produced for sale) often serve as fuels consumed onsite, as intermediates for purification and recovery of various gaseous products, or as feedstocks for isomerization and alkylation processes within a facility.

The constituents of each type of gas may be similar (with the exception of the olefin-type gases produced during thermal processes) but the variations of the amounts of these constituents cover wide ranges. Each type of gas may be analyzed by similar methods although the presence of high boiling hydrocarbons and non-hydrocarbon species such as CO_2 and H_2S may require slight modifications to the analytical test methods.

2.1 LIQUEFIED PETROLEUM GAS

Liquefied petroleum gas (*LPG*) is the term applied to certain specific hydrocarbons and their mixtures, which exist in the gaseous state under atmospheric ambient conditions but can be converted to the liquid state under conditions of moderate pressure at ambient temperature. Typically, fuel gas with four or less carbon atoms in the hydrogen-carbon combination have boiling points that are lower than room temperature and these products are gases at ambient temperature and pressure.

The constituents of LPG are produced during natural gas refining, petroleum stabilization, and petroleum refining (Hsu and Robinson, 2006; Gary et al., 2007; Speight, 2014, 2017). Thus, LPG is a hydrocarbon mixture containing propane ($CH_3CH_2CH_3$, boiling point: $-42°C$, $-44°F$), butane ($CH_3CH_2CH_2CH_3$, boiling point: $0°C$, $32°F$), *iso*-butane ($CH_3CH(CH_3)CH_3$, boiling point: $-11.7°C$, $-10.9°F$), and to a lesser extent propylene ($CH_3CH{=}CH_2$, boiling point: $-47.6°C$, $-53.7°F$) or butylene ($CH_3CH_2CH{=}CH_2$, boiling point: $-6.5°C$, $20.4°F$). The most common commercial products are propane, butane, or some mixture of the two and are generally extracted from natural gas or crude petroleum. Propylene and butylene isomers result from cracking other hydrocarbons in a petroleum refinery and are two important chemical feedstocks.

Commercial Propane consists predominantly of propane and/or propylene while *Commercial Butane* is mainly composed of butanes and/or butylenes. Both must be free from harmful amounts of toxic constituents (API, 2009) and free from mechanically entrained water (that may be further limited by specifications) (ASTM D1835). Analysis by gas chromatography is possible (IP 405).

Commercial Propane-Butane mixtures are produced to meet particular requirements such as volatility, vapor pressure, specific gravity, hydrocarbon composition, sulfur and its compounds, corrosion of copper, residues, and water content. These mixtures are used as fuels in areas and at times where low ambient temperatures are less frequently encountered. Analysis by gas chromatography is possible (ASTM D5504; ASTM D6228).

Special Duty Propane is intended for use in spark-ignition engines and the specification includes a minimum *motor octane number* to ensure satisfactory antiknock performance. Propylene ($CH_3CH{=}CH_2$) has a significantly lower octane number than propane, so there is a limit to the amount of this component that can be tolerated in the mixture. Analysis by gas chromatography is possible (ASTM D5504; ASTM D6228; IP 405). *Mixed gas* is a gas prepared by adding natural gas or LPG to a manufactured gas, giving a product of better utility and higher heat content or Btu value.

LPG and *liquefied natural gas* can share the facility of being stored and transported as a liquid and then vaporized and used as a gas. In order to achieve this, LPG must be maintained at a moderate pressure but at ambient temperature. The liquefied natural gas can be at ambient pressure but must be maintained at a temperature of roughly -1 to $60°C$ ($30–140°F$). In fact, in some applications it is actually economical and convenient to use LPG in the liquid phase. In such cases, certain aspects of gas composition (or quality such as the ratio of propane to butane and the presence of traces of heavier hydrocarbons, water, and other extraneous materials) may be of lesser importance compared to the use of the gas in the vapor phase.

For normal (gaseous) use, the contaminants of LPG are controlled at a level at which they do not corrode fittings and appliances or impede the flow of the gas. For example, H_2S and carbonyl sulfide (COS) should be absent. Organic sulfur to the level required for adequate odorization (ASTM D5305), or *stenching*, is a normal requirement in LPG, dimethyl sulfide (CH_3SCH_3), and ethyl mercaptan (C_2H_5SH); it is commonly used at a concentration of up to 50 ppm. Natural gas is similarly treated with a wider range of volatile sulfur compounds.

The presence of water in LPG (or in natural gas) is undesirable since it can produce hydrates that will cause, for example, line blockage due to the formation of hydrates under conditions where the water *dew point* is attained (ASTM D1142). If the amount of water is above acceptable levels, the addition of a small methanol will counteract any such effect.

2.2 NATURAL GAS

Natural gas, which is predominantly methane (CH_4), occurs in underground reservoirs separately or in association with crude oil (Speight, 2014). The principal types of gaseous fuels are oil (distillation) gas, reformed natural gas, and reformed propane or LPG.

The principal constituent of natural gas is CH_4. Other constituents are paraffinic hydrocarbons such as ethane (CH_3CH_3), propane ($CH_3CH_2CH_3$), and the butanes [$CH_3CH_2CH_2CH_3$ and/or $(CH_3)_3CH$]. Many natural gases contain N_2 as well as CO_2 and H_2S. Trace quantities of argon, hydrogen, and helium may also be present. Generally, the hydrocarbons having a higher molecular weight than CH_4, CO_2, and H_2S are removed from natural gas prior to its use as a fuel. Gases produced in a refinery contain CH_4, ethane, ethylene, propylene, hydrogen, carbon monoxide, CO_2, and N_2, with low concentrations of water vapor, oxygen, and other gases.

The composition of natural gas can vary so widely that no single set of specifications can cover all situations. The requirements are usually based on performances in burners and equipment, on minimum heat content, and on maximum sulfur content. Gas utilities in most states come under the supervision of state commissions or regulatory bodies and the utilities must provide a gas that is acceptable to all types of consumers and that will give satisfactory performance in all kinds of consuming equipment. However, there are specifications for LPG (ASTM D1835) which depend upon the required volatility.

The different methods for gas analysis include absorption, distillation, combustion, mass spectroscopy, infrared spectroscopy, and gas chromatography (ASTM D2163, ASTM D2650, ASTM D4424). Absorption methods involve absorbing individual constituents one at a time in suitable solvents and recording of contraction in volume measured. Distillation methods depend on the separation of constituents by fractional distillation and measurement of the volumes distilled. In combustion methods, certain combustible elements are caused to burn to CO_2 and water, and the volume changes are used to calculate composition. Infrared spectroscopy is a useful application, but for the most accurate analysis, mass spectroscopy and gas chromatography are the preferred methods.

The specific gravity of product gases, including LPG, may be determined conveniently by several methods and a variety of instruments (ASTM D1070).

The *heat value* of gases is generally determined at constant pressure in a flow calorimeter in which the heat released by the combustion of a definite quantity of gas is absorbed by a measured quantity of water or air. A continuous recording calorimeter is available for measuring heat values of natural gases (ASTM D1826).

The lower and upper limits of *flammability* of organic compounds indicate the percentage of combustible gas in air below which and above which flame will not propagate. When flame is initiated in mixtures having compositions within these limits, it will propagate and therefore the mixtures are flammable. Knowledge of flammable limits and their use in establishing safe practices in handling gaseous fuels is important, e.g., when purging equipment used in gas service, in controlling factory or mine atmospheres, or in handling liquefied gases.

2.3 SHALE GAS

Shale gas is natural gas produced from shale formations that typically function as both the reservoir and source rocks for the natural gas (Speight, 2013b). In terms of chemical makeup, shale gas is typically a dry gas composed primarily of CH_4 (60%–95% v/v), but some formations do produce wet gas. The Antrim and New Albany plays have typically produced water and gas. Gas shale formations are organic-rich shale formations that were previously regarded only as source rocks and seals for gas accumulating in the strata near sandstone and carbonate reservoirs of traditional onshore gas.

Gas produced from tight formations (such as shale gas formations) has a broad variation in compositional makeup, with some having wider component ranges, a wider span of minimum and maximum heating values, and higher levels of water vapor and other substances than pipeline tariffs or purchase contracts may typically allow. Indeed, because of these variations in gas composition, shale gas formation can have unique interactions with biotechnological agents.

Although not highly sour in the usual sense of having high H_2S content, and with considerable variation from play to resource to resource and even from well to well within the same resource (due to extremely low permeability of the shale even after fracturing) (Speight, 2013b), shale gas often contains varying amounts of H_2S with wide variability in the CO_2 content. The gas is not ready for pipelining immediately after it has exited the shale formation. The challenge in treating such gases is the low (or differing) H_2S/CO_2 ratio and the need to meet pipeline specifications. In a traditional gas processing plant, the olamine of choice for content for H_2S removal is N-methyldiethanolamine (Mokhatab et al., 2006; Speight, 2007, 2014) but whether this olamine will suffice to remove the H_2S without removal of excessive amounts of CO_2 is another issue.

2.4 REFINERY GAS

The terms *petroleum gas* or *refinery gas* are often used to identify LPG or even gas that emanates from the top of a refinery distillation column. For the purpose of this text, petroleum gas not only describes LPG but also natural gas and refinery gas (Hsu and Robinson, 2006; Mokhatab et al., 2006; Gary et al., 2007; Speight, 2014, 2017). In this chapter, each gas is, in turn, referenced by its name rather than the generic term *petroleum gas* (ASTM D4150). However, the composition of each gas varies (Table 2.2) and recognition of this is essential before testing protocols are applied (Mokhatab et al., 2006; Speight, 2014).

Refinery gas (fuel gas) is the non-condensable gas that is obtained during distillation of crude oil or treatment (cracking, thermal decomposition) of petroleum (Robinson and Faulkner, 2000; Speight, 2014). Refinery gas is produced in considerable quantities during the different refining processes and is used as fuel for the refinery itself and as an important feedstock for the production of petrochemical products. It consists mainly of hydrogen (H_2), methane (CH_4), ethane (C_2H_6), propane (C_3H_8), butane (C_4H_{10}), and olefins ($RCH=CHR^1$, where R and R^1 can be hydrogen or a methyl group) and may also include off-gases from petrochemical processes (Table 2.3). Olefins such as ethylene ($CH_2=CH_2$, boiling point: $-104°C$, $-155°F$), propene (propylene, $CH_3CH=CH_2$, boiling point: $-47°C$, $-53°F$), butene (butene-1, $CH_3CH_2CH=CH_2$, boiling point: $-5°C$, $23°F$), *iso*-butylene (($CH_3)_2C=CH_2$, boiling point: $-6°C$, $21°F$), *cis-* and *trans*-butene-2 ($CH_3CH=CHCH_3$, boiling point: ca. $1°C$, $30°F$), and butadiene ($CH_2=CHCH=CH_2$, boiling point: $-4°C$, $24°F$), as well as higher boiling olefins, are produced by various refining processes.

Still gas is a broad terminology for low-boiling hydrocarbon mixtures and is the lowest boiling fraction isolated from a distillation (*still*) unit in the refinery. If the distillation unit is separating light hydrocarbon fractions, the still gas will be almost entirely CH_4 with only traces of ethane (CH_3CH_3) and ethylene ($CH_2=CH_2$). If the distillation unit is handling higher boiling fractions, the still gas might also contain propane ($CH_3CH_2CH_3$), butane ($CH_3CH_2CH_2CH_3$), and their respective isomers. *Fuel gas* and still gas are terms that are often used interchangeably but the term *fuel gas* is intended to denote the product's destination—to be used as a fuel for boilers, furnaces, or heaters.

Table 2.2 Composition of Associated Natural Gas From a Petroleum Well

Category	Component	Amount (%)
Paraffinic	Methane (CH_4)	70–98
	Ethane (C_2H_6)	1–10
	Propane (C_3H_8)	Trace–5
	Butane (C_4H_{10})	Trace–2
	Pentane (C_5H_{12})	Trace–1
	Hexane (C_6H_{14})	Trace–0.5
	Heptane and higher (C_7^+)	None–trace
Cyclic	Cyclopropane (C_3H_6)	Traces
	Cyclohexane (C_6H_{12})	Traces
Aromatic	Benzene (C_6H_6), others	Traces
Non-hydrocarbon	Nitrogen (N_2)	Trace–15
	Carbon dioxide (CO_2)	Trace–1
	Hydrogen sulfide (H_2S)	Trace occasionally
	Helium (He)	Trace–5
	Other sulfur and nitrogen compounds	Trace occasionally
	Water (H_2O)	Trace–5

Table 2.3 Possible Constituents of Natural Gas and Refinery Gas

Gas	Molecular Weight	Boiling Point 1 atm °C (°F)	Density at 60°F (15.6°C), 1 atm	
			g/L	Relative to Air = 1
Methane	16.043	−161.5 (−258.7)	0.6786	0.5547
Ethylene	28.054	−103.7 (−154.7)	1.1949	0.9768
Ethane	30.068	−88.6 (−127.5)	1.2795	1.0460
Propylene	42.081	−47.7 (−53.9)	1.8052	1.4757
Propane	44.097	−42.1 (−43.8)	1.8917	1.5464
1,2-Butadiene	54.088	10.9 (51.6)	2.3451	1.9172
1,3-Butadiene	54.088	−4.4 (24.1)	2.3491	1.9203
1-Butene	56.108	−6.3 (20.7)	2.4442	1.9981
cis-2-Butene	56.108	3.7 (38.7)	2.4543	2.0063
trans-2-Butene	56.108	0.9 (33.6)	2.4543	2.0063
iso-Butene	56.104	−6.9 (19.6)	2.4442	1.9981
n-Butane	58.124	−0.5 (31.1)	2.5320	2.0698
iso-Butane	58.124	−11.7 (10.9)	2.5268	2.0656

In a series of processes commercialized under the names Platforming, Power-forming, Cat-forming, and Ultra-forming, paraffinic and naphthenic (cyclic nonaromatic) hydrocarbons, in the presence of hydrogen and a catalyst, are converted into aromatics, or isomerized to more highly branched hydrocarbons. Catalytic reforming processes thus not only result in the formation of a liquid product of higher octane number, but also produce substantial quantities of gases. The latter are rich in hydrogen, but also contain hydrocarbons from CH_4 to butanes, with a preponderance of propane ($CH_3CH_2CH_3$), n-butane ($CH_3CH_2CH_2CH_3$), and isobutane [$(CH_3)_3CH$]. The composition of these gases varies in accordance with reforming severity and reformer feedstock. Since all catalytic reforming processes require substantial recycling of a hydrogen stream, it is normal to separate reformer gas into a propane ($CH_3CH_2CH_3$) and/or a butane [$CH_3CH_2CH_2CH_3/(CH_3)_3CH$] stream, which becomes part of the refinery LPG production, and a lighter gas fraction, part of which is recycled. In view of the excess of hydrogen in the gas, all products of catalytic reforming are saturated, and there are usually no olefin-type gases present in either gas stream.

A second group of refining operations that contributes to gas production is that of the catalytic cracking processes (Hsu and Robinson, 2006; Gary et al., 2007; Speight, 2014, 2017). Both catalytic and thermal cracking processes, the latter being now largely used to produce chemical raw materials, result in the formation of unsaturated hydrocarbons, particularly ethylene ($CH_2\!=\!CH_2$), but also propylene (propene, $CH_3CH\!=\!CH_2$), isobutylene [isobutene, $(CH_3)_2C\!=\!CH_2$], and the n-butenes ($CH_3CH_2CH\!=\!CH_2$, and $CH_3CH\!=\!CHCH_3$) in addition, to hydrogen (H_2), CH_4 and smaller quantities of ethane (CH_3CH_3), propane ($CH_3CH_2CH_3$), and butane isomers [$CH_3CH_2CH_2CH_3$, $(CH_3)_3CH$]. Diolefins such as butadiene ($CH_2\!=\!CHCH\!=\!CH_2$) are also present. Additional gases are produced in refineries with coking or visbreaking facilities for the processing of their heaviest crude fractions. A further source of refinery gas is hydrocracking, a catalytic high-pressure pyrolysis process in the presence of fresh and recycled hydrogen.

The first and most important aspect of gaseous testing is the measurement of the volume of gas (ASTM D1071). In this test method, several techniques are described and may be employed for any purpose where it is necessary to know the quantity of gaseous fuel. In addition, the thermophysical properties of CH_4 (ASTM D3956), ethane (ASTM D3984), propane (ASTM D4362), *n*-butane (ASTM D4650), and *iso*-butane (ASTM D4651) should be available for use and consultation.

Residual refinery gases, usually in more than one stream, which allow a degree of quality control, are treated for H_2S removal and gas sales are usually on a thermal content (calorific value, heating value) basis with some adjustment for variation in the calorific value and hydrocarbon type (Speight, 2007, 2014).

3.0 LIQUID PRODUCTS

A wide variety of liquid products are produced from petroleum that varies from the high-volatile naphtha to the low-volatile lubricating oil (Speight, 2014, 2017). The liquid products are often characterized by a variety of techniques that include measurement of physical properties and fractionation into group types. The purpose of this section is to present examples of the various liquid effluents and the methods of analysis that can be applied to the analysis of various liquid effluents from refinery processes. The examples chosen are those at the lower end of the petroleum product boiling range (naphtha) and at the higher end of the petroleum product boiling range (residual fuel oil).

The exposure of liquid products to biotechnological applications are varied although in some aspects product behavior can be predicted, but only with a degree of caution since behavior is very much dependent upon composition. For this reason, reference is made to the various test methods dedicated to these products and which can be applied to the products boiling in the intermediate range. In the light of the various tests available for composition, such tests will be deemed necessary depending on the environmental situation and the requirements of the legislation, as well as at the discretion of the analyst.

3.1 NAPHTHA

Naphtha (often referred to as *naft* in the older literature) is a generic term applied to refined, partly refined, or an unrefined low-to-medium boiling petroleum distillate fraction. Naphtha resembles gasoline in terms of boiling range and carbon number, being a precursor to gasoline. In the strictest sense of the term, not less than 10% of the naphtha should distill below 175°C (345°F) and not less than 95% of the material should distill below 240°C (465°F) under standardized distillation conditions (ASTM D86). The main uses of petroleum naphtha fall into the general areas of (1) precursor to gasoline and other liquid fuels, (2) solvents (diluents) for paints, (3) dry-cleaning solvents, (4) solvents for cutback asphalts, (5) solvents in rubber industry, and (6) solvents for industrial extraction processes. Turpentine, the older and more conventional solvent for paints has now been almost completely replaced by the cheaper and more abundant petroleum naphtha.

The term *petroleum solvent* describes the liquid hydrocarbon fractions obtained from petroleum and is used in industrial processes and formulations. These fractions are also referred to as *naphtha* or *industrial naphtha*. By definition, the solvents obtained from the petrochemical industry such as alcohols, ethers, and the like are not included in this chapter. A refinery is capable of producing hydrocarbons of a high degree of purity and at the present time petroleum solvents are available covering a wide range of solvent properties including both volatile and high boiling qualities. Other petroleum products boiling within the naphtha boiling range include (1) *industrial spirit* and (2) *white spirit*.

Industrial spirit comprises liquids distilling between 30 and 200°C (−1 to 390°F), with a temperature difference between 5% volume and 90% volume distillation points, including losses, of not more than 60°C (140°F). There are several (up to eight) grades of industrial spirit, depending on the position of the cut in the distillation range defined above. On the other hand, *white spirit* is an industrial spirit with a flash point above 30°C (99°F) and has a distillation range from 135 to 200°C (275–390°F).

Naphtha contains varying amounts of paraffins, olefins, naphthene constituents, and aromatics and olefins in different proportions, in addition to potential isomers of paraffin that exist in naphtha boiling range. As a result, naphtha is divided predominantly into two main types: (1) aliphatic naphtha and (2) aromatic (naphtha). The two types differ in two ways: first, in the kind of hydrocarbons making up the solvent, and second, in the methods used for their manufacture. Aliphatic solvents are composed of paraffinic hydrocarbons and cycloparaffins (naphthenes), and may be obtained directly from crude petroleum by distillation. The second type of naphtha contains aromatics, usually alkyl-substituted benzene, and is very rarely, if at all, obtained from petroleum as straight-run materials.

Stoddard solvent is a petroleum distillate widely used as a dry cleaning solvent and as a general cleaner and degreaser. It may also be used as paint thinner, as a solvent in some types of photocopier toners, in some types of printing inks, and in some adhesives. Stoddard solvent is considered to be a form of mineral spirits, white spirits, and naphtha but not all forms of mineral spirits, white spirits, and naphtha are considered to be Stoddard solvent. Stoddard solvent consists of linear alkanes (30%–50%), branched alkanes (20%–40%), cycloalkanes (30%–40%), and aromatic hydrocarbons (10%–20%). The typical hydrocarbon chain ranges from C_7 through C_{12} in length.

Turpentine, the older more conventional solvent for paints, has now been almost completely replaced with the discovery that the cheaper and more abundant petroleum naphtha is equally satisfactory. The differences in application are slight: naphtha causes a slightly greater decrease in viscosity when added to some paints than does turpentine, and depending on the boiling range, may also show difference in evaporation rate.

Cutback asphalt is asphalt cement diluted with a petroleum distillate to make it suitable for direct application to road surfaces with little or no heating. Asphalt cement, in turn, is a combination of hard asphalt with a heavy distillate or with a viscous residuum of an asphaltic crude oil. The products are classified as rapid, medium, and slow curing, depending on the rate of evaporation of the solvent. A rapid-curing product may contain 40%–50% of material distilling up to 360°C (680°F); a slow-curing mixture may have only 25% of such material. Gasoline naphtha, kerosene, and light fuel oils boiling from 38 to 330°C (100–30°F) are used in different products and for different purposes; the use may also dictate the nature of the asphaltic residuum that can be used for the asphalt.

If spilled or discharged in the environment, naphtha represents a threat of the toxicity of the constituents to land and/or to aquatic organisms (Speight, 2005; Speight and Arjoon, 2012). A significant spill may cause long-term adverse effects in the aquatic environment. The constituents of naphtha predominantly fall in the C_5–C_{16} carbon range: alkanes, cycloalkanes, aromatics and, if they are subject to a cracking process, alkenes as well. Naphtha may also contain a preponderance of aromatic constituents (up to 65%), others contain up to 40% alkenes, while all of the others are aliphatic in composition, up to 100%.

3.2 GASOLINE

Gasoline, also called *gas* (United States and Canada) or *petrol* (Great Britain) or *benzine* (Europe), is mixture of volatile, flammable liquid hydrocarbons derived from petroleum and used as fuel for internal combustion engines. It is also used as a solvent for oils and fats. Originally a by-product of the petroleum industry (kerosene being the principal product), gasoline became the preferred automobile fuel because of its high energy of combustion and capacity to mix readily with air in a carburetor.

Gasoline is a mixture of hydrocarbons that usually boils below 180°C (355°F) or, at most, below 200°C (390°F). The hydrocarbon constituents in this boiling range are those that have 4–12 carbon atoms in their molecular structure and fall into three general types: paraffins (including the cycloparaffins and branched materials), olefins, and aromatics.

Automotive gasoline typically contains about 150–200 hydrocarbon compounds, although nearly 1000 have been identified. The relative concentration of the compounds varies considerably depending on the source of crude oil, refinery process, and product specifications. Typical hydrocarbon chain lengths range from C_4 through C_{12} with a general hydrocarbon distribution consisting of 4%–8% v/v alkane derivatives, 2%–5% v/v alkenes, 25%–40% v/v iso-alkane derivatives, 3%–7% v/v cycloalkane derivatives, 1% to 4% v/v cycloalkene derivatives, and 20%–50% v/v aromatic derivatives. However, these proportions vary greatly. Unleaded gasoline may have higher proportions of aromatic hydrocarbons than leaded gasoline.

Gasoline can vary widely in composition: even those with the same octane number may be quite different, not only in the physical makeup but also in the molecular structure of the constituents and the octane number regularly needs enhancement (Watanabe et al., 2010). The variation in aromatics content, as well as the variation in the content of normal paraffins, branched paraffin derivatives, cyclopentane derivatives, and cyclohexane derivatives, involves characteristics of any one individual crude oil and may in some instances be used for crude oil identification. Furthermore, straight-run gasoline generally shows a decrease in paraffin content with an increase in molecular weight, but the cycloparaffins (naphthenes) and aromatics increase with increasing molecular weight. Indeed, the hydrocarbon type variation may also vary markedly from process to process.

Aviation gasoline is form of motor gasoline that has been especially prepared for use for aviation piston engines. It has an octane number suited to the engine, a freezing point of −60°C (−76°F), and a distillation range usually within the limits of 30–180°C (86–356°F) compared to −1 to 200°C (30–390°F) for automobile gasoline. The narrower boiling range ensures better distribution of the vaporized fuel through the more complicated induction systems of aircraft engines. Aircraft operate at altitudes at which the prevailing pressure is less than the pressure at the surface of the earth (pressure at 17,500 feet is 7.5 psi compared to 14.7 psi at the surface of the earth). Thus, the vapor pressure of aviation gasoline must be limited to reduce boiling in the tanks, fuel lines, and carburetors. Thus, the aviation gasoline does not usually contain the gaseous hydrocarbons (butanes) that give automobile gasoline the higher vapor pressures.

Gasoline-type jet fuel includes all light hydrocarbon oils for use in aviation turbine power units that distill between 100 and 250°C (212–480°F). It is obtained by blending kerosene and gasoline or naphtha in such a way that the aromatic content does not exceed 25% v/v. Additives can be included to improve fuel stability and combustibility. To increase the proportion of higher boiling octane components, such as aviation alkylate and xylenes, the proportion of lower boiling components must also be increased to maintain the proper volatility. *Iso*-pentane and, to some extent, *iso*-hexane are the lower boiling components used. *Iso*-pentane and *iso*-hexane may be separated from selected naphtha by super-fractionators or synthesized from the normal hydrocarbons by isomerization. In general, most aviation gasoline is made by blending a selected straight-run naphtha fraction (aviation base stock) with *iso*-pentane and aviation alkylate.

The final step in gasoline manufacture is *blending* the various streams into a finished product (Speight, 2014, 2017). It is not uncommon for the finished gasoline to be made up of six or more streams and several factors make this flexibility critical: (1) the requirements of the gasoline specification and the regulatory requirements and (ii2) performance specifications that are subject to local climatic conditions and regulations. However, from the aspect of biotechnological application, gasoline is not merely a mixture of pure hydrocarbons; there are many additives that may interfere with biotechnical interactions of gasoline.

Additives are gasoline-soluble chemicals that are mixed with gasoline to enhance certain performance characteristics or to provide characteristics not inherent in the gasoline. Additives are generally derived from petroleum-based materials and their function and chemistry are highly specialized. They produce the desired effect in parts per million (ppm) concentration range.

Oxidation inhibitors (antioxidants) are aromatic amines and hindered phenols that prevent gasoline components (particularly olefins) from reacting with oxygen in the air to form *peroxides* or *gums*. Peroxides can degrade antiknock quality, cause fuel pump wear, and attack plastic or elastomeric fuel system parts; soluble gums can lead to engine deposits, and insoluble gums can plug fuel filters. Inhibiting oxidation is particularly important for fuels used in modern fuel-injected vehicles, as their fuel recirculation design may subject the fuel to more temperature and oxygen-exposure stress.

Corrosion inhibitors are carboxylic acids and carboxylates that prevent free water in the gasoline from rusting or corroding pipelines and storage tanks. *Demulsifiers* are polyglycol derivatives that improve the water-separating characteristics of gasoline by preventing the formation of stable emulsions.

Antiknock compounds are compounds (such as tetraethyl lead) that increase the antiknock quality of gasoline. Gasoline containing tetraethyl lead was first marketed in 1920s and the average concentration of lead in gasoline gradually was increased until it reached a maximum of about 2.5 g per gallon (g/gal.) in the late 1960s. After that, a series of events resulted in the use of less lead and Environment Protection Agency (EPA) regulations required the phased reduction of the lead content of gasoline beginning in 1979. The EPA completely banned the addition of lead additives to on-road gasoline in 1996 and the amount of incidental lead may not exceed 0.05 g/gal. *Anti-icing additives* are surfactants, alcohols, and glycols that prevent ice formation in the carburetor and fuel system. The need for this additive is being reduced as older-model vehicles with carburetors are replaced by vehicles with fuel injection systems.

Dyes are oil-soluble solids and liquids used to visually distinguish batches, grades, or applications of gasoline products. For example, gasoline for general aviation, which is manufactured to different and more exacting requirements, is dyed blue to distinguish it from motor gasoline for safety reasons. *Markers* are a means of distinguishing specific batches of gasoline without providing an obvious visual clue. A refiner may add a marker to its gasoline so it can be identified as it moves through the distribution system. *Drag reducers* are high-molecular-weight polymers that improve the fluid flow characteristics of low-viscosity petroleum products. Drag reducers lower pumping costs by reducing friction between the flowing gasoline and the walls of the pipe.

Oxygenates are carbon-, hydrogen-, and oxygen-containing combustible liquids that are added to gasoline to improve performance. The addition of oxygenates to gasoline is not new since ethanol (ethyl alcohol or grain alcohol) has been added to gasoline for decades. Thus, *oxygenated gasoline* is a mixture of conventional hydrocarbon-based gasoline and one or more oxygenates. The current oxygenates belong to one of two classes of organic molecules: alcohols and ethers. The most widely used oxygenates in the United States are ethanol, methyl tertiary-butyl ether, and tertiary-amyl methyl ether. Ethyl tertiary-butyl ether (ETBE) is another ether that could be used. Oxygenates may be used in areas of the United States where they are not required if concentration limits (as defined by environmental regulations) are observed.

3.3 SOLVENTS

The so-called *petroleum ether* solvents are specific boiling range naphtha as is *ligroin*. Thus, the term *petroleum solvent* describes a special liquid hydrocarbon fraction obtained from naphtha and used in industrial processes and formulations. These fractions are also referred to as *industrial naphtha*. Other solvents include *white spirit* that is subdivided into *industrial spirit* (distilling between 30 and 200°C, 86 to 392°F) and *white spirit* (light oil with a distillation range of 135–200°C (275°F to 392°F)). The special value of naphtha as a solvent lies in its stability and purity.

3.4 KEROSENE AND DIESEL FUEL

Kerosene (also spelled *kerosine*), also called paraffin or paraffin oil, is a flammable pale yellow or colorless oily liquid with a characteristic odor. It is obtained from petroleum and used for burning in lamps and domestic heaters or furnaces, as a fuel or fuel component for jet engines, and as a solvent for greases and insecticides. Kerosene is intermediate in volatility between gasoline and gas/diesel oil. It is a medium oil distilling between 150 and 300°C (300–570°F). Kerosene has a flash point about 25°C

(77°F) and is suitable for use as an illuminant when burned in a wide lamp. The term *kerosene* is also too often incorrectly applied to various fuel oils, but a fuel oil is actually any liquid or liquid petroleum product that produces heat when burned in a suitable container or that produces power when burned in an engine.

Diesel fuels originally were straight-run products obtained from the distillation of crude oil. Currently, diesel fuel may also contain varying amounts of selected cracked distillates to increase the volume available. The boiling range of diesel fuel is approximately 125–328°C (302 to 575°F) (Table 2.1). Thus, in terms of carbon number and boiling range, diesel fuel occurs predominantly in the kerosene range and this many of the test methods applied to kerosene can also be applied to diesel fuel. Diesel fuel depends upon the nature of the original crude oil and the refining processes by which the fuel is produced, and the additive (if any) used, such as the solvent red dye. Furthermore, the specification for diesel fuel can exist in various combinations of characteristics such as, for example, volatility, ignition quality, viscosity, gravity, and stability.

Stove oil, like kerosene, is always a straight-run fraction from suitable crude oils, whereas other fuel oils are usually blends of two or more fractions, one of which is usually cracked gas oil. The straight-run fractions available for blending into fuel oils are heavy naphtha, light and heavy gas oils, reduced crude, and pitch. Cracked fractions such as light and heavy gas oils, cracked coal tar, and fractionator bottoms from catalytic cracking may also be used as blends to meet the specifications of the different fuel oils.

Jet fuel is a light petroleum distillate that is available in several forms suitable for use in various types of jet engines. The major jet fuels used by the military are JP-4, JP-5, JP-6, JP-7, and JP-8. Briefly, JP-4 is a wide-cut fuel developed for broad availability. JP-6 is a higher cut than JP-4 and is characterized by fewer impurities. JP-5 is specially blended kerosene, and JP-7 is high flash point special kerosene used in advanced supersonic aircraft. JP-8 is kerosene modeled on Jet A-1 fuel (used in civilian aircraft). From what data are available, typical hydrocarbon chain lengths characterizing JP-4 range from C_4 to C_{16}. Aviation fuels consist primarily of straight and branched alkanes and cycloalkanes. Aromatic hydrocarbons are limited to 20%–25% of the total mixture because they produce smoke when burned. A maximum of 5% alkenes is specified for JP-4. The approximate distribution by chemical class is: straight chain alkanes (32%), branched alkanes (31%), cycloalkanes (16%), and aromatic hydrocarbons (21%).

Kerosene type jet fuel is a medium distillate product that is used for aviation turbine power units. It has the same distillation characteristics and flash point as kerosene (between 150 and 300°C, 300 and 570°F, but not generally above 250°C, 480°F). In addition, it has particular specifications (such as freezing point) which are established by the International Air Transport Association.

3.5 FUEL OIL

Fuel oil is classified in several ways but was formally divided into two main types: *distillate fuel oil* and *residual fuel oil*, each of which was a blend of two or more refinery streams (Speight, 2014, 2017). Distillate fuel oil is vaporized and condensed during a distillation process and thus have a definite boiling range and do not contain high-boiling constituents. A fuel oil that contains any amount of the residue from crude distillation of thermal cracking is a residual fuel oil. The terms *distillate fuel oil* and *residual fuel oil* are losing their significance, since fuel oil is now made for specific uses and may be either distillates or residuals or mixture of the two. The terms *domestic fuel oil, diesel fuel oil,* and *heavy fuel oil* are more indicative of the uses of fuel oils. All of the fuel oil classes described here are refined from crude petroleum and may be categorized as either a distillate fuel or a residual fuel depending on the method of production.

No. 1 fuel oil is a petroleum distillate that is one of the most widely used of the fuel oil types. It is used in atomizing burners that spray fuel into a combustion chamber where the tiny droplets bum while in suspension. It is also used as a carrier for pesticides, as a weed killer, as a mold release agent in the ceramic and pottery industry, and in the cleaning industry. It is found in asphalt coatings, enamels, paints, thinners, and varnishes. No. 1 fuel oil is a light petroleum distillate (straight-run kerosene) consisting primarily of hydrocarbons in the range C_9–C_{16}. No. 1 fuel oil is very similar in composition to diesel fuel; the primary difference is in the additives.

No. 2 fuel oil is a petroleum distillate that may be referred to as domestic or industrial fuel oil. The domestic fuel oil is usually lower boiling and a straight-run product. It is used primarily for home heating. Industrial distillate is a cracked product or a blend of both. It is used in smelting furnaces, ceramic kilns, and packaged boilers. No. 2 fuel oil is characterized by hydrocarbon chain lengths in the C_{11}–C_{20} range. The composition consists of aliphatic hydrocarbons (straight chain alkanes and cycloalkanes) (64%), 1–2% unsaturated hydrocarbons (alkenes) (1%–2%), and aromatic hydrocarbons (including alkyl benzenes and 2-ring, 3-ring aromatics) (35%) but contains only low amounts of the polycyclic aromatic hydrocarbons (<5%). Domestic fuel oil is fuel oil that is used primarily in the home. This category of fuel oil includes kerosene, stove oil, and furnace fuel oil; they are distillate fuel oils.

No. 6 fuel oil (also called *Bunker C oil* or *residual fuel oil*) is the residuum from crude oil after naphtha-gasoline, No. 1 fuel oil, and No. 2 fuel oil have been removed. No. 6 fuel oil can be blended directly to heavy fuel oil or made into asphalt. Residual fuel oil is more complex in composition and impurities than distillate fuels. Limited data are available on the composition of No. 6 fuel oil. Polycyclic aromatic hydrocarbons (including the alkylated derivatives) and metal-containing constituents are components of No. 6 fuel oil.

3.6 LUBRICATING OIL

Lubrication is the introduction of various substances (referred to as *lubricants*) between sliding surfaces to reduce wear and friction. Thus, *lubricating oil* is used to reduce friction and wear between bearing metallic surfaces that are moving with respect to each other, by separation of metallic surfaces by a film of the oil. Lubricating oil is distinguished from other fractions of crude oil by an extremely high boiling point (>400°C, >750°F) (Table 2.1) (Banaszewski and Blythe, 2000; Speight, 2014, 2015a; Speight and Exall, 2014).

Mineral oils are often used as lubricating oils but also have medicinal and food uses. A major type of hydraulic fluid is the mineral oil class of hydraulic fluids. The mineral-based oils are produced from heavy-end crude oil distillates. Hydrocarbon numbers ranging from C_{15} to C_{50} occur in the various types of mineral oils, with the heavier distillates having higher percentages of the higher carbon number compounds.

Lubricating oil is distinguished from other fractions of crude oil by the usually high (>400°C, >750°F) boiling point, as well as their high viscosity. Materials suitable for the production of lubricating oils are comprised principally of hydrocarbons containing from 25 to 35 or even 40 carbon atoms per molecule, whereas residual stocks may contain hydrocarbons with 50 or more (up to 80 or so) carbon atoms per molecule. The composition of lubricating oil may be substantially different from the lubricant fraction from which it was derived, since wax (normal paraffins) is removed by distillation or refining by solvent extraction and adsorption preferentially removes non-hydrocarbon constituents, as well as polynuclear aromatic compounds and the multi-ring cycloparaffins (Speight, 2014; Speight and Exall, 2014).

Mono-, di-, and trinuclear aromatic compounds appear to be the main constituents of the aromatic portion, but material with more aromatic nuclei per molecule may also be present. For the dinuclear aromatics, most of the material consists of naphthalene types. For the trinuclear aromatics, the phenanthrene type of structure predominates over the anthracene type. There are also indications that the greater part of the aromatic compounds occurs as mixed aromatic-cycloparaffin compounds.

Used lubricating oil—often referred to as *waste oil* without further qualification—is any lubricating oil, whether refined from crude or synthetic components, which has been contaminated by physical or chemical impurities as a result of use (Boughton and Horvath, 2004; Speight and Exall, 2014).

3.7 WHITE OIL, INSULATING OIL, INSECTICIDES

Lubricating oil is not the only material manufactured from the high-boiling fraction of petroleum, and there are several important types of oil that may be contained in the so-called lubricant fraction.

3.7.1 White Oil

White oils generally fall into two classes: (1) those often referred to as technical white oils, which are employed for cosmetics, textile lubrication, insecticide vehicles, paper impregnation, and so on; and (2) pharmaceutical white oils, which may be employed as laxatives or for the lubrication of food-handling machinery. The colorless character of these oils is important in some cases, as it may indicate the chemically inert nature of the hydrocarbon constituents. Textile lubricants should be colorless to prevent the staining of light-colored threads and fabrics. Insecticide oils should be free of reactive (easily oxidized) constituents so as not to injure plant tissues when applied as sprays. Laxative oils should be free of odor, taste, and also hydrocarbons, which may react during storage and produce unwanted by-products. These properties are attained by the removal of nitrogen-containing, oxygen-containing, and sulfur-containing compounds, as well as reactive hydrocarbons by, say, sulfuric acid.

3.7.2 Insulating Oil

Petroleum oil for electrical insulation fall into two general classes: (1) those used in transformers, circuit breakers, and oil-filled cables; and (2) those employed for impregnating the paper covering of wrapped cables. The first are highly refined fractions of low viscosity and comparatively high boiling range and resemble heavy burning oils, such as mineral seal oil, or the very light lubricating fractions known as non-viscous neutral oils. The second are usually highly viscous products, often naphthenic distillates, and are not usually highly refined.

The deterioration of transformer oils in service is closely connected with oxidation by air, which brings on deposition of sludge and the development of acids, resulting in overheating and corrosion, respectively. The *sludge* formed is one of three types: (1) sludge attributed to the direct oxidation of the hydrocarbon constituents to oil-insoluble products, (2) thick soap-like material resulting from the reaction of acid products of oxidation with metals in the transformer, and (3) carbonaceous material formed by any arc or corona discharge occurring in service.

3.7.3 Insecticides

Insecticides are agents of chemical or biological origin that control insects. Control may result from killing the insect or otherwise preventing it from engaging in behaviors deemed destructive. Paraffins and naphthenes are the major components of the refined spray oils, and the former appear to be the more toxic. With both naphthenic and paraffinic hydrocarbons, the insecticidal effect increases with molecular weight but becomes constant at about 350 for each; the maximum toxicity has also been attributed to that fraction boiling between 240 and 300°C (465–570°F) at 40 mm Hg pressure.

4.0 SEMI-SOLID AND SOLID PRODUCTS

The semi-solid products and the solid products from petroleum refining (although valuable) often take a back seat to the liquid products. Residua, asphalt, and coke do not receive the same amount of attention as the liquid fuels and (incorrectly) are often considered as by-products though their use makes them as important as many of the liquid fuels.

4.1 GREASE

Grease is lubricating oil to which a thickening agent has been added for the purpose of holding the oil to surfaces that must be lubricated. The development of the chemistry of grease formulations is closely linked to an understanding of the physics at the interfaces between the machinery and the grease. With this insight, it is possible to formulate greases that are capable of operating in increasingly demanding and wide-ranging conditions.

There are three basic components that contribute to the multiphase structure of lubricating grease: (1) a base fluid, (2) a thickener, and (3) very frequently, in modern grease, a group of additives. The function of the thickener is to provide a physical matrix to hold the base fluid in a solid structure until operating conditions, such as load, shear, and temperature, initiate viscoelastic flow in the grease. To achieve this matrix, a careful balance of solubility between the base fluid and the thickener is required.

Finally, the key to providing a grease matrix that is stable, both over time and under the operating shear within machine components, can be found in the thickener system. The thickeners themselves also contribute significantly to the extreme pressure and anti-wear characteristics of grease and additionally, thickeners provide a grease gel capable of carrying additives which, in turn, extends performance in these areas.

4.2 WAX

Petroleum wax is of two general types: (1) *paraffin wax* in petroleum distillates and (2) *microcrystalline wax* in petroleum residua. The melting point of wax is not directly related to its boiling point, because waxes contain hydrocarbons of different chemical nature. Nevertheless, waxes are graded according to their melting point and oil content.

Paraffin wax is a solid crystalline mixture of straight-chain (normal) hydrocarbons ranging from C_{20} to C_{30} and possibly higher, that is, $CH_3(CH_2)_nCH_3$ where $n \geq 18$. It is distinguished by its solid state at ordinary temperatures (25°C, 77°F) and low viscosity (35–45 SUS at 99°C, 210°F) when melted. However, in contrast to petroleum wax, petrolatum (*petroleum jelly*), although solid at ordinary temperatures, does in fact contain both solid and liquid hydrocarbons. It is essentially a low-melting, ductile, microcrystalline wax. The physical properties of microcrystalline waxes are greatly affected by the oil content (Kumar et al., 2007) and hence by achieving desired level of oil content waxes of desired physical properties can be obtained.

The melting point of paraffin wax has both direct and indirect significance in most wax utilization. All wax grades are commercially indicated in a range of melting temperatures rather than at a single value, and a range of 1°C (2°F) usually indicates a good degree of refinement. Other common physical properties that help to illustrate the degree of refinement of the wax are color (ASTM D156), oil content, API gravity (ASTM D287), flash point (ASTM D92), and viscosity (ASTM D88; ASTM D445) although the last three properties are not usually given by the producer unless specifically requested.

Microcrystalline wax forms approximately 1%–2% w/w of crude oil and are a valuable product having numerous applications. This type of wax is usually obtained from heavy lube distillates by solvent dewaxing and from tank bottom sludge by acid clay treatment. However, these crude waxes usually contain appreciable quantity (10%–20% w/w) of residual oil and, as such, are not suitable for many applications such as paper coating, electrical insulation, textile printing, and polishes.

4.3 ASPHALT

First a word about nomenclature: asphalt is known by many different names (some of which, of course, cannot be mentioned here). Names such as asphalt concrete, asphalt cement, asphalt binder, hot mix asphalt, plant mix, bituminous mix, and bituminous concrete are in common use. To be more specific to this text, road asphalt is a combination of two primary ingredients – asphalt and aggregate.

Asphalt production from crude petroleum as a product of refineries in the early 20th century and the increasing popularity of the automobile served to greatly expand the asphalt industry. Asphalt may be residual (straight-run) asphalt, which is made up of the nonvolatile hydrocarbons in the feedstock, along with similar materials produced by thermal alteration during the distillation sequences, or they may be produced by air blowing residua. Alternatively, asphalt may be the residuum from a vacuum distillation unit. In either case, the properties of the asphalt are, essentially, the properties of the residuum (Table 2.4) (Speight and Exall, 2014). If the properties are not suitable for the asphalt product top meet specifications, changing the properties by, for example, blowing is necessary.

The nature of the asphalt is determined by such factors as the nature of the medium (paraffinic or aromatic), as well as the nature and proportion of the asphaltenes and of the resins (Speight, 2014, 2015b). The asphaltene constituents have been suggested to be lyophobic; the resins are lyophilic, and the *interaction* of the resins with the asphaltenes is responsible for asphaltene dispersion, which seems to exercise marked control on the nature of the asphalt. The asphaltenes vary in character but are of sufficiently high molecular weight to require dispersion as micelles, which are peptized by the resins. If the asphaltenes are relatively low in molecular weight, the resins plentiful, and the medium aromatic in nature, the result may be viscous asphalt without anomalous properties. If, however, the medium is paraffinic and the resins are scarce, and the asphaltenes are high in molecular (or micellar) weight (these conditions are encouraged by vacuum, steam reduction, or air blowing), the asphalt is of the gel type and exhibits the properties that accompany such structure. A high content of resins imparts to a product desirable adhesive character and plasticity; high asphaltene content is usually responsible for the harder, more brittle, asphalt as evidenced from the structure and rheological properties of modified asphalt (Giavarini et al., 2000).

Table 2.4 Properties of Atmospheric and Vacuum Residua

Feedstock	Gravity API	Sulfur (wt%)	Nitrogen (wt%)	Nickel (ppm)	Vanadium (ppm)	Asphaltenes (Heptane) (wt%)	Carbon Residue (Conradson) (wt%)
Arabian Light >650°F	17.7	3.0	0.2	10.0	26.0	1.8	7.5
Arabian Light >1050°F	8.5	4.4	0.5	24.0	66.0	4.3	14.2
Arabian Heavy > 650°F	11.9	4.4	0.3	27.0	103.0	8.0	14.0
Arabian Heavy >105°F	7.3	5.1	0.3	40.0	174.0	10.0	19.0
Alaska, North Slope >650°F	15.2	1.6	0.4	18.0	30.0	2.0	8.5
Alaska, North Slope >1050°F	8.2	2.2	0.6	47.0	82.0	4.0	18.0
Lloydminster (Canada) >650°F	10.3	4.1	0.3	65.0	141.0	14.0	12.1
Lloydminster (Canada) >1050°F	8.5	4.4	0.6	115.0	252.0	18.0	21.4

The use of asphalt—in many cases this was natural *bitumen* or a residuum (Speight, 2014, 2015b) rather than a processed material goes back into antiquity—was in fact the first petroleum derivative that was used extensively. Nowadays, a good portion of the asphalt produced from petroleum is consumed in paving roads; the remainder is employed for roofing, paints, varnishes, insulating, rust-protective compositions, battery boxes, and compounding materials that go into rubber products, brake linings, and fuel briquettes. However, asphalt uses can be more popularly divided into use as road oils, cutback asphalt, asphalt emulsion, and solid asphalt. The properties of asphalt are defined by a variety of standard tests (Speight, 2015b) that can be used to define quality and viscosity specifications.

Asphalt is characterized by their properties at different temperatures and stages of life simulated by laboratory aging. *Consistency* is the term used to describe the degree of fluidity or plasticity of binders at any particular temperature. The consistency of binder varies with temperature. Binders are graded based on ranges of consistency at a standard temperature. When the binder is exposed to air in thin films and is subjected to prolonged heating, i.e., during mixing with aggregates, the binder tends to harden. This means that the consistency (viscosity) of the binder has increased for any given temperature. A limited increase is allowable. However, careless temperature and mixing control can cause more damage to the binder, through hardening, than many years of service on the finished roadway.

4.4 COKE

Coke is the residue left by the destructive distillation of petroleum residua in processes such as the delayed coking process (Hsu and Robinson, 2006; Gary et al., 2007; Speight, 2011, 2017). Coke is a gray to black solid carbonaceous residue that is produced from petroleum during thermal processing; characterized by having a high carbon content (95%+ by weight) and a honeycomb type of appearance and is insoluble in organic solvents. The coke formed in catalytic cracking operations is usually non-recoverable, as it is often employed as fuel for the process.

The composition of petroleum coke varies with the source of the crude oil, but in general, large amounts of high-molecular-weight complex hydrocarbons (rich in carbon but correspondingly poor in hydrogen) make up a high proportion. The solubility of petroleum *coke* in carbon disulfide has been reported to be as high as 50%–80%, but this is in fact a misnomer, since the coke is the insoluble, honeycomb material that is the end product of thermal processes.

Coke does not offer the same potential for biotechnological changes as the gaseous and liquid petroleum products. It is used predominantly as a refinery fuel unless otherwise used for the production of a high-grade coke or carbon as desired. In the former case, the constituents of the coke that will release environmentally harmful gases such as nitrogen oxides, sulfur oxides, and particulate matter should be known. In addition, stockpiling coke on a site where it awaits use or transportation can lead to leachates that result from rainfall (or acid rainfall), which are highly detrimental. In such a case, application of the toxicity characteristic leaching procedure to the coke (Speight and Arjoon, 2012) that is designed to determine the mobility of both organic and inorganic contaminants present in materials is warranted before stockpiling the coke in the open.

Three physical structures of coke can be produced by delayed coking: (1) shot coke, (2) sponge coke, or (3) needle coke.

Shot coke is an abnormal type of coke resembling small balls. Because mechanisms are not well understood, the coke from some coker feedstocks form into small, tight, non-attached clusters that look like pellets, marbles, or ball bearings. It is usually a very hard coke, i.e., with low hardgrove grindability index (Speight, 2013a). Such coke is less desirable to the end users because of difficulties in handling and grinding. It is believed that feedstocks high in asphaltene constituents and low API favor shot coke formation. Blending aromatic materials with the feedstock and/or increasing the recycle ratio reduces the yield of shot coke. Fluidization in the coke drums may cause formation of shot coke.

Sponge coke is the common type of coke produced by delayed coking units. It is in a form that resembles a sponge and has been called honeycombed. Sponge coke, mostly used for anode-grade, is dull and black, having a porous, amorphous structure.

Needle coke (*acicular coke*) is a special quality coke produced from aromatic feedstocks. It is silver-gray, having crystalline broken needle structure, and is believed to be chemically produced through cross-linking of condensed aromatic hydrocarbons during coking reactions. It has a crystalline structure with more unidirectional pores and is used in the production of electrodes for the steel and aluminum industries and is particularly valuable because the electrodes must be replaced regularly.

Petroleum coke is employed for a number of purposes, but its chief use is in the manufacture of carbon electrodes for aluminum refining, which requires a high-purity carbon, low in ash and sulfur-free; the volatile matter must be removed by calcining. In addition, to its use as a metallurgical reducing agent, petroleum coke is employed in the manufacture of carbon brushes, silicon carbide abrasives, and structural carbon (e.g., pipes and Raschig rings), as well as calcium carbide manufacture from which acetylene is produced:

$$Coke \rightarrow CaC_2$$

$$CaC_2 + H_2O \rightarrow HC \equiv CH$$

5.0 PROCESS WASTES

Petroleum refineries are complex, but integrated, unit process operations that produce a variety of products from various feedstocks and feedstock blends (Speight, 2014, 2017). During petroleum refining, refineries use and generate an enormous amount of chemicals, some of which are present in air emissions, wastewater, or solid wastes. Emissions are also created through the combustion of fuels, and as by-products of chemical reactions occurring when petroleum fractions are upgraded. A large source of air emissions is, generally, the process heaters and boilers that produce carbon monoxide, sulfur oxides, and nitrogen oxides, leading to pollution and the formation of acid rain.

$$CO_2 + H_2O \rightarrow H_2CO_3 \text{ (carbonic acid)}$$
$$SO_2 + H_2O \rightarrow H_2SO_3 \text{ (sulfurous acid)}$$
$$2SO_2 + O_2 \rightarrow 2SO_3$$
$$SO_3 + H_2O \rightarrow H_2SO_4 \text{ (sulfuric acid)}$$
$$NO + H_2O \rightarrow HNO_2 \text{ (nitrous acid)}$$
$$2NO + O_2 \rightarrow NO_2$$
$$NO_2 + H_2O \rightarrow HNO_3 \text{ (nitric acid)}$$

Hence, there is the need for gas-cleaning operations on a refinery site so that such gases are cleaned from the gas stream prior to entry into the atmosphere.

5.1 GASES AND LOWER BOILING CONSTITUENTS

Gases and lower boiling constituents (refinery gases) contain one or more organic and inorganic constituents and are mixtures of individual compounds existing in the gaseous phase at normal environmental temperatures. These constituents typically have extremely low melting and boiling points. They also have high vapor pressures and low octanol/water partition coefficients. The aqueous solubility of these components varies, and can range from low parts per million (hydrogen gas) to several hundred 1000 parts per million (ammonia). The environmental fate characteristics of refinery gases are governed by these physical-chemical attributes.

All components of these gases will partition to the air where interaction with hydroxyl radicals may be either an important fate process or have little influence, depending on the constituent. Many of the gases are chemically stable and may be lost to the atmosphere or simply become involved in the environmental recycling of their atoms. Some show substantial water solubility, but their volatility eventually causes these gases to enter the atmosphere. This leads to the recognition that air emissions from a refinery include point and non-point sources. Point sources are emissions that exit stacks and flares and, thus, can be monitored and treated. Non-point sources are "fugitive emissions" which are difficult to locate and capture. Fugitive emissions occur throughout refineries and arise from the thousands of valves, pumps, tanks, pressure relief valves, and flanges. While individual leaks are typically small, the sum of all fugitive leaks at a refinery can be one of its largest emission sources.

The numerous process heaters used in refineries to heat process streams or to generate steam (boilers) for heating or steam stripping can be potential sources of SO_x, NO_x, CO, particulates and hydrocarbons emissions. When operating properly and when burning cleaner fuels such as refinery fuel gas, fuel oil, or natural gas, these emissions are relatively low. If, however, combustion is not complete, or heaters are fired with refinery fuel pitch or residuals, emissions can be significant.

Sulfur is removed from a number of refinery process off-gas streams (sour gas) in order to meet the SO_X emission limits of the Clean Air Act and to recover saleable elemental sulfur. Process off-gas streams, or sour gas, from the coker, catalytic cracking unit, hydrotreating units, and hydroprocessing units can contain high concentrations of H_2S mixed with light refinery fuel gases. Before elemental sulfur can be recovered, the fuel gases (primarily CH_4 and ethane) need to be separated from the H_2S. This is typically accomplished by dissolving the H_2S in a chemical solvent. Solvents most commonly used are amines, such as diethanolamine (DEA). Dry adsorbents such as molecular sieves, activated carbon, iron sponge and zinc oxide are also used. In the amine solvent processes, DEA solution or another amine solvent is pumped to an absorption tower where the gases are contacted and H_2S is dissolved in the solution. The fuel gases are removed for use as fuel in process furnaces in other refinery operations. The amine-H_2S solution is then heated and steam stripped to remove the H_2S gas.

Current methods for removing sulfur from the H_2S gas streams are typically a combination of two processes: the Claus process followed by the Beavon process, SCOT process, or the Wellman-Land process. The Claus process consists of partial combustion of the H_2S-rich gas stream (with one-third the stoichiometric quantity of air) and then reacting the resulting sulfur dioxide and unburned H_2S in the presence of a bauxite catalyst to produce elemental sulfur.

Since the Claus process by itself removes only about 90% of the H_2S in the gas stream, the Beavon, SCOT, or Wellman-Land processes are often used to further recover sulfur. In the Beavon process, the H_2S in the relatively low concentration gas stream from the Claus process can be almost completely removed by absorption in quinone solution. The dissolved H_2S is oxidized to form a mixture of elemental sulfur and hydroquinone. The solution is injected with air or oxygen to oxidize the hydroquinone back to quinone. The solution is then filtered or centrifuged to remove the sulfur and the quinone is then reused. The Beavon process is also effective in removing small amounts of sulfur dioxide, COS, and carbon disulfide that are not affected by the Claus process. These compounds are first converted to H_2S at elevated temperatures in a cobalt molybdate catalyst prior to being fed to the Beavon unit. Air emissions from sulfur recovery units will consist of H_2S, SO_x and NO_x in the process tail gas, as well as fugitive emissions and releases from vents.

The SCOT process is also widely used for removing sulfur from the Claus tail gas. The sulfur compounds in the Claus tail gas are converted to H_2S by heating and passing it through a cobalt-molybdenum catalyst with the addition of a reducing gas. The gas is then cooled and contacted with a solution of diisopropanolamine (DIPA) which removes all but trace amounts of H_2S. The sulfide-rich DIPA is sent to a stripper where H_2S gas is removed and sent to the Claus plant and the cleaned DIPA is returned to the absorption column.

Many of the gaseous and liquid constituents of the lower boiling fractions of petroleum and in petroleum products fall into the class of chemicals which have one or more of the following characteristics that are declared hazardous by EPA in terms of the following properties (1) ignitability-flammability, (2) corrosivity, (3) reactivity, and (4) hazardous.

An ignitable liquid is a liquid that has a flash point of less than 60°C (140°F). Examples are: benzene, hexane, heptane, pentane, petroleum ether (low boiling), toluene, and xylene(s). An aqueous solution that has a pH of less than or equal to 2, or greater than or equal to 12.5 is considered *corrosive*. Most petroleum constituents and petroleum products are not corrosive but many of the chemicals used in refineries are corrosive. Corrosive materials also include substances such as sodium hydroxide and some other acids or bases. Chemicals that react violently with air or water are considered *reactive*. Examples are sodium metal, potassium metal, phosphorus, etc. Reactive materials also include strong oxidizers such as perchloric acid ($HClO_4$), and chemicals capable of detonation when subjected to an initiating source, such as solid, dry $< 10\%$ H_2O picric acid, benzoyl peroxide, or sodium borohydride ($NaBH_4$). Solutions of certain cyanides or sulfides that could generate toxic gases are also classified as reactive. *Hazardous chemicals* have toxic, carcinogenic, mutagenic, or teratogenic effects on humans or other life forms and are designated either as *Acutely Hazardous Waste* or *Toxic Waste* by the EPA. Substances containing any of the toxic constituents so listed are to be considered *hazardous* unless, after considering the following factors it can reasonably be concluded that the chemical (waste) is not capable of posing a substantial present or potential hazard to public health or the environment when improperly treated, stored, transported, or disposed of, or otherwise managed.

The issues to be held in consideration are (1) the nature of the toxicity presented by the constituent, (2) the concentration of the constituent in the waste, (3) the potential of the constituent or any toxic degradation product of the constituent to migrate from the waste into the environment under the types of improper management, (4) the persistence of the constituent or any toxic degradation product of the constituent, (5) the potential for the constituent or any toxic degradation product of the constituent to degrade into non-harmful constituents and the rate of degradation, (6) the degree to which the constituent or any degradation product of the constituent accumulates in an ecosystem, (7) the plausible types of improper management to which the waste could be subjected, (8) the quantities of the waste generated at individual generation sites or on a regional or national basis, (9) the nature and severity of the public health threat and environmental damage that has occurred as a result of the improper management of wastes containing the constituent, and (10) actions taken by other governmental agencies or regulatory programs based on the health or environmental hazard posed by the waste or waste constituent.

5.2 HIGHER BOILING CONSTITUENTS

Naphthalene and its homologs are less acutely toxic than benzene but are more prevalent for a longer period during oil spills. The toxicity of different crude oils and refined oils depends on not only the total concentration of hydrocarbons but also the hydrocarbon composition in the water-soluble fraction (WSF) of petroleum, water solubility, concentrations of individual components, and toxicity of the components. The WSFs prepared from different oils will vary in these parameters. WSF of refined oils (for example, No. 2 fuel oil and Bunker C oil) are more toxic than WSF of crude oil to several species of fish (killifish and salmon). Compounds with either more rings or methyl substitutions are more toxic than less substituted compounds, but tend to be less water soluble and thus less plentiful in the WSF.

Among the polynuclear aromatic hydrocarbons, the toxicity of petroleum is a function of its di- and tri-aromatic hydrocarbon content. Like the single aromatic ring variations, including benzene, toluene, and the xylenes, all are relatively volatile compounds with varying degrees of water solubility. The larger and higher molecular weight aromatic structures (with four to five aromatic rings), which are more persistent in the environment, have the potential for chronic toxicological effects. Since these compounds are nonvolatile and are relatively insoluble in water, their main routes of exposure are through ingestion and epidermal contact. Some of the compounds in this classification are considered possible human carcinogens; these include benzo(a and e)pyrene, benzo(a)anthracene, benzo(b, j, and k)fluorene, benzo(ghi)perylene, chrysene, dibenzo(ah)anthracene, and pyrene.

Mixtures of polynuclear aromatic hydrocarbons are often carcinogenic and possibly photo-toxic. One way to approach site-specific risk assessments would be to collect the complex mixture of polynuclear aromatic hydrocarbons and other lipophilic contaminants in a semipermeable membrane device (SPMD, also known as a *fat bag*), then test the mixture for carcinogenicity, toxicity, and phototoxicity.

The solubility of hydrocarbon components in petroleum products is an important property when assessing toxicity. The water solubility of a substance determines the routes of exposure that are possible. Solubility is approximately inversely proportional to molecular weight; lighter hydrocarbons are more soluble in water than higher molecular weight compounds. Lower molecular weight hydrocarbons (C4 to C8, including the aromatic compounds) are relatively soluble, up to about 2000 ppm, while the higher molecular weight hydrocarbons are nearly insoluble. Usually, the most soluble components are also the most toxic.

5.3 WASTEWATER

Process wastewater is also a significant effluent from a number of refinery processes. Atmospheric distillation units and vacuum distillation units create the largest volumes of process wastewater, about 26 gallons per barrel of oil processed. Fluid catalytic cracking and catalytic reforming also generate considerable amounts of wastewater (15 and 6 gallons per barrel of feedstock, respectively). A large portion of wastewater from these three processes is contaminated with oil and other impurities and must be subjected to primary, secondary, and sometimes tertiary water treatment processes, some of which also create hazardous waste.

The predominant process wastewater from petroleum refining consists of cooling water, process water, storm water, and sanitary sewage water. A large portion of water used in petroleum refining is used for cooling and most cooling water is recycled. Cooling water typically does not come into direct contact with process oil streams and therefore contains a lesser number of contaminants than process wastewater. However, it may contain some oil contamination due to leaks in the process equipment.

Refinery effluent water contains various hydrocarbon components including gasoline blending stocks, kerosene, diesel fuel, and heavier liquids. Also present may be suspended mineral solids, sand, salt, organic acids, and sulfur compounds. The nature of the components depends on the constituents of the inlet crude oil, as well as the processing scheme of the refinery (Speight, 2005). Most of these constituents would be undesirable in the effluent water, so it is necessary to treat the water to remove the contaminants.

Water used in processing operations account for a significant portion of the total wastewater. Process wastewater arises from desalting crude oil, steam stripping operations, pump gland cooling, product fractionator reflux drum drains, and boiler blowdown. Because process water often comes into direct contact with oil, it is usually highly contaminated. Storm water (i.e., surface water runoff) is intermittent and will contain constituents from spills to the surface, leaks in equipment, and any materials that may have collected in drains. Runoff surface water also includes water coming from crude and product storage tank roof drains.

5.4 SPENT CAUSTIC

In the petroleum refining industry, caustic solutions (i.e., NaOH) are regularly used to remove H_2S and organic sulfur compounds from hydrocarbon streams. Once H_2S is reacted with NaOH, the solution becomes known as *spent caustic* or *spent sulfidic caustic*. Spent caustics typically have a pH value (>12) and high sulfide concentrations (2%–3% w/w). Depending on the source, spent caustic may also contain phenols, mercaptans, amines, and other organic compounds that are soluble or emulsified in the caustic (Speight, 2005; Speight and Arjoon, 2012).

Most spent caustics are sent off-site for commercial recovery or reuse, e.g., in pulp and paper mills, for treatment by wet air oxidation, or for disposal by deep-well injection. The main methods that deal with caustics are chemical methods, such as neutralization and oxidation, but these need large investments and have a high operating cost. And the most important is that it will lead to a serious environmental pollution and equipment corrosion.

5.5 SOLID WASTE

Solid wastes are generated from many of the refining processes, petroleum handling operations, as well as wastewater treatment. Both hazardous and nonhazardous wastes are generated, treated, and disposed. Refinery wastes are typically in the form of sludge (including sludge from wastewater treatment), spent process catalysts, filter clay, and incinerator ash. Treatment of these wastes includes incineration, land treating off-site, land filling onsite, land filling off-site, chemical fixation, neutralization, and other treatment methods.

A significant portion of the non-petroleum product outputs of refineries is transported off-site and sold as by-products. These outputs include sulfur, acetic acid, phosphoric acid, and recovered metals. Metals from catalysts and from the crude oil that have deposited on the catalyst during the production often are recovered by third-party recovery facilities.

Storage tanks are used throughout the refining process to store crude oil and intermediate process feeds for cooling and further processing. Finished petroleum products are also kept in storage tanks before transport off-site. Storage tank bottoms are mixtures of iron rust from corrosion, sand, water, and emulsified oil and wax, which accumulate at the bottom of tanks. Liquid tank bottoms (primarily water and oil emulsions) are periodically drawn off to prevent their continued build up. Tank bottom liquids and sludge are also removed during periodic cleaning of tanks for inspection. Tank bottoms may contain amounts of tetraethyl or tetramethyl lead (although this is increasingly rare due to the phasing-out of leaded products), other metals, and phenols. Solids generated from leaded gasoline storage tank bottoms are listed as a Resource Conservation and Recovery Act hazardous waste.

Sulfonic acids are produced by when petroleum is treated with sulfuric acid. Sulfuric acid treating of petroleum distillates is generally applied to dissolve unstable or colored substances and sulfur compounds, as well as to precipitate asphaltic materials. When drastic conditions are employed, as in the treatment of lubricating fractions with large amounts of concentrated acid or when fuming acid is used in the manufacture of white oils, considerable quantities of petroleum sulfonic acids are formed. Extensive side reactions, mainly oxidation, also occur and increase with the proportion of sulfur trioxide in the acid.

Many of the lower molecular weight paraffins are physically absorbed by concentrated and fuming sulfuric acids; chemical activity increases with rise in molecular weight, and compounds containing tertiary carbons are especially responsive. *n*-Hexane, *n*-heptane, and *n*-octane are essentially inactive in cold fuming acid; but at the boiling point of the hydrocarbons rapid sulfonation takes place to give mono- and disulfonic acids:

$$RH + H_2SO_4 \rightarrow RSO_3H + H_2O$$
Paraffin   sulfonic acid

The five- and six-membered ring lower naphthene derivatives are stable to cold concentrated sulfuric acid, but fuming sulfuric acid reacts with cyclohexane to give mono- and naphthene derivatives and mono-aromatic sulfonic acids, along with products based on cyclic olefins formed through hydrogen-transfer reactions.

Acid sludge produced during the use of sulfuric acid as a treating agent is mainly of two types: (1) sludge from light oils (gasoline and kerosene) and (2) sludge from lubricating stocks, medicinal oils, and the like. In the treatment of the latter oils it appears that the action of the acid causes precipitation of asphaltene constituents and resin constituents, as well as the solution of color-bearing and sulfur compounds. Sulfonation and oxidation-reduction reactions also occur but to a lesser extent since much of the acid can be recovered. In the desulfurization of cracked distillates, however, chemical interaction is more important, and polymerization, ester formation, aromatic-olefin condensation, and sulfonation also occur. N_2 bases are neutralized, and the acid dissolves naphthenic acids; thus the composition of the sludge is complex and depends largely on the oil treated, acid strength, and the temperature.

Sulfuric acid sludge from *iso*-paraffin alkylation and lubricating oil treatment are frequently decomposed thermally to produce sulfur dioxide (which is returned to the sulfuric acid plant) and *sludge acid coke*. The coke, in the form of small pellets, is used as a substitute for charcoal in the manufacture of carbon disulfide. Sulfuric acid coke is different from other petroleum coke in that it is pyrophoric in air and also reacts directly with sulfur vapors to form carbon disulfide.

REFERENCES

API, June 10, 2009. Refinery Gases Category Analysis and Hazard Characterization. Submitted to the EPA by the American Petroleum Institute, Petroleum HPV Testing Group. HPV Consortium Registration # 1100997 United States Environmental Protection Agency, Washington, DC.

ASTM D86, 2017. Standard Test Method for Distillation of Petroleum Products at Atmospheric Pressure. Annual Book of Standards. ASTM International, West Conshohocken, Pennsylvania.

ASTM D88, 2017. Standard Test Method for Saybolt Viscosity. Annual Book of Standards. ASTM International, West Conshohocken, Pennsylvania.

ASTM D92, 2017. Standard Test Method for Flash and Fire Points by Cleveland Open Cup Tester. Annual Book of Standards. ASTM International, West Conshohocken, Pennsylvania.

ASTM D156, 2017. Standard Test Method for Saybolt Color of Petroleum Products (Saybolt Chromometer Method). Annual Book of Standards. ASTM International, West Conshohocken, Pennsylvania.

ASTM D287, 2017. Standard Test Method for API Gravity of Crude Petroleum and Petroleum Products (Hydrometer Method). Annual Book of Standards. ASTM International, West Conshohocken, Pennsylvania.

ASTM D445, 2017. Standard Test Method for Kinematic Viscosity of Transparent and Opaque Liquids (and Calculation of Dynamic Viscosity). Annual Book of Standards. ASTM International, West Conshohocken, Pennsylvania.

ASTM D1070, 2017. Standard Test Methods for Relative Density of Gaseous Fuels. Annual Book of Standards. ASTM International, West Conshohocken, Pennsylvania.

ASTM D1071, 2017. Standard Test Methods for Volumetric Measurement of Gaseous Fuel Samples. Annual Book of Standards. ASTM International, West Conshohocken, Pennsylvania.

ASTM D1142, 2017. Standard Test Method for Water Vapor Content of Gaseous Fuels by Measurement of Dew-Point Temperature. Annual Book of Standards. ASTM International, West Conshohocken, Pennsylvania.

ASTM D1319, 2017. Standard Test Method for Hydrocarbon Types in Liquid Petroleum Products by Fluorescent Indicator Adsorption. Annual Book of Standards. ASTM International, West Conshohocken, Pennsylvania.

ASTM D1826, 2017. Standard Test Method for Calorific (Heating) Value of Gases in Natural Gas Range by Continuous Recording Calorimeter. Annual Book of Standards. ASTM International, West Conshohocken, Pennsylvania.

ASTM D1835, 2017. Standard Specification for Liquefied Petroleum (LP) Gases. Annual Book of Standards. ASTM International, West Conshohocken, Pennsylvania.

ASTM D2163, 2017. Standard Test Method for Determination of Hydrocarbons in Liquefied Petroleum (LP) Gases and Propane/Propene Mixtures by Gas Chromatography. Annual Book of Standards. ASTM International, West Conshohocken, Pennsylvania.

ASTM D2425, 2017. Standard Test Method for Hydrocarbon Types in Middle Distillates by Mass Spectrometry. Annual Book of Standards. ASTM International, West Conshohocken, Pennsylvania.

ASTM D2549, 2017. Standard Test Method for Separation of Representative Aromatics and Nonaromatics Fractions of High-Boiling Oils by Elution Chromatography. Annual Book of Standards. ASTM International, West Conshohocken, Pennsylvania.

ASTM D2650, 2017. Standard Test Method for Chemical Composition of Gases by Mass Spectrometry. Annual Book of Standards. ASTM International, West Conshohocken, Pennsylvania.

ASTM D2789, 2017. Standard Test Method for Hydrocarbon Types in Low Olefinic Gasoline by Mass Spectrometry. Annual Book of Standards. ASTM International, West Conshohocken, Pennsylvania.

ASTM D3956, 2017. Standard Specification for Methane Thermophysical Property Tables. Annual Book of Standards. ASTM International, West Conshohocken, Pennsylvania.

ASTM D3984, 2017. Standard Specification for Ethane Thermophysical Property Tables. Annual Book of Standards. ASTM International, West Conshohocken, Pennsylvania.

ASTM D4150, 2017. Standard Terminology Relating to Gaseous Fuels. Annual Book of Standards. ASTM International, West Conshohocken, Pennsylvania.

ASTM D4362, 2017. Standard Specification for Propane Thermophysical Property Tables. Annual Book of Standards. ASTM International, West Conshohocken, Pennsylvania.

ASTM D4424, 2017. Standard Test Method for Butylene Analysis by Gas Chromatography. Annual Book of Standards. ASTM International, West Conshohocken, Pennsylvania.

ASTM D4650, 2017. Standard Specification for Normal Butane Thermophysical Property Tables. Annual Book of Standards. ASTM International, West Conshohocken, Pennsylvania.

ASTM D4651, 2017. Standard Specification for Isobutane Thermophysical Property Tables. Annual Book of Standards. ASTM International, West Conshohocken, Pennsylvania.

ASTM D5305, 2017. Standard Test Method for Determination of Ethyl Mercaptan in LP-Gas Vapor. Annual Book of Standards. ASTM International, West Conshohocken, Pennsylvania.

ASTM D5504, 2017. Standard Test Method for Determination of Sulfur Compounds in Natural Gas and Gaseous Fuels by Gas Chromatography and Chemiluminescence. Annual Book of Standards. ASTM International, West Conshohocken, Pennsylvania.

ASTM D6228, 2017. Standard Test Method for Determination of Sulfur Compounds in Natural Gas and Gaseous Fuels by Gas Chromatography and Flame Photometric Detection. Annual Book of Standards. ASTM International, West Conshohocken, Pennsylvania.

Banaszewski, A., Blythe, J., 2000. In: Lucas, A.G. (Ed.), Modern Petroleum Technology. Downstream, vol. 2. John Wiley & Sons Inc., New York (Chapter 30).

Boughton, R., Horvath, A., 2004. Environmental assessment of used oil management methods. Environmental Science & Technology 38 (2), 353–358.

Gary, J.H., Handwerk, G.E., Kaiser, M.J., 2007. Petroleum Refining: Technology and Economics, fifth ed. CRC Press, Taylor & Francis Group, Boca Raton, Florida.

Giavarini, C., Mastrofini, D., Scarsella, M., 2000. Macrostructure and rheological properties of chemically modified residues and bitumens. Energy & Fuels 14, 495–502.

Hsu, C.S., Robinson, P.R. (Eds.), 2006. Practical Advances in Petroleum Processing Volume 1 and Volume 2. Springer Science, New York.

Kumar, S., Nautiyal, S.P., Agrawal, K.M., 2007. Physical properties of petroleum waxes 1: effect of oil content. Petroleum Science and Technology 25, 1531–1537.

Mokhatab, S., Poe, W.A., Speight, J.G., 2006. Handbook of Natural Gas Transmission and Processing. Elsevier, Amsterdam, Netherlands.

Robinson, J.D., Faulkner, R.P., 2000. In: Lucas, A.G. (Ed.), Modern Petroleum Technology. Downstream, vol. 2. John Wiley & Sons Inc., New York. Chapter 1.

Speight, J.G., 2005. Environmental Analysis and Technology for the Refining Industry. John Wiley & Sons Inc., Hoboken, New Jersey.

Speight, J.G., 2007. Natural Gas: A Basic Handbook. GPC Books. Gulf Publishing Company, Houston, Texas.

Speight, J.G., 2011. The Refinery of the Future. Gulf Professional Publishing, Elsevier, Oxford, United Kingdom.

Speight, J.G., Arjoon, K.K., 2012. Bioremediation of Petroleum and Petroleum Products. Scrivener Publishing, Salem, Massachusetts.

Speight, J.G., 2013a. The Chemistry and Technology of Coal, third ed. CRC Press, Taylor & Francis Group, Boca Raton, Florida.

Speight, J.G., 2013b. Shale Gas Production Processes. Gulf Professional Publishing, Elsevier, Oxford, United Kingdom.

Speight, J.G., 2014. The Chemistry and Technology of Petroleum, fifth ed. CRC Press, Taylor & Francis Group, Boca Raton, Florida.

Speight, J.G., Exall, D.I., 2014. Refining Used Lubricating Oils. CRC Press, Taylor & Francis Group, Boca Raton, Florida.

Speight, J.G., 2015a. Handbook of Petroleum Product Analysis, second ed. John Wiley & Sons Inc., Hoboken, New Jersey.

Speight, J.G., 2015b. Asphalt Materials Science and Technology. Butterworth-Heinemann, Elsevier, Oxford, United Kingdom.

Speight, J.G., 2017. Handbook of Petroleum Refining. CRC Press, Taylor & Francis Group, Boca Raton, Florida.

Watanabe, K., Nagai, K., Aratani, N., Saka, Y., Chiyoda, N., Mizutani, H., December 2010. Techniques for octane number enhancement in FCC gasoline. In: Proceedings. 20th Annual Saudi-Japan Symposium on Catalysts in Petroleum Refining & Petrochemicals. Saudi Arabia, Dhahran.

INTRODUCTION TO PETROLEUM BIOTECHNOLOGY

3

1.0 INTRODUCTION

Biotechnology is the use of an environmentally friendly technique used to restore soil and water to its original state by using indigenous microbes to break down petroleum constituents into simpler usable products or to breakdown petroleum and free the environment from environmentally unfriendly contaminants into products that are removed from the environment or a benign in terms of environmental change. The implementation of biotechnological processes in the oil industry is an important aspect for the future production of fuels and valuable products.

In fact, in recent years, with more and more attention given to the study of microorganisms in oil fields to mitigate potential environmental issues (Speight and Arjoon, 2012) as well as application to enhanced oil recovery (Chapter 4), there has been a surge of interest in the potential for the application of biotechnology to other aspects of petroleum technology, such as the potential for biorefining of crude oil (El-Gendy and Speight, 2015). For example, biotechnology can be used in petroleum refining in processes such as desulfurization (Chapter 6) and denitrogenation (Chapter 7) (El-Gendy and Speight, 2015). The microorganisms used in biotechnological applications may be a collage of multiple organisms with each organism designed to perform a specific chemical task that can be integrated into the whole operation.

More particularly, the term biotechnology (as used in this text) can be broadly defined as *using living organisms or their products for commercial purposes*. As such, biotechnology has been practiced by human society since the beginning of recorded history in such activities as baking bread, brewing alcoholic beverages, or breeding food crops or domestic animals. In the current context, a more specific definition of biotechnology is *the commercial application of living organisms or their products, which involves the deliberate manipulation of their DNA molecules*, which implies that a set of laboratory techniques developed within the last 20 years that have been responsible for the current scientific and commercial interest in biotechnology, the founding of many new companies, and the redirection of research efforts and financial resources among established companies and universities.

Thus biotechnology refers to the use of living systems and organisms (or part of them) to develop or make products, or any technological application that uses biological systems, living organisms, or derivatives thereof, to make or modify products or processes for specific use. Depending on the applications, biotechnology overlaps (as an umbrella term) with the fields of bioengineering and biomanufacturing. The advantages of biocatalysts are its specificity and high selectivity; its flexibility as it can operate in a wide range of conditions, mostly, under ambient temperature and pressure; low energetic costs; low toxicity; low emissions; and no generation of undesirable by-products. Based on the aforementioned criteria, the whole microbial cell or its enzymes can transform compounds into products. Furthermore, biotechnology can be applied in different sectors of petroleum industry.

By way of definition for the current context, *petroleum biotechnology* is based on biotransformation processes to recover, refine, and transform petroleum (and petroleum products) into valuable derivatives.

Introduction to Petroleum Biotechnology. https://doi.org/10.1016/B978-0-12-805151-1.00003-5

Petroleum microbiology research is advancing on many fronts, spurred on most recently by new knowledge of cellular structure and function gained through molecular and protein engineering techniques, combined with more conventional microbial methods. *Petroleum bioremediation* refers specifically to the cleanup of spills of petroleum and petroleum products using microorganisms (Speight and Arjoon, 2012). Furthermore, *biodegradation (biotic degradation, biotic decomposition) is the chemical degradation of contaminants by bacteria or other biological means.* Organic material can be degraded aerobically (in the presence of oxygen) or anaerobically (in the absence of oxygen). Most bioremediation systems run under aerobic conditions, but a system under anaerobic conditions may permit microbial organisms to degrade chemical species that are otherwise nonresponsive to aerobic treatment and vice versa. Thus, in the current context of crude oil biotransformation, biotechnology has the potential to enable crude oil refineries to reduce waste, minimize water use, prevent pollution, and curb greenhouse gas emissions.

Generally, biological processing of petroleum feedstocks offers an attractive alternative to conventional thermochemical treatment due to the mild operating conditions and greater reaction specificity afforded by the nature of biocatalysts. Efforts in microbial screening and development have identified microorganisms capable of petroleum desulfurization, denitrogenation, and demetallization. Biological desulfurization of petroleum may occur either oxidatively or reductively. In the oxidative approach, organic sulfur is converted to sulfate and may be removed in process water. This route is attractive because it would not require further processing of the sulfur and may be amenable for use at the well head where process water may then be reinjected. In the reductive desulfurization scheme, organic sulfur is converted into hydrogen sulfide, which may then be catalytically converted into elemental sulfur, an approach of utility at the refinery. Regardless of the mode of biodesulfurization (BDS), key factors affecting the economic viability of such processes are biocatalyst activity and cost, differential in product selling price, sale or disposal of coproducts or wastes from the treatment process, and the capital and operating costs of unit operations in the treatment scheme.

However, for any petroleum biotechnology to be effective, the microorganisms must convert the petroleum constituents into harmless saleable products (or products that may require additional treatment). In the case of bioremediation, the microorganisms must convert the petroleum constituents into environmentally benign products. In addition, since bioremediation can be effective only where environmental conditions permit microbial growth and activity, its application often involves the manipulation of environmental parameters to allow microbial growth and degradation to proceed at a faster rate. However, as is the case with other technologies, bioremediation has its limitations and there are several disadvantages that must be recognized (Table 3.2).

Furthermore, the control and optimization of biotechnology processes is a complex system of many factors. These factors include (1) the existence of a microbial population capable of degrading the crude oil constituents; (2) the availability of contaminants to the microbial population; (3) the environment factors, i.e., the type of soil, the temperature, and the pH; and (4) the presence of oxygen or other electron acceptors and nutrients.

Also, the general concept of biotransformation encompasses a wide range of procedures for modifying chemicals in living organisms according to human needs. In addition, and even more pertinent to this text, bioengineering is a related field that more heavily emphasizes higher systems approaches (not necessarily the altering or using of biological materials directly) for interfacing with and utilizing microbes and is the application of the principles of engineering and natural sciences to molecular transformation of the feedstock. At the same time, it must be recognized that the environment can be affected by biotechnologies, both positively and adversely. The cleanup of environmental waste is an example of an application of environmental biotransformation of petroleum-based contaminants and care must be taken to ensure that there is no loss of containment of nonindigenous microbes that could bring harm to an ecosystem (Speight and Arjoon, 2012) are examples of environmental implications of biotechnology.

Moreover, due to the revolution in protein and genetic engineering, the study of extremophilic microorganisms, biocatalysts in nonaqueous media, and nanobiocatalysts, biotechnology found its way in petroleum refining (Fig. 3.1) as a means of offering biorefining options (Fig. 3.2) for the refinery. The major potential applications of biorefining are BDS, biodenitrogenation (BDN), biodemetallization (BDM), biotransformation of heavy crude oils into lighter crude oils, and finally biodepolymerization of asphaltene constituents (Le Borgne and Quintero, 2003; Bachmann et al., 2014). However, the major applications of petroleum biotechnology are biotreatment of waste streams (wastewater or gases), bioremediation of hydrocarbon-polluted soils and sediments, and finally, the microbial enhanced oil recovery (MEOR). But the main well-established application is related to effluent treatment and bioremediation. Although, biodegradation of resins and asphaltenes have been reported, and microorganisms are associated with the degradation of metalloporphyrins. However, there is a little clear evidence that BDM of crude oil can be achieved.

In fact, biotechnology is now accepted as an attractive means of improving the efficiency of any industrial processes and resolving serious environmental problems. One of the reasons for this is the

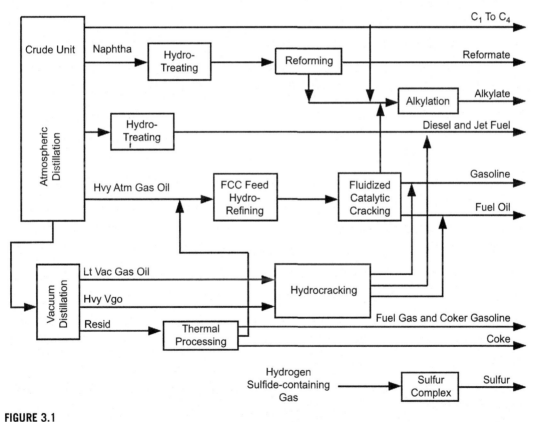

FIGURE 3.1

Refinery schematic.

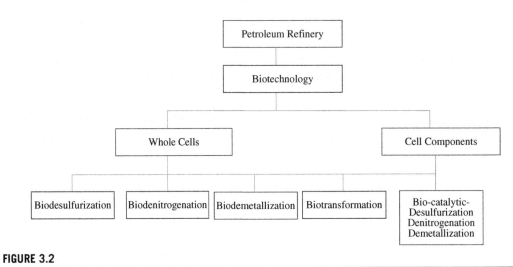

FIGURE 3.2

Potential applications of biotechnology in a petroleum refinery.

extraordinary metabolic capability that exists within the bacterial world. Microbial enzymes can bio-transform a wide range of compounds, and the worldwide increase in attention being paid to this concept can be attributed to several factors, including the presence of a wide variety of catabolic enzymes and the ability of many microbial enzymes to transform a broad range of unnatural compounds (xenobiotic compounds) as well as natural compounds. Biotransformation processes have several advantages compared with chemical processes, including the following: (1) microbial enzyme reactions are often more selective, (2) biotransformation processes are often more energy-efficient, (3) microbial enzymes are active under mild conditions, and (4) microbial enzymes are environment-friendly biocatalysts. Although many biotransformation processes have been described, only a few of these have been used as part of an industrial process and opportunities exist for biorefining of petroleum (Mohebali and Ball, 2008). Of particular interest in this context is the phenomenon of BDS (biological desulfurization, microbial desulfurization) which is used to oxidize sulfur compounds in crude oil ultimately resulting in desulfurization (El-Gendy and Speight, 2015). This represents the ability of microbial species to "desulfurize compounds that are recalcitrant to the current standard technology in the oil industry" (Abin-Fuentes et al., 2013).

Biobased processes afford the potential for substantially lower capital expenses per ton of capacity in the form of fewer unit operations since bioprocesses can often handle multiple steps of a process in a single-unit operation—fermentation. Fewer operations mean less equipment and lower costs. In addition, bioprocesses can often design organisms to produce exactly the chemical of interest, rather than the mix of hydrocarbons. In fact, biotechnology has the potential to become a must-have component of a petroleum production and refining portfolios as the bio-based processes prove their commercial reliability and economics.

Thus it is the purpose of this chapter to present to the reader an introduction to petroleum biotechnology and the potential for the commercial installation of biotransformation as a part of refinery operations.

2.0 **PRINCIPLES OF BIOTECHNOLOGY**

The principles of biotechnology as applied to crude oil are defined by the composition of crude oil. Crude oil is a complex mixture of thousands of various compounds, organic and inorganic, including aliphatic and aromatic hydrocarbons as well as higher molecular weight compounds containing sulfur, nitrogen, oxygen, and metals. Nonhydrocarbon compounds (those compounds that are not composed solely of carbon and hydrogen) include sulphur compounds in the form of hydrogen sulfide ($H2S$), mercaptans (compounds containing the –SH group), organic sulfide derivatives, organic thiophene derivatives, as well as benzothiophene derivatives and naphthothiophene derivatives (Chapter 1). These compounds are unfavorable to biotechnology processes due to their resistance to biochemical change. Nitrogen compounds represent nonhydrocarbon compounds that occur in crude oil and occurs as (1) basic and (2) nonbasic derivatives (Chapter 1). The first group includes pyridine derivatives and quinoline derivatives while the second group comprises pyrrole derivatives, indole derivatives, and carbazole derivatives (Chapter 1). Oxygen compounds such as phenol derivatives, carboxylic acid derivatives, and furan derivatives also occur in crude oil. Porphyrin derivatives often occur in crude oil and are composed of pyrrole rings connected by methine bridges.

Trace elements are present in crude oil in ppm quantities. Besides porphyrins, trace elements occur as naphthenic acids soaps (particularly compounds of Zn, Ti, Ca, and Mg), as well as metalorganic bonds (V, Cu, Ni, Fe). The highest concentration of trace elements that have been determined corresponds to vanadium, nickel, and iron as well as calcium and iron. Crude oil is naturally enriched with these elements during its migration from the source rock to the reservoir rock and even within the reservoir rock. Particularly high contents of vanadium have been found in crude oils from Venezuela.

Resin constituents and asphaltene constituents represent the high-boiling fractions of crude oil, particularly in those crude oils that are designated as naphthenic crude oils. The resin constituents and asphaltene constituents have very complex chemical structures and include most of the heteroatoms (nitrogen, oxygen, and sulfur compounds), trace elements, and polynuclear aromatic hydrocarbon derivatives (PNAs, also called polycyclic aromatic hydrocarbon derivatives, PAHs).

The conditions prevailing in a crude oil reservoir significantly differ from environmental settings typical to the occurrence of living organisms on Earth. The red-ox potential is very low, the pressure and temperature are very high, and the salt content may reach up to over 10%. Moreover, this setting lacks electron acceptors, such as oxygen, typical for most microorganisms, while sulfate and carbonate are present and the range of electron donors admissible for microorganisms is very wide.

Most hydrocarbon derivatives (especially the heteroatom-containing compounds including the trace metal constituents) occurring in crude oil have toxic effects resulting mainly from their chemical structure (Speight and Arjoon, 2012). These toxic hydrocarbons include both aliphatic and aromatic compounds, such as PAHs, whose toxicity increases proportionally to the number of carbon atoms in the compound. This is particularly in the case of the PAH derivatives with more than four-member rings in the structure. Despite the toxicity of the chemical compounds occurring in crude oil, several groups of microorganisms have been found in this setting.

The main sources of carbon for microorganisms in crude oil are hydrocarbons, both aliphatic and aromatic, but also organic compounds that are often the products of crude oil biodegradation. These organic compounds include organic acids such as acetic acid, benzoic acid, butyric acid, formic acid, propanoic acid, and naphthenic acids. The electron donors may be hydrogen and, in the case of immature oil, the resin constituents and the asphaltene constituents, whose metabolic availability is confirmed by the fact that anaerobic microorganisms may develop in cultures with crude oil without any modifications of the composition.

In spite of the complexity of the various crude oils (which is reservoir specific), industrial biotechnology, of which petroleum biotechnology is a part, is one of the most promising new approaches to pollution prevention, resource conservation, and cost reduction. It is often referred to as the third wave in biotechnology. If developed to the full potential, industrial biotechnology may have a larger impact on the world than health care and agricultural biotechnology. The concept offers businesses a way to reduce costs and create new markets while protecting the environment. Also, since many of the products do not require the lengthy review times that drug products must undergo, it is a quicker, easier pathway to the market. The application of biotechnology to industrial processes is not only transforming how products are manufactured but is also providing with new products that could not even be imagined a few years ago. However, because industrial biotechnology is so new, its benefits are still not well known or understood by industry, policymakers, or consumers.

From the initial inception of the concept, industrial biotechnology has integrated product improvements with pollution prevention. This is illustrated by the way in which industrial biotechnology solved the phosphate water pollution problems in the 1970s caused by the use of phosphates in laundry detergent. Biotechnology companies developed enzymes that removed stains from clothing better than phosphates, thus enabling replacement of a polluting material with a nonpolluting biobased additive while improving the performance of the end product. This innovation dramatically reduced phosphate-related algal blooms in surface waters around the globe, and simultaneously enabled consumers to get their clothes cleaner with lower wash water temperatures and concomitant energy savings.

Biotechnology can be used to design customized organisms that act as catalysts to efficiently convert a crude oil feedstock or a crude oil–derived product into a desired molecule, such as butadiene, which can then be send to the petrochemical section of a refinery to produce other products. Recent innovations enable biotechnologists to engineer these organisms and comprehensive end-to-end processes so they can produce a wider range of chemical products.

Within the petroleum industry, industrial biotechnology can be used to (1) create new products, such as biodegradable plastics; (2) integrate biomass with petroleum-based feedstocks by processing biomass in biorefineries to produce electricity, transport fuels, or chemicals; (3) modify existing processes and develop new processes, such as the use of biotransformation processes to reduce the amount of environmentally harsh chemical products; and (4) reduce the environmental impact of manufacturing, such as the treatment of refinery waste products on site rather than seeking off-site methods for the disposal of such wastes.

2.1 HISTORY

Historically, industrial biotechnology actually dates back to at least 7000 BC when various cultures used fermentation to produce alcoholic beverages (wine and beer) (Table 3.1). Over time, mankind's knowledge of fermentation increased, enabling the production of cheese, yogurt, vinegar, and other food products. In the 1800s, Louis Pasteur proved that fermentation was the result of microbial activity. Then in 1928, Sir Alexander Fleming extracted penicillin from mold. In the 1940s, large-scale fermentation techniques were developed to make industrial quantities of this wonder drug. Not until after World War II, however, did the biotechnology revolution begin, giving rise to modern industrial biotechnology.

Since that time, industrial biotechnology has produced enzymes for use in our daily lives and for the manufacturing sector. For instance, meat tenderizer is an enzyme and some contact lens cleaning fluids

Table 3.1 Highlights Timeline of the History of Biotechnology[a]	
Pre-Christian Era	
7000 BC	The Chinese discover fermentation (beer making)
6000 BC	Babylonians used yeast to make beer
4000 BC	The Egyptians baked leavened bread using yeast
250 BC	The Greeks fermented grapes to make wine
100 BC	Chinese use chrysanthemum as a natural insecticide
Pre-20th Century	
1663	First recorded description of living cells by Robert Hooke
1675	Antoine van Leeuwenhoek discovers and describes bacteria and protozoa
1798	Edward Jenner uses first viral vaccine to inoculate against smallpox
1862	Louis Pasteur discovers the bacterial origin of fermentation
1877	Robert Koch develops a technique for staining bacteria for identification
20th Century	
1928	Alexander Fleming discovered noticed that a mold could stop the duplication of bacteria
1942	Penicillin is mass-produced in microbes for the first time
1950	The first synthetic antibiotic is created
1953	James D. Watson and Francis Crick describe the structure of DNA
1974	Scientists invent the first biocement for industrial applications
21st Century	
2001 et seq.	Expansion of fermentation to produce biobased fuels, such as bioethanol
[a]*With reference to industrial biotechnology.*	

contain enzymes to remove sticky protein deposits. In the main, industrial biotechnology involves the microbial production of enzymes, which are specialized proteins. These enzymes have evolved in nature to be super-performing biocatalysts that facilitate and speed-up complex biochemical reactions. These enzyme catalysts are what make industrial biotechnology such a powerful new technology.

2.2 INDUSTRIAL BIOTECHNOLOGY

Industrial biotechnology involves working with nature to maximize and optimize existing biochemical pathways that can be used in manufacturing. The industrial biotechnology revolution rides on a series of related developments in three fields of study of detailed information derived from the cell: genomics, proteomics, and bioinformatics. As a result, scientists can apply new techniques to a large number of microorganisms ranging from bacteria, yeasts, and fungi to marine diatoms and protozoa.

Industrial biotechnology companies use many specialized techniques to find and improve nature's enzymes. Information from genomic studies on microorganisms is helping researchers capitalize on the wealth of genetic diversity in microbial populations. Researchers first search for enzyme-producing microorganisms in the natural environment and then use DNA probes to search at the molecular level for genes that produce enzymes with specific biocatalytic capabilities. Once isolated, such enzymes can be identified and characterized for their ability to function in specific industrial processes. If necessary, they can be improved with biotechnology techniques.

Many biocatalytic tools are rapidly becoming available for industrial applications because of the recent and dramatic advances in biotechnology techniques. In many cases, the biocatalysts or whole-cell processes are so new that many chemical engineers and product development specialists in the private sector are not yet aware that they are available for deployment. This is a good example of a "technology gap" where there is a lag between availability and widespread use of a new technology. This gap must be overcome to accelerate progress in developing more economic and sustainable manufacturing processes through the integration of biotechnology. "New Biotech Tools for a Cleaner Environment" provides dramatic illustrations of what these powerful new tools can do. The report aims to spark more interest in this powerful technology, to help close this technology gap, and facilitate progress toward a more sustainable future.

2.3 APPLICATIONS IN THE PETROLEUM INDUSTRY

Biotechnology has applications in four major industrial areas, including health care (medical); crop production and agriculture; nonfood (industrial) uses of crops and other products such as biodegradable plastics, vegetable oil, biofuels; and finally, the generation of products from petroleum-based feedstocks. In the latter case, this includes MEOR (Chapter 4), BDS (Chapter 6) and BDN (Chapter 7) (Obire, 1990, 1993; Taylor et al., 1998; Speight, 2014a; El-Gendy and Speight, 2015). Thus *industrial biotechnology* (known in some countries as *white biotechnology*) is biotechnology applied to industrial processes such as using microbial entities (including enzymes) to produce a useful chemical product. Another example is the use of enzymes as industrial catalysts to either produce valuable chemicals or destroy hazardous/polluting chemicals (Speight and Arjoon, 2012). This form of biotechnology tends to consume less in resources than traditional processes used to produce industrial goods.

The application of biotechnology to the petroleum industry is not new and has been practiced for many years as a method of bioremediation of oil spills (Speight and Arjoon, 2012) and from this much can be learned about the potential of the application of biotechnology to the refining industry (El-Gendy and Speight, 2015). Typically, bioremediation is a direct function of biotransformation (biodegradation), which may refer to complete *mineralization* of the organic contaminants into carbon dioxide, water, inorganic compounds, and cell protein or transformation of complex organic contaminants to other simpler organic compounds that are not detrimental to the environment.

In fact, unless they are overwhelmed by the amount of the spilled material or its toxicity, many indigenous microorganisms in soil and/or water are capable of degrading hydrocarbon contaminants. In fact, bioremediation is an environmentally friendly technique used to restore soil and water to its original state by using indigenous microbes to break down and eliminate contaminants. The microorganisms used for bioremediation may be indigenous to a contaminated area or they may be isolated from elsewhere and brought to the contaminated site. Contaminants are transformed by living organisms through reactions that take place as a part of their metabolic processes. Biodegradation of a compound is often a result of the actions of multiple organisms. When microorganisms are imported to a contaminated site to enhance degradation we have a process known as bioaugmentation.

Furthermore, biotechnology is now accepted as an attractive means of improving the efficiency of any industrial processes and resolving serious environmental problems. One of the reasons for this is the extraordinary metabolic capability that exists within the bacterial world. Microbial enzymes are capable of biotransforming a wide range of compounds, and the worldwide increase in attention being paid to this concept can be attributed to several factors, including the presence of a wide variety of catabolic enzymes and the ability of many microbial enzymes to transform a broad range of unnatural compounds (xenobiotic compounds) as well as natural compounds. Biotransformation processes have several advantages compared with chemical processes, including the following: (1) microbial enzyme reactions are often more selective, (2) biotransformation processes are often more energy-efficient, (3) microbial enzymes are active under mild conditions, and (4) microbial enzymes are environment-friendly biocatalysts. Although many biotransformation processes have been described, only a few of these have been used as part of an industrial process and opportunities exist for biorefining of petroleum (Mohebali and Ball, 2008). Of particular interest in this context is the phenomenon of BDS (biological desulfurization, microbial desulfurization) in which are used to oxidizes sulfur compounds in crude oil ultimately resulting in desulfurization. This represents the ability of microbial species to desulfurize compounds that are recalcitrant to the current standard technology in the oil industry (Abin-Fuentes et al., 2013; El-Gendy and Speight, 2015). From this work, it is evident that biorefining is a possible alternative to some of the current oil-refining processes (Figs. 3.1 and 3.2) (Speight, 2000; Parkash, 2003; Hsu and Robinson, 2006; Ancheyta and Speight, 2007; Gary et al., 2007; Speight, 2014a).

For biotechnology to be effective, microorganisms must convert the petroleum constituents into the necessary products. However, since biotechnology can be effective only where reaction conditions permit microbial growth and activity; the application of the technique often involves the manipulation of environmental parameters to allow microbial growth and degradation to proceed at a faster rate.

Finally, biotechnology is a key technology for the emerging biomass-based industries and is worthy of inclusion here and later in this text (Chapter 8). Biomass is biological material that has come from animal, vegetable, or plant matter and is *carbon neutral*—while the plant is growing, it uses the energy of the sun to absorb the same amount of carbon from the atmosphere as it releases into the atmosphere. By maintaining this closed carbon cycle, it is felt, with some mathematical meandering, that there is no overall increase in carbon dioxide levels through emissions to the atmosphere.

Biomass includes a wide range of materials that produce a variety of products that are dependent upon the feedstock (Balat, 2011; Demirbaş, 2011; Ramroop Singh, 2011; Speight, 2011a). For example, typical biomass wastes include wood material (bark, chips, scraps, and saw dust), pulp and paper industry residues, agricultural residues, organic municipal material, sewage, manure, and food processing by-products. Agricultural residues such as straws, nut shells, fruit shells, fruit seeds, plant stalks and stover, green leaves, and molasses are potential renewable energy resources. Many developing countries have a wide variety of agricultural residues in ample quantities. Large quantities of agricultural plant residues are produced annually worldwide and are vastly underutilized. Agricultural residues, when used a fuel, through direct combustion, only a small percentage of their potential energy is available, due to inefficient burners used. Current disposal methods for these agricultural residues have caused widespread environmental concerns. For example, disposal of rice and wheat straw by open-field burning causes air pollution. In addition, the widely varying heat content of the different types of biomass varies widely and must be taken into consideration when designing any conversion process (Jenkins and Ebeling, 1985).

Raw materials that can be used to produce biomass fuels are widely available and arise from many different sources and in numerous forms. The main basic sources of biomass material are (1) wood, including bark, logs, sawdust, wood chips, wood pellets, and briquettes; (2) high-yield energy crops, such as wheat, that are grown specifically for energy applications; (3) agricultural crop and animal residues, like straw or slurry; (4) food waste, both domestic and commercial; and (5) industrial waste, such as waste wood products or waste paper products.

Liquid biofuels (such as biodiesel, which is not typically produced through the agency of petroleum biotechnology) and biobased chemicals include a wide variety of products, some of which are well established and already commercialized to a significant extent and others that are emerging. Many products are innovative in terms of manufacturing process or raw material, particularly those that are produced using biocatalysis. These processes create biobased products, or more specifically for this investigation, liquid biofuels and biobased chemicals. The most common liquid biofuel produced in the United States is ethyl alcohol, or ethanol, which is primarily manufactured from the starch portion of corn kernels. Liquid biofuels and biobased chemicals include a wide variety of products, some of which are well established and already commercialized to a significant extent and others that are emerging. Many products are innovative in terms of manufacturing process or raw material, particularly those that are produced using biocatalysis.

The biomass industries (which use biotechnological concepts) are emerging as a response to the declining global supply of cheap, easily extracted petroleum. There is also more public pressure for cleaner industrial practices, lower greenhouse gas emissions, and transition to renewable raw materials. Biomass-based industries are an important part of the bioeconomy, which refers to sustainable production, collection, and conversion of biomass into a range of fuels and chemical products.

3.0 BIOTRANSFORMATION OF PETROLEUM CONSTITUENTS

Biotransformation is the chemical modification (or modifications) made by an organism on a chemical compound. If this modification ends in mineral compounds, such as carbon dioxide, CO_2, water, H_2O, or ammonia NH_4^+, the biotransformation is regarded as being complete and is referred to as *mineralization*.

A modern refinery accepts a variety of different crude oils for processing, which are blended prior to a variety of processing sequences (Parkash, 2003; Hsu and Robinson, 2006; Gary et al., 2007; Speight, 2011a,b, 2012, 2014a, 2017). Thus petroleum and petroleum products are mixtures of differing molecular species hydrocarbons and the constituents of these molecular categories are present in varied proportions, resulting in high variability in petroleum and petroleum products. In terms of bulk fractions (Chapter 2), the resin constituents and the asphaltene constituents are of interest (or notoriety) because these constituents generally resist degradation. During biotransformation, the constituents of petroleum and petroleum products are subjected to physical and chemical processes such as evaporation or oxidation, which produce changes in the composition of the crude oil (Speight, 2014a; Taghvaei Ganjali et al., 2007).

The biotransformation of petroleum constituents is a complex process that depends on the nature and on the amount of the hydrocarbons present. Petroleum hydrocarbons can be divided into four classes: the saturates, the aromatics, resin constituents, and asphaltene constituents (phenols, fatty acids, ketones, esters, and porphyrins), and the resins (pyridines, quinolines, carbazoles, sulfoxides, and amides) (Colwell et al., 1977). Different factors influencing hydrocarbon degradation have been reported (Cooney et al., 1985). One of the important factors that limit biotransformation of crude oil constituents is the availability to microorganisms (Barathi and Vasudevan, 2001) and hydrocarbons differ in their susceptibility to microbial attack (Speight and Arjoon, 2012). The susceptibility of hydrocarbons to microbial degradation can be generally ranked as follows: linear alkane derivatives, branched alkane derivatives, low-molecular-weight aromatic derivatives, and cyclic alkane derivatives (Perry, 1984). Some compounds, such as the high-molecular-weight PNAs or PAHs may not be degraded at all (Kanal and Harayama, 2000; Bamforth and Singleton, 2005; Atlas and Bragg, 2009).

At the time of writing, the predominant commercial practice of biotransformation of crude oil constituents has focused not on refining but primarily on the remediation and cleanup of petroleum hydrocarbons in the environment (Speight and Arjoon, 2012). Thus successful application of bioremediation technology to a contaminated ecosystem requires knowledge of the characteristics of the site and the parameters that affect the microbial biotransformation of pollutants and it is from these published works that a process for biorefining crude oil might be developed.

Moreover, it is essential to recognize that the biotransformation of petroleum is dependent on the ability of the local microbiota to adapt to the different petroleum constituents (Greenwood et al., 2009). The different structural and functional response of microbial subgroups to the different constituents confirms that the overall response of microbial entities is sensitive to petroleum composition. This suggests that the preferred response to the different constituents may be engineered by preexposure of the microbes to representative chemicals (Leuenberger, 1990). The controlled adaptation of microbes to a chemical is the basis of proactive bioremediation technology (Speight and Arjoon, 2012).

The premise being that microbial species adapted through a history of exposure to petroleum hydrocarbons is less severely impacted by microbial species with no such preexposure or adaptation (Page et al., 1996; Peters et al., 2005). Indeed, the diversity of microbes for the biotransformation of petroleum constituents may be significant but, in the absence of a previous history of exposure to petroleum constituents, the numbers of the microbes may be low due to lack of and prior stimulus and the potential for adaptation (Swannell et al., 1996).

The biotransformation of various petroleum-based pollutants (Speight and Arjoon, 2012) is a sustainable way to cleanup environments that have been contaminated by spill of crude oil and/or crude oil products. This form of biotransformation (usually referred to as bioremediation and/or biodegradation) harnesses the naturally occurring, microbial catabolic diversity to degrade, transform, or accumulate a huge range of petroleum-based compounds including hydrocarbon derivatives and PAHs. Major methodological breakthroughs in recent years have enabled insights into biotransformation pathways and the ability of organisms to adapt to changing environmental conditions. Functional approaches are increasing the understanding of the relative importance of different pathways and regulatory networks to the biotransformation of petroleum constituents in various environments and are accelerating the development and inception of biotransformation processes (Peixoto et al., 2011; Speight and Arjoon, 2012).

Chemically, the biotransformation of petroleum constituents is a complex process that depends on the nature and on the amount of the hydrocarbons present. Petroleum hydrocarbons can be divided into four classes: the saturates, the aromatics, resin constituents, and asphaltene constituents (phenols, fatty acids, ketones, esters, and porphyrins), and the resins (pyridines, quinolines, carbazoles, sulfoxides, and amides) (Colwell et al., 1977). Different factors influencing hydrocarbon degradation have been reported (Cooney et al., 1985). One of the important factors that limit biotransformation of crude oil constituents is the availability to microorganisms (Barathi and Vasudevan, 2001) and hydrocarbons differ in their susceptibility to microbial attack (Speight and Arjoon, 2012). The susceptibility of hydrocarbons to microbial degradation can be generally ranked as follows: linear alkane derivatives, branched alkane derivatives, low molecular weight aromatic derivatives, and cyclic alkane derivatives (Perry, 1984). Some compounds, such as the high molecular weight polynuclear (polycyclic) aromatic hydrocarbon derivatives (PNAs or PAHs) may not be degraded at all (Kanal and Harayama, 2000; Bamforth and Singleton, 2005; Atlas and Bragg, 2009).

A modern refinery accepts a variety of different crude oils for processing, which are blended prior to a variety of processing sequences (Parkash, 2003; Hsu and Robinson, 2006; Gary et al., 2007; Speight, 2011b, 2012, 2014a, 2017). However, crude oil and crude oil products are mixtures of differing molecular species hydrocarbons and the constituents of these molecular categories are present in varied proportions, resulting in high variability in crude oil and its products (Chapters 1 and 2). In terms of bulk fractions of crude oil and crude oil products (Chapter 1), the resin constituents and the asphaltene constituents are of interest (or notoriety) because these constituents generally resist degradation. During biotransformation, the constituents of petroleum and petroleum products are subjected to physical and chemical processes such as evaporation or oxidation which produce changes in the composition of the crude oil (Speight, 2014a; Taghvaei Ganjali et al., 2007).

Moreover, organism that chemically transform crude oil have a specific order of preference for compounds that are converted. Progressive degradation of crude oil tends to remove saturated hydrocarbons first, concentrating heavy polar and asphaltene components in the residual oil. This leads to decreasing crude oil quality by lowering the API gravity while increasing (1) the viscosity, (2) the sulfur content, and (3) the metal content. In addition to lowering reservoir recovery efficiencies, the economic value of the oil generally decreases with biodegradation, owing to a decrease in refinery distillate yields and an increase in vacuum residua yields. Furthermore, biotransformation typically leads to the formation of naphthene derivatives that increase the acidity of the oil, typically measured as total acid number (TAN). An increase in the TAN may further reduce the value of the crude oil in the reservoir and may contribute to production and downstream handling problems such as equipment corrosion and the formation of difficult-to-break emulsions.

3.1 ALKANE DERIVATIVES

Alkanes are major constituents of conventional petroleum and petroleum products. Conventional (light) petroleum contains 10%–40% w/w normal alkanes, but weathered and heavier oils may have only a fraction of a percent. Higher molecular weight alkanes constitute 5%–20% w/w of light oils and up to 60% w/w of the more viscous oils and tar sand bitumen. Of these, the normal alkane series (straight-chain alkane series) is the most abundant and the most quickly degraded. Compounds with chains of up to 44 carbon atoms can be metabolized by microorganisms, but those having 10–24 carbon atoms (C_{10}–C_{24}) are usually the easiest to metabolize. Shorter chains (up to approximately C_8) also evaporate relatively easily. Only a few species can use C_1–C_4 alkanes and C_5–C_9 alkanes are degradable by some microorganisms but toxic to others.

Branched alkane derivatives are usually more resistant to biodegradation than normal alkanes but less resistant than cycloalkanes (naphthenes)—those alkanes having carbon atoms in ringlike central structures. Branched alkanes are increasingly resistant to microbial attack as the number of branches increases. At low concentrations, cycloalkanes may be degraded at moderate rates, but some highly condensed cycloalkanes can persist for long periods after a spill.

Understanding the bacterial degradation pathway of cycloalkane derivatives, as well as that of n-alkane derivatives, is important from the standpoint of biotransformation (Fujii et al., 2004).

Pristane isomer of C19
2,6,10,14-Tetramethylpentadecane

Phytane isomer of C20
2,6,10,14-Tetramethylhexadecane

C27 Steranes

C30 Hopanes

Oleanane

Generally, with respect to the molecular composition of the aliphatic constituents of petroleum and petroleum-related products, microbial biotransformation will biotransform the n-alkane derivatives and the branched-chain alkanes. The polycyclic alkane derivatives of the sterane and triterpane type tend to be somewhat resistant to biotransformation. Since this is the case even for naphthene-type petroleum (which is originally depleted in nonring alkane derivatives), the biotransformation of petroleum constituents may be restricted to n-alkane derivatives and isoprenoid derivatives (Fujii et al., 2004; Antić et al., 2006).

Generally, with respect to the molecular composition of the aliphatic constituents of petroleum and petroleum-related products, microbial biodegradation attacks n-alkanes and isoprenoid alkanes. The polycyclic alkanes of the sterane and triterpane type tend to be somewhat resistant to biodegradation. Since this is the case even for naphthenic type petroleum (which is originally depleted in n-alkanes), it has been concluded that the biodegradation of petroleum type pollutants, under natural conditions, will be restricted to n-alkanes and isoprenoids (Antić et al., 2006).

3.2 AROMATIC HYDROCARBON DERIVATIVES

Aromatic hydrocarbon derivatives are characterized by the presence of at least one benzene (or substituted benzene) ring. The low-molecular-weight aromatic hydrocarbon derivatives are relatively easily subject to biotransformation. Light crude oil typically contains between 2% and 20% w/w low-boiling aromatic compound derivatives, whereas heavy oil contains less than 2% w/w aromatic compounds. As the molecular weight and complexity increase of the aromatic derivatives increases, biotransformation is less likely to occur. Thus the degradation rate of PNA derivatives is slower than the degradation rate of monocyclic aromatic derivatives (Gibson and Subramanian, 1984).

However, it is uncommon to find organisms that could effectively react and change both aliphatic constituents and aromatic constituents of petroleum possibly due to differences in metabolic routes and pathways for the degradation of the two classes of hydrocarbons. There are indications of the existence of bacterial species with propensities for simultaneous degradation of aliphatic hydrocarbons and aromatic hydrocarbons (Amund et al., 1987; Obayori et al., 2009). This rare ability may be as a result of long exposure of the organisms to different hydrocarbon pollutants, resulting in genetic alteration and acquisition of the appropriate degradative genes.

The biodegradation of alkyltetralins has also been studied. However, tetralin has been shown to be biodegraded by both mixed cultures of microbes (Strawinski and Stone, 1940; Soli and Bens, 1972) and by some strains able to utilize the compound as sole carbon and energy source (Schreiber and Winkler, 1983; Sikkema and Bont, 1991; Hernáez et al., 1999).

It has been demonstrated that *rhodococci* strains are able to react with alkyltetralin derivatives (Frenzel et al., 2009). The identification of such bacteria capable of the bioreactivity of alkyltetralins may be an important step toward the development of bioremediation strategies for sites contaminated by toxic aromatic hydrocarbons.

3.3 POLYNUCLEAR AROMATIC HYDROCARBON DERIVATIVES

PAHs, in the current context, are organic compounds with two or more aromatic rings in various structural configurations. PAHs constitute a large and diverse class of organic compounds. However, derivatives such as tetralin (1,2,3,4-tetrahydronaphthalene) and decalin (decahydronaphthalene, bicyclo[4.4.0]decane) are not included in this group but are included in the alkane group because of the saturated ring.

Tetralin   Decalin

The biotransformation of PAHs is typically accompanied by the accumulation of neutral and acidic oxidation products. Both neutral and acidic water-soluble fractions are also formed when various mixed bacterial cultures degrade weathered crude oil (Chapman et al., 1995; Foght and Westlake, 1988). However, most PAH derivatives occur as hybrids encompassing various structural components, such as in the PAH, benzo[*a*]pyrene.

Benzo(a)pyrene

Generally, an increase in the size and angularity of a PAH molecule results in a concomitant increase in hydrophobicity and electrochemical stability (Zander, 1983; Harvey, 1997). The molecule stability and hydrophobicity of PAHs are two primary factors that contribute to their persistence of in the environment.

PAH derivatives are present as natural constituents in fossil fuels and (through refining) in crude oil products (Pavlova and Ivanova, 2003; Speight, 2014a, 2017) and can be formed during the incomplete combustion of organic material, and are therefore present in relatively high concentrations in products of fossil fuel refining (Speight, 2014a, 2017). PAH derivatives that are released into the environment may originate from petroleum products such as including gasoline, diesel fuel, and fuel oil (Pavlova and Ivanova, 2003). The concentration of PAHs in crude oil and crude oil products varies widely, depending on (1) the crude oil and (2) the production method (Parkash, 2003; Hsu and Robinson, 2006; Gary et al., 2007; Speight, 2011b, 2012, 2014a, 2017).

The toxic, mutagenic, and carcinogenic properties of PAHs have resulted in some of these compounds (including naphthalene, phenanthrene, and anthracene) to be designated as priority pollutants. In addition, the solubility of PAHs in aqueous media is very low (Luning Prak and Pritchard, 2002), which affects degradation of these compounds and can lead to biomagnification within an ecosystem.

The chemical properties, and hence the ability of PAHs to undergo biotransformation, are dependent in part upon both molecular size (i.e., the number of aromatic rings) and the pattern of ring linkage. Ring linkage patterns (also known as molecular topology) in PAHs may occur such that the tertiary carbon atoms are centers of two or three interlinked rings, as in the linear kata-condensed PAH anthracene or the pericondensed PAH pyrene.

Interest in the biotransformation degradation mechanisms and the mechanism by which biotransformation of PAHs can be achieved is of utmost importance because of the ubiquitous distribution of these chemicals in crude oil and their potential effect on the environment (Cerniglia, 1984; Gibson and Subramanian, 1984; Cerniglia and Heitkamp, 1989, 1990; Cerniglia, 1992; Shuttleworth and Cerniglia, 1995). Evidence also suggests that in some cases, PAH-toxicity also increases with size, up to at least four or five fused benzene rings (Cerniglia, 1992). The relationship between PAH-environmental toxicity and increasing numbers of benzene rings is consistent with the results of various studies correlating environmental biotransformation rates and PAH molecule size (Shuttleworth and Cerniglia, 1995).

The biodegradation of naphthalene (the simplest PAHs) process was optimized with preliminary experiments in slurry aerobic microcosms (Bestetti et al., 2003). From soil samples collected on a contaminated site, a *Pseudomonas putida* strain (designated as M8), capable to degrade naphthalene was selected. Microcosms were prepared with M8 strain by mixing noncontaminated soil and a mineral medium. Different experimental conditions were tested varying naphthalene concentration, soil/water ratio, and inoculum density. The disappearance of hydrocarbon, the production of carbon dioxide, and the ratio of total heterotrophic and naphthalene-degrading bacteria were monitored at different incubation times. The kinetic equation that best fitted the disappearance of contaminant with time was determined. The results showed that the isolated strain enhanced the biodegradation rate with respect to the natural biodegradation.

Of the four-ring PAHs, fluoranthene, pyrene, chrysene, and benz[a]anthracene have been investigated to various degrees.

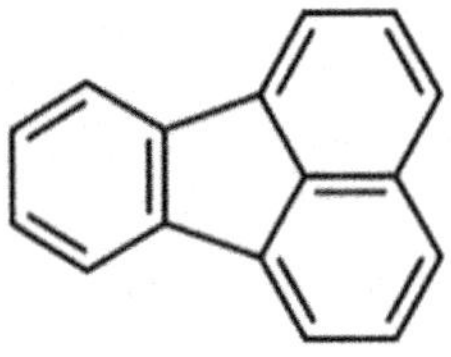

Fluoranthene

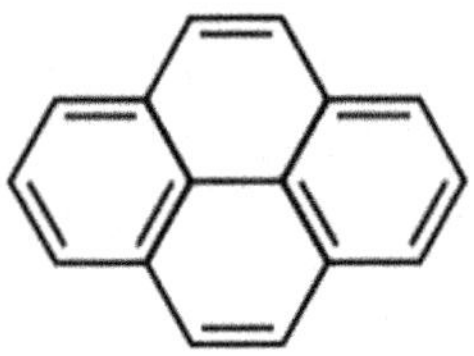

Pyrene

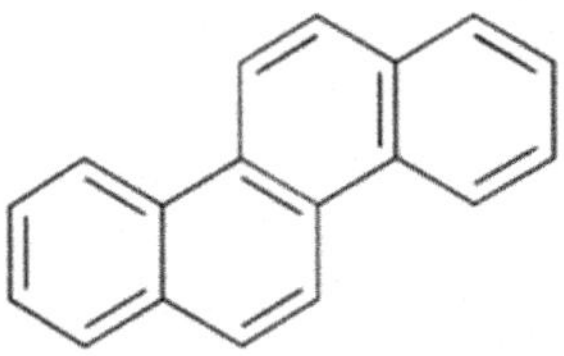

Chrysene

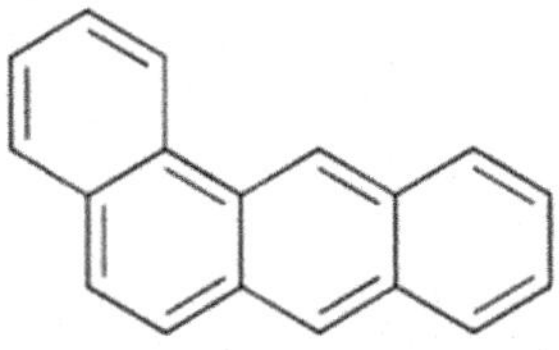

Benz(a)anthracene

Fluoranthene, a PAH, containing a five-membered ring, has been shown to be metabolized by a variety of bacteria, and pathways describing its biodegradation have been proposed (Mueller et al., 1990; Weissenfels et al., 1990, 1991; Ye et al., 1996). Fluoranthene has been used as a model compound in studies that have investigated the effects of surface-active compounds on PAH biodegradation. Comparisons of the mineralization of fluoranthene by four fluoranthene-degrading strains in the presence of the nonionic surfactants showed that responses differed between strains (Willumsen et al., 1998). In addition, the biotransformation of pyrene, a pericondensed PAH, as well as the benzo(a) pyrene, has been reported and several proposed mechanistic pathways have been suggested (Heitkamp and Cerniglia, 1988, 1989; Cerniglia and Heitkamp, 1990).

Generally, aromatic constituents with five or more rings are not easily attacked and may persist in the environment for long periods. High-molecular-weight aromatics comprise 2%–10% w/w conventional (light) petroleum and up to 35% w/w of the more viscous petroleum. But, currently, there is still a limited information regarding the bacterial biotransformation of PAHs-derivatives with five or more rings. Most studies have focused on the five-ring benzo(a)pyrene due to the potential hazards of this chemical to human.

Measuring the success of the biotransformation of petroleum-related PAH derivatives is based on several parameters. Though the lower n-alkanes are generally considered the most biodegradable compound class within crude oils (Leahy and Colwell, 1990; Atlas and Bartha, 1992; Prince, 1993), other studies point to exceptional conditions in which PAH derivatives degrade preferentially to n-alkanes (Speight, 2016).

An increase in the understanding of the biotransformation of PAH-degrading microbes and the mechanisms by which PAH biotransformation occur will prove helpful for predicting the behavior of these compounds leading to the development of practical PAH biotransformation strategies in the future (Okerentugba and Ezeronye, 2003).

3.4 HETEROCYCLIC DERIVATIVES

The biotransformation of various heterocyclic chemicals has been investigated extensively in the last several decades. The metabolic fate of such nonindigenous chemicals under aerobic conditions has been widely studied, and the predominant persistence patterns and degradation pathways are well elaborated (Young, 1984) and clearly demonstrate the importance of anaerobic microbial transformations of organic compounds in anoxic environments.

Heterocyclic aromatic compounds serve as substrates for a variety of microorganisms. For example, pyridine is a compound that can be metabolized by microorganisms under either aerobic or anaerobic conditions. Under anaerobic conditions, the initial step in pyridine metabolism can be either ring reduction or ring hydroxylation. The source of atomic oxygen for the ring hydroxylation reaction can be molecular oxygen or water; under anoxic conditions the source is water (Berry et al., 1987).

4.0 FACTORS AFFECTING BIOTRANSFORMATION

The biotransformation of any petroleum constituent is a measure of the ability of that constituent to be metabolized (or cometabolized) by bacteria or other microorganisms through a series of biological process, which include ingestion by organisms as well as microbial degradation (Payne and McNabb, 1984). The chemical characteristics of the contaminants influence biodegradability; in addition, the location and distribution of petroleum contamination in the subsurface can significantly influence the likelihood of success for bioremediation.

Moreover, the biotransformation of petroleum constituents and petroleum products is inherently influenced by the composition of the substrate (Chapter 1). For example, petroleum is quantitatively biotransformed and kerosene, which consists almost exclusively of medium chain-length alkanes, is completely biotransformed under suitable conditions but for viscous asphaltene-containing crude oil and tar sand bitumen, approximately 6%–10% w/w of the material oil may be biotransformed within a reasonable time period, even when the conditions are favorable for biotransformation (Bartha, 1986; Okoh et al., 2001, 2002; Okoh, 2003; Okoh, 2006; Okoh and Trejo-Hernandez, 2006). In addition, the biotransformation of petroleum-based constituents can be enhanced by use of a consortium of different bacteria compared to the activity of single-bacterium species (Ghazali et al., 2004; Milić et al., 2009).

When biotransformation occurs in an oil reservoir, the process dramatically affects the fluid properties (Chapter 4) and hence the value and producibility of an oil accumulation. Specifically, petroleum biotransformation typically raises viscosity of the residual material (which reduces oil producibility) and reduces the API gravity (which reduces the value of the produced oil). The process increases the asphaltene content (relative to the saturated and aromatic hydrocarbon content and the starting material), the concentration of certain metals, the sulfur content, and oil acidity.

There are indications that petroleum biotransformation involves more biological components than just the microorganisms that directly attack petroleum constituents (the primary degraders) and shows

Table 3.2 General Description of Chemical Reaction Types Catalyzed by Microbial Transformations as Might Be Applied to Petroleum and Petroleum Products

Reaction	Result
Oxidation	Hydroxylation and dehydrogenation of C—C bonds
	Oxidative degradation of alkyl chains
	Oxidative removal of substituents
	Oxidative deamination
	Oxidation of heterofunctions
	Oxidative ring fission
Reduction	Reduction of organic acids and derivatives
	Hydrogenation of olefin (C=C) bonds
	Reduction of heteroatom functions
	Dihydroxylation
	Reductive elimination of substituents
Hydrolysis	Hydrolysis of acid derivatives
	Hydration of C=C bonds and epoxides
Condensation	Dehydration
Isomerization	Migration of double bonds or oxygen functions
Bond formation	Formation of carbon-carbon (C—C) bonds
	Formation of carbon-heteroatom bonds

that the primary degraders interact with these components (Head et al., 2006). In addition, primary degraders need to compete with other microorganisms for limiting nutrients, and the nonpetroleum-degrading microorganisms can be affected by metabolites and other compounds that are released by oil-degrading bacteria and vice versa.

In a refining scenario, the rapid biotransformation of petroleum hydrocarbon derivatives is attributed to conditions favorable to biotransformation. It is possible that elevated nutrient levels possibly provide an active and capable microbial community that may be enhanced due to prior exposure to petroleum.

4.1 CONDITIONS FOR BIOTRANSFORMATION

The composition of petroleum and petroleum products (Chapters 1 and 2) is the first and most important consideration when the suitability of biotransformation in the context of the refinery is to be evaluated. Heavy crude oil (typically alkane-deficient) is generally much more difficult to biotransform than light crude oil (typically alkane-rich). Also, the amount of heavy crude oil biotransformed by some bacterial species increases with increasing concentration of the feedstock (Okoh et al., 2002; Rahman et al., 2002, 2003). In fact, an important aspect of the conditions for biotransformation is the ability of microorganisms to produce enzymes to catalyze metabolic reactions (Table 3.2), which is governed by the genetic composition of the organism(s). Enzymes produced by microorganisms in the presence of carbon sources cause initial attack on the hydrocarbon constituents while other enzymes are utilized to complete the breakdown of the hydrocarbon. Thus lack of an appropriate enzyme either prevents attack or is a barrier to complete hydrocarbon degradation. Furthermore, the biotransformation of petroleum-related constituents can occur under both aerobic (oxic) and anaerobic (anoxic) conditions (Zengler et al., 1999), usually by the action of different consortia of microorganisms.

The biotransformation of hydrocarbons (being fully reduced substrates) requires an exogenous electron sink. In the initial attack, this electron sink should be molecular oxygen. In the subsequent steps too, oxygen is the most common electron sink. In the absence of molecular oxygen, further biotransformation of partially oxygenated intermediates may be supported by nitrate or sulfate reduction.

4.2 EFFECTS OF BIOTRANSFORMATION

These early stages of the biotransformation process (loss of n-paraffins followed by loss of acyclic iso-paraffin derivatives) can be readily detected by gas chromatography analysis of the product (Speight, 2005). However, in heavily biotransformed crude oils, gas chromatographic analysis alone cannot distinguish differences in biotransformation due to interference of the unresolved complex mixture that dominates the gas chromatographic traces of heavily degraded crude oils. Among such crude oils, differences in the extent of biotransformation can be assessed using gas chromatography–mass spectrometry to quantify the concentrations of biomarkers with differing resistances to the biotransformation process (Jacquot et al., 1996).

During biotransformation, the properties of the petroleum fluid changes because different classes of compounds in petroleum have different susceptibilities to biotransformation (Goodwin et al., 1983). The early stages of biotransformation (in addition to any evaporation effects) are characterized by the loss of n-paraffins (n-alkanes or branched alkanes) followed by loss of acyclic isoprenoid derivatives (e.g., norpristane, pristane, and phytane). Compared with those compound groups, other compound classes (such as highly branched and cyclic saturated hydrocarbons as well as aromatic compounds) are more resistant to biotransformation. However, even the more-resistant compound classes are eventually destroyed as biotransformation proceeds.

4.3 EFFECT OF NUTRIENTS

Different types of nutrients (primarily nitrogen and phosphorus) have been applied to improve the biotransformation of petroleum hydrocarbon derivatives, including classic (water soluble) nutrients and oleophilic and slow-release fertilizers. Generally, hydrocarbons have low-to-poor solubility in water and, as a result, are adsorbed on to clay or humus fractions, so they pass very slowly to the aqueous phase where they are metabolized by microorganisms. Cyclodextrins are natural compounds that form soluble inclusion complexes with hydrophobic molecules and increase degradation rate of hydrocarbons in vitro.

In the perspective of an in situ application such as MEOR (Chapter 4), β-cyclodextrin does not increase eluviation (the lateral or downward movement of the suspended material in soil through the percolation of water) of hydrocarbons through the formation and consequently does not increase the risk of groundwater pollution (Jean et al., 2008; Sivaraman et al., 2010). Thus in situ bioremediation of PAH derivatives can be improved by the augmentation of degrading microbial populations and by the increase of hydrocarbon bioavailability (Bardi et al., 2007).

Generally, the addition of nutrients is necessary to enhance the biotransformation of petroleum-related pollutants (Choi et al., 2002; Kim et al., 2005; Joshi and Pandey, 2011). In fact, even in harsh sub-Arctic climates, it has been observed that the effectiveness of fertilizers for petroleum increases the chemical, microbial, and toxicological parameters compared to the used of various fertilizers in a pristine environment.

On the other hand, in an investigation of the role of the nitrogen source in the biotransformation of crude oil constituents by a defined bacterial consortium under cold, marine conditions (10°C/50°F), it was observed that nitrate did not affect the pH, whereas ammonium amendment led to progressive acidification, accompanied by an inhibition of the biotransformation of aromatic (particularly polynuclear aromatic) hydrocarbons (Foght et al., 1999). However, the aromatic systems were degraded or cometabolized in the absence of nutrients where the pH remained almost unchanged. The best overall biotransformation process was observed in the presence of nitrate without ammonium, plus high phosphate buffering—a disadvantage of nitrate is that significant emulsification of the petroleum occurs. Generally, it is worth bearing in mind that acidity/alkalinity (pH) is an important factor that requires consideration as it affects the solubility of both PAHs as well as the metabolism of the microorganisms, showing an optimal range for biotransformation between 5.5 and 7.8 (Bossert and Bartha, 1984; Wong et al., 2001).

4.4 EFFECT OF TEMPERATURE

Temperature plays an important role in the biotransformation of petroleum-related hydrocarbons not only because of the direct effect on the chemistry of the pollutants but also because of the effect on the physiology and diversity of the microbial surroundings (Atlas, 1975). In short, temperature can play the role of increasing a microbial reaction or inhibiting a microbial reaction in a similar manner to the general rules for the influence of temperature on chemical reactions.

Typically, biotransformation of petroleum and petroleum products occurs at temperatures less than 80°C (<176°F) whereas many of the microorganisms cannot exist at higher temperatures (unless the microbes are of a specific thermophilic type) (Chen and Taylor, 1995, 1997a, 1997b). The biotransformation process is terminated by lowering the temperature below 40°C (104°F). Thus the ambient temperature of the biotransformation affects both the properties of spilled crude oil or crude oil products (Speight, 2014a) and the activity or population of microorganisms (Venosa and Zhu, 2003). At low temperatures, the viscosity of the oil increases, while the volatility of toxic low-molecular-weight hydrocarbons is reduced, delaying the onset of biotransformation. Temperature also variously affects the solubility of hydrocarbon derivatives.

Although the biotransformation of hydrocarbon derivatives can occur over a wide range of temperatures, the rate of biotransformation generally decreases with decreasing temperature. The highest rates of biotransformation reactions generally occur in the range of 30–40°C (86–104°F) (Bossert and Bartha, 1984). In fact, the biotransformation of petroleum is highly dependent not only on composition but also on microbial incubation temperature—at 20°C (68°F). As expected from petroleum chemistry and composition (Chapters 1 and 2) (Speight, 2014a), the rate of biotransformation for the constituents of heavy oil is significantly lower (at 20°C, 68°F) than for conventional oil. During biotransformation, some preference is shown for removal of the paraffin constituents over the aromatic and asphaltic constituents, especially at low temperatures (Whyte et al., 1998). Branched paraffins, such as pristane, are biotransformed at both 10 and 20°C (50 and 68°F).

4.5 RATES OF BIOTRANSFORMATION

Biotransformation is a multivariable process and optimization of the reaction rates through classical methods is subject to question. To overcome the disadvantages of the process, response surface methodology has been advocated for analyzing the effects of several independent variables on the biotransformation process to assess the optimum conditions for the process (Nasrollahzadeh et al., 2007; Huang et al., 2008; Pathak et al., 2009; Vieira et al., 2009; Mohajeri et al., 2010; Zahed et al., 2010). The outcome is the suggestion that the rates of biotransformation can be increased by modifying selected physical and chemical conditions that control biotransformation in multiphase systems, namely, (1) bioavailability and (2) terminal electron acceptor availability (Sandrin et al., 2006). With respect to bioavailability, model simulations suggest that (1) increasing the interfacial area between the aqueous and solid phases, (2) increasing the rate of contaminant solubilization, and (3) minimizing the accumulation of the contaminant in nonaqueous phase liquids will result in significantly higher rates of biotransformation.

The complex array of factors that influence biotransformation of petroleum-related constituents is not realistic to expect a simple rate model or kinetic model to provide precise and accurate descriptions of concentrations during different seasons and in different environments. Therefore, it is nearly impossible to predict the rates of the biotransformation process. To give a final answer on how much time remediation processes require and what the final mineral oil concentrations will be, experiments should be continued until the biotransformation processes have stopped completely. In future, it will be necessary to use complex models to yield a more exact assessment of soil remediation to the desired level (Maletić et al., 2009).

Finally, as with all efforts at modeling the outcome of complex processes, the variable parameters used in the models must be based on (1) the properties of the material (in this case the petroleum-based contaminant), (2) data retrieved about the conditions of the actual site, and on (3) experiments performed using the original aged contaminant without any additions (model compounds or analytical *spikes*).

5.0 BIOTRANSFORMATION OF CRUDE OIL IN THE RESERVOIR

The biotransformation of crude oil in a reservoir is linked to numerous interactions between the crude oil and the reservoir environment. Bacteria influence the chemical composition of crude oil, the conditions of its in-situ exploitation through the decomposition of some oil fractions, through the production of metabolism products such as biopolymers, biosurfactants, organic acids, and gases (such as methane, CH_4, carbon dioxide, CO_2, hydrogen sulfide, H_2S, and hydrogen, H_2, among others), and by the presence of microorganisms that may change the properties of the reservoir rock (Wolicka and Borkowski, 2012).

The biotransformation of crude oil in-situ (within the crude oil reservoir) significantly affects the composition of the crude oil (Speight, 2012, 2014a). The determination of the stage of the biotransformation process may be also based on the relations of pristane and phytane (Pr + Ph) to $nC_{17} + nC_{18}$, C_{30} alpha-beta hopane to (Pr + Ph), and C_{25}-nor C_{29} alpha-beta-hopane to C_{30} alpha-beta-hopane (Wolicka and Borkowski, 2012). Furthermore, depending on the stage of the progression of the process, different proportions of hydrocarbon derivatives are contained in the crude oil. Thus the composition of the crude oil is linked to the susceptibility of the specific crude oil to microbiological degradation.

Moreover, in a crude oil reservoir, a factor complicating the determination of the degree of oil biotransformation is the influx (migration) of crude oil (or an immature crude oil precursor) from the source rock. If the initial fraction of oil was subjected to microbiological transformations and portions of unchanged oil were introduced later, then the total product will have very different composition, with biomarkers pointing to both high and low microbiological evolution of the raw material. Also, there will be differences in the physical properties of the oil such as viscosity. Complete homogenization of crude oil is not possible even in the geological time scale. Due to processes of crude oil mixing and diffusion (as well as the permeability of the reservoir rock), the composition of the crude oil may become uniform only locally.

Typically, biotransformation of crude oil takes place in the contact zone between crude oil and water. The resulting solution gradient causes influx of substances prone to degradation such as n-alkane derivatives and isoprene alkane derivatives into this zone and outflow of reaction products in the opposite direction. Mixing of biotransformed crude oil with inflowing unchanged oil takes place also in this region. In the reservoir, advanced biotransformation is favored by moderate temperatures in shallow reservoirs. In addition, the influence of biotransformation on the crude oil composition is reflected in the concentration of compound types. For example, in the case of PAH derivatives, three elements determine their susceptibility to microbiological degradation: (1) the number of rings in the compound, (2) the number of alkyl substituents, and (3) the location of the bonds of these substituents. The percentage content of hydrocarbons with large number of rings is inversely proportional to the rate of biotransformation (Huang et al., 2008).

A similar relationship takes place in the case of the number of substituents, e.g., dimethylnaphthalene is more susceptible to biotransformation than trimethylnaphthalene. In the case of methylnaphthalene derivatives, the most thermodynamically stable isomers are very quickly decomposed. This indicates that the stereochemical configuration of the hydrocarbon compounds is the dominant factor affecting biotransformation of crude oil rather than thermodynamic effects (Wolicka and Borkowski, 2012). In the case of methylphenanthrene derivatives, the highest resistance to biotransformation was observed in those with methyl groups in positions 9 and 10 in the molecule. In the case of steroid hydrocarbon derivatives such as monoaromatic derivatives and triaromatic derivatives, the monoaromatic derivatives are more resistant to microbiological degradation than the triaromatic derivatives. In addition, branched triaromatic derivatives with short side chains are more effectively biotransformed than branched triaromatic derivatives with long chains. In turn, short-chain pregnane derivatives are rather resistant to biotransformation in comparison to long-chain typical sterane derivatives.

Furthermore, the thermal maturity of the molecule is related to high values of the concentration ratios of polynuclear ring systems, while the degree of biotransformation increases when the ratio $[C_{21}/(C_{21}+C_{28})]$ decreases and when the ratio of $[C_{21}$ and C_{22}-pregnane derivatives$/(C_{27}$ and C_{29}-sterane derivatives$)]$ increases. Thus the indicators for crude oil thermal maturity might be influenced by biotransformation because bacteria generally decompose thermodynamically stable components very quickly. In addition, thermal maturity trends and biotransformation indicators may differ between oil reservoirs due to different species within the bacterial population, red-ox conditions, crude oil composition, and nutrients availability (Wolicka and Borkowski, 2012). On the other hand, in situ biotransformation of the light hydrocarbon fraction, e.g., n-paraffin derivatives, isoparaffin derivatives, cycloalkane derivatives, as well as benzene and alkylbenzene derivatives, may also take place in crude oil reservoirs (Vieth and Wilkes, 2005). All of these compounds are subject to biotransformation under anaerobic conditions that typically occur in oil reservoirs.

Typically, within the nonaromatic hydrocarbon derivatives, bacteria degrade first n-paraffin derivatives and then isoparaffin derivatives; therefore an increase of the i-C_5/n-C_5 ratio may be used as an indicator of the biotransformation progress. In the case of methylcyclohexane/n-C_7 ratio, the same trend also indicates a high degree of crude oil degradation. Another method that allows the determination of oil biotransformation degree is the study of the isotopic ratios of specific constituents. The preference of microorganisms for low-molecular-weight molecular isotopes causes the enrichment of heavier isotopes in the crude oil. This method may be, however, applied only in the case of low-molecular-weight carbon compounds. Removal of lower-boiling constituents because of biotransformation is an unfavorable event because it causes the decrease of the economic value of the crude oil (Vieth and Wilkes, 2005).

The biotransformation of oil within the reservoir rock may cause the formation of an upper gas cap, which is composed mainly of methane that is formed during the reduction of carbon dioxide by hydrogen. If the crude oil contains high quantities of sulfur, the activity of sulfur-reducing bacteria would result in the formation of hydrogen sulfide at concentrations that may reach over 10% in the gas. Similarly, the activity of other microbes may result in the transformation of primarily wet gas into dry gas containing at least 90% v/v methane. This process is linked to the preferential removal of C_3 to C_5 constituents by bacteria. At the same time, the crude oil fraction that is capable of being transformed will move to higher molecular weight (and higher boiling) (Wolicka and Borkowski, 2012).

There is also a link between the acidity of the oil, which is measured as the content of potassium hydroxide (KOH) required for neutralizing this acidity (the acid number of the crude oil) (Speight, 2014a,b), and the degree of oil biotransformation. This is observed particularly in oil with high acid value (>0.5 g KOH/g oil). This relates to the production of carboxylic acids during in situ microbiological degradation of hydrocarbons. However, other factors also influence the total acid value of the oil, because some nonbiotransformed oils have low pH, which is probably linked with high sulfur content. In oil with advanced biotransformation, the concentration of hopane acids increases; however, this increase is not enough to affect the overall acidity of the crude oil.

The controlled influence of parameters such as crude oil-water interfacial tension, crude oil viscosity, and the permeability of the reservoir rock through the use of microorganisms are the basis of MEOR, which has great potential as an enhanced oil recovery process (Chapter 4). In fact, MEOR methods can be applied in a variety of oil reservoir conditions such as low or high temperature or at high salinity concentrations. The versatility of MEOR applications is explained by the natural adaptation of bacterial species to oil reservoir conditions. Microorganisms play also a positive role in the removal of pollutants in crude oil fields or along transportation routes. Some microorganisms can biodegrade many of the chemical compounds contained in crude oils. Bioremediation processes may occur naturally (without human intervention); however, natural bioremediation process may take a long time (years) to be completed.

The complex nature of biochemical processes, which take place in various environments such as crude oil reservoirs, causes several interpretation problems during the analysis of the models used to describe the microbiological interactions in these settings. Although a complete description of the biotransformation reactions taking place in crude oil-microorganisms systems has not been fully resolved and understood, existing knowledge, even although it is incomplete, can be applied for the recovery of additional crude oil from mature oil fields as well as in bioremediation processes for the remediation of soil-water environments polluted by crude oil and oil-derived products.

The quality of crude oil and natural gas in the reservoir reflects the compositional characteristics of hydrocarbons that impact the economic viability of an exploration, development, or production opportunity. Compositions may affect the direct value of the product (e.g., crude valuation relative to a benchmark oil) or the development or facility costs (e.g., additional wells required, emulsion processing, use of special steels), or they may even cause the oil to be unrecoverable. Typical oil-quality properties include API gravity, viscosity, sulfur, asphaltene, and metals (e.g., vanadium, nickel, and iron content), residua (e.g., vacuum residua or Conradson carbon content), acidity (TAN), wax content or pour point, and sensitivity to emulsion formation upon production (Parkash, 2003; Hsu and Robinson,

2006; Gary et al., 2007; Speight, 2011b, 2012, 2014a,b, 2017). Biotransformation impacts essentially all of the properties that have an influence on crude oil quality.

The majority of the petroleum resource is partly biodegraded. This is of considerable practical significance and can limit economic exploitation of petroleum reserves and lead to problems during petroleum production. Knowledge of the microorganisms present in petroleum reservoirs, their physiological properties, and the biochemical potential for hydrocarbon degradation benefits successful petroleum exploration. Anaerobic conditions prevail in petroleum reservoirs and biological hydrocarbon degradation is apparently inhibited at temperatures above 80–90°C (176–194°F) (Rölling et al., 2003; Aitken et al., 2004; Jones et al., 2008; Parnell et al., 2017).

Reservoir gas caps and solution gases also undergo biodegradation in cool reservoirs. C_{2+} gas components, particularly propane ($CH_3CH_2CH_3$) and n-butane (n-$CH_3CH_2\ CH_2CH_3$), are preferentially removed from natural gas, making biodegraded gases drier through the enrichment of methane (CH_4). Most organisms that play a role in the biotransformation processes also generate carbon dioxide (CO_2) as a byproduct when they degrade hydrocarbons, increasing the carbon dioxide content of solution gas or gas caps. An elevated content of carbon dioxide can impact development economics negatively by necessitating the use of special steels to resist corrosion.

MEOR (Chapter 4) in which petroleum undergoes biotransformation in the reservoir before recovery operations commence is an important tertiary oil recovery method, which is cost-effective and eco-friendly technology to drive the residual oil trapped in the reservoirs. The potential of microorganisms to degrade heavy crude oil to reduce viscosity is very effective in MEOR. Earlier studies of MEOR (1950s) were based on three broad areas: injection, dispersion, and propagation of microorganisms in petroleum reservoirs; selective degradation of oil components to improve flow characteristics; and production of metabolites by microorganisms and their effects. Since thermophilic spore-forming bacteria can thrive in very extreme conditions in oil reservoirs, they are the most suitable organisms for the purpose.

5.1 RESERVOIR CHARACTER

The major controls on primary oil composition are characteristics of the source rock and the reservoir rock such as (1) organic-matter type, (2) depositional environment, (3) level of maturity, and (4) the quality of the crude oil in the reservoir. These controls exert the dominant influence on as-generated gravity, viscosity, gas/oil ratio, sulfur, and residua contents.

After emplacement in cool reservoirs, hydrocarbons are subject to biotransformation, in addition to a number of other potential alteration processes, including water washing, phase separation, gravity segregation, and deasphalting (asphaltene deposition). Some reservoirs have a complex history with multiple episodes of charge and degradation. Fresh charge to a reservoir may upgrade quality, while earlier episodes of severe transformation of the crude oil constituents (possibly when the reservoir was shallower and cooler than at present) may downgrade quality. Thus knowledge of generation and charge timing and reservoir temperature history can help improve prerecovery predictions of the quality of the crude oil (Wenger et al., 2002).

The impact of biotransformation in the reservoir on crude oil quality can be significant. Although different bacterial types and reservoir environments do have some effect on the order of compounds removed, the general order-of-preference trends are usually applicable. Thus the straight-chain n-alkanes are typically attacked before branched saturates (e.g., isoprenoid derivatives), cycloalkane derivatives, and aromatic derivatives. As a result, significant differences in the distribution of the crude oil constituents (as well as the bulk properties of the crude oil) are intimately linked to the level of biotransformation.

5.2 TEMPERATURE EFFECTS

In a reservoir where the temperature is less than approximately 80°C (176°F), oil biotransformation is common and detrimental. Crude oil from shallow, cool reservoirs tends to be progressively more bio-degraded than those in deeper, hotter reservoirs (Wenger et al., 2002). Increasing levels of biodegrada-tion generally cause a decline in oil quality, diminishing the producibility and value of the oil as API gravity and distillate yields decrease. Additionally, viscosity, sulfur, asphaltene, metals, vacuum residua, and TANs increase. For a specific hydrocarbon system (similar source type and level of matu-rity), general trends exist for oil-quality parameters versus present-day reservoir temperatures of <80°C (<176°F). However, other controls on biodegradation may also have significant effects, making predrill prediction of oil quality difficult in some areas.

An important factor influencing microorganism activity is temperature and the biotransformation reactions are not (theoretically) considered to occur at temperatures above 130–150°C (266–302°F). Such conditions would correspond to deep reservoirs at depths in excess of 13,000 feet. Thus the res-ervoir temperature range is critical to biotransformation. Above temperatures of approximately 80°C (176°F), petroleum degrading bacterial activity is significantly inhibited. At a temperature below this limit, the biotransformation reactions generally operate at a lower rate (i.e., lower efficiency) (Wenger et al., 2002).

5.3 PRESSURE EFFECTS

A major effect of reservoir pressure (through the presence of a gas cap) is the separation of a solid phase (consisting of asphaltene constituents) from the oil. This phenomenon of gas deasphalting in the reser-voir is caused by the pressure-induced dissolution of gas cap constituents (low-boiling hydrocarbon derivatives) in the crude oil. This destabilizes the asphaltene constituents and caused separation result-ing in the production of a crude oil with a lower asphaltene content. Biotransformation reactions can cause an increase in reservoir pressure, especially if low-boiling volatile (crude oil-soluble) gases such as hydrocarbons (an example is methane, CH_4), hydrogen, and carbon dioxide are the result of such reactions. In fact, the initiation of biotransformation reactions that increase the reservoir pressure could be beneficial and lead to production of a higher quality crude oil at the surface.

In general, a pressure up to approximately 2900 psi (along with temperatures up to 80°C/176°F) in crude oil reservoirs are within the limits for the operation of biotransformation mechanisms.

6.0 CHALLENGES AND OPPORTUNITIES

Biotechnology is now accepted as an attractive means of improving the efficiency of many refining processes as well as resolving serious environmental problems (Speight and Arjoon, 2012; El-Gendy and Speight, 2015). For example, microbial enzymes are capable of biotransforming a wide range of compounds, and the worldwide increase in attention being paid to this concept can be attributed to several factors, including the presence of a wide variety of catabolic enzymes and the ability of many microbial enzymes to transform a broad range of unnatural compounds (xenobiotics) as well as natural compounds. Biotransformation processes have several advantages compared with chemical processes, including the following: (1) microbial enzyme reactions are often more selective, (2) biotransformation processes are often more energy-efficient, (3) microbial enzymes are active under mild conditions, and (4) microbial enzymes are environment-friendly biocatalysts.

Biorefining is a possible alternative to some of the current oil-refining processes. The major potential applications of biorefining are BDS, BDN, BDM, and biotransformation of heavy crude oils into lighter crude oils. The most advanced area is BDS (El-Gendy and Speight, 2015) with the result that the concepts related to, and performance of, biocatalytic desulfurization (BDS) is showing promise for application to other areas of biorefining.

Although the main points of research and development in the field of in situ and ex situ petroleum biotechnology nowadays mainly concerns with MEOR, bioremediation of petroleum pollutants, and BDS, it is expected in the near future, that the use of biotechnology would be extended to other areas of petroleum refining. This would have a concern on BDN; biotransformation of heavy oils into low-molecular-weight products; more research on the applications of biocatalysts in hydrocracking, isomerization, polymerization, or alkylation; and finally, the introduction of biocatalysts in petrochemical industry.

The economic viability of in situ petroleum biotechnology such as MEOR depends mainly on the types of the indigenous microbial populations and their metabolic pathways. Not only this, but it depends also on the application of cost effective nutrients, as well as the ability to efficiently deliver nutrients, specific growth inhibitors, enzymes (free or immobilized), polymers, chemicals etc., to the point of application.

The success of ex situ petroleum biotechnology in petroleum refining and petrochemical industry will depend on the enhanced activity of the applied biocatalyst (i.e., rate and substrate range) and its long-term stability under the harsh conditions in petroleum industry under the actual process conditions, its reusability for successive cycles without losing its activity (i.e., be active for several hundred hours), and finally, its economic production. Thus it must be able to fit in-line (i.e., alternative and/or complimentary) with the existing upstream process (e.g., pH, temperature, pressure). Moreover, it should produce valuable products and avoid the production of toxic by-products. The most important point of view is to allow for easy separation of cells/biocatalyst and product. The advantages in the advances of biocatalyst applications in nonaqueous media, and the active and stable nature of the enzymes in organic and gaseous phases, as well as in supercritical fluids, would add to the industrial application of petroleum biotechnology processes, in which, the substrates are generally insoluble in water and have to be dissolved in organic solvents. In petroleum, the fuel itself can act as the required organic solvent. Potential biocatalysts can be enzymes (free or immobilized) and whole cells. Enzymes usually require less water than microorganisms to fulfill its maximum activity. Theoretically, only one film of water, covering their surface, is sufficient for its catalytic activity (Le Borgne and Quintero, 2003). However, in the case of complex transformations that require a mixture of several enzymes organized in a complex metabolic pathway and the presence of cofactors (electron-transporting molecules), the whole cells are preferable. Thus to design and develop biotransformation in organic media, microbial isolates or mutants that are able to survive in the presence of high concentrations of organic solvents are very important, to catch the wave of industrial petroleum biotechnology.

The development of commercial bioprocess depends mainly on significant improvements in a cheap and abundant production of highly active and stable biocatalysts adapted to the extreme conditions encountered in petroleum refining (such as complex hydrophobic media, high temperatures etc.): extremophilic microorganisms, such as psychrophiles: optimal growth temperature <20°C; thermophiles and hyperthermophiles: optimal growth temperature >50°C and 75°C, respectively; high pressure ones (barophiles: optimal growth at > 1 atm); extreme pH ones (acidophiles and alkalophiles: optimal growth at pH <3 and >10, respectively); and finally, the high salt concentrations ones (halophiles: optimal growth in 2–5.2 M NaCl) (Rinker et al., 1999). Many of the aforementioned conditions are similar to those in the petroleum industry. Thus extremophiles and their enzymes (i.e., extremozymes) can be recommended for petroleum industry, since they are extremely thermostable and resistant to salinity, organic solvents, and extreme pH. The recent discovery for microorganisms that are able to survive at 110°C opens a new research for their application in refineries, since at higher temperatures, the bioavailability and solubility of hydrocarbons are higher. Moreover, the enzymes from the halophiles would be very promising, particularly, in low water activity reaction media (such as organic solvents). Since, halophilic environments are characterized by low water activities. Thus extremophilic microorganisms possessing activities relevant to biorefining have to be isolated and screened, in addition to the application of genetic engineering.

The application of genetic engineering is very important to improve the biocatalyst activity. Improvement in the bioreactor design, phase contact, and separation systems should be also taken into consideration for the future commercialization of bioprocess in petroleum industry.

Thus the most important point of view is the establishment of the interdisciplinary participation of biotechnology, biochemistry, geology, refining processes, and engineering.

REFERENCES

Abin-Fuentes, A., Mohamed, M.E.S., Wang, E.I.C., Prather, K.L.J., 2013. Exploring the mechanism of biocatalyst inhibition in microbial desulfurization. Applied and Environmental Microbiology 79 (24), 7807–7817.

Aitken, C.M., Jones, D.M., Larter, S.R., 2004. Anaerobic hydrocarbon biodegradation in deep subsurface oil reservoirs. Nature 431 (7006), 291–294.

Amund, O.O., Adewale, A.A., Ugoji, E.O., 1987. Occurrence and characterization of hydrocarbon utilizing bacteria in Nigerian soils contaminated with spent motor oil. Indian Journal of Microbiology 27, 63–87.

Ancheyta, J., Speight, J.G., 2007. Hydroprocessing Heavy Oils and Residua. CRC Press, Taylor & Francis Group, Boca Raton, Florida.

Antić, M.P., Jovančićević, B.S., Ilić, M., Vrvić, M.M., Schwarzbauer, J., 2006. Petroleum pollutant degradation by surface water microorganisms. Environmental Science and Pollution Research 13 (5), 320–327.

Atlas, R.W., 1975. Effects of temperature and crude oil composition on petroleum biodegradation. Advances in Applied Microbiology 30 (3), 396–403.

Atlas, R.M., Bartha, R., 1992. Hydrocarbon biodegradation and oil spill bioremediation. Advances in Microbial Ecology 12, 287–338.

Atlas, R.M., Bragg, J., 2009. Bioremediation of marine oil spills: when and when not – the Exxon Valdez experience. Microbial Biotechnology 2 (2), 213–221.

Bachmann, R.T., Johnson, A.C., Edyvean, R.G.J., 2014. Biotechnology in the petroleum industry: an overview. International Biodeterioration & Biodegradation 86, 225–237.

Balat, M., 2011. Fuels from biomass – an overview. In: Speight, J.G. (Ed.), The Biofuels Handbook. Royal Society of Chemistry, London, United Kingdom. Part 1, (Chapter 3).

Bamforth, S.M., Singleton, I., 2005. Bioremediation of polycyclic aromatic hydrocarbons: current knowledge and future directions. Journal of Chemical Technology and Biotechnology 80, 723–736.

Barathi, S., Vasudevan, N., 2001. Utilization of petroleum hydrocarbons by *Pseudomonas fluorescens* isolated from a petroleum-contaminated soil. Environment International 26 (5–6), 413–416.

Bardi, L., Martini, C., Opsi, F., Bertolone, E., Belviso, S., Masoero, G., Marzona, M., Ajmone Marsan, F., 2007. Cyclodextrin-enhanced in situ bioremediation of polyaromatic hydrocarbons-contaminated soils and plant uptake. Journal of Inclusion Phenomena and Macrocyclic Chemistry 57, 439–444.

Bartha, R., 1986. Microbial Ecology: Fundamentals and Applications. Addisson-Wesley Publishers, Reading, Massachusetts.

Bestetti, G., Collina, E., Di Gennaro, P., Lasagni, M., Pitea, D., 2003. Kinetic study of naphthalene biodegradation in aerobic slurry phase microcosms for the optimization of the process. Water, Air, and Soil Pollution: Focus 3, 223–231.

Bossert, I., Bartha, R., 1984. The fate of petroleum in soil ecosystems. In: Atlas, R.M. (Ed.), Petroleum Microbiology. Macmillan, New York, pp. 453–473.

Cerniglia, C.E., 1984. Microbial metabolism of polycyclic aromatic hydrocarbons. Advances in Applied Microbiology 30, 31–71.

Cerniglia, C.E., 1992. Biodegradation of polycyclic aromatic hydrocarbons. Biodegradation 3, 351–368.

Cerniglia, C.E., Heitkamp, M.A., 1989. Microbial degradation of polycyclic aromatic hydrocarbons in the aquatic environment. In: Varanasi, U. (Ed.), Metabolism of Polycyclic Aromatic Hydrocarbons in the Aquatic Environment. CRC Press, Inc., Boca Raton, Florida, pp. 41–68.

Cerniglia, C.E., Heitkamp, M.A., 1990. Polycyclic aromatic hydrocarbon degradation by *Mycobacterium*. Methods in Enzymology 188, 148–153.

Chapman, P.J., Shelton, M., Grifoll, M., Selifonov, S., 1995. Fossil fuel biodegradation: laboratory studies. Environmental Health Perspectives 103 (Suppl. 5), 79–83.

Chen, C.I., Taylor, R.T., 1995. Thermophilic biodegradation of BTEX by two *Thermus* species. Biotechnology and Bioengineering 48, 614–624.

Chen, C.I., Taylor, R.T., 1997a. Batch and fed-batch bioreactor cultivations of a *Thermus* species with thermophilic BTEX-degrading activity. Applied Microbiology and Biotechnology 47, 726–733.

Chen, C.I., Taylor, R.T., 1997b. Thermophilic biodegradation of BTEX by two consortia of anaerobic bacteria. Applied Microbiology and Biotechnology 48, 121–128.

Choi, S.-C., Kwon, K.K., Sohn, J.H., Kim, S.-J., 2002. Evaluation of fertilizer additions to stimulate oil biodegradation in sand seashore mescocosms. Journal of Microbiology and Biotechnology 12, 431–436.

Colwell, R.R., Walker, J.D., Cooney, J.J., 1977. Ecological aspects of microbial degradation of petroleum in the marine environment. Critical Reviews in Microbiology 5 (4), 423–445.

Cooney, J.J., Silver, S.A., Beck, E.A., 1985. Factors influencing hydrocarbon degradation in three freshwater lakes. Microbial Ecology 11 (2), 127–137.

Demirbaş, A., 2011. Production of fuels from crops. In: Speight, J.G. (Ed.), The Biofuels Handbook. Royal Society of Chemistry, London, United Kingdom. Part 2, (Chapter 1).

El-Gendy, N.Sh., Speight, J.G., 2015. Handbook of Refinery Desulfurization. CRC Press, Taylor & Francis Group, Boca Raton, Florida.

Foght, J., Semple, K., Gauthier, C., Westlake, D.S., Blenkinsopp, S., Sergy, G., Wang, Z., Fingas, M., 1999. Effect of nitrogen source on biodegradation of crude oil by a defined bacterial consortium incubated under cold, marine conditions. Environmental Technology 20, 839–849.

Foght, J.M., Westlake, D.W., 1988. Degradation of polycyclic aromatic hydrocarbons and aromatic heterocycles by a *Pseudomonas* species. Canadian Journal of Microbiology 34 (10), 1135–1141.

Frenzel, M., James, P., Burton, S.K., Rowland, S.J., Lappin-Scott, H.M., 2009. Towards bioremediation of toxic unresolved complex mixtures of hydrocarbons: identification of bacteria capable of rapid degradation of alkyl-tetralins. Journal of Soils and Sediments 9, 129–136.

Fujii, T., Narikawa, T., Takeda, K., Kato, J., 2004. Biotransformation of various alkanes using the *Escherichia coli* expressing an alkane hydroxylase system from *Gordonia* sp. TF6. Bioscience, Biotechnology, and Biochemistry 68 (10), 2171–2177.

Gary, J.G., Handwerk, G.E., Kaiser, M.J., 2007. Petroleum Refining: Technology and Economics, fifth ed. CRC Press, Taylor & Francis Group, Boca Raton, Florida.

Ghazali, F.M., Rahman, R.N.Z.A., Salleh, A.B., Basri, M., 2004. Biodegradation of hydrocarbons in soil by microbial consortium. International Biodeterioration & Biodegradation 54, 61–67.

Gibson, D.T., Subramanian, V., 1984. Microbial degradation of aromatic hydrocarbons. In: Gibson, D.T. (Ed.), Microbial Degradation of Organic Compounds. Marcel Dekker, Inc., New York, N.Y., pp. 181–252.

Goodwin, N.S., Park, P.J.D., Rawlinson, A.P., 1983. Crude oil biodegradation under simulated and natural conditions. In: Bjorøy, M. (Ed.), Advances in Organic Geochemistry 1981. John Wiley & Sons Inc., New York, pp. 650–658.

Greenwood, P.F., Wibrow, S., George, S.J., Tibbett, M., 2009. Hydrocarbon biodegradation and soil microbial community response to repeated oil exposure. Organic Geochemistry 40, 293–300.

Harvey, R.G., 1997. Polycyclic Aromatic Hydrocarbons. John Wiley & Sons Inc.-VCH, New York.

Head, I.M., Jones, D.M., Röling, W.F.M., 2006. Marine microorganisms make a meal of oil. Nature Reviews Microbiology 4, 173–182.

Heitkamp, M.A., Cerniglia, C.E., 1988. Mineralization of polycyclic aromatic hydrocarbons by a bacterium isolated from sediment below an oil field. Applied and Environmental Microbiology 54, 1612–1614.

Heitkamp, M.A., Cerniglia, C.E., 1989. Polycyclic aromatic hydrocarbon degradation by a *Mycobacterium* sp. In: Microcosms Containing Sediment and Water from a Pristine EcosystemAppl. Environ. Microbiol., vol. 55, pp. 1968–1973.

Hernáez, M.J., Reineke, W., Santero, E., 1999. Genetic analysis of biodegradation of tetralin by a *Sphingomonas* strain. Applied and Environmental Microbiology 65 (4), 1806–1810.

Hsu, C.S., Robinson, P.R. (Eds.), 2006. Practical Advances in Petroleum Processing, vols. 1 and 2. Springer Science, New York.

Huang, L., Ma, T., Li, D., Liang, F., Liu, R., Li, G., 2008. Optimization of nutrient component for diesel oil degradation by *Rhodococcus erythropolis*. Marine Pollution Bulletin 56, 1714–1718.

Jacquot, F., Guiliano, M., Doumenq, P., Munoz, D., Mille, G., 1996. In vitro photo-oxidation of crude oil maltenic fractions: evolution of fossil biomarkers and polycyclic aromatic hydrocarbons. Chemosphere 33, 671–681.

Jean, J.M., Lee, M.K., Wang, S., Chattopadhyay, P., Maity, J., 2008. Effects of inorganic nutrient levels on the biodegradation of benzene, toluene, and xylene (BTX) by *Pseudomonas* sp in a laboratory porous media sand aquifer model. Bioresource Technology 99, 7807–7815.

Jenkins, B.M., Ebeling, J.M., (May-June) 1985. Thermochemical properties of biomass fuels. California Agriculture 14–18.

Jones, D.M., Head, I.M., Gray, N.D., Adams, J.J., Rowan, A.K., Aitken, C.M., Bennett, B., Huang, H., Brown, A., Bowler, B.F., Oldenburg, T., Erdmann, M., Larter, S.R., 2008. Crude oil biodegradation via methanogenesis in subsurface petroleum reservoirs. Nature 451 (7175), 176–180.

Joshi, P.A., Pandey, G.B., 2011. Screening of petroleum degrading bacteria from cow dung. Research Journal of Agricultural Sciences 2 (1), 69–71.

Kanal, R.A., Harayama, S., 2000. Biodegradation of high-molecular-weight polycyclic aromatic hydrocarbons by bacteria. Journal of Bacteriology 182 (8), 2059–2067.

Kim, S., Choi, D.H., Sim, D.S., Oh, Y., 2005. Evaluation of bioremediation effectiveness on crude oil-contaminated sand. Chemosphere 59, 845–852.

Le Borgne, S., Quintero, R., 2003. Biotechnological processes for the refining of petroleum. Fuel Processing Technology 81, 155–169.

Leahy, J.G., Colwell, R.R., 1990. Microbial degradation of hydrocarbons in the environment. Microbiological Reviews 54, 305–315.

Leuenberger, H.G.W., 1990. Biotransformation – a useful tool in organic chemistry. Pure & Applied Chemistry 62 (4), 753–768.

Luning Prak, D.J., Pritchard, P.H., 2002. Solubilization of polycyclic aromatic hydrocarbon mixtures in micellar nonionic surfactant solution. Water Research 36, 3463–3472.

Maletić, S., Dalmacija, B., Rončević, S., Agbaba, J., Petrović, O., 2009. Degradation kinetics of an aged hydrocarbon-contaminated soil. Water Air Soil Pollution 202, 149–159.

Milić, J.S., Beškoski, V.P., Ilić, M.V., Ali, S.A.M., Gojgić-Cvijović, G.Đ., Vrvić, M.M., 2009. Bioremediation of soil heavily contaminated with crude oil and its products: composition of the microbial consortium. Journal of the Serbian Chemical Society 74 (4), 455–460.

Mohajeri, L., Aziz, H.A., Isa, M.H., Zahed, M.A., 2010. A Statistical experiment design approach for optimizing biodegradation of weathered crude oil in coastal sediments. Bioresource Technology 101, 893–900.

Mohebali, G., Ball, A.S., 2008. Biocatalytic desulfurization (biodesulfurization) of petrodiesel fuels. Microbiology 154, 2169–2183.

Mueller, J.G., Chapman, P.J., Blattmann, B.O., Pritchard, P.H., 1990. Isolation and characterization of a fluoranthene-utilizing strain of *Pseudomonas paucimobilis*. Applied and Environmental Microbiology 56, 1079–1086.

Nasrollahzadeh, H.S., Najafpour, G.D., Aghamohammadi, N., 2007. Biodegradation of phenanthrene by mixed culture consortia in a batch bioreactor using central composite face-entered design. International Journal of Environmental Research 1, 80–87.

Obayori, O.S., Ilori, M.O., Adebusoye, S.A., Oyetibo, G.O., Omotayo, A.E., Amund, O.O., 2009. Degradation of hydrocarbons and biosurfactant production by *Pseudomonas sp.* strain LP1. World Journal of Microbiology & Biotechnology 25, 1615–1623.

Obire, O., 1990. Bacterial degradation of three different crude oils in Nigeria. Nigerian Journal of Botany 1, 81–90.

Obire, O., 1993. The suitability of various Nigerian petroleum fractions as substrate for bacterial growth. Discovery and Innovation 9, 25–32.

Okerentugba, P.O., Ezeronye, O.U., 2003. Petroleum degrading potentials of single and mixed microbial cultures isolated from rivers and refinery effluent in Nigeria. African Journal of Biotechnology 2 (9), 288–292.

Okoh, A.I., Ajisebutu, S., Babalola, G.O., Trejo-Hernandez, M.R., 2001. Potentials of *Burkholderia cepacia* strain RQ1 in the biodegradation of heavy crude oil. International Microbiology 4, 83–87.

Okoh, A.I., Ajisebutu, S., Babalola, G.O., Trejo-Hernandez, M.R., 2002. Biodegradation of Mexican heavy crude oil (Maya) by *Pseudomonas aeruginosa*. Journal of Tropical Biosciences 2 (1), 12–24.

Okoh, A.I., 2003. Biodegradation of Bonny light crude oil in soil microcosm by some bacterial strains isolated from crude oil flow stations saver pits in Nigeria. African Journal of Biotechnology 2 (5), 104–108.

Okoh, A.I., 2006. Biodegradation alternative in the cleanup of petroleum hydrocarbon pollutants. Biotechnology and Molecular Biology Review 1 (2), 38–50.

Okoh, A.I., Trejo-Hernandez, M.R., 2006. Remediation of petroleum hydrocarbon polluted systems: exploiting the bioremediation strategies. African Journal of Biotechnology 5, 2520–2525.

Page, D.S., Boehm, P.D., Douglas, G.S., Bence, A.E., Burns, W.A., Mankiewic, P.J., 1996. The natural petroleum hydrocarbon background in subtidal sediments of Prince William Sound Alaska, USA. Environmental Toxicology and Chemistry 15, 1266–1281.

Parkash, S., 2003. Refining Processes Handbook. Gulf Professional Publishing, Elsevier, Amsterdam, Netherlands.

Parnell, J., Baba, M., Bowden, S., Muirhead, D., 2017. Subsurface biodegradation of crude oil in a fractured basement reservoir, Shropshire, UK. Jounral of the Geological Society. http://jgs.geoscienceworld.org/content/jgs/early/2017/02/24/jgs2016-s2129.full.pdf.

Pathak, H., Kantharia, D., Malpani, A., Madamwar, D., 2009. Naphthalene biodegradation using *Pseudomonas* sp. HOB1: in vitro studies and assessment of naphthalene degradation efficiency in simulated microcosms. Journal of Hazardous Materials 166, 1466–1473.

Pavlova, A., Ivanova, R., 2003. Determination of petroleum hydrocarbons and polycyclic aromatic hydrocarbons in sludge from wastewater treatment basins. Journal of Environmental Monitoring 5, 319–323.

Payne, J.R., McNabb Jr., G.D., 1984. Weathering of petroleum in the marine environment. Marine Technology Society Journal 18 (3), 24.

Peixoto, R.S., Vermelho, A.B., Rosado, A.S., 2011. Petroleum-degrading enzymes: bioremediation and new prospects. Enzyme Research. :475193https://www.hindawi.com/journals/er/2011/475193/.

Perry, J.J., 1984. Microbial metabolism of cyclic alkanes. In: Atlas, R.M. (Ed.), Petroleum Microbiology. Macmillan, New York, pp. 61–98.

Peters, K.E., Walters, C.C., Moldowan, J.M., 2005. The Biomarker Guide, second ed. Cambridge University Press, Cambridge, United Kingdom.

Prince, R., 1993. Petroleum spill bioremediation in marine environments. Critical Reviews in Microbiology 19, 217–242.

Rahman, K.S.M., Thahira-Rahman, J., Lakshmanaperumalsamy, P., Banat, I.M., 2002. Towards efficient crude oil degradation by a mixed bacterial consortium. Bioresource Technology 85, 257–261.

Rahman, K.S.M., Thahira-Rahman, T., Kourkoutas, Y., Petsas, I., Marchant, R., Banat, I.M., 2003. Enhanced bioremediation of n-alkanes in petroleum sludge using bacterial consortium amended with rhamnolipid and micronutrients. Bioresource Technology 90 (2), 159–168.

Ramroop Singh, N., 2011. Biofuel. In: Speight, J.G. (Ed.), The Biofuels Handbook. Royal Society of Chemistry, London, United Kingdom. Part 1, (Chapter 5).

Rinker, K.D., Han, C.J., Adams, M.W.W., Kelly, R.M., 1999. In: Demain, A.L., Davies, J.E. (Eds.), Manual of Industerial Microbiology and Biotechnology. American Society for Microbiology, Washington, DC.

Rölling, W.F., Head, I.M., Larter, S.R., 2003. The microbiology of hydrocarbon degradation in subsurface petroleum reservoirs: perspectives and prospects. Research in Microbiology 154 (5), 321–328.

Sandrin, T.R., Kight, W.B., Maier, W.J., Maier, R.M., 2006. Influence of a non-aqueous phase liquid (NAPL) on biodegradation of phenanthrene. Biodegradation 17, 423–435.

Schreiber, A.F., Winkler, U.K., 1983. Transformation of tetralin by whole cells of *Pseudomonas stutzeri* As39. Applied Microbiology and Biotechnology 18 (1), 6–10.

Shuttleworth, K.L., Cerniglia, C.E., 1995. Environmental aspects of polynuclear aromatic hydrocarbon biodegradation. Applied Biochemistry and Biotechnology 54, 291–302.

Sikkema, J., Bont, J.A.M., 1991. Isolation and initial characterization of bacteria growing on tetralin. Biodegradation 2 (1), 15–23.

Sivaraman, C., Ganguly, A., Mutnuri, S., 2010. Biodegradation of hydrocarbons in the presence of cyclodextrins. World Journal of Microbiology & Biotechnology 26, 227–232.

Soli, G., Bens, E.M., 1972. Bacteria which attack petroleum hydrocarbons in a saline medium. Biotechnology and Bioengineering 14, 319–330.

Speight, J.G., 2000. The Desulfurization of Heavy Oil and Residua, second ed. Marcel Dekker Inc., New York.

Speight, J.G., 2005. Environmental Analysis and Technology for the Refining Industry. John Wiley & Sons Inc., Hoboken, New Jersey.

Speight, J.G., 2011a. The Biofuels Handbook. Royal Society of Chemistry, London, United Kingdom.

Speight, J.G., 2011b. The Refinery of the Future. Gulf Professional Publishing, Elsevier, Oxford, United Kingdom.

Speight, J.G., 2012. Crude Oil Assay Database. Knovel, Elsevier, New York. Online version available at: http://www.knovel.com/web/portal/browse/display?_EXT_KNOVEL_DISPLAY_bookid=5485&VerticalID=0.

Speight, J.G., Arjoon, K.K., 2012. Bioremediation of Petroleum and Petroleum Products. Scrivener Publishing, Beverly, Massachusetts.

Speight, J.G., 2014a. The Chemistry and Technology of Petroleum, fifth ed. CRC Press, Taylor & Francis Group, Boca Raton, Florida.

Speight, J.G., 2014b. High Acid Crudes. Gulf Professional Publishing, Elsevier, Oxford, United Kingdom.

Speight, J.G., 2016. Microbial Enhanced Oil Recovery. Introduction to Enhanced Recovery Methods for Heavy Oil and Tar Sands (Second Edition). Pages 323–351.

Speight, J.G., 2017. Handbook of Petroleum Refining. CRC Press, Taylor & Francis Group, Boca Raton, Florida.

Strawinski, R.J., Stone, R.W., 1940. The utilization of hydrocarbons by bacteria. Journal of Bacteriology 40 (3), 461.

Swannell, R.P.J., Lee, K., McDonagh, M., June 1996. Field Evaluation of marine oil spill bioremediation. Microbiological Reviews 342–365.

Taghvaei Ganjali, S., Nahri Niknafs, B., Khosravi, M., et al., 2007. Photo-oxidation of crude petroleum maltenic fraction I natural simulated conditions and structural elucidation of photoproducts. Iranian Journal of Environmental Health Science & Engineering 4 (1), 37–42.

Taylor, T.R., Jackson, K.J., Duba, A.G., Chen, C.I., May 19, 1998. In Situ Thermally Enhanced Biodegradation of Petroleum Fuel Hydrocarbons and Halogenated Organic Solvents. United States Patent 5753122.

Venosa, A.D., Zhu, X., 2003. Biodegradation of crude oil contaminating marine shorelines and freshwater wetlands. Spill Science & Technology Bulletin 8 (2), 163–178.

Vieira, P.A., Faria, S.R., Vieira, B., De Franca, F.P., Cardoso, V.L., 2009. Statistical analysis and optimization of nitrogen, phosphorus, and inoculum concentrations for the biodegradation of petroleum hydrocarbons by response surface methodology. Journal of Microbiology and Biotechnology 25, 427–438.

Vieth, A., Wilkes, H., 2005. Deciphering biodegradation effects on light hydrocarbons in crude oils using their stable carbon isotopic composition: a case study from the Gullfaks oil field, offshore Norway. Geochimica 70, 651–665.

Weissenfels, W.D., Beyer, M., Klein, J., 1990. Degradation of phenanthrene, fluorene and fluoranthene by pure bacterial cultures. Applied Microbiology and Biotechnology 32, 479–484.

Weissenfels, W.D., Beyer, M., Klein, J., Rehm, H.J., 1991. Microbial metabolism of fluoranthene: isolation and identification of ring fission products. Applied Microbiology and Biotechnology 34, 528–535.

Wenger, L.M., Davis, C.L., Isaksen, G.H., 2002. Multiple controls on petroleum biodegradation and impact on oil quality. SPE Reservoir Evaluation & Engineering 5 (05), 375–383.

Whyte, L.G., Hawari, J., Zhou, E., Bourbonnière, L., Inniss, W.E., Greer, C.W., 1998. Biodegradation of variable-chain-length alkanes at low temperatures by a psychrotrophic *Rhodococcus* sp. Applied and Environmental Microbiology 64, 2578–2584.

Willumsen, P.A., Karlson, U., Pritchard, P.H., 1998. Response of fluoranthene-degrading bacteria to surfactants. Applied Microbiology and Biotechnology 50, 475–483.

Wolicka, D., Borkowski, A., 2012. Microorganisms and crude oil. In: Romero-Zerón, L. (Ed.), Introduction to Enhanced Oil Recovery (EOR) Processes and Bioremediation of Oil-contaminated Sites (Chapter 5).

Wong, J.W.C., Lai, K.M., Wan, C.K., Ma, K.K., Fang, M., 2001. Isolation and optimization of PAHs-degradative bacteria from contaminated soil for PAHs bioremediation. Water, Air, and Soil Pollution 13, 1–13.

Ye, D., Siddiqi, M.A., Maccubbin, A.E., Kumar, S., Sikka, H.C., 1996. Degradation of polynuclear aromatic hydrocarbons by *Sphingomonas paucimobilis*. Environmental Science & Technology 30, 136–142.

Young, L.Y., 1984. Anaerobic degradation of aromatic compounds. In: Gibson, D.T. (Ed.), Microbial Degradation of Organic Compounds. Marcel Dekker, Inc., New York, pp. 487–523.

Zahed, M.A., Aziz, H.A., Isa, M.H., Mohajeri, L., 2010. Enhancement of biodegradation of n-alkanes from crude oil contaminated seawater. International Journal of Environmental Research 4 (4), 655–664.

Zander, M., 1983. Physical and chemical properties of polycyclic aromatic hydrocarbons. In: Bjørseth, A. (Ed.), Handbook of Polycyclic Aromatic Hydrocarbons. Marcel Dekker, Inc., New York, pp. 1–26.

Zengler, K., Richnow, H.H., Rossello-Mora, R., Michaelis, W., Widdel, F., 1999. Methane formation from long-chain alkanes by anaerobic microorganisms. Nature 401, 266–269.

FURTHER READING

Alexander, M., 1994. Biodegradation and Bioremediation. Academic Press Inc., New York.

Atlas, R.M., 1991. Microbial hydrocarbon degradation – bioremediation of oil spills. Journal of Chemical Technology and Biotechnology 52, 149–156.

Bardi, L., Ricci, R., Mario Marzona, M., 2003. *In situ* bioremediation of a hydrocarbon polluted site with cyclodextrin as a coadjuvant to increase bioavailability. Water, Air, and Soil Pollution: Focus 3, 15–23.

Berry, D.F., Francis, A.J., Bollag, J.-M., 1987. Microbial metabolism of homocyclic and heterocyclic aromatic compounds under anaerobic conditions. Microbiological Reviews 51 (1), 43–59.

Dandie, C.E., Weber, J., Aleer, S., Adetutu, E.M., Ball, A.S., Juhasz, A.L., 2010. Assessment of five bioaccessibility assays for predicting the efficacy of petroleum hydrocarbon biodegradation in aged contaminated soils. Chemosphere 81, 1061–1068.

Freijer, J.I., de Jonge, H., Bouten, W., Verstraten, J.M., 1996. Assessing mineralization rates of petroleum hydrocarbons in soils in relation to environmental factors and experimental scale. Biodegradation 7, 487–500.

Gibson, D.T., Sayler, G.S., 1992. Scientific Foundation for Bioremediation: Current Status and Future Needs. American Academy of Microbiology, Washington, DC.

Margesin, R., Schinner, F., 1997a. Efficiency of indigenous and inoculated cold-adapted soil microorganisms for biodegradation of diesel oil in alpine soils. Applied and Environmental Microbiology 63, 2660–2664.

Margesin, R., Schinner, F., 1997b. Bioremediation of diesel-oil-contaminated alpine soils at low temperatures. Applied Microbiology and Biotechnology 47, 462–468.

Sleat, R., Robinson, J.P., 1984. The bacteriology of anaerobic degradation of aromatic compounds. Journal of Applied Bacteriology 57, 381–394.

Vidali, M., 2001. Bioremediation: an overview. Pure and Applied Chemistry 73 (7), 1163–1172.

Zekri, A.Y., Abou-Kassem, J.H., Shedid, S.A., 2009. A new technique for measurement of oil biodegradation. Petroleum Science and Technology 27, 666–677.

MICROBIAL ENHANCED OIL RECOVERY

4

1.0 INTRODUCTION

Biotechnology refers to the use of living systems and organisms to develop or make products or any technological application that uses biological systems, living organisms, or derivatives thereof, to make or modify products or processes for specific use (Table 4.1). Depending on the applications of biotechnology, it overlaps (as an umbrella term) with the fields of bioengineering and biomanufacturing. But first and by way of a refresher definition in the context of this book, *petroleum biotechnology* (Chapter 3) is based on biotransformation processes such as biodegradation and bioremediation (Speight and Arjoon, 2012).

Furthermore, *biodegradation (biotic degradation, biotic decomposition)* is the chemical degradation of chemicals by bacteria or other biological means. Organic material can be degraded aerobically (in the presence of oxygen) or anaerobically (in the absence of oxygen). Most bioremediation systems operate under aerobic conditions, but a system under anaerobic conditions may permit microbial organisms to degrade chemical species that are otherwise nonresponsive to aerobic treatment, but at a slower rate; the converse also applies—chemical species that are nonresponsive to anaerobic treatment may undergo biotransformation under aerobic conditions, usually at faster rate than anaerobic process. Furthermore, the general concept of biotransformation encompasses a wide range of procedures for modifying chemicals in living organisms according to human needs. In addition, and even more pertinent to this text, bioengineering is a related field that more heavily emphasizes higher systems approaches (not necessarily the altering or using of biological materials *directly*) for interfacing with and utilizing microbes and is the application of the principles of engineering and natural sciences to molecular transformation of the feedstock.

Biotechnology has applications in four major industrial areas, including health care (medical), crop production and agriculture, nonfood (industrial) uses of crops and other products such as biodegradable plastics, vegetable oil, biofuels, and the generation of products from petroleum-based feedstocks (Chapter 3). In the latter case, this includes microbial enhanced oil recovery (MEOR), biodesulfurization (Chapter 6) and biodenitrogenation (Chapter 7) (Speight, 2014a; El-Gendy and Speight, 2016). Thus, *industrial biotechnology* (known in some countries as *white biotechnology*) is biotechnology applied to industrial processes such as using microbial entities (including enzymes) to produce a useful chemical product. Another example is the use of enzymes as industrial catalysts to either produce valuable chemicals or destroy hazardous/polluting chemicals (Speight and Arjoon, 2012). This form of biotechnology tends to consume less in resources than traditional processes used to produce industrial products.

Introduction to Petroleum Biotechnology. https://doi.org/10.1016/B978-0-12-805151-1.00004-7

Table 4.1 Potential Process Application of Biotechnology	
Storage	Reduction in corrosion
Exploration	Geological prospecting
Production	Microbial enhanced oil recovery
Refining	Desulfurization
	Denitrogenation
	Demetallization
	Transformation to lower molecular weight products
Petrochemicals	Phenol production and polymerization
	Oxidation of alkene derivatives
	Oxidation of sulfur derivatives
	Oxidation of aromatic derivatives
Process effluents	Reduction in amounts
	Reduction in pollution
	Biotransformation and bioremediation

2.0 OIL RECOVERY

Recovery, as applied in the petroleum industry, is the production of oil from a reservoir (Speight, 2014a, 2016). There are several methods by which this can be achieved that range from recovery due to reservoir energy (i.e., the oil flows from the well hole without assistance) to enhanced recovery methods in which considerable energy must be added to the reservoir to produce the oil. However, the effect of the method on the oil and on the reservoir (Table 4.2) must be considered before application. This section, for the most part, deals with those recovery methods (bio-based recovery methods) that are applied to recovery of conventional crude oil and, in some cases, to recovery of heavy oil (Speight, 2013a).

The process of recovering oil from any conventional reservoir requires (1) a pathway which connects oil in the pore space of a reservoir to the surface and (2) sufficient energy in the reservoir to drive the oil to the surface. Lack of these requirements in the environment results in oil remaining in the reservoir. In this case, it is not economical to implement incremental development activities. In addition, all of the theoretically displaceable oil cannot be recovered, even if there is a pathway and adequate reservoir energy.

Generally, crude oil reservoirs sometimes exist with an overlying *gas cap*, in communication with aquifers, or both. The oil resides together with water and free gas in very small holes (pore spaces) and fractures. The size, shape, and degree of interconnection of the pores vary considerably from place to place in an individual reservoir. Below the oil layer the sandstone is usually saturated with salt water. The oil is released from this formation by drilling a well and puncturing the limestone layer on either side of the limestone dome or fold. If the peak of the formation is tapped, only the gas is obtained. If the penetration is made too far from the center, only salt water is obtained. Oil wells may be either on land or under water. In North America, many wells are *offshore* in the shallow parts of the oceans. The crude oil or unrefined oil is typically collected from individual wells by small pipelines.

Table 4.2 Recovery Process Parameters and Their Potential Effects

Property	Comments
Carbon dioxide injection	Lowers pH; can change oil composition
Miscible flooding	Hydrocarbon-rich gases lower the solubility parameter of the oil
	Separation of asphaltene material
Organic chemicals	Can lower the solubility parameter and solubility parameter of the oil
	Separation of asphaltene material
	Blocking of flow channels
Acidizing	Interaction of crude oil constituents upsetting molecular balance
	Deposition of sludge
	Blocking of flow channels
Pressure decrease	Can change composition of oil medium
	Phase separation of asphaltene material
	Blocking of flow channels
Temperature decrease	Change in composition of oil medium leading to phase separation of asphaltene material
	Blocking of flow channels

Recovery of crude oil when a well is first opened is usually by natural flow forced by the pressure of the gas or fluids that are contained within the deposit. There are several means that serve to drive the petroleum fluids from the formation, through the well, and to the surface, and these methods are classified as either natural or applied flow.

2.1 PRIMARY PROCESSES

If the underground pressure in the oil reservoir is sufficient, then the oil will be forced to the surface under this pressure (primary recovery, natural methods). Gaseous fuels or natural gas are usually present, which also supplies needed underground pressure. In this situation, it is sufficient to place a complex arrangement of valves (the Christmas tree) at the wellhead to connect the well to a pipeline network for storage and processing. This is called primary oil recovery.

Thus, *primary oil production* (*primary oil recovery*) is the first method of producing oil from a well and depends upon natural reservoir energy to drive the oil through the complex pore network to producing wells. If the pressure on the fluid in the reservoir (reservoir energy) is great enough, the oil flows into the well and up to the surface. Such driving energy may be derived from liquid expansion and evolution of dissolved gases from the oil as reservoir pressure is lowered during production, expansion of free gas, or a gas cap, influx of natural water, gravity, or combinations of these effects.

Crude oil moves out of the reservoir into the well by one or more of three processes. These processes are: *dissolved gas drive, gas cap drive,* and *water drive*. Early recognition of the type of drive involved is essential to the efficient development of an oil field. In *dissolved gas drive*, the propulsive force is the gas in solution in the oil, which tends to come out of solution because of the pressure release at the point of penetration of a well. Dissolved gas drive is the least efficient type of natural drive as it is difficult to control the gas-oil ratio; the bottom-hole pressure drops rapidly, and the total eventual recovery of petroleum from the reservoir may be less than 20%. If gas overlies the oil beneath the top of the trap, it is compressed and can be utilized (*gas cap drive*) to drive the oil into wells situated at the bottom of the oil-bearing zone. By producing oil only from below the gas cap, it is possible to maintain a high gas-oil ratio in the reservoir until almost the very end of the life of the pool. If, however, the oil deposit is not systematically developed, so that bypassing of the gas occurs, an undue proportion of oil is left behind. The usual recovery of petroleum from a reservoir in a gas cap field is 40%–50%.

Usually the gas in a gas cap (*associated natural gas*) contains methane and other hydrocarbons that may be separated out by compressing the gas. A well-known example is *natural gasoline* that was formerly referred to as *casinghead gasoline* or *natural gas gasoline*. However, at high pressures, such as those existing in the deeper fields, the density of the gas increases and the density of the oil decreases until they form a single phase in the reservoir. These are the so-called retrograde condensate pools because a decrease (instead of an increase) in pressure brings about condensation of the liquid hydrocarbons. When this reservoir fluid is brought to the surface and the condensate is removed, a large volume of residual gas remains. The modern practice is to cycle this gas by compressing it and injecting it back into the reservoir, thus maintaining adequate pressure within the gas cap, and condensation in the reservoir is prevented. Such condensation prevents recovery of the oil, for the low percentage of liquid saturation in the reservoir precludes effective flow.

The most efficient propulsive force in driving oil into a well is natural *water drive*, in which the pressure of the water forces the lighter recoverable oil out of the reservoir into the producing wells. In anticlinal accumulations, the structurally lowest wells around the flanks of the dome are the first to come into water. Then the oil-water contact plane moves upward until only the wells at the top of the anticline are still producing oil; eventually these also must be abandoned as the water displaces the oil.

In a water drive field, it is essential that the removal rate be adjusted so that the water moves up evenly as space is made available for it by the removal of the hydrocarbons. An appreciable decline in bottom-hole pressure is necessary to provide the pressure gradient required to cause water influx. The pressure differential needed depends on the reservoir permeability; the greater the permeability, the less the difference in pressure necessary. The recovery of petroleum from the reservoir in properly operated water drive pools may run as high as 80%. The force behind the water drive may be hydrostatic pressure, the expansion of the reservoir water, or a combination of both. Water drive is also used in certain submarine fields.

Gravity drive is an important factor when oil columns of several thousands of feet exist, as they do in some North American fields. Furthermore, the last bit of recoverable oil is produced in many pools by gravity drainage of the reservoir. Another source of energy during the early stages of withdrawal from a reservoir containing under-saturated oil is the expansion of that oil as the pressure reduction brings the oil to the bubble point (the pressure and temperature at which the gas starts to come out of solution).

For primary recovery operations, no pumping equipment is required. If the reservoir energy is not sufficient to force the oil to the surface, then the well must be pumped. In either case, nothing is added to the reservoir to increase or maintain the reservoir energy or to sweep the oil toward the well. The rate of production from a flowing well tends to decline as the natural reservoir energy is expended. When a flowing well is no longer producing at an efficient rate, a pump is installed. The recovery efficiency for primary production is generally low when liquid expansion and solution gas evolution are the driving mechanisms. Much higher recoveries are associated with reservoirs with water and gas cap drives and with reservoirs in which gravity effectively promotes drainage of the oil from the rock pores. The overall recovery efficiency is related to how the reservoir is delineated by production wells. Thus, for maximum recovery by primary recovery it is often preferable to sink several wells into a reservoir, thereby bringing about recovery by a combination of the methods outlined here.

2.2 SECONDARY PROCESSES

Over the lifetime of the well the pressure will fall, and at some point, there will be insufficient underground pressure to force the oil to the surface. If economical, and it often is, the remaining oil in the well is extracted using secondary oil recovery methods. It is at this point that secondary recovery methods must be applied.

Secondary oil recovery methods use various techniques to aid in recovering oil from depleted or low-pressure reservoirs. Sometimes pumps on the surface or submerged (electrical submersible pumps, ESPs) are used to bring the oil to the surface. Other secondary recovery techniques to increase the reservoir's pressure by water injection and gas injection, which injects air or some other gas into the reservoir, are used. Together, primary recovery and secondary recovery allow 25%–35% v/v of the crude oil in the reservoir oil to be recovered.

The most common follow-up, or *secondary recovery*, operations usually involve the application of pumping operations or of injection of materials into a well to encourage movement and recovery of the remaining petroleum. The pump (generally known as the horsehead pump or the sucker rod pump) provides mechanical lift to the fluids in the reservoir (Speight, 2014a). The most commonly recognized oil-well pump is the reciprocating or plunger pumping equipment (also called a *sucker-rod pump*), which is easily recognized by the *horsehead* beam pumping jacks. A pump barrel is lowered into the well on a string of 6-inch (inner diameter) steel rods known as sucker rods. The up-and-down movement of the sucker rods forces the oil up the tubing to the surface. A walking beam powered by a nearby engine may supply this vertical movement, or it may be brought about by a pump jack, which is connected to a central power source by means of pull rods. Electrically powered centrifugal pumps and submersible pumps (both pump and motor are in the well at the bottom of the tubing) have proven their production capabilities in numerous applications.

There are also *secondary oil recovery* operations that involve the injection of water or gas into the reservoir. When water is used, the process is called a *waterflood*; with gas, a *gas flood*. Separate wells are usually used for injection and production. The injected fluids maintain reservoir pressure or re-pressure the reservoir after primary depletion and displace a portion of the remaining crude oil to production wells. In fact, the first method recommended for improving the recovery of oil was probably the re-injection of natural gas, and there are indications that gas injection was utilized for this purpose before 1900 (Craft and Hawkins, 1959; Frick, 1962).

The success of secondary recovery processes depends on the mechanism by which the injected fluid displaces the oil (displacement efficiency) and on the volume of the reservoir that the injected fluid enters (conformance or sweep efficiency). In most proposed secondary projects, water does both these things more effectively than gas. It must be decided if the use of gas offers any economic advantages because of availability and relative ease of injection. In reservoirs with high permeability and high vertical span, the injection of gas may result in high recovery factors because of gravity segregation, as described in a later section. However, if the reservoir lacks either adequate vertical permeability or the possibility for gravity segregation, a frontal drive similar to that used for water injection can be used (dispersed gas injection). Thus, dispersed gas injection is anticipated to be more effective in reservoirs that are relatively thin and have little dip. Injection into the top of the formation (or into the gas cap) is more successful in reservoirs with higher vertical permeability (200 md or more) and enough vertical relief to allow the gas cap to displace the oil downward.

2.3 TERTIARY PROCESSES

Enhanced oil recovery (EOR, *tertiary oil recovery*) (Speight, 2014a, 2016) is the incremental ultimate oil that can be recovered from a petroleum reservoir over oil that can be obtained by primary and secondary recovery methods. The residual crude oil (non-recoverable crude oil) in reservoirs represents the relative inefficiency of the primary and secondary production techniques. Extraction of this trapped oil can be achieved by injecting chemicals (polymers or surfactants), gases (carbon dioxide, hydrocarbons, or nitrogen), or steam into the reservoir. The chemicals used for EOR must be compatible with the physical and chemical environments of oil reservoirs. The varying permeability of petroleum reservoirs is also a major concern in processes involving enhanced oil recovery. When water is injected to displace the oil, it preferentially flows through areas of highest permeability and bypasses much of the oil (Speight, 2014a).

The viscosity (or the API gravity) of petroleum is an important factor that must be considered when heavy oil is recovered from a reservoir. In fact, certain reservoir types, such as those with very viscous crude oils and some low-permeability carbonate (limestone, dolomite, or chert) reservoirs, respond poorly to conventional secondary recovery techniques (Speight, 2014a).

In these reservoirs, it is desirable to initiate EOR operations as early as possible. This may mean considerably abbreviating conventional secondary recovery operations or bypassing them altogether. Thermal floods using steam and controlled in situ combustion methods are also used. Thermal methods of recovery reduce the viscosity of the crude oil by heat so that it flows more easily into the production well. Thus, tertiary techniques are usually variations of secondary methods with a goal of improving the *sweeping* action of the invading fluid.

Thus, EOR methods are designed to reduce the viscosity of the crude oil (i.e., to reduce the pour point of the crude oil relative to the temperature of the reservoir), thereby increasing oil production. EOR methods are applied starting when secondary oil recovery techniques are no longer enough to sustain production. Thermally enhanced oil recovery methods are tertiary recovery techniques that heat the oil and make it easier to extract. Steam injection is the most common form of this process and is used extensively to increase oil production. In situ combustion is another form of thermally enhanced oil recovery but instead of using steam to reduce the crude oil viscosity, some of the oil is burned to heat the surrounding oil. Detergents are also used to decrease oil viscosity.

EOR processes use *thermal, chemical,* or *fluid phase behavior* effects to reduce or eliminate the capillary forces that trap oil within pores, to thin the oil, or otherwise improve its mobility or to alter the mobility of the displacing fluids. In some cases, the effects of gravity forces, which ordinarily cause vertical segregation of fluids of different densities, can be minimized or even used to advantage. The various processes differ considerably in complexity, the physical mechanisms responsible for oil recovery, and the amount of experience that has been derived from field application. The degree to which the EOR methods are applicable in the future will depend on development of improved process technology. It will also depend on improved understanding of fluid chemistry, phase behavior, physical properties, and on the accuracy of geology and reservoir engineering in characterizing the physical nature of individual reservoirs.

Chemical methods include polymer flooding, surfactant (micellar or polymer and microemulsion) flooding, and alkaline flood processes.

Polymer flooding (Polymer augmented waterflooding) is waterflooding in which organic polymers are injected with the water to improve horizontal and vertical sweep efficiency. The process is conceptually simple and inexpensive, and its commercial use is increasing despite relatively small potential incremental oil production. Surfactant flooding is complex and requires detailed laboratory testing to support field project design. As demonstrated by field tests, it has excellent potential for improving the recovery of low-viscosity to moderate-viscosity oil. Surfactant flooding is expensive and has been used in few large-scale projects. Alkaline flooding has been used only in those reservoirs containing specific types of high-acid-number crude oils.

The terms *microemulsion* and *micellar solution* are used to describe concentrated, surfactant-stabilized dispersions of water and hydrocarbons that are used to enhance oil recovery. At concentrations above a certain critical value, the surfactant molecules in solution form aggregates called micelles. These micelles are capable of solubilizing fluids in their cores and are called swollen micelles. Spherical micelles have size ranges from 10^{-6} to 10^{-4} mm. The micellar solution or microemulsion is homogeneous, transparent or translucent, and stable to phase separation. The term *soluble oil* is often used to describe an oil external system having little or no dispersed water.

Although used to describe the process, neither *microemulsion* or *micellar solution* accurately describes all compositions that are used in emulsion flooding. In fact, many systems used do not have an identifiable external or continuous phase and these terms do not always apply.

Microemulsion flooding (micellar/emulsion flooding) refers to a fluid injection process in which a stable solution of oil, water, and one or more surfactants along with electrolytes of salts is injected into the formation and is displaced by a mobility buffer solution. Injecting water in turn displaces the mobility buffer. Depending on the reservoir environment, a pre-flood may or may not be used. The microemulsion is the key to the process. Oil and water are displaced ahead of the microemulsion slug, and a stabilized oil-water bank develops. The displacement mechanism is the same under secondary and tertiary recovery conditions. In the secondary case, water is the primary produced fluid until the oil bank reaches the well.

In *microemulsion flooding*, two approaches have developed to enhance oil recovery. In the first process, a relatively low-concentration surfactant microemulsion is injected at large pore volumes of 15%–60% to reduce the interfacial tension between water and oil, thereby increasing oil recovery. In the second process, a relatively small pore volume, from 3% to 20% of a high-concentration surfactant microemulsion, is injected. With the high concentration of surfactant in the microemulsion, the micelles solubilize the oil and water in the displacing microemulsion. Consequently, the high-concentration system may initially displace the oil in a miscible manner. However, as the high-concentration slug moves through the reservoir it is diluted by the formation fluids and the process ultimately or gradually reverts to a low-concentration flood. However, this initial displacement forms an oil bank, which is very important in establishing displacement efficiency. Low-concentration systems typically contain 2%–4% surfactant, whereas high-concentration systems contain 8%–12% w/w.

Mobility control is important to the success of the process. The mobility of the microemulsion can be matched to that of the stabilized water-oil bank by controlling the microemulsion viscosity. The mobility buffer following the microemulsion slug prevents rapid slug deterioration from the rear and thus minimizes the slug size required for efficient oil displacement. Water external emulsions and aqueous solutions of high molecular weight polymers have been used as mobility buffers.

Microemulsion flooding can be applied over a wide range of reservoir conditions. Generally, wherever a waterflood has been successful, microemulsion flooding may also be applicable. In cases in which waterflooding was a failure because of poor mobility relationships, microemulsion flooding might be technically successful because of the required mobility control. Of course, if waterflooding was a failure because of certain reservoir conditions, such as fracturing or very high permeability streaks, microemulsion flooding will most likely also fail. In microemulsion flooding, the slug must be designed for specific reservoir conditions of temperature, resident water salinity, and crude oil type. If the temperature is very high, a fluid-handling problem may result in the field because of the increased vapor pressure of the hydrocarbon in microemulsion.

In analyzing the applicability of microemulsion-polymer flooding to a given reservoir, the need for a thorough understanding of the reservoir and fluid characteristics cannot be overemphasized. As mentioned, such characteristics as the nature of the oil and water content, relative permeability, mobility ratios, formation fractures, and variations in permeability, porosity, formation continuity, and rock mineralogy can have a dramatic effect on the success or failure of the process.

Conventional waterflooding can often be improved by the addition of polymers (*polymer flooding*) to injection water to improve the mobility ratio between the injected and in-place fluids. The polymer solution affects the relative flow rates of oil and water and sweeps a larger fraction of the reservoir than water alone, thus contacting more of the oil and moving it to production wells. Polymers currently in use are produced both synthetically (polyacrylamides) and biologically (polysaccharides). The polymers may also be cross-linked in situ to form highly viscous fluids that will divert the subsequently injected water into different reservoir strata.

Polymer flooding has its greatest utility in heterogeneous reservoirs and those that contain moderately viscous oils. Oil reservoirs with adverse waterflood mobility ratios have a potential for increased oil recovery through better horizontal sweep efficiency. Heterogeneous reservoirs may respond favorably because of improved vertical sweep efficiency. Because the microscopic displacement efficiency is not affected, the increase in recovery over waterflood will likely be modest and limited to the extent that sweep efficiency is improved, but the incremental cost is also moderate. Currently, polymer flooding is being used in a significant number of commercial field projects. The process may be used to recover oils of higher viscosity than those for which a surfactant flood might be considered. Polymer solutions must be stable for a prolonged period at reservoir conditions. Mechanical, chemical, thermal, and microbial effects can degrade polymers. However, degradation can be minimized or even prevented by using specific equipment or methods (Speight, 2014a).

Stability problems may occur because of oxygen contamination of the polymer solutions. Such contamination can lower the screen factor of polyacrylamide solutions by as much as 30%. In field operations, the loss of mobility reduction due to oxygen may be more serious since control of the reservoir fluid composition can be difficult. Sodium hydrosulfite in low concentrations is an effective oxygen collector for polyacrylamide solutions. However, sodium hydrosulfite tends to catalyze polymer deterioration when free oxygen and decomposed polymers are present. Therefore, the proper use of sodium hydrosulfite is imperative to avoid severe polymer degradation. In addition, caution is necessary to prevent oxygen from reentering the system once sodium hydrosulfite has been added to the makeup water.

Surfactant flooding is a multiple-slug process involving the addition of surface-active chemicals to water. These chemicals reduce the capillary forces that trap the oil in the pores of the rock. The surfactant slug displaces most of the oil from the reservoir volume contacted, forming a flowing oil-water bank that is propagated ahead of the surfactant slug. The principal factors that influence the surfactant slug design are interfacial properties, slug mobility in relation to the mobility of the oil-water bank, the persistence of acceptable slug properties and slug integrity in the reservoir, and cost.

A slug of water containing polymer in solution follows the surfactant slug. The polymer solution is injected to preserve the integrity of the costlier surfactant slug and to improve the sweep efficiency. Both these goals are achieved by adjusting the polymer solution viscosity in relation to the viscosity of the surfactant slug to obtain a favorable mobility ratio. The polymer solution is then followed by injection of drive water, which continues until the project is completed. However, each reservoir has unique fluid and rock properties, and specific chemical systems must be designed for each individual application. The chemicals used, their concentrations in the slugs, and the slug sizes depend upon the specific properties of the fluids and the rocks involved and upon economic considerations.

Alkaline flooding adds inorganic alkaline chemicals, such as sodium hydroxide, sodium carbonate, or sodium orthosilicate derivatives, to the water to enhance oil recovery by one or more of the following mechanisms: interfacial tension reduction, spontaneous emulsification, or wettability alteration. These mechanisms rely on the in situ formation of surfactants during the neutralization of petroleum acids in the crude oil by the alkaline chemicals in the displacing fluids.

Although emulsification in alkaline flooding processes decreases injection fluid mobility to a certain degree, emulsification alone may not provide adequate sweep efficiency. Sometimes polymer is included as an ancillary mobility control chemical in an alkaline waterflood to augment any mobility ratio improvements due to alkaline-generated emulsions.

Other variations on this theme include the use of steam and the means of reducing interfacial tension using various solvents. The solvent approach has had some success when applied to bitumen recovery from mined tar sand but when applied to non-mined material losses of solvent and dissolved bitumen are always an issue (Speight, 2013b). However, this approach should not be rejected out of hand since a novel concept may arise that guarantees minimal (acceptable) losses of bitumen and solvent.

Miscible fluid displacement (*miscible displacement*) is an oil displacement process in which an alcohol, a refined hydrocarbon, a condensed petroleum gas, carbon dioxide, liquefied natural gas, or even exhaust gas is injected into an oil reservoir, at pressure levels such that the injected gas or alcohol and reservoir oil are miscible; the process may include the concurrent, alternating, or subsequent injection of water.

The procedures for miscible displacement are the same in each case and involve the injection of a slug of solvent that is miscible with the reservoir oil followed by injection of either a liquid or a gas to sweep up any remaining solvent. It must be recognized that the miscible *slug* of solvent becomes enriched with oil as it passes through the reservoir and its composition changes, thereby reducing the effective scavenging action. However, changes in the composition of the fluid can also lead to wax deposition, as well as deposition of asphaltene constituents. Therefore, caution is advised.

Microscopic observations of the leading edge of the miscible phase have shown that the displacement takes place at the boundary between the oil and the displacing phase. The small amount of oil that is bypassed is entrained and dissolved in the rest of the slug of miscible fluids; mixing and diffusion occur to permit complete recovery of the remaining oil. If a second miscible fluid is used to displace the first, another zone of displacement and mixing follows. The distance between the leading edge of the miscible slug and the bulk of pure solvent increases with the distance traveled, as mixing and reservoir heterogeneity cause the solvent to be dispersed.

Other parameters affecting the miscible displacement process are reservoir length, injection rate, porosity, and permeability of reservoir matrix, size and mobility ratio of miscible phases, gravitational effects, and chemical reactions. Miscible floods using carbon dioxide, nitrogen, or hydrocarbons as miscible solvents have their greatest potential for enhanced recovery of low-viscosity oils. Commercial hydrocarbon-miscible floods have been operated since the 1950s, but carbon dioxide-miscible flooding on a large scale is relatively recent and is expected to make the most significant contribution to miscible enhanced recovery in the future.

Carbon dioxide can displace many crude oils, thus permitting recovery of most of the oil from the reservoir rock that is contacted (*carbon dioxide-miscible flooding*). The carbon dioxide is not initially miscible with the oil. However, as the carbon dioxide contacts the in situ crude oil, it extracts some of the hydrocarbon constituents of the crude oil into the carbon dioxide and carbon dioxide is also dissolved in the oil. Miscibility is achieved at the displacement front when no interfaces exist between the hydrocarbon-enriched carbon dioxide mixture and the carbon dioxide-enriched oil. Thus, by a *dynamic* (*multiple-contact*) process involving interphase mass transfer, miscible displacement overcomes the capillary forces that otherwise trap oil in pores of the rock.

The reservoir operating pressure must be kept at a level high enough to develop and maintain a mixture of carbon dioxide and extracted hydrocarbons that, at reservoir temperature, will be miscible with the crude oil. Impurities in the carbon dioxide stream, such as nitrogen or methane, increase the pressure required for miscibility. Mixing due to reservoir heterogeneity and diffusion tends to locally alter and destroy the miscible composition, which must then be regenerated by additional extraction of hydrocarbons. In field applications both miscible and near-miscible displacements may proceed simultaneously in different parts of the reservoir.

The volume of carbon dioxide injected is specifically chosen for each application and usually range from 20% to 40% of the reservoir pore volume. In the later stages of the injection program, carbon dioxide may be driven through the reservoir by water or a lower cost inert gas. To achieve higher sweep efficiency, water and carbon dioxide are often injected in alternate cycles.

In some applications, particularly in carbonate (limestone, dolomite, and chert) reservoirs where it is likely to be used most frequently, carbon dioxide may prematurely break through to producing wells. When this occurs, remedial action using mechanical controls in injection and production wells may be taken to reduce carbon dioxide production. However, substantial carbon dioxide production is considered normal. Generally, this produced carbon dioxide is reinjected, often after processing to recover valuable light hydrocarbons.

For some reservoirs, miscibility between the carbon dioxide and the oil cannot be achieved and is dependent upon the oil properties. However, carbon dioxide can still be used to recover additional oil. The carbon dioxide swells crude oils, thus increasing the volume of pore space occupied by the oil and reducing the quantity of oil trapped in the pores. It also reduces the oil viscosity. Both effects improve the mobility of the oil. Carbon dioxide-immiscible flooding has been demonstrated in both pilot and commercial projects, but overall it is expected to make a relatively small contribution to enhanced oil recovery.

The solution gas-oil ratio for carbonated crude oil should be measured in the normal way and plotted as gas-oil ratio in volume per volume versus pressure. The greater the solubility of carbon dioxide in the oil, the larger is the increase in the solution gas-oil ratio. In fact, the increase in the gas-oil ratio usually parallels the increase in the oil formation volume factor due to swelling. It should be noted that the gas in any gas-oil ratio experiment is not carbon dioxide but contains hydrocarbons that have vaporized from the liquid phase. Consequently, whether the gas-oil ratio is measured in a pressure-volume-temperature cell or from a slim tube experiment, compositional analysis must be carried out to obtain the composition of the gas, as well as that of the equilibrium liquid phase. If actual measured values are not available, the correlation developed for crude oil containing dissolved gases can be used but give only approximate values at best. Since the density of pure gases is a function of pressure and temperature, for crude oil saturated with gases, the density in the mixing zone must be specified as a function of pressure and mixing zone composition.

Hydrocarbon gases and condensates have been used for over 100 commercial and pilot miscible floods. Depending upon the composition of the injected stream and the reservoir crude oil, the mechanism for achieving miscibility with reservoir oil can be similar to that obtained with carbon dioxide (dynamic or multiple-contact miscibility), or the miscible solvent and in situ oil may be miscible initially (first-contact miscibility). Except in special circumstances, these light hydrocarbons are generally too valuable to be used commercially.

Nitrogen and flue gases have also been used for commercial miscible floods. Minimum miscibility pressures for these gases are usually higher than for carbon dioxide, but in high-pressure, high-temperature reservoirs where miscibility can be achieved these gases may be a cost-effective alternative to carbon dioxide.

Thermal EOR processes add heat to the reservoir to reduce oil viscosity and/or to vaporize the oil. In both instances, the oil is made more mobile so that it can be more effectively driven to producing wells. In addition to adding heat, these processes provide a driving force (pressure) to move oil to producing wells.

Thermal recovery methods include cyclic steam injection, steam flooding, and in situ combustion. The steam processes are the most advanced of all EOR methods in terms of field experience and thus have the least uncertainty in estimating performance, provided that a good reservoir description is available. Steam processes are most often applied in reservoirs containing viscous oils and tars, usually in place of rather than following secondary or primary methods. Commercial application of steam processes has been underway since the early 1960s. In situ combustion has been field tested under a wide variety of reservoir conditions, but few projects have proven economical and advanced to commercial scale.

Steam drive injection (steam injection) has been commercially applied since the early 1960s. The process occurs in two steps: (1) steam stimulation of production wells, that is, direct steam stimulation, and (2) steam drive by steam injection to increase production from other wells (indirect steam stimulation).

When there is some natural reservoir energy, steam stimulation normally precedes steam drive. In steam stimulation, heat is applied to the reservoir by the injection of high-quality steam into the produce well. This cyclic process, also called *huff and puff* or *steam soak*, uses the same well for both injection and production. The period of steam injection is followed by production of reduced viscosity oil and condensed steam (water). One mechanism that aids production of the oil is the flashing of hot water (originally condensed from steam injected under high pressure) back to steam as pressure is lowered when a well is put back on production.

Cyclic steam injection is the alternating injection of steam and production of oil with condensed steam from the same well or wells. Thus, steam generated at surface is injected in a well and the same well is subsequently put back on production. A cyclic steam injection process includes three stages. In the first stage is injection, during which a measured amount of steam is introduced into the reservoir. In the second stage (*the soak period*) requires that the well be shut in for a period of time (usually several days) to allow uniform heat distribution to reduce the viscosity of the oil (alternatively, to raise the reservoir temperature above the pour point of the oil). Finally, during the third stage, the now-mobile oil is produced through the same well. The cycle is repeated until the flow of oil diminishes to a point of no returns.

Cyclic steam injection is used extensively in heavy-oil reservoirs, tar sand deposits, and in some cases to improve injectivity prior to steam flooding or in situ combustion operations. Cyclic steam injection is also called *steam soak* or the *huff 'n' puff* method.

In practice, steam is injected into the formation at greater than fracturing pressure (150–1600 psi for Athabasca sands) followed by a *soak* period after which production is commenced. The technique has also been applied to the California tar sand deposits and in some heavy oil reservoirs north of the Orinoco deposits. The steam flooding technique has been applied, with some degree of success, to the Utah tar sands and has been proposed for the San Miguel (Texas) tar sands.

In situ combustion is normally applied to reservoirs containing low-gravity oil but has been tested over perhaps the widest spectrum of conditions of any EOR process. In the process, heat is generated within the reservoir by injecting air and burning part of the crude oil. This reduces the oil viscosity and partially vaporizes the oil in place, and the oil is driven out of the reservoir by a combination of steam, hot water, and gas drive. *Forward combustion* involves movement of the hot front in the same direction as the injected air. *Reverse combustion* involves movement of the hot front opposite to the direction of the injected air.

The relatively small portion of the oil that remains after these displacement mechanisms have acted becomes the fuel for the in situ combustion process. Production is obtained from wells offsetting the injection locations. In some applications, the efficiency of the total in situ combustion operation can be improved by alternating water and air injection. The injected water tends to improve the utilization of heat by transferring heat from the rock behind the combustion zone to the rock immediately ahead of the combustion zone. The performance of in situ combustion is predominantly determined by the four following factors: (1) the quantity of oil that initially resides in the rock to be burned, (2) the quantity of air required to burn the portion of the oil that fuels the process, (3) the distance to which vigorous combustion can be sustained against heat losses, and (4) the mobility of the air or combustion product gases.

In *modified extraction* processes, combinations of in situ and mining techniques are used to access the reservoir. A portion of the reservoir rock must be removed to enable application of the in situ extraction technology. The most common method is to enter the reservoir through a large-diameter vertical shaft, excavate horizontal drifts from the bottom of the shaft, and drill injection and production wells horizontally from the drifts. Thermal extraction processes are then applied through the wells. When the horizontal wells are drilled at or near the base of the tar sand reservoir, the injected heat rises from the injection wells through the reservoir, and drainage of produced fluids to the production wells is assisted by gravity.

However, in addition to the conventional methods of crude oil recovery from the reservoir, there has arisen a strong incentive to seek new technologies in order to increase or extend oil production from areas which face production flow problems, and it has been demonstrated that it is possible to obtain a significant increase in production and in the recovery factor through the use of biological products, even when the operating mechanisms are only partially understood (Table 4.3; Saikia et al., 2013). Nevertheless, with this need in mind, microbially enhanced oil recovery presents significant advantages and can be applied via existing producing or injection wells. In fact, there is a growing awareness that microorganisms can play a much wider role than previously thought and may represent the application

Table 4.3 Products of Microbial Reactions Products—Effects in the Reservoir

Products	Effects
Acid	Increase in porosity
	Increase in permeability
	Production of carbon dioxide via reaction with carbonate minerals
Biomass	Selective and nonselective plugging
	Emulsification through adhesion to oil
	Changing in wettability of mineral surfaces
	Reduction of oil viscosity
	Reduction of pour point
	Desulfurization
Gases	Reservoir repressurization
	Viscosity reduction
	Solubilization of carbonate rocks—increase in permeability
Solvents	Dissolution of oil
Surfactants	Lowering of interfacial tension
	Emulsification
Polymers	Mobility control
	Selective or nonselective plugging

of similar processes to those involved in the formation of crude oil. The possibility of biological emulsification and demulsification of oil suggests that it would be possible to emulsify oil in order to facilitate its withdrawal from the rock matrix, and later demulsify it in order to recover it, after it is released from the reservoir. Together, these processes could allow an improvement in the production and recovery of the oil in place.

Briefly, microorganisms can synthesize useful products by fermenting low-cost substrates or raw materials. Therefore, MEOR processes are substitute chemical enhanced oil recovery (CEOR) processes. In MEOR processes, chosen microbial colonies that are specific to the crude oil type and the reservoir type are used to biotransform the crude oil constituents to products that enable enhanced flow of the crude oil to a production well (Sarkar et al., 1989; Suthar et al., 2008; Banat et al., 2010; Al-Bahry et al., 2013).

3.0 MICROBIAL ENHANCED OIL RECOVERY

A crude oil reservoir is a complex environment containing living (microorganisms) and nonliving factors (minerals) which interact with each other in a complicated dynamic network of nutrients and energy fluxes. Since the reservoir is heterogeneous, the variety of ecosystems containing diverse microbial communities are also diverse leading to properties that affect reservoir behavior and mobilization of the crude oil. Put simply, microbes are living entities that produce excretion products and new cells which can interact with each other or with the reservoir environment, positively or negatively. For example, the production of acid derivatives can interact with and dissolve carbonate minerals that lead to enhanced flow of crude oil through the reservoir. In fact, all of the entities (enzymes, extracellular polymeric substances, and the cells themselves) may participate as catalysts or reactants. Such complexity is increased by the interplay with the environment which can play an important role by affecting cellular function.

MEOR is one of the EOR techniques where bacteria and their by-products are utilized for oil mobilization in a reservoir. In principle, MEOR is a process that increases oil recovery through inoculation of microorganisms in a reservoir, aiming that bacteria and their by-products cause some beneficial effects such as the formation of stable oil-water emulsions, mobilization of residual oil as a result of reduced interfacial tension, and diverting of injection fluids through upswept areas of the reservoir by clogging high permeable zones. Microbial technologies are becoming accepted worldwide as cost-effective and environmentally friendly approaches to improve oil production (Springham, 1984; Sarkar et al., 1989, 1994; Youssef et al., 2009).

Thus, MEOR involves, as the name suggests, insertion (injection) of microbes into the reservoir to modify the properties of the oil so that it can be recovered by flow through a well. To understand the process and assist in the selection of a suitable microbe colony (or suitable microbe colonies), it is necessary to understand the types of biotransformations that occur when microbes are added to crude oil (Chapter 3).

MEOR is a technique that enables the improvement of the recovery of crude oil by injection of microorganisms and/or their products into depleted oil reservoirs. The process involves biotransformation of the crude oil constituents so that the crude oil can be recovered through a well (Beckman, 1926; ZoBell, 1947; Jobson et al., 1972; Jack, 1983). During the last several decades, the interest in MEOR has been revived throughout the world in the search for a high performance and cost-effective enhanced recovery method. Furthermore, the emergence of heavy oils as major refinery feedstocks and the recovery as well as the potential for upgrading the heavy oil as part of the recovery process.

There are some environmental factors that affect the performance of microbial enhanced recovery processes (Chapter 3). These are temperature, permeability, pH, salinity of the medium, and oxygen content. As all oil reservoirs are essentially devoid of oxygen, anaerobic bacteria are generally preferred in field applications. *Clostridium acetobutylicum* is an anaerobic microorganism that utilizes nutrients such as molasses and yeast extract in an aqueous solution. On fermenting these nutrients, this culture forms gases (carbon dioxide and hydrogen), acetic and butyric acids, and solvents such as acetone and butanol (Ballangue et al., 1987). At the first stage of the fermentation, gases and acids are produced, and at the second stage, with partial conversion of acids to solvents, acetone (CH_3COCH_3) and butanol ($CH_2CH_2CH_2CH_2OH$) are formed (LePage et al., 1987).

3.1 APPROACHES

There are several historical approaches to take advantage of the natural metabolic processes of autochthonous microorganism to boost crude oil recovery (Al-Sayegh et al., 2015). This includes the production of biosurfactants which are amphipathic molecules (molecules with a polar affinity and a nonpolar affinity, i.e., hydrophilic and hydrophobic parts) which are produced by a variety of microorganisms. Surfactants have the ability to reduce the surface and interfacial tension by accumulating at the interface of immiscible fluids and increasing the solubility and mobility of hydrophobic or insoluble organic compounds.

As biosurfactant sources (due to many microorganisms being capable of synthesizing biosurfactants) a microorganism community with such properties may be selected for the controlled production of biosurfactants for injection into the reservoir to decrease the interfacial tension between water and oil, thus releasing oil trapped in the rock by capillary forces. Furthermore, the presence of biosurfactants in the reservoir environment aids the biotransformation of hydrocarbons by autochthonous microorganisms. Similarly, biosurfactants may also become an easily accessible carbon source for autochthonous microorganisms, which produce biogenic gases such as carbon dioxide, methane, and hydrogen that would increase the reservoir pressure, which favors the displacement of crude oil toward production wells.

On the other hand, there is the phenomenon of biogenic gas production by the direct introduction of selected microorganisms that are capable of crude oil biotransformation with the goal of producing biogenic gases to increase the reservoir pressure. There are also microorganisms that develop biofilms on the reservoir rock surfaces, which cause plugging of watered pores, thus redirecting the injected water to upswept areas of higher oil saturations making waterflooding more effective in displacing oil toward the production wells. Finally, the activity of sulfate reducing bacteria can be hampered by the introduction of chemical compounds into the reservoir that would increase the activity of autochthonous microorganism groups to slow down the activity of the sulfate reducing bacteria. The main advantages of MEOR methods over conventional EOR methods are the much lower energy consumption and low or nontoxicity that makes the MEOR method an environmentally friendly process.

MEOR projects can be divided into two main groups: (1) ex situ production of the metabolites such as biosurfactants, biopolymers, and emulsifiers using exogenous or indigenous bacteria and (2) in situ production of the metabolites. Several factors concomitantly affect microbial growth and activity. In oil reservoirs, such environmental constraints permit to establish criteria as to assess and compare the suitability of microorganisms. Those constrains may not be as harsh as other environments on Earth – as an example, connate brine's salinity is higher than that of sea water but lower than that of salt lakes. In addition, pressure and temperature in oil reservoirs (Chapter 3) are generally (but not always) within the limits for the survival of other microorganisms (Saikia et al., 2013).

3.1.1 Microbes and Nutrients

To allay any misunderstanding, there are many kinds of microorganisms found within the reservoir, although anaerobic microorganisms are considered to be the indigenous true inhabitants (Magot et al., 2000). Commonly used bacterial species are *Bacillus* and *Clostridium*. The *Bacillus* species produce surfactants, acids, and some gases, and *Clostridium* produce surfactants, gases, alcohols, and solvents. Few *Bacillus* species also produce polymers. *Bacillus* and *Clostridium* are often able to bear extreme conditions existing in the oil reservoirs. The survival of these organisms originates from the ability to form spores. The spores are dormant, resistant forms of the cells (Bryant and Burchfield, 1989), which can survive in stressful environments exposing them to high temperature, drying, and acid. The duration of the dormancy can be extremely long and yet the survival rate is large (Madigan et al., 2003).

Microorganisms are complex in their way of responding to the surrounding environment. The cells change physiological state in order to have optimal chances for survival meaning that substrate consumption, growth, and metabolite production may change significantly (Van Hamme et al., 2003). Microorganisms present in an oil reservoir or other porous media are subjected to many physical (temperature, pressure, pore size/geometry), chemical (acidity, oxidation potential, salinity) and biological factors (cell processes).

Anaerobes ferment and cannot use oxygen for respiration, and for strict anaerobes, the presence of oxygen is toxic. Both aerobic and facultatively aerobic microorganisms have also been found. The aerobes can respire, while the facultative aerobes are able to grow either as aerobes or anaerobes determined by the nutrient availability and environmental conditions (Madigan et al., 2003). Regarding the presence of aerobes and facultative aerobes, the role as true inhabitants is uncertain and thus considered contaminants which are transferred to the reservoir through fluid injection or during drilling devices (Magot et al., 2000).

Furthermore, microbes can be classified in terms of their oxygen intake (Chapter 3). Thus, there are aerobes where the growth depends on a plentiful supply of oxygen to make cellular energy. On the other hand, anaerobes, by contrast, which are sensitive to even low concentration of oxygen and are found in deep oil reservoirs. These anaerobes do not contain the appropriate complement of enzymes that are necessary for growth in an aerobic environment (Lazar et al., 2007). The third group of bacteria is facultative microbes, which can grow either in the presence or reduced concentration of oxygen. However, there are many sources for bacterial species that play an important role in MEOR. On the other hand, it has been suggested that there are four main sources that are suitable for bacterial isolation: (1) formation waters, (2) sediments from formation water purification plants, (3) sludge from biogas operations, and (4) effluents from sugar refineries.

Microbial growth is determined by the presence of different nutrients. The primary nutrients consumed is carbon and nitrogen, which are the main constituent parts of the cell and enter many cell processes (Madigan et al., 2003). The substrates for growth of hydrocarbon-degrading bacteria include different crude oil components such as *n*-alkane derivatives, homocyclic aromatic compound derivatives, polycyclic aromatic compound derivatives, as well as nitrogen and sulfur heterocyclic derivatives (Van Hamme et al., 2003).

The microbes require mainly three components for growth and metabolic productions: carbon, nitrogen, and phosphorous sources, generally in the ratio of carbon-100/nitrogen-10/phosphorus-1. Media optimization is very important since the types of bio-products that are produced by different types of bacteria are highly dependent on the types, concentrations, and components of the nutrients provided. Some microbes utilize oil as the carbon source, which is excellent for heavy oil production, since it will reduce the carbon chain of heavy oil and thus increase its quality. In fact, the presence of crude oil in the media can significantly increase the production of methane and carbon dioxide but the growth rate of the microbial colony might be reduced. Thus, it is important to carefully test the nutritional preferences of the studied microbes that would maximize the production of desired metabolites if cost-effective supplies are assured.

3.1.2 Ex Situ Production of Metabolites

In the case of ex situ production of the metabolites, the microorganisms are grown using industrial fermenters or mobile plants and then injected into the oil formation as aqueous solutions. This method of delivery draws from the approach of CEOR in that this option produces the desired bio-products outside of the well after which the bio-products are injected into the wellhead to enhance oil recovery. Such a method is appealing because it allows more direct control of the process since specific compositions, compounds, and products can be selected and injected into the well. The method also uses microbes either grown or engineered in the laboratory under controlled conditions. In addition, microbial products of interest such as biosurfactants are often extracted from the laboratory-generated microbes and mixed with the water before injection, often in combination with synthetic chemicals. In another option within the ex situ system, isolated laboratory microbes may be injected into the well, with the objective of producing the desired products within the reservoir. However, the option to directly inject laboratory microbes assumes that the laboratory-generated microbial colonies will out-compete those colonies already acclimated to the relatively harsh well conditions (compared to the "more microbe-friendly" laboratory conditions) and indigenous to the reservoirs.

3.1.3 In Situ Production of Metabolites

In the case of the in situ production of metabolites, the formation of metabolites is the result of the microbiological activity that takes place directly in the reservoir. The metabolites are produced by indigenous bacteria or by exogenous bacteria that are injected into the reservoir. The in situ option can be divided into two categories depending upon the method of injection of microorganisms and nutritional media (such as molasses or chemical products) into the reservoir.

The first category consists of the in situ stimulation of the natural indigenous microflora of the reservoir by means of injecting nutrients into the reservoir. The alternative (second category) is the injection of microbial cultures (exogenous or indigenous) along with nutrients; which is the preferred mode of application in the field. However, the development of these methods is not possible without the knowledge of the physicochemical and microbiological conditions in the environment in which the crude oil exists. Oil recovery by the use of microbes is characterized at later stages by the formation of a mature biocenosis—the interaction of organisms living together in a habitat (biotope). The growth of this integrated microorganism community depends on the availability of nutrients.

The *in situ* production of metabolites is carried out in two stages. In the first stage, water and oxygen are pumped into the formation as a water-air mixture containing mineral salts, nitrogen, and phosphorous to activate the indigenous microflora. In the presence of water and air, aerobic bacteria oxidize hydrocarbons producing low molecular weight organic acids (such as acetic acid, propionic acid, and butyric acid), alcohol derivatives (such as methanol and ethanol), biosurfactants, and carbon dioxide, which increase the pressure in the reservoir.

In the second stage, oxygen-free water is injected into the reservoir to activate anaerobic indigenous bacteria that metabolize crude oil to acids and gas (such as methane and carbon dioxide). The accumulation of these biogases increases the reservoir pressure and, if the pressure in the reservoir is sufficiently high, methane can be dissolved into the liquid crude oil phase thereby reducing the viscosity of the crude oil. Similarly, carbon dioxide could also reduce the oil viscosity if the pressure in the formation allows the miscibility of carbon dioxide into the bulk oil phase. A reduction of the oil viscosity improves the crude oil-displacing properties through the reservoir increasing oil production. Furthermore, carbon dioxide can react with the reservoir minerals and dissolve carbonate minerals resulting in an increase in the permeability of the formation rock.

Thus, in contrast to the ex situ approach, the in situ approach stimulates the microbial colonies that are indigenous to the wells to produce the desired bio-products. In the in situ option, indigenous microbes of interest are often stimulated with specifically chosen substrates to produce and release compounds such as biosurfactants, bio-acids, and bio-solvents. The stimulation of bio-film production to decrease reservoir permeability has also been employed. While both ex situ and in situ approaches have potential, and while both options could be used in tandem, the choice of the various operators is typically the in situ operations.

3.2 MECHANISMS AND EFFECTS

The mechanisms by which the bacteria can improve the oil recovery are (1) biotransformation of crude oil. A proposed mechanism of MEOR utilization of bacteria that can degrade crude oil and consume the higher molecular weight constituents (particularly the resin constituents and the asphaltene constituents) (Ali et al., 2012; Lavania et al., 2012; Tabatabaee and Assadi, 2013; Kopytov et al., 2014). As a result of this process, there is an overall reduction in the density of the oil (increase in the API gravity) of the crude oil and a reduction in the molecular weight, as well as a decrease in the viscosity of the oil (Bryant and Burchfield, 1989). For example, microbes such as *Pseudomonas*, *Arthrobacter*, and other aerobic bacteria are especially effective in the degradation of crude oil (Bushnell and Haas, 1941; Bryant, 1990). However, this degradation is focused on the lower molecular weight constituents of the crude oil—especially the paraffin constituents—and bacterial treatment is beneficial for removal of paraffin derivative from the wellbore, which can seriously restrict the flow of the (modified) crude oil (Pelger, 1992).

During the MEOR process, the components are mixed at surface facilities and injected into the oil reservoir. Inside the reservoir, the bacteria are transported by the injected water and accumulate in porous zones at the oil/rock and oil/water interfaces where metabolites such as solvents, surfactants, acids, and carbon dioxide are produced. These bio-products interact with the crude oil in the reservoir to reduce the oil viscosity and interface tension at the oil/rock and oil/water interfaces and improve rock permeability by removing paraffin, mud, and other debris that plug the porous media (Lazar et al., 2007; Youssef et al., 2009). Microbial cells are continuously generated, as well as the in situ production of metabolites. The prolonged interaction of metabolites with the oil in the reservoir changes the oil properties in such a way that immobile unrecoverable oil is converted into movable oil that can flow to the production wells increasing oil output accordingly.

The main mechanisms acting on oil recovery by MEOR are (1) formation of bio-acids that could dissolve some of the minerals—such as clay minerals and carbonate minerals—in the formation rocks which leads to increases the porosity and permeability of the reservoir, (2) production of solvents and biogases, leading to lower oil viscosity that facilitates crude oil displacement from the rock and flow through the porous media, (3) formation of biosurfactants, biopolymers, and other compounds that interact with the crude oil by emulsifying the oil, reducing the oil viscosity, and reducing the interfacial tension at the oil-water interface, and (4) production of microbial biomass that could change the wettability of the reservoir rock.

The initial phase of the MEOR process is the partial oxidation of residual oil in the bottom-hole area that yields bio-products such as carbon dioxide and low molecular-weight fatty acids. Furthermore, when these bio-products are transported by the injected water into the anaerobic zones of the reservoir, methanogenic microorganisms are activated. Production of methane is important because this is an easily extractible gas that can be dissolved into the oil phase if the reservoir pressure is high enough, thereby improving the mobility of the crude oil.

After implementation of the process, continuous monitoring through bacteria population growth and the oil output rate within certain period of time is essential. The hydrocarbon-oxidizing bacteria population is kept under continuous surveillance and the population size is compared to that of the indigenous hydrocarbon oxidizing bacteria before the beginning of the process. In addition, the fluids produced by the interaction of the bacteria with the crude oil are also monitored for the presence of hydrocarbon-oxidizing bacteria in order to assess how far the microorganisms have moved into the reservoir after injection.

During the biotransformation process, the produced gases (such as carbon dioxide, hydrogen, and methane) can improve the oil recovery in two ways: (1) the gas dissolves in the crude oil and thus reduces the viscosity of the oil or (2) increases the pressure in the reservoir (Donaldson and Clark, 1983; Behlulgil et al., 1992). The source of this produced gas is in situ fermentation of carbon sources such as glucose by usually anaerobic bacteria. The most important gas-producing bacteria are *Clostridium*, *Desulfovibrio*, *Pseudomonas*, and certain methanogens (Bryant and Burchfield, 1989). In addition, the production of specific (nongaseous) chemicals during the biotransformation of crude oil can be useful in the improvement of crude oil. Chemicals such as: organic acids, alcohols, solvents, surfactants, and polymers are produced by a wide array of microorganisms (Bryant and Lockhart, 2001).

Apart from the production of gases and chemicals that assist in the recovery of crude oil from the reservoir, microbes (bacteria) can be used in selective plugging (permeability modification) operations. In this method, polymers or the bacteria themselves are used to reduce the permeability of highly permeable zones or of water channels that form in heterogeneous reservoirs. The unswept formations are invaded by the water and sweep efficiency increases. *Bacillus*, *Xanthamonas*, and *Leuconostoc* strains have been shown to be effective in such processes (Yakimov et al., 1997; Jenneman et al., 1994). Other uses of bacteria in the petroleum industry include the control of unwanted bacteria (such as sulfate-reducing bacteria) in oil fields (Hitzman, 1983; Hitzman and Sperl, 1994) and biodegradation of hazardous wastes caused by petroleum-related activities for the controlling and removal of environmental pollution (Ronchel et al., 1995; Speight and Arjoon, 2012).

The presence of autochthonous microorganisms (indigenous or native microorganisms), as well as their persistence and reproduction within crude oil reservoirs is advantageous for the implementation of MEOR methods, which are also known as microbial improved oil recovery methods (MEOR methods), allowing additional oil recovery at relatively low cost and the extension of the exploitation life of mature reservoirs, otherwise abandoned.

3.3 CHANGES IN RESERVOIR ROCK PERMEABILITY

Typically, in microbial enhanced recovery operations, the emphasis is on the environmental controls on the growth and survival of bacteria in subsurface environments. However, consideration must be given to the effects that bacterial growth can exert on the surrounding environment—particularly when the environment is a formation that contains crude oil. The significance of microbial activity in crude oil-containing reservoirs can lead to beneficial effects which include (1) the selective degradation of higher molecular weight constituents, leading to a reduction in viscosity, (2) the production of surfactants, leading to a reduction in interfacial tension, and (3) the production of gaseous products, leading to additional reservoir energy to move the crude oil to the production well. However, there are detrimental effects of microbial activity in petroleum reservoirs which include (1) corrosion of well-bore casings from by-products such as hydrogen sulfide and (2) the consumption of the hydrocarbon derivatives by the bacteria. Furthermore, permeability reduction, due to metabolic products or the bacteria themselves, can exert positive as well as negative effects, by causing secondary flow paths to become active.

A well-known and often studied method of MEOR is the selective plugging of reservoir zones with the purpose of diverting injected water toward reservoir areas of high oil saturations. The cause of this phenomenon is suspected to be the microorganisms' mass and the products of their biological activity such as biopolymers, which effectively block the flow of water within large pores and fractures. Biopolymers allow bacteria to connect and form biofilms, which improve nutrient gain and decrease their sensitivity to toxic substances.

3.4 CHANGES IN RESERVOIR ROCK WETTABILITY

Wettability is a major factor controlling residual oil saturation, and thus, it is essential to characterize reservoir wettability. Reservoir rock wettability can be altered by contact with absorbable crude oil components (such as resin constituents and asphaltene constituents), which can lead to heterogeneous forms of wettability characterized by the term fractional wettability. A fractional-wet system is where a portion of the reservoir rock is strongly oil-wet, while the rest is strongly water-wet.

Fractional-wet systems have previously been studied by packing columns with different ratios of water-wet rock and sand rendered oil-wet by treatment with an organic silane solution found that nonuniform wettability can distort the capillary pressure curve such that it no longer represents the true pore-size distribution and wettability can have a dramatic effect on residual oil entrapment.

The effect of microorganisms is often manifested on the wettability of the reservoir rock by the crude oil and approaches have been taken to understand such wettability alteration during a MEOR project. Typically, surface wettability has been quantified by placing a liquid drop on a solid surface and then measuring the resulting contact angle, known as the wetting angle. Other than directly measuring contact angle, porous media wettability is often quantified using macro-scale indices. Changes in these macro-scale indices due to microbial activity have been found to be more water-wet after a MEOR project. However, the direction in which microorganisms change wettability is not consistent.

3.5 BIOLOGICAL DEMULSIFICATION OF CRUDE OIL

MEOR results from the production of biosurfactants by the microbes. The two main effects of these surface-active compounds are (i) a reduction in the interfacial tension between oil and water, and (ii) the formation of micelles. The first of these effects reduces the hydrostatic pressure that must be applied to the liquid in the pores of the formation to overcome the capillary effect, while the second provides a physical mechanism whereby oil can be mobilized by a moving aqueous phase. Both effects result from the presence of hydrophilic and hydrophobic structural elements, with affinities for the water phase and for the oil phase, respectively.

Thus, MEOR applications also include the demulsification of crude oil. In crude oil processing, there are two types of emulsions, i.e., oil in water and water in oil. These emulsions are formed at different stages of crude oil exploitation, production, and processing, causing significant operational problems to the industry (Singh et al., 2007). The release of oil from these emulsions (demulsification) typically includes: (1) centrifuging, (2) warming, (3) application of an electric current, and (4) the addition of chemical compounds to break up the emulsions.

There are many microorganisms with favorable demulsification properties including *Acinetobacter calcoaceticus, Acinetobacter radioresistens, Aeromonas* sp., *Alcaligenes latus, Alteromonas* sp., *Bacillus subtilis, Corynebacterium petrophilium, Micrococcus* sp., *Pseudomonas aeruginosa, Pseudomonas carboxydohybrogena, Rhodococcus aurantiacus, Rhodococcus globerulus, Rhodococcus rubropertinctus, Sphingobacterium thalophilum,* and *Torulopsis bombicola*. Microorganisms benefit from the double hydrophobic-hydrophilic nature of surfactants or the hydrophobic cell surface to remove emulsifiers from the interfacial surface between oil and water. Temperature increase is favorable for the demulsification process, because it decreases viscosity, increases the density difference between the phases, attenuates the stabilizing action of the interfacial surface, and increases the drop collision rate, which leads to coalescence (Singh et al., 2007).

4.0 **UPGRADING DURING RECOVERY**

As already stated, MEOR is based on two principles: (1) the advancement of crude oil through porous media, which is expedited by modifying the interfacial properties of the oil-water minerals and (2) upgrading the crude oil. In the first principle (increased flow through porous media), the microbial activity alters fluidity. The second principle involves upgrading to the crude oil during the process in which higher molecular weight constituents are transformed into lower molecular weight constituents (heavy oils are transformed into lighter crude oils) by the microbial activity. In addition, the microbial process can also aid in the removal of sulfur from heavy oils, as well as the removal of nitrogen (Chapters 6 and 7) (El-Gendy and Speight, 2016).

Thus, MEOR is a process to enable crude oil flow through porous media which is achieved by modification of the interfacial properties of the crude oil-water minerals. In the system, microbial activity enhances (1) fluid flow, through viscosity reduction, (2) displacement efficiency, through decrease of interfacial tension, (3) sweep efficiency through mobility control and/or selective plugging, (4) and driving force, through changes in the reservoir pressure. These changes in crude oil properties are often referred to as crude oil upgrading and, in this case, the transformation of heavy oils (lower API gravity) into lighter crude oil (higher API gravity) occurs by microbial activity. Also, it can aid in the removal of sulfur from heavy oils, as well as the removal of heavy metals. Thus, MEOR can be considered to be a technology that can accomplish a degree of upgrading during the recovery process (Fujiwara et al., 2004; Al-Sulaimani et al., 2011; Speight, 2014a). However, although successful microbial enhanced oil recesses are specific for each well (sometime cited as a disadvantage), preprocess planning can make the process one of the most promising recovery method which can lead to recovery of the billions of barrels of crude oil that might otherwise remain unrecoverable by conventional recovery technologies.

Along with the growing awareness of the microbial action on the constituents of crude oil which can play an important role in oil recovery, there is also the awareness that microbial application to refining process is also a benefit (El-Gendy and Speight, 2016). For example, the selective biotransformation of long chain constituents or high molecular weight constituents, such as the resin constituents and the asphaltene constituents, into products that flow more easily, represents a possible application of biotechnology in cases where production is more difficult due to the type of oil involved. In fact, in light of the advanced stage of the development of methods for microbial enhanced recovery of crude oil, as much in relation to emulsification as to demulsification of oil, this concept is now undergoing serious evaluation. Thus, the concept relating to the selective biotransformation of crude oil constituents that would permit a significant increase in production with the biotransformation of only small superficial amounts of the crude oil within the reservoir matrix forms the basis for further development of the biotransformation concept leading to the production of partially upgraded crude oil.

In fact, bacterial degradation of oils in reservoirs has long been recognized (Chapter 3) (Wenger et al., 2002). The reservoir temperature range is critical to bacterial degradation (Chapter 3). Above temperatures of approximately 82°C (180°F), the biotransformation of crude oil constituents by bacteria is significantly inhibited. At temperatures just below this limit, the bacteria are generally operating at lower efficiency. Crude oil in lower temperature reservoirs (e.g., <50°C, <122°F) are much more likely to undergo biotransformation. The salinity of the formation water may also impact the efficiency of biotransformation which necessitates consideration of the combined temperature-salinity parameters. As a consequence, the various reservoir parameters will play an important role in the biotransformation of crude oil constituents.

Although different bacterial types and reservoir environments do have some effect on the order of biotransformation of the various constituents, the following general trends in order-of-preference reactivity are usually applicable. Thus, straight-chain *n*-alkane derivatives are typically attacked before branched-chain derivatives (such as isoprenoid derivatives), then cyclic alkane derivatives, and finally, aromatic hydrocarbon derivatives. The multi-ring biomarker compounds tend to be resistant through moderate-to-heavy biotransformation levels and these constituents provide a method for correlating and identifying biotransformed (degraded) and non-biotransformed (non-biodegraded) crude oils.

Progressive biotransformation almost invariably reduces oil quality. As the high-quality saturated hydrocarbons are removed, there is a corresponding increase of the amount of higher molecular weight, multi-ring non-hydrocarbon constituents (e.g., resin and asphaltene constituents). These changes in crude oil composition lead to (1) higher density, i.e., lower API lower gravity, (2) higher viscosity, (3) higher sulfur content, (4) higher nitrogen content, and (5) higher asphaltene content, as well as resin content of the remaining oil. These changes in composition result in lower value for the crude oil, diminished recovery efficiency, and possible additional production problems associated with handling and processing heavier oils. One quality of the crude oil that does not diminish with the degree of biotransformation is (in certain crude oils) the wax content. Because wax is composed of high molecular weight *n*-alkane derivatives, these constituents are biotransformed more readily than the higher molecular weight, multi-ring constituents, the loss of wax constituents may lead to decreased API gravity which may be offset by a lower pour point and the benefit of lower deposition of the wax constituents in reservoir channels, well pipes, and surface facilities.

In addition to the concentration of the lower quality constituents (i.e., resin and asphaltene constituents) because of biotransformation of the alkane and other hydrocarbon derivatives, the products of biotransformation reactions may have a negative impact on crude oil. In many cases, bacteria manufacture organic acid derivatives, most of which fall within the groups of constituents known as are naphthenic acids—typically the naphthenic acids and multi-ring compounds that can serious affect refinery processes (Speight, 2014a,b, 2017). Because of the differences in solubility of these acid derivatives, the low molecular weight acids (C_1–C_5 acids) occur predominantly in the aqueous phase (i.e., in the associated formation water and production water) while the higher molecular weight species (C_{6+} acids) are concentrated in the oil phase. The acid content of crude oil is typically monitored as the total acid number (TAN) that is determined by potentiometric titration (ASTM D664) and is a standard method by which oils are assayed and valued, although caution is advised in the interpretation of the data (Speight, 2014b, 2015).

Generally, the TAN increases with increasing levels of biotransformation and the activity of biodegrading organisms may be most important in determining organic acid contents because acid derivatives may dissipate rapidly owing to relatively high water-solubility and high reactivity. An elevated content of naphthenic acid derivatives (TAN > ~1 mg KOH/g oil) are detrimental to crude oil value because acids cause refinery equipment corrosion at high temperatures (Speight, 2014a,b) which can result in corrosion, as well as to the formation of emulsions during the production of crude oil that has been subject to biotransformation. Sometimes these emulsions can be broken by conventional means and the inconvenience (that is the additional expense) associated with emulsion-breaking operations, especially on production platform sites in deep water, can further challenge field economics. The low molecular weight organic acid derivatives in water may often be odiferous and can cause wastewater disposal problems in refineries processing some crude oil that has been biotransformed.

5.0 CHALLENGES AND OPPORTUNITIES

A MEOR process is not merely a matter of "throwing bugs down the hole" and waiting for the results! Prior to the application of a MEOR process, the process must be assessed to determine the compatibility of the crude oil and reservoir properties with the process by taking into account the physicochemical properties of the crude oil, reservoir production performance, and reservoir properties (i.e., temperature). At the preliminary stage of the process, reservoir fluid samples are collected and tested for compatibility with the MEOR systems. The first stage is the identification of the indigenous hydrocarbon-consuming bacteria, which is already adapted to the in situ reservoir conditions; after which the best action strategy for each process is designed and developed.

Some of the general advantages of a MEOR project are; (1) increase of the productivity of the oil fields, (2) increase in the total oil produced and more efficient operation of wells and oil fields, (3) increase of the viscosity of the formation water due to the upsurge of biomass concentration and the metabolic products produced by the microorganisms, such as soluble biopolymers, which reduces the mobility of the formation water within the formation rock, (4) low energy input requirement for microbes to produce microbial agents, and the process is environmentally friendly, because microbial products are biodegradable. Generally, the advantages are considered to outweigh the disadvantages (Tables 4.1 and 4.4) (Youssef et al., 2009).

In terms of the quality of the oil produced, some benefits are (1) an increase in the yield of low-boiling ($<C_{20}$) alkane derivatives, (2) reduction of the average content of C_{20}–C_{40} alkane derivatives, (3) biodegradation of higher molecular weight constituents of the crude oil, (4) biotransformation of sulfur-containing organic compounds, and (5) emulsification of crude oil that allows easier mobility and flow to the production well. Other effects, such as transformation of aromatic constituents and phenol constituents are also possible but very much dependent upon the transforming abilities of the microbes (She et al., 2011).

Biotechnological approaches use microorganisms that oxidize oil to break-up asphaltene-resin-paraffin sludge that have been accumulated in the wellbore for years affecting the crude oil production due to plugging of the oil production zones. In mature reservoirs, it is common to find production wells rendering water cuts of 70%–90% v/v of the total production. The microbial stimulation of aged production wells to break-up unwanted paraffin/asphaltene deposits is an economical way to reactivate old production zones. After the initial injection of microorganisms, it is only necessary to supply nutrients into the reservoir rock to enable the development of the existing bacterial population that will produce the required biosurfactants. Some of the common microorganisms used in MEOR processes include: *P. aeruginosa*, *Bacillus licheniformis*, *Xanthomonas campestris* and *Desulfovibrio desulfuricans* (Singh et al., 2007).

Table 4.4 Summary of the Advantages and Disadvantages of Microbial Enhanced Oil Recovery

Advantages

- Injected microbes and nutrients are easy to handle in the field and independent of oil prices.
- Economically attractive for mature oil fields before abandonment.
- Increases oil production.
- Existing facilities require slight modifications.
- Easy application.
- Low energy input requirement for microbes to produce active products.
- More efficient than other enhanced oil recovery methods when applied to carbonate oil reservoirs.
- Microbial activity increases with microbial growth.
- Cellular products are biodegradable and considered to be environmentally friendly.

Disadvantages

- Anaerobic processes require large amounts of nutrients, such as sugar.
- Exogenous (ex situ) microbes require facilities for their cultivation.
- Indigenous (in situ) microbes need a standardized framework for evaluating microbial activity.
- Microbial growth is favored when:
    layer permeability is greater than 50 md;
    reservoir temperature is less than to 80 C (176°F),
    salinity is below 150 g/L,
    reservoir depth is less than 2400 m (8000 feet).
- Oxygen used for aerobic processes can act as corrosive agent on equipment and down-hole piping.
- Production of hydrogen sulfide that can act as corrosive agent on equipment and down-hole piping.

In practice, the method may be difficult to implement because of the uncertainty in providing suitable conditions for the development of the microorganism groups taking part in the metabolic pathway. Furthermore, after microbial application it is essential to remove the bio-products of microorganisms' activity, as well as the bacteria themselves to maintain the composition of the crude oil (it is not desirable to give up light fractions of the crude oil to biotransformation). For instance, in situ bacterial cultivation may cause decrease of oil production. Another potential disadvantage of MEOR is the unpredictability of the process, although this can be overcome to a great extent by knowledge of the crude oil constituents, the influence of microbes on these constituents, and choosing the appropriate microbes for the task (Chapter 3). Other resolvable problems linked to MEOR include (1) plugging of the reservoir rock by the bacterial mass in undesirable locations, (2) in situ biotransformation of the applied chemical compounds, and (3) acidification of the crude oil by the bioproduction of hydrogen sulfide in the reservoir (Almeida et al., 2004; Patel et al., 2015).

Finally, there are issues that need to continually address the design and implementation of MEOR processes. These are: (1) injectivity, (2) dispersion, (3) optimization of metabolic activity, and (4) concentration of the metabolites.

1. Injectivity lost due to microbial plugging of the wellbore—to avoid wellbore plugging, some actions must be taken such as filtration before injection, avoiding biopolymers production, and minimizing microbial adsorption to rock surface by using dormant cell forms, spores, or ultra-micro-bacteria.

2. Dispersion or transportation of all necessary components to the target zone.

3. Optimization of the desired in situ metabolic activity due to the effect of variables such as pH, temperature, salinity, and pressure for any in situ MEOR operation. Isolation of microbial strains, adaptable to the extreme reservoir conditions of pH, temperatures, pressure, and salinity.

4. Low in situ concentration of bacterial metabolites; the solution to this problem might be the application of genetic engineering techniques.

Nevertheless, MEOR is a cost-effective and eco-friendly process that shows several advantages over other enhanced oil recovery processes. MEOR has great potential to become a viable alternative to the traditional enhanced oil recovery chemical methods. Although MEOR is a highly attractive method in the field of oil recovery, there are still uncertainties in meeting the engineering design criteria required by the application of microbial processes in the field, as outlined above. Therefore, a better understanding of the MEOR processes and the mechanism from an engineering standpoint are required; as well as the systematic evaluation of the major factors affecting this process such as (1) crude oil composition and properties, (2) reservoir heterogeneity, characteristics, and geology, as well as (3) selection of the appropriate microbial consortia, to improve the process efficiency which can only be accomplished by a thorough understanding of the process characteristics and process options (see, for example, Romero-Zerón, 2012).

REFERENCES

Al-Bahry, S.N., Al-Wahaibi, Y.M., Elshafie, A.E., Al-Bemani, A.S., Joshi, S.J., Al-Makhmari, H.S., Al-Sulaimani, H.S., 2013. Biosurfactant production by *Bacillus subtilis* B20 using date molasses and its possible application in enhanced oil recovery. International Biodeterioration and Biodegradation 81, 141–146.

Al-Sayegh, A., Al-Wahaibi, Y., Al-Bahry, S., Elshafie, A., Al-Bemani, A., Joshi, S., 2015. Microbial enhanced heavy crude oil recovery through biodegradation using bacterial isolates from an Omani oil field. Microbial Cell Factories (Electronic Resource) 14, 141.

Al-Sulaimani, H., Joshi, S., Al-Wahaibi, Y., Al-Bahry, S.N., Elshafie, A., Al-Bemani, A., 2011. Microbial biotechnology for enhancing oil recovery: current developments and future prospects. Biotechnology, Bioinformatics and Bioengineering Journal 1 (2), 147–158.

Ali, H.R., El-Gendy, N.Sh., Moustafa, Y.M., Roushdy, M.I., Hashem, A.I., 2012. Degradation of asphaltenic fraction by locally isolated halotolerant bacterial strains. ISRN Soil Science:435485. http://dx.doi.org/10.5402/2012/435485.

Almeida, P.F., Moreira, R.S., Almeida, R.C.C., Guimaraes, A.K., Carvalho, A.S., Quintella, C., Esperidia, M.C.A., Taft, C.A., 2004. Selection and application of microorganisms to improve oil recovery. Engineering in Life Sciences 4, 319–325.

ASTM D664, 2017. Standard Test Method for Acid Number of Petroleum Products by Potentiometric Titration. Annual Book of Standards. ASTM International, West Conshohocken, Pennsylvania.

Ballangue, J., Masion, E., Amine, J., Petitdemange, H., Gay, R., 1987. Inhibitor effect of products of metabolism on growth of *Clostridium acetobutylicum*. Applied Microbiology and Biotechnology 26, 568.

Banat, I.M., Franzetti, A., Gandolfi, I., Bestetti, G., Martinotti, M.G., Fracchia, L., Smyth, T.J., Marchant, R., 2010. Microbial biosurfactants production, applications and future potential. Applied Microbiology and Biotechnology 87 (2), 427–444.

Beckman, J.W., 1926. The action of bacteria on mineral oil. Industrial and Engineering Chemistry News Edition 4, 23–26.

Behlulgil, K., Mehmetoglu, T., Donmez, S., 1992. Application of microbial enhanced oil recovery technique to a Turkish heavy oil. Applied Microbiology and Biotechnology 36 (6), 833–835.

Bryant, R.S., 1990. Screening Criteria for Microbial EOR Processes, Topical Report. Bartlesville Project Office, Department of Energy, Bartlesville, Oklahoma.

Bryant, R.S., Burchfield, T.E., 1989. Review of microbial technology for improving oil recovery. SPE Reservoir Engineering Journal 4 (2), 151.

Bryant, S.L., Lockhart, T.P., 2001. Reservoir engineering analysis microbial enhanced oil recovery. Journal of Petroleum Technology 53 (1), 57.

Bushnell, L.D., Haas, H.F., 1941. The utilization of certain hydrocarbons by microorganisms. Journal of Bacteriology 41, 529.

Craft, B.C., Hawkins, M.F., 1959. Applied Petroleum Reservoir Engineering. Prentice-Hall, Englewood Cliffs, New Jersey.

Donaldson, E.C., Clark, J.B. (Eds.), 1983. Proceedings. 1982 International Conference on Microbial Enhancement of Oil Recovery, Afton, Okla, May 16–21. National Technical Information Service (NTIS), Springfield, Virginia.

Frick, T.C., 1962. Petroleum Production Handbook, vol. II. McGraw-Hill, New York.

Fujiwara, K., Sugai, Y., Yazawa, N., Ohno, K., Hong, C.X., Enomoto, H., 2004. Biotechnological approach for development of microbial enhanced oil recovery technique. Studies in Surface Science and Catalysis 151, 405–445.

Hitzman, D.O., 1983. Petroleum microbiology and its role in enhanced oil recovery. In: Proceedings. 1982 International Conference on Microbial Enhancement of Oil Recovery, Afton, Okla, May 16–21. National Technical Information Service (NTIS), Springfield, Virginia.

Hitzman, D.O., Sperl, G.T., 1994. A new microbial technology for enhanced oil recovery and sulfide prevention and reduction. In: Paper No. SPE/DOE 27752. Proceedings. SPE/DOE 9th Symposium on Improved Oil Recovery, Tulsa, Oklahoma. Society of Petroleum Engineers, Richardson, Texas.

Jack, T.R., 1983. Enhanced oil recovery by microbial action. In: Yen, T.F., Kawahara, F.K., Hertzberg, R. (Eds.), Chemical and Geochemical Aspects of Fossil Energy Extraction. Ann Arbor Science Publishers, Ann Arbor, Michigan.

Jenneman, G.E., Moffatt, P.D., Young, G.R., 1994. Application of a microbial selective plugging process at the North Burbank Unit: pre-pilot tests and results. In: Proceedings. Paper No. SPE/DOE 27827. SPE/DOE 9th Symposium on Improved Oil Recovery, Tulsa, Oklahoma. Society of Petroleum Engineers, Richardson, Texas.

Jobson, A., Cook, F.D., Westlake, D.W.S., 1972. Microbial utilization of crude oil. Advances in Applied Microbiology 23 (6), 1082–1089.

Kopytov, M.A., Filatov, D.A., Altunina, L.K., 2014. Biodegradation of high-molecular mass heteroatomic components of heavy oil. Petroleum Chemistry 54 (1), 58–64.

Lavania, M., Cheema, S., Sarma, P.M., Mandal, A.K., Lal, B., 2012. Biodegradation of asphalt by *Garciaella petrolearia* TERIG02 for viscosity reduction of heavy oil. Biodegradation 23 (1), 15–24.

Lazar, I., Petrisor, I.G., Yen, T.F., 2007. Microbial enhanced oil recovery. Petroleum Science and Technology 25, 1353–1366.

LePage, C., Fayolle, F., Hermann, M., Vandecasteele, J.P., 1987. Changes in membrane lipid composition of *Clostridium acetobutylicum* during acetone-butanol fermentation: effects of solvents, growth temperature, and pH. Journal of General Microbiology 133, 103.

Madigan, M., Martinko, J., Parker, J., 2003. Brock Biology of Microorganisms, thirteenth ed. Benjamin Cummings, Pearson Education, San Francisco.

Magot, M., Ollivier, B., Patel, B., 2000. Microbiology of petroleum reservoirs. Anton van Leeuwenhoek 77, 103–116.

El-Gendy, N.Sh., Speight, J.G., 2016. Handbook of Refinery Desulfurization. CRC Press, Taylor & Francis Group, Boca Raton, Florida.

Patel, J., Borgohain, S., Kumar, M., Rangarajan, V., Somasundaran, P., Sen, R., 2015. Recent developments in microbial enhanced oil recovery. Renewable and Sustainable Energy Reviews 52, 539–1558.

Pelger, J.W., 1992. Wellbore stimulation using microorganisms to control and remediate existing paraffin accumulations. In: Paper No. SPE 23813. Proceedings. SPE Intl. Symposium on Formation Damage Control, Lafayette, Louisiana. Society of Petroleum Engineers, Richardson, Texas.

Romero-Zerón, L., 2012. Introduction to Enhanced Oil Recovery (EOR) Processes and Bioremediation of Oil-Contaminated Sites. INTECH, Rijecka, Croatia.

Ronchel, M.C., Ramos, C., Jensen, L.B., Molin, S., Ramos, J.S., 1995. Construction and behavior of biologically contained bacteria for environmental applications in bioremediation. Applied and Environmental Microbiology 61 (8), 2990.

Saikia, U., Bharanidharan, R., Vendhan, E., Kumar Yadav, S., Shankar, S., 2013. A brief review on the science, mechanism and environmental constraints of microbial enhanced oil recovery (MEOR). International Journal of ChemTech Research 5 (3), 1205–1212.

Sarkar, A.K., Goursaud, J.C., Sharma, M.M., Georgiou, G., 1989. Critical evaluation of MEOR processes. In Situ 13 (4), 207–238.

Sarkar, A., Georgiou, G., Sharma, M., 1994. Transport of bacteria in porous media: II. A model for convective transport and growth. Biotechnology and Bioengineering 44, 499–508.

She, Y.H., Shu, F.C., Zhang, F., Wang, Z.L., Kong, S.Q., Yu, L.J., 2011. The enhancement of heavy crude oil recovery using bacteria degrading polycyclic aromatic hydrocarbons. Advanced Materials Research 365, 320–325.

Singh, A., Van Hamme, J.D., Ward, O.P., 2007. Surfactants in microbiology and biotechnology: Part 2. Application aspects. Biotechnology Advances 25, 99–121.

Speight, J.G., 2013a. Heavy Oil Production Processes. Gulf Professional Publishing, Elsevier, Oxford, United Kingdom.

Speight, J.G., 2013b. Oil Sand Production Processes. Gulf Professional Publishing, Elsevier, Oxford, United Kingdom.

Speight, J.G., 2014a. The Chemistry and Technology of Petroleum, fifth ed. CRC-Taylor and Francis Group, Boca Raton, Florida.

Speight, J.G., 2014b. High Acid Crudes. Gulf Professional Publishing, Elsevier, Oxford, United Kingdom.

Speight, J.G., 2015. Handbook of Petroleum Product Analysis, second ed. John Wiley & Sons Inc., Hoboken, New Jersey.

Speight, J.G., 2016. Introduction to Enhanced Recovery Methods for Heavy Oil and Tar Sands, second ed. Gulf Professional Publishing, Elsevier, Oxford, United Kingdom.

Speight, J.G., Arjoon, K.K., 2012. Bioremediation of Petroleum and Petroleum Products. Scrivener Publishing, Beverly, Massachusetts.

Springham, D.G., 1984. Microbial methods for the enhancement of oil recovery. Biotechnology & Genetic Engineering Reviews 1, 187–221.

Suthar, H., Hingurao, K., Desai, A., Nerurkar, A., 2008. Evaluation of bioemulsifier mediated microbial enhanced oil recovery using sand pack column. Journal of Microbiological Methods 75 (2), 225–230.

Tabatabaee, M.S., Assadi, M.M., 2013. Vacuum distillation residue upgrading by an indigenous *Bacillus cereus*. Journal of Environmental Health Sciences & Engineering 11, 18–24.

Van Hamme, J., Singh, A., Ward, O., 2003. Recent advances in petroleum microbiology. Microbiology and Molecular Biology Reviews 67, 503–549.

Wenger, L.M., Davis, C.L., Isaksen, G.H., 2002. Multiple controls on petroleum biodegradation and impact on oil quality. SPE Reservoir Evaluation & Engineering 5 (5), 375–383 Paper No. SPE-80168. Society of Petroleum Engineers, Richardson, Texas.

Wolicka, D., Suszek, A., Borkowski, A., Bielecka, A., 2009. Application of aerobic microorganisms in bioremediation in situ of soil contaminated by petroleum products. Bioresource Technology 100 (13), 3221–3227.

Yakimov, M.M., Amro, M.M., Bock, M., Boseker, K., Fredrickson, H.L., Kessel, D.G., Timmis, K.N., 1997. The potential of *Bacillus licheniformis* strains for in-situ enhanced oil recovery. Journal of Petroleum Science and Engineering 18, 147.

Youssef, N., Elshahed, M.S., McInerney, M.J., 2009. Microbial processes in oil fields: culprits, problems, and opportunities. In: Laskin, A.I., Sariaslani, S., Gadd, G.M. (Eds.), Advances in Applied Microbiology. Academic Press, Academic-Elsevier, Amsterdam, Netherland, pp. 141–251.

ZoBell, C.E., 1947. Bacterial release of oil from oil bearing materials. World Oil 126, 36.

FURTHER READING

Meredith, W., Kelland, S.J., Jones, D.M., 2000. Influence of biodegradation on crude oil acidity and carboxylic acid composition. Organic Geochemistry 31, 1059–1073.

Speight, J.G., 2012. Crude Oil Assay Database. Knovel, Elsevier, New York. Online version available at: http://www.knovel.com/web/portal/browse/display?_EXT_KNOVEL_DISPLAY_bookid=5485&VerticalID=0.

BIO-UPGRADING HEAVY CRUDE OIL

1.0 INTRODUCTION

The world depends mainly on fossil fuels (oil, coal and natural gas) for 81.1% of its total energy need and oil is the main supply (31.3%), while hydropower, biofuels and others (solar, wind…etc.) account for only 2.4%, 10.3%, and 1.4% respectively (International Energy Agency, 2015). The oil in the various reservoirs has taken millions of years to accumulate and may be consumed in less than two centuries at the current rates of production. The production peak was estimated to occur sometime between 2010 and 2020, and, by the end of this century, oil resources was estimated to be drastically reduced (Hall et al., 2003). The United States Energy Information Administration estimated that the total oil demand in the world is expected to grow up to 123 mm bpd (million barrels per day) by 2025. In addition, the organization of petroleum exporting countries (OPEC) has estimated that production will be approximately 61 million barrels per day by 2025, which is less than half of the demand (Hutzler et al., 2003). Crude oil can be generally classified into light crude oil (API >35 degrees API), medium crude oil (20–35 degrees API), and heavy crude oil (API ≤20 degrees API) oil by the American Petroleum Institute (API) gravity to (Table 5.1, Fig. 5.1) (Speight, 2013a,b, 2014, 2015). However, the classification of petroleum, heavy oil, extra heavy oil, and bitumen from tar sand formations based on a single property is not an accurate method of classification and lacks credibility because of the potential range of data that can be obtained by the analytical method (Speight, 2014, 2015) since the limits of experimental difference are typically on the order of ±3%. As a consequence, it should be asked "what is the difference between a heavy oil with a viscosity of 9950 cP and a tar sand bitumen with a viscosity of 10,050 cP?" The answer is "very little" (if anything, especially when the limits of experimental difference are considered) and to draw a fine line of differentiation does not serve any purpose in such cases.

More specifically, heavy oil is a *type* of petroleum that is different from conventional petroleum insofar as they are much more difficult to recover from the subsurface reservoir and more difficult to refine than conventional crude oil (Speight, 2013a,b, 2014, 2015). Heavy oil, particularly heavy oil formed by biodegradation of organic deposits, are found in shallow reservoirs, formed by unconsolidated sands. This characteristic, which causes difficulties during well drilling and completion operations, may become a production advantage due to higher permeability. In simple terms, heavy oil is a type of crude oil which is very viscous and does not flow easily. The common characteristic properties (relative to conventional crude oil) are high specific gravity (low API gravity), low hydrogen to carbon ratios, high carbon residue, and high content of asphaltene constituents, heavy metals, sulfur, and nitrogen (Speight, 2014). Specialized refining processes are required to produce more useful fractions, such as: naphtha, kerosene, and gas oil (Speight, 2013a,b, 2014).

In addition to conventional petroleum and heavy crude oil, there remains an even more viscous material that offers some relief to the potential shortfalls in supply (Speight, 2013a,b, 2014, 2017) and this is the *bitumen* found in *tar sand* (*oil sand*) deposits. *Tar sand*, also variously called *oil sands* or

Introduction to Petroleum Biotechnology. https://doi.org/10.1016/B978-0-12-805151-1.00005-9

Table 5.1 Simplified Differentiation[a] Between Conventional Crude Oil, Tight Oil, Heavy Crude Oil, Extra Heavy Crude Oil, and Tar Sand Bitumen

Type	Comment
Light crude oil	Mobile in the reservoir
	API gravity: >35 degrees
	High-permeability reservoir
	Produced by primary and secondary recovery methods
Medium crude oil	Mobile in the reservoir
	API gravity: 25–35 degrees
	High-permeability reservoir
	Produced by primary and secondary recovery methods
Tight oil	Similar to the properties of conventional crude oil
	API gravity: >25 degrees
	Immobile in the reservoir
	Low-permeability reservoir
	Produced by horizontal drilling into reservoir and fracturing to release fluids/gases
Heavy crude oil	More viscous than conventional crude oil
	API gravity: 10–20 degrees
	Mobile in the reservoir
	High-permeability reservoir
	Produced by secondary and tertiary recovery methods (enhanced oil recovery)
Extra heavy crude oil	Similar to the properties of tar sand bitumen
	API gravity: <10 degrees
	Mobile in the reservoir
	High-permeability reservoir
	High temperature reservoir or deposit
	Produced by secondary and tertiary recovery methods (enhanced oil recovery)
Tar sand bitumen	Immobile in the deposit
	API gravity: <10 degrees
	High-permeability reservoir
	Mining (often preceded by explosive fracturing)
	Innovative methods[b]

[a]*This list is not intended for use as a means of classification.*
[b]*Innovative methods are exclusive of tertiary (enhanced) recovery methods but does include methods such as steam-assisted gravity drainage (SAGD), vapor-assisted extraction (VAPEX), and variants or hybrids thereof, as well as extreme heating methods such as in-situ combustion.*

bituminous sands, are loose-to-consolidated sandstone or a porous carbonate rock, impregnated with bitumen, a heavy asphaltic crude oil with an extremely high viscosity under reservoir conditions.

The term *tar sand bitumen* (also, on occasion referred to as *extra heavy oil* and *native asphalt*, although the latter term is incorrect) includes a wide variety of reddish brown to black materials of near solid to solid character that exist in nature either with no mineral impurity or with mineral matter contents that exceed 50% by weight. Bitumen is frequently found filling pores and crevices of sandstone,

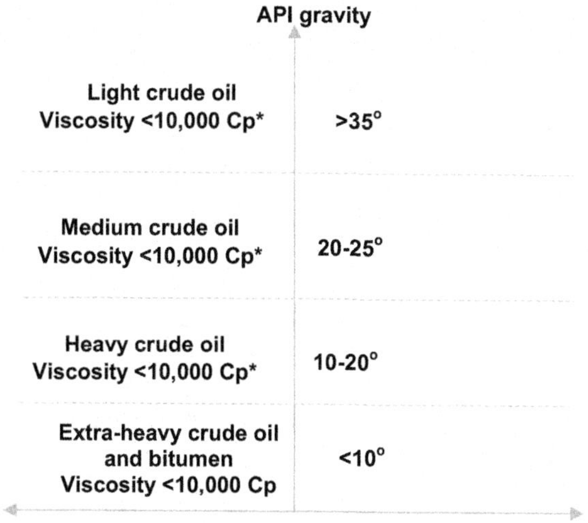

API gravity

Light crude oil **Viscosity <10,000 Cp***	**>35°**
Medium crude oil **Viscosity <10,000 Cp***	**20-25°**
Heavy crude oil **Viscosity <10,000 Cp***	**10-20°**
Extra-heavy crude oil **and bitumen** **Viscosity <10,000 Cp**	**<10°**

FIGURE 5.1

Simplified differentiation of crude oil, heavy crude oil, extra heavy crude oil, and tar sand bitumen according to API gravity and viscosity (Speight, 2014, 2015). *Different ranges of viscosity can be assigned but (like the API gravity ranges) are speculative and remain open to question (Speight, 2013a,b, 2014).

limestone, or argillaceous sediments, in which case the organic and associated mineral matrix is known as *rock asphalt*.

There have been many attempts to define tar sand deposits and the bitumen contained therein (Speight, 2013a,b, 2014, 2015, 2017). To define conventional petroleum, heavy oil, and bitumen, the use of a single physical parameter such as viscosity is not sufficient. Other properties such as API gravity, elemental analysis, composition, and, most of all, the properties of the bulk deposit must also be included in any definition of these materials (Speight, 2014, 2015). Only then will it be possible to classify petroleum and its derivatives. In fact, the most appropriate definition of *tar sands* (which is also recognized legally in the courts of the United States) is found in the writings of the US government (US Congress, 1976):

> Tar sands are the several rock types that contain an extremely viscous hydrocarbon which is not recoverable in its natural state by conventional oil well production methods including currently used enhanced recovery techniques. The hydrocarbon-bearing rocks are variously known as bitumen-rocks oil, impregnated rocks, oil sands, and rock asphalt.

This definition relies upon the character of the bitumen through the method of recovery. Thus, the bitumen found in tar sand deposits is an extremely viscous material that is *immobile under reservoir conditions* and cannot be recovered through a well by the application of secondary or enhanced recovery techniques. Mining methods match the requirements of this definition (since mining is not one of the specified recovery methods) and the bitumen can be recovered by alteration of its natural state such as thermal conversion to a product that is then recovered. In this sense, changing the natural state (the chemical composition) as occurs during several thermal processes (such as some in-situ combustion processes) also matches the requirements of the definition.

By inference, conventional petroleum and heavy oil are also included in this definition. Petroleum is the material that can be recovered by conventional oil well production methods whereas heavy oil is the material that can be recovered by enhanced recovery methods. Tar sand is currently recovered by a mining process followed by separation of the bitumen by the hot water process. The bitumen is then used to produce hydrocarbons by a conversion process.

It has been estimated that 2 trillion barrels (2×10^{12} bbls) of conventional light crude oil and 5 trillion barrels of heavy crude oil remain in reservoirs worldwide (Al-Sulaimani et al., 2011; Lavania et al., 2012; Demirbas et al., 2016). The largest heavy and extra heavy crude oil reserves in the world are found in the Orinoco oil belt of Venezuela, the Athabasca oil sands in Alberta, Canada, and the Olenik oil sands in Siberia, Russia (BGR, 2015). In Canada only, the heavy oil reserve considered to be potentially recoverable is estimated to be 280–315 bbl (billion barrels of petroleum liquids), larger than the Saudi Arabia oil reserves estimated at 264 bbl (OPEC, 2009; Government of Alberta, Canada, 2011). Heavy oils are also located in various countries and are being produced in India, Colombia, Indonesia, China, Mexico, Brazil, Trinidad, Argentina, Eastern Europe, Ecuador, Egypt, Saudi Arabia, Oman, Kuwait, Turkey, Australia, Nigeria, Angola, the North Sea, Romania, Iran, and Italy (Atkins et al., 2010; Vartivarian and Andrawis, 2006).

Heavy crude oils typically have a specific gravity approaching or even exceeding that of water and have a high viscosity, high density, as well as and high contents of sulfur, nitrogen, metal constituents, and polar high-molecular weight resin constituents and asphaltene constituents compared to conventional oil (Leòn and Kumar, 2005; Speight, 2014; Al-Bahry et al., 2016). The high viscosity, together with the flocculation and deposition of asphaltene constituents, resin constituents, and paraffin wax harden its production and refining processes and that comes with the required fulfillment of more severe worldwide environmental regulations. Where, there is a great attempt to decrease the emissions of the problematic sulfur dioxide and nitrogen oxides. For example, emissions of sulfur dioxide (SO_2) declined in the United Kingdom and the European Union by 71% and 72%, respectively, between 1986 and 2001, while nitrous oxide emissions declined by about 40% (Fowler et al., 2005).

The current sulfur limitation in diesel oil in most of the countries is to reach to ultra-low sulfur diesel (ULSD) 10–15 ppm S. The current goal set by Environmental Protection Agency (EPA) under the tier 3 program mandates 10 ppm S for gasoline (EPA, 2014) and the Euro 5/6 legislations mandates "sulfur-free" gasoline and ≤ 10 ppm S diesel (The International Council on Clean Transportation, 2016). While, tailpipe and evaporative combined emissions of non-methane organic gas (NMOG) and NOx should be reduced from 109 mg/mile in 2017 to 30 mg/mile by 2025 (EPA, 2014). Since 1993, the aromatic hydrocarbon content in diesel fuel has been set to 10% in California and 11% in Europe. A low content of aromatic hydrocarbons in fuels reduces the emission of carcinogenic substances, such as benzene and polycyclic aromatic hydrocarbons (PAH) (Ayala et al., 2007). The presence of sulfur derivatives, nitrogen derivatives, and metal-containing constituents leads to the formation of coke and poisons the refining and upgrading catalyst (Ancheyta and Speight, 2007; Ancheyta, 2016). To remove those pollutants and to obtain cleaner fuels, the traditional refineries must work under severe conditions of higher temperature and pressure and more expensive catalysts which would lead to an increase of the cost.

In addition, approximately two-thirds (c. 67% v/v) of the world oil consumption is in the transport sector (International Energy Agency, 2015). There is also a growing demand for transportation fuels (gasoline, jet fuel, and diesel), which are produced from naphtha (boiling point <175°C) and middle distillates (boiling point <370°C) (ExxonMobil, 2016). Heavy oils have lower yields of these liquid distillates compared to light oils. Thus, heavy oils require deep conversion and substantial upgrading involving deep thermal and catalytic processes and significant hydrogen inputs to obtain fractions suitable to produce transportation fuels (Ancheyta and Speight, 2007; Ancheyta, 2016).

Heavy oils, residuum, and oil wastes represent a substantial resource if a low-cost technology for their processing could be developed (Kirkwood et al., 2004). In terms of reserves, 50%–70% of original oil is still in place and is available. However, it is heavy and requires extensive secondary and tertiary recovery technology. Similarly, wastes from oil processing amount to 400 million gallons annually within the United States alone (Premuzic and Lin, 1999a).

The properties and economic value of petroleum depend on the relative amounts of the bulk fractions present in the crude oil. These are, for convenience rather than precise chemical character, called the SARA fractions (saturates, aromatics, resin constituents, asphaltene constituents) (Table 5.2, Fig. 5.2). In crude oils, the asphaltene constituents, resin constituents, the aromatic constituents, and the alkane

Table 5.2 Examples of Fractional Composition (SARA Analysis), Density and API Gravity of Various Heavy Oils (Hinkle et al., 2008; Speight, 2013a, 2013b, 2014)

Origin	Saturates, % W/W	Aromatics, % W/W	Resins, % W/W	Asphaltenes, % W/W	Density, G/mL	API Gravity
Alaska	23	22	35	18	0.997	10.4
Canada	18	27	27	15	0.991	11.3
Texas	4	17	37	43	1.119	−5
Utah	19	14	46	20	1	8.05
Venezuela	19	32	29	18	1.013	8.05

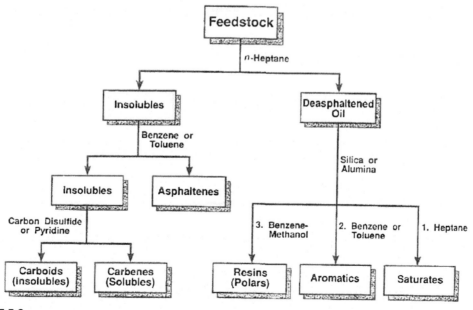

FIGURE 5.2

A typical SARA-Type (saturates, aromatics, resins, asphaltenes) separation scheme for various feedstocks.

constituents compose a dynamic stable system, in which the petroleum alkanes act as solvents, the asphaltene constituents as micelles, and the resin constituents as stabilizers (Speight, 2014).

Petroleum can be described in terms of light and heavy crude oils, where, with the transition between light to heavy crude oil, the carbon, metal, and the heteroatom content increase. The N, S, Ni, and V concentrations are reported to be inversely proportional to API gravity (Duyck et al., 2007). High viscosity significantly hampers the pumping, transportation, refining, and handling of petroleum. Common methods used to overcome problems associated with high viscosity include heating, dilution, and chemical additives. All are expensive and require specialized equipment and/or safety procedures. Industry has long recognized the need for a safe, economical, and effective method for reducing viscosity. The addition of a solvent would overcome this problem, and would allow their production and pipelining over a significant distance (Dehaghani and Badizad, 2016). But, the cost of suitable solvents is another problem to be solved. In downstream processing, heavy crude requires the conversion of the vacuum residue component into distillable oils. This upgrading has typically been accomplished with either thermal cracking or by catalytic hydroconversion. Thermal processing ranges from mild cracking (to reduce viscosity) to severe cracking (with the formation of coke). These processes are energy and cost-intensive, less selective, and environmentally reactive and require supporting infrastructure for the supply of hydrogen and treatment of hydrogen sulfide in cracked off-gases.

Asphaltene constituents are the heaviest and most polar fraction of crude oil. Asphaltene constituents are defined as the fraction of crude oil that is insoluble in n-heptane or n-pentane, but is soluble in benzene or toluene. The asphaltene concentration in a light crude oil may be less than 1%, while the concentration in heavy bitumen can be more than 20% of the total weight. The asphaltene constituents are responsible for sludge formation resulting in flow reduction by plugging downstream equipment and production of less valuable coke in current upgrading of petroleum. Moreover, the utilization of distillation residue, that is mainly composed of asphaltene constituents and entrapped heavy metals, is of a great interest in petroleum refineries, nowadays. That is because of the expected increase of utilization of heavy crudes from different sources, e.g., tar sands, oil shales, and off-shore reservoirs. The asphaltene content and the wax content typically are responsible for the high viscosity and many problems associated with recovery, separation, and processing of heavy oils and bitumen.

Crude oil is a colloidal system where the asphaltene constituents are the disperse phase. In crude oil, the interaction between resin constituents and asphaltene constituents render the latter stabilized. Waxes consist of long chain alkanes, i.e., paraffin wax (C18–C36) and naphthenic hydrocarbons or cycloparaffins (C30–C60). As in the case of waxes, the deposition of asphaltene constituents is the consequence of oil instability. In reservoirs, paraffins and asphaltene constituents remain in equilibrium, but when crude oil is extracted, this equilibrium is lost due to temperature and pressure changes, and therefore, asphaltene constituents and waxes tend to precipitate. This precipitate forms deposits during extraction in oil wells, blending of oils from different origins, storage, transportation, and refining of heavy fractions. Heavy crudes also have characteristic molecular interactions (basically the van der Waals interactions), which are small for small molecules but large for asphaltene constituents. These forces keep the asphaltene molecules together (>100 atoms) with good fitting. At low temperatures (<60°C) these close contacts are more common and result in increased crude viscosity. Other molecular interactions responsible for increasing viscosity are at free radical sites, which are associated with condensed polycyclic aromatic structures with highly reactive unpaired electrons. These sites are involved in complexation of metals, inter- and intra-molecular reactions, molecular rearrangements, and hydrogen bonding. Thus, the biotransformation or upgrading of asphaltene constituents and other aromatic compounds contained in petrochemical and other heavy hydrocarbon streams is one of the main interest in petroleum industry to maximize the usage of crude oil and minimize the waste (Gupta and Gera, 2015).

In the past, petroleum biorefining as a branch of petroleum biotechnology has been related to the production of single cell protein (SCP) from waxy *n*-alkanes. Nowadays, there is an aim to be applied in upgrading of heavy crude oils (Fig. 5.3). Petroleum biotechnology involves the use of wide range of conditions, milder temperature and pressure, cleaner and selective processes, lower emissions and non-generation of undesirable by-products. Moreover, microbial and enzymatic catalysts can be manipulated and used for more specific applications. Heavy crude oils can be subjected to biorefining to get rid of most of the sulfur, nitrogen, toxic metals, and asphaltene constituents (Le Borgne and Quintero, 2003).

Benedik et al. (1998) reported that any fuel-upgrading process would be a low-margin (probably less than US$ 1/bbl or US$ 0.02/L, value added), large volume (approximately 10^{10} L/year), commodity enterprise, where, an efficient operation is essential for economic viability. Premuzic et al. (1999) reported that the bioconversion of heavy crude oils depends on the distribution of polar compounds containing N, S, O, and trace metals.

Due to the revolution in protein and genetic engineering, the study of extremophilic microorganisms, biocatalysts in nonaqueous media and nano-biocatalysts, biotechnology found its way into petroleum refining (i.e., biorefining). For example; biodesulfurization (BDS), biodenitrogenation (BDN), biodemetallization (BDM), biotransformation of heavy crudes into lighter crudes, and finally biodepolymerization of asphaltene constituents.

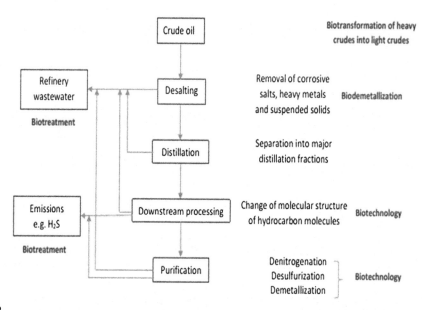

FIGURE 5.3

Simplified flow chart for possible applications of petroleum biotechnology during heavy crude oil processing.

2.0 BIODESULFURIZATION

Sulfur is the major concern for producers and refiners and has long been a key determinant of the value of crude oils. It is the third most abundant element in crude oil after carbon and hydrogen. Heavy oil and bitumen contain 3%–6% sulfur that must be removed before its usage as a refinery feedstock. The combustion of S-containing fuels would lead to the increased emissions of sulfur oxides (SOx), the main cause of acid rains, particulate matter (PM), and black smoke associated with diesel and gasoline vehicles. It has been reported that the total PM emissions from diesel engines are proportional to the diesel sulfur content (Stanislaus et al., 2010; Mohebali and Ball, 2016). If desulfurization occurs on the crude feedstocks before they ever enter the refinery system, this would minimize the downstream desulfurization costs. Crude oils with higher viscosities and higher densities usually contain higher amounts of more complex sulfur compounds. The aliphatic acyclic sulfides (thioethers) and cyclic sulfides (thiolanes) are easy to remove during a hydrodesulfurization process or by thermal treatment. On the other hand, sulfur contained in aromatic rings, such as thiophene and its benzologs (e.g., benzothiophene, dibenzothiophene, benzonaphthothiophene) are more resistant to sulfur removal by hydrodesulfurization and thermal conversion (Gray et al., 1996).

Thus, the key techno-economic challenge to the ability of BDS processes is to establish a cost-effective means of implementing the two-phase bioreactor system and de-emulsification steps, as well as the product recovery step (Kaufman et al., 1998; Pacheco et al., 1999; McFarland, 1999). Use of multiple-stage air-lift reactors can reduce mixing costs, and centrifugation approaches can facilitate de-emulsification, desulfurized oil recovery, and recycling of the cells. Most of the applications of petroleum biotechnology are at the level of laboratory research except BDS, for which, pilot plants have been established. BDS was first developed by ENCHIRA Biotechnology Corporation (formerly, Energy Biosystems Corporation), by the mid-1990s.

Due to the milder and safer process conditions of BDS, the CO_2 emissions and energy requirements of a BDS-based processes is estimated to be lower than that of hydrodesulfurization processes (Linguist and Pacheco, 1999; Singh, 2012). Moreover, the capital costs to setup a BDS-process is reported to be 50% lower than that of a hydrodesulfurization process (Pacheco et al., 1999; Linguist and Pacheco, 1999; Atlas et al., 2001).

BDS has been applied to mid-distillates (Grossman et al., 1999; Pacheco et al., 1999; El-Gendy, 2001, 2004; Li et al., 2005; El-Gendy et al., 2006; Aribike et al., 2009); partially hydrodesulfurization-treated mid-distillates (Folsom et al., 1999); extensively hydrodesulfurization-treated mid-distillates (Grossman et al., 2001); light gas oils (Chang et al., 1998; Pacheco et al., 1999; Furuya et al., 2003; Ishii et al., 2005; Dinamarca et al., 2010); cracked stocks (Pacheco et al., 1999); and crude oils (Premuzic and Lin, 1999a,b; El-Gendy, 2001, 2004; El-Gendy et al., 2006). However, *Pseudomonas* sp. is reported to be an ideal candidate for BDS in petroleum, because they are organic solvent-tolerant and have a high growth rate (El-Gendy and Speight, 2015).

A thermophilic bacterial strain, identified as *Paenibacillus*, can selectively desulfurize dibenzothiophene without degrading its hydrocarbon matrix. These strains follow the same metabolic pathway of *Rhodococcus erythropolis* IGTS8. *Paenibacillus* enzymes are homologous to *Rhodococcus* enzymes; however, they are active at higher temperatures: from 50 to 60°C. Thus, it is proposed for the development of a BDS process for crude oil at high temperatures where crude oil viscosity is lower and mass transfer limitations are reduced (Konishi et al., 2000). Baldi et al. (2003) reported the isolation of a yeast strain, *Rhodosporidium toruloides* DBVPG 6662, that can grow on a variety of sulfur compounds and can desulfurize orimulsion (a bitumen amended with an emulsifying agent and water) by 68% in 15 days. El-Gendy et al. (2006) reported a BDS of crude oil by the halotolerant yeast, *Candida parapsilosis* NSh45, isolated from Egyptian hydrocarbon polluted sea water, in a batch process of 1/3 O/W phase ratio of 7 days at 30°C and mixing rate of 200 rpm. The NSh45 reduced the sulfur content of Belayim Mix crude oil (2.74% w/w sulfur) by 75%, with a decrease in the average molecular weight of asphaltene constituents by approximately 28% and dynamic viscosity by approximately 70%, compared to that of *R. erythropolis* IGTS8 which expressed total sulfur removal of approximately 64%, and a decrease in average molecular weight of asphaltene constituents and crude dynamic viscosity of approximately 24% and 64%, respectively, under the same conditions (El-Gendy et al., 2006).

Gunam et al. (2013) reported that BDS removed most of the sulfur (59% w/w) from Liaoning crude oil within 72 h, using *Shingomonas subarctica* T7B. *Stachybotrys* sp. WS4 is reported to accomplish 76% and 65% desulfurization of heavy crude oil from the Soroush and Kunhemond oil fields within 72 and 144 h, respectively (Torkamani et al., 2009). *Mycobacterium goodie* X7B and *R. erythropolis* XP have been reported to remove 47.2%–62.3% w/w of sulfur from crude oil after 72 h treatment at 30°C (86°F) (Yu et al., 2006; Li et al., 2007). The fungus strain *Stachybotrys* sp. which has been isolated to selectively remove sulfur and nitrogen from heavy crude oil at 30°C (86°F), by the Petroleum Engineering Development Company, a subsidiary of National Iranian Oil Company, is proved to be able to remove 76% and 64.8% w/w of the sulfur from heavy crude oil of Soroush oil field and the Kuhemond oil field—with an initial sulfur content of 5% w/w and 7.6% w/w, within 72 and 144 h, respectively (Torkamani et al., 2009).

Anaerobic BDS of crude oil and its distillates has been also reported by *Desulfovibrio desulfuricans* M6 (Kim et al., 1990), extreme thermopile *Pyrococcus furiosus* (Tilstra et al., 1992), Nocardioform actinomycete FE9 (Finnerty, 1992, 1993), *Desulfomicrobium scambium* and *Desulfovibrio longreachii* (Yamada et al., 2001). The desulfurization of petroleum under anaerobic conditions would be attractive because it avoids costs associated with aeration; it has the advantage of liberating sulfur as a gas and does not liberate sulfate as a by-product that must be disposed by some appropriate treatment. Under anaerobic conditions, oxidation of hydrocarbons to undesired compounds such as colored and gum forming products is minimal (McFarland, 1999). These advantages can be counted as incentives to continue research on reductive BDS. However, maintaining an anaerobic process is extremely difficult and the specific activity of most of the isolated strains have been reported to be insignificant for dibenzothiophenes (Armstrong et al., 1995). Due to low reaction rates, safety and cost concerns, and the lack of identification of specific enzymes and genes responsible for anaerobic desulfurization, anaerobic microorganisms effective enough for practical petroleum desulfurization have not been found yet, and anaerobic BDS process has not been developed.

Agarwal and Sharma (2010) reported the 63.29% and 61.40% BDS of two crude oil samples, heavy and light crude oils with sulfur content of 1.88% and 0.378%, respectively, by *Pantoea agglomerans* D23W3. However, the use of *P. agglomerans* D23W3 under anaerobic conditions showed marginally better results than those under aerobic conditions. The use of the thermophile *Klebsiella* sp. 13T resulted in 68.08% and 62.43% sulfur removal from heavy and light crude oil, respectively (Agarwal and Sharma, 2010). Li and Jiang (2013) reported also 51.7% sulfur removal from bunker oil within 7 d of incubation using microbial mixed culture.

The conventional technologies cannot achieve the target of ULSD in a cost-effective manner. Thus, a combination of BDS and hydrodesulfurization technologies is suggested to achieve the future goals.

In fact, significant progress has been made toward the commercialization of crude oil BDS. This progress includes the isolation and characterization of crude oil candidates for the BDS process; improved biocatalyst performance that directly relates to crude oil BDS (thermotolerant, solvent tolerant, overcome the toxicity of by-products); development of analytical methodology, which led to breakthroughs in the characterization of BDS-recalcitrant compounds; development of a process concept for crude oil BDS; and construction and testing of a prototype bench unit. Technical hurdles still need to be overcome to achieve commercialization. The major obstacles to the economical BDS of crude oil include biocatalyst specificity, rate, reusability, and stability for a long time under process conditions. Work continues to modify the biocatalyst to increase its effectiveness and to screen other organisms for additional desulfurization capabilities. In addition, mass transfer and separations hurdles must be overcome in crude oils with increased oil viscosity and density.

The selective removal of sulfur atom from the hydrocarbon structure (e.g., dibenzothiophene), through the 4S-pathway (Fig. 5.4), by retaining its hydrocarbon skeleton and fuel value is recommendable. The dibenzothiophene is not degraded but transformed into 2-hydroxybiphenyl (2-HBP) or 2,2′-bihydroxybiphenyl (Monticello, 2000; Nassar et al., 2016), which, in some strains transformed to the less toxic 2-methoxybiphenyl and 2,2′-dimethoxy-1,1′-biphenyl (El-Gendy et al., 2014), and both are portioned to the hydrocarbon phase (i.e., the fuel), while the sulfur is eliminated as inorganic sulfate in the aqueous phase containing the biocatalyst (Bordoloi et al., 2014). Several microorganisms are reported retaining the 4S-pathway; the first is *Rhodococcus. erythropolis* IGTS8, followed by different new isolates belonging to the *Rhodococcus, Gordonia, Nocardia, Microbacterium, Mycobacterium* species, *Actinomycetales, Sphingomonas, P. agglomerans, Stenotrophomonas, Brevibacillus* species. Moderate thermophiles and thermotolerant species have been also reported, which are beneficial in real field operation, especially downstream the hydrodesulfurization process; *Paenibacillus, Bacillus subtilis, Mycobacterium, Thermobifida, Rubrobacter,* and *Klebsiella* (Ayala and LeBorgne, 2010; Morales and Le Borgne, 2014; El-Gendy, 2015; Ayala et al., 2016; Mohebali and Ball, 2016; El-Gendy and Speight, 2015).

It has been estimated that the BDS catalyst must have an S-removal activity within the range of 1–3 mmol dibenzothiophene/g dry cell weight/h in real petroleum fractions. This means, the activity of the current biocatalysts is needed to be improved by 500-fold to attain a commercially viable process (Kilbane, 2006). Alves et al. (2015) performed a cost analysis study, comparing two BDS process designs; upstream and downstream conventional hydrodesulfurization. The BDS costs and emission estimations were made considering the BDS of dibenzothiophene, as model for S-compounds, while, hydrodesulfurization estimations were made based on crude oil hydrodesulfurization. The BDS downstream hydrodesulfurization configuration is found to be the best alternative to be applied in oil refinery, from the point of energy consumption, greenhouse gas emissions, and operational costs, to obtain almost S-free fuels, with a much lower emission of greenhouse gases and carbon dioxide.

Certain technical issues hamper the applicability of BDS. One of them is the biphasic nature of the process. An aqueous phase is needed to maintain a viable biocatalyst in the BDS process and, thus, significant quantities of water must be added to the fuel. The values found in the literature generally indicate low oil-to-water volumetric ratios in the desulfurization of real fractions, ranging from 1% to 25% of oil in water. The highest oil-to-water ratio has been 50%–80% as reported by Yu et al. (1998) and Monot et al. (2002), respectively. It has been reported that, combining oxidative desulfurization with BDS would achieve 91% sulfur removal from heavy oil (Agarwal and Sharma, 2010).

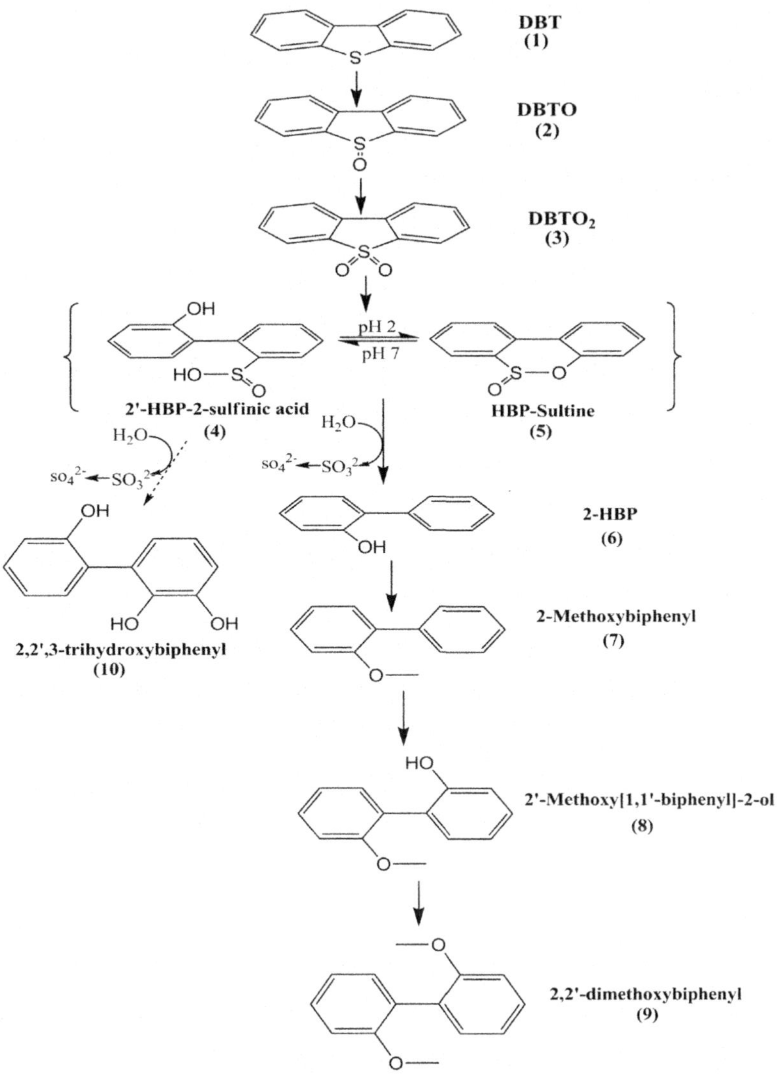

FIGURE 5.4

Selective biodesulfurization pathway (El-Gendy et al., 2014).

3.0 **BIODENITROGENATION**

Nitrogen, is like sulfur; it is considered as a petroleum contaminant. Éigenson and Ivchenko (1977) reported that the higher the sulfur content in petroleum fractions, the higher is its nitrogen content. Moreover, nitrogen containing compounds (NCCs) coexist with sulfur containing compounds in fossil fuels (Yi et al., 2014). Although, specific BDS of petroleum and its distillates has been reasonably investigated, there is little information about BDN of oil feed without affecting its calorific value. It has been estimated that BDN of petroleum would be beneficial for deep denitrogenation, where, the classical hydroprocessing methods are costly and nonselective (Vazquez-Duhalt et al., 2002). Microorganisms such as *Pseudomonas ayucida, Aneurinibacillus* sp., *Pseudomonas stutzeri, Yokenella* sp., and *Pseudomonas nitoreducens* are issued for BDN of fossil fuels (Kilbane et al., 2003). A thermophilic carbazole-degrading bacteria, *Anoxybacillus rupiensis*, that can tolerate up to 80°C, with maximum activity at temperature range 55–65°C has been reported, which would be advantageous for application in real petroleum processing (Fadhil et al., 2014). However, so far, no microorganism that selectively extracts nitrogen atom from carbazole, in a microbial pathway equivalent to the 4S-pathway, have been isolated.

Atlas and Aislabie (1992) reported the selective removal of NCCs in raw shale oil by *Pseudomonas* and *Acinetobacter* sp. where, amines, nitriles, quinolines, and pyridines were converted to the water-soluble hydroxylated compounds. Kilbane et al. (2000) reported the selective BDN of quinine without affecting its carbon-skeleton by *P. ayucida* strain IGTN9m (ATCC N° PTA-806), where, approximately 5% nitrogen, was removed by this isolate from petroleum without affecting its energetic value. Since IGTN9m converts quinine to 8-hydroxycoumarin without further degradation, no carbon is lost. Resting cells of IGTN9m selectively transform 68% of quinine from shale oil within 16 h. The low BDN efficiency in oil is attributed to the low abundance of quinine relative to other NPAHs in oil and the narrow substrate range of the quinine-degrading enzymes. Sugaya et al. (2001) reported the optimum QN-BDN conditions using *Comamonas* sp. TKV3-2-1, for its application in petroleum feed. The degradation rate reaches 1.6 mmol/g cell/h at 83% (v/v) petroleum/aqueous medium with a cell density of 28.5 g/L. Moreover, Sugaya et al. (2001) suggested the application of BDN of crude oil, during its storage period, as it would overcome the problems of extra treatment period and cost. In addition, with the decrease in petroleum nitrogen content, this would effectively improve its quality and thus increase its products' yields (such as gasoline). Kayser and Kilbane (2004) reported the decrease of the CAR content in a petroleum sample by approximately 95% in 2:10 petroleum/aqueous medium within 16 h using a genetically engineered bacterium. Anaerobic BDN has been also reported to modify the properties of heavy crude oil and enhance oil recovery (Fallon et al., 2010).

From the practical point of view, BDN and BDS should be integrated, where, sulfur and nitrogen, would be removed through specific enzymatic attack of the C—N and C—S bonds, respectively, but without C—C bond attack, to preserve the fuel value of the biotreated products. Duarte et al. (2001) in PETROBRAS, the Brazilian oil company, have isolated *Gordonia* sp. strain F.5.25.8 that can utilize dibenzothiophene through the 4S-pathway and CAR as a sole source of S and N, respectively. F.5.25.8 is the first reported strain that can simultaneously metabolize dibenzothiophene and carbazole (Santos et al., 2005). Santos et al. (2006) reported that F.5.25.8 can tolerate up to 42°C, which would add to its advantages in industrial application of BDS/BDN as complementary to hydrotreatment process. Moreover, it is reported to have a different genetic organization of the BDS (dsz) and BDN (car) gene clusters relative to *R. erythropolis* IGTS8 and *Pseudomonas* sp. IGTN9m, respectively.

BDN process suffers from the main limitations of BDS: low nitrogen removal activity and the need for an aqueous phase. Moreover, it has two more limitations that should be economically attractive: no microorganisms capable of selectively removing nitrogen without carbon loss have been developed yet, and denitrogenation is an additional step that does not produce an upgraded fuel; therefore, reaction velocities should be even higher than for BDS.

4.0 BIODEMETALLIZATION

Crude oil contains metals in the form of salts (zinc, titanium, calcium, and magnesium), petroporphyrins, and other complexes in the asphaltene constituents (vanadium, copper, nickel, and iron) (Hobson and Pohl, 1975; Ali and Abbas, 2006; Speight, 2014). The more residual the oil, the higher the metal content, those metal species possibly clustered by heavy molecular mass compounds (Speight, 2014). The higher the asphaltene content in crude oil is, the higher the heavy metal content would be (Duyck et al., 2007). Metal accumulation in the heaviest polar fractions of crude oils plays a significant role in establishing the refining procedure, since V, Fe, Ni, and Mo have both negative and positive effects on product recoveries (Panariti et al., 2000). During more than 70 years since the discovery of metalloporphyrins in petroleum, the origin and significance of heavy metals in petroleum has been poorly investigated, due to its low concentrations and the limitations of analytical methods for such complex matrices (i.e., petroporphyrins and asphaltene constituents) in crude oil and its heavy fractions (Duyck et al., 2008). The vanadium (V) and nickel (Ni) containing compounds are the most predominate and persist mainly in the resin and asphaltene fractions of crude oil. Total metal content in crude oils has an extended concentration range.

Depending on the origin of crude oil, the concentration of the vanadium varies from as low as 0.1 ppm to as high as 1200 ppm, while that of nickel commonly varies from trace to 150 ppm (Ali and Abbas, 2006). However, Ni and V porphyrins occur as high as 120–1500 ppm, respectively, in heavy oil (Hessley, 1990; Duyck et al., 2007). The V/Ni ratio is constant in crude oils of common source rocks and dependent on the geological age of the rocks, with oils from Triassic or older ages showing a value higher than unity (Ball et al., 1960), and this ratio is also used for tracing source effects (Al-Shahristani and Al-Atyia, 1972; El-Gayar et al., 2002). However, biodegradation of asphaltene constituents and resin constituents is reported to influence the Ni/V ratio in these fractions (Duyck et al., 2007). Fossil fuels with a particularly high content of organometallics include Bosean, Cerro Negro, Mayan, Wilmington, and Prudhoe Bay Crude oils (Fish et al., 1984). The metallic salts usually occur as inorganic water-soluble forms and are easily removed during the crude desalting process, in which they are concentrated in the aqueous phase. However, porphyrins are embedded in the extremely complex structure of asphaltene; thus, metal removal from petroporphyrins and complexes is a problem to be solved.

A major problem and expense during the refining of oil and other petroleum products is the continuous contamination of solid, porous catalysts by various porphyrins, metalloporphyrins, chlorins, and natural degradation products of these compounds, such as petroporphyrins, containing metals such as vanadium and nickel. The porphyrin deposits, however, consist of several different metals on the catalytic surface including vanadium (V), nickel (Ni), titanium (Ti), iron (Fe), copper (Cu), or a combination thereof with the concentrations of V and Ni varying from a few to several hundred parts per million (ppm), depending on the type of crude oil supply. Vanadium is usually present in a concentration greater than other metals with much more than half of all V being deposited on the catalyst arising from the porphyry complex. Heavy metals poison refining catalyst. During petroleum refining, heavy metals are concentrated in the residual fraction, which is usually subjected to catalytic cracking. During the catalytic cracking, metals in the oil deposit on the cracking catalysts, and decrease their selectivity and activity. Catalyst deactivation in cracking, hydrogenation, and hydrodesulfurization processes can also occur by pore clogging, metal deposition, deformation, and destruction of reactors (Salehizadeh et al., 2007), which in turn demands that the catalytic process be interrupted to either replace or clean the catalyst at a huge cost (Altgelt and Boduszynski, 1994; Callejas et al., 2001). The direct, material cost of replacing contaminated catalysts for the United States petroleum industry is estimated at more than $1 billion, annually (Paul and Smith, 2009). The presence of metal contaminants in the fluid catalytic cracking (FCC) feeds presents another and potentially more serious problem because although sulfur can be converted to gaseous forms which can be readily handled in an FCC unit, the nonvolatile metal contaminants tend to accumulate in the unit and during the cracking process they are deposited on the catalyst together with the coke.

Moreover, since both nickel and vanadium exhibit dehydrogenation activity, their presence on the catalyst particles tends to promote dehydrogenation reactions during the cracking sequence and this result in increased amounts of coke and light gases at the expense of gasoline production (Elliot, 1996; Ali and Abbas, 2006). Heavy metals can be liberated into the environment during fuel combustion in the form of ash with high concentrations of toxic metal oxides. That leads to undesirable waste disposal problem (Xu et al., 1997). Moreover, heavy metals (mostly Ni and V) are furthermore corrosive (Montiel et al., 2009). Metal porphyrins are relatively volatile and when crude oil is vacuum-distilled they tend to carry over the heavier fraction of the distillated liquids. Hence, traces of vanadium are usually found in vacuum gas oil. The well-established demetallization processes in petroleum industry are the physical deasphalting process, where the lighter oils are physically separated from heavier asphaltene constituents by mixing the heavy oil/residue with a very low boiling solvent such as propane, butane, or isobutene (Yamada et al., 1979; Speight, 1981; Farag et al., 1989).

The distillation separates crude oils into fractions according to boiling point, so that each of the processing units following will have the feedstock that meet the required specifications. The metallic constituents concentrate in the residues (Reynolds et al., 1987). The thermal processes, such as visbreaking (Rollmann and Walsh, 1980) and coking (Speight, 1981) are basically the reshuffling of the hydrogen distribution in the residue to produce lighter products containing more hydrogen while the asphaltene constituents and metals are removed in the form of coke or visbreaking residue. The filtration by porous membranes is reported to be effective for removal of N, S, Al, Cr, Cu, Ni, V, and asphaltene constituents from spent diesel, lubricating oil, crude oil, heavy oils, or bitumen (Kutowy et al., 1989). Osterhuber (1989) of Exxon developed a method for upgrading heavy oils by solvent dissolution and ultra-filtration at high pressure. The process is especially suitable for removing trace metals (mainly Ni and V) and lowering the Conradson Carbon Residue of the resulting oil.

Finally, we come to the chemical process, which is the selective removal of metals from the organic moieties with a minimal conversion of the remaining petroleum. For example; metal removal with solvent (Savastano, 1991), oxidative demetallization of petroleum asphaltene constituents and residua (Gould, 1980), and the hydrodemetallization process (Kashima Oil Co., Ltd. Japan., 1983; Adarme et al., 1990; Bartholdy and Hannerup, 1990; Piskorz et al., 1996) are usually used for demetallization of crude oils. But, these are expensive and usually produce secondary pollution in the environment (Hernandez et al., 1998).

Fedorak et al. (1993) reported the destruction of porphyrin ring systems using a chlorination reaction using *Caldariomyces fumago* CPO enzyme leading to the biocatalytic demetallization of petroporphyrins and asphaltene constituents with a reduction in nickel. For example, nickel was reduced from octaethyl porphyrin and vanadyl octaethyl porphyrin by approximately 93% and 53%, respectively. But, chlorination produces chlorinated compounds which have negative impact on the environment. Biocatalytic removal of Ni and V from petroporphyrins by chloroperoxidase (CPO) has been also reported by Mogollón et al. (1998).

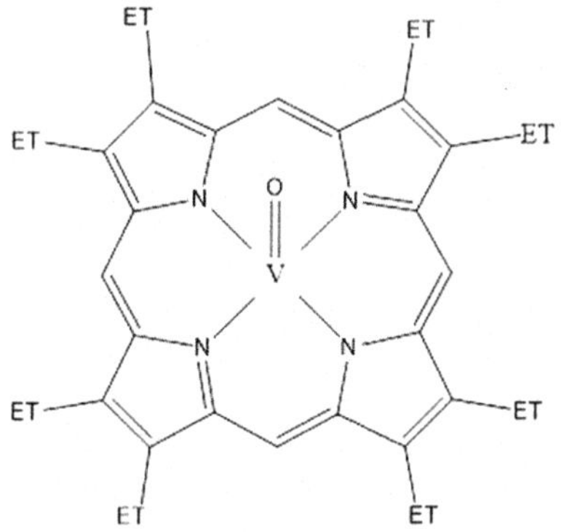

FIGURE 5.5

Example of a vanadyl porphyrin.

Xu et al. (1997, 1998) reported the BDM of petroporphyrins and crude oil, through the oxidation of prophyrinic rings using hemoproteins (cytochrome C reductases) from *Bacillus megaterium* and *Catharanthus roseuse* in the presence of the cofactor NADPH (the reduced form of nicotinamide adenine dinucleotide phosphate NADP). In contrast to peroxidases, such as CPO, oxygenases from *Escherichia coli*, animal cells (such as liver or kidney cells), plant cells (such as from mung beans or *Arabidopsis thaliana*) or yeast cells (such as *Candida tropiculis*) can degrade porphyrin molecules without subjecting the hydrocarbon to chlorine or peroxide, and the metals which can be removed by this method include nickel, vanadium, cobalt, copper, iron, magnesium, and zinc. The produced metals can be readily removed by extraction (such as a de-salt wash), distillation, ion exchange, and/or column chromatography. The main advantages of this process is it can be operated in a batch, semicontinuous or continuous mode, alone or in combination, with one or more additional biorefining processes (such as BDS), in a sealed or open vessel and in the presence or absence of light.

The removal of vanadyl porphyrins (Fig. 5.5) from the crude oils is reported to be of a great importance. Salehizadeh et al. (2007) reported the possibility of microbial demetallization of crude oil using *Aspergillus* sp. MS-100, isolated from polluted soil of Isfahan refinery for its ability to consume vanadium oxide octaethyl porphyrin as a sole carbon source. Horse myoglobin and plant peroxidase were chosen as the proteins used to synthesize the agents for removal of porphyrins from uncracked fuels (Paul and Smith, 2009). Kilbane (2005) reported that, since most of the metals in petroleum are associated with organonitrogen compounds, it would be possible to simultaneously perform the BDN and biometallization processes to reduce the nitrogen and metal content in one process instead to do it in two separate processes.

5.0 **BIOTRANSFORMATION OF HEAVY CRUDE OIL**

The efficient recovery, serration, and processing of heavy crude oils and bitumen are hampered by the presence of high concentration of asphaltene constituents. The presence of the asphaltene constituents is one of the main reasons for the increase of the crude oil viscosity (El-Gendy, 2001, 2004; El-Gendy et al., 2006; Bachman et al., 2014). Moreover, it enhances the propensity to form emulsions, polymers, and coke (Vazquez-Duhalt et al., 2002; Hernández-López et al., 2015). Recently, biological processes have emerged as a cost-effective and environmentally favorable alternative to break asphaltenic structures to obtain high-value light oils from less-value heavy oils (i.e., bio-cracking). The asphaltene constituents can be described as condensed aromatic cores containing alkyl and alicyclic moieties, as well as nitrogen, sulfur, and metal containing non- and heterocyclic groups (Vazquez-Duhalt et al., 2002; Hernández-López et al., 2015).

The asphaltene constituents have extremely complex and variable molecular structures containing sulfur (0.3%–10.3%), oxygen (0.3%–4.8%), nitrogen (0.6%–3.3%), and metal elements, such as Fe, Ni, and V in a small amount, with an average molecular weight ranging between 600 and 2,000,000 (Tavassoli et al., 2012). Asphaltene is also considered to be the product of complex heteroatomic aromatic macrocyclic structures polymerized through sulfide linkages (El-Gendy et al., 2006; Ali et al., 2012). Breaking the asphaltene constituents into smaller molecules and cutting an internal aliphatic linkage (sulfides, esters, and ethers) of an asphaltene molecule can lead to a reduction in viscosity (Peng et al., 1997; El-Gendy, 2001, 2004).

The asphaltene constituents are relatively high molecular weight and large and highly hydrophobic;thus, mass transfer limitations are expected in aqueous reactions and the biotransformation rates are limited by the mass transfer of target molecules to the biocatalyst and, in the case of whole cells, across the cell membrane (Leòn and Kumar, 2005). Despite these difficulties, there is evidence in the literature for bacterial transformation of these complex, high molecular weight substrates. This is possible because these compounds contain carbon, hydrogen, sulfur, nitrogen, and oxygen, which are necessary elements for the survival of microorganisms. Fig. 5.6 summarizes the possible biotransformation point of view in the complex structure of the asphaltene constituents.

Different extremophile bacterial genera, such as *Achromobacter*, *Leptospirillum*, *Pseudomonas*, *Sulfolobus*, *Thiobacillus* have been reported for their capabilities to transform heavy oils into lighter ones (Premuzic et al., 1997, 1999; Premuzic and Lin, 1999a,b). These genera are adapted to resist high temperatures, pressures, salt, and hydrocarbon concentrations. They interact with heteroatoms and organometallic sites in the heavy oils that serve as attachment and initiation points for the microbial activity. The involved reactions probably include oxidation, redistribution, and fragmentation of the heavy polar fractions (asphaltene constituents) into lower fractions (maltenes). These asphaltene constituents would probably decompose by rupture at active sites containing heteroatoms, allowing the liberation of the trapped, lower molecule weight constituents. The heterocyclic compounds would be possibly oxidized into more soluble compounds that migrated to the aqueous phase. This occurs with a complimentary increase in the concentration of saturated C-chains (C8–C26) and decrease in the higher molecular weight hydrocarbons, heteroatoms, and metal contents. Unfortunately, up till now, the specific microorganisms capable of performing this biotransformation are not known with certainty, and there is no available data concerning the involved biochemical reactions and metabolic pathways.

A complex set of multiple biochemical reactions between select microorganisms and heavy crude oils under controlled conditions have been reported by Premuzic et al. (1999) that led to a significant lowering (24%–40%) of the N, S, O, and trace metal contents, with a concurrent redistribution of hydrocarbons. The reactions are both biocatalyst and crude oil-dependent and, in terms of chemical mechanisms, appear to involve asphaltene and the associated polar fractions. Asphaltene constituents from a crude oil rich in heavy metals (Castilla crude oil) have been fractionated and the biocatalytic

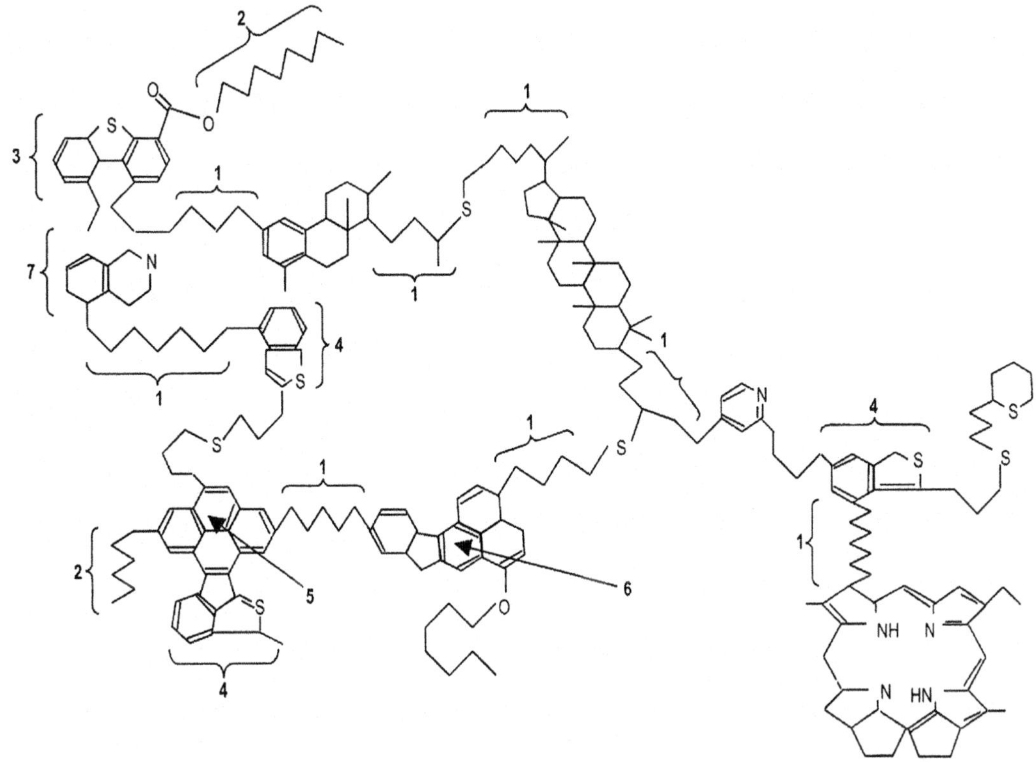

FIGURE 5.6

Asphaltene structure. Regions susceptible of fragmentation and biodegradation in an asphaltene molecule: 1. photooxidation; 2. beta-oxidation; 3. dibenzothiophene metabolic path; 4. path similar to dibenzothiophene; 5. pyrene path; 6. path similar to benzo(a)pyrene; 7. similar to carbazoles.

modifications of these fractionated asphaltene constituents by three different hemoproteins, CPO, cytochrome C peroxidase (Cit-C), and lignin peroxidase, have been evaluated in both aqueous buffer and organic solvents. However, only the CPO-mediated reactions were effective in eliminating the Soret peak in both aqueous and organic solvent systems and the CPO has been reported to be able to alter components in the heavy fractions of petroleum and remove 53% and 27% of total heavy metals (Ni and V, respectively) from petroporphyrin-rich fractions and asphaltene constituents (Mogollón et al., 1998). Premuzic and Lin (1999a) adapted and modified extremophilic microorganisms (thermophilic, thermo-adapted, barophilic, extreme pH, high salinity, and toxic metal adapted microorganisms) such as those belonging to the *Thiobacillus thiooxidans*, *Thiobacillus ferrooxidans*, *Leptospirillum ferrooxidans*, *Acinetobacter calcoaceticus*, *Sulfolobus solfataricus*, *Achromobacter* sp., *Arthrobacter* sp., and *Pseudomonas* sp., for biochemical conversion of a feedstock of heavy crude oils.

The upgraded oil feedstock produced by the process are characterized by increased lighter fractions of oils, increased content of saturated hydrocarbons, decreased content of organic sulfur containing components which have been decreased by about 20% to about 50%, decreased content of organic nitrogen containing components which have been reduced by about 15% to about 45% and, a significantly decreased concentration of trace metals by about 16% to about 60% by weight, where, the increase in the relative content of the lighter fractions of oil and in the content of saturated hydrocarbons depends on the chemistry of the starting material. The upgraded oil obtained by the process of the present invention contains an increased content of hydrocarbon surfactants such as emulsifying agents and hydrocarbon-based detergents. Additionally, the upgraded oil also has an increased content of oxygenates which are additives used by gasoline manufacturers to enhance fuel combustion.

Le Borgne and Quintero (2003) reported that the fungal depolymerization of coal asphaltene constituents, through oxidative attack by extracellular radical generating enzymes (Hofrichter et al., 1997), could be investigated for bio-upgrading of petroleum asphaltene constituents. García-Arellano et al. (2004) reported the biotransformation of porphyrins and asphaltene constituents by a semisynthetic biocatalyst (PEG-Cyt-Met), prepared by double chemical modification of cytochrome c. the chemically modified hemeprotein proved to be an active, solvent tolerable, and high temperature resistible biocatalyst. The FTIR spectra proved the oxidation of sulfur and carbon atoms in asphaltene molecules. Ayala et al. (2007) reported that the peroxidase-catalyzed reactions have the potential to be applied in biotransformation of asphaltene constituents.

Rontani et al. (1985) studied asphaltenic fraction degradation of Asthart crude, which was partially degraded by a marine mixed bacterial population with saturated hydrocarbons as a co-substrate. Pendrys (1989) isolated seven Gram-negative, aerobic, asphalt-degrading bacteria by an enrichment technique. The predominant genera of these isolates were *Pseudomonas*, *Acinetobacter*, *Alcaligenes*, *Flavimonas*, and *Flavobacterium*. A mixed culture preferentially degraded the saturate and naphthene aromatic fractions of asphalt cement-20 and utilized asphalt as the sole carbon and energy source. Hao et al. (2004) reported the biodegradation of asphaltene constituents, resins, and high molecular weight polyaromatic hydrocarbons and reduction of sulfur and nitrogen contents of heavy crude oils by a Gram-negative thermophile strain TH-2, which was isolated from Shengli oil field in China and tolerates up to 85°C. The bioconversions that occurred in the crude oil increased the lighter hydrocarbons and redistributed the saturates. The viscosity and paraffin content were also decreased by 10.1%–55.4% and 16.2%–37.7%, respectively. Pineda Flores et al. (2004) reported on the utilization of asphaltene constituents as a sole carbon and energy source by a consortium; *Corynebacterium* sp., *Bacillus* sp., *Brevibacillus* sp., and *Staphylococcus* sp., isolated from Maya crude oil. Van Hamme et al. (2004) isolated bacteria capable of cleaving subterminal C-S bonds within alkyl chains using the novel fluorinated organosulfur compound, bis-(3-pentafluorophenylpropyl)-sulfide, as a substrate. This may not only help in desulphurization of heavy crudes but also in viscosity reduction as these bacteria can cleave alkyl C-S bonds of an asphaltene molecule without reducing the carbon value of the substrate. Moustafa et al. (2006) reported that, for the biomodification of asphaltene constituents, the reactions with organosulfur moieties could be very significant, because sulfur is the most abundant element in the asphaltene fraction after carbon and hydrogen; sulfur can form up to 8% wt. of the asphaltene fraction and has an important role in its molecular structure. *Neosartorya fischeri* is a fungus isolated from natural asphalt Lake of Guanoco located in Sucre, Venezuela, that can biodegrade 15.5% of asphaltene constituents, with 13.2% mineralized to carbon dioxide within 11 days of incubation and an extracellular laccase activity induced during the asphaltene metabolism (Uribe-Alvarez et al., 2011).

It has been reported that the cytochrome P450 (CYP) monooxygenases have been reported to be involved in the fungal biotransformation of asphaltene constituents and high molecular weight polyaromatic hydrocarbons (Hernández-López et al., 2016). Ali et al. (2012) reported the 83%–96% biodegradation of 2500 mg/L asphaltene constituents as sole source of carbon and energy source, within 21 d by three halotolerant and thermotolerant bacterial isolates—*Bacillus* sp. Asph1, *Pseudomonas aeruginosa* Asph2, and *Micrococcus* sp. Asph3. The gel permeation chromatographic analysis revealed a decrease in asphaltene constituents' average molecular weights, indicating a microbial attack on the polysulfide linkages, which would lead to biodepolymerization of the asphaltene fraction. The FTIR showed significant alternations in functional groups after biotreatment, which would be due to oxidative alterations of the macromolecular structures induced by the bacteria, and exhibit distinct changes at the bands for sulfones and sulfoxides, suggesting that the bacteria can oxidize the abundant thioether linkages of macromolecular structures to sulfoxide and sulfone functions. *Pestalotiopsis* sp., a halotolerant fungus, is reported to degrade 21.4% of asphaltene constituents as a sole carbon source within 15 days of incubation (Yanto and Tachibana, 2013, 2014). Tavassoli et al. (2012) reported the isolation of 25 species from oil and polluted soil samples in Dorood oil field, in the south of Iran, that can utilize asphaltene as a sole carbon and energy source. The best five bacterial isolates; *Pseudomonas* sp. TMU2-5, *Bacillus licheniformis* TMU1-1, *Bacillus lentus* TMU5-2, *Bacillus cereus* TMU8-2, and *Bacillus firmus* TMU6-2 reported biodegradation of 40%–46%, where, the Gram-positive *B. lentus* TMU5-2 expressed the greatest capability for asphaltene degradation, and a consortium of these five isolates recorded 48% of asphaltene degradation.

Jahromi et al. (2014) reported the isolation of four bacterial consortia from oil contaminated soils and sludge; consortium 1 (*P. aeruginosa* and *Pseudomonas fluorescens*), Consortium 2 (*Citrobacter amalonaticus* and *Enterobacter cloacae*), Consortium 3 (*Staphylococcus hominis*) and consortium 4 (*B. cereus* and *Lysinibacillus fusiformis*). That showed good capabilities for biodegradation of 35 g/L asphaltene constituents 51.5%, 43%, 21.5% and 33.5%, respectively at 40 C within 2 months.

Finnerty and Singer (1983) reported that acid producing microbes can reduce the viscosity of crude oil (8–150 degrees API) with hexadecane from 25,000 cP to 275 cP. Lavania et al. (2012) reported the decrease in heavy oil viscosity from 3520 to 2029 cP, due to the anaerobic biodegradation of asphaltene by *Garciaella petrolearia* TERIG0 that was isolated from sea buried oil pipeline known as Mumbai Uran trunk line located on the western coast of India. When each of the fractions namely aliphatic, aromatic, and asphalt were treated with TERIG02 for 30 days, it was found that maximum degradation was in the case of asphalt followed by aromatic fraction, suggesting that the TERIG02 could tolerate the toxicity of these compounds and was capable of utilizing them as a carbon and energy source.

5.1 BIODEAROMATIZATION

The low-sulfur, low-aromatic, and high cetane number diesel is the fuel of the future. Hydrogenation, including a selective fragmentation and aromatic ring opening, would decrease the density and increase the cetane number of diesel (Rossini, 2003). There is an inverse proportional relation between cetane number and emissions of PM and NOx (Leliveld and Eijsbouts, 2008). Thus, the cetane number, which describes the ignition quality of a diesel fuel, is the most important property to be controlled in diesel fuel. Catalytic hydrotreatment is the main applied process for dearomatization. However, the balance between hydrogenolysis and hydrogenation reactions reduces the efficiency of the process and generates low molecular weight and volatile products, which would be lost from the diesel fraction (Rossini, 203). More hydrogen consumption is needed for the dearomatization of streams, when heavy oils are used as refinery feedstock, which have a larger volume of aromatic streams (Leliveld and Eijsbouts, 2008).

The naphthalene degradation pathway (Fig. 5.7) serves as the basis to develop a biodearomatization process. It is suggested that if step 5 (Fig. 5.7) is interrupted, a more linear, oxidized hydrocarbon with the same carbon content as naphthalene could be obtained. Thus, this would both reduce the aromaticity of the compound and generate hydrocarbons that could be hydrogenated under milder conditions. The naphthalene degradation pathway has been characterized mainly in *Pseudomonas* sp. and many other species: Gram-negative—*Sphingomonas, Burkholderia, Ralstonia, Rahnella, Polaromonas, Comamonas* sp., and Gram-positive—*Rhodoccocus, Nocardioides,* and *Mycobacterium* sp.

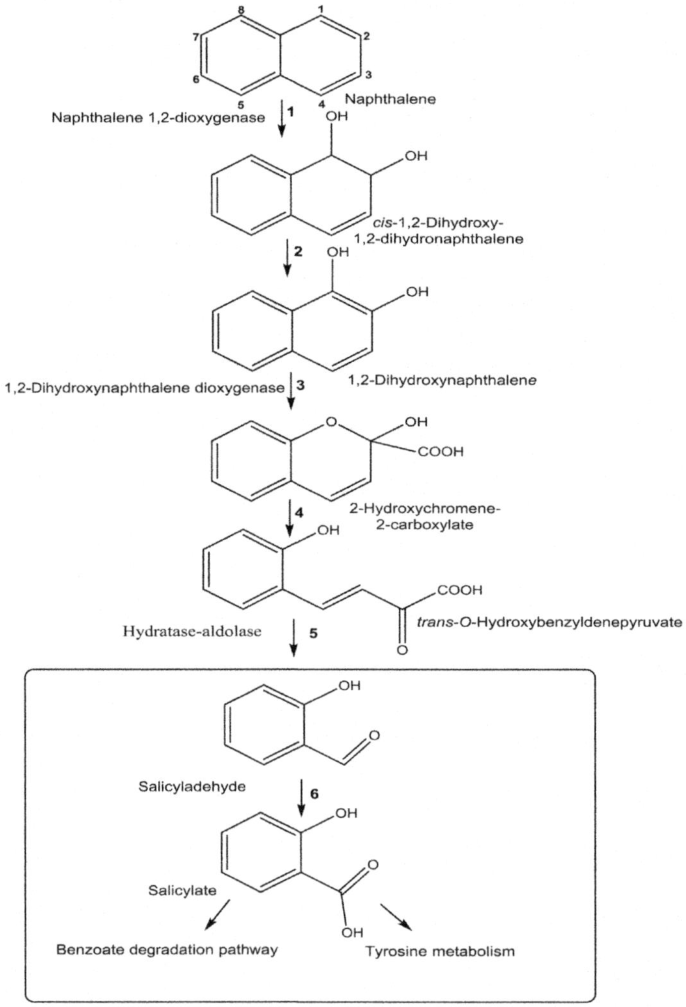

FIGURE 5.7

Naphthalene biodegradation pathway.

The gene encoding the hydratase-aldolase enzyme (step 5; Fig. 5.7) was interrupted in *P. fluorescens* LP6a and *Sphingomonas yanoikuyae* N2, to generate mutants capable of oxidatively opening the aromatic ring, without degrading the resulting hydrocarbon. The mutants are active water–oil biphasic systems and catalyzed the ring opening of model compounds such as naphthalene, methyl naphthalene, dibenzothiophene, 4-methyl dibenzothiophene, phenanthrene, and carbazole; the biocatalysts also reduced the aromatic content of distillates, such as light gas oil (Foght, 2004; Kotlar et al., 2004). The biotransformation of aromatic rings in petroleum has been reported, that involves the hydroxylation of the aromatic rings. Then, the activated molecules are then processed by conventional hydrogenation and/or hydrogenolysis processes producing cracked and ring-opened products (Coyle et al., 2000). Leòn et al. (2003) reported the isolation of some bacteria from Guanaco asphalt lake, Venezuela, capable of degrading dibenzothiophene through the Kodama-pathway, producing 3-hydroxyformyl benzothiophene. These bacteria are reported to be able to grow on bitumen and reduce the asphaltene constituents, resin, sulfur, and nitrogen contents of heavy crude oil, with a recorded reduction in viscosity.

A range of attractive diol precursors for chemical synthesis can be produced by naphthalene dioxygenase which also catalyzes a variety of other reactions including monohydroxylation, desaturation, O- and N-dealkylation and sulfoxidation (Resnick et al., 1996). Enantiospecific conversions of petrochemical substrates and their derivatives can be achieved by stereo-selective biocatalytic hydroxylation reactions using cytochrome p450-dependent monooxygenases, dioxygenases, lipoxygenases, and peroxidases (Kikuchi et al., 1999). Cytochrome p450cam monooxygenase from *Pseudomonas putida* has been successfully evolved to function more efficiently in the hydroxylation of naphthalene and dioxygenases with improved thermos-stability and substrate specificity (Furukawa et al., 2000).

5.2 BIOTRANSFORMATION OF ASPHALTENE CONSTITUENTS

The asphaltene constituents have drawn considerable attention due to problems caused by their detrimental effects in the extraction, transportation, and processing of residua because of their viscous and flocculating nature and their relative resistance to biodegradation following spills (Speight and Arjoon, 2012; Speight, 2014).

The asphaltene constituents are the highest molecular weight and most polar fraction of crude oil. Despite that the structure of asphaltene constituents has not been fully elucidated, it is widely accepted that it is constituted by interacting systems of polyaromatic sheets bearing alkyl side-chains. Asphaltene molecules have a high content of O, N, and S heteroatoms, as well as metals (V, Ni, and Fe) (Speight, 2014). The problems associated with asphaltene constituents have increased due to the need to extract heavier crude oils, as well as the trend to extract larger amounts of light fractions out of crude oil by cracking and visbreaking.

The asphaltenic fraction is recognized as the most recalcitrant oil fraction. There is no clear evidence that asphaltene constituents can be degraded or transformed by microbial activity. Microorganisms have been found associated with naturally occurring bitumen (Wyndham and Costerton, 1981; Naranjo et al., 2007), which contains high amounts of asphaltene constituents. A molecular study (Kim and Crowley, 2007) revealed a wide range of phylogenetic groups within the Archaea and Bacteria domains in natural asphalt-rich tar pits; interestingly, genes encoding novel oxygenases were also detected in such samples.

On the other hand, an extensive screening involving more than 750 strains of filamentous fungi was carried out to select strains able to modify untreated hard coal (Bublitz et al., 1994; Hofrichter et al., 1997). Only six of the 750 strains tested exhibited some activity, from which the most active fungi, *Panus tigrinus*, growing on wood shavings coated with coal asphaltene constituents led to a decrease of the average molecular weight (Hofrichter et al., 1997), although the average molecular weight of any complex mixture is not a measure of the constituents that were actually biodegraded.

Furthermore, most of studies on asphaltene constituents biodegradation should be considered cautiously as the asphaltene content was usually determined gravimetrically after *n*-alkane precipitation, and thus the reported changes may be attributed to the disruption of the asphaltenic matrix by the production of surfactants during bacterial growth, liberating trapped hydrocarbons. Other studies have reported that the asphaltene fraction does not support bacterial growth, and no changes in asphaltene content are found after bioconversion of heavy oil and asphaltene constituents (Lacotte et al., 1996; Thouand et al., 1999).

There have been claims of the biodegradation of asphaltene constituents by mixed bacteria (Bertrand et al., 1983; Rontani et al., 1985). However, none of these reports described the analytical results of extractable materials recovered from appropriate sterile controls. Therefore, most of the asphaltene losses during microbial activity could be considered abiotic losses (Lacotte et al., 1996).

A study (Pineda Flores et al., 2004) reported a bacterial consortium able to grow in the asphaltene fraction as the sole carbon source. Mineralization of the asphaltene constituents was estimated by measuring production. The authors found in two control experiments (inoculum without asphaltene constituents and non-inoculated asphaltene constituents) a carbon dioxide production equivalent to 39% and 26%, respectively, of that found in the consortium growing in the asphaltene fraction.

The microbial inoculum for consortium stabilization contained 1% of crude oil, which could serve as carbon source. Thus, it is not possible to distinguish the origin of the carbon dioxide production. The first clear experimental evidence that enzymes are able to modify asphaltene molecules has been reported (Fedorak et al., 1993). Chloroperoxidase from the fungus *C. fumago* and a chemically modified cytochrome c were able to transform petroporphyrin derivatives and asphaltene constituents in reaction mixtures containing organic solvents (Fedorak et al., 1993; Mogollón et al., 1998; García-Arellano et al., 2004). Notable spectral changes in the petroporphyrin-rich fraction of asphaltene constituents were observed and the enzymatic oxidation of petroporphyrin derivatives led to the removal of up to 74% of Ni and 95% of V.

According to FTIR spectra, the chemically modified cytochrome c catalyzed the oxidation of sulfur and carbon atoms in asphaltene molecules (García-Arellano et al., 2004). The enzymatic treatment of asphaltene constituents is an interesting alternative for the removal of heavy metals. It would result in reduced catalyst poisoning during hydrotreatment and cracking processes. On the other hand, the introduction of polar groups in asphaltene molecules could positively affect their sedimentation properties and improve their behavior.

It has also been reported (Uribe-Alvarez et al., 2011) that a fungus isolated from a natural asphalt lake is able to grow using asphaltene constituents as the sole source of carbon and energy.

Thus, a fungal strain isolated from a microbial consortium growing in a natural asphalt lake is able to grow in purified asphaltene constituents as the only source of carbon and energy. The asphaltene constituents were rigorously purified in order to avoid contamination from other petroleum fractions. In addition, most of petroporphyrin derivatives were removed. The 18S rRNA and b-tubulin genomic sequences, as well as some morphologic characteristics, indicate that the isolate is *N. fischeri*. After

11 weeks of growth, the fungus is able to metabolize 15.5% of the asphaltenic carbon, including 13.2% transformed to carbon dioxide. In a medium containing asphaltene constituents as the sole source of carbon and energy, the fungal isolate produces extracellular lactase activity, which is not detected when the fungus grows in a rich medium. The results obtained in this work clearly demonstrate that there are microorganisms able to metabolize and mineralize asphaltene constituents, which is considered the most recalcitrant petroleum fraction.

To overcome the shortcomings of conventional methods, microbial degradation of asphaltene has been accepted worldwide as the most promising environmentally sound technology for remediation of spills and discharges related to petroleum and petroleum products.

Furthermore, bacterial metabolites (especially polysaccharides) are of great value as enhancers of oil recovery due to their surfactant activity and bio-emulsifying properties (Banat, 1995). Because the conditions in oil deposits are often saline, the use of salt-resistant metabolites may be advantageous to the recovery of oil. Furthermore, hypersaline water and soil are often contaminated with crude oils, heavy metals, or other toxic compounds from anthropogenic sources. However, conventional microbiological treatment processes do not function at high salt concentrations; therefore, the use of moderately halophilic bacteria should be considered (Hao and Lu, 2009).

In the past, biodegradation of asphaltene constituents through the use of a microbial consortium or mixed cultures isolated from soil samples, sediments contaminated with hydrocarbons and oil wells have taken place but in low proportions of 0.55%–3.5% (Venkateswaran et al., 1995; Thouand et al., 1999). This is most likely due to the complex molecular structure of asphaltene constituents (Speight, 2014) which makes these molecular species resistant to biodegradation thereby causing their accumulation in ecosystem where petroleum and its refining byproducts are spilled in either accidental or purposeful ways (Guiliano et al., 2000).

The focus of many studies has generally been bioremediation of sites contaminated by total petroleum hydrocarbons (Iturbe et al., 2007; Machackova et al., 2008) and there is a general lack of detailed work on the biodegradation of asphaltene constituents. However, more recently, viscosity reduction by asphaltene degradation has been structurally characterized by FTIR. The work was focused on reduction of viscosity of heavy oil in order to improve enhanced recovery from the reservoir or deposit. The bacterium (*G. petrolearia* TERIG02) also showed an additional preference to degrade toxic asphalt and aromatics compounds first unlike the other known strains. Furthermore, these characteristics make the species *G. petrolearia* TERIG02 a potential candidate for residua and asphalt biodegradation and a solution to degrading toxic aromatic compounds (Lavania et al., 2012).

In contrast to low-molecular-weight hydrocarbons, polycyclic aromatics and hydrocarbons included in the asphaltene fraction are usually considered as being only slightly biodegradable because of their insufficient availability to microbial attack (Gibson and Subramanian, 1984; Cerniglia, 1992; Kanaly and Harayama, 2000). Among the pentacyclic triterpane derivatives, the hopane constituents are so stable that they are commonly used as ubiquitous biomarkers for the assessment of biodegradation levels of crude oil (Ourisson et al., 1979). They were shown to be only slightly biodegraded by specialized microflorae under laboratory conditions (Frontera-Suau et al., 2002).

The mechanism of the degradation is complex but is believed to be a sequential process in which *n*-alkane moieties are generally removed first, followed by the degradation of *iso*-alkane moieties, cycloalkane moieties, one-to-three ring aromatics, and finally polyaromatics (Greenwood et al., 2008). However, the typical pattern of degradation varies with different bacteria, as well as type and composition of oil (Greenwood et al., 2008; Zrafi-Nouira et al., 2009).

Microbes reduce the viscosity by degrading high molecular weight constituents into lower molecular weight constituents such as biological surface active substances, acids, and gases. In addition, anaerobic fermentation leads to the production of acids, carbon dioxide, hydrogen, and alcohols. Anaerobic bacteria produce acetate and butyrate during the initial growth phase (acidogenic phase) of the fermentation process.

As the culture moves to the second phase of fermentation, the stationary growth phase, there is a shift in the metabolism of the cells to solvent production (solvent-generation phase). These gaseous and liquid metabolites dissolve into the oil resulting in reduced viscosity (Bryant et al., 1998). Moreover, the reaction of asphalt degradation within an acidic background is preferable as the proton (H^+) effectively interacts with the polar functionalities in the asphaltene constituents and resin constituents thereby efficiently reducing the polar interactions which result in breaking the intermolecular associations that exist in the raw residuum or asphalt.

Furthermore, when each of the fractions namely aliphatic, aromatic, and asphaltic (asphaltene and resin) fractions were treated with *G. petrolearia* TERIG02 for 30 days it was found that maximum degradation was in the case of asphalt followed by aromatic fraction (Lavania et al., 2012). Indications were that *G. petrolearia* TERIG02 could tolerate the toxicity of these compounds and was capable in utilizing them as a carbon and energy source.

In addition, five asphaltene degrading bacterial strains were obtained from crude oil and polluted soil samples of Dorood oil field in the south of Iran. Maximum degradation ability of 46% and 48% were observed by *B. lentus* and the mixed culture of five selected isolates at 28 C, respectively. Statistical optimization of asphaltene biodegradation was successfully carried out and the optimum values of pH, salinity, and asphaltene concentration for asphaltene biodegradation at 40°C (104°F) were obtained for pure cultures of *B. lentus*. The kinetic study showed that a good model fit the biodegradation of asphaltene constituents.

6.0 CHALLENGES AND OPPORTUNITIES

Although the main effort in R&D of petroleum biotechnology has been directed toward bioremediation and biodegradation, there are also other new concerns about the biotransformation of heavy crude oils into lighter ones. That would take place through BDS, BDN, BDM, and biodepolymerization of asphaltene constituents. However, the most advanced area is the BDS, for which pilot plants have been established, while, the other technologies are still at the level of basic research. The main motivations toward crude oil upgrading and desulfurization prior to refining are the economics of higher oil value, lower processing costs, and reduced air pollution that causes smog, acid rain, and global warming.

The main advantages of the application of biotechnology in petroleum industry are as follows: the potential simplicity of the process, the selectivity of the biocatalysts (whatever, whole cells, free or immobilized enzymes), low pressure, low temperature, no hydrogen required, low chemical costs, and minimal equipment investment when compared to the conventional refinery operations.

Besides whole-cell catalysts, another plausible option offered by biotechnology is enzyme-based catalysis. Enzyme-based catalysts have several advantages over whole-cell catalysts. In terms of robustness, an enzyme may function in very low water content environments; in contrast with cells, an enzyme may even be less labile to thermal and organic solvent denaturation if the environment is hydrophobic; in terms of design, it is usually more straightforward to modulate the kinetic and stability properties of a single protein than the properties of an enzymatic cascade, intrinsically constrained by cell metabolism; finally, multiple strategies, such as immobilization, solvent engineering, and protein engineering, may be combined in order to enhance the desired characteristics of the enzyme. Heavy oils require temperatures of 60–100°C for processing as free-flowing fluids. Thus, there is a great effort to reach for microbes and enzymes that can work at those temperatures.

REFERENCES

Adarme, R., Sughrue, E.L., Johnson, M.M., Kidd, D.R., Phillips, M.D., Shaw, J.E., 1990. Demetallization of asphaltenes: thermal and catalytic effects with small pore catalysts. American Chemical Society 35, 614–618.

Agarwal, P., Sharma, D.K., 2010. Comparative studies on the bio-desulfurization of crude oil with other desulfurization techniques and deep desulfurization through integrated processes. Energy and Fuels 24 (1), 518–524.

Al-Bahry, S.N., Al-Wahaibi, Y.M., Balqees Al-Hinai, B., Joshi, S.J., Elshafie, A.E., Al-Bemani, A.S., Al-Sabahi, J., 2016. Potential in heavy oil biodegradation via enrichment of spore forming bacterial consortia. Journal of Petroleum Exploration and Production Technology 6, 787–799.

Ali, M.A., Abbas, S., 2006. A review of methods for the demetallization of residual fuel oils. Fuel Processing Technology 87, 573–584.

Ali, H.R., El-Gendy, N.Sh., Moustafa, Y.M., Mohamed, I., Roushdy, M.I., Hashem, A.I., 2012. Degradation of asphaltenic fraction by locally isolated halotolerant bacterial strains. ISRN Soil Science 2012:435485.

Al-Shahristani, H., Al-Atyia, M.J., 1972. Vertical migration of oil in Iraqi oil fields: evidence based on vanadium and nickel concentrations. Geochimica et Cosmochimica Acta 36, 929–938.

Al-Sulaimani, H., Joshi, S., Al-Wahaibi, Y., Al-Bahry, S., Elshafie, A., Bemani, A.A., 2011. Microbial biotechnology for enhancing oil recovery: current developments and future prospects. Biotechnology, Bioinformatics and Bioengineering 2, 147–158.

Altgelt, K.H., Boduszynski, M.M., 1994. Composition and Analysis of Heavypetroleum Fractions (Chemical Industries), first ed. Marcel Dekker Inc., NY, USA.

Alves, L., Paixão, S.M., Pacheco, R., Ferreira, A.F., Silva, C.M., 2015. Biodesulphurization of fossil fuels: energy, emissions and cost analysis. RSC Advances 5, 34047–34057.

Ancheyta, J., Speight, J.G., 2007. Hydroprocessing of Heavy Oils and Residua. CRC Press, Taylor & Francis Group, Boca Raton, Florida.

Ancheyta, J., 2016. Deactivation of Heavy Oil Hydroprocessing Catalysts: Fundamentals and Modeling. Wiley, Hoboken.

Aribike, D.S., Susu, A.A., Nwachukwu, S.C.U., Kareem, S.A., 2009. Microbial desulfurization of diesel by *Desulfobacterium* aniline. Academia Arena 1 (4), 11–17.

Armstrong, S.M., Sankey, B.M., Voodoo, G., 1995. Conversion of dibenzothiophene to biphenyl by sulfate reducing bacteria isolated from oil field production facilities. Biotechnology Letters 17, 1133–1136.

Atkins, L., Higgins, T., Barnes, C., 2010. Heavy Crude Oil Global Analysis and Outlook to 2030. Hart energy consulting report, Houston, Texas, USA).

Atlas, R.M., Aislabie, J., 1992. Process for Biotechnological Upgrading of Shale Oil. US Patent No. 5,143,827.

Atlas, R.M., Boron, D.J., Deever, W.R., Johnson, A.R., McFarland, B.L., Meyer, J.A., 2001. Method for Removing Organic Sulfur from Heterocyclic Sulfur Containing Organic Compounds. US Patent H1,986.

Ayala, M., LeBorgne, S., 2010. Microorganisms utilizing sulfur containing hydrocarbons. In: Timmis, K.N. (Ed.), Handbook of Hydrocarbon and Lipid Microbiology. Springer, Berlin.

Ayala, M., Vazquez-Duhalt, R., Morales, M., Le Borgne, S., 2016. Application of microorganisms to the processing and upgrading of crude oil and fractions. In: Lee, S.Y. (Ed.), Consequences of Microbial Interactions with Hydrocarbons, Oils, and Lipids: Production of Fuels and Chemicals. Handbook of Hydrocarbon and Lipid Microbiology. Springer International Publishing, AG.

Ayala, M., Verdin, J., Vazquez-Duhalt, R., 2007. The prospects for peroxidase-based biorefining of petroleum fuels. Biocatalysis and Biotransformation 25 (2–4), 114–129.

Bachman, R.T., Johnson, A.C., Edyvean, R.G.J., 2014. Biotechnology in the petroleum industry: an overview. International Biodeterioration and Biodegradation 86, 225–237.

Baldi, F., Pepi, M., Fava, F., 2003. Growth of *Rhodosporidium toruloides* strain DBVPG 6662 on dibenzothiophene crystals and orimulsion. Applied and Environmental Microbiology 69 (8), 4689–4696.

Ball, J.S., Wenger, W.J., Hyden, H.J., Horr, C.A., Myers, A.T., 1960. Metal content of twenty-four petroleums. Journal of Chemical and Engineering Data 5, 553–557.

Banat, I.M., 1995. Biosurfactants production and possible uses in microbial enhanced oil recovery and oil pollution remediation: a review. Bioresource Technology 51, 1–12.

Bartholdy, J., Hannerup, P.N., 1990. Hydrodemetallation in reside hydroprocessing. American Chemical Society 35, 619–625.

Benedik, M.J., Gibbs, P.R., Riddle, R.R., Willson, R.C., 1998. Microbial denitrogenation of fossil fuels. Trends in Biotechnology 16, 390–395.

Bertrand, J.C., Rambeloarisoa, E., Rontani, J.F., Giusti, G., Mattei, G., 1983. Microbial degradation of crude oil in sea water in continuous culture. Biotechnology Letters 5, 567–572.

BGR, 2015. Energy study 2015. Reserves, Resources and Availability of Energy Resources. Federal Institute of Geoscience and Natural Resources, Hannover, Germany.

Bordoloi, N.K., Rai, S.K., Chaudhuri, M.K., Mukherjee, A.K., 2014. Deep-desulfurization of dibenzothiophene and its derivatives present in diesel oil by a newly isolated bacterium *Achromobacter* sp. to reduce the environmental pollution from fossil fuel combustion. Fuel Processing Technology 119, 236–244.

Bryant, R.S., Bailey, S.A., Step, A.K., Evans, D.B., Parli, J.A., 1998. Biotechnology for heavy oil recovery. In: SPE Paper No. 36767 Proceedings. SPE/DOE Improved Oil Recovery Symposium, pp. 1–7.

Bublitz, F., Guenther, T., Fritsche, W., 1994. Screening of fungi for the biological modification of hard coal and coal derivatives. Fuel Processing Technology 40, 347–354.

Callejas, M.A., Martinez, M.T., Fierro, J.L.G., Rial, C., Jiménez-Mateos, J.M., Gómez-Garcia, F.J., 2001. Structural and morphological study of metal deposition on an aged hydrotreating catalyst. Applied Catalysis A: General 220, 93–104.

Cerniglia, C.E., 1992. Biodegradation of polycyclic aromatic hydrocarbons. Biodegradation 3, 351–368.

Chang, J.H., Rhee, S.K., Chang, Y.K., Chang, H.N., 1998. Desulfurization of diesel oils by a newly isolated dibenzothiophene-degrading *Nocardia* sp. strain CYKS2. Biotechnology Progress 14, 851–855.

Coyle, C., Siskin, M., Ferrunhelli, D.T., Logan, M.S.P., Zylstra, G., 2000. Biological Activation of Aromatics of Chemical Processing And/or Upgrading of Aromatic Compounds, Petroleum, Coal, Resin, Bitumen, and Other Petrochemical Streams. United States Patent 6,156,946.

Dehaghani, A.H.S., Badizad, M.H., 2016. Experimental study of Iranian heavy crude oil viscosity reduction by diluting with heptane, methanol, toluene, gas condensate and naphtha. Petroleum 2, 415–424.

Demirbas, A., Bafail, A., Nizami, A.-S., 2016. Heavy oil upgrading: unlocking the future fuel supply. Petroleum Science and Technology 34, 303–308.

Dinamarca, M.A., Ibacache-Quiroga, C., Baeza, P., Galvez, S., Villarroel, M., Olivero, P., Ojeda, J., 2010. Biodesulfurization of gas oil using inorganic supports biomodified with metabolically active cells immobilized by adsorption. Bioresource Technology 101, 2375–2378.

Duarte, G.F., Rosado, A.S., Seldin, L., de Araujo, W., van Elsas, J.D., 2001. Analysis of bacterial community structure in sulfurous-oil-containing soils and detection of species carrying dibenzothiophene desulfurization (dsz) genes. Applied and Environmental Microbiology 67, 1052–1062.

Duyck, C., Miekeley, N., Fonseca, T.C.O., Szatmari, P., Santos Neto, E.V.S., 2008. Trace element distributions in biodegraded crude oils and fractions from the Potiguar basin, Brazil. Journal of Brazilian Chemical Society 19 (5), 978–986.

Duyck, C., Miekeley, N., Porto da Silveira, C.L., Aucélio, R.Q., Campos, R.C., Grinberg, P., Brandão, G.P., 2007. Trace element distributions in biodegraded crude oils and fractions from the Potiguar basin. Brazilian Spectrochimica Acta Part B 62, 939–951.

Éigenson, A.S., Ivchenko, E.G., 1977. Distribution of sulfur and nitrogen in fractions from crude oil and residues. Chemistry and Technology of Fuels and Oils 13, 542–544.

El-Gayar, M., Mostafa, M.S., Abdelfattah, A.E., Barakat, A.O., 2002. Application of geochemical parameters for classification of crude oils from Egypt into source-related types. Fuel Processing Technology 79, 13–28.

El-Gendy, N.Sh., Farahat, L.A., Moustafa, Y.M., Shaker, N., El-Temtamy, S.A., 2006. Biodesulfurization of crude and diesel oil by *Candida parapsilosis* NSh45 isolated from Egyptian hydrocarbon polluted sea water. Biosciences, Biotechnology Research Asia 3 (1a), 5–16.

El-Gendy, N.Sh., 2001. Biodesulfurization of Organosulfur Compounds in Crude Oil and its Fractions MSc Thesis. Department of Chemistry, Cairo University, Cairo, Egypt.

El-Gendy, N.Sh., 2004. Biodesulfurization Potentials of Crude Oil by Bacteria Isolated from Hydrocarbon Polluted Environments in Egypt Ph.D. Thesis. Department of Chemistry, Cairo University, Cairo, Egypt.

El-Gendy, N.Sh., Nassar, H.N., Abu Amr, S.S., 2014. Factorial design and response surface optimization for enhancing a biodesulfurization process. Petroleum Science and Technology 32 (14), 1669–1679.

El-Gendy, N.Sh., Speight, J.G., 2015. Handbook of Refinery Desulfurization. CRC Press, Taylor & Francis, Boca Raton.

El-Gendy, N.Sh., 2015. Biodesulfurization of petroleum and its fraction. In: Pant, K.K., Sinha, S., Bajpai, S., Govil, J.N. (Eds.), Advances in Petroleum Engineering. Chemical Technology Series. Vol. 4 Advances in Petroleum Engineering II: Petrochemical, Studium Press LLC, Houston, USA, pp. 655–680 (Chapter 24).

Elliot, J.D., 1996. Delayed Coker Design and Operation: Recent Trends Andinnovations. Foster Wheeler USA Corporation.

EPA, 2014. Control of air pollution from motor vehicles: tier 3 motor vehicle emission and fuel standards; final rule. Federal Register. 79, 23413–23886. www.gpo.gov/fdsys/pkg/FR-2014-04-28/pdf/2014-06954.pdf.

ExxonMobil, 2016. Global transportation demand by fuel. In: The Outlook for Energy, a View to 2040, Outlook for Energy Charts. http://corporate.exxonmobil.com/en/energy/energy-outlook/charts2016/global-transportation-demand-by-fuel?parentId=d7323290-c766-440a-8e68-094d67a30841.

Fadhil, A.M.A., Al-Jailawi, M.H., Mahdi, M.S., 2014. Isolation and characterization of a new thermophilic, carbazole degrading bacterium (*Anoxybacillus rupiensis*) Strain Ir3 (JQ912241). International Journal of Advanced Research 2, 795–805.

Fallon, R.D., Hnatow, L.L., Jackson, S.C., Keeler, S.J., 2010. Method for Identification of Novel Anaerobic Denitrifying Bacteria Utilizing Petroleum Components as Sole Carbon Source. US patent 7740063 B2.

Farag, A.S., Sif El-Din, O.I., Youssef, M.H., Hassan, S.I., Farmawy, S., 1989. Solvent demetallization of heavy oil residue. Hungarian Journal of Industrial Chemistry 17 (3), 289–294.

Fedorak, P.M., Semple, K.M., Vazquez-Duhalt, R., Westlake, D.W.S., 1993. Chloroperoxidase-mediated modifications of petroporphyrins and asphaltenes. Enzyme Microbiology and Technology 15, 429–437.

Finnerty, W.R., 1993. Symposium on: bioremediation and bioprocessing presented before the division of petroleum chemistry, Inc. In: 205th National Meeting, American Chemical Society, Denver, Co, pp. 282–285.

Finnerty, W.R., 1992. Biodegradation 2, 223–226.

Finnerty, W.R., Singer, M.E., 1983. Microbial enhancement of oil recovery. Biotechnology 1, 47–54.

Fish, R.H., Komlenic, J.J., Wines, B.K., 1984. Characterization and comparison ofvanadyl and nickel compounds in heavy crude petroleums and asphaltenes by reverse-phase and size-exclusion liquid chromatography/graphite furnace atomic absorption spectrometry. Analytical Chemistry 56, 2452–2460.

Foght, J.M., 2004. Whole-cell bioprocessing of aromatic compounds in crude oil and fuels. In: Vazquez-Duhalt, R., Quintero-Ramirez, R. (Eds.), Studies in Surface Science and Catalysis: Petroleum Biotechnology: Developments and Perspectives, vol. 151. Elsevier, Amsterdam, pp. 145–175.

Folsom, B.R., Schieche, D.R., DiGrazia, P.M., Werner, J., Palmer, S., 1999. Microbial desulfurization of alkylated dibenzothiophenes from a hydrodesulfurized middle distillate by *Rhodococcus erythropolis* I-19. Applied and Environmental Microbiology 65, 4967–4972.

Fowler, D., Smith, R.I., Mullera, J.B.A., Hayman, G., Vincent, K.J., 2005. Changes in the atmospheric deposition of acidifying compounds in the UK between 1986 and 2001. Environmental Pollution 137, 15–25.

Frontera-Suau, R., Bost, F., McDonald, T., Morris, P.J., 2002. Aerobic biodegradation of hopanes and other biomarkers by crude oil degrading enrichment cultures. Environmental Science & Technology 36, 4585–4592.

Furuya, T., Ishii, Y., Noda, K., Kino, K., Kirimura, K., 2003. Thermophilic biodesulfurization of hydrodesulfurized light gas oils by *Mycobacterium phlei* WU-F1. FEMS Microbiology Letters 221 (1), 137–142.

García-Arellano, H., Buenrostro-Gonzalez, E., Vazquez-Duhalt, R., 2004. Biocatalytic transformation of petroporphyrins by chemical modified cytochrome C. Biotechnology and Bioengineering 85.

Gibson, D.T., Subramanian, V., 1984. Microbial degradation of aromatic hydrocarbons. In: Gibson, D.T. (Ed.), Microbial Degradation of Organic Compounds. McGraw–Hill, New York, pp. 181–252.

Gould, K.A., 1980. Oxidative demetallization of petroleum asphaltenes and residua. Fuel 59 (10), 733–736.

Government of Alberta, Canada, 2011. Oil Sands. Resource & Assessment. Alberta Energy [WWW document]. http://www.energy.alberta.ca/OilSands/1715.asp.

Gray, K.A., Pogrebinsky, O.S., Mrachko, G.T., Xi, L., Monticello, D.J., Squires, C.H., 1996. Molecular mechanisms of biocatalytic desulfurization of fossil fuels. Nature Biotechnology 14 (13), 1705.

Greenwood, P.F., Wibrow, S., George, S.J., Tibbett, M., 2008. Sequential hydrocarbon biodegradation in a soil from arid coastal Australia, treated with oil under laboratory-controlled conditions. Organic Geochemistry 39, 1336–1346.

Grossman, M.J., Lee, M.K., Prince, R.C., Garrett, K.K., Minak-Bernero, V., Pickering, I., 2001. Deep desulfurization of extensively hydrodesulfurized middle distillate oil by *Rhodococcus* sp. Strain ECRD-1. Applied and Environmental Microbiology 67, 1949–1952.

Grossman, M.J., Lee, M.K., Prince, R.C., Garrett, K.K., George, G.N., Pickering, I.J., 1999. Microbial desulfurization of a crude oil middle-distillate fraction: analysis of the extent of sulfur removal and the effect of removal on remaining sulfur. Applied and Environmental Microbiology 65, 181–188.

Guiliano, M., Boukir, A., Doumenq, P., Mille, G., 2000. Supercritical fluid extraction of BAL 150 crude oil asphaltenes. Energy and Fuels 14, 89–94.

Gunam, I.B.W., Yamamura, K., Sujaya, N., Antara, N.S., Aryanta, W.R., Tanaka, M., Tomita, F., Sone, T., Asano, K., 2013. Biodesulfurization of dibenzothiophene and its derivatives using resting and immobilized cells of *Sphingomonas subarctica* T7b. Journal of Microbiology and Biotechnology 23 (4), 473e482.

Gupta, R.K., Gera, P., 2015. Process for the upgradation of petroleum residue: review. International Journal of Advanced Technology in Engineering and Science 3 (2), 643–656.

Hall, C., Tharakan, P., Hallock, J., Cleveland, C., Jefferson, M., 2003. Hydrocarbons and the evolution of human culture. Nature 426, 318–322.

Hao, R., Lu, A., Zeng, Y., 2004. Effect on crude oil by thermophilic bacteria. Journal of Petroleum Science and Engineering 43, 247–258.

Hao, R., Lu, A., 2009. Biodegradation of heavy oils by halophilic bacterium. Progress in Natural Science 19, 997–1001.

Hernandez, A., Mellado, R., Martinez, J., 1998. Metal accumulation and vanadium induced multidrug resistance by environmental isolates of *Escherichia hermani* and *Enterobacter cloacae*. Applied and Environmental Microbiology 64, 4317–4320.

Hernández-López, E.L., Ayala, M., Vazquez-Duhalt, R., 2015. Microbial and enzymatic biotransformations of asphaltenes. Petroleum Science and Technology 33 (9), 1017–1029.

Hernández-López, E.L., Perezgasga, L., Huerta-Saquero, A., Mouriño-Pérez, R., Vazquez-Duhalt, R., 2016. Biotransformation of petroleum asphaltenes and high molecular weight polycyclic aromatic hydrocarbons by *Neosartorya fischeri*. Environmental Science and Pollution Research 23, 10773–10784.

Hessley, R.A., 1990. Fuel Science and Technology Handbook. In: Speight, J.G. (Ed.). Marcel Dekker, Inc., New York, NY, USA.

Hinkle, A., Shin, E.J., Liberatore, M.W., Herring, A.M., Batzle, M., 2008. Correlating the chemical and physical properties of a set of heavy oils from around the world. Fuel 87, 3065–3070.

Hobson, G.D., Pohl, W., 1975. Modern Petroleum Technology, fourth ed. Applied Science Publishing Co., London, UK.

Hofrichter, M., Bublitz, F., Fritsche, W., 1997. Fungal attack on coal: I. Modification of hard coal by fungi. Fuel Processing Technology 52, 43.

Hutzler, M.J., Sitzer, S., Holtberg, P.D., Conti, J., Kendell, J.M., Kydes, A.S., 2003. Annual Energy Outlook 2003 with Projection to 2025. The United States Energy Information Administration. US Department of Energy, Washington, DC, USA.

International Energy Agency, 2015. Key World Energy Statistics. OECD/IEA, Paris, France.

Ishii, Y., Kozaki, S., Furuya, T., Kino, K., Kirimura, K., 2005. Thermophilic biodesulfurization of various heterocyclic sulfur compounds and crude straight-run light gas oil fraction by a newly isolated strain *Mycobacterium phlei* WU-0103. Current Microbiology 50, 63–70.

Iturbe, R., Flores, C., Castro, A., Torres, L.G., 2007. Sub-soil contamination due to oil spills in zones surrounding oil pipeline-pump stations and oil pipeline right-of-ways in southwest-Mexico. Environmental Monitoring and Assessment 133, 387–398.

Jahromi, H., Fazaelipoor, M.H., Ayatollahi, S., Niazi, A., 2014. Asphaltenes biodegradation under shaking and static conditions. Fuel 117, 230–235.

Kanaly, R.A., Harayama, S., 2000. Biodegradation of high-molecular weight polycyclic aromatic hydrocarbons by bacteria. Journal of Bacteriology 182, 2059–2067.

Kashima Oil Co., Ltd. Japan, 1983. Removal of Heavy Metals from Heavy Petroleum Oil. JP Patent No. 58096681. Assigned to Kashima Oil Co., Ltd. Japan).

Kaufman, E.N., Harkins, J.B., Borole, A.P., 1998. Comparison of batch stirred and electrospray reactors for biodesulfurization of dibenzothiophene in crude oil and hydrocarbon feedstocks. Applied Biochemistry and Biotechnology 73, 127–144.

Kayser, K.J., Kilbane, J.J., 2004. Method for Metabolizing Carbazole in Petroleum. US Patent 6943006.

Kikuchi, M., Ohnishi, K., Harayama, S., 1999. Novel family shuffling methods for the in vitro evolution of enzymes. Gene 236, 159–167.

Kilbane, J.J., 2005. Biotechnological upgrading of petroleum. In: Ollivier, B., Magot, M. (Eds.), Petroleum Microbiology. ASM Press, Washington, USA, pp. 239–256.

Kilbane, J.J., Ribeiro, C.M.S., Linhares, M.M., 2003. Bacterial Cleavage of Only Organic C-n Bonds of Carbonaceous Materials to Reduce Nitrogen Content. US patent 6541240 B1.

Kilbane, J.J., Ranganathan, R., Cleveland, L., Kayser, K.J., Ribiero, C., Linhares, M.M., 2000. Selective removal of nitrogen from quinoline and petroleum by *Pseudomonas ayucida* IGTN9M. Applied and Environmental Microbiology 66 (2), 688–693.

Kilbane, J.J., 2006. Microbial biocatalyst developments to upgrade fossil fuels. Current Opinion in Biotechnology 17, 305–314.

Kim, H.Y., Kim, T.S., Kim, B.H., 1990. Degradation of organic compounds and the reduction of dibenzothiophene to biphenyl and hydrogen sulfide by *Desulfovibrio desulfuricans* M6. Biotechnology Letters 12, 761–764.

Kim, J.-S., Crowley, D.E., 2007. Microbial diversity in natural asphalts of the rancho La Brea tar pits. Applied and Environmental Microbiology 73, 4579–4591.

Kirkwood, K.M., Foght, J.M., Gray, M.R., 2004. Prospects for Biological Upgrading of Heavy Oils and Asphaltenes. Petroleum Biotechnology: Developments and Perspectives (Studies in Surface Science and Catalysis), Vol. 151. Elsevier, Netherlands, pp. 113–145 (Chapter 4).

Konishi, J., Ishii, Y., Okumura, K., Suzuki, M., 2000. High Temperature Desulfurization by Microorganisms. US Patent 6,130,081.

Kotlar, H.K., Brakstad, O.G., Markussen, S., Winnberg, A., 2004. Use of petroleum biotechnology throughout the value chain of an oil company: an integrated approach. In: Vazquez-Duhalt, R., Quintero-Ramirez, R. (Eds.), Studies in Surface Science and Catalysis: Petroleum Biotechnology: Developments and Perspectives, Vol. 151. Elsevier, Amsterdam, pp. 1–27.

Kutowy, O., Tweddle, T.A., Hazlett, J.D., 1989. Method for the Molecular Filtration of Predominantly Aliphatic Liquids. US Patent No. 4,814,088. (Assigned to National Research Council of Canada).

Lacotte, D.J., Mille, G., Acquaviva, M., Bertrand, J.C., 1996. Arabian light 150 asphaltene biotransformation with n-alkanes as co-substrate. Chemosphere 32, 1755–1761.

Lavania, M., Cheema, S., Sarma, P.M., Mandal, A.K., Lal, B., 2012. Biodegradation of asphalt by *Garciaella petrolearia* TERIG02 for viscosity reduction of heavy oil. Biodegradation 23, 15–24.

Le Borgne, S., Quintero, R., 2003. Biotechnological processes for refining of petroleum. Fuel Processing Technology 81, 155–169.

Leliveld, R.G., Eijsbouts, S.E., 2008. How a 70-year-old catalytic refinery process is still ever dependent on innovation. Catalysis Today 130, 183–190.

Leòn, V., Kumar, M., 2005. Biological upgrading of heavy crude oil. Biotechnology and Bioprocess Engineering 10, 471–481.

Leòn, V., Fuenmayor, S., DeSisto, A., Marcano, A., Munoz, S., Rivas, A., 2003. Isolation of bacteria strains capacities in cracking and desulfurization of heavy crude oil. In: Proceeding of 2nd ICPB the Development and Prospective of Biotechnology Applied to the Oil Industry. November 57. Mexico City, Mexico.

Li, F., Xu, P., Feng, J., Meng, L., Zheng, Y., Luo, L., Ma, C., 2005. Microbial desulfurization of gasoline in a *Mycobacterium goodii* X7B immobilized-cell system. Applied and Environmental Microbiology 71 (1), 276–281.

Li, W., Jiang, X., 2013. Enhancement of bunker oil biodesulfurization by adding surfactant. World Journal of Microbiology & Biotechnology 29 (1), 103–108.

Li, Y.G., Ma, J., Zhang, Q.Q., Wang, C.S., Chen, Q., 2007. Sulfur-selective desulfurization of dibenzothiophene and diesel oil by newly isolated *Rhodococcus erythropolis* NCC-1. Chinese Journal of Organic Chemistry 25, 400–405.

Linguist, L., Pacheco, M., 1999. Enzyme-based diesel desulfurization process offers energy, CO_2 advantages. Oil and Gas Journal 22, 45–48.

Machackova, J., Wittlingerova, Z., Vlk, K., Zima, J., Ales, L., 2008. Comparison of two methods for assessment of in situ jet fuel remediation efficiency. Water Air and Soil Pollution 187, 181–194.

McFarland, B.L., 1999. Biodesulfurization. Current Opinion in Microbiology 2, 257–264.

Mogollón, L., Rodríguez, R., Larrota, W., Ortiz, C., Torres, R., 1998. Biocatalytic removal of nickel and vanadium from petroporphyrins and asphaltenes. Applied Biochemistry and Biotechnology 70–72, 765–777.

Mohebali, G., Ball, A.S., 2016. Biodesulfurization of diesel fuels – past, present and future perspectives. International Biodeterioration and Biodegradation 110, 163–180.

Monot, F., Abbad-Andaloussi, S., Warzywoda, M., 2002. Biological Culture Containing *Rhodococcus erythropolis* and/or *Rhodococcus rhodnii* and Process for Desulfurization of Petroleum Fraction. United States Patent 6,337,204.

Monticello, D.J., 2000. Biodesulfurization and upgrading of petroleum distillates. Current Opinion in Biotechnology 11, 540–546.

Montiel, C., Quintero, R., Aburto, J., 2009. Petroleum biotechnology: technology trends for the future. African Journal of Biotechnology 8 (12), 2653–2666.

Morales, M., Le Borgne, S., 2014. Protocols for the isolation and preliminary characterization of bacteria for biodesulfurization and biodenitrogenation of petroleum-derived fuels. In: McGenity, T.J., Timmis, K.N., Nogales, B. (Eds.), Hydrocarbon and Lipid Microbiology Protocols. Springer, Berlin.

Moustafa, Y.M.M., El-Gendy, N.Sh., Farahat, L.A., Abo-State, M.A., El-Temtamy, S.A., 2006. Biodesulfurization of ras badran crude oil and its constituents with special emphasis on its asphaltene fraction. Egyptian Journal of Petroleum 15 (1), 21–30.

Naranjo, L., Urbina, H., De Sisto, A., León, V., 2007. Isolation of autochthonous non-white rot fungi with potential for enzymatic upgrading of Venezuelan extra-heavy crude oil. Biocatalysis and Biotransformation 25, 1–9.

Nassar, H.N., Deriase, S.F., El-Gendy, N.Sh., 2016. Modeling the relationship between microbial biomass and total viable count for a new bacterial isolate used in biodesulfurization of petroleum and its fractions. Petroleum Science and Technology 34 (11–12), 980–985.

OPEC, 2009. OPEC Annual Statistical Bulletin 2009. [WWW document]. URL http://www.opec.org/opec_web/en/.

Ourisson, G., Albrecht, P., Rohmer, M., 1979. The hopanoids. paleochemistry and biochemistry of a group of natural products. Pure and Applied Chemistry 51, 709–729.

Osterhuber, E., 1989. Upgrading Heavy Oils by Solvent Dissolution and Ultrafiltration. US Patent No. 4,7 97,200. (Assigned to Exxon R&E Company).

Pacheco, M.A., Lange, E.A., Pienkos, P.T., Yu, L.Q., Rouse, M.P., Lin, Q., Linquist, L.K., 1999. Recent advances in desulfurisation of diesel fuel. In: NPRA Annual Meeting. NPRA AM-99-27, San Antonio, pp. 1–26.

Panariti, N., Del Bianco, N.A., Del Piero, G., Marchionna, M., 2000. Petroleumresidue upgrading with dispersed catalysts. Part 1. Catalysts activity and selectivity. Applied Catalysis A: General 204 (2), 203–213.

Paul, J.A.K., Smith, M.L., 2009. Method for Purification of Uncatalyzed Natural Fuels from Metal Ions by Means of at Least One Hemeprotein and Use of the at Least on Hemeprotein. United States Patent 8,475,652.

Pendrys, J.P., 1989. Biodegradation of asphalt cement-20 by aerobic bacteria. Applied and Environmental Microbiology 55 (6), 1357–1362.

Peng, P., Morales-Izquierdo, A., Hogg, A., Strausz, O.P., 1997. Molecular structure of Athabasca asphaltene: sulfide, ether, and ester linkages. Energy Fuels 11, 1171–1187.

Pineda Flores, G., Bollarguello, G., Mestahoward, A., 2004. A microbial mixed culture isolated from a crude oil sample that uses asphaltenes as a carbon and energy source. Biodegradation 15, 145–151.

Piskorz, J., Radlein, D., Majerski, P., Scott, D., 1996. (. US Patent No. 5,496,464. Hydrotreating of Heavy Hydrocarbon Oils in Supercritical Fluids Assigned to Natural Resources Canada).

Premuzic, E.T., Lin, M.S., Lian, H., Zhou, W.M., Yablon, J., 1997. The use of chemical markers in the evaluation of crude bioconversion products, technology, and economic analysis. Fuel Processing Technology 52, 207–223.

Premuzic, E.T., Lin, M.S., 1999a. Biochemical Upgrading of Oils. US patent No. 5,858,766.

Premuzic, E.T., Lin, M.S., 1999b. Induced biochemical conversions of heavy crude oils. Journal of Petroleum Science and Engineering 22 (1–3), 171–180.

Premuzic, E.T., Lin, M.S., Bohenek, M., Zhou, W.M., 1999. Bioconversion reactions in asphaltenes and heavy crude oils. Energy and Fuels 13 (2), 297–304.

Resnick, S.M., Lee, K., Gibson, D.T., 1996. Diverse reactions catalyzed by naphthalene dioxygenase from *Pseudomonas* sp. strain NCIB 9816. Journal of Industrial Microbiology 17 (5), 438–457.

Reynolds, G.J., Biggs, W.R., Bezman, S.A., 1987. Removal of heavy metals fromresidual oils. ACS Symposium Series 344, 205–219.

Rollmann, D.L., Walsh, D.E., 1980. Visbreaking Process for Demetallation and Desulfurization of Heavy Oil. US Patent 4203830. (Assigned to Mobil Oil Corp., New York, USA).

Rontani, J.F., Bosser-Joulak, F., Rambeloarisoa, E., Bertrand, J.C., Faure, G.R., 1985. Analytical study of asphalt crude oil and asphaltenes biodegradation. Chemosphere 14, 1413–1422.

Rossini, S., 2003. The impact of catalytic materials on fuel reformulation. Catalysis Today 77, 467–484.

Salehizadeh, H., Mousavi, M., Hatamipour, S., Kermanshahi, K., 2007. Microbial demetallization of crude oil using *Aspergillus* sp.: vanadium oxide octaethyl porphyrin (VOOEP) as a model of metallic petroporphyrins. Iranian Journal of Biotechnology 5 (4), 226–231.

Santos, S.C., Alviano, D.S., Alviano, C.S., Padula, M., Leitao, A.C., Martins, O.B., Ribeiro, C.M., Sassaki, M.Y., Matta, C.P., Bevilaqua, J., Sebastian, G.V., Seldin, L., 2006. Characterization of *Gordonia* sp. strain F.5.25.8 capable of dibenzothiophene desulfurization and carbazole utilization. Applied Microbiology and Biotechnology 71, 355–362.

Santos, S.C., Alviano, D.S., Alviano, C.S., Padula, M., Leitao, A.C., Martins, O.B., Ribeiro, C.M., Sassaki, M.Y., Matta, C.P., Bevilaqua, J., 2005. Characterization of *Gordonia* sp. strain F.5.25.8 capable of dibenzothiophene desulfurization and carbazole utilization. Applied Microbiology and Biotechnology 66, 1–8.

Savastano, C.A., 1991. Solvent extraction approach to petroleum demetallation. Fuel Science and Technology International 9 (7), 833–871.

Singh, A., 2012. How specific microbial communities benefit the oil industry: biorefining and bioprocessing for upgrading petroleum oil. In: Whitby, C., Skovhus, T.L. (Eds.), Applied Microbiology and Molecular Biology in Oilfield Systems, pp. 121–178.

Speight, J.G., 1981. The Desulfurization of Heavy Oils and Residua. Marcel Dekker, Inc., New York, USA.

Speight, J.G., Arjoon, K.K., 2012. Bioremediation of Petroleum and Petroleum Products. Scrivener Publishing, Beverly, Massachusetts.

Speight, J.G., 2013a. Heavy Oil Production Processes. Gulf Professional Publishing, Elsevier, Oxford, United Kingdom.

Speight, J.G., 2013b. Heavy and Extra Heavy Oil Upgrading Technologies. Gulf Professional Publishing, Elsevier, Oxford, United Kingdom.

Speight, J.G., 2014. The Chemistry and Technology of Petroleum, fifth ed. CRC Press, Taylor & Francis Group, Boca Raton, Florida.

Speight, J.G., 2015. Handbook of Petroleum Product Analysis, second ed. John Wiley & Sons Inc., Hoboken, New Jersey.

Stanislaus, A., Marafi, A., Rana, M.S., 2010. Recent advances in the science and technology of ultra-low sulfur diesel (ULSD) production. Catalysis Today 153, 1–6.

Sugaya, K., Nakayama, O., Hinata, N., Kamekura, K., Ito, A., Yamagiwa, K., Ohkawa, A., 2001. Biodegradation of quinoline in crude oil. Journal of Chemical Technology and Biotechnology 76, 603–611.

Tavassoli, T., Mousavi, S.M., Shojaosadati, S.A., Salehizadeh, H., 2012. Asphaltene biodegradation using microorganisms isolated from oil samples. Fuel 93, 142–148.

The International Council on Clean Transportation, 2016. A Technical Summary of Euro 6/VI Vehicle Emission Standards. http://www.theicct.org/sites/default/files/publications/ICCT_Euro6-VI_briefing_jun2016.pdf.

Thouand, G., Bauda, P., Oudot, J., Kirsch, G., Sutton, C., Vidalie, J.F., 1999. Laboratory evaluation of crude oil biodegradation with commercial or natural microbial inocula. Canadian Journal of Microbiology 45, 106–115.

Tilstra, L., Eng, G., Olson, G.J., Wang, F.W., 1992. Reduction of sulphur from polysulphidic model compounds by the hyperthermophilic *Archaebacterium Pyrococcus furiosus*. Fuel 71, 779–783.

Torkamani, S., Shayegan, J., Yaghmaei, S., Alemzadeh, I., 2009. Annual Report of 'heavy Crude Oil Biodesulfurization Project' Initiated by Petroleum Engineering Development Company (PEDEC), a Subsidiary of National Iranian Oil Company.

Uribe-Alvarez, C., Ayala, M., Perezgasga, L., Naranjo, L., Urbina, H., Vazquez-Duhalt, R., 2011. First evidence of mineralization of petroleum asphaltenes by a strain of *Neosartorya fischeri*. Microbial Biotechnology 4, 663–672.

US Congress, 1976. Public Law FEA-76–4. United States Library of Congress, Washington, DC.

Van Hamme, J.D., Fedorak, P.M., Foght, J.M., Gray, M.R., Dettman, H.D., 2004. Use of a novel fluorinated organosulfur compound to isolate bacteria capable of carbon-sulfur bond cleavage. Applied and Environmental Microbiology 70, 1487–1493.

Vartivarian, D., Andrawis, H., 2006. Delayed coking schemes are most economical for heavy-oil upgrading. Oil and Gas Journal 104 (6), 52–56.

Vazquez-Duhalt, R., Torres, E., Valderrama, B., Le Borgne, S., 2002. Will biochemical catalysis impact the petroleum refining industry? Energy and Fuels 16, 1239–1250.

Venkateswaran, K., Hoaki, T., Kato, M., Maruyama, T., 1995. Microbial degradation of resins fractionated from Arabian light crude oil. Canadian Journal of Microbiology 41, 418–424.

Wyndham, R.C., Costerton, J.W., 1981. In vitro microbial degradation of bituminous hydrocarbons and in situ colonization of bitumen surfaces within the Athabasca oil sands deposit. Applied and Environmental Microbiology 41, 791–800.

Xu, G.-W., Mitchell, K.W., Moticello, D.J., 1997. Process for Demetallizing a Fossil Fuel. United States Patent 5,624,844.

Xu, G.-W., Mitchell, K.W., Moticello, D.J., 1998. Fuel Product Produced by Demetallizing a Fossil Fuel with an Enzyme. United States Patent 5,726,056.

Yamada, Y., Matsumoto, S., Kakiyama, H., Honda, H., 1979. Japanese patent No. JP patent 54110206, (Assigned to Agency of Industrial Sciences and Technology, Japan).

Yamada, K.O., Morimoto, M., Tani, Y., 2001. Degradation of dibenzothiophene by sulfate-reducing bacteria cultured in the presence of only nitrogen gas. Journal of Bioscience Bioengineering 91, 91–93.

Yanto, D.H.Y., Tachibana, S., 2013. Biodegradation of petroleum hydrocarbons by a newly isolated *Pestalotiopsis* sp. NG007. International Biodeterioration and Biodegradation 85, 438–450.

Yanto, D.H.Y., Tachibana, S., 2014. Potential of fungal co-culturing for accelerated biodegradation of petroleum hydrocarbons in soil. Journal of Hazardous Materials 278, 454–463.

Yi, N., Xue, G., HogShuai, G., XiangPing, Z., SuoJiang, Z., 2014. Simultaneous desulfurization and denitrogenation of liquid fuels using two functionalized group ionic liquids. Science China 57 (12), 1766–1773.

Yu, B., Xu, P., Shi, Q., Ma, C., 2006. Deep desulfurization of diesel oil and crude oils by a newly isolated *Rhodococcus erythropolis* strain. Applied and Environmental Microbiology 72, 54–78.

Yu, L., Meyer, T.A., Folsom, B.R., 1998. Oil/water/biocatalyst Three Phase Separation Process. US Patent No. 5,772,901.

Zrafi-Nouira, I., Guermazi, S., Chouari, R., Safi, N.M.D., Pelletier, E., Backhrouf, A., Saidane-Moshabi, D., Sghir, A., 2009. Molecular diversity analysis and bacterial population dynamics of an adapted seawater microbiota during the degradation of Tunisian Zarzantine Oil. Biodegradation 20, 467–486.

FURTHER READING

Bhatia, S., Sharma, D.K., 2010. Biodesulfurization of dibenzothiophene, its alkylated derivatives, and crude oil by a newly isolated strain of *Pantoea agglomerans* D23W3. Biochemical Engineering Journal 50 (3), 104–109.

Furukawa, Y., Ishimori, K., Morishima, I., 2000. Electron transfer reactions in Zn-substituted cytochrome P450cam. Biochemistry 39 (36), 10996–11004.

Lloyd, J.R., 2003. Microbial reduction of metals and radionuclides. FEMS Microbiology Reviews 27, 411–425.

Soleimani, M., Bassi, A., Margaritis, A., 2007. Biodesulfurization of refractory organic sulfur compounds in fossil fuels. Biotechnology Advances 25, 570–596.

BIOCATALYTIC DESULFURIZATION

1.0 INTRODUCTION

Although the percentage of energy obtained from fossil fuels decreased, over 82% of the world energy is still from fossil fuels, approximately half of which comes from crude oil. It has been reported that the largest estimated crude oil reserves are in Canada, Iran, and Kazakhstan and approximately 56% of the world's oil reserves are in the Middle East. However, there is a depletion of the high-quality low sulfur content low-density (high American Petroleum Institute [API] gravity) light crude oil coming with the increment of the production and use of high sulfur content heavy crude oil (Montiel et al., 2009; Srivastava, 2012; Alves et al., 2015). Crude oil is a complex mixture of different hydrocarbons and smaller amounts of nonhydrocarbons, such as derivatives of sulfur, nitrogen, oxygen, and traces of metallic constituents, suh as vanadium, nickel, iron, and copper, which have significant effects on crude oil refining (Table 6.1).

In the context of this chapter, the removal of sulfur compounds from petroleum-related sources is an important aspect of petroleum refining because of the contribution these contaminants make to the formation of sulfur oxides (SO_x) and hence to air pollution and acid rain. They also contribute to coke formation and to catalyst poisoning during the refining of crude oil, thus reducing process yields (Hsu and Robinson, 2006; Gary et al., 2007; Speight, 2011, 2014, 2017).

Sulfur is the most abundant element after carbon and hydrogen. The average sulfur content varies from 0.04% to 14% w/w in crude oil according to its type, geographical source, and origin (Voloshchuk et al., 2009). The sulfur content and the API gravity are the two properties that determine the quality and value of the crude oil. The sweet, i.e., the low sulfur content crude oil is characterized with sulfur content of <0.5% w/w, while the sour high sulfur content is characterized with sulfur content of >0.5%w/w. Most of the sulfur present in crude oil is organically bound sulfur and can be summarized in four groups: (1) thiol derivatives also called mercaptan derivatives, (2) sulfide derivatives, (3) disulfide derivatives, and (4) thiophene derivatives, which includes the derivatives of thiophene, benzo-thiophene, dibenzothiophene with hydrogen sulfide and elemental sulfur (Bahuguna et al., 2011; Speight, 2014). Crude oils with higher viscosity and density usually contain high amount of more complex sulfur compounds. Although, it is generally true that the sulfur content increases with the boiling point during distillation and sulfur concentration tends to increase progressively with increasing carbon number (Figs. 6.1 and 6.2). Thus, crude fractions in the fuel oil and asphalt boiling range have higher sulfur content than those in the jet and diesel boiling range, which in turn is characterized by higher sulfur content than those in the gasoline boiling range. However, the middle distillates may contain more sulfur than those of the higher boiling fractions because of decomposition of the higher molecular weight compounds during distillation (Speight, 2000, 2014).

The presence of sulfur compounds is very undesirable, because of their actual or potential corrosive nature, which would cause the corrosion of pipelines, pumping, and refining equipment.

Table 6.1 The Result of Sulfur Derivatives, Nitrogen Derivatives, and Metal Derivatives in Crude Oil (Speight, 2017)

Contaminant	Effect on Catalyst	Mitigation	Process
Sulfur	Catalyst fouling Deactivation of active sites	Hydrodesulfurization	Hydroprocessing
Nitrogen	Adsorption of basic nitrogen Destruction of active sites	Hydrodemetallization	Hydroprocessing
Metals	Fouling of active sites Fouling of catalyst pores	Demetallization	Demet, Met-X

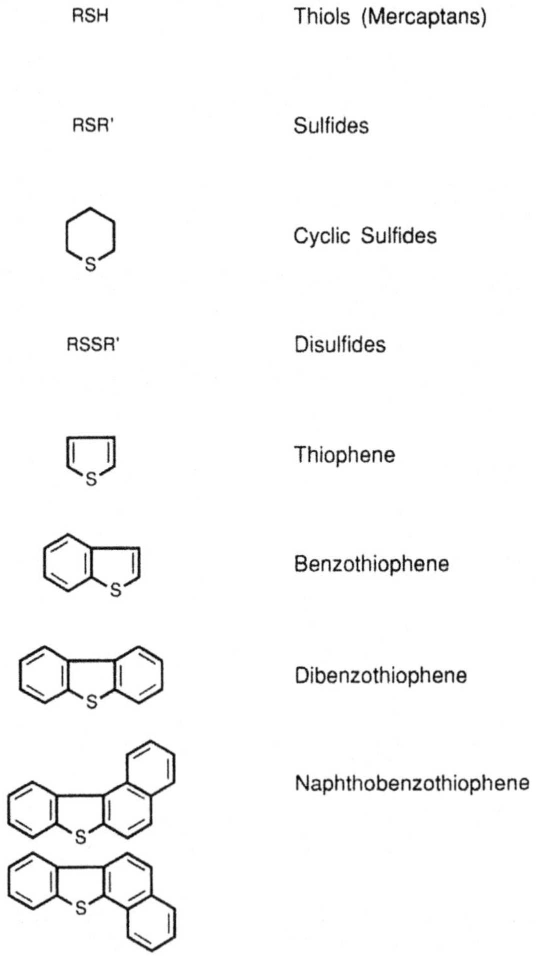

RSH   Thiols (Mercaptans)

RSR'   Sulfides

Cyclic Sulfides

RSSR'   Disulfides

Thiophene

Benzothiophene

Dibenzothiophene

Naphthobenzothiophene

FIGURE 6.1

General groups of the sulfur compounds that occur in crude oil.

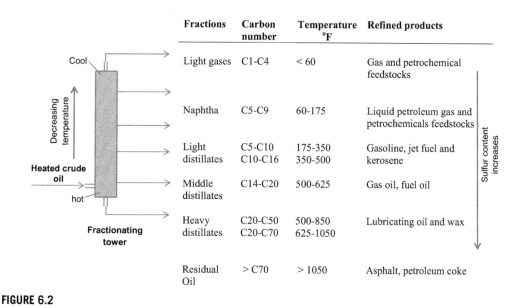

Fractions	Carbon number	Temperature °F	Refined products
Light gases	C1-C4	< 60	Gas and petrochemical feedstocks
Naphtha	C5-C9	60-175	Liquid petroleum gas and petrochemicals feedstocks
Light distillates	C5-C10 C10-C16	175-350 350-500	Gasoline, jet fuel and kerosene
Middle distillates	C14-C20	500-625	Gas oil, fuel oil
Heavy distillates	C20-C50 C20-C70	500-850 625-1050	Lubricating oil and wax
Residual Oil	> C70	> 1050	Asphalt, petroleum coke

FIGURE 6.2

Simplified schematic diagram for crude oil distillation, downstream processing, and distribution of sulfur content.

Sulfur compounds are undesirable in refining processes as they tend to deactivate some catalysts used in downstream processing and upgrading of hydrocarbons (Chengying et al., 2002). In liquid petroleum products, sulfur compounds would contribute to the formation of gummy deposits that would plug the filters of the fuel-handling systems of automobiles and other engines. Sulfur compounds in fuel oils would cause corrosion to the parts of the internal combustion engines and consequently its breakdown and premature failure (Collins et al., 1997). Moreover, hydrogen sulfide present in crude oil and its low boiling distillates is one of the most corrosive sulfur compounds that would destroy paintings and buildings. Hydrogen sulfide is a highly toxic gas, its threshold limit value is 10 ppm, and personal exposure to hydrogen sulfide level (1000–2000 ppm) would kill within 10 s. Furthermore, sulfur compounds have unfavorable influence on antiknock and oxidation characteristics. High concentration of sulfur in fuels dramatically decreases the efficiency and lifetime of emission gas treatment systems in cars (Mužic and Sertić-Bionda, 2013). Sulfur levels in automotive fuels have unfavorable poisoning effect on the catalytic converters in automotive engines, thus increasing the evolution of particulate matter, carbon monoxide (CO), carbon dioxide (CO_2), sulfur oxides (SO_x), nitrogen oxides (NO_x), and other combustion related emissions that would cause smog, global warming, and water pollution (Srivastava, 2012).

Emissions of sulfur compounds (such as sulfur dioxide, SO_2, and particulate matter continuing metal sulfates) formed during the combustion of petroleum products are the subject of environmental monitoring in all developed countries. These harmful emissions affect the stratospheric ozone, increasing the hole in the Earth's protective ozone layer (Denis, 2010). High levels of sulfur dioxide cause bronchial irritation, trigger asthma, and prolonged exposure can cause cardiopulmonary and lung cancer mortality (Mohebali and Ball, 2008). Incomplete combustion of fossil fuels causes emission of aromatic sulfur compounds. Oxidation of these compounds in the atmosphere would lead to the aerosol of sulfuric acid, which consequently would form acid rain and haziness that reduces the average temperature of affected area (Charlson et al., 1992). Thus

$$SO_2 + H_2O \rightarrow H_2SO_3 \text{ (sulfurous acid)}$$

$$2SO_2 + O_2 \rightarrow 2SO_3 \text{ (sulfur trioxide)}$$

$$SO_3 + H_2O \rightarrow H_2SO_4 \text{ (sulfuric acid)}$$

In addition, carbon dioxide and the oxides of nitrogen (Chapter 7) also contribute to the formation of acid rain:

$$CO_2 + H_2O \rightarrow H_2CO_3 \text{ (carbonic acid)}$$

$$NO + H_2O \rightarrow HNO_2 \text{ (nitrous acid)}$$

$$2NO + O_2 \rightarrow NO_2 \text{ (nitrogen dioxide)}$$

$$NO_2 + H_2O \rightarrow HNO_3 \text{ (nitric acid)}$$

Acid rain has a negative impact on living organisms, aquatic life, and agricultural crops and are responsible for corrosion of many infrastructures and destroy monuments, leather, and buildings, thus negatively affecting the economy (Grossman et al., 2001). The large-ring thiophenes exhibit some mutagenic and carcinogenic activities (Murphy et al., 1992). Observations and analyses associated with oil spills and chronic pollution have indicated that the polycyclic aromatic sulfur heterocyclic derivatives (polycyclic aromatic sulfur heterocyclic derivatives) are more resistant to microbial degradation and accumulated to a greater degree than the corresponding polycyclic aromatic hydrocarbon derivatives (Durate et al., 2001). As a result, the United States Environmental Protection Agency and other regulatory agencies worldwide have limited the amount of SO_2 released into the air and put strict regulations and limits for the sulfur content in the transportation fuels, for example, the sulfur content limits in the European Union (EU) reached 10 ppm since 2009, and 15 ppm for USA, since 2006 to ultraclean fuels <1, approaching zero-sulfur emissions upon fuel combustion (Zhang et al., 2011).

Consequently, refineries must have the capability to remove sulfur from crude oil and refinery streams to the extent needed to mitigate these unwanted effects and reach the restricted regulatory sulfur limits. The higher the sulfur content of the crude oil, the greater the required degree of sulfur control and the higher the associated cost. Nowadays, the applicable desulfurization methods in petroleum industry have its own deficiencies. From the economic point of view, it is important to maximize the sulfur removal keeping the energy content of the fuel unaltered and turn the waste into valuable products to be reused, reduce the low-value products, and increase the high-value ones, with the full utilization of the existing facilities, to provide quick cost-effective returns. Therefore, petroleum industry moves towards the application of bioprocesses as complementary technologies on diverse platforms to reduce investment and maintenance costs. Since, biotechnology involves many advantages, including milder temperature and pressure conditions, cleaner and selective processes, low emission, and no production of undesirable by-products.

The major potential applications of biorefining are biodesulfurization, biodenitrogenation, bio-demetallization, and biotransformation of heavy crude oils into lighter (lower-density, higher-API gravity) crude oils. The most advanced area is biodesulfurization for which pilot plants exist (Le Borgne and Quintero, 2003). Biodesulfurization was developed in the United States by the Enchira Biotechnology Corporation (ENBC; formerly Energy Biosystems Corporation) and patented in 1990s (Kilbane, 1994; Monticello and Kilbane, 1994; Monticello, 1996; Monticello et al., 1996). After that, a lot of patents have been developed; the top five patent holders are ENBC (21 patents), the Japanese Petroleum Energy Center (4 patents), US-based Institute of Gas Technology (4 patents), the Korean Advanced Institute of Science and Technology (3 patents), and Exxon Research and Engineering Company (2 patents) (Bachmann et al., 2014).

This chapter provides a state-of-the-art as well as the challenges and opportunities for research and development in biodesulfurization technology with special emphasis on application of nanotechnology to enhance the commercialization of biodesulfurization process in the petroleum industry.

2.0 DESULFURIZATION

There are various reported desulfurization methods to remove sulfur from fossil fuels. Among these, hydrodesulfurization (HDS) is currently considered as the most important one. Desulfurization by ionic liquids, selective adsorption or reactive adsorption, oxidative desulfurization, and biological methods have also shown good potential to be a possible substitution for hydrodesulfurization technology or to be used in line with hydrodesulfurization in future refining systems (Srivastava, 2012; Mužic and Sertić-Bionda, 2013).

2.1 HYDRODESULFURIZATION

The hydrodesulfurization process in a refinery is a process designed to remove sulfur from the feedstocks, which can, depending upon the structure of the sulfur-containing and nitrogen-containing constituents, occur concurrently with removal of nitrogen from the feedstock:

$$[S]_{feedstock} + H_2 \rightharpoonup Hydrocarbon\ product + H_2S$$

$$[N]_{feedstock} + H_2 \rightharpoonup Hydrocarbon\ product + NH_3$$

Thus, HDS is the most common technology used by refineries to remove sulfur from crude oil and its distillates (gasoline, kerosene, and diesel oil)—it is a technology that converts organic sulfur compounds to hydrogen sulfide (hydrogen sulfide) under high temperature (290–455°C; 555 to 850°F) and high pressure (150–3000 psi) and uses hydrogen gas in presence of metal catalysts, e.g., $CoMo/Al_2O_3$ or $NiMo/Al_2O_3$. The produced hydrogen sulfide is then catalytically air oxidized to elemental sulfur (Speight, 2014). Oil refiners depend on such a costly extreme chemical process to treat approximately 20 million barrels of crude oil per day (Gupta et al., 2005). The reactivity of organosulfur compounds varies widely depending on their structure and local sulfur atom environment. The low-boiling crude oil fraction contains mainly the aliphatic organosulfur compounds: mercaptans, sulfides, and disulfides. They are very reactive in a conventional hydrotreating process and can easily be completely removed from the fuel. For higher boiling crude oil fractions such as heavy straight run naphtha, straight run diesel and low-boiling naphtha from the fluid catalytic cracking unit (light FCC naphtha), the organo-sulfur compounds predominantly contain thiophene rings. These compounds include thiophene derivatives and benzothiophene (BT) derivatives. Compounds containing the thiophene moiety are more difficult to convert via hydrotreating than mercaptan derivatives and sulfide derivatives. The higher molecular weight fractions blended to the gasoline and diesel pools—bottom FCC naphtha, coker naphtha, FCC and coker diesel—contain mainly alkyl benzothiophene derivatives, dibenzothiophene derivatives, as well as polycyclic aromatic sulfur heterocyclic derivatives, i.e., the least reactive sulfur-containing molecules in the hydrodesulfurization reaction (Fig. 6.3). Deep desulfurization of the fuels implies that more and more of the least reactive sulfur compounds must be converted.

Recently, the refining industry has made a great deal of progress towards developing more active catalysts and more economical processes to remove sulfur from gasoline and diesel oil (La Paz Zavala and Rodriguez, 2004). For example, California refineries are producing gasoline that contains approximately 29 ppm sulfur, despite the high sulfur content of California and Alaska crude oils used as feed stocks up to 11,000 ppm (US EPA, 2000). Due to incentives and regulations, 10 ppm sulfur diesel fuel has been commercially available in Sweden for several years. But although the hydrodesulfurization process tends to improve diesel quality by raising its cetane number, it decreases gasoline quality by lowering its octane number (Song and Ma, 2003).

Additional investments are required to achieve low or near-zero sulfur products to meet the international environmental regulations for producing ultraclean transportation fuels, which will need higher temperatures and pressures with larger consumption of energy as well as new catalysts to desulfurize the most recalcitrant molecules, leading to increased operations and capital costs as well as more carbon dioxide emissions (Castorena et al., 2002; Babich and Moulijn, 2003). In addition to desulfurization, this will result in demetallization, carbon residue reduction, some denitrogenation, hydrocracking, and cocking (Song, 2003). Consequently, there is a significant interest in low-cost desulfurization technologies that might complement hydrodesulfurization. Biocatalytic desulfurization (biodesulfurization) is the one of the most promising technologies as it operates at ambient temperature and pressure with high selectivity, resulting in decreased energy costs, low emission, and no generation of undesirable side products.

2.2 ADSORPTIVE DESULFURIZATION

Adsorption has become a key separation technique in industry and particularly in the oil and gas industry and the process can be used for desulfurization. There are two types of adsorptive desulfurization (ADS): ADS during which physical and/or chemical adsorption of organosulfur compounds takes place on the surface of the adsorbent throughout the van der Waals and electrostatic interactions, and the reactive ADS through which organosulfur compounds react with certain chemical species on adsorbent surface, via chemisorption π-complexation and then the sulfur reacts with adsorbent producing sulfide and the hydrocarbons released free of sulfur into the steam (Xu et al., 2014). Adsorbents used industrially are generally synthetic microporous solids such as (1) activated carbon, (2) molecular sieve carbon, (3) activated alumina, (4) silica gel, (5) zeolite derivatives, (6) metal oxides, and (7) clay. They are usually agglomerated with binders in the form of beads, extrudates, and pellets of a size consistent with the application that is considered (Song and Ma, 2003). The efficiency of ADS is mainly influenced by the adsorption properties: capacity, selectivity, stability, and ability to regeneration (Babich and Moulijn, 2003).

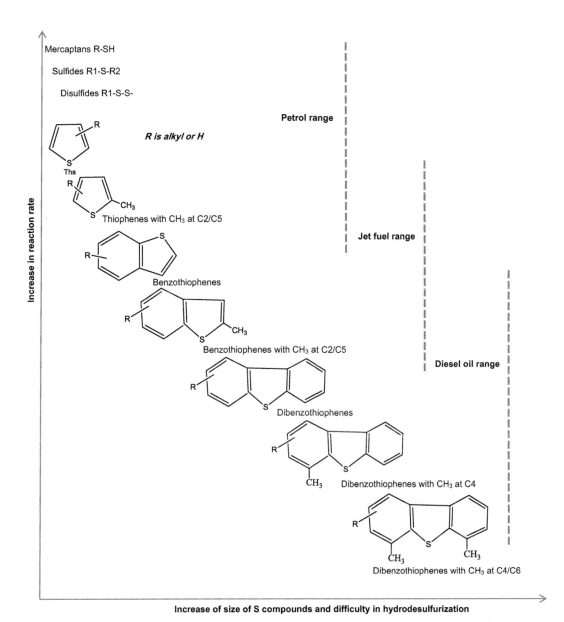

FIGURE 6.3

Reactivity of various organosulfur compounds in hydrodesulfurization process versus the molecular weight and positions of alkyl substitutions.

Activated carbon, carbon fibers, and carbon nanotube can be used in ADS processes where the adsorption would be physical that depend on the size and volume of the pores or chemical that depends on the chemisorption surface properties (Bagreev et al., 2004). Activated carbon is widely used, due to its good porosity, high surface area, high efficiency for adsorption, and can be used in batch vessel reactor, continuous stirred reactor, and fixed or fluidized bed reactor. Lee et al. (2002) investigated the ADS of diesel oil on 10 different activated carbons and reported that coconut shell–based carbon activated by high temperature was more effective for ADS than coal-based or wood-based carbons. The use of rice husk activated carbon for adsorptive removal of sulfur compounds, e.g., dibenzothiophenes from kerosene, was reported (Shimizu et al., 2007). Mužic et al. (2011) reported the desulfurization of diesel fuel in fixed bed adsorption process using activated carbon SOL-CARB C3 (Chemviron Carbon, Belgium) that yielded an output with sulfur concentration < 0.7 ppm. The commercial activated carbon and carbonized date palm kernel powder was also reported for adsorptive removal of sulfur compound from diesel fuel (Al Zubaidy et al., 2013).

Zeolite derivatives are efficiently used as adsorbents for selective adsorption of polar compounds and heterogenic organic compounds; they are hydrophilic and have large voids inside their structures (Mužic et al., 2010). The modified Y-type zeolite is popularly used as adsorbents to remove sulfur from fuels via π-complexation. It has been reported that cuprous-exchanged (Cu^+) and silver-exchanged (Ag^+) Y-typed zeolites are effective to remove sulfur compounds from gasoline (Yang et al., 2001; Takahashi et al., 2002; Hernández-Maldonado and Yang, 2004a, 2004b). McKinley and Angelici (2003) reported selective removal of dibenzothiophene and 4,6-dimethyldibenzothiophene from simulated feedstock with Ag^+/SBA-15 and Ag^+/SiO_2 as adsorbents. It has also reported (Velu et al., 2005) the selective adsorption process for removing sulfur at ambient temperature to achieve ultraclean diesel and gasoline with Ni-exchanged Y zeolite. In the USA, the S-Zorb process for reactive adsorption desulfurization was developed. This process is based on fluidized bed technology, where it can decrease sulfur in feedstock containing greater than 2 mg/mL sulfur to very low level less than 0.005 mg/mL in presence of hydrogen within operating temperature and pressure of 300–400°C (570–750°F) and 1.9–3.45 MPa, respectively. The sulfur compounds adsorbed on the adsorbent, reacted with reduced metals, producing metal sulfides and newly formed hydrocarbon that would be released into the main stream. The spent adsorbent is continuously removed from the reactor and transported into the regeneration chamber. The sulfur is removed from the surface of the adsorbent by burning and the formed sulfur dioxide is sent to the sulfur plant. The adsorbent is then reduced with hydrogen and recycled back to the reactor (Mužic and Sertić-Bionda, 2013). However, depending upon the process parameters, it is likely that the S-Zorb process also affects the hydrocarbon skeleton of dibenzothiophene and its derivatives (Stanislaus et al., 2010).

The preparation of nanoferrites (e.g., Fe_3O_4, $NiFe_2O_4$, $CuFe_2O_4$, and $MnFe_2O_4$) by a reverse (water/oil) microemulsion method, containing cetyltrimethylammonium bromide, 1-butanol, cyclohexane, and a metal salt solution, using ammonium hydroxide as a coprecipitating agent has also been reported Zaki et al., 2013). The complex nickel-iron oxide ($NiFe_2O_4$) expressed the highest adsorption capacity towards dibenzothiophene (166.3 μmol dibenzothiophene/g adsorbent), due to the high specific surface area of the oxide (233.4 m^2/g), large pore volume of approximately 0.3 cm^3/g, mesoporous framework, and strong acid sites.

However, the adsorption desulfurization process has issues that need to be addressed. There are large amounts of aromatics compared to the amounts of sulfur compounds in the fuel streams feedstock. Therefore, aromatics can also be adsorbed on the desulfurization adsorbents. Adsorbents should be well designed to achieve suitable selectivity; when the selectivity is low, the adsorbents are easy to be regenerated. But this can lead to the heat loss because of the comparative adsorption. As the selectivity increases, the spent adsorbents become more and more difficult to be regenerated (Hernández-Maldonado and Yang, 2004a, 2004b).

Solvent extraction and oxidation in the air are two methods to regenerate the desulfurization adsorbents. There are some disadvantages in these two methods. For solvent extraction method, it is difficult to separate sulfur compounds from the organic solvents and reuses these solvents. For calcinations method, sulfur compounds and aromatics are burned out, which can lose heat value of fuels. Regeneration can be also done by biological treatment (Li et al., 2008a).

Thermal processes such as visbreaking (or even hydrovisbreaking, i.e., visbreaking in an atmosphere of hydrogen or in the presence of a hydrogen donor material)—the long-ignored step-child of the refining industry—may see a surge in use as a pretreatment process (Radovanović and Speight, 2011; Speight, 2011, 2012, 2014, 2017). Management of the process to produce a liquid product that has been freed of the high potential for coke deposition (by taking the process parameters into the region where sediment forms) either in the absence or presence of (for example) a metal oxide scavenger could be valuable ally to catalyst cracking or hydrocracking units.

Scavenger additives such as metal oxides may also see a surge in use. As a simple example, a metal oxide (such as calcium oxide) has the ability to react with sulfur-containing feedstock to produce a hydrocarbon (and calcium sulfide):

$$\text{Feedstock[S]} + \text{CaO} \rightharpoonup \text{hydrocarbon product} + \text{CaS} + \text{H}_2\text{O}$$

A similar concept may also be applied to denitrogenation and demetallization processes (Chapter 7).

2.3 OXIDATIVE DESULFURIZATION

Oxidative desulfurization is an innovative technology that can be used to reduce the cost of producing ultralow sulfur diesel (ULSD) and is a process during which refractory organosulfur compounds are oxidized to sulfoxides and sulfone at relatively low temperature and at atmospheric pressure, in presence of oxidizing reagent, such as aqueous hydrogen peroxide, tert-butyl hydroperoxide, oxygen/aldehyde, and potassium ferrate. The reaction requires the presence of homogenous catalyst, such as formic acid, cobalt-manganese-nickel acetates, and acetic acid, or heterogeneous catalyst, such as tungsten/zirconia, titanium/mesoporous silica, peroxy-carboxylic acid, functionalized hexagonal mesoporous silica, and molybdenum-vanadium oxides supported on alumina, titania, ceria, niobia, silica, and cobalt-manganese-nickel oxides supported on alumina (Javadli and De Klerk, 2012). The produced sulfoxide and/or sulfones are subsequently removed from the treated feedstock by a separation method. Sulfone derivatives have higher polarity and molecular weight compared to the initial sulfur compounds and, thus, the sulfone derivatives are more easily removed from petroleum fractions by extraction, adsorption, distillation, and thermal decomposition (Mužic and Sertić-Bionda, 2013).

The oxidative desulfurization process has some technological and economic problems. Solvent extraction is commonly used; the solvent can be recovered and reused through a distillation process. However, an additional problem of product loss occurs. In addition, the oxidative desulfurization process is a source of sulfonic waste that requires special treatment. The oxidative desulfurization of dibenzothiophene using hydrogen peroxide (H_2O_2) by a recyclable amphiphilic catalyst has been reported Cui et al. (2012).

Magnetic silica nanospheres covered with complexes between 3-(trimethoxysilyl)-propyl dimethyl octa-decyl ammonium chloride and phosphotungestic acid, under mild reaction conditions 600 rpm and 50°C (122°F), and the amount of sulfur was decreased from 487 ppm to less than 0.8 ppm. The catalyst with oxidized products have been separated by external magnetic field and simply recycled by acetone eluent then successfully reused for three times, with almost constant activity.

The introduction of regulations stipulating ultralow sulfur content in fuels caused the peroxides such as hydrogen peroxide (H_2O_2) to become the most used oxidizing agents (Joskić et al., 2014) and molecular oxygen (Murata et al., 2003). The use of molecular oxygen may be appealing to refineries that already have the infrastructure for an oxidation facility to prepare blown asphalt. Other oxidizing agents such as in situ–formed per-acids; organic acids; phosphate acid and heteropolyphosphate acids; Fe-tetra amido macrocyclic ligand; Fenton and Fenton-like compounds; as well as solid catalysts such as those based on titanium-silica (tungsten-vanadium-titania, $W-V-TiO_2$); solid bases such as magnesium and lanthanum metal oxides; or hydrotalcite compounds, iron oxides, and oxidizing catalysts based on monoliths; and tert-butyl-hydroperoxide with catalyst support can effectively oxidize organic sulfur compounds into sulfones with less residue formation (Abdul Jalil and Falah Hassan, 2012; Mužic and Sertić-Bionda, 2013). Unipure Inc. has developed an oxidative desulfurization process (called ASR-2) based on peroxide (H_2O_2) oxidation in presence of formic acid (HCO_2H), within a short residence time (5 min), mild temperature (120°C, 250°F), and at approximately atmospheric pressure; the sulfones products can be separated by extraction or adsorption. However, it is known that homogeneous catalysts are difficult to separate from the reaction products and this limits their recycling. Thus, the preparation of new supported catalysts is the most desirable improvement of the oxidative desulfurization process (García-Gutiérrez et al., 2008). Currently, there are different catalysts (Wang and Yu, 2013; Kadijani et al., 2014; Mjalli et al., 2014) and ultrasound oxidative desulfurization (Wu and Ondruschka, 2010; Zhao and Wang, 2013; Liu et al., 2014; Wittanadecha et al., 2014) that are applied to improve reaction efficiency.

For example, ultrasonic-assisted catalytic ozonation combined with extraction process exhibits high catalytic efficiency for the removal of dibenzothiophene from simulated diesel oil (Zhao and Wang, 2013). SulphCO Inc. has developed oxidative desulfurization using ultrasound power during sulfur oxidation by hydrogen peroxide using tungsten phosphoric acid at 70–80°C (158–176°F) and atmospheric pressure, within residence time of 1 min; a desulfurization efficiency on the order of 80% and 90% occurred for crude oil and diesel oil, respectively (Dai et al., 2008). Lyondell Chemical Co and UOP LLC in cooperation with EniChem S.P.A. have developed oxidative desulfurization using *t*-butyl-hydroperoxide (TBHP), followed by adsorption or extraction for sulfone separation. This single liquid system is beneficial for simple reactor engineering, thus enabling the application of the fixed bed column. The major drawback of this process is the high price of TBHP and the waste treatment of tert-butyl alcohol and sulfone. The tert-butyl alcohols can be used as a potential octane-improving compounds for gasoline (Wang et al., 2003).

Compared to hydrodesulfurization, oxidative desulfurization has several advantages: (1) the refractory sulfur compounds, such as alkylated dibenzothiophene derivative, are easily oxidized under low operating temperature and pressure; (2) there is no use of expensive hydrogen, so the process is safer and can be applied in small and medium size refineries, isolated ones, and those located away from hydrogen pipelines. The overall capital cost and requirements for an oxidative desulfurization unit is significantly less than the capital cost for a deep hydrodesulfurization-unit (Gore, 2001; Guo et al., 2011).

Solvent extraction of the sulfoxide-sulfone derivatives using γ-butyrolactone, n-methyl pyrrolidone, methanol, dimethylformamide, acetonitrile, and furfural has been reported (Liu et al., 2008). In general, one of the major drawbacks of the solvent extraction method is the appreciable solubility of hydrocarbon fuels in polar solvents, which leads to significant losses of usable hydrocarbon fuel. Such a loss is completely unacceptable on a commercial basis. Beside this, sulfone derivatives are polar compounds and form strong bonding with polar solvents and it is difficult to remove them from the solvents to below 10 ppm w/w. Hence there will be build-up of sulfone derivatives in the solvent during solvent recovery (Nanoti et al., 2009).

On the other hand, the removal of sulfoxide-sulfone derivatives could be achieved using an adsorption technique (using adsorbents such as silica gel, activated carbon, bauxite, clay, coke, alumina, silicalite (polymorph of silica), ZSM-5, zeolite β, zeolite x, and zeolite y) in addition to the mesoporous oxide-based materials. These have attracted much attention in the recent years due to their large pore sizes and controlled pore size distribution that may be beneficial in allowing accessibility of large–molecular-size sulfone derivatives to the surface-active sites. However, one of the major drawbacks of the adsorption technique is that the amount of oil treated per unit weight of adsorbent is low (Babich and Moulijn, 2003).

In 1996, PetroStar Inc. combined conversion and extraction desulfurization to remove sulfur from diesel fuel (Chapados et al., 2000); briefly, the diesel fuel is oxidized by mixing with peroxyacetic acid (i.e., H_2O_2/acetic acid) at low temperature <100°C (<212°F) and under atmospheric pressure. Then liquid-liquid extraction takes place, producing low sulfur content diesel oil followed by adsorption treatment to yield ultralow sulfur diesel fuel. Recycling of extract solvent for reuse takes place and the concentrated extract is further processed to remove sulfur. The Lyondell Chemical Company used TBHP in a fixed-bed reactor for a cost-effective oxidative desulfurization process (Liotta and Han, 2003).

In recent years, the oxidative desulfurization concept has gained importance and is regarded as an excellent option after the hydrodesulfurization process, since the advantages of oxidative desulfurization—in comparison to conventional hydrodesulfurization—include the following: (1) the requirement of rather moderate reaction conditions, (2) there is no need for expensive hydrogen, and (3) the higher reactivity of aromatic sulfur compounds in oxidation reactions since the electrophilic reaction with sulfur atom is enhanced by higher electron density of aromatic rings (Javadli and De Klerk, 2012). Alkyl groups attached to the aromatic ring increase the electron density on the sulfur atom even more but the reactivity of the feedstock constituents to the process is also dependent on the type of catalyst. Thus, the alkyl-substituted dibenzothiophene derivatives are easily oxidized under low temperature and pressure conditions to form the corresponding sulfoxide derivatives and sulfone derivatives. While sulfur compounds such as disulfide derivatives that are easy to be hydrodesulfurized oxidized slowly. British Petroleum designed a process to avoid the use of organic or inorganic acids, where the oxidation of dibenzothiophene was studied with hydrogen peroxide using phosphotungstic acid as the catalyst and tetraoctylammonium bromide as the phase transfer agent in a mixture of water and toluene. The catalyzed decomposition of hydrogen peroxide competes with dibenzothiophene oxidation and, therefore, the major cost involved in treating gas oil fractions by oxidative desulfurization is the huge amount of hydrogen peroxide consumption; the sulfur-containing compounds that remain after hydrodesulfurization are particularly reactive toward oxidation and the consumption of hydrogen peroxide during this oxidation is proportional to the amount of sulfur to be oxidized (Collins et al., 1997). For this reason, oxidative desulfurization can be utilized as a second stage after existing hydrodesulfurization units, taking a low sulfur diesel (500 ppm) down to ultralow sulfur diesel (<10 ppm) levels. The efficiency and economics of an oxidative desulfurization process is strongly dependent on the methods used for oxidizing the sulfur compounds and successively the methods used for separating the sulfoxide and sulfone derivatives from the oxidized fuels. Thus, the key to successful implementation of oxidative desulfurization technology in most refinery applications is effectively integrating the oxidative desulfurization unit with the existing diesel hydrotreating unit in a revamp situation (Gatan et al., 2004).

Recently, the photo-oxidative desulfurization using TiO_2 (Tao et al., 2009) and ferric oxide (Li et al., 2012) has been reported. The order of reactivity of recalcitrant sulfur compounds in photo-oxidative desulfurization is the opposite of hydrodesulfurization, viz

dibenzothiophene < 4-methyl dibenzothiophene <4,6-dimethyl dibenzothiophene

This represents a beneficial advantage for application of photo-oxidative desulfurization, but there are some problems that should be solved to make it technically and economically feasible: the need for better solvents to increase S-compounds solubility and aromatic rejection and appropriate commercial techniques for solvent recovery. Thus, to increase the photo-transformation of S-compounds, a combination of a solvent and a photosensitizer should be optimized.

3.0 BIOCATALYTIC DESULFURIZATION

Although the concentrations of benzothiophene and dibenzothiophene are considerably decreased by hydrodesulfurization and it has been commercially used for a long time, it has several disadvantages (Hernández Maldonado and Yang, 2004c): (1) hydrodesulfurization of diesel feedstock for a low-sulfur product requires a larger reactor volume, longer processing times, and substantial hydrogen and energy inputs; (2) for refractory sulfur compounds, hydrodesulfurization requires higher temperature, pressure, and longer residence time. This makes the process costly due to the requirement of stronger reaction vessels and facilities; (3) the application of extreme conditions to desulfurize refractory compounds results in the deposition of carbonaceous coke on the catalysts; (4) exposure of crude oil fractions to severe conditions including temperatures above about 360°C decreases the fuel value of treated product; (5) deep hydrodesulfurization processes need large new capital investments and/or have higher operating costs; (6) the hydrogen sulfide that is generated poisons the catalysts and shortens their useful life; (7) deep hydrodesulfurization is affected by components in the reaction mixture such as organic hetero-compounds and polyaromatic hydrocarbons; (8) for older units, which are not competent to meet the new sulfur removal levels, erection of new hydrodesulfurization facilities and heavy load of capital cost is inevitable; (9) hydrodesulfurization removes paraffinic sulfur compound such as thiols, sulfides, and disulfides effectively. However, some aromatic sulfur-containing compounds such as 4- and 4,6-substituted dibenzothiophene and polycyclic aromatic sulfur heterocyclic derivatives are resistant to hydrodesulfurization and form the most abundant organosulfur compounds after hydrodesulfurization; (10) hydrogen atmosphere in hydrodesulfurization results in the hydrogenation of olefinic compounds and reduces the calorific value of the fuel. To increase the calorific value, the hydrodesulfurization-treated stream is sent to the FCC unit, which adds to the cost; and (11) although hydrodesulfurization is considered a cost-effective method for fossil fuel desulfurization, the cost of sulfur removal from refractory compounds by hydrodesulfurization is high because of the cost of hydrogen.

As already mentioned, hydrodesulfurization is not equally effective in desulfurizing all classes of sulfur compounds present in fossil fuels. On the other hand, the biodesulfurization process is effective regardless of the position of alkyl substituents (Pacheco, 1999). However, the hydrodesulfurization process conditions are sufficient not only to desulfurize sensitive (labile) organosulfur compounds, but also to (1) remove nitrogen and metals from organic compounds, (2) induce saturation of at least some carbon–carbon double bonds, (3) remove substances having an unpleasant smell or color, (4) clarify the product by drying it, and (5) improve the cracking characteristics of the material (Swaty, 2005). Therefore, in consideration of such advantages, placing a biodesulfurization-unit downstream of a hydrodesulfurization-unit as a complementary technology (rather than as a replacement technology) to achieve ultradeep desulfurization is a realistic consideration giving a multistage process for desulfurization of fossil fuels (Fang et al., 2006).

Biodesulfurization is a biological method in which microbes or enzymes are used as a catalyst to remove organosulfur compounds, especially the recalcitrant ones, e.g., dibenzothiophene and its derivatives. It can be performed aerobically or anaerobically. Its main disadvantage is that its conversion rate is much slower than hydrodesulfurization, since all the biological reactions are generally slower than the chemical reactions. There are three main routes for aerobic biodesulfurization (Fig. 6.4), complete mineralization where the end products are carbon dioxide (CO_2) and water (H_2O). In this process (the Kodama pathway) carbon-carbon bonds are cleaved and water-soluble by-products are produced, which would significantly inhibit microbial growth dibenzothiophene oxidation (Kilbane and Jackowski, 1992); in the 4S-pathway, the carbon skeleton is not destroyed and only sulfur is removed. The first two pathways are not recommended for desulfurization of fuels, as the efficiency of biodesulfurization depends on the biocatalyst capabilities to remove sulfur without altering the carbon skeleton or reducing the value of the fuel. But they are recommendable in bioremediation of oil spills and soil or sediments polluted with petroleum hydrocarbons (Gupta et al., 2005).

The anaerobic biodesulfurization of crude oil using the hydrogenase enzyme activity of sulfate reducing bacteria *Desulfovibrio desulfuricans* has been reported (Zobell, 1953). Anaerobic biodesulfurization proceeds more slowly than aerobic reactions, but generate the same products as the conventional hydrodesulfurization technology, hydrogen sulfide and desulfurized oil (Fig. 6.5). Furthermore, that mixed cultures containing sulfate-reducing bacteria will desulfurize a variety of model compounds such as dibenzothiophene, benzothiophene, dibenzothiophene, and dibenzothiophene (Kohler et al., 1984). A mixed methanogenic culture that was derived from a methanogenic sewage digester was used to reductively biodesulfurize benzothiophene derivatives to toluene ($C_6H_5CH_3$) with benzyl mercaptan ($C_6H_5CH_2SH$) as an intermediate in sulfur-limited medium (Miller, 1992).

In the specific desulfurization by *Desulfovibrio desulfuricans* M6, there is a desulfurization of 96% and 42% of benzothiophene and dibenzothiophene, respectively (Kim et al., 1990). Metabolite analyses proved that the strain could convert dibenzothiophene to biphenyl and hydrogen sulfide. M6 also removed 21% w/w of the sulfur in Kuwait crude oil and up to 17% w/w of sulfur from other crude oils and their distillate products. The extreme thermopile *Pyrococcus furiosus*, growing at 98°C (210°F), was shown to reduce organic polysulfide to hydrogen sulfide using the reduction of sulfur as a source of energy (Tilstra et al., 1992). A *Nocardioform actinomycete* FE9, grown on n-hexadecane with dibenzothiophene as the sole sulfur source, formed biphenyl, traces of 2-HBP, dihydroxybiphenyl, and sulfate under air atmosphere, but biphenyl, and hydrogen sulfide under nitrogen or hydrogen atmospheres, but the highest conversion rates were under a hydrogen atmosphere. This strain could remove both the

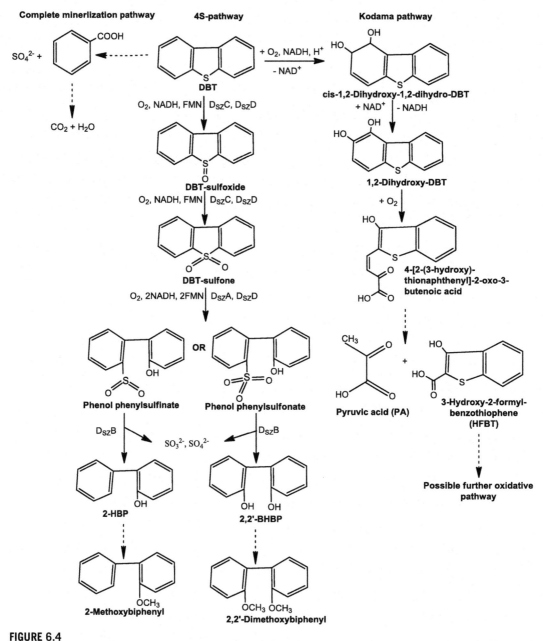

FIGURE 6.4

Aerobic pathways for biodesulfurization.

FIGURE 6.5

Anaerobic pathways for biodesulfurization.

sulfur heteroatoms in thianthrene, converting it to benzene and sulfate under oxygen atmosphere. Terthiophene (an oligomer of thiophene, $C_4H_3SC_4H_2SC_4H_3S$)

was transformed to hydrogen sulfide and unsaturated product tentatively identified as 1,3,5,7,9,11-dodecahexaene (Finnerty, 1993). Some anaerobic microorganisms, such as *Desulfomicrobium scambium* and *Desulfovibrio longreachii*, have been reported to desulfurize only about 10% of dibenzothiophene dissolved in kerosene. The gas-chromatographic analysis of samples showed unknown metabolites, indicating that the bacteria had possibly followed a pathway different from common anaerobic pathways (Yamada et al., 2001). An 81.5% anaerobic biodesulfurization of diesel oil with initial sulfur concentration of 166.034 ppm using an isolated bacterial strain, *Desulfobacterium indolicum* within 72 h has been reported (Kareem et al., 2012) and the ability of hyperthermophelic *Sulfolobus solfataricus* P2 to desulfurize dibenzothiophene and its derivatives has also been noted (Gun et al., 2015) but it was not possible to desulfurize benzothiophene at 78°C (172°F).

The desulfurization of petroleum under anaerobic conditions would be attractive because it avoids costs associated with aeration, it has the advantage of liberating sulfur. As a gas, hydrogen sulfide is treated with existing refinery desulfurization plants (e.g., Claus process) and does not liberate sulfate as a by-product that must be disposed by some appropriate treatment (Setti et al., 1997). Under anaerobic conditions, oxidation of hydrocarbons to undesired compounds such as colored and gum-forming products is minimal (McFarland, 1999). Moreover, anaerobic microorganisms use approximately 10% of total produced energy, while aerobic ones use 50% of the total produced energy (Marcelis, 2002). These advantages can be counted as incentives to continue research on reductive biodesulfurization. However, maintaining an anaerobic process is extremely difficult and the specific activity of most of the isolated strains have been reported to be insignificant for dibenzothiophenes (Armstrong et al., 1995). Due to low reaction rates, safety and cost concerns, and the lack of identification of specific enzymes and genes responsible for anaerobic desulfurization, anaerobic microorganisms that are sufficiently effective for commercial petroleum desulfurization have not been found yet, and anaerobic biodesulfurization process has not been developed. Consequently, aerobic biodesulfurization has been the focus of most the research in biodesulfurization (Le Borgne and Quintero, 2003).

The oxidation of dibenzothiophene by the Kodama pathway (Fig. 6.4) usually produces 3-hydroxy-2-formyl benzothiophene and pyruvic acid, up till now the steps after production of 3-hydroxy-2-formyl benzothiophene are not fully known, as it is not chemically stable and probably mineralized in nature (Bressler and Fedorak, 2001; El-Gendy and Abo State, 2008; Nassar, 2009). It has been reported that when methyl-dibenzothiophene is subjected to carbon-carbon bond cleavage by the Kodama pathway, the oxidation occurred in the benzene ring that has no substitutions, producing methyl-3-hydroxy-2-formyl benzothiophene (Soleimani et al., 2007). Many bacteria can oxidize mesophilically complex aromatic sulfur compounds with the formation of water-soluble sulfur-containing products via the Kodama pathway such as *Bacillus* sp., *Micrococcus* sp., *Pseudomonas* sp., *Rhizobium* sp., *Acinetobacter* sp., *Arthrobacter* sp., and a fungus *Cunninghamella elegans* (Kim et al., 1996). On the other hand, *Sulfolobus acidocaldarius* was reported to degrade dibenzothiophene via Kodama pathway at high temperature, 70°C (Kargi and Robinson, 1984).

The complete mineralization pathway (Fig. 6.4) was first reported by van Affferden et al. (1990), using *Brevibacterium* sp. DO. Sato and Clark (1995) reported mineralization of dibenzothiophene-sulfoxide and dibenzothiophene-sulfone but not dibenzothiophene. The complete mineralization of dibenzothiophene has been reported (El-Gendy, 2006) using the Gram-positive *Aureobacterium* sp. NShB1 and Gram-negative *Enterobacter* sp. NShB2, through the production of 2-hydroxybiphenyl (2-HBP) and 2,2′-bihydroxybiphenyl, respectively, as has reported the complete mineralization of dibenzothiophene using the halo-tolerant Gram-positive *Corynebacterium variabilis* sp. Sh42 (El-Gendy et al., 2010).

The 4S-pathway that removes the sulfur without altering the octane value of the fuels has been reported (Malik, 1978) but it did not take much concern because of (1) the high reserves of the lower-density (higher-API gravity) crude oil, (2) the flexibility of environmental rules, and (3) the profitability of conventional processes at that time. By the 1990s, due to the depletion of light (lower-density, higher-API gravity) crude oil reserves, increasing worldwide fuel demand, the more restrict environmental emission regulations, led to the exploration of new reservoirs of sulfur rich oils, e.g., heavy crude oil, tar sands, and oil shale. Consequently, petroleum industry shared efforts with academia to research and developed the applicability of biodesulfurization throughout the specific oxidative biodesulfurization (i.e., the 4S-pathway, Fig. 6.4). This pathway is so-called because four S-containing intermediates are involved in this pathway; sulfoxide, sulfone, sulfinate or sulfonate, and finally sulfate. This mechanism involves four enzymes: flavin-dependent dibenzothiophene monooxygenase ($D_{SZ}C$), dibenzothiophene-sulfone monooxygenase ($D_{SZ}A$), and a third enzyme required for the activity of $D_{SZ}C$ and $D_{SZ}A$, the flavin reductase ($D_{SZ}D$). The fourth enzyme is the desulfinase ($D_{SZ}B$), which completes the reaction, producing the phenolic product. There is an important fact that should be taken into consideration, that the desulfurization genes $D_{SZ}A$, $D_{SZ}B$, and $D_{SZ}C$ that encode the enzymes for desulfurization reaction are found in one operon in the order $D_{SZ}ABC$, while the $D_{SZ}D$ gene that encodes the flavin reductase is found unlinked in all the organisms reported so far. The produced hydroxybiphenyl is very soluble in oil and can easily find its way back to the petroleum fraction, conserving the calorific value of the fuel. Recently, some microorganisms expressed an extended 4S-pathway called 4SM-pathway, where 2-hydroxybiphenyl is methoxylated to 2-methoxybiphenyl (Xu et al., 2007). The production of 2-methoxybiphenyl and 2,2′-dimethoxy-1,1′-biphenyl as final end product of biodesulfurization of dibenzothiophene through the 4S-pathway using *Mycobacterium* sp. and *Rhodococcus erythropolis* HN2, respectively, has been reported (Chen et al., 2009; El-Gendy et al., 2014). The methoxylation pathway from 2-hydroxybiphenyl to 2-methoxybiphenyl has been examined (Ohshiro et al., 1996; Monticello, 2000; Kim et al., 2004) to investigate the potential to overcome the toxicity of 2-hydroxybiphenyl, since the production of 2-methoxybiphenyl would make less inhibitory effect on the microbe due to the loss of the hydroxy structure of phenolic compounds; this expresses lower inhibitory effect on growth and biodesulfurization efficiencies of microorganisms.

Several aerobic microorganisms were isolated for their sulfur-specific biodesulfurization capabilities (Table 6.2) out of these strains; *Rhodococcus rhodochrous* IGTS8, isolated and patented by Enchira Biotechnology Corporation, has been studied in detail and its metabolic route has been established through the identification of the enzymes and genes responsible for its desulfurizing activity (Denome et al., 1994; Piddington et al., 1995; Gray et al., 1996, 2003). Almost all sulfur-specific desulfurizing bacteria are Gram-positive; *Rhodococcus*, *Paenibacillus*, *Gordonia*, *Corynebacterium*, *Mycobacterium*, and *Lysinibacillus* species. However, there are few reports for Gram-negative sulfur-specific desulfurizing bacteria: *Desulfovibrio*, *Pseudomonas*, *Sphingomonas* species. Petroleum viscosity would decrease at elevated temperatures, which would consequently increase the biodesulfurization rate, due to the higher mass transfer (Torkamani et al., 2008), so researchers focus recently on isolating thermo-philic microorganisms (Table 6.2) to make biodesulfurization more applicable as a complimentary process to hydrodesulfurization. Since distillate fractions are often treated at high temperatures during hydrodesulfurization, there would be cost savings using thermophilic microorganisms, as there would be no need for cooling of the stock to 30°C (86°F).

One of the most important rate-limiting steps, and sometimes the only rate-limiting step, is the transfer of the polycyclic aromatic sulfur heterocyclic derivative compounds from the oil phase into the cell. The other limitation that might decrease or hinder the overall rate of biodesulfurization process, including substrate acquisition, is the need for reducing equivalents and the preference of the enzymes for specific substrates (Folsom et al., 1999; Gray et al., 2003). Generally, dibenzothiophene and

Table 6.2 List of Some of the Bacterial Isolates Capable of Selectively Desulfurizing Dibenzothiophene and Its Derivatives via the 4S-Pathway

Bacterium	References	Bacterium	References
Mesophilic Microorganisms			
Gordona strai CYSK1	Rhee et al. (1998)	*Rhodococcus erythropolis* Xp	Yu et al. (2006)
Lysinibacillus sphaericus DMT-7	Bahuguna et al. (2011)	*Rhodococcus* sp. *Iawq*	Ma et al. (2006)
Pseudomonas delafieldii R8	Huang et al. (2012)	*R. erythropolis* NCC1	Li et al. (2007a)
Pseudomonas putida CECT5279	Calzada et al. (2011)	*R. erythropolis* LSSE81	Xiong et al. (2007)
Rhodococcus sp. UM3	Purdy et al. (1993)	*R. erythropolis* FSD2	Zhang et al. (2007)
Rhodococcus sp. UM9	Purdy et al. (1993)	*Shewanella putrefaciens* NCIMB	Ansari et al. (2007)
Rhodococcus sp. ECRD1	Lee et al. (1995) Grossman et al. (2001)	*Sphingomonas* sp. AD109	Darzins and Mrachko (2000)
R. erythropolis H-2	Ohshiro et al. (1995)	*Sphingomonas subarctica* T7B	Gunam et al. (2006)
Rhodococcus sp. SY1	Omori et al. (1995)	*Xanthomonas* sp.	Constanti et al. (1994)
Rhodococcus sp. X309	Denis Larose et al. (1997)	Several unidentified bacteria	Abbad-Andaloussi et al. (2003)
Rhodococcus sp. B1	Denis Larose et al. (1997)	Thermophilic microorganisms	
R. erythropolis I-19	Folsom et al. (1999)	*Bacillus subtilis* WU-S2B	Kirimura et al. (2001)
R. erythropolis KA2-5-1	Kobayashi et al. (2001)	*Mycobacterium phlei* WU-F1	Furuya et al. (2001)
Rhodococcus sp. *P32C1*	Maghsoudi et al. (2001)	*M. phlei* WU-0103	Ishii et al. (2005)
Rhodococcus sp. *T09*	Matsui et al. (2001)	*Paenibacillus* sp. A11-2	Onaka et al. (2001)
Rhodococcus sp. *IMBSO2*	Castorena et al. (2002)	*Sulfolobus acidocaldarius*	Kardinahl et al. (1996)
Rhodococcus sp. *FMF*	Akbarzadeh et al. (2003)		
R. erythropolis N1-36	Wang et al. (1996)		
Rhodococcus sp. DS3	Ma et al. (2006)		

C1-dibenzothiophene are preferentially attacked followed by the highly alkylated molecules (Cx-dibenzothiophene). The position of alkylation also influences the biodesulfurization rate. Usually, alkylation near the sulfur (e.g., 4-methyl-dibenzothiophene, 4,6-dimethy-dibenzothiophene) would lower the biodesulfurization rate. However, *Sphingomonas* strain expressed the opposite trend (Monticello et al., 1985).

For commercialization of biodesulfurization process, the microorganisms should have a broad versatility to desulfurize different sulfur compounds, especially those reported recalcitrant to hydrodesulfurization. The sulfur-specific biodesulfurization of thiophene, benzothiophene dibenzothiophene, 4-methyl dibenzothiophene, and 4,6-dimethyl dibenzothiophene by the Gram-positive bacterial isolate *Brevibacillus invocatus* C19 has been reported (Nassar et al., 2013). On the other hand, the biodesulfurization of dibenzothiophene, 4,6-dimethyl dibenzothiophene, 4,6-dipropyl dibenzothiophene, 4,6-dibutyl dibenzothiophene, 4,6-dipentyl dibenzothiophene, 4-hexyl dibenzothiophene, 2,3-dihexyl dibenzothiophene, and 3-propyl-4,8-dimethyl dibenzothiophene occurs through the 4S-pathway using the Gram-negative bacterial isolate *Sphingomonas subarctica* T7b (Wayan et al., 2013). The required biodesulfurization activity for commercialization is reported to be 1.2–3 mM/g dry cell/h (Kilbane, 2006). Genetic engineering has been applied to increase the biodesulfurization rate. It has been reported that the desulfurization enzymes are repressed in presence of inorganic sulfate, when the native promoter of $D_{SZ}ABC$ operon is replaced with a nonrepressible promoter for enhancement of biodesulfurization activity (Noda et al., 2002). There is also high biodesulfurization of dibenzothiophene (50–250 μmol/g DCW/h) using genetically modified *R. erythropolis* KA2-5-1 (Konishi et al. (2005) and there is a 200-fold increase in the biodesulfurization activity of *R. erythropolis* IGTS8 increased by genetic engineering, and the dibenzothiophene desulfurization activity reached 20 μmol/g DCW/min (Le Borgne and Quintero (2003). Introduction of dsz genes into *Rhodococcus* sp. T09, a strain capable of desulfurizing benzothiophene, improved the uptake of alkylated sulfur compounds (Matsui et al., 2001). The resulting recombinant strain grew with both dibenzothiophene and BT as the sole sulfur sources and the recombinant cells desulfurized not only alkylated benzothiophene derivatives with the production of alkylated hydroxyphenyl derivatives as the end products.

It has been reported that the desulfinase $D_{SZ}B$ catalyzes the rate-limiting step in the 4S-pathway. Thus, increasing its level, either by removing the overlap between $D_{SZ}A$ and $D_{SZ}B$ or by rearrangement of the genes in the operon, such that $D_{SZ}B$ becomes the first gene in the operon, would result in increased biodesulfurization activity (Liu et al., 2008). Deletion of $D_{SZ}B$ gene could allow the accumulation of hydroxyphenyl benzene sulfinate, which is a valuable product for surfactant industry (Yu et al., 2005). Genetic engineering can be also applied on $D_{SZ}C$, which catalyzes the first step in the 4S-pathway to improve the substrate range (Arsendorf et al., 2002). To overcome the mass transfer limitation, the microorganism should be able to tolerate high concentration of hydrocarbons (i.e., oil phase or solvents) so they would function optimally. It is well known that the solvent tolerance of *Rhodococcus* is lower than *Pseudomonas*, although of the hydrophobicity of its membrane, which makes it easily uptake very hydrophobic C_X-dibenzothiophene from oil phase (McFarland, 1999). Gene cloning of dszABC genes from *R. erythropolis* XP into *P. putida* constructing a solvent tolerant resulted in the desulfurization by the *P. putida* A4 Tao et al., 2006). This strain, when contacted with sulfur refractory compounds dissolved in hydrocarbon solvent, maintained the same substrate desulfurization traits as observed in *R. erythropolis* XP. Resting cells of *P. putida* A4 could desulfurize 86% of dibenzothiophene in 10% (v/v) p-xylene in 6h. Within the first 2h, the desulfurization occurred with a rate of 1.29 mM dibenzothiophene/g DCW/h. Further mutagenesis is attempted on $D_{SZ}B$ to increase the biocatalytic activity and thermostability of the enzyme (Ohshiro et al., 2007). The design of a recombinant microorganism resulted in the removal of the highest amount of sulfur compounds in fossil fuels (Raheb et al., 2009). The three genes (DszA,B,C) from the desulfurization operon in *R. erythropolis* IGTS8 were inserted into the chromosome of a novel indigenous *Pseudomonas putida*. As mentioned before, the reaction catalyzed by products of DszA,B,C genes require $FMNH_2$ supplied by DszD enzyme. Thus pVLT31 vector harboring dszD gene was transferred into this recombinant strain. This new indigenous bacterium is an ideal biocatalyst for desulfurizing enzyme system due to the solvent tolerant characteristic and optimum growth temperature at 40°C (104°F), which is suitable for industrial biodesulfurization process. In addition, this strain produces Rhamonolipid biosurfactant, which accelerates two-phase separation step in the biodesulfurization process through increasing emulsification. Moreover, the strain has a high growth rate, which cause to remove sulfur compounds faster than *R. erythropolis* IGTS8 and expresses higher biodesulfurization activity in shorter time.

There are few reports about biodesulfurization using fungal isolates such as the ability of *Rhodosporidium toruloides* strain DBVPG6662 to utilize dibenzothiophene as a sulfur source producing 2,2′-dihydroxybiphenyl (Baldi et al., 2003). When this strain was grown on glucose in the presence of commercial emulsion of bitumen (Orimulsion), 68% of the benzothiophene derivatives and dibenzothiophene derivatives were removed after 15 days of incubation. Not only this but it was also able to utilize the organic sulfur in a large variety of thiophene derivatives that occur extensively in commercial fuel oils by physically adhering to the organic sulfur source.

4.0 BIODESULFURIZATION OF CRUDE OIL AND ITS FRACTIONS

Biodesulfurization with high hydrocarbon phase tolerance is considered an advantage because less amount of water is required for biodesulfurization. However, very few reports are available on the biodesulfurization of real oil feed, i.e., crude oil and its fractions. The biodesulfurization of single organosulfur compound dissolved in hydrocarbons (often termed a *synthetic model oil*) has been extensively examined (Finnerty, 1993; Rhee et al., 1998; Ohshiro and Izumi, 1999; Jia et al., 2006; Caro et al., 2007; Li et al., 2007a, 2007b). However, caution is advised when using such studies as a further base from which to work since, in industry, organosulfur compounds are found in mixture and not individually and there is a need to use different crude oil feedstocks to determine biodesulfurization efficiency (Table 6.3).

The oil/water (O/W) ratio of 1:9, i.e., 10% has been used by various investigators to study the desulfurization of different fuel oils (Rhee et al., 1998; Li et al., 2003; 2007a). *R. globerulus* DAQ3, which was isolated from oil-contaminated soil, demonstrated high specific dibenzothiophene desulfurization activity and stability in oil/water systems. A model diesel with 1452 μM dibenzothiophene/hexadecane was contacted with the same volume of 10 g DCW/L of aqueous phase. It was reported that resting cells of DAQ3 could transform 92% and 100% of dibenzothiophene in 9 and 24 h, respectively. Desulfurization with IGTS8 at identical condition resulted in 29% and 32% desulfurization. The dibenzothiophene desulfurization by DAQ3 proved to be significantly higher than the one reported by IGTS8 (Yang and Marison, 2005). *Rhodococcus* sp. IMP-S02 could remove 60% of the sulfur from diesel oil when incubated for 7 d at 30°C (Castorena et al., 2002). *Pantoea agglomerans* D23W3 were found to remove 26.38%–71.42% of sulfur from different petroleum oils with highest sulfur removal from light (lower-density, higher-API gravity) crude oil at 1/9 O/W phase ratio (Bhatia and Sharma, 2010).

Table 6.3 List of Some Aerobic Biodesulfurization for Crude Oil Fractions

Bacteria	Crude Oil Fraction (S-Content)	Biodesulfurization (%)	References
Gordona sp. CYKS1 (RC[a])	Middle distillate unit feed (1500 ppm) Light gas oil (3000 ppm)	70 50	Rhee et al. (1998)
Gordonia sp. SYKS1 (RC)	Light gas oil (3000 ppm) Middle distillate unit feed (1500 ppm) Diesel oil (250 ppm)	35 60 76	Chang et al. (2000)
Mycobacterium pheli WU-F1 (RC)	B-light gas oil (350 ppm) F-light gas oil (120 ppm) X-light gas oil (34 ppm)	74.4 65 55.9	Furuya et al. (2003)
Mycobacterium sp. X7B (RC)	Hydrodesulfurized diesel oil (535 ppm)	86	Li et al. (2003)
M. pheli WU-0103 (GC[b])	Light gas oil (1000 ppm)	52	Ishii et al. (2005)
Nocardia globerula R-9 (RC)	Straight run diesel oil (1807 ppm)	59	Mingfang et al. (2003)
Pseudomonas delafieldii R-8 (RC)	Hydrodesulfurized diesel oil (591 ppm)	90.5	Guobin et al. (2005)
P. delafieldii R-8 (GC)	Hydrodesulfurized diesel oil (591 ppm)	47	Guobin et al. (2006)
Rhodococcus erythropolis XP	Hydrodesulfurized diesel oil (259 ppm)	94.5	Yu et al. (2006)
Rhodococcus sp. ECRD-1 (GC)	Middle distillate fraction of Oregon Basin crude oil (20,000 ppm)	8.1	Grossman et al. (1999)
Rhodococcus sp. ECRD-1 (GC)	Catalytic cracker middle distillate light cycle oil (669 ppm)	92	Grossman et al. (2001)
Rhodococcus sp. P32C1 (RC)	Hydrodesulfurized light diesel oil (303 ppm)	48.5	Maghsoudi et al. (2001)
R. erythropolis I-19 (RC)	Hydrodesulfurized middle distillate (1850 ppm)	67	Folsom et al. (1999)
Sphingomonas subarctica T7b (GC)	Hydrodesulfurized light gas oil (280 ppm)	59	Gunam et al. (2006)

[a]*Resting cells.*
[b]*Growing cells.*

The biodesulfurization of bunker oil MFO380 using a mixed culture enriched from an oil sludge in a two-phase system 1:50 (w/w) oil/water in presence of Triton X-100 surfactant has been reported (Li and Jiang, 2013). Also, there is a report (Aribike et al., 2008) that the anaerobic *Desulfobacterium indolicum* isolated from oil-contaminated soil exhibited very high desulfurizing ability towards kerosene at 30°C in 1:9°/w phase ratio, resulting in reduction of sulfur from 48.68 to 13.76 ppm over a period of 72 h with significant decrease in thiophene and 2,5-dimethyl thiophene.

Aribike et al. (2009) reported also the anaerobic, *Desulfobacterium anilini*, which was isolated from petroleum products–polluted soil, to express a significant decrease of benzothiophene and dibenzothiophene in diesel with 82% removal of total sulfur after 72h, at 30°C with 1/9 O/W phase ratio. Arabian et al. (2014) reported 33.26% biodesulfurization of kerosene with initial sulfur content of 2333 ppm, in a biphasic system (1/9 O/W) using the native bacterial isolate *Bacillus cereus* HN within 72 h incubation at 40°C.

The isolation of yeast, *Candida parapsilosis* NSh45 that showed in batch flasks of 1:3 O/W phase ratio brought about complete desulfurization of 1000 ppm dibenzothiophene dissolved in n-hexadecane (El-Gendy et al., 2006). However, in batch flasks of 1:3 O/W phase ratio after incubation period of 7 d at 30°C (200 rpm), 82% of 12,400 ppm sulfur content of diesel oil was desulfarized, keeping its calorific value, compared to that of *R. erythropolis* IGTS8 that expressed only 46%. NSh45 also reduced the sulfur content of Belayim Mix crude oil (2.74 wt.% sulfur) by 75%, decreased the average molecular weight of asphaltenes by ≈28%, and the crude oil dynamic viscosity by ≈70%, compared to that of *R. erythropolis* IGTS8, which expressed total sulfur removal of ≈64%, decrease in average molecular weight of asphaltenes, and crude dynamic viscosity of ≈24% and 64%, respectively, under the same conditions.

The isolation of the native fungus, which has been identified as *Stachybotrys* sp., is able to remove sulfur and nitrogen from heavy crude oil selectively at 30°C (86°F) (Torkamani et al., 2009). This fungus can desulfurize 76% and 64.8% of the sulfur content of heavy crude oil of Soroush oil field and Kuhemond oil field in Iran, with the initial sulfur contents of 5 wt.% and 7.6 wt.% within 72 and 144 h, respectively. This fungus strain has been isolated as a part of the heavy crude oil biodesulfurization project initiated by Petroleum Engineering Development Company (PEDEC), a subsidiary of National Iranian Oil Company.

4.1 ENZYMATIC OXIDATION OF ORGANOSULFUR COMPOUNDS

One of the major problems in biodesulfurization of petroleum and its fractions using whole cells is the need of aqueous phase, for viability of the cells. Thus, a two-phase system reactor is required to metabolize the hydrophobic substrate, and a microorganism with a broad substrate specificity for the various organosulfur compounds present in oil is also required. This problem can be solved by the application of enzymes with broad specificity instead of whole cells, in anhydrous organic phase, or at very low water content, so the mass transfer limitations would be reduced. But as has mentioned before, the use of enzymes is disadvantageous since extraction and purification of the enzyme is costly and, frequently, enzyme-catalyzing reactions require cofactors that must be regenerated (Setti et al., 1997). However, some studies have been carried out on biooxidation of organosulfur compounds, using isolated enzymes, in the free or immobilized form.

The hemoprotein enzymes like cytochrome P450 (Alvarez and Ortiz de Montellano, 1992), lignin peroxidase from the white rot fungus *Phanerochaete chrysosporium* (Vazquez-Duhalt et al., 1994), lactoproxidase (Doerge et al., 1991), horseradish peroxidase (Kobayashi et al., 1986), and chloroperoxidase from the fungus *Caldariomyces fumago* (Pasta et al., 1994) can be applied for biotransformation of thiophene, dibenzothiophene, and other organosulfur compounds (Ayala et al., 2007). The nonenzymatic hemoproteins are also able to perform the dibenzothiophene oxidation in vitro, such as hemoglobin (Ortiz-Leon et al., 1995), cytochrome *c* (Vazquez-Duhaltet al., 1993), and microperoxidase (Colonna et al., 1994). An enzymatic oxidation process for the two-step desulfurization of fossil fuels (Vazquez-Duhalt et al., 2002a) involves (1) first oxidation of diesel oil by chloroperoxidase, lignin peroxidase, manganese peroxidase, or cytochrome c, producing sulfoxide and sulfone, with higher boiling point, leaving the majority of the hydrocarbons in their original form and (2) distillation to remove the oxidized organosulfur compounds. Other physicochemical processes can be used for the separation of the oxidized organosulfur compounds from the main hydrocarbon mixture such as column chromatography, precipitation, and complexation with a solid support.

The biooxidation of organosulfur compounds can be performed in a batch, semicontinuous or continuous methods alone, or in a combination with one or more additional refining process. The reaction can be carried out in open or closed vessel. Cytochrome c is a biocatalyst able to oxidize thiophenes and organ sulfides and has several advantages when compared with other hemoenzymes. It is active in a pH range from 2 to 11, has the hemoprosthetic group covalently bond, exhibiting activity at high concentrations of organic solvents, and is not expensive (Vazquez-Duhalt et al., 1993). Immobilized cytochrome c in poly(ethylene)glycol (PEG-Cyt) has been used to oxidize high sulfur content diesel oil (Zeynalov and Nagiev, 2015). The oxidation using chloroperoxidase (CPO) in the presence of H_2O_2, followed by distillation at 50°C (122°F), decreased the sulfur content of straight run diesel fuel from 1.6% to 0.27%, keeping 71% of the original hydrocarbons (Vazquez-Duhalt et al., 2002a). A simple estimate of the cost of this technology has been reported and a turnover number (i.e., number of substrate molecules that can be converted per molecule of enzyme before inactivation) is on the order of 500,000. Thus 1 g of enzyme could reduce the sulfur content of 0.81 ton of fuel from 500 to 30 ppm (Ayala et al., 2007).

The activity of CPO towards oxidation of 4,6-dimethyl dibenzothiophene in water/acetonitrile mixture using free-form and immobilized into SBA-16 mesoporous material has been investigated, and the thermal stability of the immobilized enzyme was three times higher than that of free enzyme (Terres et al., 2008). In addition, the use of lipase NOVOZYM LC, in presence of H_2O_2 and carboxylic acid in absence of water or any cofactor, followed by furfural extraction of produced sulfoxide and sulfone (Singh et al., 2009). The biocatalyst is active up to 70°C (158°F) and preferably in the temperature range 35–60°C (95–140°F). Three types of diesel oil with different sulfur contents (6400, 1000, and 500 ppm) were used. The sulfur content decreased to 2300, 115, and 29 ppm, respectively.

Taking into consideration the current high cost of enzymes, the desulfurization of one barrel of fuel would be too expensive to be implemented on a large scale. It is estimated that the enzyme stability should be increased by two orders of magnitude to be economically attractive. Nevertheless, it should be kept in mind that a biotechnological process would have lower costs in terms of capital investment and energy consumption (Ayala et al., 1998; Linguist and Pacheco, 1999).

4.2 LONG-TERM REPEATED SPECIFIC BIODESULFURIZATION BY IMMOBILIZATION

All the biodesulfurization processes reported hitherto are triphasic systems composed of cells, water, and oil or in another word biphasic system (oil and cells in water). There are numerous reports on the treatment of diesel oils or model oil mixtures by using suspensions of growing or resting cells (Chang et al., 2001; Noda et al., 2003; Labana et al., 2005; Yu et al., 2006). Treatment of oils using free cells has some limitations such as high cost of the biocatalyst and low volumetric ratio between the organic phase and the aqueous one. In these cases, oil is mixed together with the cells as a suspension, which produces a sort of surfactant, emulsifying the oil. It seems to be very difficult to separate oil and water from the emulsified oil. Furthermore, the recovery of the cells is also difficult (Yu et al., 1998).

Cell immobilization was one of the most promising approaches. Compared with cell suspension, biodesulfurization with immobilized cells has some advantages: ease of biocatalyst separation from the treated fuels, low risk of contamination, relatively high oil/water volumetric ratios, high stability, and long life-time of the biocatalyst (Chang et al., 2000; Hou et al., 2005). Nevertheless, physical interactions between bacterial cells and sulfured substrates require further studies to resolve problems associated with the limited access of microorganisms to organic substrates to upscale biodesulfurization (Tao et al., 2006; Yang et al., 2007). In this context, surfactants and immobilized cells are considered promising solutions to the problem of low solubility. Biomodification of inorganic supports using cell immobilization increases the interaction between reactants present in two-phase systems, thus avoiding the need to use expensive surfactants (Feng et al., 2006). So far, very few published papers are available on biodesulfurization by immobilized cells.

Generally, entrapment and adsorption are preferred methods for cell immobilization. In entrapment, living cells are enclosed in a polymeric matrix, which is porous enough to allow diffusion of substrates to the cells and of products away from the cells. The materials used for entrapment of cells are mainly natural polymers, such as alginate, carrageenan, gelatin, and chitosan. They may also be synthetic polymers such as polysaccharides, photo-cross-linkable resins, polyurethane, polyvinyl alcohol (PVA), polyacrylamide, and so on. Major drawbacks of an entrapment technique are diffusional limitations and steric hindrance, especially when diffusion of macromolecular substrates, such as starch and proteins, is involved. Mass transfer involved in diffusion of a substrate to a reaction site and in removal of inhibitory or toxic products from the environment may be impeded. Cell immobilization by adsorption is currently gaining considerable importance because of a major advantage, namely, reducing or eliminating the mass transfer problems associated with the common entrapment methods. However, the adsorption technique is generally limited by biomass loading, strength of adhesion, biocatalytic activity, and operational stability. This is because immobilization by adsorption involves attachment of cells to the surface of an adsorbent like Celite. Adsorption is a simple physical process in which the forces involved in cell attachment are so weak that cells that are several micrometers across are not strongly adsorbed and are readily lost from the surface of the adsorbent (Shan et al., 2005a).

The first application in biodesulfurization using entrapment technique for immobilization of free cells is reported by Naito et al. (2001), where a photo-cross-linkable resin prepolymer; ENT-4000, was used as a suitable gel material and succeeded in constructing a biphasic biodesulfurization system (immobilized *R. erythropolis* KA2-5-1 cells and oil) with good desulfurization activity and without leakage of cells from the support. Furthermore, ENT-4000–immobilized cells catalyzed biodesulfurization repeatedly in this system for more than 900 h with reactivation; and recovery of both the biocatalyst and the desulfurized model oil was easy. This study would give a solution to the problems in biodesulfurization, such as the troublesome process of recovering desulfurized oil and the short life of biodesulfurization biocatalysts. However, the expensive cost of the ENT-4000 and the loss of part of the cell activity due to the toxicity of the used chemicals withdraw its further application.

Currently, one of the most widely used entrapment carriers for immobilization of enzymes and whole cells is calcium alginate (CA), that is for its good biocompatibility, cheapness, and simplicity (Tang et al., 2012). Immobilization of *Mycobacterium goodii* X7B cells by entrapment with CA, carrageenan, agar, PVA, polyacrylamide, and gelatin-glutaraldehyde has been investigated (Li et al., 2005). It was found that CA-immobilized cells had the highest dibenzothiophene desulfurization activity. When immobilized *M. goodii* X7B cells were incubated at 40°C (104°F) with Dushanzi straight-run gasoline (DSRG227) in a reaction mixture containing 10% (v/v) oil for 24 h, the sulfur content of the gasoline decreased from 275 to 121 ppm; after a 24 h reaction, the desulfurization reaction was repeated by exchanging the used immobilized cells for fresh ones. The sulfur content further decreased to 54 ppm, corresponding to a reduction of 81%.

A small microbead size is important for minimizing the mass transfer resistance problem normally associated with immobilized cell culture involved in diffusion of a substrate to a reaction site and removal of inhibitory or toxic products from the environment may be impeded (Klein et al. (1993). To minimize these effects, it is essential to minimize the diffusional distance through a reduction in bead size (Cassidy et al., 1996). Immobilization of resting cells of *Pseudomonas delafieldii* R-8 by entrapment in CA using a new technique "gas jet extrusion technique" is a rapid and simple method (Li et al., 2008b). The resultant slurry was extruded through a cone-shaped needle into a stirred 0.1 M $CaCl_2$ gelling solution. The slurry is intruded as discrete droplets to form CA beads with normal size (2.5 and 4 mm in diameter). To prepare smaller beads than 2 mm diameter, nitrogen gas was introduced around the tip of the needle to blow off the droplets. By adjusting the gas flow rate to 0.5 L/min, the size of the beads can be controlled at 1.5 mm in diameter (Fig. 6.6). The specific desulfurization rate of 1.5 mm diameter beads reported to be 1.4-fold higher than that of 4 mm. Tween 20 (polyoxyethylene sorbitan monolaurate) has been employed to improve the permeability of the entrapment–encapsulation hybrid

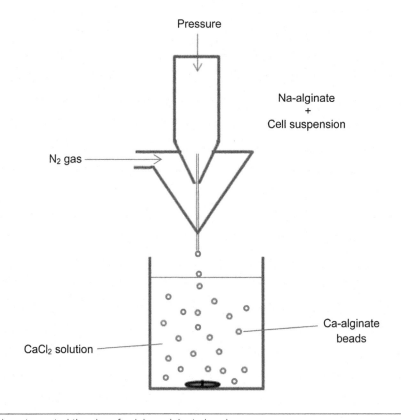

FIGURE 6.6

Simple technique to control the size of calcium alginate beads.

membrane (Song et al., 2005). The immobilized beads without Tween 20 would rupture because of the formation of carbon dioxide and nitrogen as a result of respiration and denitrification. During cell immobilization (Li et al., 2008b), the process the addition of non toxic and nonionic surfactants, including Span 20 (sorbitan monolaurate), Span 80 (sorbitan monooleate), Tween 20, and Tween 80 (polyoxyethylene sorbitan monostearate), greatly enhanced the desulfurization rate compared to the control. Span 80 showed the highest effect on desulfurization activity. Within 24 h, the desulfurization rate with the addition of 0.5% Span 80 reported to be 1.8-fold higher than that of without Span 80.

The resting cells of *Gordonia* sp. WQ-01 A, a dibenzothiophene-desulfurizing strain, can be immobilized (Peng and Wen, 2010). Batch dibenzothiophene-biodesulfurization experiments using immobilized cells and n-dodecane as the oil phase were conducted in fermenter under varying operating conditions such as initial dibenzothiophene concentration, bead loading, and the oil phase volume fraction. When the initial dibenzothiophene concentration is 0.5, 1, and 5 mmol/L, the dibenzothiophene concentration dropped almost to zero after 40, 60, and 100 h, respectively. The influence of bead loading and the oil-phase volume fraction was small to the dibenzothiophene-biodesulfurization. Furthermore, a mathematical model was proposed to simulate the batch dibenzothiophene-biodesulfurization process in an oil-water-immobilization system, which considered the internal and external mass transfer resistances of dibenzothiophene and oxygen and the intrinsic kinetics of bacteria. As with most immobilization systems, the diffusion rate of substrates and products within the bead often limits productivity. It was concluded that (1) the rate-limiting step in the oil-water-immobilization system is not mass transfer resistance but bioconversion and that (2) compared to the effect of dibenzothiophene, oxygen concentration is not an important factor affecting the dibenzothiophene-biodesulfurization in the immobilized system. That work leads to a greater understanding of the dynamic behavior of the immobilized system and may be generally applicable to other area of biocatalytic and biotransformation processes.

In the biodesulfurization of model oil with initial sulfur concentration of 300 ppm (1:1 thiophene: dibenzothiophene in n-octane) using CA-immobilized *P. delafieldii* R-8 cell beads, a triporate injector designed for cell immobilization and production of 4 mm beads was used (Huang et al., 2012). The rate of the biodesulfurization was higher than that of dibenzothiophene. The decrease of water volume is beneficial for real industry as it will allow more oil to be processed by immobilized cells. The immobilized cells showed sufficient biodesulfurization efficiency in absence of water, due to the presence of water in the CA beads enough for viability of the immobilized cells. However, the optimum ratio is reported to be 5:1 oil/water, due to the poor dispersion and adhesion of CA beads that decrease the biodesulfurization efficiency in absence of water phase. Addition of water increases the dispersion of immobilized cells in oil phase and increases the mass transfer rate. But excess of water should be avoided to improve the bioreactor efficiency. The immobilized cells are easily separated and successfully reused for 15 cycles each of 24 h, with efficient biodesulfurization activity. The last batch efficiency is only 25% lower than that of the first batch. Adegunlola et al. (2012) reported biodesulfurization of crude oil by CA-immobilized spores of *Aspergillus flavus* that recorded 94.7% biodesulfurization efficiency.

The key of entrapment method is the selection of materials and methods for preparation. The use of PVA for cells entrapment has been investigated (Lozinsky and Plieva, 1998). PVA is biologically compatible, readily available, low cost, and nontoxic. It may not have adverse effects upon cells and is becoming one of the most promising materials for entrapment. Hashimoto and Furukawa (1987) used the PVA–boric acid technique to immobilize cells in PVA. Although easy to be operated, the saturated boric acid solution used to cross-link the PVA, is highly acidic (pH is approx. 4), and could cause difficulty in maintaining cell viability. In addition, the sphering speed of the PVA is very slow and it results in agglomeration of the PVA beads. Moreover, the spherical bead rigidity is weak.

A successful attempt to quicken the sphering speed of the PVA, prevent agglomeration problem of the PVA beads, and increase the beads rigidity was adding a small amount of CA to the saturated boric acid (Wu and Wisecarver, 1992). However, the CA may be easily damaged by a salt of phosphoric acid and tend to be eroded or dissolved when used in reaction systems (Fernandes et al., 2002). To overcome the difficulties, a freezing–thawing technique was used to prepare PVA beads (Giuliano et al., 2003). The beads have excellent water resistance, elasticity and flexibility and shows high safety to living bodies because no chemical agent is used for gel formation. In addition, its high water content and porous structure make it ideal for use as a carrier suitable for the culture and propagation of the immobilized microorganisms (Freeman and Lilly, 1998). The complete biodesulfurization of model oil (100 mg/L dibenzothiophene in n-tetradecane) can be achieved with water/oil phase ratio of 7:1 using immobilized cells of *Sphingomonas subarctica* T7b in PVA, CA, or a mixture of both, within 72 h at 27°C and 160 rpm (Wayan et al., 2013). Cell-immobilized PVA beads were reported to be used for eight batch cycles of 24 h and the biodesulfurization remained stable. While the biodesulfurization efficiency using the other two immobilizing matrices was not stable and decreased. Not only was this, but the cell-immobilized PVA beads expressed higher biodesulfurization efficiency than free resting cells, over a wide range of pH 4.5–8.5 with maximum biodesulfurization 80.9 mg/kg DCW/h at pH7, and remained stable at higher pH values up to 8.5. The biodesulfurization efficiency of cell-immobilized PVA beads at elevated temperature (37°C) was higher than that of free resting cells.

Bacterial immobilization by adsorption is an improvement over the cell entrapment method that reduces mass transference and the steric effect. Bacterial cell adsorption involves the use of inorganic compounds as ideal biosupports. These materials must have controlled porosity and a high specific area. They must also be inert to biological attack, insoluble in the growth media, and nontoxic to microbial cells. Moreover, adsorbed cells on inorganic supports should be able to maintain the metabolic activity required for the biodesulfurization process. Few studies have evaluated the influence of inorganic supports, with different physicochemical properties, on the biodesulfurization activity of dibenzothiophene or gas oil (Hwan et al., 2000; Zhang et al., 2007), alumina and Celite being the most common supports used in these bioprocesses.

Celite beads are used as filter aids in pharmaceutical and beverage processing, and as bulk filters for food and planes, Celite is virtually inert for biological attack, insoluble in culture media, and nontoxic to microbial cells. These properties justified the choice of celite beads as a support material for immobilization. The desulfurization of a model oil (hexadecane containing dibenzothiophene) and diesel oil has been reported (Chang et al., 2000) by immobilized dibenzothiophene-desulfurizing bacterial strains, *Gordona* sp. CYKS1 and *Nocardia* sp. CYKS2. Celite bead was used as a biosupport for cell immobilization. Immobilized cells were used for eight cycles each of 24 h, good desulfurization was obtained, and the desulfurization rate of diesel oil was about 4–7 times higher than that of model oil since model oil contained only dibenzothiophene, a recalcitrant compound, while diesel oil contained various readily desulfurizable compounds, such as thiols and sulfides.

The efficiency of adsorption of *Pseudomonas stutzeri* on Silica (Si), Alumina (Al), sepiolite (a complex magnesium silicate $Mg_4Si_6O_{15}(OH)$), and titania (TiO_2) and their influence on the biodesulfurization of gas oil, with emphasis on the interaction of inorganic supports on metabolic activity has been investigated (Dinamarca et al., 2010). The highest interaction was observed in the *P. stutzeri*/Si and

P. stutzeri/Sep biocatalysts. A direct relation between biodesulfurization activity and the adsorption capacity of the bacterial cells was observed at the adsorption/desorption equilibrium level. The biomodification of inorganic supports generates dynamic biostructures that facilitate the interaction with insoluble organic substrates, improving the biodesulfurization of gas oil in comparison to whole non-adsorbed cells. This study concluded that immobilization by adsorption of bacterial cells is a simple and effective methodology that can be applied in biodesulfurization reactions of gas oil. It has also been reported (Dinamarca et al., 2014a) that the addition of biosurfactant or Tween 80 improves the biodesulfurization efficiency of dibenzothiophene and gas oil in free and immobilized cell systems of *R. rhodochrous* IGTS8, using adsorption on silica, alumina, and sepiolite as immobilizing technique. The biodesulfurization capacity in immobilized cell system was much higher than free cell system and the biodesulfurization with *R. rhodochrous*/sepiolite in the presence of biosurfactant expressed the highest biodesulfurization capacity and attributed this to the adsorption efficiency of the cells on the supports where IGTS8 expressed the highest adsorption on sepiolite (8.9×10^8 cells/g), and the more bioavailability of sulfur compounds to the cells is due to the formation and stabilization of sulfur-compound micelles due to the presence of biosurfactants in the catalytic media. The mechanism of immobilized cell system action with the studied surfactants were through increased mobility of the solid particles in the reaction medium, thus increasing the bioavailability of the S-compounds to the cells by reducing the interfacial tension of the aqueous medium (due to the presence of Tween 80) or/and increase the solubilization of the S-compounds by formation of micelles (i.e., biosurfactant).

As has been mentioned before, due to the slow kinetics of enzymatic reactions and mass transfer limitations, the overall rate of biodesulfurization is much slower than the conventional oxidative desulfurization. Recently, the application of ultrasonic enhancement sulfur-specific biodesulfurization has been reported (Bhasarkar et al., 2015), through the physical and chemical effects of ultrasonic by the generation of intense microturbulence in the medium and the generation of highly reactive oxidizing radicals ($O^•$ and $^•OH$) from the thermal dissociation of transient collapse that would oxidize dibenzothiophene to dibenzothiophene sulfoxide and dibenzothiophene sulfone, respectively. The utilized *R. rhodochrous* MTCC 3552 was immobilized throughout the cross-linking of the cells on commercial polyurethane foam; sonication using an ultrasound bath was used in the immobilization process to release all proteins inside the immobilized cells. The biodesulfurization reaction mixture was a liquid-liquid heterogeneous phase; organic (toluene containing different concentration of dibenzothiophene) and aqueous (water) and surfactant β-cyclodextrin, as a phase transfer agent (to enhance interphase dibenzothiophene transportation).

Immobilized cell system expressed higher biodesulfurization than that of free cell system. For free cell system, sonication showed better biodesulfurization efficiency than mechanical shaking. Additionally, β-cyclodextrin with sonication improved the biodesulfurization capacity of free cells by approximately 77%. But the increment of biodesulfurization efficiency in immobilized cell system by sonication and/or β-cyclodextrin is relatively smaller than that occurred in free cell system; recoded increment of addition of β-cyclodextrin in mechanical shaking was 23.8%, while the application of sonication alone expressed biodesulfurization increment with approximately 51% relative to mechanical shaking. However, application of sonication and β-cyclodextrin recorded biodesulfurization increment of 59.8%. The intense microconvection generated by the cavitation bubbles would result in good emulsification of the reaction mixture with high interfacial for interphase transport of dibenzothiophene from the organic phase to the aqueous phase in addition to the produced microturbulence, which enhances the diffusion of dibenzothiophene and the 2-hydroxy biphenyl sulfinate across the cell membrane. Moreover, it also enhances the back diffusion of 2-hydroxybiphenyl into the organic phase. This avoids its accumulation in the aqueous phase that consequently would decrease its inhibition effect on the cells. The microconvection enhances the movement of the microbial cells in the reaction mixture, which would cause an increase in their intercollisions and/or their collisions with the walls of the reaction reactor. Thus, kinetic energy gained by the cells through such rapid motion would help in accelerating the enzymatic reaction inside the cells. Sonication produces active oxygen radicals (O•) and hydroxyl radicals (•OH) that oxidize dibenzothiophene to its sulfoxide and sulfone, which would diffuse into the cell and undergo further transformation to 2-hydroxybiphenyl. Thus, the rate of biodesulfurization would be enhanced and the reduction of substrate inhibition would occur.

4.3 NANOBIOCATALYTIC DESULFURIZATION

The biodesulfurization of fuels takes place regularly in a three-phase system oil/water/biocatalyst. It has been reported that the transfer of polycyclic aromatic sulfur heterocyclic derivatives from the oil phase to the water phase and then from water to the cells, where oxidation reaction occurs into the cytoplasm, limits the metabolism of polycyclic aromatic sulfur heterocyclic derivatives. The more the hydrophobicity of the microorganism, the higher the rate of biodesulfurization as it would assimilate polycyclic aromatic sulfur heterocyclic derivatives directly from the oil (Monticello, 2000). The availability of production of biosurfactant would also help in increasing the rate of biodesulfurization, due to the formation of oil/water emulsion that retains the microbial viability and activity (Han et al., 2001). Biocatalyst stability, lifetime, and separation are crucial factors for commercialization of biodesulfurization. Immobilized biocatalyst has some advantages: ease of separation, high stability, low risk of contamination, and long lifetime (Hou et al., 2005). Compared to biodesulfurization, ADS has a much faster reaction rate (Song, 2003). Adsorbent preparation is the key of ADS. Recently, most adsorbents for desulfurization were based on π-complexation (Shan et al., 2005a) or formations of metal-sulfur bonds (such as nickel-sulfur and lanthanum-sulfur bonds) (Tian et al., 2006). Adsorbents based on π-complexation are easy to regenerate, but their selectivity is very low, resulting in a loss of fuel quality. Meanwhile, adsorbents that form metal-sulfur bonds with sulfur have high selectivity but are difficult to be regenerated. Hence, ADS technology also has a long way to go before being industrialized.

If a desulfurization technology has both the high reaction rate of ADS and the high selectivity of biodesulfurization, it can increase the desulfurization rate without damaging fuel quality. Adsorbent is an important factor of this coupling technology because different adsorbents have different interaction to organic sulfur compounds and cells, which would affect their assembly onto the cell surfaces and desorption behavior of organic sulfur from them. Because the adsorbents are assembled on the cells' surfaces, the property of cell surface is another factor, which impacts the coupling technology. Moreover, desulfurization conditions, such as temperature and volume ratio of oil to water phase, would also affect in situ coupling technology.

In the last 2 decades, numerous studies have been carried out on biodesulfurization using whole cells (Maghsoudi et al., 2001; Tao et al., 2006; Yang et al., 2007; Caro et al., 2008) or isolated enzymes (Monticello and Kilbane, 1994) in the free or immobilized form. As has been mentioned before, the biodesulfurization of dibenzothiophene occurs via a multienzyme system that requires cofactors (e.g., NADH—the reduced form of nicotinamide adenine dinucleotide). The use of enzymes is disadvantageous since extraction and purification of the enzyme is costly and, frequently, enzyme-catalyzing reactions require cofactors that must be regenerated (Setti et al., 1997). Therefore, biodesulfurization can often be designed by using whole cell biotransformation rather than that of the enzyme. However, there are still some bottlenecks limiting the commercialization of the biodesulfurization process.

One of the challenges is to improve the current biodesulfurization rate by about 500-fold assuming the target industrial process is 1.2–3 mmol/g dry cell weight (DCW)/h (Kilbane, 2006). When free cells are used for petroleum biodesulfurization, deactivation of the biocatalyst and troublesome oil–water–biocatalyst separation are significant barriers (Konishi et al., 2005; Yang et al., 2007). Cell immobilization may give a solution to the problems, providing advantages such as repeated or continuous use, enhanced stability, and easy separation. However, the biodesulfurization rate in immobilized cell system, especially applying entrapment technique, would be slightly better, or the same and sometimes slightly lower than that in free cell system, due to the diffusional resistance of reactants and products to and from the cells within the support (Shan et al., 2003)

For biodesulfurization, cells need to be harvested from the culture medium, and several separation schemes had been evaluated, including settling tanks (Schilling et al., 2002), hydrocyclones (Yu et al., 1998), and centrifuges (Monticello, 2000). But these procedures are time-consuming and costly. Magnetic separation technology provides a quick, easy, and convenient alternative over traditional methods in biological systems (Haukanes and Kvam, 1993). Super-paramagnetic nanoparticles are increasingly used to achieve affinity separation of high value cells and biomolecules (Molday et al., 1977). Magnetic supports for cells immobilization offer several advantages, such as the ease of magnetic collection. The magnetic supports present further options in continuous reactor systems when used in a magnetically stabilized, fluidized bed. In addition, the mass transfer resistance can be reduced by the spinning of magnetic beads under revolving magnetic field (Sada et al., 1981).

There is a report (Shan et al., 2003) of the modification of magnetite (Fe^{2+}-Fe^{3+}, Fe_3O_4) magnetic fluid preparation with a coprecipitation method, in presence of N_2 gas, to prevent oxidation of ferrous ion, solution blackening, and control the particle size (Liu et al., 2003) to produce a hydrophilic magnetic fluid. Magnetic particles prepared by coprecipitation method would have many hydroxyl groups on their surface in contact with the aqueous phase. The hydroxyl groups (OH groups) on the surface of mixed iron oxide nanoparticles (with an average particle diameter on the order of 8 nm) react readily with carboxylic acid head groups of the added oleic acid after which the excess oleic acid will be adsorbed to the first layer of oleic acid to form a hydrophilic shell. When this magnetic fluid was added to aqueous ammonium hydroxide, the outer layer of oleic acid on the magnetite surface is transformed into ammonium oleate and the hydrophilic magnetic fluid is produced. Magnetic fluid directly mixed into hydrophilic support liquids such as PVA and sodium alginate with dried cells of *P. delafieldii* R-8. Immobilized beads were formed by extruding the mixture through a syringe into a gelling solution of 0.1 M calcium chloride saturated with boric acid and solidified for 24 h. The immobilized beads formed were washed with saline, and then freeze dried for 48 h under vacuum. The magnetic fluid is mainly composed of magnetite particles, which not only provide the magnetic property of support but also improve mechanical strength of the supports, which are super-paramagnetic. Therefore, the magnetic immobilized supports could be easily separated and recycled by external magnetic field and the recovered magnetic supports could be redispersed by gentle shaking with the removal of the external magnetic field. Compared with nonmagnetic immobilized cells, the beads of magnetic immobilized cells showed higher reaction activity of desulfurization and with higher strength against swelling, longer term stability and can be reused for seven cycles of reaction while the nonmagnetic immobilized cells could only be used for five cycles. Also, the magnetic immobilized cells can be easily separated from the reaction medium, stored, and reused to give consistent results. The support is relatively cheap, easy to prepare, and good for large-scale industrial applications. Dai et al. (2014) reported biodesulfurization of dibenzothiophene in 1:9 oil/water biphasic system using magnetic Fe_3O_4/CA–immobilized cells of *Brevibacterium lutescens* CCZU12-1. The immobilized biocatalyst can be reused successfully for four times. The biodesulfurization of dibenzothiophene using nano γ-Al_2O_3/CA–immobilized *R. erythropolis* R1 has been reported Derikvand and Etemadifar, 2014). Taguchi optimization of dibenzothiophene-biodesulfurization in biphasic system and the statistical analysis showed that the concentration of γ-Al_2O_3 nanoparticles is the most significant factor in the biodesulfurization process and the optimum ratio of cells/nano gamma-alumina (γ-Al_2O_3) recorded to be 20% w/w. The biodesulfurization efficiency increased with decreasing bead size for the increment of its specific surface area, and the increase of its surface volume ratio thus reduces the mass transfer limitation, the optimum diameter reported to be 1.5 mm. The increase of alginate concentration improves the stability of the beads and its reusability. The optimum Na-alginate concentration for efficient biodesulfurization, bead stability, and reusability reported to be 2% (w/v), as a better option for multiple applications. The addition of surfactants (0.5% v/v) during preparation of immobilized cells enhances the biodesulfurization efficiency of the immobilized biocatalyst. Span 80 has better impact on biodesulfurization of dibenzothiophene as compared to Tween 80, due to the greater reduction in water surface tension in presence of Span 80. The molecules of Span 80 and Tween 80 contain oleate-chain of 18 carbons with an unsaturated bond, which can improve the stability of the O/W interface layer. But the average droplet size formed by Span 80 is smaller than that of Tween 80 (Schmidts et al., 2009). Thus, this would enhance the dibenzothiophene removal efficiency. The biodesulfurization with nano γ-Al_2O_3/CA–immobilized *R. erythropolis* R1 was more than two-folds greater than that of CA-immobilized cells. Since, the γ-Al_2O_3 nanoparticles can create pores in the cell membranes, which would facilitate the transfer of dibenzothiophene to the cells and 2-HBP from the cells. The nano γ-Al_2O_3/CA–immobilized cells can be reused for four successive cycles, each of 20 h, and the viability of the cells was retained but the biodesulfurization activity was decreased by approximately 59% than that of the first cycle. This is probably due to the reduction of the 4S-pathway cofactor levels of hydroxylamine oxidoreductase ($NADH_2$) and dihydroflavin mononuclitide ($FMNH_2$, i.e., 1,5-dihydroriboflavin 5'-(dihydrogen phosphate)). However, the addition of 10 mmol/L and 4 mmol/L of nicotinamide and riboflavin to biodesulfurization media enhanced the biodesulfurization efficiency more than 30% after the fourth cycle. Yan et al. (2008) reported that nicotinamide and riboflavin are the precursors of hydroxylamine oxidoreductase and dihydroflavin mononuclitide, respectively. Thus, their addition to biodesulfurization medium would enhance biodesulfurization rate and is very applicable in multiple use of the immobilized cells.

Other researchers have attempted to increase the efficiency of cells and to decrease the cost of operations in a biodesulfurization process (Shan et al., 2005b). In the work, magnetic PVA beads were prepared by a freezing–thawing technique under liquid nitrogen and the beads have distinct superparamagnetic properties. The desulfurization rate of the immobilized cells could reach 40.2 mmol/kg/h twice that of free cells. The heat resistance of the cells apparently increased when the cells were entrapped in magnetic PVA beads. The cells immobilized in magnetic PVA beads could be stably stored and be repeatedly used over 12 times for biodesulfurization. The immobilized cells could be easily separated by magnetic field. In order to understand cells distribution in magnetic PVA beads, the sections of the beads after being repeatedly used for six times were observed by scanning electron microscopy. A highly macroporous structure is found in the beads in favor of diffusion of substrates and dissolved gas. On average the size of the beads was about 3 mm. It is evident that the R8 cells mainly covered the edges and submarginal sections of the bead while no cell in the center of the bead because of insufficiencies of oxygen and nutrients and gradually autolysis.

In situ coupling of ADS and biodesulfurization is a new desulfurization technology for fossil oil. It has the merits of high selectivity of biodesulfurization and high rate of ADS. It is carried out by assembling nanoadsorbents onto surfaces of the microbial cells. For example, the combination of ADS and biodesulfurization and a proposed in situ coupling technology of them have been described, which increase the adsorption of dibenzothiophene from the oil phase and transfer it quickly into the cell for biodesulfurization, thus increasing the overall rate of desulfurization by about 2.5-fold (Shan et al., 2005c). The gamma-alumina (γ-Al$_2$O$_3$) nanosorbents, which can selectively adsorb dibenzothiophene from organic phase, were assembled on the surfaces of *Pseudomonas delafieldii* R-8 cell, a desulfurization strain. The γ-Al$_2$O$_3$ nanosorbents can adsorb dibenzothiophene from oil phase, and the rate of adsorption was significantly higher than that of biodesulfurization (Shan et al., 2005d). Thus, dibenzothiophene can be quickly transferred to the biocatalyst surface where nanosorbents were located, which quickened dibenzothiophene transfer from organic phase to biocatalyst surface and resulted in the increase of biodesulfurization rate. The desulfurization rate of the cells assembled with nanosorbents was approximately 2.5-fold higher than that of original cells. In order to increase the applicability of nanoparticles of γ-Al$_2$O$_3$, the problem of ease aggregation during and after its synthesis should be solved, taking into consideration the biocompatibility of dispersants. An improved in situ coupling technology of ADS and biodesulfurization by *Pseudomonas delafieldii* R-8 cells has also been reported (Zhang et al., 2007), where, γ-Al$_2$O$_3$ nanoparticles were synthesized and modified using gum Arabic (a natural gum that has good rheological properties and emulsion stability) to avoid the agglomeration in aqueous solutions and its effect on ADS and biodesulfurization was also evaluated. Results showed that γ-Al$_2$O$_3$ nanoparticlesdispersed well in aqueous solutions after modification with gum Arabic, and the ADS capacity of modified γ-Al$_2$O$_3$ nanoparticles increased by 1.12-fold than that of unmodified one. The good dispersion might be attributed to the chemical binding between the negatively charged groups of gum Arabic and the positive sites on the surface of γ-Al$_2$O$_3$ nanoparticles, giving rise to the non-DLVO (Derjaguin-Landau-Verwey-Overbeek) surface steric force, which prevents the agglomeration of nanoparticles in the aqueous solution (Leong et al., 2001). The better the dispersion of the adsorbents, the more the specific surface area the adsorbents would have, and thus, the more the adsorbents can adsorb sulfur; taking into consideration that excess gum Arabic (>1 wt.%) would take up from the adsorption site on the γ-Al$_2$O$_3$ nanoparticles that would decrease the ADS capacity. The adsorption of γ-Al$_2$O$_3$ nanoparticles onto *Pseudomonas delafieldii* R-8 increased the biodesulfurization efficiency from 14.5 mmol/kg/h to 17.8 mmol/kg/h. Compared with the unmodified γ-Al$_2$O$_3$ nanoparticles the biodesulfurization rate by adsorbing the gum Arabic-modified γ-Al$_2$O$_3$ nanoparticles onto the surfaces of R-8 cells increased to 25.7 mmol/kg/h (i.e., 1.44-fold), which may be due to the improvement in the dispersion and biocompatibility of γ-alumina nanoparticles after modification with gum Arabic, due to the stronger affinity of gum Arabic-modified γ-Al$_2$O$_3$ nanoparticles to the cells than the unmodified ones and its lower toxicity to the cells.

Other researchers (Zhang et al., 2008) applied different kinds of widely used adsorbents (alumina, molecular sieves, and active carbon) throughout in situ coupling technology of ADS and biodesulfurization. The procedure was carried out by assembling nanoadsorbents onto surfaces of *Pseudomonas delafieldii* R-8 cells. The data showed that Na–Y molecular sieves restrain the activity of R-8 cells and active carbon cannot desorb the substrate, dibenzothiophene. Thus, they are not applicable to in situ coupling desulfurization technology. γ-Al$_2$O$_3$ can adsorb dibenzothiophene from oil phase quickly, and then desorb it and transfer it to R-8 cells for biodegradation, thus increasing desulfurization rate. It was also found that nanosized γ-Al$_2$O$_3$ increases desulfurization rate more than regular-sized γ-Al$_2$O$_3$. Therefore, nano-γ-Al$_2$O$_3$ is regarded as better adsorbent for this in-situ coupling desulfurization technology. However, the ecofriendly MCM-41 mesoporous silica, with significant number of pores, ordered porosity and large specific surface area would enhance the adsorption of molecules to microbial cells relative to zeolites. Nasab et al. (2015) reported the fabrication of MCM-41 mesoporous silica nanosorbents at room temperature, using a quaternary ammonium template and cetyl trimethyl ammonium bromide surfactant (pore size 3.54 nm and specific surface area 1106 m^2/g), then it was physically well assembled on *R. erythropolis* IGTS8 cells by electrostatic and Van der Waals interactions. The dibenzothiophene-biodesulfurization efficiency of the modified cells was examined in a biphasic system (dibenzothiophene dissolved in dodecane/water 1:3) using modified cells with MCM-41/cells of 0.3:0.5 (w/w). The biodesulfurization efficiency of the modified cells (MCM-41/cells) was higher than that of free cells, due to the mass transfer improvement. The maximum specific desulfurization activity in terms of dibenzothiophene consumption and 2-HBP production rates recorded 0.34 μmol dibenzothiophene/g DCW/min and 0.126 μmol 2-HBP/g DCW/min, which were 19% and 16% higher than those of free cells, respectively.

A new technique in which magnetic nanoparticles are used to coat the cells has been developed; it could successfully overcome the difficulties of conventional cell immobilization, such as mass transfer problems, cell loss, and separation of carrier with adsorbed cells from the reaction mixture at the end of a desulfurization treatment. Coating layer of nanoparticles does not change the hydrophilic nature of the cell surface. The coating layer has negligible effect on mass transfer because the structure of the layer is looser than that of the cell wall and does not interfere with mass transfer of dibenzothiophene. The coated cells have good stability and can be reused. This new technique has the advantage of magnetic separation and is convenient as well as easy to perform so offers the potential to be suitable for large-scale industrial applications. There is a technique (Shan et al., 2005c) in which microbial cells of *Pseudomonas delafieldii* R-8 were coated with magnetic nanoparticles such as the complex iron oxide magnetite (Fe$_3$O$_4$) and then immobilized by external application of a magnetic field. The magnetic nanoparticles were strongly adsorbed on the cell surfaces because of their high specific surface area and high surface energy. It was possible to concentrate the dispersed coated cells by application of a magnetic field for reuse, and when dispersed, the coated cells experienced minimal mass transfer problems. Thus, this technique has advantages over conventional immobilization by adsorption to carrier materials such as celite. Furthermore, the method can overcome drawbacks such as limitations in biomass loading and in the loss of cells from the carrier associated with conventional immobilization by adsorption. The magnetic nanoparticles were synthesized by a coprecipitation method followed by modification with ammonium oleate. The surface-modified magnetite nanoparticles were monodispersed in an aqueous solution and did not precipitate over 18 months. Using transmission electron microscopy (TEM), the average size of the magnetic particles was found to be in the range from 10 to 15 nm. Transmission electron microscopic cross section analysis of the cells showed further that the magnetite nanoparticles were for the most part strongly absorbed by the surfaces of the cells and coated the cells. The coated cells had distinct super-paramagnetic properties. The coated cells not only had the same desulfurizing activity as free cells but could also be reused more than five times while the free cells could be used only once. Compared to cells immobilized on celite, the cells coated with magnetite nanoparticles had greater desulfurizing activity and operational stability in so far as the coated cells did not experience a mass transfer problem. But the process has a complication, because centrifugation was necessary in preparation of resting cells.

The in situ cell separation and immobilization of bacterial cells for biodesulfurization, which were developed by using super-paramagnetic magnetite nanoparticles, has also been reported (Li et al., 2009a). The magnetite nanoparticles were synthesized by coprecipitation followed by modification with ammonium oleate. The surface-modified nanoparticles were monodispersed and the particle size was on the order of 13 nm. After adding the magnetic fluids to the culture broth, *R. erythropolis* LSSE8-1 cells were immobilized by adsorption and then separated with an externally magnetic field. Analysis showed that the nanoparticles were strongly adsorbed to the surface and coated the cells. Compared to free cells, the coated cells not only had the same desulfurizing activity but could also be easily separated from fermentation broth by magnetic force. It was believed that oleate modified magnetite nanoparticles adsorbed on the bacterial cells mainly because of the nanosize effect and hydrophobic interaction. Ansari et al. (2009) reported dibenzothiophene biodesulfurization using decorated *R. erythropolis* IGT8 with magnetite nanoparticles (the average size of the prepared nanoparticle was on the order of 45–50 nm and the ratio of nanoparticles/cells was 1.78 w/w). The biodesulfurization efficiency of decorated cells was 56% higher than that of nondecorated cells.

From the point of commercial application, the *Rhodococcus* strains possess several properties favorable for desulfurization over *Pseudomonas* in an oil–water system. First, the hydrophobic nature of *Rhodococcus* makes them access preferentially Cx-dibenzothiophene derivatives from the oil, resulting in little mass transfer limitation (Le Borgne and Quintero, 2003). Moreover, the *Rhodococcus* bacteria are more resistant to solvents than *Pseudomonas* (Bouchez-Naitali et al., 2004). Therefore, there has been an attempt to develop a simple and effective technique by integrating the advantages of magnetic separation and cell immobilization for biodesulfurization process with Gram positive *R. erythropolis* LSSE8-1 and Gram –ve *P. delafieldii* R-8 (Li et al., 2009b). Cells were grown to the late exponential phase and the culture was transferred into Erlenmeyer flasks. A volume of magnetic fluids was added and mixed thoroughly—the microbial cells were coated by adsorbing the magnetic nanoparticles. The ammonium oleate–modified magnetite nanoparticles formed a stable suspension in distilled water, and the magnetic fluid did not settle during 8 months of storage at room temperature. The transmission electron microscope image of magnetite nanoparticles showed that the particles have an approximately spherical morphology with an average diameter of about 13 nm. The magnetite nanoparticles on the cell surface were not washed out by deionized water, ethanol, saline water (0.85 wt.%), or phosphate buffer (0.1 M, pH 7). Thus, there is little cell loss or decrease in biomass loading when cells are coated with magnetite nanoparticles. This outcome was different from that obtained with cells immobilized by traditional adsorption to a carrier. The cells coated with magnetite nanoparticles were super-paramagnetic. Therefore, the cell-nanoparticle aggregates in aqueous suspension could be easily separated with an externally magnetic field and redispersed by gentle shaking after the removal of magnetic field. For magnetic separation, a permanent magnet can be placed at the side of the vessel. After several minutes (3–5 min), the coated cells can be concentrated and separated from the suspension medium by decantation. This one-step technology; namely in situ magnetic separation and immobilization of bacteria, optimized dramatically the biodesulfurization process flow, and much less of magnetite nanoparticles was needed.

An adsorption mechanism between the magnetite nanoparticles and desulfurizing cells has also been proposed (Li et al., 2009b). The large specific surface area and the high surface energy of the magnetite nanoparticles (i.e., the nanosize effect) ensure that the magnetite nanoparticles are strongly adsorbed on the surfaces of microbial cells. Furthermore, the hydrophobic interaction between the bacterial cell wall and the hydrophobic tail of oleate modified magnetite nanoparticles may play another important role in cell adsorption. The suspension of oleate-modified magnetite nanoparticles was considered bilayer surfactant stabilized aqueous magnetic fluids (Liu et al., 2006). The iron oxide nanocrystals were first chemically coated with oleic acid molecule after which the excess oleic was weakly adsorbed on the primary layer through the hydrophobic interaction between the subsequent molecule and the hydrophobic tail of oleate. Since the bacterial cell wall is composed of proteins, carbohydrates, and other substances (such as peptidoglycan, lipopolysaccharide, and mycolic acid) the extracellular matrix can form hydrophobic interaction with the hydrophobic tail of oleate of nanoparticles.

A microbial method has also been used to regenerate desulfurization adsorbents (Li et al., 2006). Most of the sulfur compounds can be desorbed and removed and heat losses during the bioregeneration process are markedly reduced. The particle size of cells is similar to that of desulfurization adsorbents, which is about several microns. Therefore, it is difficult to separate regenerated adsorbents and cells. Super-paramagnetism is an efficient method to separate small particles. To solve the problem of separation of cells and adsorbents, magnetite nanoparticles modified *P. delafieldii* R-8 cells were used in the bioregeneration of adsorbents. Biodesulfurization with *P. delafieldii* R-8 strains coated with magnetite nanoparticles has been reported previously (Shan et al., 2005c) and cells coated with magnetite nanoparticles can be separated and reused for several times.

There is also a report (Li et al., 2008c) of the bioregeneration of desulfurization adsorbents AgY zeolite with magnetic cells. Super-paramagnetic magnetite nanoparticles are prepared by the coprecipitation method followed by modification with ammonia oleate. Magnetic *P. delafieldii* R-8 cells can be prepared by mixing the cells with magnetite nanoparticles. The biodesulfurization activity of the magnetic cells is similar to that of free cells. When the magnetic cells were used in the bioregeneration of desulfurization adsorbent AgY, the concentration of dibenzothiophene and 2-hydroxybiphenyl with free cells is a little higher than that with magnetic cells. The adsorption capacity of the regenerated adsorbent after being desorbed with magnetic *P. delafieldii* R-8, dried at 100°C for 24 h, and calcined in the air at 500°C for 4 h is 93% that of the fresh one. The magnetic cells can be separated from adsorbent bioregeneration system after desulfurization with external magnetic field and, thus, can be reused.

Zhang et al. (2011) reported the assembling of γ-Al$_2$O$_3$ nanoparticles onto magnetic immobilized *R. erythropolis* LSSE8-1-vgb. The optimum ratio of cells/magnetite nanoparticles was 50:1 (g/g) for better biodesulfurization efficiency and separation of magnetic immobilized cells, taking into consideration that the optimum ratio between magnetite nanoparticles/γ-Al$_2$O$_3$ nanoparticles is 1:5 (g/g). Assembling of γ-Al$_2$O$_3$ nanoparticles, which is characterized by small particle size, large pore volume, large specific surface area, and strongest electrostatics interaction with desulfurization cells, onto magnetic immobilized cells enhanced the biodesulfurization efficiency, as it improved the adsorption of dibenzothiophene from the oil phase. The γ-Al$_2$O$_3$ nanoparticles/magnetic immobilized cells have been used in three successive biodesulfurization cycles each of 9 h and its biodesulfurization efficiency was 20% higher than that of magnetic immobilized cells. The activity decreased less than 10% throughout the three successive cycles. So integrating γ-Al$_2$O$_3$ nanoparticle with magnetite nanoparticles in biocatalyst preparation would enhance the biodesulfurization efficiency and the cells can be collected and reused conveniently by an external magnetic field.

The decoration of *R. erythropolis* R1 with oleate-modified magnetite nanoparticles (average particles size < 10 nm) has been reported (Etemadifar et al., 2014) where response surface methodology was used to optimize the biodesulfurization conditions of dibenzothiophene in a biphasic system (dibenzothiophene in tetradecane/water). The dibenzothiophene concentration, incubation temperature, and the interaction of these two factors have significant effect on the biodesulfurization efficiency of the decorated cells. The hydrophobic nature of the *Rhodococcus* strains, its resistance to high concentration of dibenzothiophene and solvents, and the decrease of dibenzothiophene toxicity as it was dissolved in tetradecane would enhance the adsorption of dibenzothiophene from the oil phase and consequently the biodesulfurization efficiency (Monticello, 2000; Bouchez-Naitali et al., 2004). The first and third enzymes of the 4S-pathway ($D_{SZ}C$ and $D_{SZ}B$) are reported to be more sensitive to temperature changes relative to the other enzymes involved in the 4S-pathway and they are biodesulfurization rate-limiting (Furuya et al., 2001). The biodesulfurization efficiency of the free and coated cells was nearly the same and the optimum dibenzothiophene concentration, temperature, and pH for maximum biodesulfurization were reported to be 6.67 mM, 29.63°C, and 6.84, respectively. In addition, the enhancement the anaerobic kerosene biodesulfurization using nanoparticles decorated *Desulfobacterium indolicum* at 30°C and atmospheric pressure brought about a decrease in the sulfur content from 48.68 to 13.76 ppm within 72 h (Kareem, 2014).

The immobilization of the haloalkaliphilic S-oxidizing bacteria *Thioalkalivibrio versutus* D301 using super-paramagnetic magnetite nanoparticles under haloalkaliphilic conditions (pH 9.5) to remove sulfide, thiosulfate, and polysulfide from wastewater has been studied (Xu et al., 2015). The magnetite nanoparticle–coated cells expressed similar biodesulfurization activity like that of free cells and can be reused for six batch cycles.

Mesoporous materials in the nanoscale (2–50 nm) are characterized by large surface area, ordered pore structures with suitable sizes, high chemical and physical stability, and can be chemically modified with functional groups for the covalent binding of molecules, which make them very applicable in adsorption, separation, catalytic reactions, sensors, and immobilization of huge biomolecules like enzymes (Li and Zhao, 2013; Juarez-Moreno et al., 2014). Enzymatic immobilization in nanomaterials has some advantages: increase in catalytic activity, reduction in protein misfolding, thermal stability, increase in solvent tolerance, and reduction of denaturation caused by the presence of organic solvents and mixtures of water-organic solvent molecules (Carlsson et al., 2014). Also, the immobilization (i.e., conjugation) of CPO enzyme with latex nanoparticles after its functionalization with chloromethyl groups can react with the amine groups of the protein (i.e., the enzyme) to form stable covalent bonds. The average size of the conjugated nanoparticles reported to be 180 nm that can be applied in biooxidative of fuel streams, followed by physical or chemical remove of the produced sulfoxide and sulfone to produce ultralow sulfur fuels. Immobilization of CPO (as a good candidate for oxidative desulfurization in nonaqueous media) and myoglobin (which, for its availability, represents a good candidate for industrial scale) with La-nanoparticles (40 nm) was modified with chloromethyl groups (Vertegel, 2010). The conjugated and free enzymes were used for dibenzothiophene oxidation in presence of hydrogen peroxide in nonaqueous system (dibenzothiophene in hexane) and two-phase system (dibenzothiophene in acetonitrile/water; 20% v/v). For CPO, in two-phase system and nonaqueous one, free and conjugated enzymes expressed nearly the same activity. But the stability of free enzymes was much lower than conjugated ones, where its activity deteriorates much faster. The activity of conjugated CPO was approximately five times higher than that of free cells after 2 d storage. However, the free myoglobin expressed better performance than the immobilized ones in two-phase systems, while the immobilized myoglobin expressed twice the activity in nonaqueous system, which is an encouraging result, since this protein as well as similar heme-containing protein, hemoglobin, can be readily available in large quantities from animal processing industries. The nanoimmobilized enzymes are easily separated, stable over time, and can be taken as a good prospect for a recyclable desulfurization agent.

Juarez-Moreno et al. (2015) reported the immobilization of CPO into (1D) γ-Al$_2$O$_3$ nano-rods, through physical adsorption. The dimensions of CPO are reported to be $3.1 \times 5.3 \times 5.5$ nm (Sundaramoorthy et al., 1995), the pore size needed for CPO immobilization is > 5.5 nm, and the pore size of the calcined prepared γ-Al$_2$O$_3$ nanorods at 700°C (>6 nm) and 900°C (21.3 nm) is big enough for CPO immobilization. There is a high affinity between negatively charged CPO and the positively charged prepared γ-Al$_2$O$_3$ nanorods caused by favorable electrostatic interactions between the enzyme and the support. Immobilized enzyme expressed better oxidation of dibenzothiophene to dibenzothiophene sulfoxide than that of free enzymes, and γ-Al$_2$O$_3$ nanorods at 700°C expressed the highest activity, which encourages its application in biooxidative desulfurization of fuel streams.

5.0 BIODESULFURIZATION REACTORS

In all fossil fuel bioprocessing schemes, in order to realize commercial success, there is a need to contact the biocatalyst that is in aqueous phase with an immiscible or partially miscible organic substrate. The reactors must be designed to allow a sufficient liquid-liquid and/or gas-liquid mass transfer, amenability for continuous operation and high throughput, and ability for biocatalyst recovery and reuse; also emulsion breaking is a significant issue. All of this come with simultaneously reducing operating and capital costs.

Generally, batch-stirred tank reactors have been used because of the absence of immobilization technologies. A continuous two-phase (organic/aqueous) bioreactor can affect with 12% sulfur removal from diesel oil by *R. globerulus* DAQ3 (Yang et al., 2007). Mass et al. (2014) reported dibenzothiophene-biodesulfurization by the *R. erythropolis* ATCC 4277 in a batch reactor using two-phase system; dibenzothiophene dissolved in n-dodecane/water of different ratios 20%, 80%, and 100% (v/v), recording biodesulfurization efficiency of 93.3, 98.0, and 95.5%. The highest value for the specific dibenzothiophene-biodesulfurization rate was 44 mmol dibenzothiophene/kg DCW/h and was attained in the reactor containing 80% dibenzothiophene. Multistaged airlift reactors can also be used to overcome poor reaction kinetics at low sulfur concentrations and to reduce mixing costs. This would enhance the concept of continuous growth and regeneration of the biocatalyst in the reactions system rather than in separate, external tanks. A typical biodesulfurization process consists of charging the biocatalyst, oil, air, and a small amount of water into a batch reactor (Fig. 6.7). In the reactor, as the polynuclear aromatic sulfur species are oxidized to water-soluble products, the sulfur segregates into the aqueous phase. The oil–water–biocatalyst–sulfur–by-product emulsion from the reactor effluent is separated into two streams, namely, the oil (which is further processed and returned to the refinery) and the water–biocatalyst–sulfur–by-product stream. A second separation is needed to allow most of the water and biocatalyst to return to the reactor for reuse. Airlift bioreactors with different microorganisms and operating conditions have also been employed and achieved 50%–100% w/w sulfur removal (Nandi, 2010; Irani et al., 2011).

Effective oil–cell–water contact and mixing is essential for good mass transfer. Unfortunately, a tight emulsion is usually formed, and it must be broken in order to recover the desulfurized oil, recycle the cells, and separate the byproducts. The phases are usually separated by liquid-liquid hydroclones (Yu et al., 1998). Another approach is to separate two immiscible liquids of varying densities by using a settling tank, where the liquid mixture is given enough residence time for them to form two layers, which are then drained.

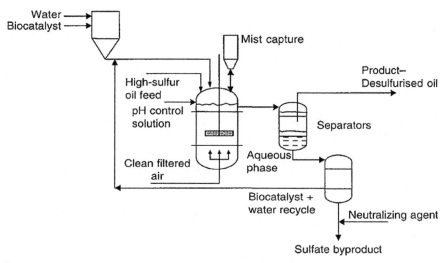

FIGURE 6.7

General flowsheet for the biodesulfurization process.

Bacteria usually partition to the oil–water interface and move with the discontinuous phase into a two-phase emulsion. In a water-in-oil emulsion, cells associate with the water droplets. A small amount of fresh oil can then be added to create an oil-in-water emulsion so that the cells will stick to the oil droplets. Passage of the emulsion through a hydrocyclone will yield a clean water phase and a concentrated cell and oil mixture that can be recycled to the reactor. By manipulating the nature of these emulsions, relatively clean oil and water can be separated from the mixture without resorting to high energy separations. This causes a significant reduction in capital and operating costs (US Department of Energy, 2003).

Throughout the effort of Ecopetrol—Instituto Colombiano del Petróleo (ICP) in Colombia towards reaching the new Colombian requirements set by the Colombian Ministry of the Environment, through Resolution 0068 issued in Jan. 18, 2001, of low sulfur content transportation fuels (gasoline and diesel oil sulfur content, 0.03 and 0.05 wt.%, by 2005, respectively), Acero et al. (2003) reported the design of a membrane bioreactor prototype and its capability for the application in biodesulfurization process of diesel oil using *Gordona rubropertinctus* ICP172. Among the nonconventional reactors that have been used at industrial levels, membrane reactors have attracted considerable interest due to its possibility of integrating the biocatalytic and the separation processes in one. Membrane bioreactors combine selective mass transportation featuring chemical reactions with the selective removal of inhibitory products from the process, thus increasing reaction conversion (Giorno and Drioli, 2000). Through this system, it is possible to immobilize cells or enzymes, and to simultaneously reach bioconversion, product separation, and enrichment in the same operating unit (Cass et al., 2000). The system was also used as a separation mechanism for the emulsion developed during the biocatalytic reaction. The reactor operates as a tube and carcass-type interchanger, where it is possible to separate the organic phase, using hydrophobic membranes on the tubing coating through which the emulsion is loaded into the carcass; the organic phase is recovered through the lumen. Two sets of experiments were carried out. In the first one, the emulsion separation was evaluated at different cellular concentrations, keeping the organic/aqueous phase relation of 25/75. In the second set of trials, the separation of phases at different organic/aqueous ratios (60/40, 50/50, 40/60 y 25/75) was evaluated, with a cellular concentration of 3 and 7 g/L (Berdugo et al., 2002). Trials carried out with a cellular concentration of 3 g/L and at different organic phase ratios showed that a greater organic/aqueous ratio allows a more effective phase separation. On the other hand, in trials reproduced at 60/40 and 40/60 ratios with a cellular concentration of 7 g/L, the phase separation was similar for 60/40 and 50/50 phase ratios. However, a smaller separation capacity was observed for organic/aqueous phase ratio of 40/60. This preliminary study showed a great potential of applying the membrane bioreactor prototype for the improvement of biodesulfurization reactions and the development of new catalytic/separation systems.

Electrospray bioreactors were investigated for use as desulfurization reactors because of their reported operational cost savings relative to mechanically agitated reactors. Unlike batch-stirred reactors, which mix the biocatalyst containing aqueous phase with the organic feedstock by imparting momentum to the entire bulk solution, electrospray reactors have the potential for tremendous cost savings, creating an emulsion <5 μm in diameter, at a cost of only 3 Watt/L. Power law relationships indicate that mechanically stirred reactors would require 100--1000-fold more energy to create such a fine emulsion, but these relationships generally do not account for the effect of endogenously produced surfactant in the system. Kaufman et al. (1998) investigated the dibenzothiophene-biodesulfurization by *Rhodococcus* sp IGTS8 in batch-stirred and electrospray reactors in a two-phase system process (dibenzothiophene in hexadecane/water). The desulfurization rates ranged between 1 and 5 mg 2-HBP/DCW/h, independent of the employed reactor, because the two reactors were capable of forming a very fine emulsion. Since, IGTS8 produces its own biosurfactant. However, the emulsion phase contactor would be advantageous for systems that do not produce its own biosurfactant that would suffer from mass transfer limitation.

There are very few reports on biodesulfurization process designs and cost analysis. In order to ensure that capital and operating costs for biodesulfurization will be lower than for hydrodesulfurization it is necessary to design a suitable biocatalytic process (Monticello, 2000). The cost of building a bioreactor can be reduced by changing from a mechanically agitated reactor to air-lift designs. An air-lift reactor was used at Enchira Biotechnology Corporation ENBC (formerly EBC) to minimize energy costs (Pacheco, 1999; Monticello, 2000). However, specific details about the process and the results achieved were not published (Kilbane and Le Borgne, 2004). A new type of air-lift reactor with immobilized *Gordonia nitida* CYKS1 cells on a fibrous support was designed (Lee et al., 2005) and used for the biodesulfurization of diesel oil. It was shown that cells immobilized on nylon fibers well sustained their growth accompanied by desulfurization activity during a series of repeated batch runs over extended period of time. Advantages of easy separation of biocatalyst from the treated fuels, high stability, and long lifetime of the biocatalyst using immobilized cells were concluded. Sanchez et al. (2008) reported biodesulfurization of 50:50 water-kerosene emulsions were carried out at 100 mL scale and in a 0.01 m³ airlift reactor with resting cells of the reference strain ATCC 39327 and *Pseudomonas* native strains No 02, 05, and 06. The reactor conditions were 30°C, pH 8.0, and 0.34 m³/h air flow. After seven culture days, the mean sulfur removal for the strains Nº 06 and ATCC 39327 was 64% and 53%, respectively, with a mean calorific power loss of 4.5% for both strains. The use of the native strain Nº 06 and the designed airlift reactor is shown as an alternative for biodesulfurization process and constitute a first step for its scale-up to pilot plant.

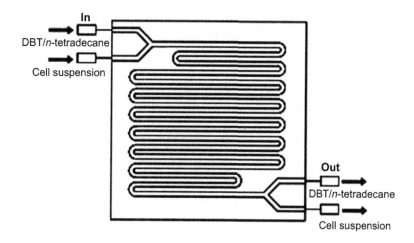

FIGURE 6.8

The Double-Y channel microchannel reactor (Noda et al., 2008).

Enhanced dibenzothiophene biodesulfurization has been accomplished using a microchannel reactor (Noda et al., 2008). The bacterial cell suspension and n-tetradecane containing 1 mM dibenzothiophene were introduced separately into the double-Y channel microfluidic device (Fig. 6.8) both at 0.2–4 µL/min. It was confirmed that a stable n-tetradecane/cell suspension interface was formed in the microchannel and the two liquids were separated almost completely at the exit of the microchannel. An emulsion was generated in the n-tetradecane/cell batch reaction, which decreased the recovery rate in the n-tetradecane phase. By contrast, the emulsion was not observed in the microchannel, and the n-tetradecane phase and cell suspension separated completely. The rate of biodesulfurization in the oil/water phase of the microchannel reaction was more than nine-fold that in a batch (control) reaction. In addition, the microchannel reaction system using a bacterial cell suspension degraded alkylated dibenzothiophene that was not degraded by the batch reaction system. This work provides a foundation for the application of a microchannel reactor system consisting of biological catalysts using an oil/water phase reaction.

Rhodococcus sp. NCIM 2891 has shown high activity to reduce sulfur level in diesel (Mukhopadhyay et al., 2006). The initial sulfur concentration was varied over the range 200–540 ppm. A trickle bed reactor (diameter 0.066 m and height 0.6 m) under continuous mode was studied with the liquid flow rate and inlet sulfur concentration as parameters. Pith balls have been used as the immobilization matrix for the microorganisms, with a constant bed porosity of 0.6. Sulfur conversion up to 99% achieved.

Li et al. (2009b) proposed a method to produce ultralow sulfur diesel by ADS and biodesulfurization in which the adsorbents were regenerated by microbial cells. The adsorption and bioregeneration properties of the adsorbents were studied with *P. delafieldii* R-8 and different types of adsorbents. The regeneration system contained n-octane, aqueous phase, lyophilized cells, and spent adsorbents. All reactions were carried out in 100 mL flasks at 30°C on a rotary shaker operated at 200 rpm. Adsorption–bioregeneration properties were tested in in-situ adsorption–bioregeneration system, which can

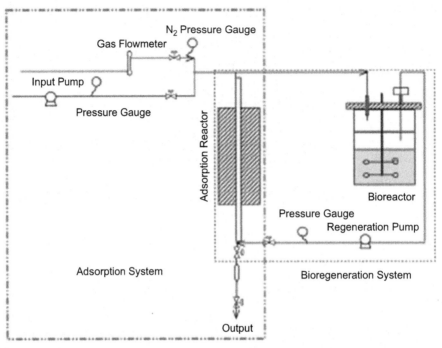

FIGURE 6.9

The in-situ adsorption-bioregeneration system (Li et al., 2009b).

conveniently be divided into two parts: adsorption and bioregeneration (Fig. 6.9). After the saturation of adsorbents, the adsorption system is shut up and adsorption reactor was connected with bioreactor. Then the desorbed sulfur compounds were converted by R-8 cells. Finally, the desorbed adsorbents were treated with air at 550°C to remove water from adsorption system. The integrated system is able to efficiently desulfurize dibenzothiophene and adsorption property of bioregenerated adsorbents is similar to fresh ones.

Immobilized cells have been reported for increased volumetric reaction rate and lower operating costs for biodesulfurization (Lee et al., 2005; Dinamarca et al., 2010) due to the utilization of high cell concentrations of the biocatalyst and increased transport rate of organosulfur compounds to the biocatalyst. In most of the biodesulfurization studies, two separate stages are employed: one for cell growth and the second for induction and biodesulfurization activity. In order to simplify the process and lower the operating cost, a single vertical rotating immobilized cell reactor (VRICR) using *R. erythropolis* IGTS8, glucose as a carbon source, and polyurethane foam for immobilization to desulfurize model oil contains 11.68 mM dibenzothiophene/hexadecane in a ratio of 1:6 oil/water, with a fed of 320 mL/h was reported (Amin, 2011). The highest biodesulfurization activity, 132 mM 2-HBP/kg DCW/h, was recorded within 15 h and then decreased dramatically. Applying 25 g/L of each of glucose and ethanol as carbon source and 11.68 mM dibenzothiophene in hexadecane as sulfur source after 35 h, activity of the reactor was restored and sulfate concentration was decreased and 2-hydroxybiphenyl was detected in bioreactor effluent at 40 h and steadily increased reaching its steady state at 55 h with complete removal of sulfur with specific desulfurization of 166.86 mM 2-hydroxybiphenyl/kg DCW/h for 120 h. The VRICR has longevity of operation, final biomass concentration, and volumetric biodesulfurization activity of 1.4, 3.5, and 7.3 times higher, respectively. This is an important aspect for applying such process in industrial implementation.

Dinamarca et al. (2014b) immobilized *R. rhodochrous* ATCC 53968 on silica with a specific surface area of 80 m^2/g, and investigated its biodesulfurization efficiency in a packed trickle bed bioreactor. It was evidenced that the low substrate flow rate, larger bed length, and larger size of immobilization particles, enabling a sufficient contact time between the microorganism and the organosulfur compounds for expression of microbial metabolism, thus, enhance the biodesulfurization efficiency. Large particle size of the adsorbent improve aeration in packed bed reactor, by offering a greater interparticle volume that can be occupied by both liquid and air. In packed bed reactor, the large particle size dominates over the higher specific surface for the cell immobilization, which is offered by the smaller particle size. Not only is this, but the importance of the meso- and macroporosity of the catalytic bed is also significant in the configuration of packed bed bioreactors for biodesulfurization. The biodesulfurization efficiency of this packed trickle bed reactor was 18% for high sulfur content gas oil (4.7 g sulfur/L). The immobilized cells maintained its biodesulfurization activity over three successive cycles and retained approximately 84% of its initial biodesulfurization efficiency. In contrast to processes using two-phase systems oil/water, the biodesulfurization operating with packed trickle bed reactors containing immobilized bacteria on inorganic support continuously fed with S-containing fuel streams advantageous, simple, effective, and ease of in-situ regeneration of the biocatalytic bed for cyclic operation of biodesulfurization of high sulfur content fuel streams.

6.0 CASE STUDIES

Transportation fuels, such as gasoline, jet fuel, and diesel, are ideal fuels due to their high-energy density, ease of storage and transportation, and established distribution network. However, their sulfur concentration must be less than 10 ppm to protect the deactivation of catalysts in reforming process and electrodes in fuel cell system (de Wild et al., 2006). Accordingly, refineries started to establish new complementary routes in addition to the hydrodesulfurization process. However, hydrodesulfurization has several disadvantages, in that it is energy intensive, costly to install and to operate, and does not work well on refractory organosulfur compounds. Recent research has therefore focused on improving hydrodesulfurization catalysts and processes and on the development of alternative technologies.

Kleshchev (1989) invented the hydrodesulfurization-biodesulfurization process. Where hydrodesulfurization of crude oil is carried out as a first step to remove labile organic sulfides, then biodesulfurization of recalcitrant sulfur compounds, e.g., dibenzothiophene, is performed by *Rhodococcus rhodochrous*.

Agarwal and Sharma (2010) reported the application of three-steps integrated desulfurization process biodesulfurization—oxidative desulfurization—reactive ADS (biodesulfurization-oxidative desulfurization-RADS). That was performed through aerobic and anaerobic biodesulfurization, using higher-density, lower-API gravity crude oil (heavy crude oil, HCO, 1.88% S) and lower-density, higher-API gravity crude oil (light crude oil, LCO, 0.38% S). Biodesulfurization of lower-density, higher-API and higher-density, lower-API gravity crude oil using *Pantoea agglomerans* D23W3 resulted in 61.40%

Table 6.4 Summary of Energy Requirements and Carbon Dioxide Generation From Biodesulfurization Compared to Hydrodesulfurization of Diesel Oil (Linguist and Pacheco, 1999)

Desulfurization Technology	Desulfurization of Diesel Oil From 0.2% to 0.005%		Desulfurization of Diesel Oil From 0.05% to 0.005%	
	Energy Consumption MBTU/bbl	CO_2 Emission lb/bbl	Energy Consumption MBTU/bbl	CO_2 Emission lb/bbl
Hydrodesulfurization	260	29.9	138	15.8
Biodesulfurization	56	6.1	37	4.3
	Decrease from that of Hydrodesulfurization Alone			
Hydrodesulfurization/ biodesulfurization	21.5%	20.4%	26.9%	26.9%

and 63.29% S removal under aerobic conditions, respectively, while, under anaerobic conditions, it showed marginally better results than those under aerobic conditions, recording 63.40% and 69.14%, respectively. The oxidative desulfurization as a second step treatment after anaerobic biodesulfurization of heavy crude oil recoded total desulfurization of approximately 94.15% followed by RADS integration resulted in maximum removal of approximately 95.21%. Meanwhile, oxidative desulfurization as a second step treatment after aerobic biodesulfurization of LCO recoded 78.84% with the RADS integration as a third step recorded total desulfurization of approximately 94.3%.

A comparison study was performed by Linguist and Pacheco (1999) to estimate the energy consumed and the CO_2 generated for biodesulfurization and hydrodesulfurization of diesel oil feedstock. The conducted two cases were biodesulfurization stand-alone or application of biodesulfurization downstream of hydrodesulfurization process (Table 6.4).

From a cost perspective, biodesulfurization has favorable features: (1) operation at low temperature and pressure; (2) biodesulfurization is estimated to have 70%–80% lower carbon dioxide emissions and energy consumption; (3) in the case of reaching adequate biodesulfurization efficiency level, the capital cost required for an industrial biodesulfurization process is predicted to be two-third of the one for a hydrodesulfurization process; and (4) cost effective: for biodesulfurization, the capital and operating costs are 50% and 10%–15% less than hydrodesulfurization, respectively.

However, the application of large-scale biodesulfurization technology is still only at the pilot scale. By the year 2000, the US Department of Energy has awarded a $900,000 grant to Enchira Biotechnology Corp. (EBC) to conduct research and development on the biodesulfurization of gasoline feed stocks, through gene-shuffling technology in an attempt to develop a biocatalyst to remove sulfur from gasoline and diesel fuel (Grisham, 2000). A pilot process has been proposed (Monticello, 2000) as a beginning of the industrial process of biodesulfurization, where three bioreactors are required to achieve very low sulfur concentration (Fig. 6.10). Since the degradation rate and metabolism of currently available microorganisms are still low, three bioreactors are essentially needed to reach low sulfur concentration. Renewing microbial biomass during the biodesulfurization process would make the process commercially viable. US-EBC has developed a pilot with working capacity of five barrels of oil per day (bpd). In the process, the biocatalyst and fossil fuel are mixed in a bioreactor where biodesulfurization takes place, then passes through series of filters to a container for

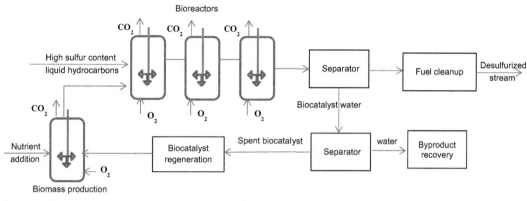

FIGURE 6.10

Biodesulfurization pilot plant.

Adapted from Monticello, D.J., 2000. Biodesulfurization and the upgrading of petroleum distillates. Current Opinion in Biotechnology 11, 540–546.

disposal of sewage. Finally, it is added to a basic aqueous solution for neutralization and removal of sodium sulfate in wastewater treatment station (Boniek et al., 2015).

Enchira Biotechnology Corporation (2001) reported that the process description and the total estimation of capital and operating costs for biodesulfurization of gasoline with initial sulfur content of 350 ppmw to reach 35 ppmw, using the gasoline resistance *P. putida* PpG1, lead to a prediction of $1.77 per barrel of gasoline. The proposed gasoline biodesulfurization process was compared to five existing chemical technologies [the licensors include UOP (ISAL process), ExxonMobil (Octagain), CD Tech (catalytic distillation), IFP (Prime-G), Phillips Petroleum (S-Zorb)], and it was found that the biodesulfurization process is competitive to the existing technologies, where the capital and operating costs for the CD Tech, IFP, and Phillips processes are very competitive and estimated in the range of $2 per barrel.

To be economically viable, a biodesulfurization process must be competitive with other commercially proven desulfurization routes. The route most chosen and best known to refiners for diesel is hydrodesulfurization. A new biodesulfurization facility must be less costly than a comparable new hydrodesulfurization facility. Or, in the various combination scenarios, using the biodesulfurization facility as a pre- or post-treatment facility combined with an existing hydrodesulfurization unit must be less costly than modifications to an existing hydrodesulfurization unit that would be required to achieve the lower sulfur requirement. As part of the study of potential biodesulfurization economics, PetroStar Inc. (Nunn et al., 2006) provided process description and total installed cost of a desulfurization process producing 6000 bpd of highway ULSD at 10 ppmw sulfur, using a straight run diesel containing 5000 ppmw sulfur from PetroStar's Valdez Refinery and *Rhodococcus* sp. The results of the study are summarized in Table 6.5. From the data, the following conclusions can be made regarding the viability of a biodesulfurization process to produce ULSD at PetroStar's Valdez Refinery: (1) a hydrodesulfurization unit has a lower installed cost than a comparable biodesulfurization unit (2) a hydrodesulfurization unit has substantially lower operating costs than a comparable biodesulfurization unit, and (3) the combination of a biodesulfurization unit and a hydrodesulfurization unit is not economically viable when compared to either of the standalone units.

Table 6.5 Estimated Annual Operating Cost of the PetroStar Inc. Desulfurization Process (Nunn et al., 2006)

Desulfurization Process	Annual Operating Cost	
	$	Cent/Gallon ULSD
Biodesulfurization alone	9,232,900	10.7
Hydrodesulfurization alone	5,800,000	6.8
Biodesulfurization followed by hydrodesulfurization	9,400,000	11.1
Hydrodesulfurization followed by biodesulfurization	8,000,000	9.4

In an attempt of Egyptian Petroleum Research Institute to catch the wave of application of nano-technology in the field of petroleum biotechnology, Zaki et al., (2013) reported the preparation of the super-paramagnetic magnetite nanoparticles (9 nm) with good pore size, volume, and high specific surface area, 3.2 nm, 0.198 cm ^{3}/g, and 110.47 m^2/g, respectively, using a reverse water/oil microemulsion method, which showed a good assembling on the Gram positive bacterial isolates, *Brevibacillus invocatus* C19, *Micrococcus* lutes RM1, and *Bacillus clausii* BS1 (Fig. 6.11), with remarkable adsorption capacity to different polyaromatic compounds: dibenzothiophene (69 μmol/g), Pyrene (7.66 μmol/g), and carbazole (95 μmol/g) through π complexation bonding. The coated cells are characterized by higher dibenzothiophene-biodesulfurization, Pyrene-biodegradation, and carbazole-biodenitrogenation rates than the free cells, respectively. Moreover, these magnetite nanoparticle–coated cells are characterized by higher storage and operational stabilities, and low sensitivity toward toxic by-products, and can be reused for four successive cycles without losing its efficiency and have the advantage of magnetic separation, which would resolve many operational problems in petroleum refinery (Saed et al., 2014; Nassar, 2015; Zakaria et al., 2015).

The application of the magnetite nanoparticle–coated *B. clausii* BS1 in biodesulfurization of diesel oil (1:4 O/W) with initial sulfur concentration of 8600 ppm showed better efficiency than that of free cells, recording approximately complete removal and 91% within 4 and 7 days of incubation at 35°C (95°F) and 200 rpm, respectively. A preliminary bench-scale process design of high-sulfur content diesel oil biodesulfurization using magnetic nanoparticles–coated BS1 can be designed (Fig. 6.12). The estimated capital cost for preparation of magnetic nanoparticles is 3.17$/g cell (Table 6.6). Taking into consideration that the free cells lose its activity after the first biodesulfurization cycle of 7 days, so for four cycles of biodesulfurization, the estimated operating and capital cost will be 197.64 and 734.08 $/gallon diesel, respectively with a lower biodesulfurization efficiency. While for the magnetic nanoparticle–immobilized cell system, the immobilized cells can be used in four successive cycles, each of 4 days, without losing its biodesulfurization activity and additionally with higher biodesulfurization rate (approximately complete removal of sulfur). Thus, the estimated operating and capital cost will be 163.42 and 682.87 $/gallon diesel for the four successive cycles. The main cost in this process comes from the media cost; preparation of magnetite nanoparticle–coated cells, 4.4 $/g cell; and the use of centrifugation for preparation of resting cells and separation of diesel oil from aqueous phase. However, there is an overall decrease in the operating and capital cost of immobilized cell system by approximately 18% and 7%, respectively. Further work is needed to decrease the cost of media by using low cost and readily available source of nutrients, and carbon as a cosubstrate, replaces the centrifugation by cyclones and decreases the cost of magnetite nanoparticles preparation and cell-coating process.

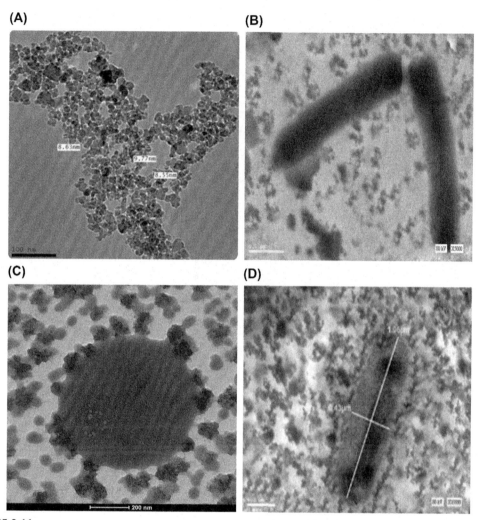

FIGURE 6.11

Transmission electron microscopy micrographs of magnetic Fe_3O_4 (A) and coated bacterial isolates; *Brevibacillus invocatus* C19 (B), *Micrococcus luteus* RM1, (C) and *Bacillus clausii* BS1 (D).

7.0 CHALLENGES AND THE FUTURE PERSPECTIVES

Petroleum refiners are a major source of CO_2 emissions in the industrial sector, to reach the new regulations for ultralow sulfur fuels; applying the conventional hydrodesulfurization technique it will consume extra energy with higher capital, energy, and maintenance costs and extra emissions of CO_2. Thus, the oil industry currently faces one of its greatest challenges: to optimize hydrocarbon production and transformation processes, in order to obtain more efficient and cleaner products that meet international quality regulations and CO_2, SOx, and NOx emission controls. The biodesulfurization technique

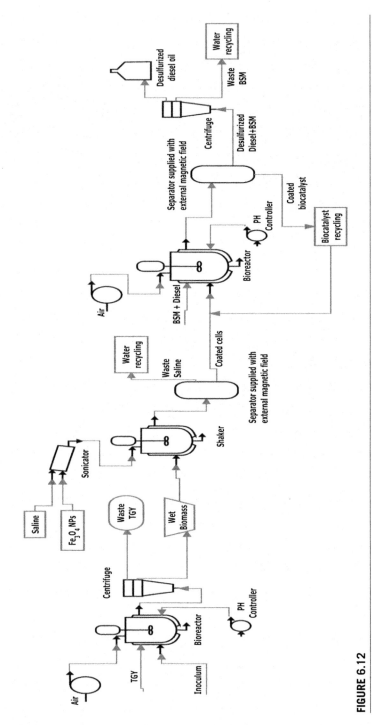

FIGURE 6.12

Biodesulfurization by magnetic nanoparticle–coated bacterial cells.

Table 6.6 Estimated Operating Cost for Biodesulfurization of High Sulfur Content Diesel Oil Using Free and Magnetic Nanoparticle–Coated Bacterial Cells

Fe_3O_4 Nanoparticle–coated cells decrease the S-content from 8600 to 112.66 ppm	
Operating cost $/gallon diesel	Capital cost $/gallon diesel
93.22	376.63
Free cells decrease the S-content from 8600 to 747.34 ppm	
Operating cost $/gallon diesel	Capital cost $/gallon diesel
49.41	183.52

was expected as a cost-effective method of meeting the new lower sulfur standards. However, there are a few reports on biodesulfurization process designs and cost analysis. Based on the advantages of providing lower energy consumption and thus, lower CO_2, mild operating conditions and moreover, with biocatalysts, there will be no need for the unit to operate 4–5 years between shutdowns, since generation of fresh biocatalyst is a continuous process. But, still, the critical factor for commercialization of biodesulfurization is to be a cost-effective one and its ability to be integrated as seamlessly as possible into the existing petrochemical operations. There are a lot of ongoing researches to improve the biodesulfurization process to be competitive with the conventional chemical and physical applied desulfurization methods in petroleum industry.

To apply biodesulfurization, it is important to improve its efficiency and turn the waste into treasure that should be reused. The desulfurization efficiency and economic benefits should be compatible, and the existing facilities should be used to provide quick returns.

Currently, it is thought that two major steps should be made soon to make biodesulfurization worthwhile: a rapid progress in understanding and finding optimal ways to implement biotechnology in refineries (Mohebali and Ball, 2008). At the moment, the process is still too slow, too expensive, and not suited for large-scale implementation. It might be used in an additional refinery step (Soleimani et al., 2007), but further development of hydrodesulfurization, ADS, and oxidative desulfurization might limit its future (Seeberger and Jess, 2010).

Biodesulfurization in nature is different from other more common biotechnology processes and the process has several limitations that prevent it from being applied in a modern refinery (Gupta et al., 2005). The metabolism of sulfur compounds is typically slow compared to chemical reactions employed in a refinery and generally the rate of metabolism is rate limiting in the process, though mass transfer resistance from the oil/water interface to the microbe is also slow compared to the rate of transfer of the sulfur compound to the oil-water interface. Large amounts of biomass are needed (typically 2.5 g biomass per g sulfur), and biological systems must be kept alive to function, which can be difficult under the variable input conditions found in refineries. The rate of desulfurization depends strongly on pH, temperature, and dissolved oxygen concentration. Separation of the cells from the oil can also be difficult; limited lifetime of the biocatalyst and immobilized cells often have lower activity. Furthermore, biodesulfurization involves a multistep metabolic pathway. Thus, whole cells rather than isolated enzyme reactions are needed, since the free or immobilized enzymes would exhibit lower activity than that of whole cell, 0.01 g dibenzothiophene/g protein/h, 0.4 g dibenzothiophene/g aerobic DCW/h, and 0.1 g dibenzothiophene/g anaerobic DCW/h, respectively and also to overcome the problem of expensive cofactors (coenzymes) (Setti et al., 1997).

The efficiency of the biocatalyst and environmental adaptability are the most important factors in the commercialization of biodesulfurization technology. Many researches have been done to isolate or genetically design and clone new strains with high sulfur specificity, broader range of target sulfur compounds, high solvent tolerance, high resistance to inhibitory by-products, and expressing higher biodesulfurization rate. Any small success that provides the possibility to remove sulfur at higher temperature, with higher rate or longer stability of desulfurization activity is considered a significant step toward industry level biodesulfurization. Both directed evolution and gene shuffling can increase biodesulfurization rates and widen the S-substrate range. The recombination techniques can be also applied to create a new recombinant strain with new hybrid enzymes with high activities that meet the needs of refinery operations and go far beyond the needs of the bacteria to remove sulfur from oil feed. Increasing the number of desulfurizing genes copies, manipulation of desulfurizing genes to raise the products with more activity and efficiency and increasing the rate and amount of gene expression can help researchers improve biodesulfurization efficiency (Li et al., 2008a; Sohrabi et al., 2012). Moreover, process development and any unexpected problems that might occur as a result of upscaling the operations must also be considered, for example, the reactivity of the biocatalyst in heavy feedstock, where viscosity and density play an important role in feedstock processing (Bachmann et al., 2014).

Other critical aspects of the process include mass production of biocatalyst, reactor design, oil/water separation, and biocatalyst separation and reusability. The research reported in the area of bioreactor design for biodesulfurization has employed stirred-tank reactor in most biodesulfurization studies. However, the multi-stage airlift reactors and emulsion-phase contractors with free cells reduce mixing costs and promote mass transfer (Mehrnia et al., 2005). The fluidized bed reactors with immobilized whole-cell are often advantageous over the conventional continuous stirred-tank bioreactors: better and easier biocatalyst recovery, increase oil/water volumetric ratio to 90% with respect to 30%–50% in continuous stirred-tank bioreactors, ease of water separation and disposal of a large amount of exhausted medium after the biodesulfurization process, ease of separation and recovery of the biocatalyst and treated oil, good storage stability of immobilized cells, reusability of the biocatalyst, and low risk of microbial pollution (Dinamarca et al. (2014b). Other limitations for commercialization of biodesulfurization include the logistics of stationary handling, shipment, storage, and the sufficient longevity of the biocatalyst. Although immobilization can solve most of these aforementioned limitations, biodesulfurization is not a cost-effective process for heavy or middle distillates of petroleum. In middle distillate fraction, dibenzothiophene and its alkylated derivatives Cx-dibenzothiophenes may reach to 70% of the sulfur compounds present. Application of mixtures of biocatalysts would be necessary to desulfurize a wide range of organosulfur compounds (Marcelis, 2002).

EBC reported that biodesulfurization could be competitive with conventional desulfurization techniques, only on producing valuable by-products, for example, the hydroxybiphenyl sulfinates or sulfonates applying the truncated 4S-pathway and the commercialization of these bioderived by-products in surfactant industry or as starting materials for the synthesis of other useful chemicals (Lange and Lin, 2001).

Environmentally benign and less-energy intensive option of the oxidative desulfurization is bioassisted system. The application of enzymatic-oxidative desulfurization in complementary with the conventional techniques is not far off, throughout the emergence of new in vitro tools for mutation and genetic rearrangement. The random chimeragenesis on transient templates method has applied enzymes with a higher substrate affinity and/or broader substrate range for complex alkylated derivatives of dibenzothiophene (Arsendorf et al., 2002; Song and Ma, 2003). Recent studies on immobilization of enzymes on nanoparticles have achieved good improvement in biodesulfurization efficiency (Derikvand et al., 2014).

One of the major limitations of applying immobilized biocatalysts is the mass transfer problem and the cost of biocatalyst separation and cell loss in some of the applied immobilizing techniques. So far, the new technique in which assembling nanoadsorbents that has selective adsorption to S-compounds from the organic phase on the surfaces of microbial cells can overcome this problem and increase the biodesulfurization rate (Shan et al., 2005d). Moreover, when magnetic nanoparticles are used to coat the cells, it not only has the advantage over conventional cell immobilization by adsorption, but it also successfully overcome the difficulties of mass transfer, cell loss, and separation of the carrier with adsorbed cells from the reaction mixture at the end of biodesulfurization process and does not change the hydrophilicity of the cell surface. The coated layer is looser than the cell wall, thus having a negligible mass transfer effect. Moreover, the coated cells have good stability, can be reused, and are characterized by the advantage of magnetic separation and present further options in continuous reactor systems when used in a magnetically stabilized, fluidized bed. Applying this new convenient and easy-to-perform technique may be promising for large-scale industrial application (Shan et al., 2005c; Li et al., 2009a). The most important factor in this technique is the ratio of nanoparticles to cell biomass. Berry et al. (2005) reported that if nanoparticles on a cell occupy more than two-third of the cell surface area, this will have a negative impact on the cell activity.

Total desulfurization of fossil fuel by microbial approach is not expected to be on-stream in the near future and more research is needed in this field. Since, these challenges of biodesulfurization are not only technical in nature but include the economics of introducing and implementing this new technology and the competitive threats posed by competing, established technologies. In addition, new regulatory challenges have been imposed to attain levels of sulfur in ULSD, which may prove extremely difficult in achieving using biodesulfurization alone. However, adsorptive-, oxidative-, hydro-, and bio-desulfurization are more likely to be employed complimentary to each other (Vazquez-Duhalt et al., 2002b). Thus, in order to develop the biodesulfurization process as a complementary process, a multidisciplinary approach is essential and the participation of scientists and engineers from the fields of biotechnology, biochemistry, refining processes, and engineering is essential. In the long run, applying nanotechnology to desulfurization of petroleum and petroleum products will lead to a process capable of producing ultralow sulfur to no-sulfur products on an economical and cost-effective basis.

With the increase of consumption of light crude oil, the time when available petroleum will predominantly or exclusively be heavy crude oil with high sulfur content is not so far. Here comes an aspect of biodesulfurization that should also be explored, which is the application of the process at the wellhead. Since, biodesulfurization can also play role in reducing the viscosity of crude oil, which is largely due to some superstructures that are linked due to aliphatic sulfide linkages (Srivastava and Kumar, 2008). Many of the biocatalysts that can be applied to crude oil upgrading and desulfurization need water, and in an oil field operation, water is coproduced with the oil so the process of water and oil separation is a routine field process. The only added process step would be agitation of the oil and water mixture with the biocatalyst. Thus, selective biodesulfurization at the wellhead could be very useful and if done just before doing the dewatering and desalting steps in crude oil extraction, will not require any modification in the refinery set up. The process would generate a wastewater stream, which must be handled whether it is generated in the oil field or at the refinery. In a refinery, this new waste stream becomes an added problem. However, in the field, the water containing the formed salts can be diluted and reinjected as part of the field water flood program, which, if allowable under environmental regulations, may have minimal effect on oil field operation. One advantage of such a concept is the expected potential simplicity of the process—the crude oil that is mixed with water and soluble biocatalyst (either the microorganism or the enzyme) and with air. After the reaction, the formed water/oil emulsion is separated to recover the upgraded oil—the biocatalyst remains with the water and is (potentially) available for reuse. The only added feature (that may not always be available at the wellhead) is a mixing reactor prior to separation (El-Gendy and Speight, 2015).

REFERENCES

Abbad-Andaloussi, S., Lagnel, C., Warzywoda, M., Monot, F., 2003. Multi-criteria comparison of resting cell activities of bacterial strains selected for biodesulfurization of petroleum compounds. Enzyme and Microbial Technology 32, 446–454.

Abdul Jalil, T., Falah Hasan, L., 2012. Oxidative desulfurization of gas oil using improving selectivity for active carbon from rice husk. Diyala Journal for Pure Science 8 (3), 68–81.

Acero, J., Berdugo, C., Mogollón, L., 2003. Biodesulfurization process evaluation with a *Gordona rubropertinctus* strain. CT&F - Ciencia, Tecnología y Futuro 2 (4), 43–54.

Adegunlola, G.A., Oloke, J.K., Majolagbe, O.N., Adewoyin, A.G., Adebayo, E.A., Adegunlola, F.O., 2012. Microbial desulphurization of crude oil using *Aspergillus flavus*. European Journal of Experimental Biology 2 (2), 400–403.

Agarwal, P., Sharma, D.K., 2010. Comparative studies on the bio-desulfurization of crude oil with other desulfurization techniques and deep desulfurization through integrated process. Energy & Fuels 24 (1), 518–524.

Akbarzadeh, S., Raheb, J., Aghaei, A., Karkhane, A.A., 2003. Study of desulfurization rate in *Rhodococcus* FMF native bacterium. Iranian Journal of Biotechnology 1 (1), 36–40.

Al Zubaidy, I.A.H., Tarsh, F.B., Darwish, N.N., Abdul Majeed, B.S.S., Al Sharafi, A., Abu Chacra, L., 2013. Adsorption process of sulfur removal from diesel oil using sorbent materials. Journal of Clean Energy Technologies 1 (1), 66–68.

Alvarez, J.C., Ortiz de Montellano, P.R., 1992. Thianthrene 5-oxide as a probe of the electrophilicity of hemoprotein oxidizing species. Biochemistry 31, 8315–8322.

Alves, L., Paixao, S.M., Pacheco, R., Ferreira, A.F., Silva, C.M., 2015. Biodesulfurization of fossil fuels: energy, emissions and cost analysis. RSC Advances 5, 34047–34057.

Amin, G.A., 2011. Integrated two-stage process for biodesulfurization of model oil by vertical rotating immobilized cell reactor with the bacterium *Rhodococcus erythropolis*. Journal of Petroleum and Environmental Biotechnology 2, 107.

Ansari, F., Pavel, G., Libor, S., Tothill, I.E., Ramsden, J.J., 2009. Dibenzothiophene degradation enhancement by decorating Rhodococcus erythropolis IGTS8 with magnetic Fe_3O_4 nanoparticles. Biotechnology and Bioengineering 102 (5), 1505–1512.

Ansari, F., Prayuenyong, P., Tothill, I., 2007. Biodesulfurization of dibenzothiophene by *Shewanella putrefaciens* NCIMB 8768. The Journal of Biological Physics and Chemistry 7, 75–78.

Arabian, D., Najafi, H., Farhadi, F., Dehkordi, A.M., 2014. Biodesulfurization of simulated light fuel oil by a native isolated bacteria *Bacillus cereus* HN. Journal of Petroleum Science and Engineering 4 (1), 31–40.

Aribike, D.S., Susu, A.A., Nwachukwu, S.C.U., Kareem, S.A., 2008. Biodesulfurization of kerosene by *Desulfobacterium indolicum*. Natural Sciences 1 (4), 55–63.

Aribike, D.S., Susu, A.A., Nwachukwu, S.C.U., Kareem, S.A., 2009. Microbial desulfurization of diesel by *Desulfobacterium aniline*. Academia Arena 1 (4), 11–17.

Armstrong, S.M., Sankey, B.M., Voordouw, G., 1995. Conversion of dibenzothiophene to biphenyl by sulfate reducing bacteria isolate from oil field production facilities. Biotechnology Letters 17 (10), 1133–1136.

Arsendorf, J.J., Loomis, A.K., Digrazia, P.M., Monticello, D.J., Pienkos, P.T., 2002. Chemostat approach for the directed evolution of biodesulfurization gain of unction mutants. Applied and Environmental Microbiology 68 (2), 691–698.

Ayala, M., Tinoco, R., Hernandez, V., Bremauntz, P., Vazquez-Duhalt, P., 1998. Biocatalytic oxidation of fuel as an alternative to biodesulfurization. Fuel Processing Technology 57, 101–111.

Ayala, M., Verdin, J., Vazquez-Duhalt, P., 2007. The prospects for peroxidase-based biorefining of petroleum fuels. Biocatalysis and Biotransformation 25 (2–4), 114129.

Babich, I.V., Moulijn, J.A., 2003. Science and technology of novel processes for deep desulfurization of oil refinery streams: a review. Fuel 82 (6), 607–631.

Bachmann, R.T., Johnson, A.C., Edyvean, R.G.J., 2014. Biotechnology in the petroleum industry: an overview. International Biodeterioration & Biodegradation 86, 225–237.

Bagreev, A., Menendez, J.A., Dukhno, I., Tarasenko, Y., Bandosz, T.J., 2004. Bituminous coal-based activated carbons modified with nitrogen as adsorbents of hydrogen sulfide. Carbon 42 (3), 469–476.

Bahuguna, A., Lily, M.K., Munjal, A., Singh, R., Dangwal, K., 2011. Desulfurization of dibenzothiophene (dibenzothiophene) by a novel strain *Lysinibacillus sphaericus* DMT-7 isolated from diesel contaminated soil. Environmental Science & Technology 23, 975–982.

Baldi, F., Pepi, M., Fava, F., 2003. Growth of *Rhodosporidium toruloides* strain DBVPG 6662 on dibenzothiophene crystals and orimulsion. Applied and Environmental Microbiology 69 (8), 4689–4696.

Berdugo, C., Caballero, C., Godoy, R.D., 2002. Aqueous-organic phases separation by membrane reactors in biodesulfurization reactions. CT&F-Ciencia, Tecnología and Futuro 2 (3), 97–112.

Berry, V., Gole, A., Kundu, S., Murphy, C.J., Saraf, R.F., 2005. Deposition of CTAB-terminated nanorods on bacteria to form highly conducting hybrid systems. Journal of the American Chemical Society 127 (50), 17600–17601.

Bhasarkar, J.B., Dikshit, P.K., Moholkar, V.S., 2015. Ultrasound assisted biodesulfurization of liquid fuel using free and immobilized cells of *Rhodococcus rhodochrous* MTCC 3552: a mechanistic investigation. Bioresource Technology 187, 369–378.

Bhatia, S., Sharma, D.K., 2010. Biodesulfurization of dibenzothiophene, its alkylated derivatives, and crude oil by a newly isolated strain of *Pantoea Agglomerans* D23W3. Biochemical Engineering Journal 50 (3), 104–109.

Boniek, D., Figueiredo, D., Santos, A.F.B., Stoianoff, M.A.R., 2015. Biodesulfurization: a mini review about the immediate search for the future technology. Clean Technologies and Environmental Policy 17 (1), 29–37.

Bouchez-Naitali, M., Abbad-Andaloussi, S., Warzywoda, M., Monot, F., 2004. Relation between bacterial strain resistance to solvents and biodesulfurization activity in organic medium. Applied Microbiology and Biotechnology 65, 440–445.

Bressler, D.C., Fedorak, P.M., 2001. Identification of disulfides from the biodegradation of dibenzothiophene. Applied and Environmental Microbiology 67 (11), 5084–5093.

Calzada, J., Alcon, A., Santos, V.E., Garcia-Ochoa, F., 2011. Mixtures of *Pseudomonas putida* CECT 5279 cells of different ages: optimization as biodesulfurization catalyst. Process Biochemistry 46, 1323–1328.

Carlsson, N., Gustafsson, H., Thorn, C., Olsson, L., Holmberg, K., Akerman, B., 2014. Enzymes immobilized in mesoporous silica: a physical–chemical perspective. Advances in Colloid and Interface Science 205, 339–360.

Caro, A., Boltes, K., Leton, P., Garcia-Calvo, E., 2007. Dibenzothiophene biodesulfurization in resting cell conditions by aerobic bacteria. Biochemical Engineering Journal 35, 191–197.

Caro, A., Boltes, K., Letn, P., Garcia-Calvo, E., 2008. Biodesulfurization of dibenzothiophene by growing cells of *Pseudomonas putida* CECT 5279 in biphasic media. Chemosphere 73 (5), 663–669.

Cass, B.J., Schade, F., Robinson, C.W., Thompson, J.E., Legge, R.L., 2000. Production of tomato flavor volatiles from a crude enzyme preparation using a hollow-fiber reactor. Biotechnology and Bioengineering 67 (3), 372–377.

Cassidy, M.B., Lee, H., Trevors, J.T., 1996. Environmental applications of immobilized cells: a review. Journal of Industrial Microbiology & Biotechnology 16, 79–101.

Castorena, G., Suarez, C., Valdez, I., Amador, G., Fernandez, L., Le Borgne, S., 2002. Sulfur-selective desulfurization of dibenzothiophene and diesel oil by newly isolated *Rhodococcus* sp. strains. FEMS Microbiology Letters 215 (1), 157–161.

Chang, J.H., Chang, Y.K., Cho, K.S., Chang, H.N., 2000. Desulfurization of model and diesel oils by resting cells of *Gordona* sp. Biotechnology Letters 22, 193–196.

Chang, J.H., Kim, Y.J., Lee, B.H., Cho, K.S., Ryu, H.W., Chang, Y.K., Chang, H.N., 2001. Production of a desulfurization biocatalyst by two-stage fermentation and its application for the treatment of model and diesel oils. Biotechnology Progress 17 (5), 876–880.

Chapados, D., Bonde, S.E., Chapados, D., Gore, W.L., Dolbear, G., Skov, E., 2000. Desulfurization by selective oxidation and extract of sulfur-containing compounds to economically achieve ultra-low proposed diesel fuel sulfur requirements. In: Proceedings. NPRA Annual Meeting AM-00-25, San Antonio, Texas. March 26–28.

Charlson, R.J., Schwartz, S.E., Hales, J.M., Cess, R.D., Coakley, J.A., Hansen, J.E., Hofmann, D.J., 1992. Climate forcing by anthropogenic aerosols. Science 255, 423–430.

Chen, H., Cai, Y., Zhang, W., Li, W., 2009. Methoxylation pathway in biodesulfurization of model organosulfur compounds with Mycobacterium sp. Bioresource Technology 100, 2085–2087.

Chengying, J., Huizhou, L., Yuchun, X., Jiayong, C., 2002. Isolation of soil bacteria species for degrading dibenzothiophene. Chinese Journal of Chemical Engineering 10 (4), 420–426.

Collins, F.M., Lucy, A.R., Sharp, C.J., 1997. Oxidation desulfurization of oils via hydrogen peroxide and heteropolyanion catalysis. Journal of Molecular Catalysis A: Chemical 117, 397–403.

Colonna, S., Gaggero, N., Carrea, G., Pasta, P., 1994. The microperoxidase-11 catalyzed oxidation of sulfides is enantioselective. Tetrahedron Letters 35, 9103–9104.

Constanti, M., Giralt, J., Bordons, A., 1994. Desulfurization of dibenzothiophene by bacteria. World Journal of Microbiology & Biotechnology 10, 510–516.

Cui, X., Yao, D., Li, H., Yang, J., Hu, D., 2012. Non-magnetic particles as multifunctional microreactor for deep desulfurization. Journal of Hazardous Materials 205–206, 17–23.

Dai, W., Zhou, Y., Wang, S., Su, W., Sun, Y., Zhou, L., 2008. Desulfurization of transportation fuels targeting at removal of thiophene/benzothiophene. Fuel Processing Technology 89 (8), 749–755.

Dai, Y., Shao, R., Qi, G., Ding, B.-B., 2014. Enhanced dibenzothiophene biodesulfurization by immobilized cells of *Brevibacterium lutescens* in n-octane-water biphasic system. Applied Biochemistry and Biotechnology 174 (6), 2236–2244.

Darzins, A., Mrachko, G.T., 2000. Sphingomonas Biodesulfurization Catalyst. US Patent 6,133,016.

de Wild, P.J., Nygnist, R.G., de Brujin, F.A., Stobbe, E.R., 2006. Removal of sulfur-containing odorants from fuel gases for fuel cell-based combined heat and power applications. Journal of Power Sources 159 (2), 995–1004.

Denis, P.A., 2010. On the enthalpy of formation of thiophene. Theoretical Chemistry Accounts 127, 621–626.

Denis-Larose, C., Labbe, D., Nergeron, H., Jones, A.M., Greer, C.W., Al-Hawari, J., Grossman, M.J., Sankey, B.M., Lau, P.C.K., 1997. Conservation of plasmid-encoded dibenzothiophene desulfurization genes in several *rhodococci*. Applied and Environmental Microbiology 63, 2915–2919.

Denome, S.A., Oldfield, C., Nash, L.J., Young, K.D., 1994. Characterization of the desulfurization genes from *Rhodococcus* sp. strain IGTS8. Bacteriology 176 (21), 6707–6716.

Derikvand, P., Etemadifar, Z., 2014. Improvement of biodesulfurization rate of alginate immobilized *Rhodococcus erythropolis* R1. Jundishapur Journal of Microbiology 7 (3), e9123. http://dx.doi.org/10.5812/jjm.9123.

Derikvand, P., Etemadifar, Z., Biria, D., 2014. Taguchi optimization of dibenzothiophene biodesulfurization by *Rhodococcus erythropolis* R1 immobilized cells in a biphasic system. International Biodeterioration & Biodegradation 86, 343–348.

Dinamarca, M.A., Ibacache-Quiroga, C., Baeza, P., Galvez, S., Villarroel, M., Olivero, P., Ojeda, J., 2010. Biodesulfurization of gas oil using inorganic supports biomodified with metabolically active cells immobilized by adsorption. Bioresource Technology 101, 2375–2378.

Dinamarca, M.A., Rojas, A., Baeza, P., Espinoza, G., Ibacache-Quiroga, C.I., Ojeda, J., 2014a. Optimization the biodesulfurization of gas oil by adding surfactants to immobilized cell systems. Fuel 116, 237–241.

Dinamarca, M.A., Orellana, L., Aguirre, J., Baeza, P., Espinoza, G., Canales, C., Ojeda, J., 2014b. Biodesulfurization of dibenzothiophene and gas oil using a bioreactor containing a catalytic bed with *Rhodococcus rhodochrous* immobilized on silica. Biotechnology Letters 36, 1649–1652.

Doerge, D.R., Cooray, N.M., Brewster, M.E., 1991. Peroxidase-catalyzed S-oxygenation: mechanism of oxygen transfer for lactoperoxidase. Biochemistry 30, 8960–8964.

Duarte, G.F., Rosado, A.S., Seldin, L., Araujo, W., Van Elsa, J., 2001. Analysis of bacterial community structure in sulfurous-oil-containing soils and detection of species carrying dibenzothiophene desulfurization (dsz) Genes. Applied and Environmental Microbiology 67 (3), 1052–1062.

El-Gendy, N.Sh., 2006. Biodegradation potentials of dibenzothiophene by new bacteria isolated from hydrocarbon polluted soil in Egypt. Biosciences Biotechnology Research Asia 3 (1a), 95–106.

El-Gendy, N.Sh., Moustafa, Y.M., Habib, S.A., Ali, S., 2010. Evaluation of *Corynebacterium variabilis* Sh42 as a degrader for different poly aromatic compounds. Journal of American Science 6 (11), 343–356.

El-Gendy, N.Sh., Abo-State, M.A., 2008. Isolation, characterization and evaluation of *Staphylococcus gallinarum* NK1 as a degrader for dibenzothiophene, phenanthrene and naphthalene. Egyptian Journal of Petroleum 17 (2), 75–91.

El-Gendy, N.Sh., Farahat, L.A., Moustafa, Y.M., Shaker, N., El-Temtamy, S.A., 2006. Biodesulfurization of crude and diesel oil by *Candida parapsilosis* NSh45 isolated from Egyptian hydrocarbon polluted sea water. Biosciences Biotechnology Research Asi 3 (1a), 5–16.

El-Gendy, N.Sh., Nassar, H.N., Abu Amr, S.S., 2014. Factorial design and response surface optimization for enhancing a biodesulfurization process. Petroleum Science and Technology 32 (14), 1669–1679.

El-Gendy, N.Sh., Speight, J.G., 2015. Handbook of Refinery Desulfurization. CRC Press, Taylor & Francis Group, Boca Raton, Florida.

Enchira Biotechnology Corporation, 2001. Gasoline Desulfurization Program. DE-FC07–97ID13570. Final Report.

Etemadifar, Z., Derikvand, P., Emtiazi, G., Habibi, M.H., 2014. Response surface methodology optimization of dibenzothiophene biodesulfurization in model oil by nanomagnet immobilized *Rhodococcus erythropolis* R1. Journal of Materials Science and Engineering: B 4 (10), 322–330.

Fang, X.X., Zhang, Y.L., Luo, L.L., Xu, P., Chen, Y.L., Zhou, H., Hai, L., 2006. Organic sulfur removal from catalytic diesel oil by hydrodesulfurization combined with biodesulfurization. Mod. Chem. Ind. 26, 234–238.

Feng, J., Zeng, Y., Ma, C., Cai, X., Zhang, Q., Tong, M., Yu, B., Xu, P., 2006. The surfactant tween 80 enhances biodesulfurization. Applied and Environmental Microbiology 72 (11), 7390–7393.

Fernandes, P., Vidinha, P., Ferreira, T., Silvestre, H., Cabral, J.M.S., Prazeres, D.M.F., 2002. Use of free and immobilized *Pseudomonas putida* cells for the reduction of a thiophene derivative in organic media. Journal of Molecular Catalysis B: Enzymatic 19–20, 353–361.

Finnerty, W.R., 1993. Organic sulfur biodesulfurization in non-aqueous media. Fuel 72, 1631–1634.

Folsom, B., Schieche, D., DiGarazia, P., Werner, J., Palmer, S., 1999. Microbial desulfurization of alkylated dibenzothiophenes from a hydrodesulfurized middle distillate by *Rhodococcus erythropolis* I-19. Applied and Environmental Microbiology 65 (11), 4967–4972.

Freeman, A., Lilly, M.D., 1998. Effect of processing parameters on the feasibility and operational stability of immobilized viable microbial cells. Enzyme and Microbial Technology 23, 335–345.

Furuya, T., Ishii, Y., Noda, K., Kino, K., Kirimura, K., 2003. Thermophilic biodesulfurization of hydrodesulfurized light gas oils by *Mycobacterium phlei* WU-F1. FEMS Microbiology Letters 221 (1), 137–142.

Furuya, T., Kirimura, K., Kino, K., Usami, S., 2001. Thermophilic biodesulfurization of dibenzothiophene and its derivatives by *Mycobacterium phlei* WU-F1. FEMS Microbiology Letters 204, 129–133.

García-Gutiérrez, J.L., Fuentes, G.A., Hernandez-Teran, M.E., Garcıa, P., Murrieta-Guevara, F., Jimenez-Cruz, F., 2008. Ultra-deep oxidative desulfurization of diesel fuel by the Mo/Al$_2$O$_3$-H$_2$O$_2$ system: the effect of system parameters on catalytic activity. Applied Catalysis A: General 334, 366–373.

Gary, J.H., Handwerk, G.E., Kaiser, M.J., 2007. Petroleum Refining: Technology and Economics, fifth ed. CRC Press, Taylor & Francis Group, Boca Raton, Florida.

Gatan, R., Barger, P., Gembicki, V., Cavanna, A., Molinari, D., 2004. Oxidative desulfurization: a new technology for ULSD. Preprints Division of Energy and Fuels - American Chemical Society 49 (2), 577–579.

Giorno, L., Drioli, E., 2000. Biocatalytic membrane reactors: applications and perspectives. Trends in Biotechnology 18, 339–349.

Giuliano, M., Schiraldi, C., Maresca, C., Esposito, V., Rosa, M.D., 2003. Immobilized *Proteus mirabilis* in poly(vinyl alcohol) cryogels for L(–)-carnitine production. Enzyme and Microbial Technology 32, 507–512.

Gore, W., 2001. Method of Desulfurization of Hydrocarbons. US Patent 6,274,785.

Gray, K.A., Pogrebinsky, O., Mrachko, G., Xi, L., Monticello, D., Squires, C.H., 1996. Molecular mechanisms of biocatalytic desulfurization of fossil fuels. Nature Biotechnology 14, 1705–1709.

Gray, K.A., Machkoyz, T., Squires, C.H., 2003. Biodesulfurization of fossil fuels. Current Opinion in Microbiology 6, 229–235.

Grisham, J., 2000. Biocatalytic success. Nature Biotechnology 18, 701.

Grossman, M.J., Lee, M.K., Prince, R.C., Garrett, K.K., George, G.N., Pickering, I.J., 1999. Microbial desulfurization of a crude oil middle-distillate fraction: analysis of the extent of sulfur removal and the effect of removal on remaining sulfur. Applied and Environmental Microbiology 65, 181–188.

Grossman, M.J., Lee, M.K., Prince, R.C., Minak-Bernero, V., George, G.N., Pickering, I.J., 2001. Deep Desulfurization of extensively hydrodesulfurized middle distillate oil by *Rhodococcus* sp. strain ECRD-1. Applied and Environmental Microbiology 67 (4), 1949–1952.

Gün, G., Yürüm, Y., Doğanay, G.D., 2015. Revisiting the biodesulfurization capability of hyperthermophilic archaeon Sulfolobus solfataricus P2 revealed dibenzothiophene consumption by the organism in an oil/water two-phase liquid system at high temperatures. Turkish Journal of Chemistry 39 (2), 255–266.

Gunam, I.B.W., Yaku, Y., Hirano, M., Yamamura, K., Tomita, F., Sone, T., Asano, K., 2006. Biodesulfurization of alkylated forms of dibenzothiophene and benzothiophene by *Sphingomonas subarctica* T7b. Journal of Bioscience and Bioengineering 101, 322–327.

Guo, W., Wang, C., Lin, P., Lu, X., 2011. Oxidative desulfurization of diesel with TBHP/Isobutyl aldehyde/air oxidation system. Applied Energy 88, 175–179.

Guobin, S., Jianmin, X., Huaiying, Z., Huizhou, L., 2005. Deep desulfurization of hydrodesulfurized diesel oil by *Pseudomonas delafieldii* R-8. Journal of Chemical Technology and Biotechnology 80, 420–424.

Guobin, S., Huaiying, Z., Jianmin, X., Guo, C., Wangliang, L., Huizhou, L., 2006. Biodesulfurization of hydrodesulfurized diesel oil with *Pseudomonas delafieldii* R-8 from high density culture. Biochemical Engineering Journal 27, 305–309.

Gupta, N., Roychoudhury, P.K., Dep, J.K., 2005. Biotechnology of desulfurization of diesel: prospects and challenges. Applied Microbiology and Biotechnology 66, 356–366.

Han, J.W., Park, H.S., Kim, B.H., Shin, P.G., Park, S.K., Lim, J.C., 2001. Potential use of nonionic surfactants in the biodesulfurization of bunker-C oil. Energy & Fuels 15 (1), 189–196.

Hashimoto, S., Furukawa, K., 1987. Immobilization of activated sludge by the PVA–boric acid method. Biotechnology and Bioengineering 30 (1), 52–59.

Haukanes, B.I., Kvam, C., 1993. Application of magnetic beads in bioassays. Nature Biotechnology 11, 60–63.

Hernández-Maldonado, A.J., Yang, R.T., 2004a. New sorbents for desulfurization of diesel fuels via π-complexation. AIChE J 50 (4), 791–801.

Hernández-Maldonado, A.J., Yang, R.T., 2004b. Desulfurization of diesel fuels by adsorption via p-complexation with vapor phase exchanged (VPIE) Cu(I)-Y zeolites. Journal of the American Chemical Society 126, 992–993.

Hernández-Maldonadoa, A.J., Yang, R.T., 2004. Desulfurization of transportation fuels by adsorption. Catalysis Reviews - Science and Engineering 46, 111–150.

Hou, Y., Kong, Y., Yang, J., Zhang, J., Shi, D., Xin, W., 2005. Biodesulfurization of dibenzothiophene by immobilized cells of *Pseudomonas stutzeri* UP-1. Fuel 84, 1975–1979.

Hsu, C.S., Robinson, P.R., 2006. Practical Advances in Petroleum Processing Volume 1 and Volume 2. Springer Science, New York.

Huang, T., Qiang, L., Zelong, W., Daojiang, Y., Jianmin, X., 2012. Simultaneous removal of thiophene and dibenzothiophene by immobilized *Pseudomonas delafieldii* R-8 cells. Chinese Journal of Chemical Engineering 20 (1), 47–51.

Hwan, J., Keun, Y., Wook, H., Nam, H., 2000. Desulfurization of light gas oil in immobilized-cell systems of Gordona sp. CYKS1 and Nocardia sp. CYKS2. FEMS Microbiology Letters 182, 309–312.

Irani, Z.A., Mehrnia, M.R., Yazdian, F., Soheily, M., Mohebali, G., Rasekh, B., 2011. Analysis of petroleum biodesulfurization in an airlift bioreactor using response surface methodology. Bioresource Technology 102, 10585–10591.

Ishii, Y., Kozaki, S., Furuya, T., Kino, K., Kirimura, K., 2005. Thermophilic biodesulfurization of various heterocyclic sulfur compounds and crude straight-run light gas oil fraction by a newly isolated strain *Mycobacterium phlei* WU-0103. Current Microbiology 50, 63–70.

Javadli, J., De Klerk, A., 2012. Desulfurization of heavy oil. Applied Petrochemical Research 1 (1–4), 3–19.

Jia, X., Wen, J., Sun, Z., Caiyin, Q., Xie, S., 2006. Modeling of dibenzothiophene biodegradation behaviors by resting cells of *Gordonia* sp. WQ-01 and its mutant in oil–water dispersions. Chemical Engineering Science 61, 1987–2000.

Joskić, R., Dunja, M., Sertić-Bionda, K., 2014. Oxidative desulfurization of model diesel fuel with hydrogen peroxide. Goriva I Maziva 53 (1), 11–18.

Juarez-Moreno, K., Diaz de Leon, J.N., Zepeda, T.A., Vazquez-Duhalt, R., Fuentes, S., 2015. Oxidative transformation of dibenzothiophene by chloroperoxidas enzyme immobilized on (1D)- γ-Al$_2$O$_3$ nanorods. Journal of Molecular Catalysis B: Enzymatic 115, 90–95.

Juarez-Moreno, K., Pestryakov, A., Petranovskii, V., 2014. Engineering of supported nanomaterials. Procedia Chemistry 10, 25–30.

Kadijani, J.A., Narimani, E., Kadijani, H.A., 2014. Oxidative desulfurization of organic sulfur compounds in the presence of molybdenum complex and acetone as catalysts. Petroleum and Coal 56 (1), 116–123.

Kardinahl, S., Schmidt, C., Petersen, C.L., Schafer, G., 1996. Isolation, characterization and crystallization of an iron-superoxide dismutase from the crenarchaeon *Sulfolobus acidocaldarius*. FEMS Microbiology Letters 138 (1), 65–70.

Kareem, S.A., 2014. Anaerobic microbial desulfurization of kerosene. Journal of Nanomedicine & Nanotechnology 5, 5.

Kareem, S.A., Aribike, D.S., Nwachukwu, S.C., Latinwo, G.K., 2012. Microbial desulfurization of diesel by *Desulfobacterium indolicum*. Journal of Environmental Science & Engineering 54 (1), 98–103.

Kargi, F., Robinson, J.M., 1984. Microbial oxidation of dibenzothiophene by the thermophilic organism *Sulfolobus acidocaldarius*. Biotechnology and Bioengineering 26 (7), 687–690.

Kaufman, E.N., Harkins, J.B., Borole, A.P., 1998. Comparison of batch-stirred and electrospray reactors for biodesulfurization of dibenzothiophene in crude oil and hydrocarbon feedstocks. Applied Biochemistry and Biotechnology 73 (2), 127–144.

Kilbane, J.J., Le Borgne, S., 2004. Petroleum biorefining: the selective removal of sulfur, nitrogen, and metals. In: Vazquez-Duhalt, R., Quintero-Ramirez, R. (Eds.), Petroleum Biotechnology, Developments and Perspectives. Elsevier, Amsterdam, Netherlands, pp. 29–65.

Kilbane, J.J., 1994. Microbial Cleavage of Organic C-s Bonds. US Patent 5,358,869.

Kilbane, J.J., 2006. Microbial biocatalysts developments to upgrade fossil fuels. Current Opinion in Biotechnology 17, 305–314.

Kilbane, J.J., Jackoswki, F., 1992. Biodesulfurization of water soluble coal-derived material by *Rhodococcus rhodochrous* IGTS8. Biotechnology and Bioengineering 40 (9), 1107–1114.

Kim, B.H., Shin, P.K., Na, J.U., Park, D.H., Bang, S.H., 1996. Microbial petroleum desulfurization. Journal of Microbiology and Biotechnology 6 (5), 299–308.

Kim, T.S., Kim, H.Y., Kim, B.H., 1990. Petroleum desulfurization by *Desulfovibrio desulfuricans* M6 using electrochemically supplied reducing equivalent. Biotechnology Letters 12 (10), 757–760.

Kim, Y.J., Chang, J.H., Cho, K.S., Ryu, H.W., Chang, Y.K., 2004. A physiological study on growth and dibenzothiophene (dibenzothiophene) desulfurization characteristics of *Gordonia* sp. CYKS1. Korean Journal of Chemical Engineering 21, 436–441.

Kirimura, K., Furuya, T., Nishii, Y., Yoshitaka, I., Kino, K., Usami, S., 2001. Biodesulfurization of dibenzothiophene and its derivatives through the selective cleavage of carbon-sulfur bonds by a moderately thermophilic bacterium *Bacillus subtilis* WU-S2B. Journal of Bioscience and Bioengineering 91 (3), 262–266.

Klein, J., Stock, J., Vorlop, D.K., 1993. Pore size and properties of spherical calcium alginate biocatalysts. European Journal of Applied Microbiology and Biotechnology 18, 86–91.

Kleshchev, S.M., 1989. Method of Removing Organic Sulfur Compounds from Petroleum Products. RU-patent: 1,505,960.

Kobayashi, M., Onaka, T., Ishii, Y., Konishi, J., Takaki, M., Okada, H., Ohta, Y., Koizumi, K., Suzuki, M., 2001. Desulfurization of alkylated forms of both dibenzothiophene and benzothiophene by a single bacterial strain. FEMS Microbiology Letters 197, 123–126.

Kobayashi, S., Nakano, M., Goto, T., Kimura, T., Schaap, A.P., 1986. An evidence of the peroxidase-dependent oxygen transfer from hydrogen peroxide to sulfides. Biochemical and Biophysical Research Communications 135, 166–171.

Kohler, M., Genz, I.L., Schicht, B., Eckart V, 1984. Microbial desulfurization of petroleum and heavy petroleum fractions. 4. Anaerobic degradation of organic sulfur compounds. Zentralblatt für Mikrobiologie 139, 239–247.

Konishi, M., Kishimoto, M., Tamesui, N., Omasa, I., Shioya, S., Ohtake, H., 2005. The separation of oil from an oil–water–bacteria mixture using a hydrophobic tubular membrane. Biochemical Engineering Journal 24, 49–54.

La Paz Zavala, C., Rodriguez, J.E., 2004. Practical applications of a process simulator of middle distillates hydrodesulfurization. Petroleum Science and Technology 22 (1/2), 61–71.

Labana, S., Pandey, G., Jain, R.K., 2005. Desulphurization of dibenzothiophene and diesel oils by bacteria. Letters in Applied Microbiology 40 (3), 159–163.

Lange, E.A., Lin, Q., 2001. Compositions Comprising 2-(2-hydroxyphenyl) Benzene Sulfinate and Alkyl Substituted Derivatives Thereof. US Patent 6,303,562.

Le Borgne, S., Quintero, R., 2003. Biotechnological processes for the refining of petroleum. Fuel Processing Technology 81, 155–169.

Lee, I.S., Bae, H., Ryu, H.W., Cho, K., Chang, Y.K., 2005. Biocatalytic desulfurization of diesel oil in an air-lift reactor with immobilized *Gordonia nitida* CYKS1 Cells. Biotechnology Progress 21, 781–785.

Lee, M.K., Senius, J.D., Grossman, M.J., 1995. Sulfur-specific microbial desulfurization of sterically hindered analogs of dibenzothiophene. Applied and Environmental Microbiology 61, 4362–4366.

Lee, S.H.D., Kumar, R., Krumpelt, M., 2002. Sulfur removal from diesel fuel contaminated methanol. Separation and Purification Technology 26, 247–258.

Leong, Y.K., Seah, U., Chu, S.Y., Ovy, B.V., 2001. Effects of gum Arabic macromolecules on surface forces in oxide dispersions. Colloids and Surfaces A: Physicochemical and Engineering 182, 263–268.

Li, F.L., Xu, P., Ma, C.Q., Luo, L.L., Wang, X.S., 2003. Deep desulfurization of hydrodesulfurization-treated diesel oil by a facultative thermophilic bacterium *Mycobacterium* sp. X7B. FEMS Microbiology Letters 223, 301–307.

Li, F., Xu, P., Feng, J., Meng, L., Zheng, Y., Luo, L., Ma, C., 2005. Microbial desulfurization of gasoline in a *Mycobacterium goodii* X7B immobilized-cell system. Applied and Environmental Microbiology 71 (1), 276–281.

Li, W., Xing, J., Li, Y., Xiong, X., Li, X., Liu, H., 2006. Feasibility study on the integration of adsorption/bioregeneration of π-complexation adsorbent for desulfurization. Industrial & Engineering Chemistry Research 45 (8), 2845–2849.

Li, Y.G., Ma, J., Zhang, Q.Q., Wang, C.S., Chen, Q., 2007a. Sulfur-selective desulfurization of dibenzothiophene and diesel oil by newly isolated *Rhodococcus erythropolis* NCC-1. Chinese Journal of Organic Chemistry 25, 400–405.

Li, F., Zhang, Z., Feng, J., Cai, X., Xu, P., 2007b. Biodesulfurization of dibenzothiophene in tetradecane and crude oil by a facultative thermophilic bacterium *Mycobacterium goodii* X7B. Journal of Biotechnology 127, 222–228.

Li, G.-G., Li, S.-S., Zhang, M.-J., Wang, J., Zhu, L., Liang, F.-L., Liu, R.-J., Ma, T., 2008a. Genetic rearrangement strategy for optimizing the dibenzothiophene biodesulfurization pathway in *Rhodococcus erythropolis*. Applied and Environmental Microbiology 74 (4), 971–976.

Li, Y.G., Xing, J.M., Xiong, X.C., Li, W.L., Gao, S., Liu, H.Z., 2008b. Improvement of biodesulfurization activity of alginate immobilized cells in biphasic systems. Journal of Industrial Microbiology & Biotechnology 35, 145–150.

Li, W., Xing, J., Li, Y., Xiong, X., Li, X., Liu, H., 2008c. Desulfurization and bio-regeneration of adsorbents with magnetic *P. delafieldii* R-8 cells. Catalysis Communications 9, 376–380.

Li, Y.G., Gao, H.S., Li, W.L., Xing, J.M., Liu, H.Z., 2009a. In situ magnetic separation and immobilization of dibenzothiophene-desulfurizing bacteria. Bioresource Technology 100, 5092–5096.

Li, W., Tang, H., Liu, Q., Xing, J., Li, Q., Wang, D., Yang, M., Li, X., Liu, H., 2009b. Deep desulfurization of diesel by integrating adsorption and microbial method. Biochemical Engineering Journal 44, 297–301.

Li, F.-T., Liu, Y., Sun, Z.-M., Zhao, Y., Liu, R.-H., Chen, L.-G., Zhao, D.-S., 2012. Photocatalytic oxidative desulfurization of dibenzothiophene under simulated sunlight irradiation with mixed-phase Fe_2O_3 prepared by solution combustion. Catalysis Science & Technology 2, 1455–1462.

Li, W., Jiang, X., 2013. Enhancement of bunker oil biodesulfurization by adding. World Journal of Microbiology & Biotechnology 29, 103–108.

Li, W., Zhao, D., 2013. An overview of the synthesis of ordered mesoporous materials. Chemical Communications 49, 943–946.

Linguist, L., Pacheco, M., 1999. Enzyme-based diesel desulfurization process offers energy, CO_2 advantages. Oil Gas J 97, 45–48.

Liotta, F.J., Han, Y.Z., 2003. Production of ultra-low sulfur fuels by selective hydroperoxide oxidation. In: Proceedings. NPRA Annual Meeting AM-03-23, San Antonio, Texas. March 26–28.

Liu, L., Zhang, Y., Tan, W., 2014. Ultrasound-Assisted oxidation of dibenzothiophene with phosphotungstic acid supported on activated carbon. Ultrasonics Sonochemistry 21 (3), 970–974.

Liu, S., Wang, B., Cui, B., Sun, L., 2008. Deep desulfurization of diesel oil oxidized by Fe(vi) systems. Fuel 87, 422–428.

Liu, X., Kaminski, M.D., Guan, Y., Chen, H., Liu, H., Rosengart, A.J., 2006. Preparation and characterization of hydrophobic superparamagnetic gel. Journal of Magnetism and Magnetic Materials 306, 248–253.

Liu, X.Q., Liu, H.Z., Xing, J.M., Guan, Y.P., Ma, Z.Y., Shan, G.B., Yang, C.L., 2003. Preparation and characterization of superparamagnetic functional polymeric microparticles. China Particuology 1, 76–79.

Lozinsky, V.I., Plieva, F.M., 1998. Poly (vinyl alcohol) cryogels employed as matrices for cell immobilization. 3. Overview of recent research and development. Enzyme and Microbial Technology 23, 227–242.

Ma, T., Li, G., Li, J., Liang, F., Liu, R., 2006. Desulfurization of dibenzothiophene by *Bacillus subtilis* recombinants carrying dszABC and dszD Genes. Biotechnology Letters 28, 1095–1100.

Maghsoudi, S., Vossoughi, M., Kheirolomoom, A., Tanaka, E., Katoh, S., 2001. Biodesulfurization of hydrocarbons and diesel fuels by *Rhodococcus* sp. strain P32C1. Biochemical Engineering Journal 8, 151–156.

Malik, K.A., 1978. Microbial removal of organic sulphur from crude oil and the environment: some new perspectives. Process Biochemistry 13, 10–13.

Marcelis, C., 2002. Anaerobic Biodesulfurization of Thiophenes (PhD Thesis). Wageningen University, Wageningen, Netherlands.

Mass, D., de Oliveira, D., de Souza, A.A., Souza, S.M., 2014. Biodesulfurization of a system containing synthetic fuel using *Rhodococcus erythropolis* ATCC 4277. Applied Biochemistry and Biotechnology 174 (6), 2079–2085.

Matsui, T., Hirasawa, K., Koizumi, K., Maruhashi, K., Kurane, R., 2001. Effect of dsz D gene expression on benzothiophene degradation of *Rhodococcus* sp. strain T09. Process Biochemistry 37 (1), 31–34.

McFarland, B.L., 1999. Biodesulfurization. Current Opinion in Microbiology 2, 257–264.

McKinley, S.G., Angelici, R.J., 2003. Deep Desulfurization by selective adsorption of dibenzothiophenes on Ag$^+$/SBA-15 and Ag$^+$/SiO$_2$. Chemical Communications 20, 2620–2621.

Mehrnia, M.R., Towfighi, J., Bonakdarpour, B., Akbarnejad, M.M., 2005. Gas hold-up and oxygen transfer in a draft-tube airlift bioreactor with petroleum-based liquids. Journal of Biochemical Engineering 22, 105–110.

Miller, K.W., 1992. Reductive desulfurization of dibenzyl disulfide. Applied and Environmental Microbiology 58 (7), 2176–2179.

Mingfang, L., Zhongxuan, G., Jianmin, X., Huizhou, L., Jiayong, C., 2003. Microbial desulfurization of model and straight-run oils. Journal of Chemical Technology and Biotechnology 78, 873–876.

Mjalli, F.S., Ahmed, A.U., Al-Wahaibi, T., Al-Wahaibi, Y., Al Nashef, I.M., 2014. Deep oxidative desulfurization of liquid fuels. Reviews in Chemical Engineering 30 (4), 337–378.

Mohebali, G., Ball, A.S., 2008. Biocatalytic desulfurisation (biodesulfurization) of petrodiesel fuels. Microbiology 154, 2169–2183.

Molday, R.S., Yen, S.P., Rembaum, A., 1977. Application of magnetic microspheres in labelling and separation of cell. Nature 268, 437–438.

Monticello, D.J., 2000. Biodesulfurization and the upgrading of petroleum distillates. Current Opinion in Biotechnology 11, 540–546.

Monticello, D.J., Bakker, D., Finnerty, W.R., 1985. Plasmid-mediated degradation of dibenzothiophene by Pseudomonas species. Applied and Environmental Microbiology 49 (4), 756–760.

Monticello, D.J., Haney, I.I.I., William, M., 1996. Biocatalytic Process for Reduction of Petroleum Viscosity. US Patent 5,529,930.

Monticello, D.J., Kilbane, J.J., 1994. Microemulsion Process for Direct Biocatalytic Desulfurization of Organosulfur Molecules. US Patent 5,358,870.

Monticello, D.J., 1996. Multistage Process for Deep Desulfurization of a Fossil Fuel. US Patent 5,510,265.

Montiel, C., Quintero, R., Aburto, J., 2009. Petroleum biotechnology: technology trends for the future. African Journal of Biotechnology 8 (12), 2653–2666.

Mukhopadhyay, M., Chowdhury, R., Bhattacharya, P., 2006. Biodesulfurization of hydrodesulfurized diesel in a tickle bed reactor – experiments and modelling. Journal of Scientific & Industrial Research 65, 432–436.

Murata, S., Murata, K., Kidena, K., Nomura, M., 2003. Oxidative desulfurization of diesel fuels by molecular oxygen. Preprints of Papers American Chemical Society, Division of Fuel Chemistry 48 (2), 531.

Murphy, S.E., Amin, S., Coletta, K., Hoffmann, D., 1992. Rat liver metabolism of benzo[*b*]naphtho[2,1-*d*]thiophene. Chemical Research in Toxicology 5, 491–495.

Mužic, M., Sertić-Bionda, K., 2013. Alternative processes for removing organic sulfur compounds from petroleum fractions. Chemical and Biochemical Engineering Quarterly 27 (1), 101–108.

Mužic, M., Sertić-Bionda, K., Adžamić, T., 2011. Desulfurization of diesel fuel in a fixed bed adsorption column: experimental study and simulation. Petroleum Science and Technology 29 (22), 2361–2371.

Mužic, M., Sertić-Bionda, K., Adžamić, T., Gomiz, Z., 2010. A design of experiments investigation of adsorptive desulfurization of diesel fuel. Chemical and Biochemical Engineering Quarterly 24 (3), 253–264.

Naito, M., Kawamoto, T., Fujino, K., Kobayashi, M., Maruhashi, K., Tanaka, A., 2001. Long term repeated biodesulfurization by immobilized *Rhodococcus*. Applied Microbiology and Biotechnology 55, 374–378.

Nandi, S., 2010. Biodesulfurization of hydro-desulfurized diesel in airlift reactor. Journal of Scientific & Industrial Research 69, 543–547.

Nanoti, A., Dasgupta, S., Goswami, A.N., Nautiyal, B.R., Rao, T.V., Sain, B., Sharma, Y.K., Nanoti, S.M., Garg, M.O., Gupta, P., 2009. Mesoporous silica as selective sorbents for removal of sulfones from oxidized diesel fuel. Microporous and Mesoporous Materials 124, 94–99.

Nasab, N.A., Kumleh, H.H., Kazemzad, M., Panjeh, F.G., Davoodi-Dehaghani, F., 2015. Improvement of desulfurization performance of *Rhodochrous erythropolis* IGTS8 by assembling spherical mesoporous silica nanosorbents on the surface of the bacteria cells. Journal of Applied Chemical Research 9 (2), 81–91.

Nassar, H.M., 2009. Potentials of Microorganisms Isolated from Egyptian Hydrocarbon Polluted Sites on Degradation of Polycyclic Aromatic Sulfur Heterocycles (Polycyclic Aromatic Sulfur Heterocyclic Derivatives) Compounds (M.Sc. Thesis). Al-Azhar University, Cairo, Egypt.

Nassar, H.N., El-Gendy, N.Sh, Abostate, M.A., Moustafa, Y.M., Mahdy, H.M., El-Temtamy, S.A., 2013. Desulfurization of dibenzothiophene by a novel strain *Brevibacillus invocatus* C19 isolated from Egyptian coke. Biosciences Biotechnology Research Asia 10 (1), 29–46.

Nassar, H.N., 2015. Development of biodesulfurization process for petroleum fractions using nano-immobilized catalyst (Ph.D. degree). Al-Azhar University, Cairo, Egypt.

Noda, K., Kogure, T., Irisa, S., Murakami, Y., Sakata, M., Kuroda, A., 2008. Enhanced dibenzothiophene biodesulfurization in a microchannel reactor. Biotechnology Letters 30, 451–454.

Noda, K., Watanabe, K., Maruhashi, K., 2002. Cloning of a rhodococcal promoter using a transposon for dibenzothiophene biodesulfurization. Biotechnology Letters 24 (22), 1875–1882.

Noda, K.I., Watanabe, K., Maruhashi, K., 2003. Isolation of a recombinant desulfurizing 4,6-dipropyl dibenzothiophene in n-tetradecane. Biosciences and Bioengineering 95 (4), 354–360.

Nunn, D., Boltz, J., DiGrazia, P.M., Nace, L., 2006. The biocatalytic desulfurization project. In: Final Technical Progress Report. Prepared for National Energy Technology Laboratory, U.S. Department of Energy.

Ohshiro, T., Hirata, T., Izumi, Y., 1995. Microbial desulfurization of dibenzothiophene in the presence of hydrocarbon. Applied Microbiology and Biotechnology 44, 249–252.

Ohshiro, T., Izumi, Y., 1999. Microbial desulfurization of organic sulfur compounds in petroleum. Bioscience, Biotechnology, and Biochemistry 63, 1–9.

Ohshiro, T., Ohkita, R., Takikawa, T., Manabe, M., Lee, W.C., Tanokura, M., Izumi, Y., 2007. Improvement of 2'-hydroxybiphenyl-2-sulfinate desulfinase, an enzyme involved in the dibenzothiophene desulfurization pathway from *Rhodococcus erythropolis* KA2-5-1 by site-directed mutagenesis. Bioscience, Biotechnology, and Biochemistry 71 (11), 2815–2821.

Ohshiro, T., Suzuki, K., Izumi, Y., 1996. Regulation of dibenzothiophene degrading enzyme activity of *Rhodococcus erythropolis* D-1. Journal of Fermentation and Bioengineering 82, 121–124.

Omori, T., Saiki, Y., Kasuga, K., Kodama, T., 1995. Desulfurization of alkyl and aromatic sulfides and sulfonates by dibenzothiophene desulphurising *Rhodococcus* sp. strain SY1. Bioscience, Biotechnology, and Biochemistry 59, 1195–1198.

Onaka, T., Kobayashi, M., Ishii, Y., Konishi, J., Maruhashi, K., 2001. Selective cleavage of the two C unknown; S bonds in asymmetrically alkylated dibenzothiophenes by *Rhodococcus erythropolis* KA2-5-1. Journal of Bioscience and Bioengineering 92, 80–82.

Ortiz-Leon, M., Velasco, L., Vazquez-Duhalt, R., 1995. Biocatalytic oxidation of polycyclic aromatic hydrocarbons by hemoglobin and hydrogen peroxide. Biochemical and Biophysical Research Communications 215, 968–973.

Pacheco, M.A., 1999. Recent advances in biodesulfurization (biodesulfurization) of diesel fuel. In: Paper Presented at the NPRA Annual Meeting, San Antonio, TX, 21–23 March 1999.

Pasta, P., Carrea, G., Colonna, S., Gaggero, N., 1994. Effects of chloride on the kinetics and stereochemistry of chloroperoxidase catalyzed oxidation of sulfides. Biochimica et Biophysica Acta 1209 (2), 203–208.

Peng, Y., Wen, J., 2010. Modeling of dibenzothiophene biodesulfurization by resting cells of *Gordonia* Sp. Chemical and Biochemical Engineering 24 (1), 85–94.

Piddington, C.S., Kovacevich, B.R., Rambosek, J., 1995. Sequence and molecular characterization of a DNA region encoding the dibenzothiophene desulfurization operon of *Rhodococcus* sp. strain IGTS8. Applied and Environmental Microbiology 61, 468–475.

Purdy, R.F., Lepo, J.E., Ward, B., 1993. Biodesulphurisation of organic sulphur compounds. Current Microbiology 27, 219–222.

Radovanović, L., Speight, J.G., 2011. Visbreaking: A Technology of the Future. In: Proceedings. First International Conference – Process Technology and Environmental Protection (PTEP 2011). University of Novi Sad, Technical Faculty "Mihajlo Pupin," Zrenjanin, Republic of Serbia. December 7, pp. 335–338.

Raheb, J., Hajipour, M.J., Saadati, M., Rasekh, B., Memari, B., 2009. The enhancement of biodesulfurization activity in a novel indigenous engineered *Pseudomonas putida* Iran. Biomedical Journal 13 (4), 207–213.

Rhee, S.K., Chang, J.H., Chang, Y.K., Chang, H.N., 1998. Desulfurization of dibenzothiophene and diesel oils by a newly isolated *Gordona* strain CYKS. Applied and Environmental Microbiology 64 (6), 2327–2331.

Sada, E., Katon, S., Terashima, M., 1981. Enhancement of oxygen absorption by magnetite-containing beads of immobilized glucose oxidase. Biotechnology and Bioengineering 21, 1037–1044.

Saed, D., Nassar, H.N., N.Sh., El-G., Zaki, T., Moustafa, Y.M., Badr, I.H.A., 2014. The enhancement of pyrene biodegradation by assembling MFe_3O_4 nano-sorbents on the surface of microbial cells. Energy Sources, Part A 36 (17), 1931–1937.

Sanchez, O.F., Almeciga-Diaz, C.J., Silva, E., Cruz, J.C., Valderrama, J.D., Caicedo, L.A., 2008. Reduction of sulfur levels in kerosene by Pseudomonas sp. Strain in an airlift reactor. Latin American Applied Research 38, 329–335.

Sato, H., Clark, D.P., 1995. Degradation of dibenzothiophene sulphoxide and sulphone by *Arthrobacter* strain dibenzothiopheneS2. Microbios 83, 145–159.

Schilling, B.M., Alvarez, L.M., Wang, D.I.C., Cooney, C.L., 2002. Continuous desulfurization of dibenzothiophene with *Rhodococcus rhodochrous* IGTS8 (ATCC 53968). Biotechnology Progress 18, 1207–1213.

Schmidts, T., Dobler, D., Nissing, C., Runkel, F., 2009. Influence of hydrophilic surfactants on the properties of multiple W/O/W emulsions. Journal of Colloid and Interface Science 338, 184–192.

Seeberger, A., Jess, A., 2010. Desulfurization of diesel oil by selective oxidation and extraction of sulfur compounds by ionic liquids-a contribution to a competitive process design. Green Chemistry 12 (4), 602–608.

Setti, L., Lanzarini, G., Pifferi, P.G., 1997. Whole cell biocatalysis for an oil desulfurization process. Fuel Processing Technology 52, 145–153.

Shan, G.B., Xing, J.M., Luo, M.F., Liu, H.Z., Chen, J.Y., 2003. Immobilization of *Pseudomonas delafieldii* with magnetic polyvinyl alcohol beads and its application in biodesulfurization. Biotechnology Letters 25, 1977–1983.

Shan, G.B., Zhang, H., Liu, H., Xing, J.M., 2005a. π-Complexation studied by fluorescence technique: application in desulfurization of petroleum product using magnetic π-complexation sorbents. Separation Science and Technology 40 (14), 2987–2999.

Shan, G.B., Xing, J.M., Guo, C., Liu, H.Z., Chen, J.Y., 2005b. Biodesulfurization using *Pseudomonas delafieldii* in magnetic polyvinyl alcohol beads. Letters in Applied Microbiology 40, 30–36.

Shan, G.B., Xing, J.M., Zhang, H., Liu, H.Z., 2005c. Biodesulfurization of dibenzothiophene by microbial cells coated with magnetite nanoparticles. Applied and Environmental Microbiology 71 (8), 4497–4502.

Shan, G.B., Xing, J.M., Zhang, H., Liu, H.Z., 2005d. Improvement of biodesulfurization rate by assembling nano-sorbents on the surface of microbial cells. Biophysical Journal 89 (6), L58–L60.

Shimizu, Y., Kumagai, S., Takeda, K., Enda, Y., 2007. Adsorptive Removal of Sulfur Compounds in Kerosene by Using Rice Husk Activated Carbon.

Singh, M.P., Kumar, M., Kalsi, W.R., Pulikottil, A.C., Sarin, R., Tuli, D.K., Malhutra, R.K., Verma, R.P., Bansal, B.M., 2009. Method for Bio-oxidative Desulfurization of Liquid Hydrocarbon Fuels and Product Thereof. US Patent Appl. 2009/0217571 A1.

Sohrabi, M., Kamyab, H., Janalizadeh, N., Huyop, F.Z., 2012. Bacterial desulfurization of organic sulfur compounds exists in fossil fuels. Journal of Pure and Applied Microbiology 6 (2), 717–729.

Soleimani, M., Bassi, A., Margaritis, A., 2007. Biodesulfurization of refractory organic sulfur compounds in fossil fuels. Biotechnology Advances 25 (6), 570–596.

Song, C., Ma, X., 2003. New design approaches to ultra-clean diesel fuels by deep desulfurization and deep de-aromatization. Applied Catalysis B: Environmental 41 (1–2), 207–238.

Song, C.S., 2003. An overview of new approaches to deep desulfurization for ultra-clean gasoline, diesel fuel, and jet fuel. Catalysis Today 86, 211–263.

Song, S.H., Choi, S.S., Park, K., Yoo, Y.J., 2005. Novel hybrid immobilization of microorganisms and its applications to biological denitrification. Enzyme and Microbial Technology 37, 567–573.

Speight, J.G., 2000. The Desulfurization of Heavy Oils and Residua, second ed. Marcel Dekker, New York.

Speight, J.G., 2011. The refinery of the future. Gulf Professional Publishing. Elsevier, Oxford, United Kingdom.

Speight, J.G., 2012. Visbreaking: a technology of the Past and the future. Scientia Iranica C 19 (3), 569–573.

Speight, J.G., 2014. The Chemistry and Technology of Petroleum, fifth ed. CRC Press, Taylor & Francis Group, Boca Raton, Florida.

Speight, J.G., 2017. Handbook of Petroleum Refining. CRC Press, Taylor & Francis Group, Boca Raton, Florida.

Srivastava, C.C., 2012. An evaluation of desulfurization technologies for sulfur removal from liquid fuels. RSC Advances 2, 759–783.

Srivastava, P., Kumar, A., 2008. Biodesulfurization of petroleum fractions for a cleaner environment. Toxicology Materials Research 28 (3), 8–12.

Stanislaus, A., Marafi, A., Rana, M.S., 2010. Recent advances in the science and technology of ultra-low sulfur diesel (ULSD) production. Catalysis Today 153 (1), 1–68.

Sundaramoorthy, M., Terner, J., Poulos, T.L., 1995. The crystal structure of chloroperoxidase: a heme peroxidase-cytochrome P450 functional hybrid. Structure 3 (12), 1367–1377.

Swaty, T.E., September 2005. Global refining industry trends: the present and future. Hydrocarbon Processing 35–46.

Takahashi, A., Yang, F.H., Yang, R.T., 2002. New sorbents for desulfurization by -complexation: thiophene/benzene adsorption. Industrial & Engineering Chemistry Research 41, 2487–2496.

Tang, H., Li, Q., Wang, Z.L., Yan, D.J., Xing, J.M., 2012. Simultaneous removal of thiophene and dibenzothiophene by immobilized *Pseudomonas delafieldii* R-8 cells. Chinese Journal of Chemical Engineering 20 (1), 47–51.

Tao, F., Yu, B., Xu, P., Ma, C.O., 2006. Biodesulfurization in biphasic systems containing organic solvents. Applied and Environmental Microbiology 72 (7), 4604–4609.

Tao, H., Nakazato, T., Sato, S., 2009. Energy-efficient ultra-deep desulfurization of kerosene based on selective photooxidation and adsorption. Fuel 88, 1961–1969.

Terres, E., Montiel, M., Le-Borge, S., Torres, E., 2008. Immobilization of chloroperoxidase on mesoporous materials for the oxidation of 4,6-dimethyldibenzothiophene, a recalcitrant organic sulfur compound present in petroleum fractions. Biotechnology Letters 30, 173–179.

Tian, F., Wu, J.Z., Liang, C., Yang, Y., Ying, P., Sun, X., Cai, T., Li, C., 2006. The study of thiophene adsorption on to La(III)-exchanged zeolite nay by FTIR spectroscopy. Journal of Colloid and Interface Science 301, 395–401.

Tilstra, L., Eng, G., Olsonn, G.J., Wang, F.W., 1992. Reduction of sulphur from polysulphidic model compounds by the hyperthermophilic archaebacterium *Pyrococcus furiosus*. Fuel 7 (7), 779–783.

Torkamani, S., Shayegan, J., Yaghemaei, S., Alemzadeh, I., 2008. Study of a newly isolated thermophilic bacterium capable of kuhemond heavy crude oil and dibenzothiophene biodesulfurization following 4S pathway at 60°C. Journal of Chemical Technology and Biotechnology 83, 1689–1693.

Torkamani, S., Shayegan, J., Yaghmaei, S., Alemzadeh, I., 2009. Heavy crude oil biodesulfurization project. In: Annual Report, Petroleum Engineering Development Company (PEDEC). National Iranian Oil Company, Tehran, Iran.

US Department of Energy, 2003. Gasoline biodesulfurization. In: Office of Energy Efficiency and Renewable Energy, May. United States Department of Energy, Washington, DC.

US EPA, 2000. Heavy-duty engine and vehicle standards and highway diesel fuel sulfur control requirements. In: Report No. EPA 420-F-00–057. US Environmental Protection Agency, Washington, DC.

Van Afferden, M., Schacht, S., Klein, J., Trüper, H.G., 1990. Degradation of dibenzothiophene by *Brevibacterium*sp. DO. Archives of Microbiology 153, 324–328.

Vazquez-Duhalt, R., Semple, K.M., Westlake, D.W.S., Fedorak, P.M., 1993. Enzyme Microb. Technol 15, 936.

Vazquez-Duhalt, R., Westlake, D.W.S., Fedorak, P.M., 1994. Lignin peroxidase oxidation of aromatic compounds in systems containing organic solvents. Applied and Environmental Microbiology 60, 459.

Vazquez-Duhalt, R.V., Bremauntz, M.D., Barzana, E., Tinoco, R., 2002a. Enzymatic Oxidation Process for Desulfurization of Fossil Fuels.US Patent 6,461,859.

Vazquez-Duhalt, R., Torres, E., Valderrama, B., Le Borgne, S., 2002b. Will biochemical catalysis impact the petroleum refining industry? Energy & Fuels 16, 1239–1250.

Velu, S., Ma, X., Song, C., 2005. Desulfurization of JP-8 jet fuel by selective adsorption over a Ni-Based adsorbent for micro solid oxide fuel cells. Energy & Fuels 19, 1116–1125.

Vertegel, A., 2010. 55th Annual report on research 2010. In: Under Sponsorship of the American Chemical Society Petroleum Research Fund. https://acswebcontent.acs.org/prfar/2010/reports/P10734.html.

Voloshchuk, A., Karsilnkova, O., Serbrykova, N., Artamonova, S., Aliev, A., Khozina, E., Kiselev, V., 2009. Selective adsorption of organic sulfur-containing compounds from diesel fuel using type-Y zeolite and γ-alumina oxide. Protection of Metals and Physical Chemistry of Surfaces 25, 512–517.

Wang, D., Qian, E.W., Amano, H., Okata, K., Ishihara, A., Kabe, T., 2003. Oxidation of dibenzothiophenes using tert-butyl hydroperoxide. Applied Catalysis A 253, 91–99.

Wang, P., Humphrey, A.E., Krawiec, S., 1996. Kinetic analyses of desulfurization of dibenzothiophene by *Rhodococcus erythropolis* in continuous cultures. Applied and Environmental Microbiology 62, 3066–3068.

Wang, R., Yu, F., 2013. Deep oxidative desulfurization of dibenzothiophene in simulated oil and real diesel using heteropolyanion-substituted hydrotalcite-like compounds as catalysts. Molecules: A Journal of Synthetic Chemistry and Natural Product Chemistry 18 (11), 13691–13704.

Wayan, G.I.B., Yamamura, K., Sujaya, I.N., Antara, N.S., Aryanta, W.R., Tanaka, M., Tomita, F., Sone, T., Asano, K., 2013. Biodesulfurization of dibenzothiophene and its derivatives using resting and immobilized cells of *Sphingomonas subarctica* T7b. Journal of Microbiology and Biotechnology 23 (4), 473–482.

Wittanadecha, W., Laosiripojana, N., Ketcong, A., Ningnuek, N., Praserthdam, P., Assabumrungrat, S., 2014. Synthesis of Au/C catalysts by ultrasonic-assisted technique for vinyl chloride monomer production. Engineering Journal 8 (3), 65–71.

Wu, K.Y.A., Wisecarver, K.D., 1992. Cell immobilization using PVA cross-linked with boric acid. Biotechnology and Bioengineering 39 (4), 447–449.

Wu, Z., Ondruschka, B., 2010. Ultrasound-assisted oxidative desulfurization of liquid fuels and its industrial application. Ultrasonics Sonochemistry 17 (6), 1027–1032.

Xiong, X., Xing, J., Li, X., Bai, X., Li, W., Li, Y., Liu, H., 2007. Enhancement of biodesulfurization in two-liquid systems by heterogeneous expression of vitreoscilla hemoglobin. Applied and Environmental Microbiology 37 (7), 2394–2397.

Xu, P., Yu, B., Li, F.L., Cai, X.F., Ma, C.Q., 2007. Microbial degradation of sulfur, nitrogen and oxygen heterocycles. Trends in Microbiology 14, 398–405.

Xu, X., Cai, Y., Song, Z., Qiu, X., Zhou, J., Liu, Y., Mu, T., Wu, D., Guan, Y., Xing, J., 2015. Desulfurization of immobilized sulfur-oxidizing bacteria, Thialkalivibrio versutus, by magnetic nanoparticles under haloalkaliphilic conditions. Biotechnology Letters 37 (8), 1631–1635.

Xu, X., Zhang, S., Li, P., Shen, Y., 2014. Adsorptive desulfurization of liquid Jet-A fuel at ambient conditions with an improved adsorbent for on-board fuel treatment for SOFC applications. Fuel Processing Technology 124, 140–146.

Yamada, K.O., Morimoto, M., Tani, Y., 2001. Degradation of dibenzothiophene by sulfate-reducing bacteria cultured in the presence of only nitrogen gas. Journal of Bioscience and Bioengineering 91, 91–93.

Yan, H., Sun, X., Xu, Q., Ma, Z., Xiao, C., Jun, N., 2008. Effects of nicotinamide and riboflavin on the biodesulfurization activity of dibenzothiophene by *Rhodococcus erythropolis* USTB-03. Journal of Environmental Sciences 20, 613–618.

Yang, J., Marison, I.W., 2005. two-stage process design for the biodesulphurisation of a model diesel by a newly isolated Rhodococcus globerulus DAQ3. Biochemical Engineering Journal 27, 77–82.

Yang, J., Hu, Y., Zhao, D., Wang, S., Lau, P.C.K., Marison, I.W., 2007. Two-layer continuous-process design for the biodesulfurization of diesel oils under bacterial growth conditions. Biochemical Engineering Journal 37 (2), 212–218.

Yang, R.T., Hernández-Maldonado, A.J., Yang, F.H., 2001. New sorbents for desulfurization of liquid fuels by π-complexation. Industrial & Engineering Chemistry Research 40, 6236–6239.

Yu, B., Huang, J., Cai, X.F., Liu, X.Y., Ma, C.Q., Li, F.L., Xu, P., 2005. A novel method for conversion of valuable biodesulfurization intermediates HPBS/Cx-HPBS from dibenzothiophene/Cx-dibenzothiophene. Chinese Chemical Letters 16 (7), 935–938.

Yu, B., Xu, P., Shi, Q., Ma, C., 2006. Microbial desulfurization of gasoline by free whole-cells of *Rhodococcus erythropolis* XP. Applied and Environmental Microbiology 72, 54–78.

Yu, L., Meyer, T.A., Folsom, B.R., 1998. Oil/water/biocatalyst Three Phase Separation Process. US Patent 5772901.

Zakaria, B.S., Nassar, H.N., Saed, D., El-Gendy, N.Sh., 2015. Enhancement of carbazole denitrogenation rate using magnetically decorated *Bacillus clausii* BS1. Petroleum Science and Technology 33 (7), 802–811.

Zaki, T., Saed, D., Aman, D., Younis, S.A., Moustafa, Y.M., 2013. Synthesis and characterization of MFe_3O_4 sulfur nanoadsorbants. Journal of Sol-Gel Science and Technology 65, 269–276.

Zeynalov, E., Nagiev, T., 2015. Enzymatic catalysis of hydrocarbons oxidation "in vitro" (review). Chemistry & Chemical Technology 9 (2), 157–164.

Zhang, H., Liu, Q.F., Li, Y., Li, W., Xiong, X., Xing, J., Liu, H., 2008. Selection of adsorbents for in-situ coupling technology of adsorptive desulfurization and biodesulfurization. Science in China Series B-Chemistry 51 (1), 69–77.

Zhang, H., Shan, G., Liu, H., Xing, J., 2007. Surface Modification of γ-Al_2O_3 Nanoparticles with gum Arabic and its applications in adsorption and biodesulfurization. Surface and Coatings Technology 201, 6917–6921.

Zhang, T.M., Li, W.L., Chen, X.X., Tang, H., Li, Q., Xing, J.M., Liu, H.Z., 2011. Enhanced biodesulfurization by magnetic immobilized *Rhodococcus erythropolis* LSSE8-1-vgb assembled with nano-gamma Al_2O_3. World Journal of Microbiology & Biotechnology 27 (2), 299–305.

Zhao, Y., Wang, R., 2013. Deep desulfurization of diesel oil by ultrasound-assisted catalytic ozonation combined with extraction process. Petroleum and Coal 55 (1), 62–67.

Zobell, C.E., 1953. Process of Removing Sulfur from Petroleum Hydrocarbons and Apparatus. US Patent 2,641,564.

FURTHER READING

Parkash, S., 2003. Refining processes handbook. In: Gulf Professional Publishing. Elsevier, Amsterdam, Netherlands.

BIOCATALYTIC DENITROGENATION

1.0 INTRODUCTION

Sulfur, nitrogen, and metals in crude oil are major concerns for producers and refiners and have long been key determinants of the value of crude oils for several reasons (Table 7.1; Hsu and Robinson, 2006; Gary et al., 2007; Speight, 2014, 2011, 2017). The amount of sulfur in many finished products is limited by law, and desulfurization offers refiners the opportunity to reduce the sulfur of their crude feedstocks before they ever enter the refinery system, minimizing downstream desulfurization costs. The regulations restricting allowable levels of sulfur in the end products continue to become increasingly stringent. This creates an ever more challenging technical and economic situation for refiners as the sulfur levels in available crude oils continue to rise and create a market disadvantage for producers of high-sulfur crudes. Lower-sulfur crudes continue to command a premium price in the market, while higher sulfur crude oils sell at a discount. Desulfurization would offer producers the opportunity to economically upgrade their resources. In a similar manner, the presence of nitrogen in crude oil and crude oil products also caused problems for refiners.

In the context of this chapter, the degradation of nitrogen compounds from petroleum-related sources is an important aspect of petroleum refining because of the contribution these contaminants make to the formation of nitrogen oxides (NO_x) and hence to air pollution and acid rain (Benedik et al., 1998). They also contribute to coke formation and catalyst poisoning during the refining of crude oil, thus reducing process yields (Hsu and Robinson, 2006; Gary et al., 2007; Speight, 2011, 2014, 2017).

Nitrogen compounds occur in crude oil and represent non-hydrocarbon compounds that occur in crude oil at the level of 0.01%–2% w/w, although over 10% w/w concentrations have been noted. The microbial degradation (biotransformation) of nitrogen compounds from fossil fuels is important because of the above stated reasons. The removal of aromatic nitrogen contaminants from petroleum is important for many reasons of which the two most important are (1) combustion leads directly to the formation of nitrogen oxides (NO_x); emissions of nitrogen oxides contribute to acid rain, as rendered in the below equations:

$$NO + H_2O \rightarrow HNO_2 \text{ (nitrous acid)}$$

$$2NO + O_2 \rightarrow NO_2 \text{ (nitrogen dioxide)}$$

$$NO_2 + H_2O \rightarrow HNO_3 \text{ (nitric acid)}$$

These gases are under increasingly stringent control by environmental regulation, (2) the presence of aromatic nitrogen compounds can lead to significant poisoning of refining catalysts, resulting in a

Introduction to Petroleum Biotechnology. https://doi.org/10.1016/B978-0-12-805151-1.00007-2

Table 7.1 The Potential Effects of Sulfur Derivatives, Nitrogen Derivatives, and Metal Derivatives in Crude Oil (Speight, 2017)

Contaminant	Effect on Catalyst	Mitigation	Process
Sulfur	Catalyst fouling Deactivation of active sites	Hydrodesulfurization	Hydroprocessing
Nitrogen	Adsorption of basic nitrogen Destruction of active sites	Hydrodemetallization	Hydroprocessing
Metals	Fouling of active sites Fouling of catalyst pores	Demetallization	Demet, Met-X

decrease in yield. In addition, carbon dioxide and the oxides of sulfur (Chapter 6) also contribute to the formation of acid rain:

$$CO_2 + H_2O \rightarrow H_2CO_3 \text{ (carbonic acid)}$$

$$SO_2 + H_2O \rightarrow H_2SO_3 \text{ (sulfurous acid)}$$

$$2SO_2 + O_2 \rightarrow 2SO_3 \text{ (sulfur trioxide)}$$

$$SO_3 + H_2O \rightarrow H_2SO_4 \text{ (sulfuric acid)}$$

Carbazole, the major nonbasic species, directly impacts the refining process in two ways: (1) it is converted during the cracking process into basic derivatives that can adsorb to the active sites of the cracking catalyst, and (2) it is unexpectedly, potent as a direct inhibitor of hydrodesulfurization (HDS), which is commonly included in the refining process to meet sulfur content criteria. The practical consequence of catalyst poisoning is that the removal of carbazole and other nitrogen species can significantly increase the extent of catalytic-cracking conversion and the yield of gasoline, with a 90% reduction in nitrogen content, an increase in gasoline yields of up to 20% w/w may be achievable, which would represent a major economic improvement in low-margin, high volume refining processes. Finally, the presence of nitrogen compounds promotes the corrosion of refining equipment such as storage tanks and piping, which adds to the refining cost.

Of interest are the polycyclic systems and the nitrogen polyaromatic heterocyclic compounds (polycyclic aromatic hydrocarbon derivatives) where one or more carbon atoms in the fused ring structure of the polycyclic aromatic hydrocarbons are replaced by nitrogen atom(s) (Table 7.2). The nitrogen polyaromatic heterocyclic compounds tend to occur in a strong association with polycyclic aromatic hydrocarbons in the environment because they come from the same sources. They can be emitted due to the incomplete combustion of organic matter, including wood, waste, and fossil fuels (such as gasoline, diesel, and coal) (Furlong and Carpenter, 1982; Chuang et al., 1991; Osborne et al., 1997; Wu et al., 2014). Nitrogen polyaromatic heterocyclic compounds are also released into the environment from spills, wastes, and effluents of several industrial activities such as oil drilling, refining and storage, coal tar processing, chemical manufacturing, and wood preservation (Lopes and Furlong, 2001). Both soil and groundwater are frequently contaminated by nitrogen, sulfur, and oxygen heterocycles at sites contaminated with petroleum and wood preservation wastes such as creosote.

Table 7.2 Illustration of the Nitrogen Types That Occur in Crude Oil

Nonbasic		
Pyrrole	C_4H_5N	
Indole	C_8H_7N	
Carbazole	$C_{12}H_9N$	
Benzo(a)carbazole	$C_{16}H_{11}N$	
Basic		
Pyridine	C_5H_5N	
Quinoline	C_9H_7N	
Indoline	C_8H_9N	
Benzo(f)quinoline	$C_{13}H_9N$	

2.0 NITROGEN IN CRUDE OIL

The nitrogen constituents in petroleum may be classified arbitrarily as basic and non-basic (Table 7.2) and tend to exist in the higher boiling fractions and residues (Speight, 2014, 2017). The basic nitrogen compounds, which are composed mainly of pyridine homologues and occur throughout the boiling ranges, have a decided tendency to exist in the higher boiling fractions and residua. The nonbasic nitrogen compounds, which are usually of the pyrrole, indole, and carbazole types, also occur in the higher boiling fractions and nonvolatile residua. It has been estimated that neutral nitrogen compounds (which include carbazoles, indoles, and pyrroles) account for less than 30% w/w of all organic nitrogen compounds. The basic nitrogen compounds include, for example, pyridine and quinoline derivatives (Fig. 7.2).

The nitrogen constituents of crude oil are problematic in refining processes because they lead to catalyst deactivation and cause fuel instability during transportation or storage (Speight, 2014, 2017).

In general, the nitrogen content of crude oil is low and generally falls within the range 0.1%–0.9% w/w, although some crude oil may contain up to 2% w/w nitrogen (Speight, 2012a, 2014). However, crude oils with no detectable nitrogen or even trace amounts are not uncommon, but in general the more asphaltic the oil, the higher its nitrogen content. Insofar as an approximate correlation exists between the sulfur content and API gravity of crude oils (Speight, 2012a, 2014), there also exists a correlation between nitrogen content and the API gravity of crude oil. It also follows that there is an approximate correlation between the nitrogen content and the carbon residue: the higher the nitrogen content, the higher the carbon residue. The presence of nitrogen in petroleum is of much greater significance in refinery operations than might be expected from the small amounts present. Nitrogen compounds can be responsible for the poisoning of cracking catalysts, and they also contribute to gum formation in such products as domestic fuel oil. The trend in recent years toward cutting deeper into the crude to obtain stocks for catalytic cracking has accentuated the harmful effects of the nitrogen compounds, which are concentrated largely in the higher boiling portions.

Basic nitrogen compounds with a relatively low molecular weight can be extracted with dilute mineral acids; equally strong bases of higher molecular weight remain unextracted because of unfavorable partitioning between the oil and aqueous phases. A method has been developed in which the nitrogen compounds are classified as basic or nonbasic, depending on whether they can be titrated with perchloric acid in a 50:50 solution of glacial acetic acid and benzene. Application of this method has shown that the ratio of basic to total nitrogen is approximately constant (0–30 0.05) irrespective of the source of the crude. Indeed, the ratio of basic to total nitrogen was found to be approximately constant throughout the entire range of distillate and residual fractions. Nitrogen compounds extractable with dilute mineral acids from petroleum distillates were found to consist of alkyl pyridine derivatives, alkyl quinoline derivatives, and alkyl isoquinoline derivatives carrying alkyl substituents, as well as pyridine derivatives in which the substituent was a cyclopentyl or cyclohexyl group. The compounds that cannot be extracted with dilute mineral acids contain the greater part of the nitrogen in petroleum and are generally of the carbazole, indole, and pyrrole types. Moreover, from an examination of various functional subfractions, it was shown that amphoteric species and basic nitrogen species contain polycyclic aromatic systems having two-to-six rings per system (Speight, 2014).

Carbazole derivatives are the major nitrogen heteroaromatics in coal-tar creosote (Speight, 2014) and, although it is useful as an industrial raw material for dyes, medicines, insecticides, and plastics, carbazole is also known to be an environmental pollutant. Despite its widespread use, little is known about the fate of carbazole in the environment (Arcos and Argus, 1968; Singh et al., 2011a,b). The nitrogen polyaromatic heterocyclic compounds are hazardous to the environment, since they are more polar, water-soluble, and consequently more mobile and bioavailable in the environment than their corresponding polycyclic aromatic hydrocarbon analogs (Bleeker et al., 2003; WHO, 2004). Nitrogen polyaromatic heterocyclic compounds, including carbazole and its derivatives, have been detected in contaminated atmospheric samples, river sediments, and groundwater sites. This causes concern because carbazole is known to be both mutagenic and toxic; even though it is not highly toxic itself, it readily undergoes radical chemistry to generate the genotoxic hydroxynitrocarbazole (Arcos and Argus, 1968; Zakaria et al., 2015a,b, 2016). Moreover, the mutagenic potential of nitrogen polyaromatic heterocyclic compounds increases with increasing the number of rings; thus, quinoline (Qn) is known to be more mutagenic than pyridine (Mohammed and Hopfinger, 1983). Growth impairment is also reported to increase with the increase in the number of rings per compound and with the increase in nitrogen content within the ring (Millemann et al., 1984).

Nitrogen compounds interfere with the refining processes, causing equipment corrosion and catalyst poisoning, reducing the processes yields, and adding to the refining costs. Upon combustion of N-compounds in fossil fuels, it produces NO_x, which contributes to acid rain, atmospheric contamination, and destruction of ozone layer (Montil et al., 2009; Larentis et al., 2011; Liu et al., 2013). The presence of nitrogen containing compounds promotes tank corrosion and oil degradation during storage (Singh et al., 2010). The nitrogen polyaromatic heterocyclic compounds found in crude oils as listed before fall into two classes—the "non-basic" molecules include pyrroles and indoles, but are predominantly mixed alkyl derivatives of carbazole, while the "basic" molecules are largely derivatives of pyridine and quinolone (Benedik et al., 1998; Bachman et al., 2014). Fig. 7.1 represents examples of nitrogen polyaromatic heterocyclic compounds that can be found in petroleum and its fractions. The total nitrogen content of crude oils averages around 0.3%– 2% w/w, of which the nonbasic compounds comprise approximately 70%–75% (Benedik et al., 1998; Bachman et al., 2014) and in some other reports range from 0.1% to 0.9% w/w (Speight, 2014). In Egyptian crude oils, carbazole and its derivatives concentrations range from 0.3% to 1% w/w (Bakr, 2009). Carbazole as an example for the nonbasic nitrogen polyaromatic heterocyclic compounds can directly impact the refining processes in two ways: (1) during the cracking process, carbazole can be converted into basic derivatives, which can be adsorbed to the active sites of the cracking catalysts. (2) It directly inhibits the function of the catalysts in HDS processes. Thus, removal of carbazole and other nitrogen-compounds would significantly increase the extent of catalytic cracking and consequently the gasoline yield. It has been reported that by

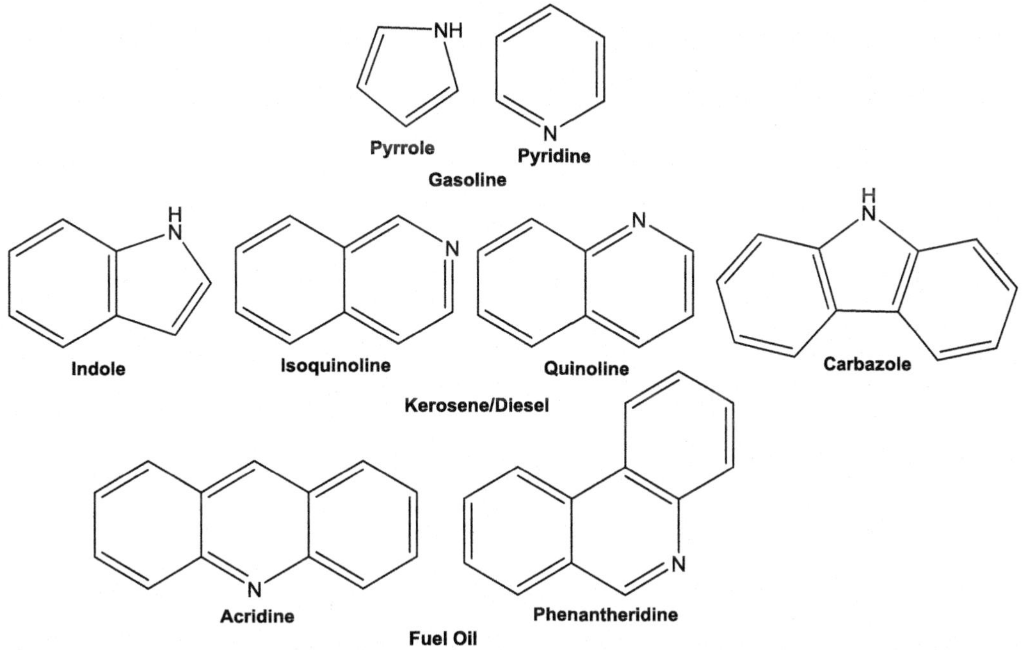

FIGURE 7.1

Different nitrogen polyaromatic heterocyclic compounds that occur in crude oil and crude oil fractions.

90% reduction in nitrogen content a 20% w/w increase in gasoline yield occurs. That has a major economic improvement in low-margin, high volume refining processes (Benedik et al., 1998). Basic nitrogen compounds are more inhibitory for catalysts than the nonbasic ones. But, they can potentially be converted to basic compounds during the refining/catalytic cracking process. Thus, they are also inhibitory to catalysts. Moreover, metals like nickel and vanadium are potent inhibitors for catalysts and in petroleum, metals are typically associated with nitrogen compounds (Hegedus and McCabe, 1981; Mogollon et al., 1998).

Although only trace amounts (usually ppm levels) of nitrogen are found in the middle distillate fractions, both neutral and basic nitrogen compounds have been isolated and identified in fractions boiling below 343°C (650°F) (Hirsch et al., 1974). Pyrrole derivatives and indole derivatives account for about two-thirds of the nitrogen while the remainder is found in the basic alkylated pyridine and alkylated Qn compounds (Table 7.2).

In the vacuum gas oil range, the nitrogen-containing compounds include higher molecular weight pyridines, Qns, benzoquinoline derivatives, amides, indoles, carbazole, and molecules with two nitrogen atoms (diaza compounds) with three and four aromatic rings are especially prevalent (Green et al., 1989). Typically, about one-third of the compounds are basic, i.e., pyridine and its benzologs, while the remainder is present as neutral species (amides and carbazoles). Although benzo- and dibenzo-quinolines found in petroleum are rich in sterically hindered structures, hindered and unhindered structures have been found to be present at equivalent concentrations in source rocks.

The residuum (the collection of constituents that typically boil above 510°C, 950°F) is the most complex fraction of crude oil and contains the majority of the heteroatoms originally in the crude oil and molecular weights of the constituents fall into a wide range that extends, as near as can be determined and subject to method dependence, up to several thousand (Speight, 2014). The fraction is so complex that the characterization of individual species is virtually impossible, no matter what claims have been made or will be made. Separation of the residuum by group type can be difficult and confusing because of the multi-substitution of aromatic and naphthenic species, as well as by the presence of multiple functionalities in single molecules. In this fraction, the amount of nitrogen may begin to approach the concentration of sulfur. The nitrogen species consistently concentrate in the most polar fractions to the extent that every molecule contains more than one heteroatom. At this point, structural identification is somewhat fruitless and characterization techniques are used to confirm the presence of the functionalities found in lower boiling fractions such as, for example, nonbasic (carbazole-type) nitrogen and basic (Qn -type) nitrogen (Table 7.2).

Porphyrins are a naturally occurring chemical species that exist in petroleum and usually occur in the nonbasic portion of the nitrogen-containing concentrate. They are not usually considered among the usual nitrogen-containing constituents of petroleum, nor are they considered as organic material containing metallo-compounds that also occur in some crude oils. Because of these early investigations, there arose the concept of porphyrins as biomarkers that could establish a link between compounds found in the geosphere and their corresponding biological precursors.

3.0 DENITROGENATION

Crude oil contains many thousands of different compounds that vary in molecular weight from methane (CH_4, molecular weight: 16) to more than 2000 (Speight, 2014). This broad range in molecular weights results in boiling points that range from 160°C (−288°F) to temperatures on the order of nearly 1100°C

(2000°F). The organic nitrogen is spread throughout crude oil fractions but exists mainly in the higher molecular weight constituents which invariably undergo thermal reaction to concentrate the nitrogen in the nonvolatile coke (Speight, 2014).

The selective removal of nitrogen-containing compounds from crude oil and from crude oil products is of interest because of the potential deleterious impact of such compounds on products and processes. Problems caused by nitrogen-containing compounds include gum formation, acid catalyst inhibition and deactivation, acid–base pair-related corrosion, and metal complexation (Speight, 2014, 2017).

As for desulfurization (Chapter 6), thermal processes such as visbreaking (or even hydrovisbreaking—visbreaking in an atmosphere of hydrogen or in the presence of a hydrogen donor material) the long-ignored step-child of the refining industry—may see a surge in use as a pretreatment process (Radovanović and Speight, 2011; Speight, 2011, 2012b, 2014, 2017). Management of the process to produce a liquid product that has been freed of the high potential for coke deposition (by taking the process parameters into the region where sediment forms) either in the absence or presence of (for example) a metal oxide scavenger could be a valuable ally to catalyst cracking or hydrocracking units. Scavenger additives such as metal oxides may also see a surge in use.

3.1 HYDRODENITROGENATION

The hydrodenitrogenation process in a refinery is a process designed to remove nitrogen from the feedstocks, which often occurs concurrently with removal of sulfur from the feedstock:

$$[S]_{feedstock} + H_2 \rightarrow \text{Hydrocarbon product} + H_2S$$

$$[N]_{feedstock} + H_2 \rightarrow \text{Hydrocarbon product} + NH_3$$

Typically, the process used is the hydrocracking process (hydroconversion process) which is used (principally) to convert waxy distillate and deasphalted oil into kerosene and gas oil by thermal decomposition of some of the constituents (Hsu and Robinson, 2006; Gary et al., 2007; Speight, 2011, 2014, 2017). The process is carried out in two stages, the first to reduce the amount of nitrogen, sulfur, and oxygen impurities that may reach the second stage catalyst, and the second to continue the process of cracking, hydrogenating, and isomerizing the compounds in the oil. In addition to denitrogenation, other reactions such as desulfurization, deoxygenation, hydrogenation, hydrocracking, and isomerization occur, all of which are exothermic and, except for isomerization, consume hydrogen.

The catalysts used for hydrodenitrogenation catalysts are typically nickel-promoted molybdenum sulfide (MoS_2), supported on alumina (Al_2O_3). These nickel-molybdenum catalysts are more active for hydrogenation than the corresponding cobalt catalysts. Nickel-molybdenum are generally good HDS catalysts (El-Gendy and Speight, 2015; Speight, 2017).

Furthermore, the high concentration of nitrogen in the higher molecular weight constituents has an adverse effect on catalysts. Therefore, process choice often favors thermal process but catalytic processes can be used if catalyst replacement and catalyst regeneration is practiced. Alternatively, catalyst poisoning can be minimized by mild hydrogenation to remove nitrogen from feedstocks in the presence of more resistant catalysts, such as cobalt-molybdenum-alumina ($Co-Mo-Al_2O_3$). The reactions involved in nitrogen removal are somewhat analogous to those of the sulfur compounds and follow a stepwise mechanism to produce ammonia and the relevant substituted aromatic compound.

The heat released in the process is absorbed by injecting cold hydrogen quench gas between the catalyst beds. Without the quench step, the heat released would generate high temperatures and rapid reactions leading to greater heat release and an eventual runaway. All the reactions, except for denitrogenation, desulfurization, and deoxygenation, which only occur in the first stage, happen in both stages.

3.2 THERMAL DENITROGENATION

The organic nitrogen originally in the asphaltene constituents invariably undergoes thermal reaction to concentrate in the nonvolatile coke (Chapter 10). Thus, although asphaltenes produce high yields of thermal coke, little is known of the actual chemistry of coke formation. In a more general scheme, the chemistry of asphaltene coking has been suggested to involve the thermolysis of thermally labile bonds to form reactive species that then react with each other (*condensation*) to form coke. In addition, the highly aromatic and highly polar (refractory) products separate from the surrounding oil medium as an insoluble phase and proceed to form coke.

Nitrogen species contribute to the pattern of the thermolysis insofar as the carbon-carbon bonds adjacent to ring nitrogen undergo thermolysis quite readily (Speight, 2014). Thus, in the visbreaking delayed coking process (which uses temperature on the order of 500°C/930°F), the initial reactions of the nitrogen-containing heterocyclic species involve thermolysis of aromatic-alkyl bonds that are enhanced by the presence of the heterocyclic nitrogen. In fact, the higher molecular weight constituents fraction, which contain nitrogen and other heteroatoms (and have lower volatility than the pure hydrocarbons), are the prime movers in the production of coke. Such species, containing various polynuclear aromatic systems, can be denuded of the attendant hydrocarbon moieties and are insoluble. While the overall goal of crude oil refining is to mitigate coke formation by elimination or modification of the primary chemical reactions, the early formation of high-nitrogen coke can be used to an advantage thereby reducing the potential for the occurrence of nitrogen in the majority of the nitrogen in the distillate (volatile gases and liquids) reaction products.

In summary, the two main routes for nitrogen removal from crude oil (and crude oil products) are (1) hydrotreating, and (2) thermal denitrogenation with hydrodenitrogenation being the most viable process for nitrogen removal from oils with high nitrogen content (Speight, 2014, 2017; Prado et al., 2017). Other processes such as liquid-liquid phase partitioning, solvent deasphalting, adsorption, chemical conversion followed by separation, and microbial conversion have received lesser attention with the last process (microbial conversion) receiving considerable interest. Chemical conversion processes include oxidative denitrogenation, N-alkylation, complexation with metal salts, and conversion in high-temperature water. Adsorption denitrogenation in which the more polar nitrogen constituents are adsorbed on to a solid adsorbent (such as alumina, Al_2O_3, or a more complex adsorbent) also offers an option for crude oil denitrogenation (Shiraishi et al., 2004). In fact, there are several processes for denitrogenation by separation of the nitrogen-rich products from oil without removing the nitrogen group from the nitrogen-containing compounds. Consequently, most of these processes are viable mainly for removal of nitrogen from low-nitrogen-content oils, typically with <0.1% w/w nitrogen.

4.0 BIOCATALYTIC DENITROGENATION

Although the number of nitrogen-containing constituents in crude oil is lower than the number of sulfur-containing constituents, the existence of the nitrogen-containing constituents is sufficient to affect the invariability of the product products. The nitrogen-containing constituents also contribute to catalyst poisoning during the refining of crude oil, thus reducing the active life of the catalyst rate of the catalyst and increasing process costs. Furthermore, many nitrogen-containing polycyclic aromatic compounds possess mutagenic and toxic activities. In addition, the combustion of these contaminants form nitrogen oxides (NO_x), releasing of which to the air will cause the formation of acid rain and hence to air pollution. The classical hydroprocessing methods of nitrogen removal are costly and complicated, and, as a result, there is increased microbial denitrogenation (biotransformation).

The biotransformation of nitrogen heteroaromatics can be used to alleviate catalyst inhibition in several ways (Bachmann et al., 2014). Carbazole, for example, can be completely metabolized to carbon dioxide and biomass, or (using appropriate blocked mutant strains) converted to anthranilic acid or other intermediates. These appear likely to cause less catalyst inhibition than their parent compound, and many polar intermediates could be readily extracted from petroleum streams into water. For example, carbazole-enrichment cultures can degrade a wide range of alkyl carbazole derivatives present in crude oil, generally yielding water-soluble, nontoxic metabolites.

As a result, much research has been undertaken for the microbial transformation of nitrogen polyaromatic heterocyclic compounds, to alleviate the catalyst inhibition. Several microorganisms capable of degrading nitrogen polyaromatic heterocyclic compounds have been isolated from wastewater sludge, hydrocarbon-contaminated soil and water, industrial effluents, and coal- and shale-liquefaction sites. Most of the attack is aerobic, but anaerobic degradation has been also noted. Several species of the genus *Pseudomonas, known* for their solvent tolerance, have been isolated that degrade carbazole and its alkyl derivatives. Moreover, other microorganisms have been reported to mineralize nonbasic nitrogen compounds, including species of *Bacillus, Sphingomonas, Xanthomonas, Gordonia, Klebsiella, Burkholderia, Arthrobacter,* and *Novosphingobium* (Singh et al., 2011a,b; Zakaria et al., 2016).

The initial attack on nitrogen polyaromatic heterocyclic compounds is normally by oxygenase enzymes that activate the dioxygen molecule and use it for region-selective and stereo-selective oxidation by the insertion of molecular oxygen into the organic substrate. These enzymes function under mild conditions of pH and temperature, allowing high yields of hydroxylated products. Dioxygenases incorporate two atoms of molecular oxygen into one molecule of substrate, while monooxygenase derivatives add only one atom of oxygen (the other atom being reduced to water). Although there are different types of monooxygenases, most add a single hydroxyl group to an already hydroxylated substrate to generate a dihydroxy product. Dioxygenases that form *cis*-diols are composed of two or three components, forming *cis*-dihydrodiol derivatives or *cis*-diol carboxylic acid derivatives, respectively. The three-component dioxygenases are composed of a flavoprotein, a ferredoxin, and a terminal oxygenase. Dioxygenases often have broad substrate specificity and require only a minimal characteristic structure for substrate recognition (Marcelis et al., 2003).

Among the nonbasic nitrogen compounds (Benedik et al., 1998), pyrrole and indole have been found to be readily degraded, while carbazole is more recalcitrant to microbial degradation. Although carbazole is relatively resistant to microbial attack, several researchers have reported the isolation of carbazole-degrading bacteria (Kimura et al., 1996; Sato et al., 1997a,b; Bressler and Gray, 2002; Singh et al., 2013; Zakaria et al., 2016). Carbazole is often chosen as a model compound in biodenitrogenation of petroleum and biodegradation of recalcitrant polycyclic aromatic hydrocarbon derivatives. Carbazole can be metabolized to give free amines, alcohols, phenols, ketones, aldehydes, and carboxylic acids (Ouchiyama et al., 1993). The functional groups often are combined in a single metabolite. The results obtained for carbazole degradation by *Pseudomonas* strain LD2 indicate that carbazole is oxidized initially by angular oxygenation at the 2 and 3 positions (Gieg et al., 1996) to form 2,9-aminobiphenyl-2,3-diol (via an unstable intermediate), which is further degraded by meta-cleavage of the diol ring to form 2-hydroxy-6-oxo-6-(29-aminophenyl)-hexa-2,4-dienoic acid (Ouchiyama et al., 1993). The second step is the breakage of the first C–N bond, creating an amino group. In this step, the aromatic ring undergoes hydrolysis to form 2-hydroxy-4-pentenoate and anthranilic acid, which then enters the tricarboxylic acid cycle after conversion to catechol. Formation of catechol is the step where nitrogen is removed; however, since it is degraded further, the whole molecule is lost and mineralized (Fig. 7.2). CARDO (1.9a-dioxygenase), which is responsible for the first step in carbazole biodegradation to 2'-aminobiphenyl-2,3-diol, is reported to have a broad substrate range and catalyzes diverse oxygenation; angular deoxygenation, *cis*-dihydroxylation, and mono-oxygenation (Nojiri et al., 1999; Takagi et al., 2002). Yoon et al. (2002) reported that *Pseudomonas rhodesiae* strain KK1 mineralizes carbazole much faster than naphthalene.

Moreover, the production of different metabolites via the biodegradation of carbazole by *Arthrobacter* sp. P1-1 has been reported (Seo et al., 2006) (Fig. 7.3). Guo et al. (2008) reported the carbazole mineralization with the production of NH_4^+-N as inorganic nitrogen, by a stable bacterial consortium isolated from a refinery wastewater sample— *Chryseobacterium* sp. NCY and *Achromobacter* sp. NCW. However, none of them can singly utilize carbazole. Sing et al. (2010) reported the isolation of a Gram negative bacterial isolate GBS.5 with high carbazole-biodegradation capability (11.36 µmol/min/g dry cell weight). Larentis et al. (2011) reported the isolation of *Pseudomonas stutzeri* (ATCC 31258) for biodenitrogenation of high concentration of carbazole (1000 mg/L). Singh et al. (2011a) reported the isolation of the Gram negative *Enterobacter* sp. from hydrocarbon contaminated soil. Zakaria et al. (2016) reported the isolation of a Gram positive *Bacillus clausii* BS1 with a higher carbazole-degradation efficiency relevant to the well-known biodenitrogenating bacterium strain *Pseudomonas resinovorans* CA10, recording 77.15% and 60.66% removal of 1000 ppm carbazole with the production of 119.79 and 102.43 ppm anthranilic acid, and 121.19 and 90.33 ppm catechol, as by-products, respectively. There are many more reported microorganisms capable for biodegradation of carbazole (Table 7.3). However, although complete mineralization of carbazole is recommendable for bioremediation of polluted environment, it is not recommendable for biodenitrogenation of crude oil and its distillates, as it will reduce the carbon value of the fuel. Lobastova et al. (2004) reported the hydroxylation of carbazole by *Aspergillus flavous* VKM F-1024 to 3-hydroxycarbazole. Moreover, naphthalene 1,2-dioxygenase from *Pseudomonas* sp. NCIB 9816-4 and biphenyl dioxygenase from *Beijerinckia* sp. B8/36 also oxidizes carbazole to 3-hydroxycarbazole (Resnick et al., 1993). The 3-hydroxy-carbazole is less toxic to the refinery catalysts and has many pharmaceutical applications as it has a strong antioxidant activity and has also some wide applications in therapies for encephalopathy, cardiopathy, hepatopathy, and arteriosclerosis (Lobastova et al., 2004). The capability of some bacterial isolates, for example, *P. resinovorans* CA10 and *Sphingomonas wittichii* RW1 to degrade not only carbazole but other persistent toxic substances or persistent organic pollutants— polyaromatic heterocyclic compounds and polycyclic aromatic hydrocarbons such as some poisonous oxygen compounds and their chloro-derivatives (e.g., the highly toxic polychlorinated dibenzo-*p*-dioxins) have been also reported (Habe et al., 2001; Widada et al., 2003; Saiki et al., 2003; Furukawa, 2003; Nam et al., 2006; Xu et al., 2006). That will have a great potential on the bioremediation of polluted environment.

Indole is another example of the nitrogen polyaromatic heterocyclic compounds, where the observed microbial degradation pathways differ significantly between microorganisms. Indole biodegradation pathways can be generally summarized into three pathways (Arora et al., 2015). In the first pathway, the degradation of indole occurs via anthranilate, which can be further metabolized to denitrogenated

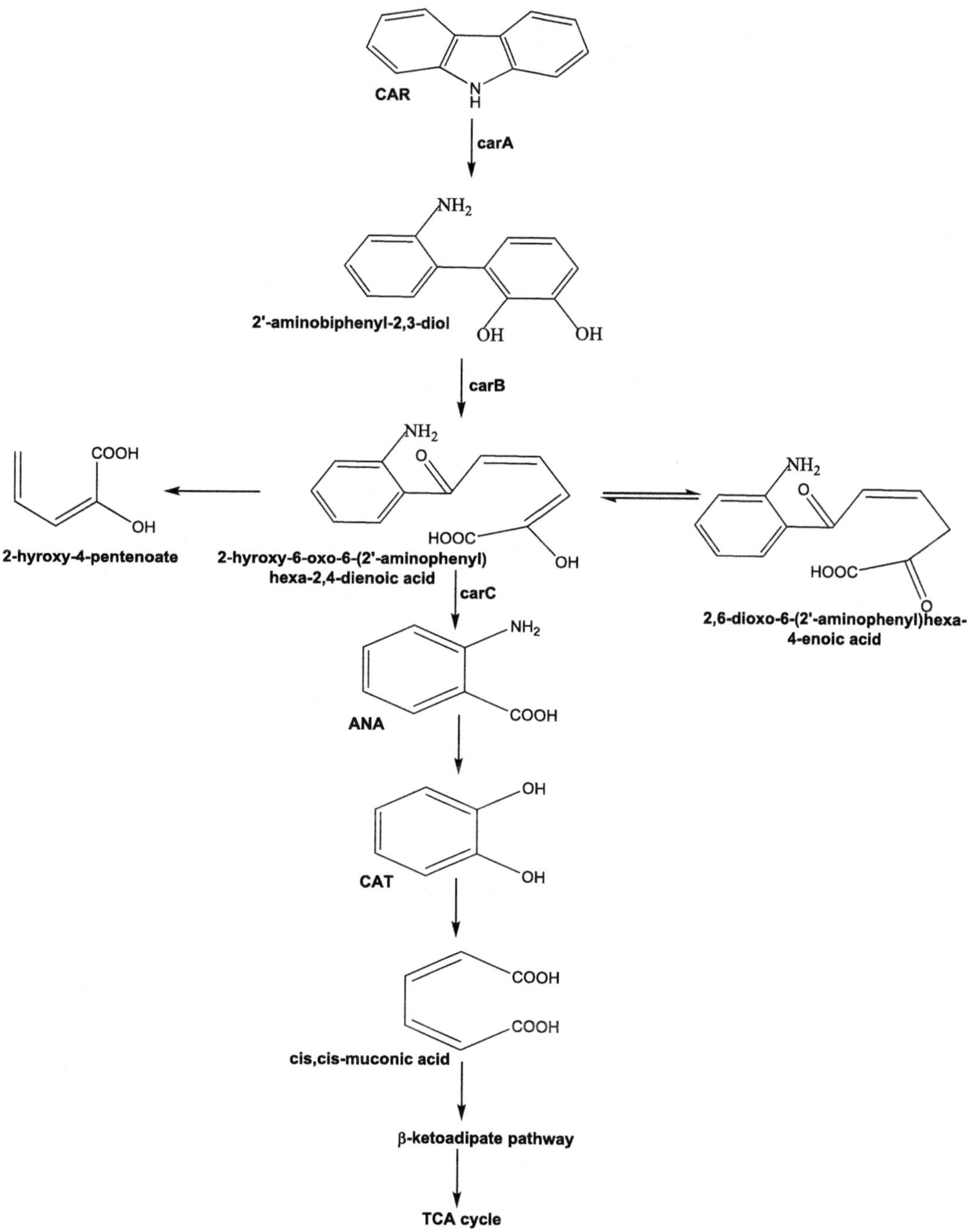

FIGURE 7.2

The biodegradation pathway for carbazole (CAR-Biodegradation pathway).

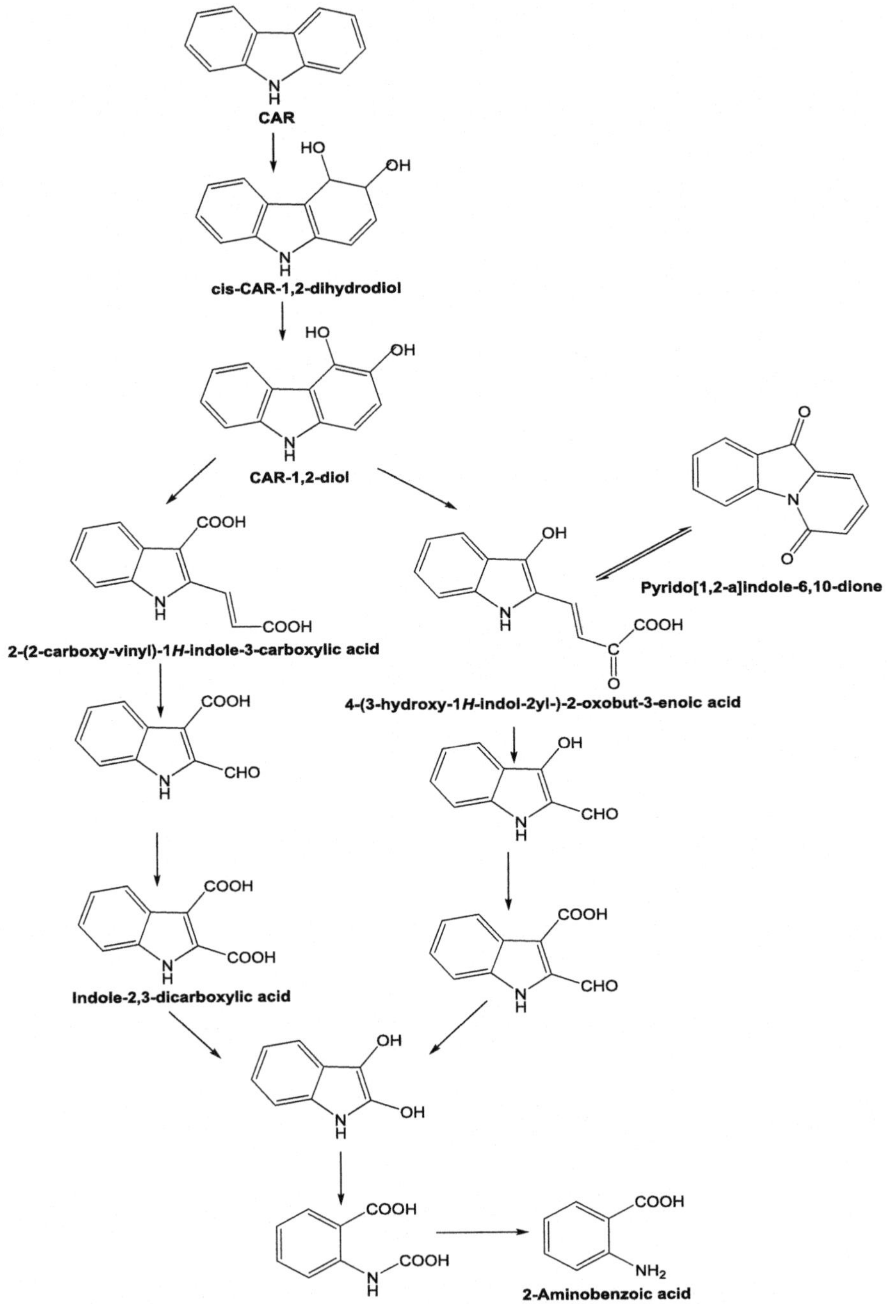

FIGURE 7.3

The biodegradation pathway of carbazole by *Arthrobacter* sp. P1-1 (Seo et al., 2006).

Table 7.3 Carbazole Biodegrading Microorganisms

Microorganisms	References
Pseudomonas sp. CA06 and CA10	Ouchiyama et al. (1993)
Pseudomonas stutzeri ATCC 31258	Hisatsuka and Sato (1994)
Ralstonia sp. RJGII.123	Grosser et al. (1991) and Schneider et al. (2000)
Pseudomonas cepacia F297	Grifoll et al. (1995)
Pseudomonas sp. LD2	Gieg et al. (1996)
Sphingomonas CB3	Shotbolt-Brown et al. (1996)
Pseudomonas stutzeri OM1	Ouchiyama et al. (1998)
Sphingomonas sp. CDH-7	Kirimura et al. (1999)
Pseudomonas putida ATCC 17484	Loh and Yu (2000)
Ralstonia sp. RGII.123	Schneider et al. (2000)
Novosphingobium sp. KA1	Habe et al. (2002)
Sphingomonas sp. KA1	Inoue et al. (2004)
Pseudomonas rhodesiae KK1	Yoon et al. (2002)
Sphingomonas sp. CDH-7	Nakagawa et al. (2002)
Sphingomonas sp. GTIN11	Kilbane et al. (2002)
Pseudomonas sp. C3211	Jensen et al. (2003)
Neptuniibacter sp. CAR-SF	Fuse et al. (2003) and Nagashima et al. (2010)
Sphingomonas sp. CP19	Bressler et al. (2003)
Burkholderia sp. strain IMP5GC	Castorena et al. (2005)
Pseudomonas sp. XLDN4-9	Li et al. (2006)
Nocardioides aromaticivorans IC177	Inoue et al. (2006)
Klebsiella sp. strain LSSE-H2 (CGMCC No. 1624)	Li et al. (2008)
Acinetobacter sp.	Singh et al. (2011b)
Pseudomonas sp. strain GBS.5	Singh et al. (2013)
Achromobacter sp. strain CAR1389	Farajzadeh and Karbalaei-Heidari (2012)
Bacillus clausii BS1	Zakaria et al. (2016)

products (Fig. 7.4A and B). In the second pathway, indole has been found to be degraded via catechol (Fig. 7.4C). The third pathway is a tryptophan independent pathway (Arora and Bae, 2014). The anaerobic biodegradation of indole has been reported under denitrifying, sulfate-reducing, or methanogenic conditions (Bak and Widdel, 1986; Madsen and Bollag, 1988; Madsen et al., 1988; Shanker and Bollag, 1990; Gu and Berry, 1991; Liu et al., 1994; Gu et al., 2002). The sulfate reducing bacteria *Desulfobacterium indolicum* was reported to metabolize indole by indole oxygenase to give anthranilate (Johansen et al., 1997). Claus and Kutzner (1983) reported indole degradation by *Alcaligenes* sp. via gentisate pathway (Fig. 7.4A). Doukyu and Aono (1997) reported the mineralization of indole via isatin and isatoic acid in *Pseudomonas* sp. strain ST-200. Kamatht and Vaidyanathan (1990) reported indole degradation via catechol by *Aspergillus niger*. However, *Phomopsis liquidambari* the endophytic fungus initially oxidizes indole into oxindole and isatin, which is then transformed to 2-dioxindole. That is further converted to 2-aminobenzoic acid via a pyridine ring cleavage (Chen et al., 2013). The *Arthrobacter* sp. SPG is reported to biotransform indole to indole-3-acetic acid via the tryptophan pathway (Arora and Bae, 2014).

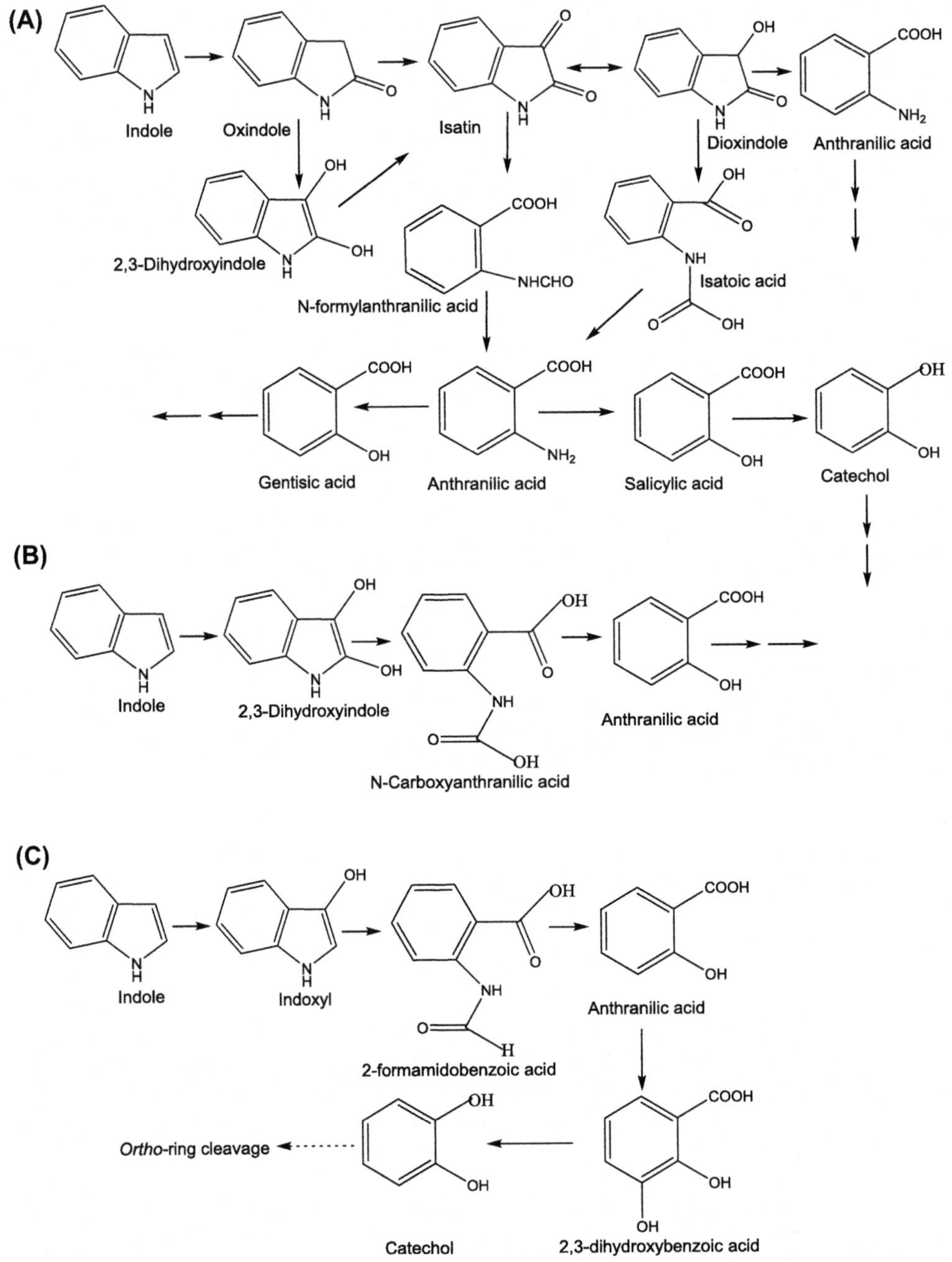

FIGURE 7.4

The different biotransformation pathways of indole.

Qn as one of the major basic nitrogen polyaromatic heterocyclic compounds is usually used as a typical N-containing compound in crude oil (Sugaya et al., 2001; Sona et al., 2010). Many aerobic and anaerobic metabolisms of Qn and its derivatives have been reported (Atlas and Rothenburger, 1993; Solomon et al., 1995; O'Loughlin et al., 1996). The selective transformation of Qn and methyl quinoline in shale oil by *Pseudomonas aeruginosa*, keeping the calorific value of the hydrocarbons within the fuel range, has been reported (Aislabie et al., 1990). However, Qn has a general suggested pathway by many researchers; the initial step is the hydroxylation at position 2 of the heterocyclic aromatic ring, with the formation of 2-hydroxyquinoline (Shukla, 1986). *P. stutzeri* is reported to degrade quinolone by the pathway involving 2-hydroxyquinoline, 8-hydroxycoumarin, and 2,8-dihydroxyquinoline. The 8-hydroxycoumarin is further metabolized to 2,3-dihydroxyphenylpropionic acid, where the rate limiting step is the oxidation of the 2,8-dihydroxyquinoline (Shukla, 1989). Sona et al. (2010) reported the isolation of the Gram positive *Bacillus licheniformis* strain CRC-75 from petroleum-contaminated soil for its capability to selectively metabolize Qn as an N-source but not as a C-source. However, the isolation of the Gram negative *P. aeruginosa* KDQ4 from a Qn enrichment culture obtained from the activated sludge of a coking wastewater treatment plant which has high biodegradation capabilities for high concentrations of Qn and phenol has been reported (Zhang et al., 2016).

Pseudomonas putida is reported to transform both the homocyclic and the heterocyclic moieties of Qns (Boyd et al., 1987). The attack on the homocyclic ring yielded the corresponding *cis*-hydrodiol derivatives and the monohydroxylated derivatives (e.g., 8-hydroxyquinoline or 5-hydroxyisoquinoline). Miethling et al. (1993) reported the biodegradation of Qn by *Comamonas acidovorans* DSM6426 to 2-hydroxyquinoline, which further degraded to 8-hydroxycoumarin. Kilbane et al. (2000) reported the selective biodenitrogenation of Qn without affecting its carbon skeleton by *Pseudomonas ayucida* strain (IGTN9m), producing 8-hydroxycoumarin and ammonia, where, approximately 5% of the nitrogen was removed by this isolate from petroleum without affecting its energetic value. Sugaya et al. (2001) reported the aerobic biodegradation of Qn by *Comamonas* sp. TKV3-2-1 to 2-hydroxyquinoline and then to ammonia and water-soluble intermediates. The rate-limiting step controlling the overall biodenitrogenation process of Qn in crude oil was found to be the degradation reaction of 2-hydroxyquinoline (Miethling et al., 1993; Sugaya et al., 2001). Thus, the rate of degradation of 2-hydroxyquinoline, is used as an index for evaluating the overall performance of Qn biodenitrogenation process in crude oil. The optimum operating conditions were reported to be crude oil concentration of 83% (v/v), cell concentration of 28.5 g/dm^3, and mixing rate of 11.7 s^{-1}, with maximum biodenitrogenation rate of 211 mg 2-hydroxyquinoline/g cell. The treated crude oil and cell suspension were efficiently separated by centrifugation (Sugaya et al., 2001). Fig. 7.5 represents the metabolic pathways of Qn and *iso*-Qn.

There are two general pathways reported for the biodegradation of pyridine. One involves a complete metabolic pathway for the degradation of pyridine by *Bacillus* strain through hydroxylation reactions, followed by reduction (Fig. 7.6A; Watson and Cain, 1975). While, the other, reported for *Nocardia* strain Z1 throughout a reductive pathway was not initiated by hydroxylation (Fig. 7.6B; Rhee et al., 1997).

5.0 CHALLENGES AND OPPORTUNITIES

While significant progress has been made toward the commercialization of crude oil biodenitrogenation (biodenitrogenation), technical hurdles still need to be overcome to achieve commercialization. The major obstacles to the economical biodenitrogenation of crude oil include catalyst specificity and

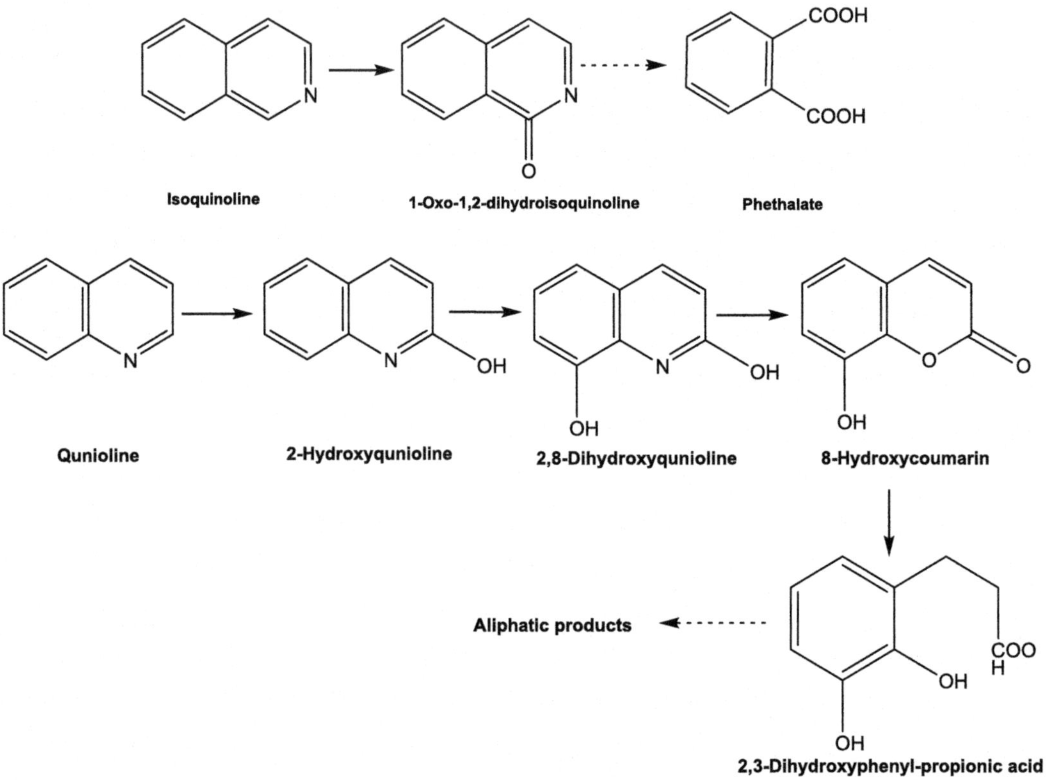

FIGURE 7.5

Metabolic pathways for the biodegradation of quinoline and isoquinoline.

rate. Work continues to modify the catalyst to increase its effectiveness and to screen other organisms for additional desulfurization capabilities. In addition, mass transfer and separations hurdles must be overcome in crude oils with increased oil viscosity and density.

Research on using microorganisms to upgrade or improve crude oil properties is a high-risk but necessary venture because the potential rewards of achieving such a process are significant both environmentally and economically. The main drivers for this work are: (1) tightening of environmental restrictions on total sulfur in refined oil products and lower carbon dioxide emission from refineries, (2) the diminishing availability of high gravity sweet crudes, (3) the rising cost of coking and hydrotreating operations, and (4) the increasing cost of meeting CO_x and sulfur restrictions as future emission allowances are reduced. To this end, it has been reported (Benedik et al., 1998) that any fuel-upgrading process would be a low-margin (probably less than US\$ 1/bbl or US\$ 0.02/L, value added), large volume (approximately 10^{10} L/ year), commodity enterprise, where, an efficient operation is essential for economic viability.

Bio-upgrading of petroleum and its fractions have not yet been applied to any large extent on an industrial scale. But, with the worldwide restrictions and regulations for ultra-low nitrogen and sulfur fuels, it can be achieved by the high-cost deep hydrotreating process that also non-selectively modifies other components in the treated petroleum fuels. That creates a demand for new technologies capable of reducing the hetero-atomic concentrations in fuels and raises the interest toward the microbial approaches.

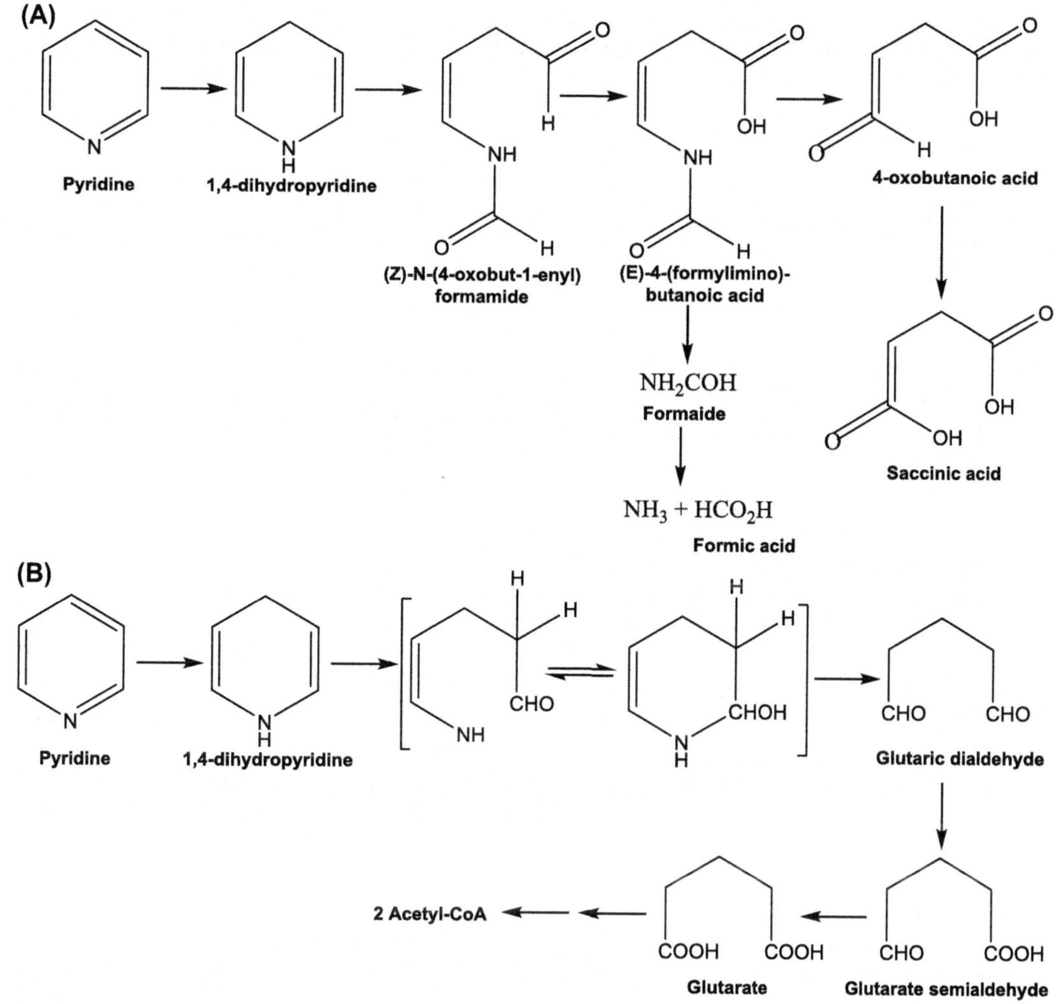

FIGURE 7.6

Metabolic pathways for pyridine biodegradation.

The use of microorganisms to specifically biodesulfurize polycyclic aromatic hydrocarbon derivatives containing sulfur in petroleum and in petroleum fractions will reduce the emissions of sulfur oxides that consequently solve a serious environmental pollution problem. Many researches have been conducted on genetically engineered microorganisms that are required to treat heavy crude oils with more complex structures of polycyclic aromatic hydrocarbon derivative containing sulfur, in addition to cultures with wide substrate ranges. However, unlike biodesulfurization, biodenitrogenation of petroleum and its distillates has not received much attention.

Microbial metabolism of petroleum-nitrogen leads to nitrogen removal and alleviation of the poisoning of the refining catalysts. It will also eliminate the contribution of fuel nitrogen to NO_x emissions. However, the economics of nitrogen-removal processes are affected by the amount of associated hydrocarbon lost from the fuel, during the denitrogenation process. Generally, the currently well-established carbazole-biodenitrogenation pathway resembles that of dibenzothiophene-Kodama pathway, which results in the loss of the fuel value. This would consequently make the biodenitrogenation of fuel streams economically unfeasible. Moreover, most of the carbazole degrading microorganisms, produce; 2′aminobiphenyl-2,3-diol as the first step in carbazole-biodenitrogenation pathway. However, recovering of carbazole-nitrogen as anthranilic acid or 2′aminobiphenyl-2,3-diol, the less inhibitory to refining catalysts, would solve part of that problem. Since, the entire carbon-content of the fuel is preserved. This can be performed by mutant or recombinant strains.

Other pathways would liberate nitrogen from carbazole in the form of ammonia (Rhee et al., 1997). Most of the biodenitrogenation research is carried out in an aqueous phase. However, more research is required for biodenitrogenation in oil/water biphasic systems. Castorena et al. (2006) managed to isolate a thermo- and solvent-tolerant bacterial isolate— *Burkholderia* sp. strain IMP5GC—which is capable of metabolizing high concentrations of carbazole in biphasic-system *n*-hexadecane/water (11.9 μmol/min g wet cell) and gas oil. Li et al. (2008) have isolated the Gram negative *Klebsiella* sp. LSSE-H2 from a mixture of dye-contaminated soil samples, for its ability to biodenitrogenate high concentrations of carbazole (19 mmol/L), with a specific activity toward carbazole in biphasic system (3.3 μmol/min g wet cell) *n*-dodecane/water. Kayser and Kilbane (2004) reported a genetically engineered bacterium that can be applied for the transformation of carbazole to the less toxic 2-aminobiphenyl-2,3-diol in shale oil, petroleum products, and coal tar, where, the carbazole content in a petroleum sample is reported to be reduced by 95% in 2:10 petroleum/aqueous medium within 16 h using that genetically engineered bacteria.

Moreover, a dual microbial process for both selective biodesulfurization and biodenitrogenation, with the overcome of the significant technical hurdles, such as; tolerance against solvents, high concentration of nitrogenous compounds, high oil to water ratio, would make microbial refining processes and bio-upgrading of petroleum and its fractions feasible on a large scale (Kilbane, 2006). Yu et al. (2006) introduced carbazole dioxygenase gene, which was amplified from *Pseudomonas* sp. strain XLDN4-9 into the excellent 4S-dibenzothiophene-biodesulfurization bacteria *Rhodococcus erythropolis* XP and designated the recombinant as SN8. The recombinant *R. erythropolis* SN8 expressed good biodesulfurization and biodenitrogenation activities toward a wide range of recalcitrant alkyl carbazoles and dibenzothiophene derivatives in crude oil, in just a one-step bioprocess. Sugaya et al. (2001) reported that, it is expected that, if biodenitrogenation of crude oil is applied, during its storage period, it would effectively improve its quality and thus increases its products' yields (such as gasoline), since its nitrogen concentration would be decreased, as it would overcome the problems of extra treatment period and cost. Sona et al. (2010) reported the biodenitrogenation of crude oil (5% v/v oil/water ratio) using resting cells of *Bacillus lichiformis* strain CRC-75. Maass et al. (2015) reported biodesulfurization/biodenitrogenation of heavy gas oil (HGO, is an intermediate fraction obtained from vacuum distillation used in the production of diesel and some lubricants) by *R. erythropolis* ATCC 4277 in a batch reactor. That reached maximum desulfurization and denitrogenation rate of 148 mg S/kg HGO/h and 162 mg N/kg HGO/h at 40% (v/v) HGO/water, respectively. The advantage of that biodesulfurization/biodenitrogenation process is that N and S were removed directly from the petroleum fraction without the need for addition of a chemical reagent, or surfactant, or immobilization of the cells to increase the bioavailability; moreover, ATCC 4277 tolerates high concentration of oil feed. That would promote the applicability of biodesulfurization and biodenitrogenation processes as economically and environmentally/industrially viable techniques. The use of enzymes for biodenitrogenation has also been reported. Laccase from *Coriolopsis gallica* has been reported for biodenitrogenation of carbazole in a medium with 15% w/w acetonitrile (Bressler et al., 2000).

Moreover, immobilization of bacteria is one of the recommendable methods for enhancing biodeni-trogenation rate (Ulonska et al., 2004; Balasubramaniyan and Swaminathan, 2007). Since, immobi-lized microorganisms would overcome the problem of the cost of biocatalyst preparation and high ratio of water-to-oil (v/v), it enhances the stability of the biocatalyst, facilitates its recovery and reusability, and decreases the risk of contamination. Immobilization by entrapment technique has some mass-transfer limitations such as steric hindrance and limited diffusion. Thus, immobilization by adsorption technique reduces/eliminates these limitations.

Castorena et al. (2008) reported the biofilm immobilization of *Burkholderia* sp. IMP5GC and its application in a packed reactor (Fig. 7.7) for a semicontinuous carbazole-biodegradation process in a biphasic-system mixture of 85% gas oil and 15% light cycle oil/water of 0.1 (i.e., 90% oil/water). The cells of IMP5GC were immobilized on the surface of porous glass cylinders 95 mm diameter, 6 mm height, pore size 100–160 µm. The reactor was packed with 170 cylinders covered with the IMP5GC-biofilm. The fuel mixture was replaced every 12 h, while, the aqueous phase was recycled to the reactor. A small volume of the aqueous phase should be freshly added to restore its original volume that is mainly lost by evaporation. The specific activity of the immobilized cells was lower than the free cells recording 74 and 153.3 mg carbazole/g protein/h, respectively. This might be attributed to mass transfer

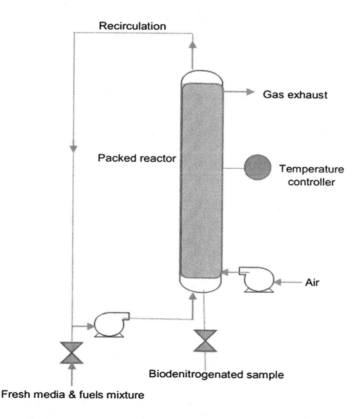

FIGURE 7.7

Schematic diagram of a packed-bed reactor for the semi-continuous fuel-biodegradation.

limitation of carbazole from medium to the cell cytoplasm through the biofilm instead of just the aqueous phase, as well as other regulatory and metabolic changes that might have been occurred due to the switching from planktonic to biofilm growth. However, the specific carbazole-degrading activity of the immobilized cells was stable in the reactor for a period of 60 days and 120 batches of the fuel mixture. Most of the water was recycled back to the reactor for reuse and thus reduced the ratio of water/oil in the process (during the 60 days, barely 1 L of water was added, i.e., 10.81 fuel/1 L aqueous media). Complete removal of carbazole occurred within 12 h, while 40 h were needed to remove 70% of 2-methycarbazole and 3-methylcarbazole and 50% 4-methylcarbazole and 1-methyl-carbazole from the oil mixture. That might be attributed to the depletion of required nutrient components in aqueous medium and/or the accumulation of inhibitory metabolites during the long incubation period. The hydrocarbons profile of the treated oil did not change. Thus, for deeper biodenitrogenation, it is suggested that, the fuel mixtures would be circulated through a sequence of successive semicontinuous packed reactors coupled with the addition of fresh aqueous phase.

Applying nanoparticles (NPs) with large specific surface area and high surface energy is very recommendable, since, nanotechnology represents a new generation of environmental-remediation technologies that could provide cost-effective solutions to some of the most challenging environmental clean-up problems. Magnetic nanoparticles are of special interest for their unique magnetic properties due to their reduced size and potential use in many technological applications (Hyeon, 2003). Magnetite is used in a wide range of applications, including data storage (Hyeon, 2003), magnetic fluids (Chikazumi et al., 1987), biotechnology (Gupta and Gupta, 2004), catalysis (Lu et al., 2004), magnetic resonance imaging (MRI) (Mornet et al., 2006), and environmental remediation (Elliot and Zhang, 2001; Takafuji et al., 2004). The magnetic Fe_3O_4 is one of the common iron oxides which has many important technological applications. The importance of magnetic nanoparticles is its applications in the separation of biomolecules for characterization or purification and they are a well-established alternative to centrifugal separation of biological solutions (Pankhurst et al., 2003). They can be easily manipulated by permanent magnets or electromagnets, independent of normal microfluidic or biological processes. Therefore, the best advantage of the magnetic separation in biotechnology is ease of manipulation of biomolecules that are coated by magnetic particles. Upon placing the magnetic nanoparticles in solution, any target cells can be captured by the functionalized surfaces. Using a magnet at the side of the solution, a magnetic moment is induced in each of the freely floating particles and sets up a field gradient across the solution. The magnetized particles will move along the field lines and aggregate toward the permanent magnet, separating their bound target from the solution. Magnetic nanoparticles can resolve many separation problems in industry and are being investigated for several different chemical separations applications. The catalysts are easily separated by utilizing the magnetic interaction between the magnetic nanoparticle and an external applied magnetic field that can be easily conjugated with biomolecules (Fig. 7.8). Biocompatible magnetic nanoparticles have a wide range of applications in bioscience and they are also able to solve many separation problems in industry; small size enables the NPs to penetrate or diffuse into contaminated area where micro-sized particles fail to reach and nano-sized particles have higher reactivity to redox-amenable contaminants. The nanoscale particles (1–100 nm) afford very high surface areas without the use of porous absorbents and can be recovered for reuse. There have been wide studies of magnetic separation techniques of cells, proteins, viruses, bacteria, and other biomolecules which achieved enormous success (Olsvik et al., 1994; Prestvik et al., 1997; Arshady, 2001; Neuberger et al., 2005).

Wang et al. (2007) stated that *Sphingomonas* sp. XLDN2-5, as a carbazole-degrading strain, was entrapped in the mixture of Fe_3O_4 nanoparticles and gellan gum using modified traditional entrapment method. The magnetically immobilized XLDN2-5 expresses higher carbazole-biodegradation (3479 μg

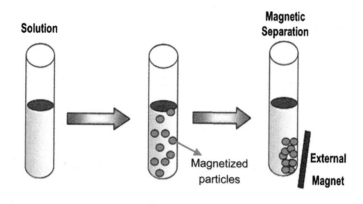

FIGURE 7.8

Magnetic separation of magnetized particles.

carbazole/g wet weight cell/h) than the non-magnetically immobilized (1761 μg carbazole/g wet weight cell/h) and free (3092 μg carbazole/g wet weight cell/h XLDN2-5) and can be used for eight successive cycles. Where, the specific degradation rate increased from 3479 to 4638 μg carbazole/g wet weight cell/h in the eighth cycle, due to the good growth of cells in the magnetic gellan gel beads. The observed decrease of carbazole biodegradation rate in the non-magnetically immobilized matrix is attributed to the mass transfer limitation and steric hindrance. The increase in the magnetically immobilized matrix is due to the presence of magnetic nanoparticles which loosen the binding of the sheets of the gellan gum matrix and the existence of many pores between the sheets of gellan gum matrix. The highest carbazole-biodegradation activity occurred at concentration of 9 mg ferric oxide per milliliter (9 mg Fe_2O_3/mL) and saturation magnetization of the magnetically immobilized cells of 11.08 emu/g.

Li et al. (2013) reported the direct assembling of *Sphingomonas* sp. XLDN2-5 by Fe_2O_3 NPs, where the ratio of cells/magnetic nanoparticles was 1:1. The average diameter of Fe_2O_3 was 20 nm, with superparamagnetic properties and saturation magnetization of 45.5 emu/g. The resulting microbial cell/Fe_3O_4 biocomposite and free cells exhibited the same carbazole-biodegradation efficiency. Thus, the magnetic nanoparticles did not exhibit any negative impact on XLDN2-5. That was attributed to the biocompatibility of magnetic nanoparticles, i.e., the coating layer itself, as it does not change the hydrophilicity of the cell surface. Moreover, this coating layer has a negligible effect on mass transfer, as its structure is looser than that of the cell wall. Thus, microbial cell/Fe_3O_4 biocomposite produces a system that is not limited by diffusional limitations. The activity of microbial cell/Fe_3O_4 biocomposite increased gradually during the recycling process. Where complete removal of 3500 μg carbazole occurred within 9 h for sixth successive cycles, but the same amount was completely removed in only 2 h within the 7thto the 10th cycles.

Zakaria et al. (2015b) reported the good assembling of magnetic Fe_3O_4 NPs (8–10 nm) on the Gram positive long bacilli bacterial isolate, *B. claussi* BS1 (0.495 μm×2.06 μm). The rate of carbazole biodegradation was doubled by coated cells, recording $t_{1/2}$ values of 31.36 and 64.78 h for coated and free-cells, respectively. The coated cells do not experience any mass transfer problem. Magnetic nanoparticles express good adsorption capacity toward carbazole (9.51 mmol/g), thus increasing the adsorption of carbazole to the cells for biodegradation. The adsorption between magnetite (Fe_3O_4) nanoparticles and carbazole is electrostatic, so the adsorption is reversible, i.e., it can be easily desorbed to the cells for biodenitrogenation. The coated cells are characterized by high storage and operational stability and reusability. It can be used for four successive cycles without losing its activity. It has lower sensitivity toward high concentrations of toxic by-products and metabolites (anthranilic acid and catechol) than the free cells. It has also the advantages of magnetic separation, which would resolve many operational problems in petroleum refinery. That would encourage the application of magnetite nanoparticles/*B. claussi* BS1 biocomposite in biodenitrogenation or biotransformation of nitrogen polyaromatic heterocyclic compounds in petroleum industry.

REFERENCES

Aislabie, J., Bej, A.K., Hurst, H., Rothenburger, S., Atlas, R.M., 1990. Microbial degradation of quinoline and methylquinolines. Applied and Environmental Microbiology 56 (2), 345–351.

Arcos, J.C., Argus, M.F., 1968. Molecular geometry and carcinogenic activity of aromatic compounds: new perspectives. Advances in Cancer Research 11, 305–471.

Arora, P.K., Bae, H., 2014. Identification of new metabolites of bacterial transformation of indole by gas chromatography-mass spectrometry and high performance liquid chromatography. International Journal of Analytical Chemistry. 239641. https://doi.org/10.1155/2014/239641.

Arora, P.K., Sharma, A., Bae, H., 2015. Microbial degradation of indole and its derivatives. Journal of Chemistry. 129159. https://doi.org/10.1155/2015/129159.

Arshady, R., 2001. Microspheres Microcapsules and Liposomes. Citus Books, London, United Kingdom. ISBN: 9780953218769.

Atlas, M.R., Rothenburger, S., 1993. Hydroxylation and biodegradation of 6-methylquinoline by *Pseudomonas* in aqueous and no aqueous immobilized-cell bioreactors. Applied and Environmental Microbiology 59 (7), 2139–2144.

Bachman, R.T., Johnson, A.C., Edyvean, R.G.J., 2014. Biotechnology in the petroleum industry: an overview. International Biodeterioration and Biodegradation 86, 225–237.

Bak, F., Widdel, F., 1986. Anaerobic degradation of indolic compounds by sulfate-reducing enrichment cultures, and description of *Desulfobacterium indolicum* gen. nov., sp. nov. Archives of Microbiology 146 (2), 170–176.

Bakr, M.M.Y., 2009. Occurrence and geochemical significance of carbazoles and xanthones in crude oil from the Western Desert, Egypt. Journal of King Abdulaziz University Earth Sciences 20 (2), 127–159.

Balasubramaniyan, S., Swaminathan, M., 2007. Enhanced degradation of quinoline by immobilized *Bacillus brevis*. Journal of the Korean Chemical Society 51 (2), 154–159.

Benedik, M.J., Gibbs, P.R., Riddle, R.R., Willson, R.C., 1998. Microbial denitrogenation of fossil fuels. Trends in Biotechnology 16, 390–395.

Bleeker, E.A.J., Wiegman, A.S., Droge, S.T.J., Kraak, M.H.S., van Gestel, C.A.M., 2003. Towards an Improvement of the Risk Assessment of Polycyclic (Hetero) Aromatic Hydrocarbon, Report 2003-01 of the Department of Aquatic Ecology and Ecotoxicology. Vrije Universiteit, Amsterdam, The Netherlands, pp. 27–32.

Boyd, D.R., McMordie, R.A.S., Porter, H.P., 1987. Metabolism of bicyclic aza-arenes by *Pseudomonas putida* to yield vicinal cis-dihydrodiols and phenols. Journal of the Chemical Society, Chemical Communications 22, 1722–1724.

Bressler, D.C., Gray, M.R., 2002. Hydrotreating chemistry of model products from bioprocessing of CARs. Energy and Fuels 16 (5), 1076–1086.

Bressler, D.C., Fedorak, P.M., Pickard, M.A., 2000. Oxidation of carbazole, p-ethylcarbazole, fluorine and dibenzothiophene by laccase of *Coriolopsis gallica*. Biotechnology Letters 22, 1119–1125.

Bressler, D.C., Kirkpatrick, L.A., Foght, J.M., Fedorak, P.M., Gray, M.R., 2003. Denitrogenation of carbazole by combined biological and catalytic treatment. American Chemical Society. Petroleum Chemistry Division Preprints 48 (1), 44.

Castorena, G., Acuña, M.E., Aburto, J., Bustos-Jaimes, I., 2008. Semi-continuous biodegradation of carbazole in fuels by biofilm-immobilized cells of *Burkholeria* sp. strain IMP5GC. Process Biochemistry 43, 1318–1321.

Castorena, G., Mugica, V., Le Borgne, S., Acuña, M.E., Bustos-Jaimes, I., Aburto, J., 2005. Carbazole biodegradation in gas oil/water biphasic media by a new isolated bacterium *Burkholderia* sp. strain IMP5GC. Journal of Applied Microbiology. 100 (4), 739–745. https://doi.org/10.1111/j.1365-2672.2005.

Castorena, G., Mugica, V., Le Borgne, S., Acuña, M.E., Bustos-Jaimes, I., Aburto, J., 2006. Carbazole biodegradation in gas oil/water biphasic media by a new isolated bacterium *Burkholderia* sp. Strain IMP5GC. Journal of Applied Microbiology 100, 739–745.

Chen, Y., Xie, X.-G., Ren, C.-G., Dai, C.-C., 2013. Degradation of N-heterocyclic indole by a novel endophytic fungus *Phomopsis liquidambari*. Bioresource Technology 129, 568–574.

Chikazumi, S., Taketomi, S., Ukita, S., Mizukami, M., Miyajima, M., Setogawa, H., Kurihara, Y., 1987. Physics of magnetic fluid. Journal of Magnetism and Magnetic Materials 65, 245–251.

Chuang, J.C., Mack, G.A., Kuhlman, M.R., Wilson, N.K., 1991. Polycyclic aromatic hydrocarbons and their derivatives in indoor and outdoor air in an 8-home study. Atmospheric Environment B-urban Atmosphere 25, 369–380.

Claus, G., Kutzner, H.J., 1983. Degradation of indole by *Alcaligenes* spec. Systematic and Applied Microbiology 4 (2), 169–180.

Doukyu, N., Aono, R., 1997. Biodegradation of indole at high concentration by persolvent fermentation with *Pseudomonas* sp. ST-200. Extremophiles: Life under Extreme Conditions 1 (2), 100–105.

El-Gendy, N.Sh., Speight, J.G., 2015. Handbook of Refinery Desulfurization. CRC Press, Taylor & Francis Group, Boca Raton, Florida.

Elliott, D.W., Zhang, W.X., 2001. Field assessment of nanoscale biometallic particles for groundwater treatment. Environmental Science & Technology 35, 4922–4926.

Farajzadeh, Z., Karbalaei-Heidari, H.R., 2012. Isolation and characterization of a new *Achromobacter* sp. strain CAR1389 as a carbazole-degrading bacterium. World Journal of Microbiology & Biotechnology 28, 3075–3080.

Fuse, H., Takimura, O., Murakami, K., Inoue, H., Yamaoka, Y., 2003. Degradation of chlorinated biphenyl, dibenzofuran, and dibenzo-p-dioxin by marine bacteria that degrade biphenyl, carbazole, or dibenzofuran. Bioscience, Biotechnology and Biochemistry 67 (5), 1121–1225.

Furlong, E.T., Carpenter, R., 1982. Azaarenes in Puget sound sediments. Geochimica et Cosmochimica Acta 46, 1385–1396.

Furukawa, K., 2003. 'Super bugs' for bioremediation. Trends in Biotechnology 21, 187–190.

Gary, J.H., Handwerk, G.E., Kaiser, M.J., 2007. Petroleum Refining: Technology and Economics, fifth ed. CRC Press, Taylor & Francis Group, Boca Raton, Florida.

Gieg, L.M., Otter, A., Fedorak, P.M., 1996. Carbazole degradation by *Pseudomonas* sp. LD2: metabolic characteristics and the identification of some metabolites. Environmental Science & Technology 30, 575–585.

Green, J.A., Green, J.B., Grigsby, R.D., Pearson, C.D., Reynolds, J.W., Sbay, I.Y., Sturm Jr., O.P., Thomson, J.S., Vogh, J.W., Vrana, R.P., Yu, S.K.Y., Diem, B.H., Grizzle, P.L., Hirsch, D.E., Hornung, K.W., Tang, S.Y., Carbongnani, L., Hazos, M., Sanchez, V., 1989. Analysis of Heavy Oils: Method Development and Application to Cerro Negro Heavy Petroleum, NIPER-452 (DE90000200. Volumes I and II. Research Institute, National Institute for Petroleum and Energy Research (NIPER), Bartlesville, Oklahoma.

Grifoll, M., Selifonov, S.A., Gatlin, C.V., Chapman, P.J., 1995. Actions of a versatile fluorine degrading bacterial isolate on polycyclic aromatic compounds. Applied and Environmental Microbiology 61 (10), 3711–3723.

Grosser, R.J., Warshawsky, D., Robie Vestal, J., 1991. Indigenous and enhanced mineralization of pyrene, benzo[a] pyrene, and carbazole in soil. Applied and Environmental Microbiology 57, 3462–3469.

Gu, J.-D., Berry, D.F., 1991. Degradation of substituted indoles by an indole-degrading methanogenic consortium. Applied and Environmental Microbiology 57 (9), 2622–2627.

Gu, J.-D., Fan, Y., Shi, H., 2002. Relationship between structures of substituted indolic compounds and their degradation by marine anaerobic microorganisms. Marine Pollution Bulletin 45, 379–384.

Guo, W., Li, D., Tao, Y., Gao, P., Hu, J., 2008. Isolation and description of a stable carbazole-degrading microbial consortium consisting of Chryseobacterium sp. NCY and Achromobacter sp. NCW. Current Microbiology 57, 251–257.

Gupta, A.K., Gupta, M., 2004. Synthesis and surface engineering of iron oxide nanoparticles for biomedical applications. Biomaterials 26, 3995–4021.

Habe, H., Ide, K., Yotsumoto, M., Tsuji, H., Hirano, H., Widada, J., Yoshida, T., Nojiri, H., Omori, T., 2001. Preliminary examinations for applying a carbazole-degrader, *Pseudomonas* sp. strain CA10, to dioxin-contaminated soil remediation. Applied Microbiology and Biotechnology 56 (5), 788–795.

Habe, H., Ide, K., Yotsumoto, M., Tsuji, H., Hirano, H., Widada, J., Yoshida, T., Nojiri, H., Omori, T., 2001. Preliminary examinations for applying a carbazole-degrader, *Pseudomonas* sp. strain CA10, to dioxin-contaminated soil remediation. Applied Microbiology and Biotechnology 56 (5), 788–795.

Hegedus, L.L., McCabe, R.W., 1981. Catalyst poisoning. Catalysis Review-Science and Engineering 23, 377–476.

Hirsch, D.E., Cooley, J.E., Coleman, H.J., Thompson, C.J., 1974. Qualitative Characterization of Aromatic Concentrates of Crude Oils from GPC Analysis. Report 7974. Bureau of Mines, U.S. Department of the Interior, Washington, DC.

Hisatsuka, K., Sato, M., 1994. Microbial transformation of carbazole to anthranilic acid by Pseudomonas stutzeri. Bioscience, Biotechnology and Biochemistry 58, 213–214.

Hsu, C.S., Robinson, P.R., 2006. Practical Advances in Petroleum Processing Volume 1 and Volume 2. Springer Science, New York.

Hyeon, T., 2003. Chemical synthesis of magnetic nanoparticles. Chemical Communications 8, 927–935.

Inoue, K., Widada, J., Nakai, S., Endoh, T., Urata, M., Ashikawa, Y., Shintani, M., Saiki, Y., Yoshida, T., Habe, H., Omori, T., Nojiri, H., 2004. Divergent structures of carbazole degradative car operons isolated from gram-negative bacteria. Bioscience, Biotechnology, and Biochemistry 68, 1467–1480.

Inoue, K., Habe, H., Yamane, H., Nojiri, H., 2006. Characterization of novel carbazole catabolism genes from gram-positive carbazole degrader *Nocardioides aromaticivorans* IC177. Applied and Environmental Microbiology 72, 3321–3329.

Jensen, A.-M., Finster, K.W., Karlson, U., 2003. Degradation of carbazole, dibenzothiophene, and dibenzofuran at low temperature by *Pseudomonas* sp. strain C3211. Environmental Toxicology and Chemistry. 22 (4), 730–735. https://doi.org/10.1002/etc.5620220408.

Johansen, S.S., Licht, D., Arvin, E., 1997. Metabolic pathways of quinoline, indole and their methylated analogs by *Desulfobacterium Indolicum* (DSM 3383). Applied Microbiology and Biotechnology 47 (3), 292–300.

Kamatht, A.V., Vaidyanathan, C.S., 1990. New pathway for the biodegradation of indole in *Aspergillus Niger*. Applied and Environmental Microbiology 56 (1), 275–280.

Kayser, K.J., Kilbane, J.J., 2004. Method for Metabolizing Carbazole in Petroleum. US Patent 6,943,006.

Kilbane, J.J., 2006. Microbial biocatalyst developments to upgrade fossil fuels. Current Opinion in Biotechnology 17, 305–314.

Kilbane, J.J., Daram, A., Abbasian, J., Kayser, K., 2002. Isolation and characterization of *Sphingomonas* sp. GTIN11 capable of carbazole metabolism in petroleum. Biochemical and Biophysical Research Communications 297, 242–248.

Kilbane, J.J., Ranganathan, R., Cleveland, L., Kayser, K.J., Ribiero, C., Linhares, M.M., 2000. Selective removal of nitrogen from quinoline and petroleum by *Pseudomonas ayucida* IGTN9M. Applied and Environmental Microbiology 66 (2), 688–693.

Kimura, T., Yan, Z., Kodama, T., Omori, T., 1996. Isolation and characterization of Tn 5-induced mutants deficient in CAR catabolism. FEMS Microbiology Letters 135 (1), 65–70.

Kirimura, K., Nakagawa, H., Tsuji, K., Matsuda, K., Kurane, R., Usami, S., 1999. Selective and continuous degradation of carbazole contained in petroleum oil by resting cells of *Sphingomonas* sp. CDH-7. Bioscience, Biotechnology and Biochemistry 63, 1563–1568.

Larentis, A.L., Sampaio, H.C.C., Carneiro, C.C., Martins, O.B., Alves, T.L.M., 2011. Evaluation of growth, carbazole biodegradation and anthranilic acid production by *Pseudomonas stutzeri*. Brazilian Journal of Chemical Engineering 28 (1), 37–44.

Li, L., Li, Q., Li, F., Shi, Q., Yu, B., Liu, F., Xu, P., 2006. Degradation of carbazole and its derivatives by a *Pseudomonas* sp. Applied Microbiology and Biotechnology 73, 941–948.

Li, Y., Du, X., Wu, C., Liu, X., Wang, X., Xu, P., 2013. An efficient magnetically modified microbial cell biocomposite for carbazole biodegradation. Nanoscale Research Letters 8, 522.

Li, Y.-G., Li, W.-L., Huang, J.X., Xiong, X.-C., Gao, H.-S., Xing, J.-M., Liu, H.-Z., 2008. Biodegradation of carbazole in oil/water biphasic system by a newly isolated bacterium *Klebsiella* sp. LSSE-H2. Biochemical Engineering Journal 41, 166–170.

Liu, S.M., Jones, W.J., Rogers, J.E., 1994. Influence of redox potential on the anaerobic biotransformation of nitrogen heterocyclic compounds in anoxic freshwater sediments. Applied Microbiology and Biotechnology 41 (6), 717–724.

Liu, X., Zhang, Y., Han, W., Tang, A., Shen, J., Cui, Z., Vitousek, P., Erisman, J.W., Goulding, K., Christie, P., Fangmeier, A., Zhang, F., 2013. Enhanced nitrogen deposition over China. Nature 494, 459–462.

Lobastova, T.G., Sukhodolskaya, G.V., Nikolayeva, V.M., Baskunov, B.P., Turchin, K.F., Donova, M., 2004. Hydroxylation of carbazoles by *Aspergillus flavus* VKM F-1024. FEMS Microbiology Letters 235 (1), 51–56.

Loh, K.-C., Yu, Y.-G., 2000. Kinetics of carbazole degradation by Pseudomonas putida in presence of sodium salicylate. Water Research 34 (17), 4131–4138.

Lopes, T.J., Furlong, E.T., 2001. Occurrence and potential adverse effects of semi volatile organic compounds in streambed sediment. United States. 1992–1995. Environmental Toxicology and Chemistry 20, 727–737.

Lu, A., Schmidt, W., Matoussevitch, N., Pnnermann, H., Bspliethoff, B., Tesche, B., Bill, E., Kiefer, W., Schvth, F., 2004. Nanoengineering of a magnetically separable hydrogenation catalyst. Angewandte Chemie International Edition 43, 4303–4305.

Maass, D., Todescato, D., Moritz, D.E., Vladimir Oliveira, J., Oliveira, D., Ulson de Souza, A.A., Guelli Souza, S.M.A., 2015. Desulfurization and denitrogenation of heavy gas oil by *Rhodococcus erythropolis* ATCC 4277. Bioprocess and Biosystems Engineering 38 (8), 1447–1453.

Madsen, E.L., Bollag, J.M., 1988. Pathway of indole metabolism by a denitrifying microbial community. Archives of Microbiology 151 (1), 71–76.

Madsen, E.L., Francis, A.J., Bollag, J.M., 1988. Environmental factors affecting indole metabolism under anaerobic conditions. Applied and Environmental Microbiology 54 (1), 74–78.

Marcelis, C.L.M., Ivanova, A.E., Janssen, A.J.H., Stams, A.J.M., 2003. Anaerobic desulfurization of thiophenes by mixed microbial communities from oilfields. Biodegradation 14, 173–182.

Miethling, R., Hecht, V., Deckwer, W.D., 1993. Microbial degradation of quinoline: kinetic studies with *Comamonas acidovorans* DSM6426. Biotechnology and Bioengineering 42, 589–595.

Millemann, R.E., Birge, W.J., Black, J.A., Cushman, R.M., Daniels, K.L., Franco, P.J., Giddings, J.M., McCarthy, J.F., Stewart, A.J., 1984. Comparative acute toxicity to aquatic organisms of components of coal-derived synthetic fuels. Transactions of the American Fisheries Society 113 (1), 74–85.

Mogollon, L., Rodriguez, R., Larrota, W., Ortiz, C., Torres, R., 1998. Biocatalytic removal of nickel and vanadium from petroporphyrins and asphaltenes. Applied Biochemistry and Biotechnology 70/72, 765–777.

Mohammed, S.N., Hopfinger, A.J., 1983. Intrinsic mutagenicity of polycyclic aromatic hydrocarbons: a quantitative structure activity study based upon molecular shape analysis. Journal of Theoretical Biology 102, 323–331.

Montil, C., Aquintero, R., Aburto, J., 2009. Petroleum biotechnology: technology trends for the future. African Journal of Biotechnology 8 (2), 2653–2666.

Mornet, S., Vasseur, S., Grasset, F., Veverka, P., Goglo, G., Demourgues, A., Portlier, J., Pollert, E., Duguet, E., 2006. Magnetic nanoparticle design for medical applications. Progress in Solid State Chemistry 4, 237–247.

Nakagawa, H., Kirimura, K., Nitta, T., Kino, K., Kurane, R., Usami, S., 2002. Recycle use of *Sphingomonas* sp. CDH-7 cells for continuous degradation of carbazole in the presence of $MgCl_2$. Current Microbiology 44, 251–256.

Nagashima, H., Zulkharnain, A.B., Maeda, R., Fuse, H., Iwata, K., Omori, T., 2010. Cloning and nucleotide sequences of carbazole degradation genes from marine bacterium *Neptuniibacter* sp. strain CAR-SF. Current Microbiology 61 (1), 50–56.

Nam, I., Kim, Y.-M., Schmidt, S., Chang, Y.-S., 2006. Biotransformation of 1,2,3-tri- and 1,2,3,4,7,8-hexa-chlorodibenzo-p-dioxin by *Sphingomonas wittichii* strain RW1. Applied and Environmental Microbiology 72 (1), 112–116.

Neuberger, T., Schopf, B., Hofman, H., Hofman, M., Rechenberg, B., 2005. Superparamagnetic nanoparticles for biomedical applications: possibilities and limitations of a new drug delivery system. Journal of Magnetism and Magnetic Materials 293, 483–496.

Nojiri, H., Nam, J.-W., Kosaka, M., Morii, K.-I., Takemura, T., Furihata, K., Yamane, H., Omori, T., 1999. Diverse oxygenations catalyzed by carbazole 1,9a-dioxigenase from *Pseudomonas* sp. strain CA10. Journal of Bacteriology 181 (10), 3105–3113.

O'Loughlin, E.J., Kehrmeyer, S.R., Sims, G.K., 1996. Isolation, characterization and substrate utilization of a quinoline-degrading bacterium. International Biodeterioration and Biodegradation 32 (2), 107–118.

Olsvik, O., Popovic, T., Skjerve, E., Cudjoe, K.S., Horns, E., Ugelstad, J., Uhlen, M., 1994. Magnetic separation techniques in diagnostic microbiology. Clinical Microbiology Reviews 7, 43–54.

Osborne, P.J., Preston, M.R., Chen, H.Y., 1997. Azaarenes in sediments, suspended particles and aerosol associated with the River Mersey estuary. Marine Chemistry 58, 73–83.

Ouchiyama, N., Zhang, Y., Omori, T., Kodama, T., 1993. Biodegradation of CAR by *Pseudomonas* spp. CA06 and CA10. Bioscience, Biotechnology and Biochemistry 57, 455–460.

Ouchiyama, N., Miyachi, S., Omori, T., 1998. Cloning and nucleotide sequence of carbazole catabolic genes from *Pseudomonas stutzeri* strain OM1, isolated from activated sludge. The Journal of General and Applied Microbiology 44, 57–63.

Pankhurst, Q.A., Connolly, J., Jones, S.K., Dobson, J., 2003. Applications of magnetic nanoparticles in biomedicine. Journal of Physics D: Applied Physics 36, 167–181.

Prado, G.H.C., Rao, Y., De Klerk, A., 2017. Nitrogen removal from oil: a review. Energy and Fuels 31 (1), 14–36.

Prestvik, W.S., Berge Mork, P.C., Stenstad, P.M., Ugelstad, J., 1997. Preparation and application of monosized magnetic particles in selective cell separation. In: Hafeli, U., Schütt, W., Teller, J., Zborowsk, M. (Eds.), Scientific and Clinical Applications of Magnetic Carriers. Plenum Press, New York, pp. 11–36.

Radovanović, L., Speight, J.G., 2011. Visbreaking: a technology of the future. In: Proceedings. First International Conference – Process Technology and Environmental Protection (PTEP 2011). University of Novi Sad, Technical Faculty "Mihajlo Pupin," Zrenjanin, Republic of Serbia. December 7, pp. 335–338.

Resnick, S.M., Torok, D.S., Gibson, D.T., 1993. Oxidation of carbazole to 3-hydroxycarbazole by naphthalene 1,2-dioxygenase and biphenyl 2,3-dioxygenase. FEMS Microbiology Letters 113, 297–302.

Rhee, S.K., Lee, K.S., Chung, J.C., Lee, S.T., 1997. Degradation of pyridine by *Nocardioides* sp. strain OS4 isolated from the oxic zone of a spent shale column. Canadian Journal of Microbiology 43 (2), 205–209.

Saiki, Y., Habe, H., Yuuki, T., Ikeda, M., Yoshida, T., Nojiri, H., Omori, T., 2003. Rhizoremediation of dioxin-like compounds by a recombinant *Rhizobium tropici* strain expressing carbazole 1,9a-dioxygenase constitutively. Bioscience, Biotechnology and Biochemistry 67, 1144–1148.

Sato, S.I., Ouchiyama, N., Kimura, T., Nojiri, H., Yamane, H., Omori, T., 1997a. Cloning of genes involved in CAR degradation of *Pseudomonas* sp. strain CA10: nucleotide sequences of genes and characterization of meta-cleavage enzymes and hydrolase. Bacteriology 179, 4841–4849.

Sato, S.I., Nam, J.W., Kasuga, K., Nojiri, H., Yamane, H., Omori, T., 1997b. Identification and characterization of genes encoding CAR 1,9α-dioxygenase in *Pseudomonas* sp. strain CA10. Bacteriology 179, 4850–4858.

Schneider, J., Grosser, R.J., Jayasimhulu, K., Xue, W., Kinkle, B., Warshawsky, D., 2000. Biodegradation of carbazole by *Ralstonia* sp. RJGII.123 isolated from a hydrocarbon contaminated soil. Canadian Journal of Microbiology 46, 269–277.

Seo, J.-S., Keum, Y.-S., Cho II, K., Li, Q.X., 2006. Degradation of dibenzothiophene and carbazole by *Arthrobacter* sp. P1-1. International Journal and Biodegradation 58, 36–43.

Shanker, R., Bollag, J.-M., 1990. Transformation of indole by methanogenic and sulfate-reducing microorganisms isolated from digested sludge. Microbial Ecology 20 (2), 171–183.

Shiraishi, Y., Yamada, A., Hirai, T., 2004. Desulfurization and denitrogenation of light oils by methyl viologen-modified aluminosilicate adsorbent. Energy and Fuels 18 (5), 1400–1404.

Shotbolt-Brown, J., Hunter, D.W.F., Aislabie, J., 1996. Isolation and description of carbazole-degrading bacteria. Canadian Journal of Microbiology 42, 79–82.

Shukla, O.P., 1989. Microbiological degradation of quinoline by *Pseudomonas-stutzeri*–the coumarin pathway of quinoline catabolism. Microbios 59 (238), 47–63.

Shukla, O.P., 1986. Microbial transformation of quinoline by a *Pseudomonas* Sp. Applied and Environmental Microbiology 51 (6), 1332–1342.

Singh, G.B., Srivastava, S., Gupta, N., 2010. Biodegradation of carbazole by a promising gram-negative bacterium. World Academy of Science, Engineering and Technology 4, 681–684.

Singh, G.B., Gupta, S., Srivastava, S., Gupta, N., 2011a. Biodegradation of carbazole by newly isolated *Acinetobacter* spp. Bulletin of Environmental Contamination and Toxicology 87, 522–526.

Singh, G.B., Srivastava, S., Gupta, S., Gupta, N., 2011b. Evaluation of carbazole degradation by *Enterobacter* sp. isolated from hydrocarbon contaminated soil. Recent Research in Science and Technology 3 (11), 44–48.

Singh, G.B., Gupta, S., Gupta, N., 2013. Carbazole degradation and biosurfactant production by newly isolated *Pseudomonas* sp. strain GBS.5. International Journal of Biodeterioration and Biodegradation 84, 35–43.

Solomon, B.O., Hecht, V., Posten, C., Deckwer, W.D., 1995. Estimation of the energetic parameters associated with aerobic degradation of quinoline by *Comamonas acidovorance* in continuous culture. Journal of Chemical Technology and Biotechnology 62, 94–97.

Sona, K., Maryam, M., Esmat, F., 2010. Quinoline biodegradation by *Bacillus licheniformis* strain CRC-75. Iranian Journal of Chemistry and Chemical Engineering 29 (2), 151–158.

Speight, J.G., 2011. The Refinery of the Future. Gulf Professional Publishing, Elsevier, Oxford, United Kingdom.

Speight, J.G., 2012a. Crude Oil Assay Database. Knovel, Elsevier, New York. Online version available at: http://www.knovel.com/web/portal/browse/display?_EXT_KNOVEL_DISPLAY_bookid=5485&VerticalID=0.

Speight, J.G., 2012b. Visbreaking: a technology of the past and the future. Scientia Iranica C 19 (3), 569–573.

Speight, J.G., 2014. The Chemistry and Technology of Petroleum, fifth ed. CRC Press, Taylor & Francis Group, Boca Raton, Florida.

Speight, J.G., 2017. Handbook of Petroleum Refining. CRC Press, Taylor & Francis Group, Boca Raton, Florida.

Sugaya, K., Nakayama, O., Hinata, N., Kamekura, K., Ito, A., Yamagiwa, K., Ohkawa, A., 2001. Biodegradation of quinoline in crude oil. Journal of Chemical Technology and Biotechnology 76, 603–611.

Takafuji, M., Ide, S., Ihara, H., Xu, Z., 2004. Preparation of poly-(1-vinylimidazole)-grafted magnetic nanoparticles and their application for removal of metal ions. Chemistry of Materials 16, 1977–1983.

Takagi, T., Nojiri, H., Yoshida, T., Habe, H., Omori, T., 2002. Detailed comparison between the substrate specificities of two angular dioxygenases, dibenzofuran 4,4a-dioxygenase from *Terrabacter* sp. and carbazole 1,9a-dioxygenase from *Pseudomonas resinovorans*. Biotechnology Letters 24 (8), 2099–2106.

Ulonska, A., Deckwer, W.-D., Hecht, V., 2004. Degradation of quinoline by immobilized *Comamonas acidovorans* in a three-phase airlift reactor. Journal of Biotechnology and Bioengineering 46 (1), 80–87.

Wang, X., Gai, Z., Yu, B., Feng, J., Xu, C., Yuan, Y., Deng, Z., Xu, P., 2007. Degradation of CAR by microbial cells immobilized in magnetic gellan gum gel beads. Applied and Environmental Microbiology 73, 6421–6428.

Watson, G.K., Cain, R.B., 1975. Microbial metabolism of the pyridine ring. Biochemical Journal 146, 157–172.

WHO, 2004. Coal Tar Creosote. Concise International Chemical Assessment Document 62. World Health Organization, Geneva, Switzerland.

Widada, J., Nojiri, H., Yoshida, T., Habe, H., Omori, T., 2003. Enhanced degradation of carbazole and 2,3-dichlorodibenzo-p-dioxin in soils by *Pseudomonas resinovorans* strain CA10. Chemosphere 49, 485–491.

Wu, G., Chen, D., Tang, H., Ren, Y., Chen, Q., Lv, Y., Zhang, Z., Zhao, Y.-L., Yao, Y., Xu, P., 2014. Structural insights into the specific recognition of N-heterocycle biodenitrogenation-derived substrates by microbial amide hydrolases. Molecular Microbiology 9 (15), 1009–1021.

Xu, P., Yu, Bo., Li, F.L., Cai, X.F., Ma, C.Q., 2006. Microbial degradation of sulfur, nitrogen and oxygen heterocycles. Trends in Microbiology 14 (9), 398–405.

Yoon, B.-J., Lee, D.-H., Kang, Y.-S., Oh, D.-C., Kim II, A., Oh, K.-H., Kahng, H.-Y., 2002. Evaluation of carbazole degradation by *Pseudomonas rhodesiae* KK1 isolated from soil contaminated with coal tar. Journal of Basic Microbiology 42 (6), 434–443.

Yu, B., Xu, P., Zhu, S., Cai, X., Wang, Y., Li, L., Li, F., Liu, X., Ma, C., 2006. Selective biodegradation of S and N heterocycles by recombinant *Rhodococcus erythropolis* strain containing carbazole dioxygenase. Applied and Environmental Microbiology 72 (3), 2235–2238.

Zakaria, B.S., Nassar, H.N., Abu Amr, S.S., El-Gendy, N.Sh., 2015a. Applying factorial design and response surface methodology to enhance microbial denitrogenation by tween 80 and yeast extract. Petroleum Science and Technology 33 (8), 880–892.

Zakaria, B.S., Nassar, H.N., Saed, D., El-Gendy, N.Sh., 2015b. Enhancement of carbazole denitrogenation rate using magnetically decorated *Bacillus clausii* BS1. Petroleum Science and Technology 33 (7), 802–811.

Zakaria, B.S., Nassar, H.N., El-Gendy, N.Sh., El-Temtamy, S.A., Sherif, S.M., 2016. Denitrogenation of carbazole by a novel strain *Bacillus clausii* BS1 isolated from Egyptian coke. Energy Sources, A: Recovery, Utilization, and Environmental Effects 38 (13), 1840–1851.

Zhang, P., Jia, R., Zhang, Y., Shi, P., Chai, T., 2016. Quinoline-degrading strain *Pseudomonas aeruginosa* KDQ4 isolated from coking activated sludge is capable of the simultaneous removal of phenol in a dual substrate system. Journal of Environmental Science and Health, Part A. Toxic/hazardous Substances and Environmental Engineering 51 (13), 1139–1148.

FURTHER READING

An, S., Tang, K., Nemati, M., 2010. Simultaneous biodesulphurization and denitrification using an oil reservoir microbial culture: effects of sulphide loading rate and sulphide to nitrate loading ratio. Water Research 44, 1531–1541.

Ayala, M., Vazquez-Duhalt, R., Morales, M., Borgne, Le, 2016. Application of microorganisms to the processing and upgrading of crude oil and fractions. In: Lee, S.Y. (Ed.), Consequences of Microbial Interactions with Hydrocarbons, Oils, and Lipids: Production of Fuels and chemicals. Handbook of hydrocarbon and lipid microbiology. Springer International Publishing, Berlin, Germany.

Bei, S.K., Dalai, A.K., Adjay, J., 2001. Comparison of hydrogenation of basic and nonbasic nitrogen compounds present in oil sand derived heavy gas oil. Energy and Fuels 15, 377–383.

Berry, D.F., Madsen, E.L., Bollag, J.-M., 1987. Conversion of indole to oxindole under methanogenic conditions. Applied and Environmental Microbiology 53 (1), 180–182.

Choi, K., Korai, Y., Mochida, I., Ryu, J.-W., Min, W., 2004. Impact or removal extent of nitrogen species in gas oil on its HDS performance: an efficient approach to its ultra-deep desulfurization. Applied Catalysis B: Environmental 50, 9–16.

Choudhary, T.V., Parrott, S., Johnson, B., 2008. Unraveling heavy oil desulfurization chemistry: targeting clean fuels. Environmental Science & Technology 42, 1944–1947.

Diaz, E., Garcia, J.L., 2010. Genetics engineering for removal of sulfur and nitrogen from fuel heterocycles. In: Timmis, K.N., McGenity, T.J., van der Meer, J.R., de Lorenzo, V. (Eds.), Handbook of Hydrocarbon and Lipid Microbiology. Springer-Verlag GmbH, Berlin Heidelberg, Germany, pp. 2787–2801.

Divakar, N.G., Subramanian, V., Sugumaran, M., Vaidyanathan, C.S., 1979. Indole oxygenase from the leaves of *Jasminum grandiflorum*. Plant Science Letters 15, 177–181.

Duarte, G.F., Rosado, A.S., Seldin, L., de Araujo, W., van Elsas, J.D., 2001. Analysis of bacterial community structure in sulfurous-oil-containing soils and detection of species carrying dibenzothiophene desulfurization (dsz) genes. Applied and Environmental Microbiology 67, 1052–1062.

Eaton, R.W., Chapman, P.J., 1992. Bacterial metabolism of naphthalene: construction and use of recombinant bacteria to study ring cleavage of 1,2-dihydroxynaphthalene and subsequent reactions. Journal of Bacteriology 174, 7542–7554.

Éigenson, A.S., Ivchenko, E.G., 1977. Distribution of sulfur and nitrogen in fractions from crude oil and residues. Chemistry and Technology of Fuels and Oils 13, 542–544.

Fallon, R.D., Hnatow, L.L., Jackson, S.C., Keeler, S.J., 2010. Method for Identification of Novel Anaerobic Denitrifying Bacteria Utilizing Petroleum Components as Sole Carbon Source. US patent 7,740,063.

Fedorak, P.M., Westlake, D.W.S., 1984. Microbial degradation of alkylcarbazols in Norm Wells crude oil. Applied and Environmental Microbiology 47 (4), 858–862.

Ishihara, A., Dumeignil, F., Aoyagi, T., Ishikawa, M., Hosomi, M., Qian, E.W., Kabe, Y., 2008. Degradation of carbazole by *Novosphingobium* sp. strain NIY3. Journal of the Japan Petroleum Institute 51 (3), 174–179.

Jha, A.M., Bharti, M.K., 2002. Mutagenic profiles of carbazole in the male germ cells of Swiss albino mice. Mutation Research 500, 97–101.

Jian, I.W., Xiangchun, Q., Liping, H., Yi, Q., Hegemann, W., 2002. Microbial degradation of quinoline by immobilized cells of *Burkholeria pickettii*. Water Research 36 (9), 2288–2296.

Kilbane, J.J., Ribeiro, C.M.S., Linhares, M.M., 2001. *Pseudomonas ayucida* Useful for Cleavage of Organic C-n Bonds. US patent 6,221,651.

Kilbane, J.J., Ribeiro, C.M.S., Linhares, M.M., 2003. Bacterial Cleavage of Only Organic C-n Bonds of Carbonaceous Materials to Reduce Nitrogen Content. US patent 6,541,240.

Kobayashi, T., Kurane, R., Nakajima, K., Nakamura, Y., Kirimura, K., Usami, S., 1995. Isolation of bacteria degrading carbazole under microaerobic conditions, i.e. nitrogen gas substituted conditions. Bioscience, Biotechnology and Biochemistry 59 (5), 932–933.

Laredo, G., De Los Reyes, A., Cano, J., Castillo, J., 2001. Inhibition effects of nitrogen compounds on the hydrodesulfurization of dibenzothiophene. Applied Catalysis A: General 207, 103–112.

Laredo, G.C., Leva, S., Alvarez, R., Mares, M.T., Castillo, J.J., Cano, J.L., 2002. Nitrogen compounds characterization in atmospheric gas oil and light cycle oil from a blend of Mexico crudes. Fuel 81, 1341–1350.

Le Borgne, S., Quintero, R., 2003. Biotechnological processes for the refining of petroleum. Fuel Processing Technology 81, 155–169.

Li, L., Xu, P., Blankespoor, H.D., 2004. Degradation of carbazole in the presence of non-aqueous phase liquids by *Pseudomonas* sp. Biotechnology Letters 26, 581–584.

Licht, D., Johansen, S.S., Arvin, E., 1997. Transformation of indole and quinoline by *Desulfobacterium indolicum* (DSM 3383). Applied Microbiology and Biotechnology 47 (2), 167–172.

Maeda, K., Nojiri, H., Shintani, M., Yoshida, T., Habe, H., Omori, T., 2003. Complete nuclitide sequence of carbazole/dioxin-degrading plasmid pCAR1 in *Pseudomonas resinovorans* strain CA10 indicates its mosaicity and the presence of large catabolic transport Tn4676. Journal of Molecular Biology 326, 21–33.

Min, W.-S., Choi, K., Khang, S.-Y., Min, D.-S., Ryu, J.-W., Yoo, K.-S., Kim, J.-H., 2001. Method for Manufacturing Cleaner Fuels. US Patent No. 6,248,230.

Monticello, D.J., Finnerty, W.R., 1985. Microbial desulfurization of fossil fuels. Annual Review of Microbiology 39, 371–389.

Morales, M., Le Borgne, S., 2010. Microorganisms utilizing nitrogen-containing hydrocarbons. In: Timmis, K.N. (Ed.), Handbook of Hydrocarbon and Lipid Microbiology. Springer, Berlin, Germany.

Morales, M., Le Borgne, S., 2014. Protocols for the isolation and preliminary characterization of bacteria for biodesulfurization and biodenitrogenation of petroleum-derived fuels. In: McGenity, T.J., Timmis, K.N., Nogales, B. (Eds.), Hydrocarbon and Lipid Microbiology Protocols. Springer, Berlin, Germany.

Mushrush, G.W., Beal, E.J., Hardy, D.R., Hughes, J.M., 1999. Nitrogen compound distribution in middle distillate fuels derived from petroleum, oil shale, and tar sand sources. Fuel Processing Technology 61, 197–210.

Nam, J., Nojiri, H., Noguchi, H., Uchimura, H., Yoshida, T., Habe, H., Yamane, H., Omori, T., 2002. Purification and characterization of carbazole 1,9a-dioxygenase, a three-component dioxygenase system of *Pseudomonas resinovorans* strain CA10. Applied and Environmental Microbiology 68, 5882–5890.

Nielsen, L.E., Kaday, D.R., Rajagopal, S., Drijber, R., Nickerson, K.W., 2005. Survey of extreme solvent tolerance in gram-positive cocci: membrane fatty acid changes in *Staphylococcus haemolyticus* grown in toluene. Applied and Environmental Microbiology 71, 5171–5176.

Nojiri, H., 2012. Structural and molecular genetic analyses of bacterial carbazole degradation system. Bioscience, Biotechnology and Biochemistry 76, 1–18.

Nojiri, H., Omori, T., 2007. Carbazole metabolism by *Pseudomonas*. In: Ramos, J.-L., Filloux, A. (Eds.), Pseudomonas. Springer, NY, USA, pp. 107–145.

Nojiri, H., Ashikawa, Y., Noguchi, H., Nam, J.-W., Urata, M., Fujimoto, Z., Uchimura, H., Terada, T., Nakamura, S., Shimizu, K., Yoshida, T., Habe, H., Omori, T., 2005. Structure of the terminal oxygenase component of angular dioxygenase, carbazole, 1,9a-dioxygenase. Journal of Molecular Biology 351 (2), 355–370.

Parkash, S., 2003. Refining Processes Handbook. Gulf Professional Publishing, Elsevier, Amsterdam, Netherlands.

Richard, J.J., Junk, G.A., 1984. Steam distillation, solvent extraction, and ion exchange for determining polar organics in shale process waters. Analytical Chemistry 56 (9), 1625–1628.

Riddle, R.R., Gibbs, P.R., Willson, R.C., Benedik, M.J., 2003a. Purification and properties of 2-hydroxy-6-oxo-6-(2′-aminophenyl)-hexa-2,4-dienoic acid hydrolase involved in microbial degradation of carbazole. Protein Expression and Purification 28 (1), 182–189.

Riddle, R.R., Gibbs, P.R., Willson, R.C., Benedik, M.J., 2003b. Recombinant carbazole-degrading strains for enhanced petroleum processing. Journal of Industrial Microbiology & Biotechnology 30, 6–12.

Santos, S.C.C., Alvinao, D.S., Alviano, C.S., Pádula, M., Leitão, A.C., Martins, O.B., Ribeiro, C.M.S., Sassaki, M.Y.M., Matta, C.P.S., Bevilaqua, J., Sebasian, G.V., Seldin, L., 2001. Characterization of Gordonia sp. strain F.5.25.8 capable of dibenzothiophene desulfurization and carbazole utilization. Applied Microbiology and Biotechnology 71, 355–362.

Shepherd, J.M., Lloyd-Jones, G., 1998. Novel carbazole degradation genes of *Shingomonas* CB3: sequence analysis, transcription, and molecular ecology. Biochemical and Biophysical Research Communications 247, 129–135.

Shin, S., Sakanishi, K., Mochida, I., 2000. Identification and reactivity of nitrogen molecular species in gas oils. Energy and Fuels 14, 539–544.

Sikkema, J., de Bont, J.A., Polman, B., 1995. Mechanisms of membrane toxicity of hydrocarbons. Microbiological Reviews 59, 201–202.

Sutton, S.D., Pfaller, S.L., Shann, J.R., Warshawsky, D., Kinkle, B.K., Vestal, J.R., 1996. Aerobic biodegradation of 4-methylquinoline by a soil bacterium. Applied and Environmental Microbiology 62 (8), 2910–2914.

Szymanska, A., Lewandowski, M., Sayag, C., Djéga-Mariadassou, G., 2003. Kinetic study of the hydrodenitrogenation of carbazole over bulk molybdenum carbide. Journal of Catalysis 218, 23–31.

Tissot, B.B., Welte, D.H., 1984. Petroleum Formation and Occurrence 2nd Revised and Enlarged Edition. Springer-Verlag, Berlin Heidelberg New York Tokyo.

US Environmental Protection Agency (EPA), 1986. Health and Environmental Effects Profile for Carbazole. US Environmental Protection Agency, Washington, DC. EPA/600/X-86/334 (NTIS PB88218789).

Von der Weid, I., Marques, J.M., Cunha, C.D., Lippi, R.K., Santos, S.C.C., Rosado, A.S., Lins, U., Seldin, L., 2007. Identification and biodegradation potential of a novel strain of *Dietzia cinnamea* isolated from a petroleum contaminated tropical soil. Systematic and Applied Microbiology 30, 331–339.

Wang, Y.-T., Suidan, M.T., Pfeffer, J.T., 1984. Anaerobic biodegradation of indole to methane. Applied and Environmental Microbiology 48 (5), 1058–1060.

Williams, P.T., Chishti, H.M., 2001. Reaction of nitrogen and sulphur compounds during catalytic hydrotreatment of shale oil. Fuel 80, 957–963.

Yamazoe, A., Yagi, O., Oyaizu, H., 2004. Biotransformation of fluorine, biphenyl ether, dibenzo-*p*-dioxin and carbazole by *Janibacter* sp. Biotechnology Letters 26, 479–486.

Zeuthen, P., Knudsen, K.G., Whitehurst, D.D., 2001. Organic nitrogen compounds in gasoil blends, their hydrotreated products and the importance to hydrotreatment. Catalysis Today 65, 307–314.

Zhao, C., Zhang, Y., Li, X., Wen, D., Tang, X., 2011. Biodegradation of carbazole by the seven *pseudomonas* sp. strains and their denitrification potential. Journal of Hazardous Materials 190, 253–259.

BIOTRANSFORMATION IN THE ENVIRONMENT

1.0 INTRODUCTION

One of the major and continuing environmental problems is hydrocarbon contamination resulting from the activities related to petroleum and petroleum products. It is estimated that between 1.7 and 1.8 million metric tons of oil are released into the aquatic and soil environments every year, where, approximately 30% the spilled oil reaches freshwater ecosystems (Dadrasnia and Agamuthu, 2013). Soil contamination with hydrocarbons causes extensive damage of local systems since accumulation of pollutants in animals and plant tissue may cause death or mutations. However, not all petroleum products are harmful to health and the environment. There are records of the use of *petroleum spirit* for medicinal purposes. This was probably a higher boiling fraction of or than *naphtha* or a low boiling fraction of *gas oil* that closely resembled the modern-day *liquid paraffin*, used for medicinal purposes. In fact, the so-called liquid paraffin has continued to be prescribed up to modern times as a means for miners to take in prescribed doses to lubricate the alimentary tract and assist coal dust, taken in during the working hours, to pass through the body.

There are however, those constituents of petroleum that are extremely harmful to health and the environment. Indeed, petroleum constituents either in the pure form or as the components of a fraction have been known to belong to the various families of carcinogens and neurotoxins. Whatever the name given to these compounds, they are extremely toxic.

As a result, once a spill has occurred, every effort must be made to rid the environment of the toxins. The chemicals of known toxicity range in degree of toxicity from low to high and represent considerable danger to human health and must be removed (Frenzel et al., 2009). Many of these chemicals substances come in contact with soil and/or water and are sequestered in these media. While conventional methods to remove, reduce, or mitigate the effects of toxic chemical in nature include (1) pump and treat systems, (2) soil vapor extraction, (3) incineration, and (4) containment, each of these conventional methods of treatment of contaminated soil and/or water suffers from recognizable drawbacks and may involve some level of risk. In short, these methods, depending upon the chemical constituents of the spilled material, may limited effectiveness and can be expensive (Speight, 1996, 2005; Speight and Lee, 2000).

Although the effects of bacteria (microbes) on hydrocarbons has been known for decades, this technology has shown promise and, in some cases, high degrees of effectiveness for the treatment of these contaminated sites, since it is cost-effective and will lead to complete mineralization. Environmental biotransformation also refers to complete *mineralization* of the organic contaminants into carbon dioxide, water, inorganic compounds, and cell protein or transformation of complex organic contaminants to other simpler organic compounds that are not detrimental to the environment. In fact, unless they are overwhelmed by the amount of the spilled material or it is toxic, many indigenous microorganisms in soil and/or water are capable of degrading hydrocarbon contaminants.

Introduction to Petroleum Biotechnology. https://doi.org/10.1016/B978-0-12-805151-1.00008-4

The United States Environmental Protection Agency (US EPA, 2006) uses environmental biotransformation because it takes advantage of natural processes and relies on microbes that occur naturally or can be laboratory cultivated; these consist of bacteria, fungi, actinomycetes, cyanobacteria, and to a lesser extent, plants. These microorganisms either consume and convert the contaminants or assimilate within them all harmful compounds from the surrounding area, thereby, rendering the region virtually contaminant-free. Generally, the substances that are consumed as an energy source are organic compounds, while those, which are assimilated within the organism, are heavy metals. Environmental biotransformation harnesses this natural process by promoting the growth and/or rapid multiplication of these organisms that can effectively degrade specific contaminants and convert them to nontoxic by-products.

The capabilities of microorganisms and plants to degrade and transform contaminants provide benefits in the cleanup of pollutants from spills and storage sites. These remediation ideas have provided the foundation for many ex situ waste treatment processes (including sewage treatment) and a host of in situ environmental biotransformation methods that are currently in practice.

2.0 BIOTRANSFORMATION

The concept of biotransformation exploits various naturally occurring mitigation processes: (1) *natural attenuation*, (2) *biostimulation*, and (3) *bioaugmentation*. Biotransformation which occurs without human intervention other than monitoring is often called *natural attenuation*. This natural attenuation relies on natural conditions and behavior of soil microorganisms that are indigenous to soil. *Biostimulation* also utilizes indigenous microbial populations to remediate contaminated soils and consists of adding nutrients and other substances to soil to catalyze natural attenuation processes. *Bioaugmentation* involves introduction of exogenic microorganisms (sourced from outside the soil environment) that can detoxify a specific contaminant. On occasion, it may be necessary to employ genetically altered microorganisms.

Briefly and by means of clarification, *biotransformation (biotic degradation, biotic decomposition) is the chemical degradation of contaminants by bacteria or other biological means.* Organic material can be degraded aerobically (in the presence of oxygen) or anaerobically (in the absence of oxygen). Most environmental biotransformation systems operate under aerobic conditions, but a system under anaerobic conditions may permit microbial organisms to degrade chemical species that are otherwise nonresponsive to aerobic treatment and vice versa. Thus, biotransformation is a natural process (or a series of processes) by which spilled petroleum hydrocarbons or other organic waste material, are broken down (degraded) into nutrients that can be used by other organisms. As a result, the ability of a chemical to be biodegraded is an indispensable element in the understanding the risk posed by that chemical on the environment.

Environmental biotransformation—*the use of living organisms to reduce or eliminate environmental hazards resulting from accumulations of toxic chemicals and other hazardous wastes*—is an option that offers the possibility to destroy or render harmless various contaminants using natural biological activity (Gibson and Sayler, 1992). In addition, environmental biotransformation can also be used in conjunction with a wide range of traditional physical and chemical technology to enhance their effectiveness (Vidali, 2001).

Biotransformation is a key process in the natural attenuation (reduction or disposal) of chemical compounds at hazardous waste sites. There are constituents of concern in crude oil that can be degraded under appropriate conditions but the success of the process depends on the ability to determine these conditions and establish them in the contaminated environment. Thus, through biotransformation processes, living microorganisms (primarily bacteria, but also yeasts, molds, and filamentous fungi) can alter and/or metabolize various classes of compounds present in petroleum. Furthermore, biotransformation also alters subsurface oil accumulations of petroleum (Winters and Williams, 1969; Speight, 2014). Shallow oil accumulations, such as heavy oil reservoirs and tar sand deposits, where the reservoir temperature is low-to-moderate (<80°C, <176°F) are commonly found to have undergone some degree of biotransformation (Winters and Williams, 1969; Speight, 2014). The process allows for the breakdown of a compound to either fully oxidized or reduced simple molecules such as carbon dioxide/methane, nitrate/ammonium, and water. However, in some cases, where the process is not complete, the products of biotransformation can be more harmful than the substance degraded.

In the current context, environmental biotransformation of petroleum and petroleum fractions (or products) is the cleanup of petroleum spills or petroleum product spills using microbes to breakdown the petroleum constituents (or other organic contaminants) into less harmful (usually lower molecular weight) and easier-to-remove products (biotransformation). The microbes transform the contaminants through metabolic or enzymatic processes, which vary greatly, but the final product is usually harmless and includes carbon dioxide, water, and cell biomass. Thus, the emerging science and technology of environmental biotransformation offers an alternative method to detoxify petroleum-related soil and water contaminants.

Since many of the contaminants of concern in petroleum and petroleum-related products oil are readily biodegradable under the appropriate conditions, the success of oil-spill environmental biotransformation depends mainly on the ability to establish these conditions in the contaminated environment using the technology to optimize the total efficiency of the microorganisms.

Over the past two decades, opportunities for applying environmental biotransformation to a much broader set of contaminants have been identified. Indigenous and enhanced organisms have been shown to degrade industrial solvents, polychlorinated biphenyls (PCBs), explosives, and many different agricultural chemicals. Pilot demonstration and full-scale applications of environmental biotransformation have been carried out on a limited basis. However, the full benefits of environmental biotransformation have not been realized because processes and organisms that are effective in controlled laboratory tests are not always equally effective in full-scale applications. The failure to perform optimally in the field setting stems from a lack of predictability due, in part, to inadequacies in the fundamental scientific understanding of how and why these processes work.

In order to enhance and make favorable the parameters presented above to ensure microbial activity, there are two other environmental biotransformation technologies that offer useful options for cleanup of spills of petroleum and petroleum products: (1) fertilization and (2) seeding.

Fertilization (*nutrient enrichment*) is the method of adding nutrients such as phosphorus and nitrogen to a contaminated environment to stimulate the growth of the microorganisms capable of biotransformation. Limited supplies of these nutrients in nature usually control the growth of native microorganism populations. When more nutrients are added, the native microorganism population can grow rapidly, potentially increasing the rate of biotransformation. *Seeding* is the addition of microorganisms to the existing native oil-degrading population. Some species of bacteria that do not naturally exist in an area will be added to the native population. As with fertilization, the purpose of seeding is to increase the population of microorganisms that can biodegrade the spilled oil.

Intrinsic environmental biotransformation is the combined effect of natural destructive and nondestructive processes to reduce the mobility, mass, and associated risk of a contaminant. Non-destructive mechanisms include sorption, dilution, and volatilization. Destructive processes are aerobic and anaerobic biotransformation. *Intrinsic aerobic biotransformation* is well documented as a means of remediating soil and groundwater contaminated with fuel hydrocarbons. In fact, intrinsic aerobic degradation should be considered an integral part of the remediation process (McAllister et al., 1995; Barker et al., 1995). There is growing evidence that natural processes influence the immobilization and biotransformation of chemicals such as aromatic hydrocarbons, mixed hydrocarbons, and chlorinated organic compounds (Ginn et al., 1995; King et al., 1995).

Phytoremediation is the use of living green plants for the removal of contaminants and metals from soil and is, essentially, the in situ treatment of pollutant contaminated soils, sediments, and water—terrestrial, aquatic and wetland plants, and algae can be used for the phytoremediation process under specific cases and conditions of hydrocarbon contamination (Brown, 1995; Nedunuri et al., 2000; Radwan et al., 2000; Magdalene et al., 2009). It is best applied at sites with relatively shallow contamination of pollutants that are amenable to the various subcategories of phytoremediation: (1) phytotransformation—the breakdown of organic contaminants sequestered by plants, (2) rhizosphere environmental biotransformation—the use of rhizosphere microorganisms to degrade organic pollutants, (3) phytostabilization—a containment process using plants, often in combination with soil additives to assist plant installation, to mechanically stabilize the site, and reduce pollutant transfer to other ecosystem compartments and the food chain, (4) phytoextraction—the ability of some plants to accumulate metals/metalloids in their shoots, (5) rhizofiltration, and/or (6) phytovolatilization/rhizovolatilization—processes employing metabolic capabilities of plants and associated rhizosphere microorganisms to transform pollutants into volatile compounds that are released to the atmosphere (Korade and Fulekar, 2009).

These technologies are especially valuable where the contaminated soils are fragile, and prone to erosion. The establishment of a stable vegetation community stabilizes the soil system and prevents erosion. This aspect is especially relevant to certain types of soil where removal of large volumes of soil destabilizes the soil system, which leads to extensive erosion. However, when the above parameters are not conducive to bacterial activity, the bacteria (1) grow too slowly, (2) die, or (3) create more harmful chemicals.

Phytotransformation and *rhizosphere environmental biotransformation* are applicable to sites that have been contaminated with organic pollutants, including pesticides. It is a technology that should be considered for remediation of contaminated sites because of its cost-effectiveness, aesthetic advantages, and long-term applicability (Brown, 1995). Plants have shown the capacity to withstand relatively high concentrations of organic chemicals without toxic effects, and they can uptake and convert chemicals quickly to less toxic metabolites in some cases. In addition, they stimulate the degradation of organic chemicals in the rhizosphere by the release of root exudates, enzymes, and the buildup of organic carbon in the soil.

In recent years, in situ biotransformation concepts have been applied in treating contaminated soil and ground water. Removal rates and extent vary based on the contaminant of concern and site-specific characteristics. Removal rates also are affected by variables such as contaminant distribution and concentration; co-contaminant concentrations; indigenous microbial populations and reaction kinetics; and parameters such as pH, moisture content, nutrient supply, and temperature. Many of these factors are a function of the site and the indigenous microbial community and, thus, are difficult to manipulate. Specific technologies may have the capacity to manipulate some variables and may be affected by other variables as well (US EPA, 2006).

During biotransformation, microbes utilize chemical contaminants in the soil as an energy source and, through oxidation-reduction reactions, metabolize the target contaminant into useable energy for microbes. By-products (metabolites) released back into the environment are typically in a less toxic form than the parent contaminants. For example, petroleum hydrocarbons can be degraded by microorganisms in the presence of oxygen through aerobic respiration. The hydrocarbon loses electrons and is oxidized while oxygen gains electrons and is reduced. The result is formation of carbon dioxide and water (Nester et al., 2001). When oxygen is limited in supply or absent, as in saturated or anaerobic soils or lake sediment, anaerobic (without oxygen) respiration prevails. Generally, inorganic compounds such as nitrate, sulfate, ferric iron, manganese, or carbon dioxide serve as terminal electron acceptors to facilitate biotransformation.

Generally, a contaminant is more easily and quickly degraded if it is a naturally occurring compound in the environment, or chemically similar to a naturally occurring compound, because microorganisms capable of its biotransformation are more likely to have evolved. Petroleum hydrocarbons are naturally occurring chemicals; therefore, microorganisms which are capable of attenuating or degrading hydrocarbons exist in the environment. Development of biotransformation technologies of synthetic chemicals such as chlorocarbons or chlorohydrocarbons is dependent on outcomes of research that search for natural or genetically improved strains of microorganisms to degrade such contaminants into less toxic forms.

In summary, biotransformation is increasingly viewed as an appropriate remediation technology for hydrocarbon-contaminated polar soils. As for all soils, the successful application of biotransformation depends on appropriate biodegradative microbes and environmental conditions in situ. Laboratory studies have confirmed that hydrocarbon-degrading bacteria typically assigned to the genera *Rhodococcus, Sphingomonas,* or *Pseudomonas* are present in contaminated polar soils. However, as indicated by the persistence of spilled hydrocarbons, environmental conditions in situ are suboptimal for biotransformation in polar soils. Therefore, it is likely that ex situ biotransformation will be the method of choice for ameliorating and controlling the factors limiting microbial activity, i.e., low and fluctuating soil temperatures, low levels of nutrients, and possible alkalinity and low moisture (Aislabie et al., 2006; Nugroho et al., 2010).

2.1 METHOD PARAMETERS

Several factors that affect the decision of which method is chosen are (1) the nature of the contaminants, (2) the location of contaminated site, cost of cleanup, (3) the time allotted to the cleanup, (4) effects on humans, animals, and plants, and last but by no means least (5) the cost of the cleanup. Sometimes when one method is no longer effective and efficient, another remediation method can be introduced into the contaminated soil.

Conventional biotransformation methods used are biopiling, composting, land farming, and bioslurry reactors, but there are limitations affecting the applicability and effectiveness of these methods (Speight and Lee, 2000). With the application of the composting technique, if the operation is unsuccessful, it will result in a greater quantity of contaminated materials. Land farming is only effective if the contamination is near the soil surface or else bed preparation needs to take place. The main drawback with slurry bioreactors is that high-energy inputs are required to maintain suspension and the potential needed for volatilization.

For a biotransformation method to be successful in soil and water cleanup, the physical, chemical, and biological environment must be feasible. Parameters that affect the biotransformation process are

(1) low temperatures, (2) preferential growth of microbes obstructive to biotransformation, (3) high concentrations of chlorinated organics, heavy metals, and heavy oils poisoning the microorganisms, (4) preferential flow paths severely decreasing contact between injected fluids and contaminants throughout the contaminated zones, and (5) the soil matrix prohibiting contaminant–microorganism contact. Since most of the contaminants of concern in crude oil are readily biodegradable under the appropriate conditions, the success of oil-spill biotransformation depends mainly on the ability to establish these conditions in the contaminated environment using the above new developing technologies to optimize the microorganisms' total efficiency. The technologies used at various polluted sites depend on the limiting factor present at the location. For example, where there is insufficient dissolved oxygen, bioventing or sparging is applied; biostimulation or bioaugmentation is suitable for instances where the biological count is low.

The bioventing process combines an increased oxygen supply with vapor extraction. A vacuum is applied at some depth in the contaminated soil which draws air down into the soil from holes drilled around the site and sweeps out any volatile organic compounds. The development and application of venting and bioventing for in situ removal of petroleum from soil have been shown to remediate hydrocarbons by venting and biotransformation (Van Eyk, 1994).

Even though a technology may have reports of improving biotransformation efficiency (for example, surfactant addition), this may not be the case at times depending on the sample.

2.2 MONITORED NATURAL ATTENUATION

The term *monitored natural attenuation* refers to the reliance on natural attenuation to achieve site-specific remedial objectives within a time frame that is reasonable compared to that offered by other more active methods.

The *natural attenuation processes* that are at work in such a remediation approach include a variety of physical, chemical, or biological processes that, under favorable conditions, act without human intervention to reduce the mass, toxicity, mobility, volume, or concentration of contaminants in soil or groundwater. These in situ processes include biotransformation, dispersion, dilution, sorption, volatilization, and chemical or biological stabilization, transformation, or destruction of contaminants. A study of any contaminated site must first be performed to decide whether natural attenuation would make a positive input, and through it has degraded lighter chain hydrocarbons quite extensively; the heavier chain hydrocarbons are less susceptible.

As with any technique, there are disadvantages—the disadvantages of *monitored natural attenuation* method are the need for longer time frames to achieve remediation objectives, compared to active remediation, the site characterization may be more complex and costly and long-term monitoring will generally be necessary.

2.3 USE OF BIOSURFACTANTS

Another common emerging technology is the use of *biosurfactants*, which are microbially produced surface-active compounds. They are amphiphilic molecules with both hydrophilic and hydrophobic regions, causing them to aggregate at interfaces between fluids with different polarities found in oil spills. Many of the known biosurfactant producers are hydrocarbon-degrading organisms.

Biosurfactants have comparable solubilization properties to synthetic surfactants but have several additional advantages that make them superior candidates in biotransformation schemes. First, biosurfactants are biodegradable and are not a pollution threat. Furthermore, most studies indicate that they are nontoxic to microorganisms and therefore are unlikely to inhibit biotransformation of nonpolar organic contaminants. Biosurfactant production is less expensive, can be easily achieved ex situ at the contaminated site, and has the potential of occurring in situ.

Biosurfactants are also effective in many diverse geologic formations and are compatible with many existing remedial technologies (such as pump and treat rehabilitation, air sparging, and soil flushing), and significantly accelerate innovative approaches including microbial, natural attenuation enhanced soil flushing, and bio-slurping.

2.4 BIOENGINEERING IN BIOTRANSFORMATION

In many cases, after an oil spill, the natural microbial systems for degrading the oil are overwhelmed. Therefore, molecular engineers are constructing starvation promoters to express heterologous genes needed in the field for survival and adding additional biotransformation genes that code for enzymes able to degrade a broader range of compounds present in the contaminated environments. Various bacterial strains are also being developed, where each strain is specific for a certain organic compound present in oil spills. This will help increase the speed of biotransformation and allow detailed cleanup to take place where no organic contamination remains in the environment.

Thus, the decision to bioremediate a site is dependent on cleanup, restoration, and habitat protection objectives and the factors that are present that would have an impact on success (Speight and Arjoon, 2012). If the circumstances are such that no amount of nutrients will accelerate biotransformation, then the decision should be made on the need to accelerate oil disappearance to protect a vital living resource or simply to speed up restoration of the ecosystem. These decisions are clearly influenced by the circumstances of the spill.

3.0 ENVIRONMENTAL BIOTRANSFORMATION

In the current context, environmental biotransformation is a method for dealing with contamination by petroleum, petroleum products, and petroleum waste streams (Speight, 2005; Speight and Arjoon, 2012). The process typically occurs through the degradation of petroleum or a petroleum product through the action of microorganisms (biotransformation). The method typically relies upon indigenous bacteria (microbes) compared to the customary (physical and chemical) remediation methods. Also, the microorganisms engaged can perform a variety of biotransformation (detoxification) reactions. Furthermore, biotransformation studies provide information on the fate of a chemical or mixture of petroleum-derived chemicals (such as oil spills and process wastes) in the environment thereby opening the scientific doorway to develop further methods of cleanup by (1) analyzing the contaminated sites, (2) determining the best method suited for the environment, and (3) optimizing the cleanup techniques which lead to the emergence of new processes.

3.1 NATURAL BIOTRANSFORMATION

Natural biotransformation typically involves the use of molecular oxygen (O_2), where oxygen (the *terminal electron acceptor*) receives electrons transferred from an organic contaminant:

$$\text{Organic substrate} + O_2 \rightarrow \text{biomass} + CO_2 + H_2O + \text{other products}$$

In the absence of oxygen, some microorganisms obtain energy from fermentation and anaerobic oxidation of organic carbon. Many anaerobic organisms (*anaerobes*) use nitrate, sulfate, and salts of iron(III) as practical alternates to oxygen acceptor as, for example, in the anaerobic reduction process of nitrates, sulfates, and salts of iron(III):

$$2NO_3^- + 10e^- + 12H^+ \rightarrow N_2 + 6H_2O$$

$$SO_4^{2-} + 8e^- + 10H^+ \rightarrow H_2S + 4H_2O$$

$$Fe(OH)_3 + e^- + 3H^+ \rightarrow Fe^{2+} + 3H_2O$$

3.2 METHODS OF BIOTRANSFORMATION

Methods for the cleanup of pollutants have usually involved removal of the polluted materials, and their subsequent disposal by land filling or incineration (so called *dig, haul, bury, or burn* methods) (Speight, 1996, 2005; Speight and Lee, 2000). Furthermore, available space for landfills and incinerators is declining. Perhaps one of the greatest limitations to traditional cleanup methods is the fact that (despite their high costs) there is not always a guarantee that the contaminants will be completely destroyed. Conventional biotransformation methods that have been, and are still, used are (1) composting, (2) land farming, (3) biopiling and (4) use of a bioslurry reactor (Speight, 1996; Speight and Lee, 2000; Semple et al., 2001).

Composting is a technique that involves combining contaminated soil with nonhazardous organic materials such as manure or agricultural wastes; the presence of the organic materials allows the development of a rich microbial population and elevated temperature characteristic of composting. *Land farming* is a simple technique in which contaminated soil is excavated and spread over a prepared bed and periodically tilled until pollutants are degraded. *Biopiling* is a hybrid of land farming and composting; it is essentially engineered cells that are constructed as aerated composted piles. A *bioslurry reactor* can provide rapid biotransformation of contaminants due to enhanced mass transfer rates and increased contaminant-to-microorganism contact. These units are capable of aerobically biodegrading aqueous slurries created through the mixing of soils or sludge with water. The most common state of bioslurry treatment is batch-wise; however, continuous-flow operation is also possible.

The technology selected for a specific site (remembering that most cleanup technologies are site-specific) will depend on the limiting factors present at the location. For example, where there is insufficient dissolved oxygen, bioventing or sparging is applied, biostimulation or bioaugmentation is suitable for instances where the biological count is low. On the other hand, application of the composting technique, if the operation is unsuccessful, will result in a greater quantity of contaminated materials. Land farming is only effective if the contamination is near the soil surface or else bed preparation is required. The main drawback with slurry bioreactors is that high-energy input is required to maintain suspension and the potential needed for volatilization.

Other techniques are also being developed to improve the microbe-contaminant interactions at treatment sites to use biotransformation technologies at their fullest potential. These biotransformation technologies consist of monitored natural attenuation, bioaugmentation, biostimulation, surfactant addition, anaerobic bioventing, sequential anaerobic/aerobic treatment, soil vapor extraction, air sparging, enhanced anaerobic dechlorination, and bioengineering (Speight, 1996; Speight and Lee, 2000).

The use of traditional methods of biotransformation continues but there is also method evolution, which may involve the following steps:

1. Isolating and characterizing naturally occurring microorganisms with biotransformation potential.
2. Laboratory cultivation to develop viable populations.
3. Studying the catabolic activity of these microorganisms in contaminated material through bench-scale experiments.
4. Monitoring and measuring the progress of biotransformation through chemical analysis and toxicity testing in chemically contaminated media.
5. Field applications of biotransformation techniques using either/both steps: (1) *in situ* stimulation of microbial activity by the addition of microorganisms and nutrients and the optimization of environmental factors at the contaminated site itself (2) *ex situ* restoration of contaminated material in specifically designated areas by land-farming and composting method.

3.3 ENHANCED BIOTRANSFORMATION

Enhanced biotransformation is a process in which indigenous or inoculated microorganisms (e.g., fungi, bacteria, and other microbes) degrade (metabolize) organic contaminants found in soil and/or ground water and convert the contaminants to innocuous end products. The process relies on general availability of naturally occurring microbes to consume contaminants as a food source (petroleum hydrocarbons in aerobic processes) or as an electron acceptor (chlorinated solvents, which may be waste materials from petroleum processing). In addition to microbes being present, to be successful, these processes require nutrients such as carbon, nitrogen, and phosphorus.

Enhanced biotransformation involves the addition of microorganisms (e.g., fungi, bacteria, and other microbes) or nutrients (e.g., oxygen, nitrates) to the subsurface environment to accelerate the natural biotransformation process.

3.4 BIOSTIMULATION AND BIOAUGMENTATION

Biostimulation is the method of adding nutrients such as phosphorus and nitrogen to a contaminated environment to stimulate the growth of the microorganisms that break down oil. Additives are usually added to the subsurface through injection wells, although injection well technology for biostimulation purposes is still emerging. Limited supplies of these necessary nutrients usually control the growth of native microorganism populations. Thus, addition of nutrients causes rapid growth of the indigenous microorganism population thereby increasing the rate of biotransformation.

It is to be anticipated that the success of biostimulation is case-specific and site-specific, depending on oil properties, the nature of the nutrient products, and the characteristics of the contaminated environments. When oxygen is not a limiting factor, one of keys for the success of oil biostimulation is to maintain an optimal nutrient level in the interstitial pore water. Several types of commercial biostimulation agents are available for use in biotransformation (Zhu et al., 2004).

Bioaugmentation is the addition of pre-grown microbial cultures to enhance microbial populations at a site to improve contaminant clean up and reduce clean up time and cost. Indigenous or native microbes are usually present in very small quantities and may not be able to prevent the spread of the contaminant. In some cases, native microbes do not have the ability to degrade a specific contaminant. Therefore, bioaugmentation offers a way to provide specific microbes in sufficient numbers to complete the biotransformation (Atlas, 1991).

Mixed cultures have been most commonly used as inocula for seeding because of the relative ease with which microorganisms with different and complementary biodegradative capabilities can be isolated (Atlas, 1977). Different commercial cultures were reported to degrade petroleum hydrocarbons (Compeau et al., 1991; Leavitt and Brown, 1994; Chhatre et al., 1996; Mishra et al., 2001; Vasudevan and Rajaram, 2001).

Microbial inocula (the microbial materials used in an inoculation) are prepared in the laboratory from soil or groundwater either from the site where they are to be used or from another site where the biotransformation of the chemicals of interest is known to be occurring. Microbes from the soil or groundwater are isolated and are added to media containing the chemicals to be degraded. Only microbes capable of metabolizing the chemicals will grow on the media. This process isolates the microbial population of interest. One of the main environmental applications for bioaugmentation is at sites with chlorinated solvents. Microbes called *Dehalococcoides nethenogenes* usually perform reductive dechlorination of solvents such as perchloroethylene and trichloroethylene.

Bioaugmentation adds highly concentrated and specialized populations of specific microbes to the contaminated area while biostimulation is dependent on appropriate indigenous microbial population and organic material being present at the site.

3.5 IN SITU AND EX SITU BIOTRANSFORMATION

Biotransformation can be used as a cleanup method for both contaminated soil and water. Its applications fall into two broad categories: in situ or ex situ. In situ biotransformation treats the contaminated soil or groundwater in the location in which it was found while ex situ biotransformation processes require excavation of contaminated soil or pumping of groundwater before they can be treated.

In situ technologies do not require excavation of the contaminated soils and so may be less expensive, create less dust, and cause less release of contaminants than ex situ techniques. Also, it is possible to treat a large volume of soil at once. In situ techniques, however, may be slower than ex situ techniques, may be difficult to manage, and are only most effective at sites with permeable soil.

The most effective means of implementing in situ biotransformation depends on the hydrology of the subsurface area, the extent of the contaminated area, and the nature (type) of the contamination. In general, this method is effective only when the subsurface soils are highly permeable, the soil horizon to be treated falls within a depth of 8–10 m, and shallow groundwater is present at 10 m or less below ground surface. The depth of contamination plays an important role in determining if an in situ biotransformation project should be employed. If the contamination is near the groundwater but the groundwater is not yet contaminated, then it would be unwise to set up a hydrostatic system. It would be safer to excavate the contaminated soil and apply an on-site method of treatment away from the groundwater.

The typical time frame for an in situ biotransformation project can be in the order of 12–24 months depending on the levels of contamination and depth of contaminated soil. Due to the poor mixing in this system, it becomes necessary to treat for long periods of time to ensure that all the pockets of contamination have been treated. In addition, in situ biotransformation is a very site-specific technology that involves establishing a hydrostatic gradient through the contaminated area by flooding it with water carrying nutrients and possibly organisms adapted to the contaminants. Water is continuously circulated through the site until it is determined to be clean.

In situ biotransformation of groundwater speeds the natural biotransformation processes that take place in the water-soaked underground region that lies below the water table. One limitation of this technology is that differences in underground soil layering and density may cause reinjected conditioned groundwater to follow certain preferred flow paths. On the other hand, ex situ techniques can be faster, easier to control, and used to treat a wider range of contaminants and soil types than in situ techniques. However, they require excavation and treatment of the contaminated soil before and, sometimes, after the actual biotransformation step.

In situ biotransformation is the preferred method for large sites and is used when physical and chemical methods of remediation may not completely remove the contaminants, leaving residual concentrations that are above regulatory guidelines. This method has the potential to provide advantages such as complete destruction of the contaminant(s), lower risk to site workers, and lower equipment/operating costs. In situ biotransformation can be used as a cost-effective secondary treatment scheme to decrease the concentration of contaminants to acceptable levels or as a primary treatment method, which is followed by physical or chemical methods for final site closure.

Finally, evidence for the effectiveness of petroleum biotransformation and petroleum product biotransformation should include: (1) faster disappearance of oil in treated areas than in untreated areas, and (2) a demonstration that biotransformation was the main reason for the increased rate of oil disappearance. To obtain such evidence, the analytical procedures must be chosen carefully and careful data interpretation is essential but there are disadvantages and errors when the method is not applied correctly (Chapter 9) (Speight, 2005).

4.0 PRINCIPLES OF ENVIRONMENTAL BIOTRANSFORMATION

Environmental biotransformation (bioremediation, biodegradation) is an environmentally natural or induced process used to restore soil and water to its original state by using indigenous microbes to break down and eliminate contaminants. Biological technologies are often used as a substitute to chemical or physical cleanup of oil spills because environmental biotransformation does not require as much equipment or labor as other methods, therefore it is usually cheaper. It also allows cleanup workers to avoid contact with polluted soil and water. For environmental biotransformation to be effective, microorganisms must convert the pollutants to harmless products. As environmental biotransformation can be effective only where environmental conditions permit microbial growth and activity, its application often involves the manipulation of environmental parameters to allow microbial growth and degradation of pollutants to proceed at a faster rate. However, as is the case with other technologies, environmental biotransformation has its limitations and there are several disadvantages that must be recognized (Table 8.1).

Table 8.1 Advantages and Disadvantages of Environmental Biotransformation

Advantages	Disadvantages
Remediates contaminants that are adsorbed onto or trapped within the geologic materials of which the aquifer is composed along with contaminants dissolved in groundwater.	Injection wells and/or infiltration galleries may become plugged by microbial growth or mineral precipitation.
Application involves equipment that is widely available and easy to install.	High concentrations (TPH greater than 50,000 ppm) of low solubility constituents may be toxic and/or not bioavailable.
Creates minimal disruption and/or disturbance to on-going site activities.	Difficult to implement in low-permeability aquifers.
Time required for subsurface remediation may be shorter than other approaches (e.g., pump-and-treat).	Reinjection wells or infiltration galleries may require permits or may be prohibited. Some states require permit for air injection.
Generally recognized as being less costly than other remedial options.	May require continuous monitoring and maintenance.
Can be combined with other technologies (e.g., bioventing, SVE) to enhance site remediation.	Remediation may only occur in more permeable layer or channels within the aquifer.
In many cases this technique does not produce waste products that must be disposed.	

Biodegradation by natural populations of microorganisms is the most reliable mechanism by which thousands of xenobiotic pollutants, including crude oil, are eliminated from the environment. On the other hand, *bioremediation* is the conversion of pollutants (hydrocarbons) by microorganisms (bacteria, yeast, or fungi) into energy, cell mass, and some biological products such as biosurfactants.

Bioremediation is often based on: (1) natural attenuation, through the use of naturally occurring microorganisms in the contaminated environment without stimulating their development in any way, (2) in situ stimulation of the microbial community, also known as *biostimulation*, or (3) amending the microbial community with an inoculum of hydrocarbon-degrading bacteria (bioaugmentation). In both cases, the successful result of bioremediation depends on appropriate hydrocarbon-degrading consortia and environmental conditions. Nevertheless, the rates of microbial uptake and mineralization of organic pollutants in contaminated environment, especially seawater, is limited due to the poor availability of nitrogen and phosphorus. For that reason, in the application of biostimulation techniques the growth of oil-degrading bacteria can be strongly enhanced by fertilization with inorganic nutrients (Nikolopoulou and Kalogerakis, 2010; Santisi et al., 2015). The main advantage of bioremediation is its low investment cost, since, complex and expensive technology is not required. It is a natural process and the final products of the microbiological degradation are carbon dioxide and water. It does not involve the use of chemical compounds that may negatively influence the biocoenosis (the interaction of organisms living together in a habitat —the biotope) to the soil; moreover, the soil can be reused after its bioremediation. It can be applied in situ at the polluted site, throughout; soil cultivation, bioventilation, bioextraction, and in situ biodegradation, or ex situ throughout the removal of the polluted soil from the site and placing it in a specially prepared location. Ex situ bioremediation can be done by soil cultivation, composting, biostacks, and bioreactors (Vidali, 2001). The main disadvantage of bioremediation process is that it takes a very long period

Table 8.2 Essential Factors for Bioremediation

Factor	Optimal Conditions
Microbial population	Suitable kinds of organisms that can biodegrade all of the contaminants
Oxygen	Enough to support aerobic biodegradation (about 2% oxygen in the gas phase or 0.4 mg/L in the soil water)
Water	Soil moisture should be from 50% to 70% of the water holding capacity of the soil
Nutrients	Nitrogen, phosphorus, sulfur, and other nutrients to support good microbial growth
Temperature	Appropriate temperatures for microbial growth (0–40°C)
pH	Best range is from 6.5 to 7.5

for remediation of a polluted environment. In situ bioremediation methods are not always easy to monitor, whereas ex situ bioremediation methods are more expensive due to transportation costs and soil storage (Vidali, 2001).

4.1 FACTORS AFFECTING THE RATE OF BIOREMEDIATION

The main factors influencing the biodegradation of oil pollutants in soil are chemical structure, pollutant concentration and toxicity of hydrocarbons to the microflora, microbiological soil potential (biomass concentration, population variability, enzyme activity), physical-chemical environmental parameters (e.g., reaction, temperature, organic matter content, humidity, presence of oxygen or other electron acceptors, and nutrients, the availability of the oil-water interface), the concentration of the pollutant, its composition, availability of hydrocarbons for microorganism cells and finally, the inherent biodegradability of the petroleum hydrocarbon pollutant (Table 8.2).

Several studies have reported the optimum operating condition for sufficient soil bioremediation to be (1) biomass content over 105 cells/g dry mass, (2) relative humidity 20%–30%, (3) temperature 20–30°C/68 to 86°F, (4) pH on the order of 6.5–7.5, (5) oxygen content at least 0.2 mg/L hydrocarbons, and (6) carbon-nitrogen-phosphorous ratio (C:N:P) 100:10:1, respectively (Sztompka, 1999; Farahat and El-Gendy, 2008; Wolicka and Borkowski, 2012; Soliman et al., 2014).

4.2 TEMPERATURE

Temperature not only affects the physical property and chemistry of the pollutant, but it also affects the physiology and diversity of the microbial flora (Atlas, 1975). The decrease in temperature would result in decrease of oil viscosity that hampers microbial metabolism. Moreover, the evaporation of light compounds with toxic properties is restricted, which prolongs the time required by microorganisms to adapt to the toxic conditions, thus, delaying the onset of biodegradation. The rate of biodegradation generally decreases with the decreasing temperature. If a substance is maintained in the frozen state or below the optimal operating temperature for microbial species, such conditions can prevent biotransformation—most biotransformation occurs at temperatures between 10 and 35°C (50 and 95°F), polar regions suffer from low temperature and deficiency of nutrients and biogenic elements in soil and high pH value, especially in seaside areas. However, significant biodegradation of hydrocarbons has been reported in psychrophilic environments in temperate regions (Delille et al., 2004; Pelletier et al., 2004; Aislabie et al., 2006). Temperature influences rate of biotransformation by controlling rate of enzymatic reactions within microorganisms. Generally, the rate of an enzymatic reaction approximately doubles for each 10°C (18°F) rise in temperature (Nester et al., 2001). However, there is an upper limit to the temperature that microorganisms can withstand.

Most bacteria found in soil, including many bacteria that degrade petroleum hydrocarbons, are mesophile organisms which have an optimum working temperature range on the order of 25–45°C (77–113°F) (Nester et al., 2001). Thermophilic bacteria (those which survive and thrive at relatively high temperatures) which are normally found in hot springs and compost heaps exist indigenously in cool soil environments and can be activated to degrade hydrocarbons with an increase in temperature to 60°C (140°F). This indicates the potential for natural attenuation in cool soils through thermally enhanced environmental biotransformation techniques (Perfumo et al., 2007).

4.3 ACIDITY-ALKALINITY

Soil acidity-alkalinity (pH) is extremely important because most microbial species can survive only within a certain pH range—generally the biotransformation of petroleum hydrocarbons is optimal at a pH 7 (neutral) and the *acceptable* (or optimal) pH range is on the order of 6–8. Furthermore, soil (or water) pH can affect availability of nutrients.

Nutrients—this is a very important factor to guarantee a successful biodegradation of hydrocarbon pollutants. Upon pollution, the carbon concentration increases and the availability of nitrogen and phosphorus become the limiting factor for oil degradation. The nitrogen, phosphorous, potassium, and in some cases iron, are the most essential nutrients to enhance a bioremediation process (Cooney, 1984; Atlas, 1985; Choi et al., 2002; Kim et al., 2005). However, excessive nutrient concentration showed negative impact on the bioremediation process (Chaîneau et al., 2005; Chaillan et al., 2006). Readily available and cost-effective source of nutrients are always recommendable, for example, poultry manure (Okolo et al., 2005), corn steep liquor (El-Gendy and Farah, 2011; Soliman et al., 2014), molasses (Farahat and El-Gendy, 2007).

4.4 EFFECT OF SALT

Salt is a common co-contaminant that can adversely affect the biotransformation potential at sites such as flare pits and drilling sites (*upstream sites*) contaminated with saline produced formation water, or at oil and gas processing facilities contaminated by refinery wastes containing potassium chloride (KCl) and sodium chloride (NaCl) salts (Pollard et al., 1994). Because of increasing emphasis and interest in the viability of intrinsic biotransformation as a remedial alternative, the impact of salt on these processes is of interest.

The effect of salinity on microbial cells varies from disrupted tertiary protein structures and denatured enzymes to cell dehydration (Pollard et al., 1994), with different species having different sensitivities to salt (Tibbett et al., 2011). A range of organic pollutants, including hydrocarbons, has been shown to be mineralized by marine or salt-adapted terrestrial microorganisms that are able to grow in the presence of salt (Margesin and Schinner, 2001; Oren et al., 1992; Nicholson and Fathepure, 2004). In naturally saline soils, it has been shown that biotransformation of diesel fuel is possible at salinities up to 17.5% w/v (Riis et al., 2003; Kleinsteuber et al., 2006).

However, an inverse relationship between salinity and the biotransformation of petroleum hydrocarbons by halophilic enrichment cultures from the Great Salt Lake (Utah) has been observed (Ward and Brock, 1978). These cultures were unable to metabolize petroleum hydrocarbons at salt concentrations above 20% (w/v) in this hyper-saline environment. An inhibitory effect of salinity at concentrations above 2.4% (w/v) NaCl was found to be greater for the biotransformation of aromatic and polar fractions than for the saturated fraction of petroleum hydrocarbons in crude oil incubated with marine sediment (Mille et al., 1991). This represents ex situ petroleum hydrocarbon degradation by salt-adapted terrestrial microorganisms.

Furthermore, the effects of salt as a co-contaminant on hydrocarbon degradation in naturally non-saline systems has been described (deCarvalho and daFonseca, 2005). The results showed that in the degradation of C_5 to C_{16} hydrocarbons at 28°C (82°F) in the presence of 1.0, 2.0, or 2.5% (w/v) NaCl by the isolate *Rhodococcus erythropolis* DCL14 the lag phase of the cultures increased and growth rates decreased with increasing concentrations of sodium chloride. In a similar study (Rhykerd et al., 1995), soils were fertilized with inorganic nitrogen and phosphorus, and amended with sodium chloride at 0.4, 1.2, or 2% (w/w). After 80 days at 25°C (77°F), the highest salt concentration had inhibited motor oil mineralization.

However, investigation of the combinations of factors limiting biotransformation of petroleum hydrocarbon contamination at upstream oil and gas production facilities have received relatively little attention. A laboratory solid-phase biotransformation study reported that high salinity levels reduced the degradation rate of flare pit hydrocarbons (Amatya et al., 2002), and more recently it has been observed that addition of sodium chloride to a petroleum-contaminated Arctic soil decreased hexadecane mineralization rates in the initial stages of biotransformation and increased lag times, but that the final extent of mineralization was comparable over a narrow range of salinity from 0% to 0.4% w/w (Børresen and Rike, 2007).

Continuing investigations are necessary to determine whether the effects observed in the laboratory are site-specific or contaminant-specific, or are applicable more broadly to subsurface hydrocarbon biotransformation. Further research using more sites, including those previously having been impacted by sodium chloride, may allow inference of salt tolerance at upstream oil and gas sites. Particularly important is the impact of sodium chloride on anaerobic hydrocarbon degradation. It should be investigated.

Field evidence is sparse with respect to anaerobic biotransformation at salt contaminated upstream oil and gas sites. Before embarking on anaerobic microcosm tests, field evidence of indicators of anaerobic biotransformation, including changes in terminal electron acceptors, presence of metabolites, and isotopic analysis, would be a reasonable way to initiate the research (Ulrich et al., 2009).

4.5 BIOAVAILABILITY OF THE CONTAMINANT

One of the important factors in biological removal of hydrocarbons from a contaminated environment is their bioavailability to an active microbial population, which is *the degree of interaction of chemicals with living organisms* or the degree to which a contaminant can be readily taken up and metabolized by a bacterium (Harms et al., 2010). Moreover, the bioavailability of a contaminant is controlled by factors such as the physical state of the hydrocarbon in situ, its hydrophobicity, water solubility, sorption to environmental matrices such as soil, and diffusion out of the soil matrix. When contaminants have very low solubility in water, as in the case of *n*-alkanes and polynuclear aromatic hydrocarbons, the organic phase components will not partition efficiently into the aqueous phase supporting the microbes.

In the case of soil, the limited availability of pollutants to microorganisms is one of the important factors that limits the biodegradation, since petroleum hydrocarbon compounds bind to soil components, and they are difficult to be removed or degraded. Two-phase bioreactors containing an aqueous phase and a nonaqueous phase liquid (NAPL) have been developed and used for environmental biotransformation of hydrocarbon-contaminated soil to address this very problem, but the adherence of microbes to the NAPL-water interface can still be an important factor in reaction kinetics. Similarly, two-phase bioreactors, sometimes with silicone oil as the nonaqueous phase, have been proposed for biocatalytic conversion of hydrocarbons like styrene (Osswald et al., 1996) to make the substrate more bioavailable to microbes in the aqueous phase. When the carbon source is in limited supply, then its availability will control the rate of metabolism and hence biotransformation, rather than catabolic capacity of the cells or availability of oxygen or other nutrients. In the case of the biotransformation of chemicals in waterways, similar principles apply.

The addition to surfactants, surface-active substances (SAS) would increase the accessibility of the microorganisms to the pollutants, which would consequently enhance the rate of biodegradation. But excess SAS would be toxic to microorganisms and may hamper their growth. Biosurfactants would overcome this drawback of SAS. Biosurfactants which belong to different groups, including lipopeptides, glycolipids, neutral lipids, and fatty acids have the advantages of high surface activity, biodegradability, and low toxicity.

It has been reported that the use of biosurfactants accelerates the bioremediation rate and renders five-times increase in bioremediation effectiveness in comparison to the application of chemical surfactants (Stroud et al., 2007). The nontoxic cyclodextrin derivatives can form soluble complexes with hydrophobic substances, accelerating bioremediation by increasing bioavailability (Garon et al., 2004; Bardi et al., 2007; Landy et al., 2011). It has been reported that the greater the oil-water interface, the faster the oil degradation by microbes and the oil droplets size distribution can play a critical role in determining the biodegradation kinetics of oil spills (Abdallah et al., 2005; Vilcáez et al., 2013). Oil droplets size has been reported to vary from $2.5\,\mu m$ to $2\,mm$, where, the rate of biodegradation of small oil droplets is faster than that of large ones, since the total water-oil interface is larger with small oil droplets (Vilcáez et al., 2013).

4.6 CONTAMINANT COMPOSITION

Oil is a complex mixture of various organic compounds, including chained and aromatic hydrocarbons, which can differ significantly in their biodegradation kinetics (Auffret et al., 2009; El-Gendy and Farah, 2011). Under enhanced conditions (1) certain fuel hydrocarbons can be removed preferentially over others, but the order of preference is dependent upon the geochemical conditions, and (2) augmentation and enhancement via electron acceptors to accelerate the biotransformation process.

The n-alkane derivatives are readily degraded than the branched ones and biodegradation of cycloalkanes, with high number of rings in the compound, is much slower (Stroud et al., 2007). In fact, the susceptibility of hydrocarbons to microbial degradation can be generally ranked as follows:

linear alkanes > branched alkanes branched alkanes > small aromatics
small aromatics > cyclic alkanes
cyclic alkanes > high molecular weight polynuclear aromatic hydrocarbons (PNAs, PAHs)

The rate of biodegradation of polynuclear aromatic hydrocarbons is generally affected by the number of rings, the number of alkyl substituents, and the location of the bonds of these substituents. The percentage content of hydrocarbons with large number of rings is inversely proportional to the rate of biodegradation. The higher the number of substituents, the lower is the degradation rate. However, the stereochemical configuration of the hydrocarbon compounds is the dominant factor affecting biodegradation of crude oil rather than thermodynamic effects, where the most thermodynamically stable isomers are very quickly decomposed. The thermally mature crude oils show higher concentrations of thermodynamically stable components. High thermal maturity is related to high values of the concentration ratios of the short chain pregnane derivatives and typical long chain sterane derivatives, while the degree of biodegradation increases when the ratio $[C_{21}/(C_{21}+C_{28})]$ decreases and when the ratio of C_{21-22}-pregnane derivatives/C_{27}–C_{29} sterane derivatives increases. The dimethylnaphthalene derivatives are reported to be more susceptible for biodegradation than trimethylnaphthalene derivatives while for methylphenanthrene derivatives the highest resistance to biodegradation was observed in those with methyl groups in positions 9 and 10. But in the case of steroid hydrocarbons, for example, mono-aromatic derivatives (MSHs) are more resistant to microbiological degradation than tri-aromatic derivatives (TSHs).

Some contaminants, such as chlorinated organic or high aromatic hydrocarbons, are generally resistant to microbial attack. They are degraded either slowly or not at all; hence, it is not easy to predict the rates of clean-up for an environmental biotransformation exercise; there are no rules to predict if a contaminant can be degraded. With regard to the aromatic benzene-toluene-ethylbenzene-xylenes: (1) toluene can be preferentially removed under intrinsic environmental biotransformation conditions, (2) biotransformation of benzene is relatively slow, (3) augmentation with sulfate can preferentially stimulate biotransformation of o-xylene, and (4) ethylbenzene may be recalcitrant under sulfate-reducing conditions but readily degradable under denitrifying conditions (Cunningham et al., 2000).

The rate of biodegradation of two- and three-aromatic ring compounds is higher than that of the high molecular weight PAHs. The four-, five- and six-aromatic rings compounds are characterized by low water solubility and tend to be adsorbed on soil particles. Branched TSHs with short side chains are reported to be more effectively biodegraded than branched TSHs with long chains, while the short chain pregnane derivatives are more resistant to biodegradation relative to long chain typical steranes. But up till now no trend has been identified for the ratios of short to long chains of mono-aromatics. The alkane degradation is mainly conducted by the genus *Rhodococcus* and *Pseudomonas*, while PAHs mainly by *Pseudomonas* and *Sphingomonas*. The biodegradation of high molecular weight PAHs has been reported by different bacteria; *Pseudomonas putida*, *Pseudomonas aeruginosa*, *Pseudomonas saccharophila*, *Flavobacterium* sp., *Burkholderia cepacia*, *Rhodococcus* sp., *Stenotrophomonas* sp., *Mycobacterium* sp., *Corynebacterium* sp., *Sphingomons* sp., and *Micrococcus* sp. It has been also reported by ligninolytic fungi, such as: *Phanaerochaete chrysosporium*, *Trametes versicolor*, *Bjerkandera* sp., *Pleurotus ostreatus*, and nonlygninolytic fungi, such as *Cunninghanella elegant*, *Penicillium janthinellum*, and *Syncephalastrum* sp. (Austin et al., 1977; Kirk and Gordon, 1988; Wolicka et al., 2009; El-Gendy et al., 2010; Soliman et al., 2014).

5.0 MECHANISM OF BIOTRANSFORMATION

Biotransformation involves chemical transformations mediated by microorganisms that: (1) satisfy nutritional requirements, (2) satisfy energy requirements, (3) detoxify the immediate environment, (4) or occur fortuitously such that the organism receives no nutritional or energy benefit (Stoner, 1994).

Mineralization is the complete biotransformation of organic materials to inorganic products, and often occurs through the combined activities of microbial consortia rather than through a single microorganism (Shelton and Tiedje, 1984). *Co-metabolism* is the partial biotransformation of organic compounds that occurs fortuitously and that does not provide energy or cell biomass to the microorganisms. Co-metabolism can result in partial transformation to an intermediate that can serve as a carbon and energy substrate for microorganisms, as with some hydrocarbons, or can result in an intermediate that is toxic to the transforming microbial cell, as with trichloroethylene and methanotrophs.

5.1 CHEMICAL REACTIONS

Biotransformation of petroleum constituents can occur under both aerobic (oxic) and anaerobic (anoxic) conditions (Zengler et al., 1999), albeit by the action of different consortia of organisms. In the subsurface, oil biotransformation occurs primarily under anoxic conditions, mediated by sulfate reducing bacteria (e.g., Holba et al., 1996) or other anaerobes using a variety of other electron acceptors as the oxidant. Thus, two classes of biotransformation reactions are: (1) aerobic biotransformation and (2) anaerobic biotransformation. *Aerobic biotransformation* involves the use of molecular oxygen (O_2), where oxygen (the "terminal electron acceptor") receives electrons transferred from an organic contaminant:

$$\text{Organic substrate} + O_2 \rightarrow \text{biomass} + CO_2 + H_2O + \text{other inorganic products}$$

Thus, the organic substrate is oxidized (addition of oxygen), and the oxygen is reduced (addition of electrons and hydrogen) to water (H_2O). In this case, the organic substrate serves as the sources of energy (electrons) and the source of cell carbon is used to build microbial cells (biomass). Some microorganisms (chemo-autotrophic aerobes or litho-trophic aerobes) oxidize reduced inorganic compounds (NH_3, Fe^{2+}, or H_2S) to gain energy and fix carbon dioxide to build cell carbon:

$$NH_3 \text{ (or } Fe^{2+} \text{ or } H_2S) + CO_2 + H_2 + O_2 \rightarrow \text{biomass} + NO_3 \text{ (or Fe or SO}_4) + H_2O$$

At some contaminated sites, because of consumption of oxygen by aerobic microorganisms and slow recharge of oxygen, the environment becomes anaerobic (lacking oxygen), and mineralization, transformation, and co-metabolism depend upon microbial utilization of electron acceptors other than oxygen (anaerobic biotransformation). Nitrate (NO_3), iron (Fe^{3+}), manganese (Mn^{4+}), sulfate (SO_4), and carbon dioxide (CO_2) can act as electron acceptors if the organisms present have the appropriate enzymes (Sims, 1990).

Anaerobic biotransformation is the microbial degradation of organic substances in the absence of free oxygen. While oxygen serves as the electron acceptor in aerobic biotransformation processes forming water as the final product, degradation processes in anaerobic systems depend on alternative acceptors such as sulfate, nitrate, or carbonate yielding, in the end, hydrogen sulfide, molecular nitrogen, and/or ammonia and methane (CH_4), respectively.

In the absence of oxygen, some microorganisms obtain energy from fermentation and anaerobic oxidation of organic carbon. Many anaerobes use nitrate, sulfate, and salts of iron(III) as practical alternates to oxygen acceptor. The anaerobic reduction process of nitrates, sulfates, and salts of iron is an example:

$$2NO_3{}^- + 10e^- + 12H^+ \rightarrow N_2 + 6H_2O$$

$$SO_4{}^{2-} + 8e^- + 10H^+ \rightarrow H_2S + 4H_2O$$

$$Fe(OH)_3 + e^- + 3H^+ \rightarrow Fe^{2+} + 3H_2O$$

Anaerobic biotransformation is a multistep process performed by different bacterial groups and involves hydrolysis of polymeric substances, like proteins or carbohydrates, to monomers and the subsequent decomposition to soluble acids, alcohols, molecular hydrogen, and carbon dioxide. Depending on the prevailing environmental conditions, the final steps of ultimate anaerobic biotransformation are performed by denitrifying, sulfate-reducing, or methanogenic bacteria.

In contrast to the strictly anaerobic sulfate-reducing and methanogenic bacteria, the nitrate-reducing microorganisms, as well as many other decomposing bacteria, are mostly facultative anaerobic insofar as these microorganisms are able to grow and to degrade organic substances under aerobic, as well as anaerobic, conditions. Thus, aerobic and anaerobic environments represent the two extremes of a continuous spectrum of environmental habitats which are populated by a wide variety of microorganisms with specific biotransformation abilities.

Anaerobic conditions occur where vigorous decomposition of organic matter and restricted aeration result in the depletion of oxygen. Anoxic conditions may represent an intermediate stage where oxygen supply is limited, still allowing a slow (aerobic) degradation of organic compounds. In a digester, the various bacteria also have different requirements to the surrounding environment. For example, acidogenic bacteria need pH values from 4 to 6, whilst methanogenic bacteria from 7 to 7.5. In batch tests, the dynamic equilibrium is often interrupted because of an enrichment of acidogenic bacteria because of lacking substrate in- and outflow.

On a structural basis, the hydrocarbons in crude oil are classified as alkanes (*normal* or *iso*), cycloalkanes, and aromatics. Alkene derivatives are rare in unrefined crude oil but do occur in many refined crude oil products as a consequence of the cracking process (Speight, 2014, 2017). Increasing carbon numbers of alkanes (homology), variations in carbon chain branching (*iso*-alkanes), ring condensations, and interclass combinations, such as phenyl alkanes, account for the high numbers of hydrocarbons that occur in crude oil.

In addition, smaller amounts of oxygen-containing compounds (phenol derivatives, naphthenic acids), nitrogen-containing compounds (pyridine derivatives, pyrrole derivatives, indole derivatives), sulfur-containing compounds (thiophene derivatives), and the high molecular weight polar asphalt fraction also occur in petroleum but not in refined petroleum products (Speight, 2014).

The inherent biodegradability of these individual components is a reflection of their chemical structure, but is also strongly influenced by the physical state and toxicity of the compounds. As an example, while *n*-alkanes as a structural group are the most biodegradable petroleum hydrocarbons, the C_5 to C_{10} homologs have been shown to be inhibitory to the majority of hydrocarbon degraders. As solvents, these homologs tend to disrupt lipid membrane structures of microorganisms. Similarly, alkanes in the C_{20} to C_{40} range are hydrophobic solids at physiological temperatures. Apparently, it is this physical state that strongly influences their biotransformation (Bartha and Atlas, 1977).

Primary attack on intact hydrocarbons requires the action of oxygenase derivatives and, therefore, requires the presence of free oxygen. In the case of alkanes, mono-oxygenase attack results in the production of alcohol. Most microorganisms attack alkanes terminally whereas some perform sub-terminal oxidation. The alcohol product is oxidized finally into an aldehyde. Extensive methyl branching interferes with the beta-oxidation process and necessitates terminal attack or other bypass mechanisms. Therefore, *n*-alkanes are degraded more readily than *iso*-alkanes.

Cycloalkanes are transformed by an oxidase system to a corresponding cyclic alcohol, which is dehydrated to ketone after which a mono-oxygenase system lactonizes the ring, which is subsequently opened by a lactone hydrolase. These two oxygenase systems usually never occur in the same organisms and hence, the frustrated attempts to isolate pure cultures that grow on cycloalkanes (Bartha, 1986b). However, synergistic actions of microbial communities are capable of dealing with degradation of various cycloalkanes quite effectively.

As in the case of alkanes, the monocyclic compounds, cyclopentane, cyclohexane, and cycloheptane, have a strong solvent effect on lipid membranes, and are toxic to the majority of hydrocarbon degrading microorganisms. Highly condensed cycloalkane compounds resist biotransformation due to their relatively complex structure and physical state (Bartha, 1986a).

Condensed polycyclic aromatics are degraded, one ring at a time, by a similar mechanism, but biodegradability tends to decline with the increasing number of rings and degree of condensation (Atlas and Bartha, 1998). Aromatics with more than four condensed rings are generally not suitable as substrates for microbial growth, though, they may undergo metabolic transformations. The biotransformation process also declines with the increasing number of alkyl substituents on the aromatic nucleus.

Asphaltic constituents of petroleum (Speight, 2014) tend to increase during biotransformation in relative and sometimes absolute amounts. This would suggest that they not only tend to resist biotransformation but may also be formed de novo by condensation reactions of biotransformation and photo-degradation intermediates.

In crude petroleum, as well as in refined products, petroleum hydrocarbons occur in complex mixtures and the constituents of the mixture can exert an influence on the biotransformation of other constituents; the effects may go in negative, as well as positive, directions. Some *iso*-alkane derivatives, apparently, remain unreactive as long as *n*-alkane derivatives are available as substrates, while some condensed aromatics are metabolized only in the presence of more easily utilizable petroleum hydrocarbons, a process referred to as co-metabolism (Wackett, 1996).

Finally, a word on the issue of adhesion as it affects biotransformation and, hence, biotransformation. Adhesion to hydrophobic surfaces is a common strategy used by microorganisms to overcome limited bioavailability of hydrocarbons (Bouchez-Naïtali et al., 1999). Intuitively, it may be assumed that adherence of cells to a hydrocarbon would correlate with the ability to utilize it as a growth substrate and conversely that cells able to utilize hydrocarbons would be expected to be able to adhere to them. However, species like *Staphylococcus aureus* and *Serratia marcescens*, which are unable to grow on hydrocarbons, adhere to them (Rosenberg et al., 1980). Thus, adherence to hydrocarbons does not necessarily predict utilization (Abbasnezhad et al., 2011).

Biotransformation of poorly water-soluble, liquid hydrocarbons is often limited by low availability of the substrate to microbes. Adhesion of microorganisms to an oil–water interface can enhance this availability, whereas detaching cells from the interface can reduce the rate of biotransformation. The capability of microbes to adhere to the interface is not limited to hydrocarbon degraders, nor is it the only mechanism to enable rapid uptake of hydrocarbons, but it represents a common strategy. The general indications are that microbial adhesion can benefit growth on and biotransformation of very poorly water-soluble hydrocarbons such as *n*-alkanes and large polycyclic aromatic hydrocarbons dissolved in a nonaqueous phase. Adhesion is particularly important when the hydrocarbons are not emulsified thereby giving limited interfacial area between the two liquid phases.

5.2 KINETIC ASPECTS

The kinetics for modeling the biotransformation of contaminated soils can be extremely complicated. This is largely because the primary function of microbial metabolism is not for the remediation of environmental contaminants. Instead the primary metabolic function, whether bacterial or fungal in nature, is to grow and sustain more of the microorganisms. Therefore, the formulation of a kinetic model must start with the active biomass and factors, such as supplemental nutrients and oxygen source that are necessary for subsequent biomass growth (Cutright, 1995; Rončević et al., 2005; Pala et al., 2006).

Studies of the kinetics of the biotransformation process proceed in two directions: (1) the first is concerned with the factors influencing the amount of transformed compounds with time and (2) the other approach seeks the types of curves describing the transformation and determines which of them fits the degradation of the given compounds by the microbiologic culture in the laboratory microcosm and sometimes, in the field. However, studies of biotransformation kinetics in the natural environment are often empiric, reflecting only a basic level of knowledge about the microbiologic population and its activity in each environment (Maletić et al., 2009).

One such example of the empirical approach is the simple (perhaps oversimplified) model:

$$dC/dt = kC^n$$

C is the concentration of the substrate, t is time, k is the degradation rate constant of the compound, and n is a fitting parameter (most often taken to be unity) (Hamaker, 1972; Wethasinghe et al., 2006).

Using this model, it is possible to fit the curve of substrate removal by varying n and k until a satisfactory fit is obtained. It is evident from this equation that the rate is proportional to the exponent of substrate concentration. First-order kinetics are the most often used equation for representation of the degradation kinetics (Heitkamp et al., 1987; Heitkamp and Cerniglia, 1987; Venosa et al., 1996; Seabra et al., 1999; Holder et al., 1999; Winningham et al., 1999; Namkoonga et al., 2002; Grossi et al., 2002; Hohener et al., 2003; Collina et al., 2005; Rončević et al., 2005; Pala et al., 2006).

However, researchers involved in kinetic studies do not always report whether the model they used was based on theory or experience and whether the constants in the equation have a physical meaning or if they just serve as fitting parameters (Rončević et al., 2005).

6.0 TEST METHODS FOR BIOTRANSFORMATION

Various methods exist for the testing of biodegradability of substances. Biodegradability is assessed by following specific parameters which are indicative of the consumption of the test substance by microorganisms, or the production of simple basic compounds which indicate the mineralization of the test substance. With respect to crude oil and crude oil products there is a variety of text methods that can be applied on a before-and-after basis to determine if changes to the substrate have occurred (Speight, 2005, 2015).

Hence there are various biodegradability testing methods which measure the amount of carbon dioxide (or methane, for anaerobic cases) produced during a specified period; there are those which measure the loss of dissolved organic carbon for substances which are water-soluble; those that measure the loss of hydrocarbon infrared bands and there are yet others which measure the uptake of oxygen by the activities of microorganisms (biochemical oxygen demand, BOD).

However, when the reference is specifically to lubricants, there are two test methods that apply to testing for biodegradability of a substrate. For example, standard test method ASTM D5864 determines lubricant biotransformation. This test determines the rate and extent of aerobic aquatic biotransformation of lubricants when exposed to an inoculum under laboratory conditions. The inoculum may be the activated sewage-sludge from a domestic sewage-treatment plant, or it may be derived from soil or natural surface waters, or any combination of the three sources. The degree of biodegradability is measured by calculating the rate of conversion of the lubricant to carbon dioxide. A lubricant, hydraulic fluid or grease is classified as readily biodegradable when 60% or more of the test material carbon is converted to carbon dioxide in 28 days, as determined using this test method.

In addition, the most established test methods used by the lubricant industry for evaluating the biodegradability of their products are Method CEC-L-33-A-94 developed by the Coordinating European Council (CEC); Method OEC D 301B, the Modified Sturm Test, developed by the Organization for Economic Cooperation and Development (OECD); and Method EPA 560/6-82-003, number CG-2000, the Shake Flask Test, adapted by the U.S. Environmental Protection Agency (EPA). These tests also determine the rate and extent of aerobic aquatic biotransformation under laboratory conditions. The Modified Sturm Test and Shake Flask Test also calculate the rate of conversion of the lubricant to carbon dioxide (CO_2). The CEC test measures the disappearance of the lubricant by analyzing test material at various incubation times through infrared spectroscopy. Laboratory tests have shown that the degradation rates may vary widely among the various test methods indicated above (US Army Corps of Engineers, 1999).

Biodegradability tests based on the test method favored by the Coordinating European Council has certain trends which indicate that mineral oils, along with alkylated benzenes and polyalkylene glycol derivatives among others, generally have poor biodegradability (i.e., from 0% to 40% w/w). Specifically, mineral oil biodegradability varies from 15% to 35% w/w as conducted by the Coordinating European Council biodegradability test method.

REFERENCES

Abbasnezhad, H., Gray, M., Foght, J.M., 2011. Influence of adhesion on aerobic biodegradation and bioremediation of liquid hydrocarbons. Applied Microbiology and Biotechnology 92, 653–675.

Abdallah, R.I., Mohamed, S.Z., Ahmed, F.M., 2005. Effect of biological and chemical dispersants on oil spills. Petroleum Science and Technology 23, 463–474.

Aislabie, J., Saul, D.J., Foght, J.M., 2006. Bioremediation of hydrocarbon-contaminated polar soils. Extremophiles: Life Under Extreme Conditions 10, 171–179.

Amatya, P.L., Hettiaratchi, J.P.A., Joshi, R.C., 2002. Biotreatment of flare pit waste. Journal of Canadian Petroleum Technology 41, 30–36.

Atlas, R.M., 1975. Effects of temperature and crude oil composition on petroleum biodegradation. Journal of Applied Microbiology 30 (3), 396–403.

Atlas, R.M., 1977. Stimulated petroleum biodegradation. Critical Reviews in Microbiology 5, 371–386.

Atlas, R.M., 1985. Effects of hydrocarbons on micro-organisms and biodegradation in Arctic ecosystems. In: Engelhardt, F.R. (Ed.), Petroleum Effects in the Arctic Environment. Elsevier, London, UK, pp. 63–99.

Atlas, R.M., 1991. Bioremediation: using nature's helpers-microbes and enzymes to remedy mankind's pollutants. In: Lyons, T.P., Jacques, K.A. (Eds.), Proceedings. Biotechnology in the Feed Industry. Alltech's Thirteenth Annual Symposium. Alltech Technical Publications, Nicholasville, Kentucky, pp. 255–264.

Atlas, R.M., Bartha, R., 1998. Fundamentals and applications. In: Microbial Ecology, fourth ed. Benjamin/ Cummings Publishing Company Inc., California, pp. 523–530.

Auffret, M., Labbé, D., Thouand, G., Greer, C.W., Fayolle-Guichard, F., 2009. Degradation of a mixture of hydrocarbons, gasoline, and diesel oil additives by *Rhodococcus aetherivorans* and *Rhodococcus wratislaviensis*. Applied and Environmental Microbiology 75, 7774–7782.

Austin, B., Calomiris, J.J., Walker, J.D., Colwell, R.R., 1977. Numerical taxonomy and ecology of petroleum-degrading bacteria. Applied and Environmental Microbiology 34, 60–68.

Bardi, L., Martini, C., Opsi, F., Bertolone, E., Belviso, S., Masoero, G., Marzona, M., Ajmone Marsan, F., 2007. Cyclodextrin-enhanced in situ bioremediation of polyaromatic hydrocarbons-contaminated soils and plant uptake. Journal of Inclusion Phenomena and Macrocyclic Chemistry 57 (1), 439–444.

Barker, G.W., Raterman, K.T., Fisher, J.B., Corgan, J.M., Trent, G.L., Brown, D.R., Sublette, G.L., 1995. Assessment of natural hydrocarbon bioremediation at two gas condensate production sites. In: Hinchee, R.E., Wilson, J.T., Downey, D.C. (Eds.), Intrinsic Bioremediation. Battelle Press, Columbus, Ohio, pp. 181–188.

Bartha, R., 1986a. Microbial Ecology: Fundamentals and Applications. Addisson-Wesley Publishers, Reading, Massachusetts.

Bartha, R., 1986b. Biotechnology of petroleum pollutant biodegradation. Microbial Ecology 12, 155–172.

Bartha, R., Atlas, R.M., 1977. The microbiology of aquatic oil spills. Advances in Applied Microbiology 22, 225–266.

Børresen, M.H., Rike, A.G., 2007. Effects of nutrient content, moisture content and salinity on mineralization of hexadecane in an Arctic soil. Cold Regions Science and Technology 48, 129–138.

Bouchez-Naïtali, M., Rakatozafy, H., Marchal, R., Leveau, J.Y., Vandecasteele, J.P., 1999. Diversity of bacterial strains degrading hexadecane in relation to the Mode of substrate uptake. Journal of Applied Microbiology 86, 421–428.

Brown, K.S., 1995. The green clean: the emerging field of phytoremediation takes root. BioScience 45, 579–582.

Chaillan, F., Chaîneau, C.H., Point, V., Saliot, A., Oudot, J., 2006. Factors inhibiting bioremediation of soil contaminated with weathered oils and drill cuttings. Environmental Pollution 144 (1), 255–265.

Chaîneau, C.H., Rougeux, G., Yeprémian, C., Oudot, J., 2005. Effects of nutrient concentration on the biodegradation of crude oil and associated microbial populations in the soil. Soil Biology and Biochemistry 37 (8), 1490–1497.

Chhatre, S., Purohit, H., Shanker, R., Khanna, P., 1996. Bacterial consortia for crude oil spill remediation. Water Science and Technology: A Journal of the International Association on Water Pollution Research 34, 187–193.

Choi, S.-C., Kwon, K.K., Sohn, J.H., Kim, S.-J., 2002. Evaluation of fertilizer additions to stimulate oil biodegradation in sand seashore mesocosms. Journal of Microbiology and Biotechnology 12 (3), 431–436.

Collina, E., Bestetti, G., Di Gennaro, P., Franzetti, A., Gugliersi, F., Lasagni, M., Pitea, D., 2005. Naphthalene biodegradation kinetics in an aerobic slurry-phase bioreactor. Environment International 31 (2), 167–171.

Compeau, G.C., Mahaffey, W.D., Patras, L., 1991. Full-scale bioremediation of a contaminated soil and water site. In: Sayler, G.S., Fox, R., Blackburn, J.W. (Eds.), Environmental Biotechnology for Waste Treatment. Plenum Press, New York, pp. 91–110.

Cooney, J.J., 1984. The fate of petroleum pollutants in fresh water ecosystems. In: Atlas, R.M. (Ed.), Petroleum Microbiology. Macmillan, New York, NY, USA, pp. 399–434.

Cunningham, J.A., Hopkins, G.D., Lebron, C.A., Reinhard, M., 2000. Enhanced anaerobic bioremediation of groundwater contaminated by fuel hydrocarbons at seal Beach, California. Biodegradation 11, 159–170.

Cutright, T.J., 1995. Polycyclic aromatic hydrocarbon biodegradation and kinetics using *Cunninghamella echinulatu var. elegans*. International Biodeterioration & Biodegradation 35 (4), 397–408.

Dadrasnia, A., Agamuthu, P., 2013. Dynamics of diesel fuel degradation in contaminated soil using organic wastes. International Journal of Environmental Science and Technology 10, 769–778.

deCarvalho, C.C.C.R., daFonseca, M.M.R., 2005. Degradation of hydrocarbons and alcohols at different temperatures and salinities by *Rhodococcus erythropolis* DCL14. FEMS Microbiology Ecology 51, 389–399.

Delille, D., Coulon, F., Pelletier, E., 2004. Effects of temperature warming during a bioremediation study of natural and nutrient-amended hydrocarbon-contaminated subAntarctic soils. Cold Regions Science and Technology 40 (1/2), 61–70.

El-Gendy, N.Sh., Moustafa, Y.M., Habibi, S.A., Ali, S., 2010. Evaluation of *Corynebacterium variabilis* Sh42 as a degrader for different poly aromatic compounds. Journal of American Science 6 (11), 343–356.

El-Gendy, N.Sh., Farah, J.Y., 2011. Kinetic modeling and error analysis for decontamination of different petroleum hydrocarbon components in biostimulation of oily soil microcosm. Soil and Sediment Contamination 20, 432–446.

Farahat, L.A., El-Gendy, N.Sh., 2007. Comparative kinetic study of different bioremediation processes for soil contaminated with petroleum hydrocarbons. Material Science Research India 4 (2), 269–278.

Farahat, L.A., El-Gendy, N.Sh., 2008. Biodegradation of Baleym Mix crude oil in soil microcosm by some locally isolated Egyptian bacterial strains. Soil and Sediment Contamination 17, 150–162.

Frenzel, M., James, P., Burton, S.K., Rowland, S.J., Lappin-Scott, H.M., 2009. Towards bioremediation of toxic unresolved complex mixtures of hydrocarbons: identification of bacteria capable of rapid degradation of alkyl-tetralins. Journal of Soils and Sediments 9, 129–136.

Garon, D., Sage, L., Seigle-Murandi, F., 2004. Effects of fungal bioaugmentation and cyclodextrin amendment on fluorene degradation in soil slurry. Biodegradation 15, 1–8.

Gibson, D.T., Sayler, G.S., 1992. Scientific Foundation for Bioremediation: Current Status and Future Needs. American Academy of Microbiology, Washington, DC.

Ginn, J.S., Sims, R.C., Murarka, I.P., 1995. In situ bioremediation (natural attenuation) at a gas plant waste site. In: Hinchee, R.E., Wilson, J.T., Downey, D.C. (Eds.), Intrinsic Bioremediation. Battelle Press, Columbus, Ohio, pp. 153–162.

Grossi, V., Massias, D., Stora, G., Bertrand Burial, J.C., 2002. Exportation and degradation of acyclic petroleum hydrocarbons following simulated oil spill in bioturbated Mediterranean coastal sediments. Chemosphere 48 (9), 947–954.

Hamaker, W., 1972. Decomposition: quantitative aspects. In: Goring, C.A.I., Hamaker, J.W., Thomson, J. (Eds.), Organic Chemicals in the Soil Environment. Marcel Dekker Inc., New York.

Harms, H., Smith, K.E.C., Wick, L.Y., 2010. Problems of hydrophobicity/bioavailability. In: Timmis, K.N. (Ed.), Handbook of Hydrocarbon and Lipid Microbiology. Springer, Berlin, pp. 1439–1450 (Chapter 42).

Heitkamp, M.A., Cerniglia, C.E., 1987. Effects of chemical structure and exposure on the microbial degradation of polycyclic aromatic hydrocarbons in freshwater and Estuarine ecosystems. Environmental Toxicology and Chemistry 6 (7), 535–546.

Heitkamp, M.A., Freeman, J.P., Cerniglia, C.E., 1987. Naphthalene biodegradation in environmental microcosms: estimates of degradation rates and characterization of metabolites. Applied and Environmental Microbiology 53 (1), 129–136.

Hohener, P., Duwig, C., Pasteris, G., Kaufmann, K., Dakhel, N., Harms, H., 2003. Biodegradation of petroleum hydrocarbon vapors: laboratory studies on rates and kinetics in unsaturated alluvial sand. Journal of Contaminant Hydrology 66 (1–2), 93–115.

Holba, A.G., Dzou, I.L., Hickey, J.J., Franks, S.G., May, S.J., Lenney, T., 1996. Reservoir geochemistry of South Pass 61 field, Gulf of Mexico: compositional heterogeneities reflecting filling history and biodegradation. Organic Geochemistry 24, 1179–1198.

Holder, E.L., Miller, K.M., Haines, J.R., 1999. Crude oil component biodegradation kinetics by marine and freshwater consortia. In: Alleman, B.C., Leeson, A. (Eds.), In Situ Bioremediation of Polycyclic Hydrocarbons and Other Organic Compounds. Battelle, Columbus, Ohio, pp. 245–250.

Kim, S.-J., Choi, D.H., Sim, D.S., Oh, Y.-S., 2005. Evaluation of bioremediation effectiveness on crude oil-contaminated sand. Chemosphere 59 (6), 845–852.

King, M.W.G., Barker, J.F., Hamilton, L.K., 1995. Natural attenuation of coal tar organics in groundwater. In: Hinchee, R.E., Wilson, J.T., Downey, D.C. (Eds.), Intrinsic Bioremediation. Battelle Press, Columbus, Ohio, pp. 171–180.

Kirk, P.W., Gordon, A.S., 1988. Hydrocarbon degradation by filamentous marine higher fungi. Mycologia 80, 776–782.

Kleinsteuber, S., Riis, V., Fetzer, I., Harms, H., Müller, S., 2006. Population dynamics within a microbial consortium during growth on diesel fuel in saline environments. Applied and Environmental Microbiology 72, 3531–3542.

Korade, D.L., Fulekar, M.H., 2009. Development and evaluation of Mycorrhiza for rhizosphere bioremediation. Journal of Applied Biosciences 17, 922–929.

Landy, D., Mallard, I., Ponchel, A., Monflier, E., 2011. Cyclodextrins for remediation technologies. In: Lichtfouse, E., Schwarzbauer, J., Robert, D. (Eds.), Environmental Chemistry for a Sustainable World. Nanotechnology and Health Risk, vol. 1. Springer, London, United Kingdom, pp. 47–81.

Leavitt, M.E., Brown, K.L., 1994. Bioremediation versus bioaugmentation – three case studies. In: Hinchee, R.E., Alleman, B.C., Hoeppel, R.E., Miller, R.N. (Eds.), Hydrocarbon Bioremediation. CRC Press, Inc., Boca Raton, Florida, pp. 72–79.

Magdalene, O.E., Ufuoma, A., Gloria, O., 2009. Screening of four common Nigerian weeds for use in phytoremediation of soil contaminated with spent lubricating oil. African Journal of Plant Science 3 (5), 102–106.

Maletić, S., Dalmacija, B., Rončević, S., Agbaba, J., Petrović, O., 2009. Degradation kinetics of an aged hydrocarbon-contaminated soil. Water, Air, and Soil Pollution 202, 149–159.

Margesin, R., Schinner, F., 2001. Biodegradation and bioremediation of hydrocarbons in extreme environments. Applied Microbiology and Biotechnology 56, 650–663.

McAllister, P.M., Chiang, C.Y., Salanitro, J.P., Dortch, I.J., Williams, P., 1995. Enhanced aerobic bioremediation of residual hydrocarbon sources. In: Hinchee, R.E., Wilson, J.T., Downey, D.C. (Eds.), Intrinsic Bioremediation. Battelle Press, Columbus, Ohio, pp. 67–76.

Mille, G., Almallah, M., Bianchi, M., Van Wambeke, F., Bertrand, J.C., 1991. Effect of salinity on petroleum biodegradation. Fresenius' Journal of Analytical Chemistry 339, 788–791.

Mishra, S., Jyot, J., Kuhad, R.C., Lal, B., 2001. In situ bioremediation potential of an oily sludge-degrading bacterial consortium. Current Microbiology 43, 328–335.

Namkoonga, W., Hwangb, E.Y., Parka, J.S., Choic, J.Y., 2002. Bioremediation of diesel-contaminated soil with composting. Environmental Pollution 119 (1), 23–31.

Nedunuri, K.V., Govundaraju, R.S., Banks, M.K., Schwab, A.P., Chen, Z., 2000. Evaluation of phytoremediation for field scale degradation of total petroleum hydrocarbons. Journal of Environmental Engineering 126, 483–490.

Nester, E.W., Anderson, D.G., Roberts Jr., C.E., Pearsall, N.N., Nester, M.T., 2001. Microbiology: A Human Perspective, third ed. McGraw-Hill, New York.

Nicholson, C.A., Fathepure, B.Z., 2004. Biodegradation of benzene by halophilic and halotolerant bacteria under aerobic conditions. Applied and Environmental Microbiology 70, 1222–1225.

Nikolopoulou, M., Kalogerakis, N., 2010. Biostimulation strategies for enhanced bioremediation of marine oil spills including chronic pollution. In: Timmis, K.N. (Ed.), Handbook of Hydrocarbon and Lipid Microbiology. Springer-Verlag, Berlin, pp. 2521–2529.

Nugroho, A., Effendi, E., Karonta, Y., 2010. Petroleum degradation in soil by thermophilic bacteria with biopile reactor. Makara Journal of Teknologi 14 (1), 43–46.

Okolo, J.C., Amadi, E.N., Odu, C.T.I., 2005. Effects of soil treatments containing poultry manure on crude oil degradation in a sandy loam soil. Applied Ecology and Environmental Research 3 (1), 47–53.

Oren, A., Gurevich, P., Azachi, M., Henis, Y., 1992. Microbial degradation of pollutants at high salt concentrations. Biodegradation 3, 387–398.

Osswald, P., Baveye, P., Block, J.C., 1996. Bacterial influence on partitioning rate during the biodegradation of Styrene in A Biphasic aqueous-organic system. Biodegradation 7, 297–302.

Pala, D.M., de Carvalho, D.D., Pinto, J.C., Santa Anna Jr., G.L., 2006. A suitable model to describe bioremediation of a petroleum-contaminated soil. International Biodeterioration & Biodegradation 58 (3–4), 254–260.

Pelletier, E., Delille, D., Delille, B., 2004. Crude oil bioremediation in sub-Antarctic intertidal sediments: chemistry and toxicity of oiled residues. Marine Environmental Research 57 (4), 311–327.

Perfumo, A., Banat, I.M., Marchant, R., Vezzulli, L., 2007. Thermally enhanced approaches for bioremediation of hydrocarbon-contaminated soils. Chemosphere 66, 179–184.

Pollard, S.J.T., Hrudey, S.E., Fedorak, P.M., 1994. Bioremediation of petroleum- and creosote-contaminated soils: a review of constraints. Waste Management Research 12, 173–194.

Radwan, S.S., Al-Mailem, D., El-Nemr, I., Salamah, S., 2000. Enhanced remediation of hydrocarbon contaminated desert soil fertilized with organic carbons. International Biodeterioration and Biodegradation 46, 129–132.

Rhykerd, R.L., Weaver, R.W., McInnes, K.J., 1995. Influence of salinity on bioremediation of oil in soil. Environmental Pollution 90, 127–130.

Riis, V., Kleinsteuber, S., Babel, W., 2003. Influence of high salinities on the degradation of diesel fuel by bacterial consortia. Canadian Journal of Microbiology 49, 713–721.

Rončević, S., Dalmacija, B., Ivančev-Tumbas, I., Petrović, O., Klašnja, M., Agbaba, J., 2005. Kinetics of degradation of hydrocarbons in the contaminated soil layer. Archives of Environmental Contamination and Toxicology 49 (1), 27–36.

Rosenberg, M., Gutnick, D., Rosenberg, E., 1980. Adherence of bacteria to hydrocarbons: a simple method for measuring cell-surface hydrophobicity. FEMS Microbiology Letters 9, 29–33.

Santisi, S., Cappello, S., Catalfamo, M., Mancini, G., Hassanshahian, M., Genovese, L., Giuliano, L., Yakimov, M.M., 2015. Biodegradation of crude oil by individual bacterial strains and a mixed bacterial consortium. Brazilian Journal of Microbiology 46 (2), 377–387.

Seabra, P.N., Linhares, M.M., Santa Anna, L.M., 1999. Laboratory study of crude oil remediation by bioaugmentation. In: Alleman, B.C., Leeson, A. (Eds.), In Situ Bioremediation of Polycyclic Hydrocarbons and Other Organic Compounds. Battelle, Columbus, Ohio, pp. 421–426.

Semple, K.T., Reid, B.J., Fermor, T.R., 2001. Impact of composting strategies on the treatment of soils contaminated with organic pollutants. Environmental Pollution 112, 269–283.

Shelton, D.R., Tiedje, J.M., 1984. Isolation and partial characterization of bacteria in an anaerobic consortium that mineralizes 3-chlorobenzoic acid. Applied and Environmental Microbiology 48, 840–848.

Sims, R.C., 1990. Soil remediation techniques at uncontrolled hazardous waste sites. Journal of the Air & Waste Management Association 40 (5), 703–732.

Soliman, R.M., El-Gendy, N.Sh., Deriase, S.F., Farahat, L.A., Mohamed, A.S., 2014. The Evaluation of different bioremediation processes for Egyptian oily sludge polluted soil on a microcosm level. Energy Sources, Part A 36 (3), 231–241.

Speight, J.G., 1996. Environmental Technology Handbook. Taylor & Francis, Washington, DC.

Speight, J.G., 2005. Environmental Analysis and Technology for the Refining Industry. John Wiley & Sons Inc., Hoboken, New Jersey.

Speight, J.G., 2014. The Chemistry and Technology of Petroleum, fourth ed. CRC Press, Taylor & Francis Group, Boca Raton, Florida.

Speight, J.G., 2015. Handbook of Petroleum Product Analysis, second ed. John Wiley & Sons Inc., Hoboken, New Jersey.

Speight, J.G., 2017. Handbook of Petroleum Refining. CRC Press, Taylor & Francis Group, Boca Raton, Florida.

Speight, J.G., Arjoon, K.K., 2012. Bioremediation of Petroleum and Petroleum Products. Scrivener Publishing, Beverly, Massachusetts.

Speight, J.G., Lee, S., 2000. Environmental Technology Handbook, second ed. Taylor & Francis, New York.

Stoner, D.L., 1994. Biotechnology for the Treatment of Hazardous Waste. CRC Press, Boca Raton, Florida.

Stroud, J.L., Paton, G.I., Semple, K.T., 2007. Microbe-aliphatic hydrocarbon interactions in soil implication for biodegradation and bioremediation. Journal of Applied Microbiology 102, 1239–1253.

Sztompka, E., 1999. Biodegradation of engine oil in soil. Acta Microbiologica Polonica 48, 185–196.

Tibbett, M., George, S.J., Davie, A., Barron, A., Milton, N., Greenwood, P.F., 2011. Just add water and salt: the optimization of petrogenic hydrocarbon biodegradation in soils from semi-arid Barrow Island, Western Australia. Water, Air, and Soil Pollution 216, 513–525.

Ulrich, A.C., Guigard, S.E., Foght, J.M., Semple, K.M., Pooley, K., Armstrong, J.E., Biggar, K.W., 2009. Effect of salt on aerobic biodegradation of petroleum hydrocarbons in contaminated groundwater. Biodegradation 20, 27–38.

US Army Corps of Engineers, 1999. US Army Manual EM1110-2-1424. (Chapter 8) www.usace.army.mil/usace-docs/engmanuals/em1110-2-1424/c-8.pdf.

US EPA, 2006. In Situ and Ex Situ Biodegradation Technologies for Remediation of Contaminated Sites Report No. EPA/625/R-06/015. Office of Research and Development National Risk Management Research Laboratory, United States Environmental Protection Agency, Cincinnati, Ohio.

Van Eyk, J., 1994. Venting and bioventing for the in situ removal of petroleum from soil. In: Hinchee, R.E., Alleman, B.C., Hoeppel, R.E., Miller, R.N. (Eds.), Hydrocarbon Bioremediation. CRC Press, Boca Raton, Florida, pp. 234–251.

Vasudevan, N., Rajaram, P., 2001. Bioremediation of oil sludge-contaminated soil. Environment International 26, 409–411.

Venosa, A., Suidan, M., Wrenn, B., Strohmeier, K., Haines, J., Eberhart, B., 1996. Bioremediation of an experimental oil spill on the Shoreline of Delaware Bay. Environmental Science & Technology 30 (5), 1764–1775.

Vidali, M., 2001. Bioremediation. An overview. Pure and Applied Chemistry 73, 1163–1172.

Vilcáez, J., Li, L., Hubbard, S.S., 2013. A new model for the biodegradation kinetics of oil droplets: application to the Deepwater Horizon oil spill in the Gulf of Mexico. Geochemical Transactions 14 (4) http://www.geo-chemicaltransactions.com/content/14/1/4.

Wackett, L.P., 1996. Co-metabolism: is the emperor wearing any clothes? Current Opinion in Biotechnology 7, 321–325.

Ward, D.M., Brock, T.D., 1978. Hydrocarbon biodegradation in hypersaline environments. Applied and Environmental Microbiology 35, 353–359.

Wethasinghe, C., Yuen, S.T.S., Kaluarachchi, J.J., Hughes, R., 2006. Uncertainty in biokinetic parameters on bioremediation: health risks and economic implications. Environment International 32 (3), 312–323.

Winningham, J., Britto, R., Patel, M., McInturff, F., 1999. A land farming field study of creosote-contaminated soil. In: Alleman, B.C., Leeson, A. (Eds.), Bioremediation Technologies for PAH Compounds. Battelle, Columbus, Ohio, pp. 421–426.

Winters, J.C., Williams, J.A., 1969. Microbiological alteration of crude oil in the reservoir. Preprints. Division of petroleum chemistry. American Chemical Society 14 (4), E22–E31.

Wolicka, D., Borkowski, A., 2012. In: Romero-Zerón, L. (Ed.), . Microorganisms and Crude Oil, Introduction to Enhanced Oil Recovery (EOR) Processes and Bioremediation of Oil-contaminated Sites. InTech. ISBN: 978-953-51-0629-6. http://www.intechopen.com/books/introduction-toenhanced-oil-recovery-eor-processes-and-bioremediation-of-oil-contaminated-sites/microorganisms-andcrude-oil.

Wolicka, D., Suszek, A., Borkowski, A., Bielecka, A., 2009. Application of aerobic microorganisms in bioremediation in situ of soil contaminated by petroleum products. Bioresource Technology 100 (13), 3221–3227.

Zengler, K., Richnow, H.H., Rossello-Mora, R., Michaelis, W., Widdel, F., 1999. Methane formation from long-chain alkanes by anaerobic microorganisms. Nature 401, 266–269.

Zhu, X., Venosa, A.D., Suidan, M.T., 2004. Literature Review on the Use of Commercial Bioremediation Agents for Clean-up of Oil Contaminated Estuarine Environments Report No. EPA/600/R-04/075. National Risk Management Research Laboratory, Environmental Protection Agency, Cincinnati, Ohio.

FURTHER READING

Baker, R.S., 1999. Bioventing systems: a critical review. In: Adriana, D.C., Bollag, J.M., Frankenberger, W.T., Sims, R.C. (Eds.), Bioremediation of Contaminated Soils. Agronomy Monograph, vol. 37. Madison, Wisconsin, pp. 595–630. Amer. Soc. Agron., Crop Sci. Soc. Amer., Soil Sci. Soc. Amer.

Brown, R.A., Norris, R.D., 1994. The evolution of a technology: hydrogen peroxide. In: Hinchee, R.E., Alleman, B.C., Hoeppel, R.E., Miller, R.N. (Eds.), In Situ Bioremediation. Hydrocarbon Bioremediation, CRC Press, Boca Raton, Florida, pp. 148–162.

Flathman, P.E., Carson Jr., J.H., Whitenhead, S.J., Khan, K.A., Barnes, D.M., Evans, J.S., 1991. Laboratory evaluation of the utilization of hydrogen peroxide for enhanced biological treatment of petroleum hydrocarbon contaminants in soil. In: Hinchee, R.E., Olfenbuttel, R.F. (Eds.), In Situ Bioreclamation: Applications and Investigations for Hydrocarbon and Contaminated Site Remediation. Butterworth-Heinemann, Stoneham, Massachusetts, pp. 125–142.

Lee, M.D., Thomas, J.M., Borden, R.C., Bedient, P.B., Ward, C.H., 1988. Biorestoration of aquifers contaminated with organic compounds. Critical Reviews in Environmental Control 18, 29–89.

Lu, C.J., 1994. Effects of hydrogen peroxide on the in situ biodegradation of organic chemicals in a simulated groundwater system. In: Hinchee, R.E., Alleman, B.C., Hoeppel, R.E., Miller, R.N. (Eds.), Hydrocarbon Bioremediation. CRC Press, Inc., Boca Raton, Florida, pp. 140–147.

Lu, C.J., Hwang, M.C., 1992. Effects of hydrogen peroxide on the in situ biodegradation of chlorinated phenols in groundwater. In: Proceedings. Water Environ. Federation 65th Annual Conference, New Orleans, Louisiana. September 20–24.

Obire, O., Anyanwu, E.C., 2009. Impact of various concentrations of crude oil on fungal populations of soil. International Journal of Environmental Science and Technology 6 (2), 211–218.

Obire, O., Nwaubeta, O., 2001. Biodegradation of refined petroleum hydrocarbons in soil. Journal of Applied Sciences & Environmental Management 5 (1), 43–46.

Pardieck, D.L., Bouwer, E.J., Stone, A.T., 1992. Hydrogen peroxide use to increase oxidant capacity for in situ bioremediation of contaminated soils and aquifers: a review. Journal of Contaminant Hydrology 9, 221–242.

Schlegel, H.G., 1977. Aeration without air: oxygen supply by hydrogen peroxide. Biotechnology and Bioengineering 19, 413.

Scragg, A., 1999. Environmental Biotechnology. Pearson Education Limited, Harlow, Essex, England.

Suflita, J.M., 1989. Microbiological principles influencing the biorestoration of aquifers. In: Kerr, R.S. (Ed.), Transport and Fate of Contaminants in the Subsurface. EPA/625/4–89/019. Environmental Research Laboratory, US Environmental Protection Agency, Ada, Oklahoma.

Wenzel, W., 2009. Rhizosphere processes and management in plant-assisted bioremediation (phytoremediation) of soils. Plant and Soil 321 (1–2), 385–408.

CHEMISTRY OF BIOTRANSFORMATION

1.0 INTRODUCTION

Petroleum biotechnology is based on biotransformation processes such as biodegradation and bioremediation (Speight and Arjoon, 2012). *Biodegradation (biotic degradation, biotic decomposition) is the chemical degradation of chemicals by bacteria or other biological means. Organic material can be degraded aerobically (in the presence of oxygen) or anaerobically (in the absence of oxygen). Most bioremediation systems run under aerobic conditions, but a system under anaerobic conditions may permit microbial organisms to degrade chemical species that are otherwise nonresponsive to aerobic treatment; the converse also applies—chemical species that are nonresponsive to anaerobic may undergo biotransformation under aerobic conditions (Speight and Arjoon, 2012). Furthermore, the general concept of biotransformation encompasses a wide range of procedures for modifying chemicals living organisms according to human needs. In addition, and even more pertinent to this text, bioengineering is a related field that more heavily emphasizes higher systems approaches (not necessarily the altering or using of biological materials *directly*) for interfacing with and utilizing microbes and is the application of the principles of engineering and natural sciences to molecular transformation of the feedstock.

Pollution by the spills of crude oil and crude oil products is a severe global environmental problem causing a number of adverse negative impacts on human health, fishers, aquacultures, tourism, ecosystem, and eventually the national income. Based on the biological response values that might accompany or follow oil pollution, lethal effects occur in the range of 1–10 mg/L, while sublethal effects range is as low as 1 mg/L (Egaas and Varanis, 1982). The aliphatic and polycyclic aromatic hydrocarbon (polynuclear aromatic hydrocarbon derivatives, also called polynuclear aromatic hydrocarbons) fractions of petroleum are readily absorbed by most aquatic organisms because of their high lipid solubility and bioaccumulation in fish and shellfish (Viguri et al., 2002; El-Gendy and Moustafa, 2007). Phenols have also known to have large toxic effects on the environment and are of great danger to human health (Ojumu et al., 2005). Phenols and phenolic compounds are hazardous pollutants that can be found in waste waters from oil refineries, petrochemical plants, coal gasification plants, coking plants, and dyes industry (Martínková et al., 2009).

Thus, it is the purpose of this chapter to present a review of the chemistry involved in biodegradation process as it can be applied in bioremediation of petroleum hydrocarbon–polluted environment.

2.0 BIODEGRADING MICROORGANISMS

Phenolic compounds are known to affect microbial growth and degradation activities even in low concentrations (Vincenza and Liliana, 2007). Aliphatic derivatives and polynuclear aromatic hydrocarbon derivatives have received a lot of attention due to their ubiquitous distribution in marine sediments

Introduction to Petroleum Biotechnology. https://doi.org/10.1016/B978-0-12-805151-1.00009-6

since hydrocarbons from spilled petroleum can persist in the sedimentary environment for a substantial period of time that may be on the order of several years (Wang et al., 2006). Moller et al. (1989) reported that fish can be tainted very rapidly on exposure, within few hours at concentrations above 1 mg/L hydrocarbons in the ambient water. Méndez et al. (2001) reported that directly or indirectly the presence of the pollutants in fish tissues may be a risk for potential bioaccumulation in the food and humans can taste petroleum hydrocarbons in fishes at concentrations between 5 and 20 mg/kg.

Microorganisms that can degrade organic pollutants can be isolated from many different environments, including both contaminated and uncontaminated sites and even marine sediments (Harayama et al., 2004). The microorganisms used for environmental biotransformation may be indigenous to a contaminated area (biostimulation process) or they may be isolated from elsewhere and brought to the contaminated site (bioaugmentation process). Contaminants are transformed by living organisms through reactions that take place as a part of their metabolic processes. Biotransformation of a compound is often a result of the actions of multiple organisms. Moreover, a complex mixture of pollutants or organic compounds requires microbial communities (i.e., consortia) to work together to efficiently degrade the pollutant.

The main sources of carbon for microorganisms in crude oil are hydrocarbons, both aliphatic and aromatic, and organic compounds such as organic acids such as acetic, benzoic, butyric, formic, propanoic, and naphthenic acids that would result from the biodegradation of crude oil (Dandie et al., 2004). Polynuclear aromatic hydrocarbons are generally described as molecules that consist of three or more fused aromatic rings in various structural configurations. Despite the limited water solubility of polynuclear aromatic hydrocarbon derivatives, it has been shown that contact of water with polynuclear aromatic hydrocarbon derivatives–contaminated sediments can lead to teratogenicity and toxicity of the water. Polynuclear aromatic hydrocarbon derivatives are ubiquitous in the environment. They are known or suspected to be genotoxic or carcinogenic and have been classified as priority pollutants (Hegazi et al., 2007). Contaminated sediment in surface waters also represents a continuing source of contamination in the aquatic food chain. The toxicity of polynuclear aromatic hydrocarbon derivatives to fish is of interest since fish occupy an elemental position in relation to man and his food chain. Even minor concentration of many polynuclear aromatic hydrocarbon derivatives could be accentuated through fish, posing a potential threat to man, being at the top of the atrophic hierarchy (Ali et al., 2006).

The United States Environmental Protection Agency (US EPA) currently regulates 16 polynuclear aromatic hydrocarbon derivative compounds as priority pollutants in water and generally considers these same compounds as "total polynuclear aromatic hydrocarbon derivatives" in contaminated soils and sediments (Liu et al., 2001; Moustafa, 2004). These polynuclear aromatic hydrocarbon derivatives can be divided according to the number of rings: low-molecular-weight and high-molecular-weight polynuclear aromatic hydrocarbon derivatives. The low-molecular-weight polynuclear aromatic hydrocarbon derivatives consist of two and three aromatic rings, whereas the high-molecular-weight polynuclear aromatic hydrocarbon derivatives consist of tetra-, penta-, and hexa-aromatic rings (Viguri et al., 2002). It is well known that the most important anthropogenic sources of polynuclear aromatic hydrocarbon derivatives are petrogenic and pyrolytic, the latter type is usually prevalent in aquatic environments (Zakaria et al., 2002; Stout et al., 2004). Contamination by polynuclear aromatic hydrocarbon derivatives may result from either pyrogenic source (incomplete combustion of organic matter, emission sources, and exhausts) or from the release of petroleum into the environment—petrogenic source (Moustafa, 2004). Microbial degradation is the major route through which polynuclear aromatic hydrocarbon derivatives are removed from contaminated environments; it is considered as an effective and environmentally benign cleanup technology as it involves the partial or complete bioconversion of these pollutants to microbial biomass, carbon dioxide, and water (Radwan et al., 2005; Hassanshahian et al., 2010, 2012a). However, biodegradation of polynuclear aromatic hydrocarbon derivatives is limited by their highly hydrophobic nature and sorption to sediments and soil particles, which decrease the rate of bioremediation (Ke et al., 2009). The biodegradation rate decreases with the increase of the number of benzene rings, thus, their recalcitrance and persistence in the environment increase with aromatic ring numbers. The salinity also inversely affects the biodegradation rate. Thus, additional cosubstrates (e.g., yeast extract, glucose, sodium citrate) and nutrients (i.e., nitrogen, e.g., urea and phosphorous sources) are required to enhance the rate of biodegradation at high salinity as high as 60 g/L NaCl (Kargi and Dincer, 1996; Arulazhagan and Vasudevan, 2011a,b).

A wide variety of prokaryotes, bacteria and eukaryotes, yeasts, and filamentous fungi are capable of utilizing hydrocarbons (Lebkowska et al., 2011). Bacteria are the most active agents in petroleum degradation, and they are reported as primary degraders of spilled oil in environment. *Arthrobacter*, *Burkholderia*, *Mycobacterium*, *Pseudomonas*, *Sphingomonas*, and *Rhodococcus* were found to be involved for alkyl-aromatic degradation. Moreover, *Pseudomonas fluorescens*, *Pseudomonas aeruginosa*, *Bacillus subtilis*, *Bacillus* sp., *Alcaligenes* sp., *Acinetobacter lwoffi*, *Flavobacterium* sp., *Micrococcus roseus*, *Corynebacterium* sp., *Gordonia*, *Brevibacterium*, *Aeromicrobium*, *Dietzia*, *Burkholderia*, and *Mycobacterium* were isolated from different polluted stream for their ability to degrade crude oil (Daugulis and McCracken, 2003; Chaillan et al., 2004; Das and Mukherjee, 2007a,b; Throne-Holst et al., 2007; Yakimov et al., 2007; Brooijmans et al., 2009). Moreover, polynuclear aromatic hydrocarbon derivative–degrading members of the marine bacterial genera *Cycloclasticus* (Dyksterhouse et al., 1995; Geiselbrecht et al., 1998; Chung and King, 2001), *Neptunomonas* (Hedlund et al., 1999), *Vibrio* (Hedlund and Staley, 2001), *Marinobacter* (Hedlund et al., 2001), *Pseudoalteromonas* (Hedlund and Staley, 2006) and *Lutibacterium* (Chung and King, 2001) have been reported in addition to the polynuclear aromatic hydrocarbon derivative–degrading members that are commonly found in terrestrial habitats such as *Pseudomonas*, *Paenibacillus*, *Rhodococcus*, *Tsukamurella*, *Arthrobacter*, and *Sphingomonas* (Daane et al., 2001).

Pseudomonads, as a biosurfactant producer (e.g., rhamnolipids and glycolipids), are the best known bacteria capable of utilizing hydrocarbons as carbon and energy sources. Biosurfactants increase the surface area of oil available for the bacteria to metabolize (Nikolopoulou and Kalogerakis, 2009). Biosurfactants can act as emulsifying agents, decreasing the surface tension and forming micelles, where, the microdroplets encapsulated in the hydrophobic microbial cell surface are taken inside and degraded (Kumar et al., 2008; Mahmound et al., 2008; Ali et al., 2014; El-Gendy et al., 2014a,b). Barabas et al. (2001) reported three *Streptomyces* strains (*Streptomyces griseoflavus*, *Streptomyces parvus*, and *Streptomyces plicatus*) from the Kuwait Burgan oil field for their ability to utilize n-hexadecane, n-octadecane, kerosene, and crude oil as sole carbon and energy sources.

Ben Said et al. (2008) reported the isolation of naphthalene-degrading bacteria belonging to *Alcanivorax*, *Pseudomonas*, *Sphingomonas*, and *Nephtomonas* from Bizerte lagoon sediments, Tunisia. Tarhriz et al. (2014) isolated a *Shewanella* genus as naphthalene-degrading bacteria from Qurugol Lake located at Azerbaijan. Tebyanian et al. (2013) reported that the outer membrane of Gram-negative bacteria can better tolerate and uptake polynuclear aromatic hydrocarbon derivatives than Gram-positive bacteria. Hassanshahian and Boroujeni (2016) reported the isolation of Gram-negative naphthalene-degrading bacteria belonging to *Shewanella*, *Salegentibacter*, *Halomonas*, *Marinobacter*, *Oceanicola*,

Idiomarina and *Thalassospira* from polynuclear aromatic hydrocarbon derivative–contaminated marine sediments and seawaters from Persian Gulf. They are reported also to produce biosurfactant and bioemulsifier that can uptake and dissolve the component of crude oil especially polynuclear aromatic hydrocarbon derivatives.

A Gram-positive, thermophilic and halotolerant bacterial strain VP3 closely related to *Aeribacillus pallidus* was isolated from a geothermal oil field, located in Sfax, Tunisia, after enrichment on vanillic acid, the lignin-related aromatic acid as one of the intermediates formed by the biodegradation of ferulic acid, and abundant compound in nature. It is reported to be able to degrade a wide range of other aromatic compounds, including benzoic, *p*-hydroxybenzoic, protocatechuic, *p*-hydroxyphenylacetic, cinnamic, *p*-coumaric, caffeic, and ferulic acids, phenol, and *m*-cresol, and it can grow also on crude oil and diesel as sole carbon and energy sources (Mnif et al., 2014).

Many polynuclear aromatic hydrocarbon derivative degraders have been reported: *Mycobacterium vanbaalenii* PYR-1 (Kweon et al., 2011), *Mycobacterium* sp. JS14 (Lee et al., 2007), *Sinorhizobium* sp. C4 (Keum et al., 2006), *Sphingomonas* sp. LB126 (Van Herwijnen et al., 2003a), *Rhodococcus* spp. (Akhtar et al., 2009), *Pasteurella* sp. IFA (Šepič et al., 1998), *Staphylococcus* sp. PN/Y (Mallick et al., 2007), *Burkholderia fungorum* LB400 (Marx et al., 2004), and *Pseudomonas* sp. PP2 (Prabhu and Phale, 2003), among others (Zhong et al., 2011; Janbandhu and Fulekar, 2011; Moscoso et al., 2012a,b).

Daane et al. (2001) reported that bacteria belonging to the genus *Paenibacillus*, isolated from petroleum-contaminated sediment and salt marsh rhizosphere can use naphthalene or phenanthrene as sole carbon source and can degrade the polynuclear aromatic hydrocarbon derivatives. Whereas in a polynuclear aromatic hydrocarbon derivative–polluted sediment sample, a complete degradation of naphthalene and phenanthrene was reported by *Paenibacillus* group-PR-P3 and 71% of fluorene was reported by *Paenibacillus* group-PR-P1. Hedlund et al. (1999) reported the capabilities of *Neptunomonas naphthovorans* (NAG-2N-113 and NAG-2N-126) strains, isolated from creosote-contaminated sediments to completely degrade naphthalene. Several *Pseudomonas* strains have been reported for naphthalene biodegradation; *Pseudomonas putida* NCIB 9816 (Yang et al., 1994), *P. putida* G7 (Menn et al., 1993), *P. putida* NCIB9816-4 (Simon et al., 1993), *Pseudomonas* sp. ND6 (Li et al., 2004), *P. putida* BS202 (Balashova et al., 2001), *P. putida* sp. (Ono et al., 2007), *Pseudomonas* sp. C18 (Denome et al., 1993), *P. putida* OUS82 (Kiyohara et al., 1994), *P. putida* KF715 (Lee et al., 1996), *P. aeruginosa* Pak1 (Takizawa et al., 1999), and *Pseudomonas stutzeri* AN10 (Bosch et al., 2000; Lanfranconi et al., 2009). Biphenyl, fluorene, and dibenzofuran are reported to comprise a class of substrates that may substitute for each other as a carbon source for certain strains (Trenz et al., 1994). El-Gendy and Abostate (2008) reported the isolation of *Staphylococcus gallinarum* NK1 from a drain receiving oil refinery wastewater for its ability to degrade dibenzothiophene, naphthalene, and phenanthrene. It also showed good capabilities for producing biosurfactant and degrading n-hexadecane.

High-molecular-weight polynuclear aromatic hydrocarbon derivatives (four and higher numbers of aromatic ring compounds) are formed by pyrogenic processes during the incomplete combustion of organic material through pyrolysis and pyrosynthesis; they are generally present in relatively high concentrations in fossil fuels and in products of fossil fuel refining (Mastral and Callén, 2000). Heitkamp and Cerniglia (1988) published the first study on the isolation of a bacterium from an oil-polluted sediment sample that could extensively co-metabolically degrade four-aromatic ring polynuclear aromatic hydrocarbon derivatives, including fluoranthene, pyrene, 1-nitropyrene, 3-methylcholanthrene, 6-nitrochrysene, and five-membered ones, benzo(a)pyrene (BaP). *Cycloclasticus* strains isolated from marine sediments are reported to be co-metabolically capable of partially degrading pyrene or fluoranthene in presence of phenanthrene (Geiselbrecht et al., 1998). Mueller et al. (1989) published the first study for bacterial isolate, *Pseudomonas paucimobilis* or *Sphingomonas paucimobilis* EPA505, capable of utilization of four- or more aromatic ring polynuclear aromatic hydrocarbon derivatives a sole source of carbon and energy, where a seven-member bacterial community isolated from creosote-contaminated soil was capable of utilizing fluoranthene. *Mycobacterium sp. strain* PYR-1 is reported to completely mineralize fluoranthene (Kelley et al., 1991). Many bacterial genera have been reported as fluoranthene degraders, for example, bacteria in the genera *Sphingomonas* (Baboshin et al., 2008), *Alcaligenes* (Weissenfels et al., 1990), *Burkholderia* (Juhasz et al., 1997), *Pseudomonas* (Gordon and Dobson, 2001), *Mycobacterium* (Rehmann et al., 2001; López et al., 2006), and *Rhodococcus* (Walter et al., 1991). However, only a few fluoranthene degraders have been isolated from the marine environment that belong to the genera *Ochrobactrum* (Wu et al., 2009), *Novosphingobium* (Yuan et al., 2009), *Cycloclasticus* (Geiselbrecht et al., 1998), and *Celeribacter* (Lai et al., 2014). *S. paucimobilis* strain EPA 505 has been reported for degradation of BaP Ye et al. (1996). BaP has been also reported to be degraded by different bacteria, such as *Agrobacterium* sp., *Bacillus* sp., *Burkholderia* sp., *Rhodococcus* sp., *Mycobacterium*, and *Flavobacterium* sp. (Walter et al., 1991; Trzesicka-Mlynarz and Ward, 1995; Schneider et al., 1996; Aitken et al., 1998). Grosser et al. (1991) and Dean-Ross and Cerniglia (1996) reported the isolation of *Mycobacterium* sp. strain RJGII-135 and *Mycobacterium flavescens* for their capabilities to mineralize pyrene. *Rhodococcus* sp. strain UW1 isolated from contaminated soil is capable of utilizing pyrene and chrysene as sole sources of carbon and energy (Walter et al., 1991). Fritzsche (1994) reported the isolation *Mycobacterium* sp. strain BB1 from a former coal gasification site. BB1 has the capability to use flourantherene (0.056 h^{-1}) and pyrene (0.04 h^{-1}) as sole carbon an energy source. *Gordona* sp. strain BP9 and *Mycobacterium* sp. strain VF1 were also isolated from hydrocarbon-contaminated soil for their capabilities to use fluoranthene and pyrene as sole carbon and energy sources (Kästner et al., 1998). Three *Burkholderia cepacia* strains isolated from soil are reported to metabolize high concentrations of pyrene and utilize fluoranthene and benz[a]anthracene as sole carbon and energy sources (Juhasz et al., 1997). Rehmann et al. (1998) isolated a *Mycobacterium* spp. strain KR2 from a polynuclear aromatic hydrocarbon derivative–contaminated soil of a gas work plant, which is able to utilize pyrene as sole source of carbon and energy. Churchill et al. (1999) reported the isolation of *Mycobacterium* sp. strain CH1 from polynuclear aromatic hydrocarbon derivative–contaminated freshwater sediments for its capability to mineralize fluoranthene and pyrene, and it has also good capability to use a wide range of branched alkanes and n-alkanes as sole carbon and energy sources. El-Gendy et al. (2010) reported the isolation of *Corynebacterium variabilis* Sh42 from hydrocarbon polluted water sample collected from El-Lessan Area of Damietta River Nile Branch in Egypt, as a degrader for different poly aromatic compounds: naphthalene, anthracene, phenanthrene, pyrene, 2-hydroxybiphenyl, 2,2′-bihydroxybiphenyl, dibenzothiophene, 4-methyldibenzothiophene, and 4,6-dimethyldibenzothiphene. Soliman et al. (2011) reported the isolation of *Micrococcus lutes* RM1 from the oily sludge contaminated soil for its ability to degrade naphthalene, anthracene, and pyrene. Willison (2004) reported the biodegradation of chrysene as sole carbon and energy source by *Sphingomonas* sp. CHY-1 isolated from polynuclear aromatic hydrocarbon derivative–contaminated soil. Dave et al. (2015) reported the biodegradation of chrysene by a bacterial consortium consisting of *Achromobacter xylosoxidans*, *Pseudomonas* sp., and *Sphingomonas* sp., isolated from crude oil saline sites at Bhavnager coast, Gujarat, india. Benzo[a] anthracene is classified as a group 2A carcinogen by the International Agency for Research on Cancer and is included in the US EPA's Priority Pollutant List. Thus, there is a much concern about the environmental fate of this recalcitrant pollutant. Although of the recalcitrance and persistence of the kata-annelated high-molecular-weight polynuclear aromatic hydrocarbon derivatives, however, Sisler and ZoBell (1947) have reported the biodegradation of benz[*a*]anthracene and dibenz[*a,h*]anthracene by marine bacterial consortium. *P. fluorescens* strain P2a is reported to utilize chrysene and benz[a]anthracene as sole carbon sources (Caldini et al., 1995). *Mycobacterium* sp. strain RJGII-135 (Schneider et al., 1996), *Sphingobium yanoikuyae* strains B8/36 and B1 (Boyd et al., 2006), and *Sphingomonas* sp. strain CHY-1 (Jouanneau et al., 2006) have been reported for their capabilities to degrade benzo[a]anthracene. The biodegradation of the recalcitrant five- or more aromatic ringed polynuclear aromatic hydrocarbon derivatives, which have mutagenic, teratogenic, and potentially carcinogenic effects, have been also reported. BaP is usually used as a model compound for high-molecular-weight polynuclear aromatic hydrocarbon derivatives, due to its potential hazardous to human and its carcinogenic properties and its recalcitrance in polluted soil and sediments. BaP is characterized by low water solubility (0.0023 mg/L) and a high octanol/water partition coefficient (LogKow, 6.06), which is related to high recalcitrance to microbial degradation. However, the cometabolic mineralization of BaP in soil has been reported (Kanaly and Bartha, 1999). The cometabolic biotransformation of BaP has been also reported by *Pseudomonas* strain NCIB 9816 (Gibson et al., 1975) and the mutant *Beijerinckia* sp. strain B8/36 (Barnsley, 1975) in presence of succinate plus salicylate and succinate plus biphenyl, respectively. Trenz et al. (1994) reported that *Brevibacterium* sp. strain DPO 1361 utilizes biphenyl and fluorene as sole sources of carbon and energy. *Mycobacterium* sp. strain PYR-1 is reported to biotransform BaP in a six-component polynuclear aromatic hydrocarbon derivative mixture (Kelley and Cerniglia, 1995). *S. paucimobilis* EPA505 is reported to mineralize the five-membered ring compounds, polynuclear aromatic hydrocarbon derivatives BaP, benzo[*b*]fluoranthene, and dibenz[*a,h*]anthracene, to 33.3%, 12.5%, and 7.8% ^{14}C carbon dioxide from initial concentration of 10 mg/L, respectively, in a single substrate system within 16 h of incubation (Ye et al., 1996). The degradation of 11.65 mg/L dibenz[*a,h*]anthracene as a sole carbon and energy source after 56 days by *B. cepacia* strain VUN 10003 has been reported (Juhasz et al., 1997). However, the rate of the cometabolic degradation of B(a)P is reported to be higher than that of dibenz[*a,h*]anthracene by *B. cepacia* strains in the presence of phenanthrene (Juhasz et al., 1997). The halotolerant *Halomonas eurihalina* strain H-28A is reported to degrade naphthalene (95%), phenanthrene (50%), fluoranthene (50%), and pyrene (58%) in the presence of glucose, yeast extract, malt extract, and proteose-peptone at 51.3 g/L NaCl (Martínez-Checa et al., 2002). Zeinali et al. (2008a) reported the biodegradation of naphthalene by the moderately thermophilic bacterium *Nocardia otitidiscaviarum* strain TSH1. Pathak et al. (2009) also reported the biodegradation of naphthalene by *Pseudomonas* sp. HOB1. Arulazhagan and Vasudevan (2009) reported the degradation of the low-molecular-weight polynuclear aromatic hydrocarbon derivatives phenanthrene and fluorene and the high-molecular-weight polynuclear aromatic hydrocarbon derivatives pyrene and benzo(e)pyrene as sole carbon and energy sources by a moderately halophilic bacterial consortium: *Ochrobactrum* sp. (GenBank accession no. EU722312), *Enterobacter cloacae* (GenBank accession no. EU722313), and *Stenotrophomonas maltophilia* (GenBank accession no. EU722314), which were isolated from a mixture of marine water samples collected from seven different sites in Chennai, India were selected for the degradation study. Moreover, the consortium was also able to degrade polynuclear aromatic hydrocarbon derivatives present in crude oil–contaminated saline wastewater that had polynuclear aromatic hydrocarbon derivatives such as phenanthrene (4368 mg/L), fluorene (2615 mg/L), pyrene (3902 mg/L), and benzo(e)pyrene (114 mg/L), where approximately 80% of high-molecular-weight polynuclear aromatic hydrocarbon derivatives and 100% of low-molecular-weight polynuclear aromatic hydrocarbon derivatives were degraded. Lily et al. (2009) have isolated *B. subtilis* BMT4i (MTCC 9447) from automobile contaminated soil for its ability to degrade BaP as the sole source of carbon and energy. Moreover, it can degrade naphthalene, anthracene, and dibenzothiophene. The halotolerant *Ochrobactrum* sp. strain VA1 (EU722312) is then reported to degrade low- and high-molecular-weight polynuclear aromatic hydrocarbon derivatives, such as anthracene (88%), phenanthrene (98%), naphthalene (90%), fluorene (97%), pyrene (84%), benzo(k)fluoranthene (57%), and benzo(e)pyrene (50%) at a 30 g/L NaCl concentration (Arulazhagan and Vasudevan, 2011a). Balachandran et al. (2012) reported the isolation of *Streptomyces* sp. ERI-CPDA-1 from oil-contaminated soil in Chennai, India, for its ability to degrade petroleum, naphthalene, phenanthrene, diesel oil, petrol, kerosene, benzene, pyridine, methanol, ethanol, cyclohexane, tween-80, xylene, DMSO, and toluene as sole carbon and energy sources. *Mycobacterium frederiksbergense* LB501T has been reported to utilize anthracene as a sole carbon and energy source (Wick et al., 2003). Aitken et al. (2005) reported the mineralization of benzo[a]anthracene, chrysene, and BaP by *Pseudomonas saccharaphila* p15. *N. otitidiscaviarum* strain TSH1 has been reported for biodegradation of phenanthrene and anthracene (Zeinali et al., 2008b). The alkaliphilic *Bacillus badius* D1, isolated from Pristine Crater Lake of Lonar, Buldhana, MS, India is also reported to utilize anthracene as a sole carbon and energy source (Ahmed et al., 2012). Moscoso et al. (2012a) reported the isolation of polynuclear aromatic hydrocarbon derivatives degrading bacterial consortium species; *Staphylococcus warneri* and *Bacillus pumilus* were isolated from lab soils polluted with polynuclear aromatic hydrocarbon derivatives and heavy metals, by using an enrichment culture technique. The rate of biodegradation of the studied polynuclear aromatic hydrocarbon derivatives in a bench-stirred bioreactor decreased as the molecular weight increased: phenanthrene (0.191 h^{-1}) > pyrene (0.120 h^{-1}) > benzo[a]anthracene (0.109 h^{-1}). The observed differences in biodegradation rates of pyrene and benzo[a]anthracene although they have the same molecular weight might be explained by the differences in molecular structure features that govern reactivity and binding affinity for the enzymes. Although a lower specific degradation rate occurred in mixed polynuclear aromatic hydrocarbon derivative system, equal or higher specific degradation rates are obtained for the studied three polynuclear aromatic hydrocarbon derivatives, when working in cometabolic conditions, a fact that is advantageous in terms of bioprocess economy. Kumaria et al. (2014) reported the isolation of *Ochrobactrum* sp. P2 (Accession No. KC493414) and *Pseudomonas* sp. BP10 (Accession No. KC493413) from crude oil–polluted soil sample collected from Barauni Oil Refinery, India, for their ability to degrade pyrene. They have the ability to produce proteolipid and glycolipid biosurfactants, respectively. They expressed 96% and 80% biodegradation of 100 μg pyrene/g spiked soil after 14 days of incubation, respectively and both achieved nearly complete biodegradation of pyrene after 28 days of incubation. Salam and Obayori (2014) reported the isolation of two Gram-positive bacteria *B. subtilis* BM1 and *Bacillus amyloliquefaciens* BR1, from a highly polluted soil samples in Lagos, Nigeria, for their ability to degrade fluorene. *Novosphingobium pentaromativorans* US6-1, a marine bacterium isolated from muddy sediments of Ulsan Bay, Republic of Korea, is reported to degrade phenanthrene, pyrene, and BaP (Lyu et al., 2014). *Sphingobium* sp. KK22 is reported to be capable to use phenanthrene as sole carbon and energy source and cometabolically oxidize the high-molecular-weight polynuclear aromatic hydrocarbon derivatives fluoranthene, benzo(a)anthracene, and benzo(k)fluoranthene in different positions (Vila et al., 2015). Folwell et al. (2016) reported that the main bacteria enriched by high-molecular-weight polynuclear aromatic hydrocarbon derivatives are *Pseudomonas*, *Bacillus*, and *Microbacterium* species. *Panebacillus* sp. strain HD1 polynuclear aromatic hydrocarbon derivatives, isolated from crude oil–contaminated soil sample, has been reported to degrade benzo[a] anthracene as a sole carbon and energy source (Deka and Lahkar, 2017). Biodegradation of anthracene, benzo[a]anthracene, and dibenzo[a,h]anthracene by a consortium of aerobic heterotrophic bacteria and cyanobacteria has been reported (Tersagh et al., 2017).

Yeasts that are capable of degrading hydrocarbons mainly include genera of *Yarrowia*, *Candida*, *Pichia*, and *Debaryomyces* (Chaillan et al., 2004; Das and Chandran, 2011). However, the rate of degradation of phenanthrene by a yeast isolated from contaminated stream *Rhodotorula glutinis* is reported to be almost equal to the degradation by bacteria *P. aeruginosa* (Romero et al., 1998). *Candida albicans* and *Candida maltose* were isolated from lagoon water in Nigeria as hydrocarbon degrader (Matthew et al., 2008). Two yeast strains *Yarrowia lipolytica* PG-20 and PG-32 have been isolated from an oil-polluted area in the Persian Gulf for their ability to degrade crude oil, with preference of alkanes than polynuclear aromatic hydrocarbon derivatives and are characterized by high emulsifying activity and cell hydrophobicity (Hassanshahian et al., 2012b). Crude oil was also reported to be degraded by *Saccharomyces cerevisiae* isolated from a commercial beverage "Zobo" prepared from *Hibiscus* flower in Nigeria (Abioye et al., 2013). The capability of several yeast species to use n-alkanes and other aliphatic hydrocarbons as a sole source of carbon and energy is mediated by the existence of multiple microsomal Cytochrome P450 enzyme systems. These cytochrome P450 enzymes had been isolated from yeast species such as *Candida maltosa*, *Candida tropicalis*, and *Candida apicola* and catalyze the first step in alkane degradation through the terminal hydroxylation and ω-hydroxylation of fatty acids that form alcohols, which are then processed by fatty alcohol oxidase and fatty aldehyde dehydrogenase or by P450 monooxygenase to yield fatty acids and α, ω-dioic acids (Fig. 9.1) (Scheuer et al., 1998).

Fungi were thought to transform or cooxidize aromatic hydrocarbons, but not utilize them as carbon and energy sources. However, some filamentous fungi, basidiomycetes, white-rot fungi, and deuteromycetes have been reported to remove polynuclear aromatic hydrocarbon derivatives more competently than bacteria (Peng et al., 2008). It has been reported that there are at least two mechanisms involved in the mycodegradation of polynuclear aromatic hydrocarbon derivatives; one utilizes the cytochrome P-450 system (Yadav et al., 2006) and the other uses the soluble extracellular enzymes of lignin catabolism, including lignin peroxidase, manganese peroxidase (Steffen et al., 2003), and

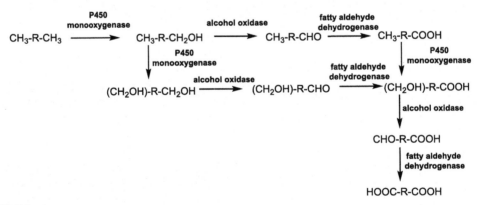

FIGURE 9.1

Aerobic biodegradation of n-alkanes by yeast.

laccase (Gianfreda et al., 1999). For example, polycyclic aromatic hydrocarbons (polynuclear aromatic hydrocarbon derivatives) are oxidized by cytochrome P-450 enzymes in fungi and other eukaryotic organisms to form arene oxides, which are then converted to trans-dihydrodiols or phenols, as dead-end products that are not further degraded further (Cerniglia, 1997). But, *Phanerochaete chrysosporium* is reported to degrade all the six components of benzene, toluene, ethylbenzene, xylene (BTEX), but does not use any of them as carbon or energy source (Yadav and Reddy, 1993). However, white-rot and litter-decomposing fungi such as *P. chrysosporium*, *Bjerkandera adusta*, and *Pleurotus ostreatus* as good producers of ligninolytic enzymes (lignin peroxidase, versatile peroxidase, Mn-peroxidase and laccase) are reported to be good polynuclear aromatic hydrocarbon derivative degraders (Pozdnyakova, 2012). Lignin peroxidase (LiP) and manganese peroxidase (MnP) can catalyze the oxidation of the recalcitrant nonphenolic lignin to form a high redox potential oxo-ferryl intermediate during the reaction of the heme cofactor with hydrogen peroxide (H_2O_2).

The heme group, found in the molecular structure of the aforementioned peroxidases, confers high redox potential to the oxo-ferryl complex. Also, the existence of specific-binding sites in their molecular structure for oxidation of their characteristic substrates, including nonphenolic aromatics in the cases of LiP and manganous iron in the case of MnP, confer their distinctive catalytic properties (Reddy et al., 2003; Martínez et al., 2005). A correlation has been elucidated between the ionization potential (IP) of polynuclear aromatic hydrocarbon derivatives and the specific activity of MnP and LiP. The lower the IP of aromatic compounds, the faster the oxidation rate by the two ligninolytic enzymes. A threshold value of IP is reported for each enzyme: LiP oxidizes polynuclear aromatic hydrocarbon derivatives with an IP 7.55 eV (Vazquez-Duhalt et al., 1994), while MnP oxidizes polynuclear aromatic hydrocarbon derivatives with an IP up to 8.2 eV (Bogan and Lamar, 1995). MnP isolated from *Irpex lacteus* is reported to be able to efficiently degrade three- and four-ringed polynuclear aromatic hydrocarbon derivatives with an IP higher than 7.8 eV (Bogan et al., 1996a,b). Laccases catalyze a one-electron oxidation concomitantly with the four-electron reduction of molecular oxygen to water using a range of phenolic compounds as hydrogen donors.

The catalysis is carried out by the presence of different copper centers, which were arranged in a trinuclear cluster with one type-1 (T1), one type-2 (T2), and two type-3 (T3) copper ions. The presence of the four cupric ions, each coordinated to a single polypeptide chain, is an absolute requirement for optimal activity (Solomon et al., 2001; Baldrian, 2006). The activity of laccases mainly depends on their redox potential, which are divided into three groups, according to their redox potential (E) at the T1 copper site. To enhance the activity of laccases, mediator should be added, since polynuclear aromatic hydrocarbon derivatives have redox potential of 500–800 mV, which is higher than that of laccase. The most common mediators used are 2,2'-azino-bis-(3-ethylbenzothiazoline-6-sulfonic acid) (ABTS) and −N−OH-type mediators such as 1-hydroxybenzotriazole (HBT) and violuric acid, with E^o values 1.09, 1.12, and 0.91 V, respectively (Xu et al., 2000, 2001; Han et al., 2004). However, the major drawbacks of those chemically synthesized mediators are the high added costs and the possible generation of toxic oxidized species. The natural compound p-coumaric acid strongly promotes the removal of polynuclear aromatic hydrocarbon derivatives by laccase. It is also reported to be better than mediator like ABTS and with similar activity like that of HBT (Caňas et al., 2007). *Cladosporium sphaerospermum*, which is reclassified as *Cladophialophora* sp. CBS 114326, is reported to completely mineralize toluene, through the initial oxidation of toluene by a cytochrome P-450-dependent toluene monooxygenase. The produced protocatechuate can be further oxidized via two simultaneous pathways: through ortho-cleavage by protocatechuate 3,4-dioxygenase or through decarboxylation by protocatechuate decarboxylase producing catechol, which is then oxidized through the other branch of the β-ketoadipate pathway (Fig. 9.2) (Weber et al., 1995; Luykx et al., 2003; Prenafeta-Boldú et al., 2006).

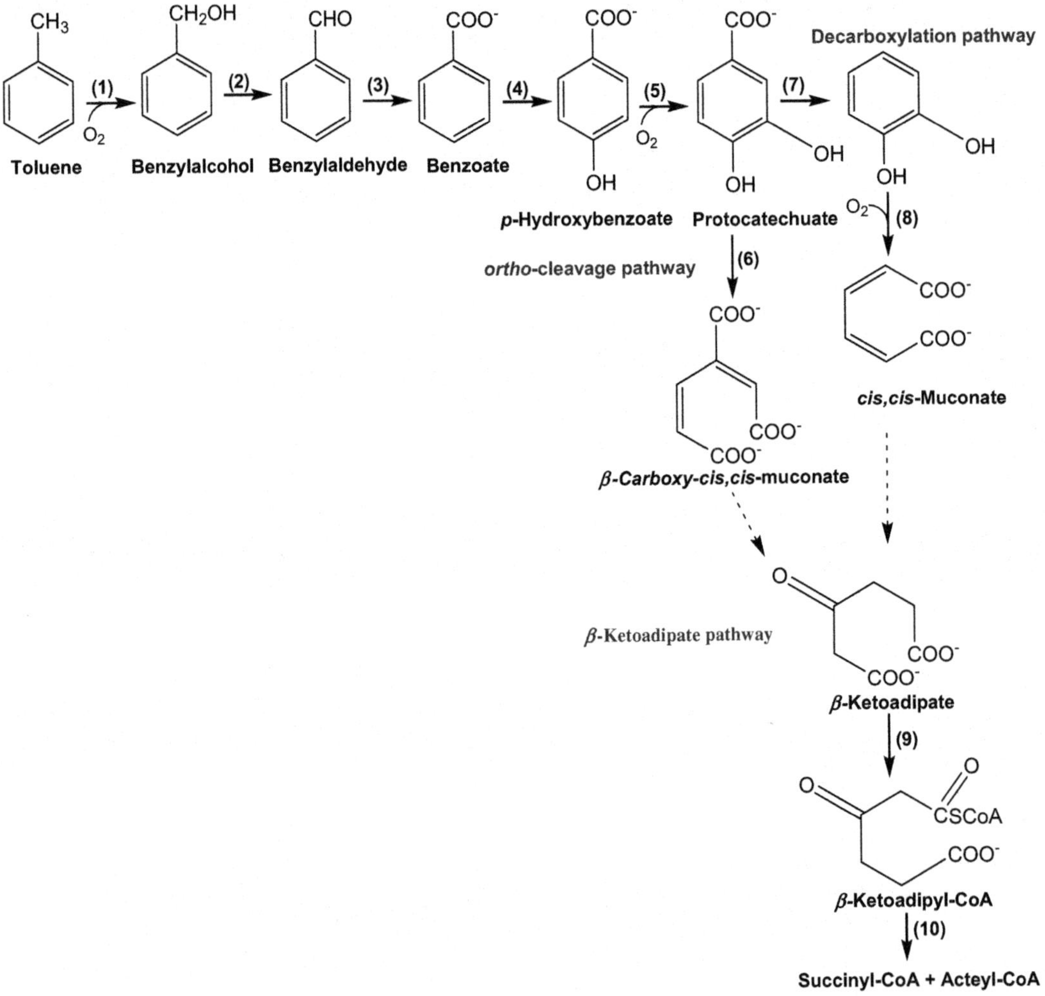

FIGURE 9.2

Aerobic biodegradation of toluene by fungi. The involved enzymes are (1) cytochrome; (2) benzyl alcohol dehydrogenase; (3) benzaldehyde dehydrogenase; (4) benzoate hydroxylase; (5) p-hydroxybenzoate hydroxylase; (6) protocatechuate 3,4-dioxygenase; (7) protocatechuate decarboxylase; (8) catechol 1,2-dioxygenase; (9) β-ketoadipate succinyl-CoA transferase; and (10) β-ketoadipyl-CoA thiolase. The biodegradation pathways of cis,cis-muconate and β-carboxy- cis,cis-muconate are illucidated in Figure 9.15.

Although CBS 114326 can grow on ethylbenzene, propylbenzene, and styrene, it cannot grow on benzene, o-xylene, or phenol. The major fungal phyla/subphyla involved in the biodegradation of oil include the *Ascomycota*, *Basidiomycota*, and *Mucoromycotina*. There are common known terrestrial strains, *Fusarium*, *Trichoderma*, and *Verticillium* spp., and others in marine environment that can live under extreme conditions, the majority of which are *Ascomycota*, mainly belonging to *Halosphaeriaceae* (*Microascales*, *Sordariomycetes*) and *Lulworthiales* (*Sordariomycetes*), and also *Mucoromycotina* is reported such as *Rhizopus stolonife*, which is reported for mycodegradation of different petroleum hydrocarbons (Saraswathy and Hallberg, 2002; Atagana et al., 2006; Adekunle and Adebambo, 2007; Gesinde et al., 2008; Husaini et al., 2008; Romerom et al., 2010). Fungi are capable of degrading this structurally diverse range of oil-derived compounds by employing a variety of mechanisms including intracellular enzymes such as cytochrome P450 monooxygenases, nitroreductases, and transferases as well as extracellular enzymes such as laccases and fungal peroxidases (Harms et al., 2011). However, nitrogen as a nutrient in bioremediation process can inhibit the lignin-degrading system of the white-rot fungi. *Phanerochaete laevis* is faster and more extensive in biodegradation of polynuclear aromatic hydrocarbon derivatives than *P. chrysosporium* (Peng et al., 2008).

Penicillium species have been reported for their ability to degrade polynuclear aromatic hydrocarbon derivatives through cytochrome P450 monooxygenase enzyme systems (Leitão, 2009). The metabolism of pyrene by *Penicillium janthinellum* SFU403, a strain isolated from petroleum-contaminated soils, has been reported to proceed via hydroxylation to 1-pyrenol, followed by 1,6- and 1,8-pyrenequinones (Fig. 9.3) (Launen et al., 1995, 1999). However, the metabolism of pyrene by *Penicillium glabrum* TW 9424, a strain isolated from a site contaminated with polynuclear aromatic hydrocarbon derivatives, showed two novel metabolites in the fungal metabolism of polycyclic aromatic hydrocarbons (1-methoxypyrene and 1,6- imethoxypyrene) (Wunder et al., 1997). *Penicillium simplicissimum* SK9117 (Marr et al., 1989), *Penicillium frequentans* Bi 7/2 (Hofrichter et al., 1992), *Penicillium chrysogenum* Thom ERK1 (Wolski et al., 2012), and the halotolerant *P. chrysogenum* CLONA 2 (Guedes et al., 2010) are reported to degrade phenol, chlorophenol, nitrophenol, and resorcinol. The biodegradation of phenols by *P. simplicissimum* SK9117 and *P. chrysogenum* Thom ERK1 was reported to occur via the *beta*-ketoadipate pathway and the *ortho* fission of catechol, respectively (Marr et al., 1989; Wolski et al., 2012). In a study performed by Cajthaml et al. (2002) on biodegradation of different polynuclear aromatic hydrocarbon derivatives by lignolytic fungus *I. lacteus*, the major degradation products of anthracene and phenanthrene were reported to be anthraquinone and phenanthrene-9,10-dihydrodiol, respectively (Fig. 9.3). Two major products were identified from fluoranthene biodegradation: 1,8-naphthalic anhydride and 2-formyl-acenaphthen-1-carboxylic acid methylester. Whereas, the major product for pyrene was the lactone of 4-hydroxy-5- phenanthrenecarboxylic acid. *P. chrysosporium* is reported to partially degrade BaP to carbon dioxide (Bumpus et al., 1985). *Armillaria* sp. F022, a white-rot fungus isolated from a tropical rain forest in Samarinda, Indonesia, is reported to biodegrade BaP, via 1,6-dihydroxy-BaP to BaP-1,6-quinone, which is then degraded to 1-hydroxy-2-naphthoic acid and benzoic acid. Laccase is reported also to play a key role in the BaP biodegradation by the white-rot fungi *Armillaria* sp. F022. The biodegradation efficiency is observed to be increased by a 2.5-fold in presence of glucose, as a cosubstrate, since glucose significantly stimulate the laccase production. Cajthaml et al. (2006) reported also the biodegradation of benz[a]anthracene by the lignolytic fungus *I. lacteus*, where the two major identified products were benzo[a]anthraquinone and 1-tetralone (Fig. 9.4). The methylating system in this group of fungi, as, elucidated in polynuclear aromatic hydrocarbon derivative biodegradation pathways (Figs. 9.3 and 9.4A), prevents futile the redox cycling of

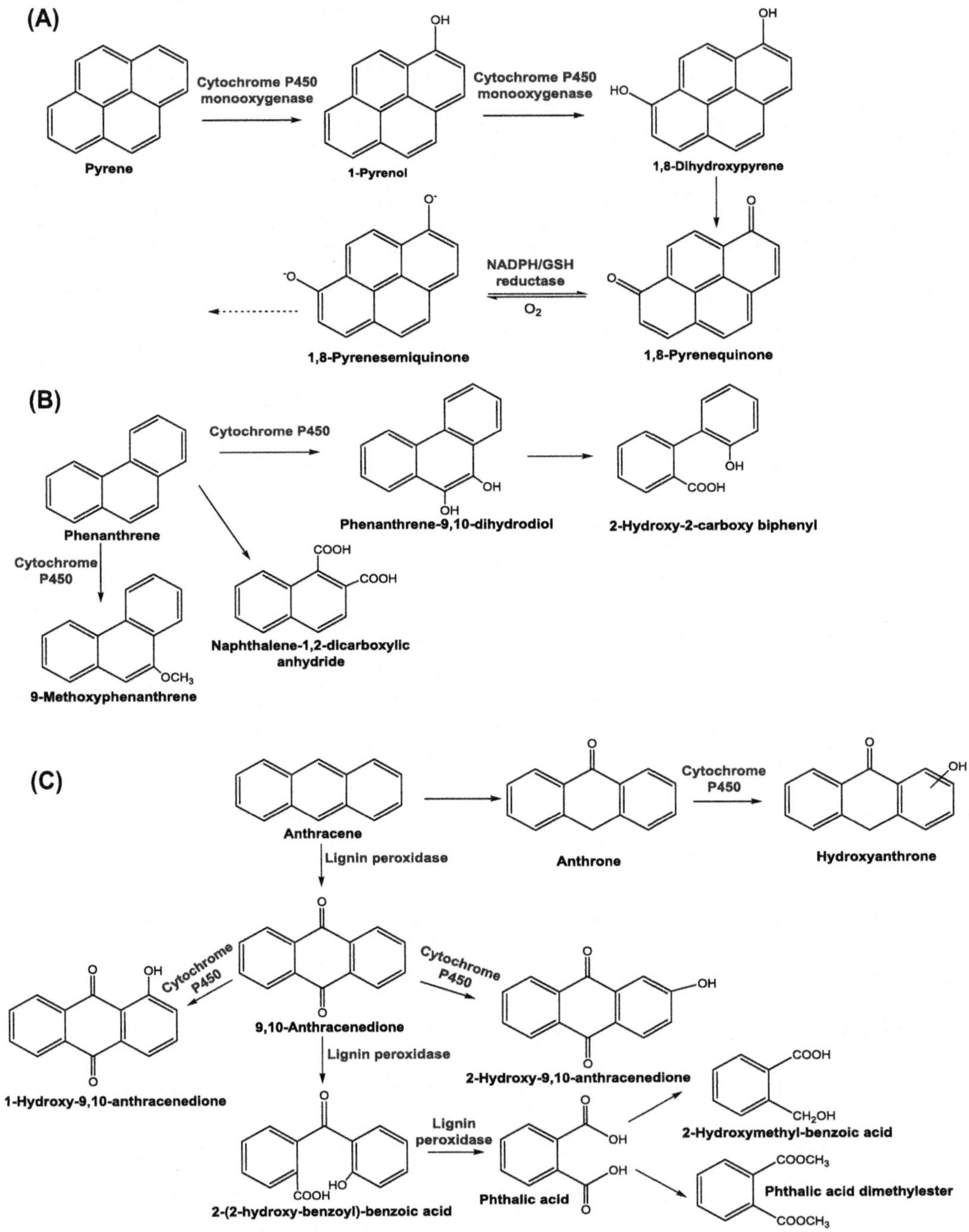

FIGURE 9.3

Suggested mycodegradation of (A) pyrene by *Penicillium janthinellum* SFU403 (Launen et al., 1995, 1999); (B) phenanthrene; and (C) anthracene by *Irpex lacteus* Cajthaml et al. (2006).

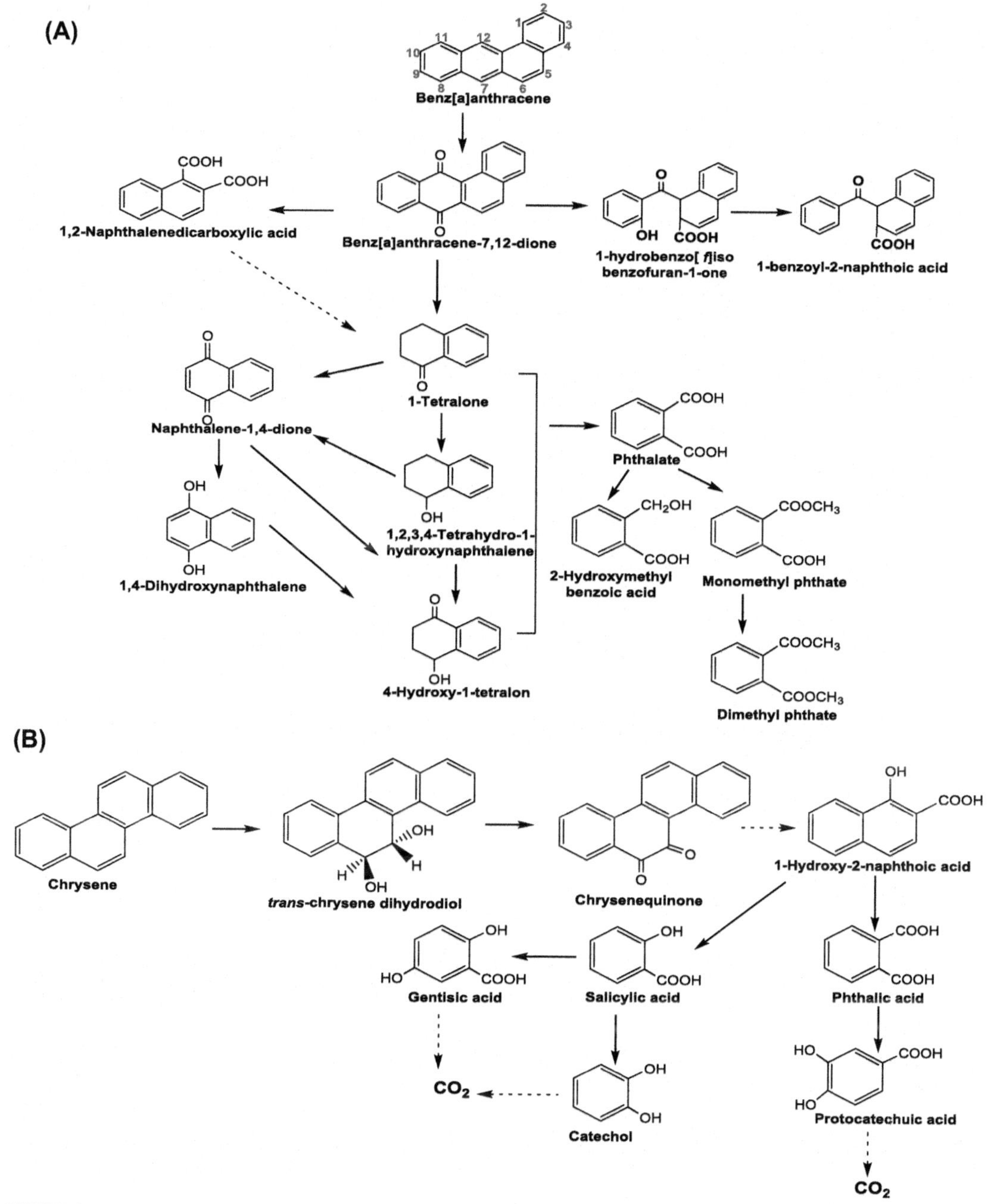

FIGURE 9.4

Proposed aerobic mycodegradation of (A) benzo[a]anthracene and (B) chrysene.

peroxidases and allows oxidation of the aromatic ring and methylation decreases the water solubility of potentially toxic compounds; thus, it can be also assumed as an auto-protection mechanism of the organism. *Polyporus* sp. S133 isolated from a petroleum-contaminated soil in Matsuyama city, Ehime, Japan has been reported for biodegradation of chrysene using MnP, LiP, laccase, 1,2-dioxygenase, and 2,3-dioxygenase throughout the production of chrysenequinone, 1-hydroxy-2-naphthoic acid, phthalic acid, salicylic acid, protocatechuic acid, catechol acid, and gentisic acid, which can enter into the tri-carboxylic acid (TCA) cycle (Fig. 9.4B). The addition of Tween 80 as a surfactant increased the rate of biodegradation by two-fold. Moreover, the addition of 10% glucose and polypeptone as a cosubstrate and nitrogen source, respectively, enhanced the rate of chrysene biodegradation, as they enhance the laccase production and activity (Hadibarata et al., 2009). In a study performed by Silva et al. (2009) under low-oxygen conditions, low-molecular-weight polynuclear aromatic hydrocarbon derivatives (2–3 rings) are reported to be degraded most extensively by *Aspergillus* sp., *Trichocladium canadense*, and *Fusarium oxysporum*. While, for high-molecular-weight polynuclear aromatic hydrocarbon derivatives (4–7 rings), maximum degradation has been observed with *T. canadense*, *Aspergillus* sp., *Verticillium* sp., and *Achremonium* sp. Four fungal strains; *Aspergillus niger*, *Penicillium documbens*, *Cochliobolus lutanus*, and *Fusarium solani*, were isolated from oil polluted beach sand collected from Pensacola beach (Gulf of Mexico) for their ability to degrade crude oil (Al-Nasrawi, 2012). *Corollospora maritima*, *Lulwoana* sp., *Gibberella stilboides*, *Hematonectria haematococca*, *Penicillium expansum*, *Alternaria tenuissima*, *Niesslia exilis*, *Capronia kleinmondensis*, *Gibellulopsis nigrescens*, and *Cadophora luteo-olivacea* are reported to grow on crude oil as sole source of carbon and energy (Garzoli et al., 2015). The halotolerant *P. chrysogenum* var. *halophenolicum* is reported to efficiently degrade high concentrations of hydroquinone under hypersaline conditions (Pereira et al., 2014). *Aspergillus flavus*, *Aspergillus fumigatus*, *A. niger*, *Aspergillus wentii*, *Collectotricum* sp., *Mortierella* sp., *Penicillium* sp., and *Trichoderma* sp. isolated from leaf surfaces of selected plants in Nigeria, such as *Alcornea cordifolia*, *Lantana camara*, *Mangnifera indica*, *Manihot esculenta*, and *Panicum maximum*, were found to be capable to degrade crude oil and some other petroleum products, such as diesel, kerosene, spent, and unspent engine oil (Adeogun and Adekunle, 2015).

The white-rot fungus *Pleurotus eryngii* F032 is reported to degrade fluorene via 9-fluorenone, phthalic acid, and benzoic acid (Arata and Kristanti, 2014). *Ascomycota* phylum fungi *Fusarium* sp. GS_I1, *Scopulariopsis* sp. GS_I2, and *Aspergillus* sp. DI_I1, isolated from oil-soaked sand patties collected from polluted beaches by the Deepwater Horizon oil spill that was occurred on 2010, are reported to express good biodegradation capabilities for crude oil over wide temperature range 20–40°C, with the preference of n-alkanes over iso-alkanes, short chain alkanes over long ones, and low-molecular-weight polynuclear aromatic hydrocarbon derivatives over the high-molecular-weight polynuclear aromatic hydrocarbon derivatives (Simister et al., 2015). The basidiomycete *P. ostreatus* D1 is reported to change its polynu-clear aromatic hydrocarbon derivative metabolic system according to the culture medium composition. For example, fluorene and fluoranthene degradation are transformed in Kirk's medium (under conditions of laccase production) to 9-fluorenone and quinone metabolite, respectively. However, complete degrada-tion with the formation of an intermediate metabolite, phthalic acid, which undergoes subsequent metab-olism, occurs in basidiomycete-rich medium (under the production of both laccase and versatile peroxidase). Both extracellular laccase and laccase on the mycelium surface can participate in the initial stages of polynuclear aromatic hydrocarbon derivative metabolism, while versatile peroxidase is respon-sible for the oxidation of the formed metabolites (Pozdnyakova et al., 2016). Guntupalli et al. (2016)

reported the biodegradation of BaP and Benzo[k]floranthene by a microbial consortium of *Acinetobacter calcoaceticus* (MTCC2409, 2289), *Serratia morcescens* (MTCC2645), *Pseudomonas* species (MTCC 2445), *S. maltophilia* (MTCC2446), and *Aspergillus terricola* var. *americanus* (MTCC2739).

The presence of *A. calcoaceticus* and *S. morcescens* is important for biosurfactant production, which enhances the rate of polynuclear aromatic hydrocarbon derivative degradation, although both strains did not show high biodegrading activity for the studied polynuclear aromatic hydrocarbon derivatives. The consortium utilized three metabolic pathways; cytochrome P450 monooxygenase with the detection of epoxy derivatives leading to phenols, *trans*-dihydrodiols, *cis*-dihyrodiols leading to salycyclic acid and catechol by bacterial dioxygenases and via quinones of ligninolytic pathway. *Fusarium* sp. F092 has been reported to degrade chrysene as carbon and energy source (Hidayat and Tachibana, 2015). The white rot fungi *Pleurotus sajor-caju mycelia* better tolerate pyrene than chrysene, and complete mineralization of pyrene is achieved by *P. sajor-caju* supplemented with Tween 80 and copper sulfate (Saiua et al., 2016).

3.0 ANAEROBIC BIOTRANSFORMATION

Anaerobic microorganisms use nitrate, iron, sulfate, manganese, and, more recently, chlorate as electron acceptors, during the anaerobic degradation, which is coupled to methanogenesis, fermentation, and phototrophic metabolism but growth of these microorganisms and their biodegradation rates are significantly lower than the aerobic degraders (Widdel and Rabus, 2001). Crude oil is a setting characterized by the presence of many microorganisms, for example, sulfate reducing bacteria that cause complete oxidation of organic compounds to CARBON DIOXIDE or incomplete oxidation of hydrocarbon compounds to acetate groups as well as iron reducing bacteria and methanogenic archaea. In-situ biodegradation can take place also in crude oil reservoirs for the light hydrocarbon fraction, e.g., n-paraffins, iso-paraffins, cycloalkanes, as well as benzene and alkylbenzens under anaerobic conditions (Fig. 9.5) (Vieth and Wilkes, 2005). Sulfate-reducing bacteria are some of the oldest microorganisms on earth. Their initial development and activity goes back to the Proterozoic Era (Rabus et al., 2000). They are Gram negative, except the *Desulfonema* sp. (Madigan et al., 2006), and are heterotrophic organisms and absolute anaerobes that can use sulfate and oxygenated sulfur compounds (sulfites, thiosulfites, trithionate, tetrathionate, and elemental sulfur) as the terminal electron acceptor in their energy metabolism and respiration process. They can be psychro-, meso-, thermophilic, halo-, and barophilic. Some species of sulfate-reducing bacteria, such as *Desulfosporosinus orientis*, *Desulfotomaculum halophilum* sp. nov., and *Desulfosporosinus meridiei* sp. nov. (Tardy-Jacquenod et al., 1998; Robertson et al., 2001) are reported to be spore-formers.

Sulfate-reducing bacteria isolated from crude oil and formation waters are characterized by a wide tolerance range in relation to salt content (0%–17%) and temperature (4–85°C) (Magot et al., 2000). Conventional sulfate-reducing microbes fall into the following phylogenetic lineages: (1) the families *Desulfoarculaceae*, *Desulfobacteraceae*, *Desulfobulbaceae*, *Desulfohalobiaceae*, *Desulfomicrobiaceae*, *Desulfonatronumaceae*, *Desulfovibrionaceae*, *Syntrophaceae*, *Syntrophobacteraceae*, and *Syntrophorhabdaceae* are within the *Deltaproteobacteria*; (2) the genera *Desulfotomaculum*, *Desulfosporomusa*, *Desulfosporosinus* (Class *Clostridia*) and *Thermodesulfobium* (Family *Thermodesulfobiaceae*) are within the phylum *Firmicutes*; and (3) the genus *Thermodesulfovibrio* is

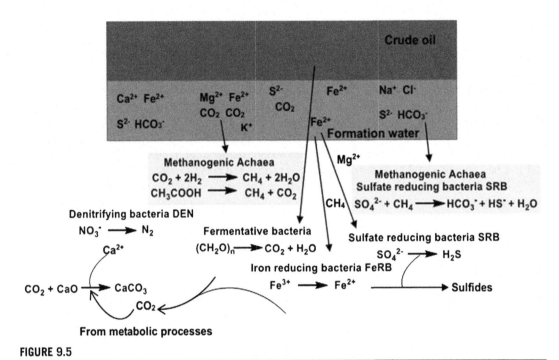

FIGURE 9.5

Possible biodegradation in a crude oil reservoir.

within the phylum *Nitrospirae*. Moreover, there is an additional group of thermophilic sulfate–reducing bacteria, which were given their own phylum, the *Thermodesulfobacteria*. Three more genera of *Archaea* are also known sulfate-reducers: *Archaeoglobus*, *Thermocladium*, and *Caldivirga* (Mori et al., 2003; Thauer et al., 2007; Muyzer and Stams, 2008; Barton and Fauque, 2009). It is important to have an idea about pathways involved in alkane metabolism coupled to sulfate reduction. Sulfate-reducing microbes are known to obtain its energy by oxidizing organic compounds or molecular hydrogen while reducing sulfate to hydrogen sulfide (Sherry et al., 2013):

$$C_{16}H_{34} + 12.25SO_4^{2-} + 8.5H^+ \rightarrow 16HCO_3^- + 12.25H_2S + H_2O$$

During the sulfate reduction, the hydrogen sulfide produced can be oxidized to thiosulfate, which can enrich the thiosulfate-reducing hydrocarbon degraders in biotreatment microcosms. The produced thiosulfate can be reduced back to HS^-, oxidized to sulfate, and disproportionated to hydrogen sulfide$^-$ and sulfate in a thiosulfate shunt characteristic of anoxic sediments (Jorgensen, 1990).

$$C_{16}H_{34} + 12.25S_2O_3^{2-} + 11.25H_2O \rightarrow 16HCO_3^- + 24.5S^{2-} + 40.5H^+$$

Sulfate-reducing microbes create problems when metal structures and pipelines are exposed to sulfate-containing water, since, the interaction between the metal and the water creates a layer of molecular hydrogen on the metal surface that is subsequently oxidized with sulfate-producing hydrogen sulfide, which leads to corrosion throughout cracking and pitting, in addition to the health and safety concern of this toxic gas (Neria-Gonzalez et al., 2006). Not only this, but souring of crude oil is mainly caused due to the presence of sulfate reducing microbes. This would occur during the secondary oil recovery when water, which would contain sulfate, is injected to repressurize the reservoir. Thus, the sulfate-reducing microbes will reduce the sulfate to sulfide whilst oxidizing the organic electron donors present in the crude oil (Hubert and Voordouw, 2007).

The *Desulfovibrionaceae* and *Desulfobacteriaceae* are reported to be abundant in Canadian oil fields, when secondary oil recovery is prompted by water flooding (Voordouw et al., 1996). Sulfate-reducing microbes have been reported to anaerobically oxidize n-alkanes completely to carbon dioxide (Aeckersberg et al., 1991, 1998; Reuter et al., 2002; Davidova et al., 2006). *Desulfococcus oleovorans* strain Hxd3 and strain Pnd3, isolated from an oil tank and from petroleum contaminated marine sediment respectively, are reported to be able to grow on C_{12}–C_{20} n-alkanes whilst reducing sulfate to sulfide (Aeckersberg et al., 1991, 1998). *Desulfoglaeba alkanexedens* strain ALDC^T and strain Lake, isolated from a naval oily wastewater storage facility and oil field production water respectively, are also n-alkane degraders (Davidova et al., 2006). A marine sulfate-reducing bacterium, *Desulfatibacillum aliphaticivorans* strain CV2803^T, isolated from a hydrocarbon-polluted sediment, is also reported to be to degrade n-alkanes nC_{13}–nC_{18} and also n-alkenes nC_7–nC_{23} (Cravo-Laureau et al., 2004).

Also, *Desulfatibacillum alkenivorans* AK-01 has been reported as alkane degrader (Callaghan et al., 2006, 2012). Under anaerobic conditions and in sulfate-containing environments, other microorganisms rather than the sulfate-reducing bacteria would play a role in the degradation of crude oil. Bioremediation of oil-polluted sediments on microcosm level under anaerobic conditions and sulfate-reducing reduction were studied and revealed good degradation of linear-alkanes (nC_7–nC_{34}) over a 686-day period. During the first 176 days of incubation, the predominant microbial population was from *Gammaproteobacteria* that were most closely related to *Marinobacterium* sp. and members of the family *Peptostreptococcaceae* within the *Firmicutes*. However, after 302 days, when sulphate was depleted and the majority of n-alkane degradation had already occurred, a shift in community composition towards *Chloroflexi* (family Anaerolineaceae), together with *Firmicutes* and the more conventional sulfate-reducing microorganisms *Deltaproteobacteria* occurred (Sherry et al., 2013). Within the depletion of sulfate and other chemical electron acceptors, such as other oxygenated sulfur compounds, the transfer of electrons on a biological acceptor may take place by using hydrogen as is the case for methanogenic archaea, which helps in the persistence of sulfate-reducing bacteria that depleted availability of electron acceptors, in a process known as syntrophic growth (Nazina et al., 2007).

Thus, the biodegradation of oil within the reservoir rock may cause the formation of an upper gas cap, which is composed mainly of methane that is formed during the reduction of carbon dioxide with hydrofen or methyl acetate. If the oil contains high quantities of sulfur, the activity of sulfate-reducing bacteria would result in the formation of hydrogen sulfide at concentrations that may reach over 10% in the gas. Methanogenes are mesophilic organisms, for example *Methanobacterium thermoautotrophicum*, that are very sensitive to even very low concentrations of oxygen and their activity is measured by the methane production rate or by the volume of produced methane; they use hydrogen as the electron donor and carbon dioxide as the electron acceptor, with low energy yield. However, extremophiles are also present, which can survive up to 110°C, for example *Methanopyrus kandleri* and *Methanococcus vulcanicus* and *Methanococcoides euhalobius* or *Methanohalophilus euhalobius*, which survived in high salt concentrations up to 140 NaCl/L, which has the ability to utilize methyamines, which are attributed to amine degradation that would occur during osmoregulation. Other examples of hydrogen oxidizing methanogenic species are the disc-shaped *Methanococcus termolitotrophicus*, *Methanoplanus petrolearius*, and the halotolerant *Methanocalculus halotolerans*, which can tolerate up to 12% salt concentration, and the rod-like *M. thermoautotrophicum*, *Methanobacterium bryantii*, and *Methanobacterium ivanovii* (Ollivier et al., 1988). Similarly, the activity of other microbes may result in the transformation of primarily wet gas into dry gas containing over 90% methane. This process is linked to the preferential removal of C3–C5 components by bacteria; thus the residual methane will be isotopically heavier and the iso-alkane to n-alkane ratio will be increased. However, the degradation of the light oil fractions is unfavorable, as it affects physical and chemical properties of the petroleum, resulting in a decrease of its hydrocarbon content and an increase in oil density, sulfur content, acidity, and viscosity. That leads to a negative impact on oil production and refining operations, thus decreasing the economic value of the crude oil (Roling, 2003; Vieth and Wilkes, 2005). *Shewanella putrefaciens* is an iron-reducing bacterium, which reduces elemental sulfur, sulfites, and thiosulfates to sulfides and can tolerate the harsh conditions of oil reservoirs. The electron donor may be hydrogen or formate, and the acceptors are the iron oxides and hydroxides.

Deferribacter thermophilus is a bacterium that reduces iron, manganese, and nitrates using yeast extract or peptone, where hydrogen and numerous organic acids are the source of energy (Greene et al., 1997). Within the nonaromatic hydrocarbons, bacteria degrade first n-paraffins and then iso-paraffins; therefore an increase of the iso-C5/n-C5 ratio can be used as an indicator of the biodegradation progress. In the case of methylcyclohexane/n-C7 ratio, the same trend also indicates a larger degree of oil degradation. Also, the increase of hopane acids in crude oil would indicate its advanced biodegradation. Degree of oil biodegradation can be followed also by the isotopic ratios of particular compounds. Also, the higher the in-situ degradation, the higher the acidity of the crude oil that would be >0.5 g KOH/g oil. However, high acidity is also recorded with nondegraded crude oil that is characterized by high sulfur content (Meredith et al., 2000; Vieth and Wilkes, 2005).

Anaerobic microorganisms utilize monocyclic aromatic hydrocarbons, such as BTEX, hexadecane, and naphthalene as the sole carbon source. Strains RCB and JJ of *Dechloromonas* (β-*Proteobacteria*) have been reported for benzene anaerobic degradation using nitrate as the electron acceptor. *Geobacter metallidurans* and *Geobacter grbicium* are reported to be capable of anaerobic toluene oxidation to carbon dioxide with the reduction of Fe(III). Some other organisms are reported for anaerobic toluene degradation through nitrate respiration: *Thauera aromatica* strains K172 and I1, *Azoarcus* sp. strain T, *A. tolulyticus* strains To14 and Td15, *Dechloromonas* strains RCB and JJ, through perchlorate respiration; *Dechloromonas* strains RCB and JJ and sulfate respiration *Desulfobacterium cetonicum*, *Desulfobacula toluolica* (Chakraborty and Coates, 2004). A *methanogenic* consortium composed of two archaea species related to *Methanosaeta* and *Methanospirillum* and two bacterial species, of *Desulfotomaculum*, are reported to degrade toluene (Beller et al., 1992).

Sulfate-reducing bacteria are reported to utilize benzene, toluene, ethylbenzene, and xylene as the sole carbon source (Coates and Chakraborty, 2002; Kniemeyer et al., 2003; Ribeiro de Nardi et al., 2007). Most of the anaerobic biodegradation were reported to the diaromatic rings (e.g., naphthalene and 2-methylnaphthalene) and a little about the three polyaromatic rings compounds (e.g., phenanthrene), which would be degraded through an initiating carboxylation reaction. It could also occur through cometabolism; for instance, benzothiophene is biodegraded in the presence of naphthalene as the subsidiary substrate. Anaerobic degradation for heterocyclic compounds such as indole or quinoline has been also reported (Meckenstock et al., 2004; Widdel and Rabus, 2001). Cycloalkanes account for 20%–40% of the total hydrocarbon fractions in crude oils. Cyclohexane is a relatively volatile hydrocarbon (boiling point 80.7°C), characterized by a low solubility in water (0.68 mM at 25°C) and is one of the abundant cycloalkanes in crude oil and usually used as a model compound for the biodegradation of cycloalkanes (Tissot and Welte, 1984; Dean, 1992). *Desulfosarcina-Desulfococcus* cluster of the *Deltaproteobacteria* is reported for biodegradation of cyclohexane and other cyclic and n-alkanes, including the gaseous alkane n-butane, with stoichiometric reduction of sulfate to sulfide (Jaekel et al., 2015).

The degradation pathway of cyclohexane under anaerobic conditions is reported to be analogous to that of n-alkanes using glycyl radical enzymes (Fig. 9.6) (Jarling et al., 2012). The most common initial reaction during the anaerobic degradation of saturated and aromatic hydrocarbons that has been detected in several physiological types of anaerobes is a radical reaction of hydrocarbons with fumarate, yielding substituted succinates (Fig. 9.7). However, *Azoarcus* strains are reported to utilize ethylbenzene, through dehydrogenation to 1-phenylethanol (Fig. 9.7) and acetophenone, carboxylation and activation to yield 3-oxo-3-phenylpropionyl-CoA, and thiolytic cleavage to acetyl-CoA and benzoyl-CoA (Rabus and Widdel, 1995; Ball et al., 1996; Champion et al., 1999; Spormann and Widdel, 2000), and sulfate-reducing bacteria are reported to degrade naphthalene and phenanthrene, through the initial carboxylation to 2-naphthoate and phenanthrene carboxylate, respectively. The 2-naphthoate (presumably as activated acid) would be further metabolized through subsequent reduction of the two rings to yield decalin-2-carboxylate (Zhang and Young, 1997; Galushko et al., 1999; Meckenstock et al., 2000), whereas, alkylnaphthalenes and alkylbenzens are reported to be

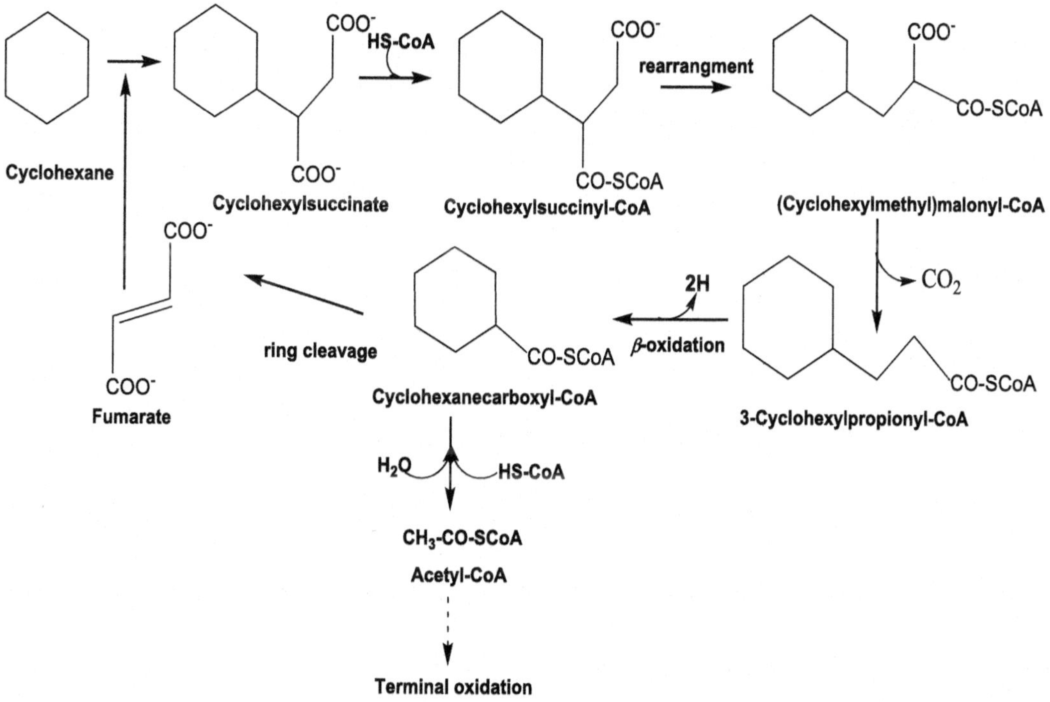

FIGURE 9.6

Proposed pathway for the anaerobic degradation of cyclohexane (Jaekel et al., 2015).

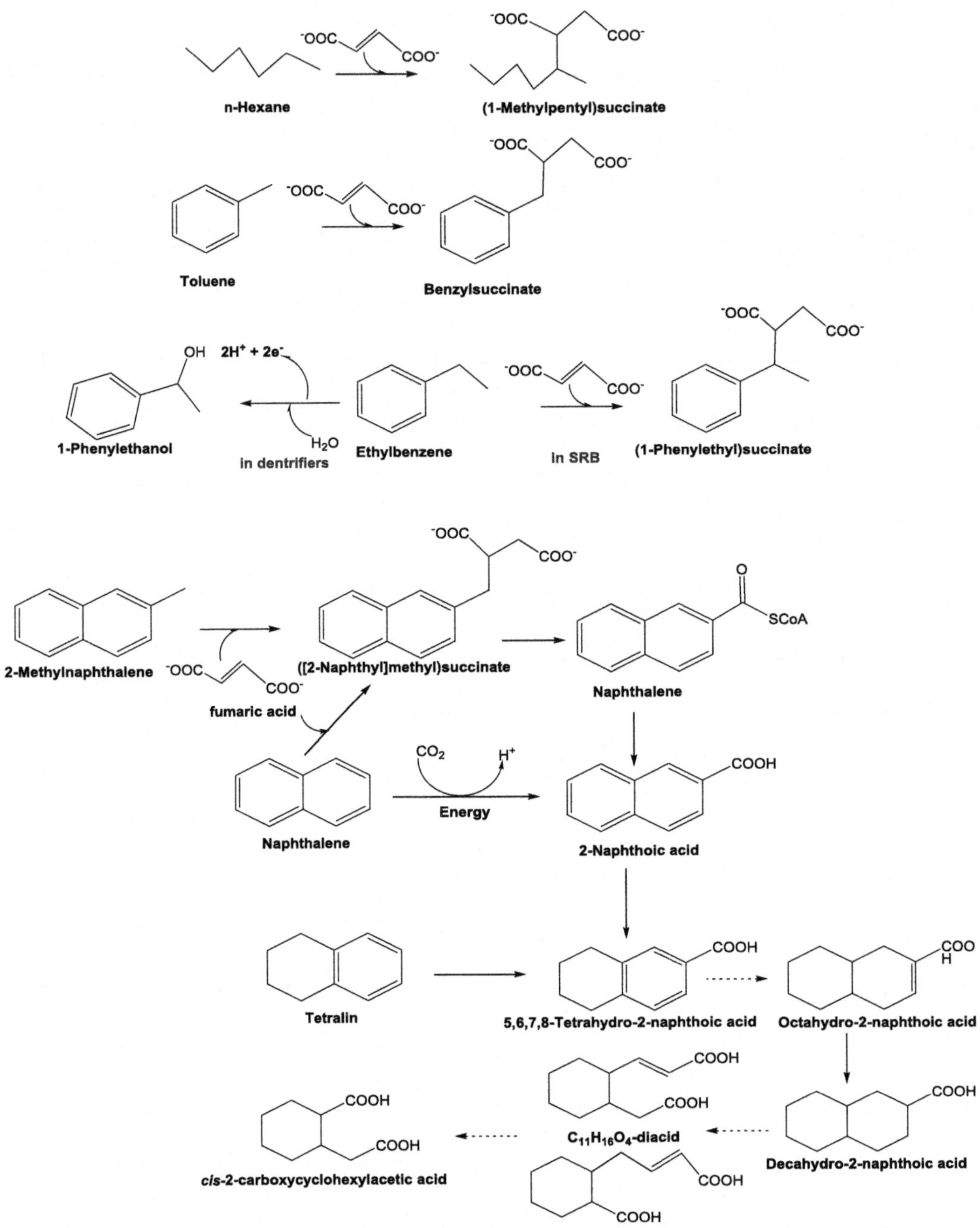

FIGURE 9.7

The proposed initial reactions during the anaerobic degradation of saturated and aromatic hydrocarbons.

easily activated by sulfate-reducing bacteria than their nonalkylated analogs. For example, 2-methylnaphthalene is reported to be metabolized to naphthyl-2-methylsuccinic, through an activation mechanism analogous to that of toluene (Annweiler et al., 2000a).

$$2C_6H_{12} + 9SO_4^{2-} \rightarrow 12HCO_3^- + 9HS^- + 3H^+$$

Anaerobic biodegradation of low-molecular-weight polynuclear aromatic hydrocarbon derivatives (two and three aromatic rings) under nitrate- and sulfate-reducing conditions has been reported (Haritash and Kaushik, 2009). Fumarate addition occurs for the metabolism of 2-methylnaphthalene under sulfate-reducing conditions, where, naphthyl-2-methylsuccinate (-succinyl-CoA) is formed and further metabolized to 2-naphthoate (2-naphthoyl-CoA) in a manner similar to that elucidated for anaerobic toluene metabolism. 2-Naphthoate can be further metabolized through ring reduction, forming, 5, 6, 7, 8-tetrahydro-2-naphthoate, then further cleaved and mineralized (Selesi et al., 2010). Annweiler et al. (2002) reported the anaerobic degradation of naphthalene, 2-methylnaphthalene, and tetralin (1,2,3,4-tetrahydronaphthalene) by a sulfate-reducing enrichment culture obtained from a contaminated aquifer, with the production of cyclohexane ring, with two carboxylic acid side chains, $C_{11}H_{16}O_4$-diacid. The side chain is elucidated to be either an acetic acid and a propenic acid or a carboxy group and a butenic acid side chain. The β-oxidation of one of those side chains led to the production of 2-carboxycyclohexylacetic acid (Fig. 9.7). Whereas, Berdugo-Clavijo et al. (2012) reported the methanogenic metabolism of 2-methylnaphthalene and 2, 6-dimethylnaphthalene with the formation of 2-naphthoic acid and 6-methyl-2-naphthoic acid, respectively.

Anaerobic biodegradation of flourene and phenathrene by sulfate-reducing bacteria have been reported through a sequence of hydration and hydrolysis reactions followed by decarboxylation with the formation of p-cresol (only in the phenanthrene system) and phenol and then mineralization (Fig. 9.8) (Tsai et al., 2009). The rate of anaerobic biodegradation is low but it is reported to be enhanced by the addition of acetate, lactate, pyruvate, sodium chloride, cellulose, or zerovalent iron (Chang et al., 2008). Schmitt et al. (1996) hypothetically elucidated the mechanism of anaerobic biodegradation into three major steps; initially high concentration of aromatic hydrocarbons is partly degraded under nitrate- and sulfate-reducing conditions to form low-molecular-weight organic acids as metabolic intermediates that can act as ligands complexing insoluble Fe (III) oxides in the aquifer and mobilizing Fe (III). Then the mobilized Fe (III) become available for iron-reducing bacteria and intensifies the degradation of aromatic hydrocarbons. The major intermediates of the anaerobic biodegradation of polynuclear aromatic hydrocarbon derivatives are benzoate (or benzoyl-CoA) and, to a lesser extent, resorcinol and phloroglucinol. Carboxylations, decarboxylations, hydroxylations, reductions, reductive dehydroxylations, deaminations, dechtorinations, aryl ether cleavages, and lyase are the common reactions involved in anaerobic biodegradation of polynuclear aromatic hydrocarbon derivatives producing central intermediates. The aromatic central intermediates are reported to be reductively attacked and cleaved by hydrolysis to noncyclic products to be further transformed by p-oxidation to central metabolites (Evans in 1977). Anaerobic polynuclear aromatic hydrocarbon derivatives degradation in soil by a mixed bacterial consortium under denitrifying conditions has been reported, where, the anaerobic biodegradation of fluorene, phenanthrene, and pyrene, occurred through fermentative and respiratory metabolism (Ambrosoli et al., 2005).

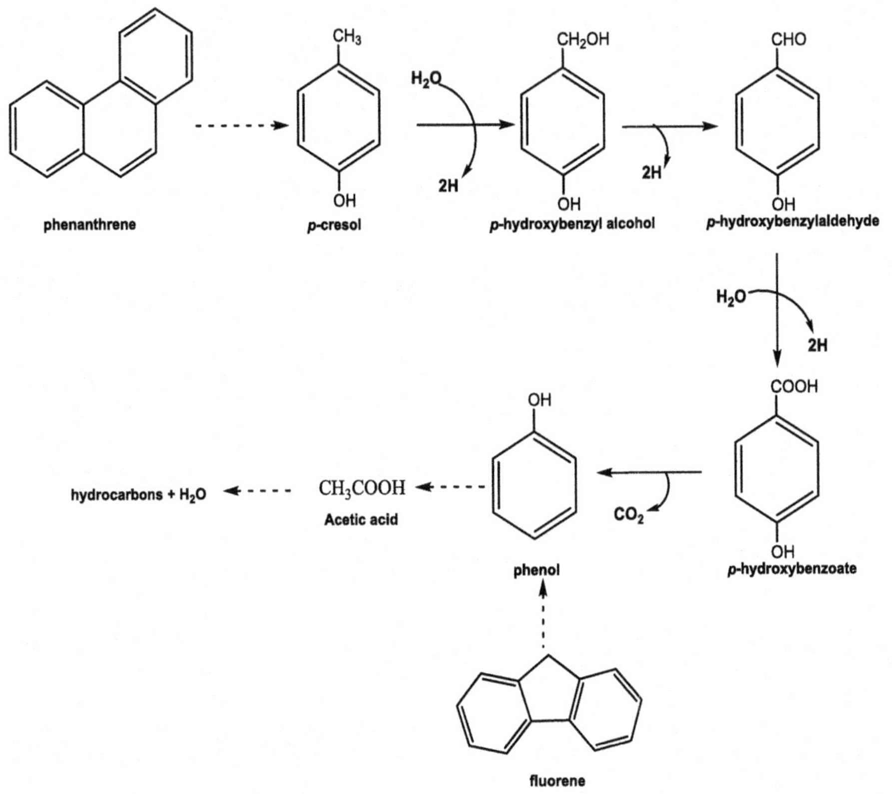

FIGURE 9.8

Anaerobic biodegradation of polynuclear aromatic hydrocarbon derivatives by sulfate-reducing bacteria (Tsai et al., 2009).

4.0 AEROBIC BIOTRANSFORMATION

Aerobic biodegradation is more rapid than anaerobic one. It is mainly initiated by oxygenases (mono- or di-) and peroxidases, depending on the type of alkyl side chain of the aromatic hydrocarbons or the type of microorganism (Fig. 9.9), followed by peripheral degradation pathways that convert organic pollutants step by step into intermediates that enters the TCA cycle (Fig. 9.10). Microorganisms degrade or transform contaminants by a variety of mechanisms.

Aalkanes—linear (n-alkanes), cyclic (cyclo-alkanes), or branched (isoalkanes)—constitute up to 50% of crude oil, depending on the oil source, and can also be produced by many living organisms such as plants, green algae, bacteria, or animals. Although, they are characterized by very low water solubility and it decreases with the increase in molecular weight, for example, the solubility of hexane and hexadecane is 1.4×10^{-4} M and 2×10^{-10} M, respectively, which would hamper their microbial uptake. But, it is reported that they can be biotransformed to carbon dioxide and water and can be used as a primary food source by the bacteria, which use the energy to generate new cells. Thus

$$2C_{12}H_{26} + 37O_2 \rightarrow 24CO_2 + 26H_2O$$

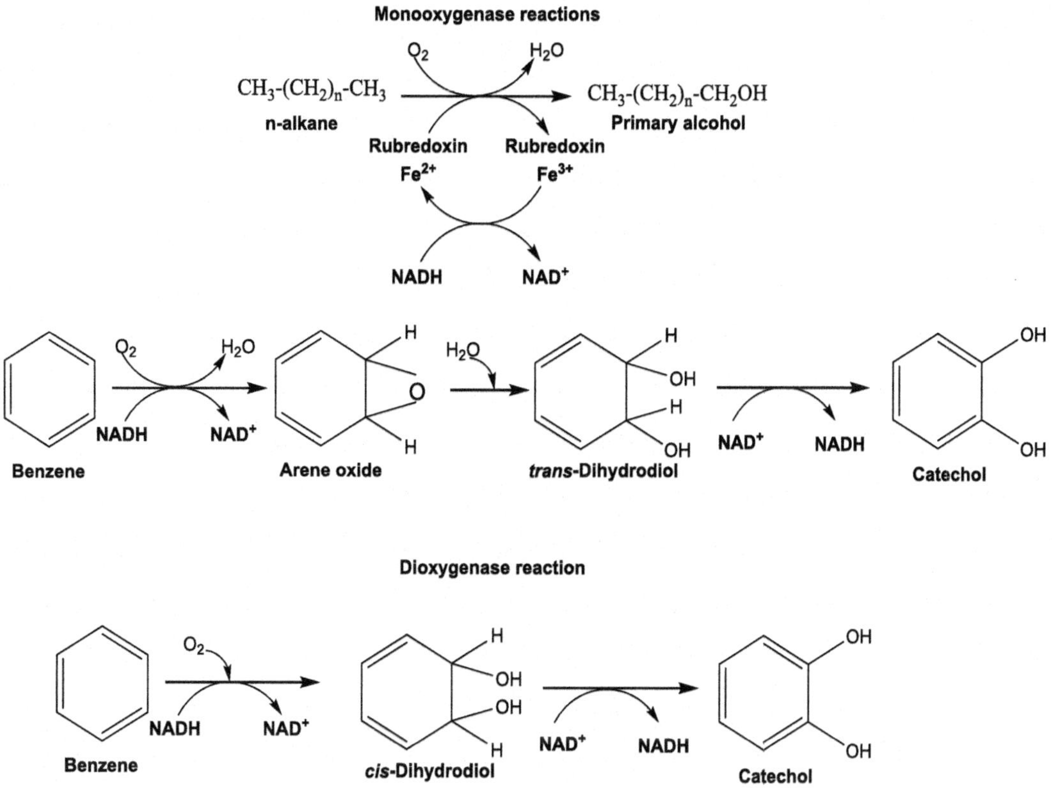

Monooxygenase reactions

FIGURE 9.9

Possible oxygenases reactions involved in the aerobic biodegradation of hydrocarbons.

The central precursor metabolites, acetyl-CoA, succinate, and pyruvate, enter in the biosynthesis of the cell biomass. Sugars are also required for various biosynthesis and microbial growth occurs through gluconeogenesis. Other processes are involved in the biodegradation, such as attachment of microbial cells to the substrates (e.g., the oil droplets) and production of biosurfactants, which enhances the rate of biodegradation as it increases the availably of the hydrocarbons pollutants to the microbial cells. Most alkane-degrading bacteria secrete diverse surfactants that facilitate emulsification of the hydrocarbons (Ron and Rosenberg, 2002; Caiazza et al., 2005; Ali et al., 2014).

Hydrocarbonoclastic bacteria, for example, *Alcanivorax*, *Thalassolituus*, *Oleispira*, and *Oleiphilus* species, are highly specialized in degrading hydrocarbons (linear and branched alkanes) but are unable to metabolize aromatic hydrocarbons, sugars, amino acids, fatty acids, and most other common carbon sources (Golyshin et al., 2002; Yakimov et al., 2003, 2004; Schneiker et al., 2006; Yakimov et al., 2007).

Cytochrome P450 enzymes, integral membrane di-iron alkane hydroxylases (e.g., alkB), soluble di-iron methane monooxygenases, and membrane-bound copper-containing methane monooxygenases are known to be the alkane oxygenase systems in prokaryotes and eukaryotes, which participate in the

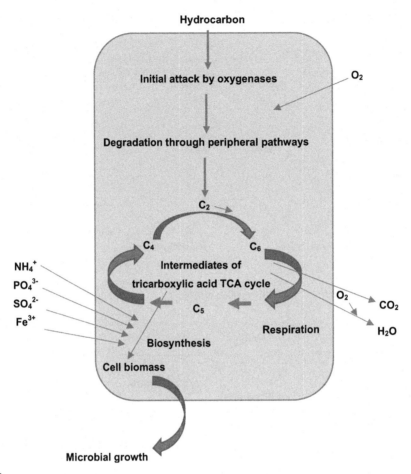

FIGURE 9.10

The main criteria in aerobic biodegradation process.

aerobic degradation of alkanes (Van Beilen and Funhoff, 2005). Cytochromes P450 are hemoproteins that hydroxylate a large number of compounds; they require a ferredoxin and a ferredoxin reductase that transfer electrons from NAD(P)H to the cytochrome.

Methane can be initially oxidized to methanol that is subsequently transformed to formaldehyde and then to formic acid, which can be further mineralized to carbon dioxide. Formaldehyde can be also assimilated for biosynthesis of multicarbon compounds either by the ribulose monophosphate pathway or by the serine pathway (Fig. 9.11). *P. putida* OCT has been reported to degrade n-alkanes using alkane hydroxylase, which composed of the membrane-bound oxygenase and two soluble components called rubredoxin and rubredoxin reductase. It produces n-alkanols that are further oxidized by a membrane-bound alcohol dehydrogenase to n-alkanals. The n-alkanals are subsequently transformed to fatty acids and then to acyl-CoA by aldehyde dehydrogenase and acyl-CoA synthetase, respectively (Fig. 9.11) (Van Beilen et al., 1994). An *Acinetobacter* strain M-1 has been reported to degrade

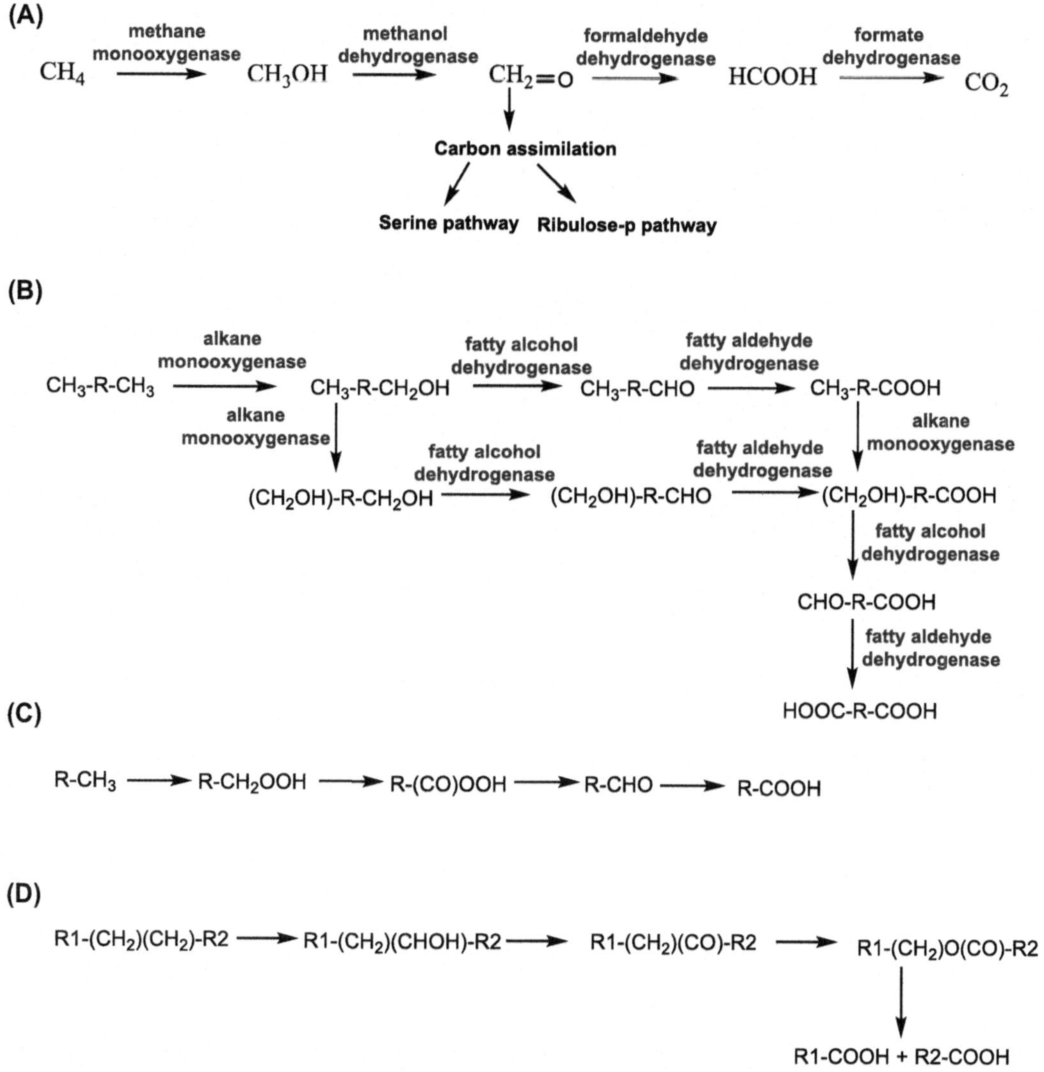

(A)

$$CH_4 \xrightarrow[\text{monooxygenase}]{\text{methane}} CH_3OH \xrightarrow[\text{dehydrogenase}]{\text{methanol}} CH_2{=}O \xrightarrow[\text{dehydrogenase}]{\text{formaldehyde}} HCOOH \xrightarrow[\text{dehydrogenase}]{\text{formate}} CO_2$$

Carbon assimilation

Serine pathway   Ribulose-p pathway

(B)

$$CH_3\text{-}R\text{-}CH_3 \xrightarrow[\text{monooxygenase}]{\text{alkane}} CH_3\text{-}R\text{-}CH_2OH \xrightarrow[\text{dehydrogenase}]{\text{fatty alcohol}} CH_3\text{-}R\text{-}CHO \xrightarrow[\text{dehydrogenase}]{\text{fatty aldehyde}} CH_3\text{-}R\text{-}COOH$$

alkane monooxygenase ↓   alkane monooxygenase ↓

$$(CH_2OH)\text{-}R\text{-}CH_2OH \xrightarrow[\text{dehydrogenase}]{\text{fatty alcohol}} (CH_2OH)\text{-}R\text{-}CHO \xrightarrow[\text{dehydrogenase}]{\text{fatty aldehyde}} (CH_2OH)\text{-}R\text{-}COOH$$

fatty alcohol dehydrogenase ↓

CHO-R-COOH

fatty aldehyde dehydrogenase ↓

HOOC-R-COOH

(C)

$$R\text{-}CH_3 \longrightarrow R\text{-}CH_2OOH \longrightarrow R\text{-}(CO)OOH \longrightarrow R\text{-}CHO \longrightarrow R\text{-}COOH$$

(D)

$$R1\text{-}(CH_2)(CH_2)\text{-}R2 \longrightarrow R1\text{-}(CH_2)(CHOH)\text{-}R2 \longrightarrow R1\text{-}(CH_2)(CO)\text{-}R2 \longrightarrow R1\text{-}(CH_2)O(CO)\text{-}R2$$

R1-COOH + R2-COOH

FIGURE 9.11

Aerobic biodegradation of n-alkanes. (A) Methane aerobic biodegradation. (B) Terminal oxidation of n-alkanes or α- and ω-hydroxylation. (C) n-Alkane degradation via alkyl hydroperoxides. (D) Subterminal oxidation of n-alkanes.

n-alkanes, through its transformation to n-alkyl peroxides, in presence of FAD^+ and Cu^{2+} as prosthetic groups, then the peroxides are further metabolized to their corresponding aldehydes (Fig. 9.11) (Maeng et al., 1996). Whyte et al. (1998) have reported that n-alkanes can be also oxidized by monooxygenase to secondary alcohols, then to ketones, and finally to fatty acids (Fig. 9.11)

The biodegradation of iso-alkanes can be occurred vis β-oxidation. *Rhodococcus* strain BPM 1613 has been reported for its capability to degrade phytane (2,6,10,14-tetramethylhexadecane), norpristane

(2,6,10-trimethylpentadecane), and farnesane (2,6,10-trimethyldodecane) via β-oxidation (Nakajima et al., 1985). *Mycobacterium* sp. HXN-1500 is reported to hydroxylate C_6–C_{11} alkanes to 1-alkanol derivatives by cytochrome P450 (Funhoff et al., 2006). Several bacterial strains can degrade $>C_{20}$ alkanes using enzyme systems. *Acinetobacter* sp. M1 is reported to grow on C_{13}–C_{44} alkanes and contains a soluble, Cu^{2+}-dependent alkane hydroxylase, dioxygenase that produces n-alkyl hydroperoxides to be further oxidized to the corresponding aldehydes (Tani et al., 2001). *Acinetobacter borkumensis* SK2 and *Acinetobacter* strain, DSM 17874 have been reported to degrade long-chain alkane by flavin-binding monooxygenase, named AlmA (Rojo, 2009). *Geobacillus thermodenitrificans* NG80-2 has been reported to degrade long-chain alkanes by long-chain alkane hydroxylase, named LadA (Feng et al., 2007).

However, when the hydrocarbons are chlorinated (as might occur in several additives to improve the performance of petroleum products) the degradation takes place as a secondary or cometabolic process rather than a primary metabolic process. In such a case, enzymes, which are produced during aerobic utilization of carbon sources such as methane, degrade the chlorinated compounds. Under aerobic conditions, a chlorinated solvent such as trichloroethylene ($ClCH$=CCl_2), which may have been mixed with the petroleum product during processing or during use, can be degraded through a sequence of metabolic steps, where some of the intermediary by-products may be more hazardous than the parent compound (e.g., vinyl chloride, CH_2=$CHCl$).

Under aerobic conditions, *Actinobacteria* and *Proteobacteria* degrade cycloalkanes such as cyclohexane by monooxygenase forming cyclohexanol, which is further oxidized to cyclohexanone, caprolactone, and adipate (Fig. 9.12) (Cheng et al., 2002). However, cycloalkanes are reported to be degraded by a co-oxidation mechanism and the substituted cycloalkanes are easily degraded than the unsubstituted analogs (Morgan and Watkinson, 1994; Harayama et al., 1999).

Polycyclic aromatic hydrocarbons (polynuclear aromatic hydrocarbon derivatives) are ubiquitous and recalcitrant organic contaminant; they are carcinogenic, teratogenic, and mutagenic, and threatening the marine biota. Cyclic aromatic compounds are known to be more stable since they have benzene rings and as the number of benzene rings increases, their microbial decomposition becomes more difficult. Many polynuclear aromatic hydrocarbon derivatives contain a "bay-region" and a "K-region," which are susceptible for initial oxidative attack to produce the corresponding dihydrodiols. However, non-Bay and non- K-region dihydrodiols, such as, 1,2-dihyroxypyrene, are known to be toxic since they form *o*-quinone through rapid auto-oxidation and subsequently dihydrodiol and *o*-quinone reacts to generate reactive oxygen species. This would lead to a decrease in culture growth and polynuclear aromatic hydrocarbon derivative degradation, due to the toxicity of those

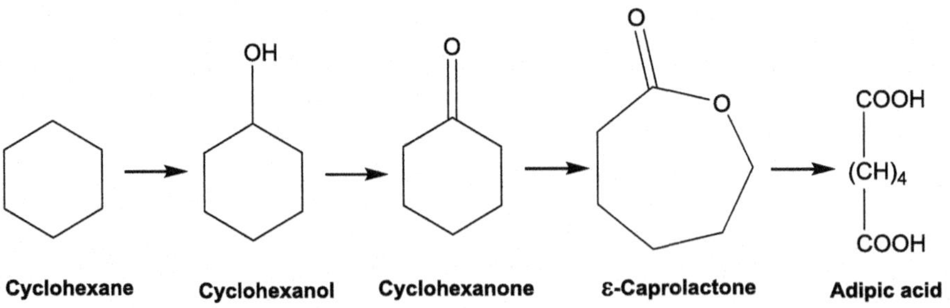

FIGURE 9.12

Aerobic biodegradation of cyclohexane.

intermediates that would accumulate in the aqueous phase. Methylation of non-K-region dihydrodiols prevents the redox cycling of catechols and quinone and is responsible for generation of reactive oxygen species (Ghosh et al., 2014).

Aromatic hydrocarbons have different elucidated aerobic biodegradation pathways. Generally, the initial step is the hydroxylation of the aromatic ring by either ring-hydroxylating dioxygenase or mono-oxygenase, producing an intermediate with two hydroxyl groups, that is subsequently cleaved by a ring-cleaving dioxygenase (Fuentes et al., 2014).

Toluene as a model compound for mono-aromatics has been reported to have different biodegradation pathways (Parales et al., 2008) (Fig. 9.13 and 9.14). The TOL plasmid-pathway (Fig. 9.13), where toluene is successively degraded to benzyl alcohol, benzaldehyde, and benzoate, is further transformed to the TCA cycle intermediates. *P. putida* F1 is reported to degrade toluene via the introduction of two hydroxyl groups, producing cis-toluene dihydrodiol, which is then converted to 3-methylcatechol, to be metabolized through the well-established meta pathway for catechol degradation. The first four steps in the pathway involve the sequential action of toluene dioxygenase (*todABClC2*), cis-toluene dihydrodiol dehydrogenase (*todD*), 3-methylcatechol 2,3-dioxygenase (*todE*), and 2-hydroxy-6-oxo-2,4-heptadienoate hydrolase (*todF*) (Zylstra et al., 1988). While, *Pseudomonas mendocina* KR1 is reported to biodegrade toluene throughout the toluene-4-monooxygenase catabolic pathway by to *p*-cresol, which is oxidized to *p*-hydroxybenzaldehyde by *p*-cresol methylhydroxylase and then oxidized to *p*-hydroxybenzoate by *p*-hydroxybenzaldehyde dehydrogenase that is subsequently hydroxylated by *p*-hydroxybenzoate hydroxylase to form protocatechuate. The protocatechuate is then oxidized by *ortho*-ring cleavage using protocatechuate-3,4-dioxygenase to produce β-carboxy-cis-cis-muconate (Fig. 8.14) (Wrightt and Olsen, 1994). *Pseudomonas pickettii* PKO1 is reported to oxidize toluene by toluene 3-monooxygease to *m*-cresol, which is further oxidized to 3-methylcatechol by another monooxygenase (Fig. 9.14) (Olsen et al., 1994). *B. cepacia* G4 is reported to metabolize toluene to *o*-cresol by toluene 2-monooxygenase, which is then transformed by another monooxygenase to 3-methylcatechol (Fig. 9.14) (Mars et al., 1996).

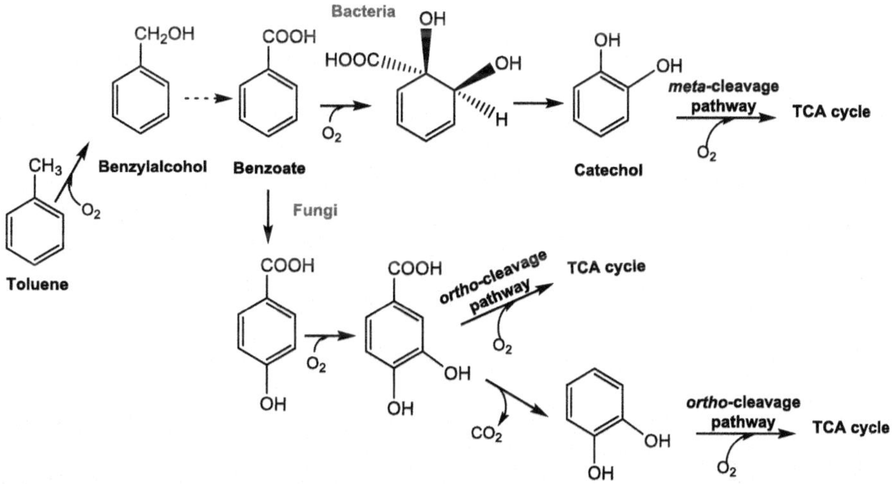

FIGURE 9.13

Toluene biodegradation by bacteria and fungi throughout the TOL-pathway.

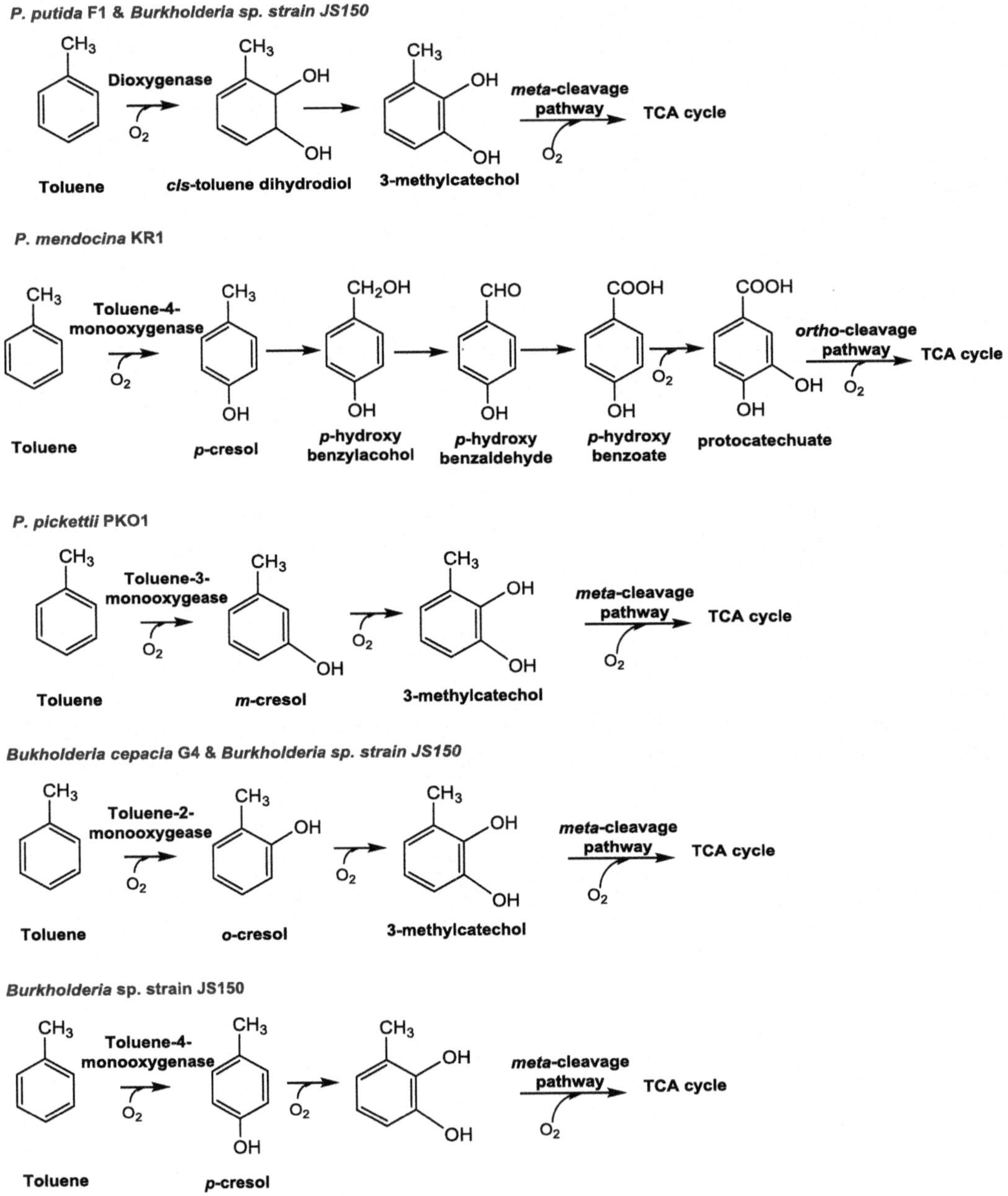

FIGURE 9.14

Aerobic biodegradation pathways for toluene by different bacterial isolates.

Burkholderia sp. strain JS150 has been reported to have at least three toluene/benzene monooxygenases to initiate toluene metabolism in addition to the toluene dioxygenase, thus it has multiple pathways for the metabolism of toluene (Fig. 9.14) (Johnson and Olsen, 1997). Sutherland et al. (1983) reported the capability of *Streptomyces setonii* strain 75Vi2 to degrade *trans*-cinnamic, p-coumaric, ferulic acids, and vanillin, using the inducible ring-cleavage dioxygenases, catechol 1,2-dioxygenase and protocatechuate 3,4-dioxygenase. The *trans*-cinnamic acid was catabolized via benzaldehyde, benzoic acid, and catechol; p-coumaric acid was catabolized via p-hydroxybenzaldehyde, p-hydroxybenzoic acid, and protocatechuic acid; ferulic acid was catabolized via vanillin, vanillic acid, and protocatechuic acid. When vanillin was used as the initial growth substrate, it was catabolized via vanillic acid, guaiacol, and catechol.

The fungal pathway for toluene degradation is reported to have some major differences from bacterial pathways; particularly the initial attack is undertaken by a cytochrome P-450 enzyme. However, there are also some similarities, including the conversion of benzyl alcohol to benzoate (like in the TOL pathway, Fig. 9.13) and the use of the *ortho*-cleavage pathway (as in the later steps in the toluene 4-monooxygenase pathway, Fig. 9.13). Previous studies of the *ortho*-cleavage pathway demonstrated that there is one clear difference between the bacterial and eukaryotic protocatechuate branches. The conversion of β-carboxymuconate to β-ketoadipate involves the participation of three enzymes in bacteria and only two in fungi (Fig. 9.15).

The biodegradation pathway of polyaromatic ring compounds has been elucidated. Dioxygenase and monooxygenase enzymes are reported as the major degrading enzymes in the oxidizing degradation of polynuclear aromatic hydrocarbon derivatives (Nievas et al., 2006). Boyd and Sheldrake (1998) categorized the dioxygenases to toluene-dioxygenase (TDO), naphthalene-dioxygenase (NDO), and biphenyl-dioxygenase (BPDO). The TDO is an extremely wide-ranging substrate enzyme. It is most suitable for cis-dihydroxylation of substituted benzene substrates and bicyclic arenes. But it has a size limitation for the accessibility of substrates, where, bulky and bendy molecules fit poorly into its active sites. However, for the larger polynuclear aromatic hydrocarbon derivatives, both NDO and BPDO are more recommendable for catalyzing dihydroxylation, with only the latter enzyme capable of metabolizing tetracyclic or larger examples and the NDO is reported to be generally incapable of catalyzing the cis-dihydroxylation of monocyclic aromatic rings. But it is particularly useful for oxidizing bi- and tricyclic polynuclear aromatic hydrocarbon derivatives substrates, such as naphthalene, phenanthrene, and anthracene. The biodegradation of polynuclear aromatic hydrocarbon derivatives is generally initiated by dihydroxylation of one of the polynuclear aromatic rings, with the cleavage of the dihydroxylated ring. These dihydroxylated intermediates may then be processed through either an *ortho*-cleavage or a *meta*-cleavage pathway, depending on the intradiol or extradiol ring-cleaving, leading to central intermediates such as protocatechuates and catechols, which are further converted to TCA cycle intermediates. The hydroxylation is catalyzed by a multicomponent dioxygenase, which consists of a reductase, a ferredoxin, and an iron sulfur protein (ISPNAR), while the ring cleavage is catalyzed by iron-containing meta-cleavage enzyme (Harayama et al., 1999; Deveryshetty and Phale, 2009). Dioxygenases are two groups: extradiols and intradiols. The extradiols cleave the hydroxylated aromatic ring at a bond proximal to one of the two hydroxylated carbon atoms, while the intradiols cleave the aromatic ring between the two hydroxylated carbon atoms, and non-heme iron is essential for the activity of these ring-cleaving dioxygenases.

Naphthalene is a ubiquitous two-ring aromatic pollutant, thus it is reported to be easily and quickly biodegraded than the higher polynuclear aromatic hydrocarbon derivatives. However, it is the simplest polynuclear aromatic hydrocarbon derivative and has long been used as a model compound in polynuclear aromatic hydrocarbon derivative biodegradation studies (Zeinali et al., 2008a). It has been

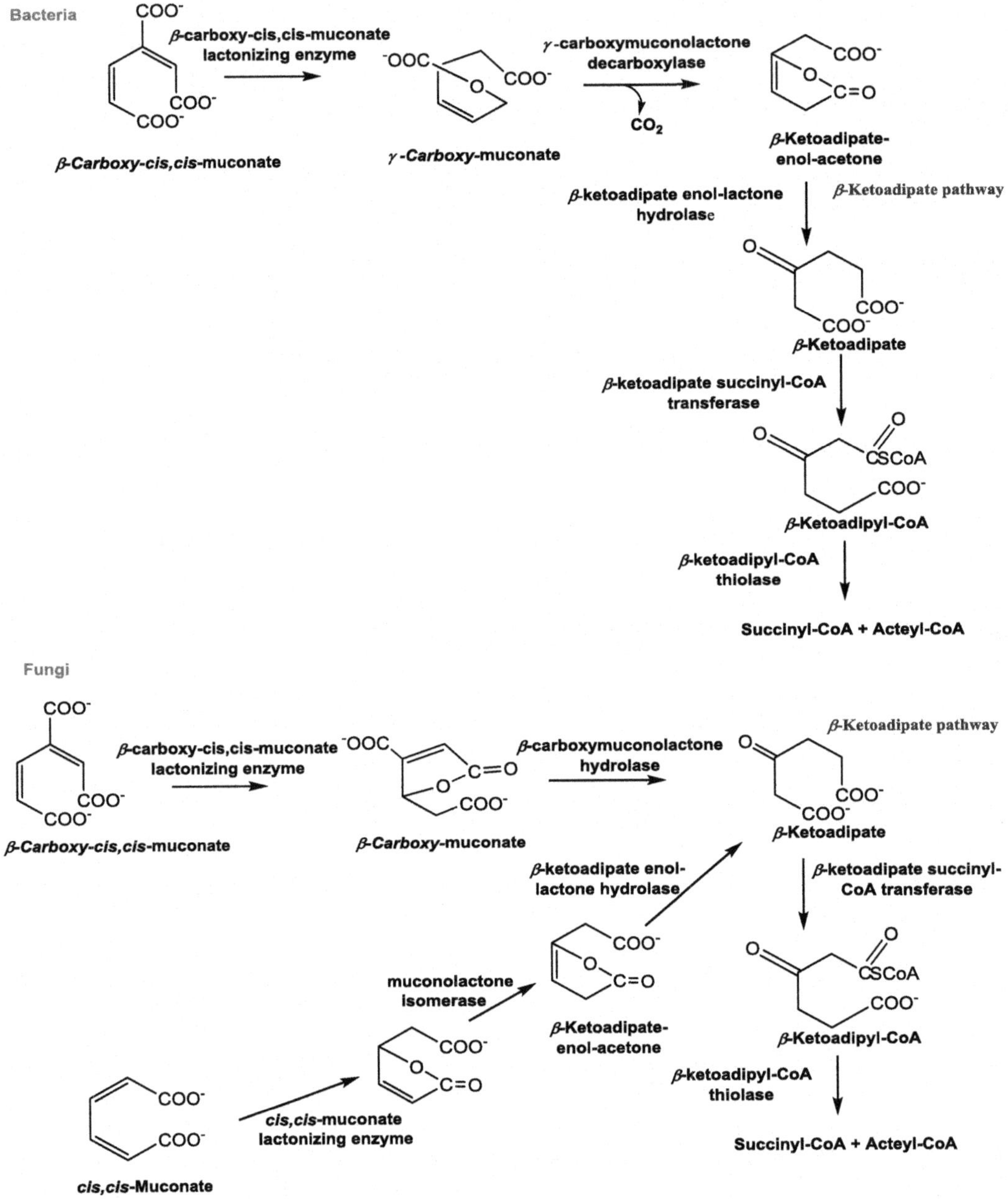

FIGURE 9.15

β-carboxymuconate metabolism in bacteria and fungi.

reported that, the enzymes involved in the conversion of naphthalene to salicylate can also degrade phenanthrene to naphthalene-1,2-diol (Peng et al., 2008). The naphthalene catabolic genes in biodegrading bacteria are reported to be organized into two operons: the *nal* operon encoding the upper pathway enzymes involved in conversion of naphthalene to salicylate, and the *sal* operon encoding the lower pathway enzymes involved in the conversion of salicylate to pyruvate and acetyl coenzyme A (Peng et al., 2008). In most of the naphthalene/phenanthrene biodegrading bacteria, the nucleotide sequences of the upper pathway genes are more than 90% identical. All these genes are arranged in a similar order: *nahAa* (ferredoxin reductase), nahAb (NDO ferredoxin), nahAc (the α-subunit of NDO), *nahAd* (the β-subunit of NDO), nahB (naphthalene-*cis*-dihydrodiol dehydrogenase), *nahF* (salicylaldehyde dehydrogenase), *nahC* (1,2-dihydroxynaphthalene dioxygenase), *nahQ* (unknown gene), *nahE* (*trans-o*-hydrobenzylidenepyruvate hydratase aldolase), and *nahD* (2-hydroxychromene-2-carboxylate isomerase) (Habe et al., 2003). However, there are some reported exceptions, for example, the naphthalene dioxygenase genes (*nag* gene) of *Comamonas testeroni* strain GZ42 (Goyal and Zylstra, 1997), and *Ralstonia* sp. U2 (Zhou et al., 2001) reported to contain all of the genes corresponding to the classical nah gene of *Pseudomonas* strains in the same order, with the exception of two extra genes inserted between the ferredoxin reductase gene and ferredoxin gene, *nagG* and *nagH*, which are structural subunits of salicylate-5-hydroxylase that can convert naphthalene to gentisate. But the pathway genes in *C. testeroni* strain GZ39 are reported to be quite different from those of the typical nah genes, where, several genes (such as extradiol dioxygenase) are not within the cluster, while, a glutathione-*S*-transferase gene is within the cluster (Zylstra et al., 1997). The naphthalene catabolic genes (*phn* gene) in *Burkholderia* sp. RP007 is quite different, encode ISPNAR α- and β-subunits of a polynuclear aromatic hydrocarbon derivative initial dioxygenase, and lack both ferredoxin and reductase components. However, the ferredoxin and reductase genes for electron transport functions of the *phn* gene is reported to be located elsewhere in the RP007 genome or could be supplied by cellular housekeeping genes (Laurie and Lloyd-Jones, 2000). Moreover, the lower pathway of *P. putida* G7, NCIB9816-4, ND6, and *P. stutzeri* AN10 strains is reported to consist of 11 genes organized in the order; *nahGTHINLOMKJY*, where, the *nahY* is not a catabolic gene, but a naphthalene chemotaxis gene (Boronin, 2001). In addition, the strains AN10 and ND6 are reported to have another salicylate hydroxylase gene (*nahW*) outside but close to the *sal* operon. It has been reported also that the iso-functional enzyme is beneficial to the bacteria to adjust its metabolism to extreme conditions (Li et al., 2004; Lanfranconi et al., 2009).

Different pathways for naphthalene biodegradation have been elucidated (Fig. 9.16). The ring of the produced gentisate (Fig. 9.16) can be cleaved between carbon 1 and carbon 2 by gentisate 1,2-dioxygenase, with further ring-fission mechanism resembleing to that of the extradiol-type fission. The extradiol dioxygenases and gentisate 1,2-dioxygenase contain non-heme ferrous iron, whereas the intradiol dioxygenases contain nonheme ferric iron (Iwabuchi and Harayama, 1998a,b). Yen and Serdar (1988) reported the metabolism of naphthalene by *P. putida* PpG7 throughout the production of salicylic acid. There is another reported naphthalene biodegradation pathway by *N. otitidiscaviarum* strain TSH1 Zeinali et al. (2008a) and *Streptomyces* sp. ERI-CPDA-1 (Balachandran et al., 2012). This pathway is initiated, as usual, by naphthalene hydroxylation by 1,2-dioxygenase, giving *cis* 1,2-naphthalenedihydrodiol, that further dehydrogenated to produce 1,2-dihydroxynaphthalene, which undergoes extradiol cleavage to 4-(-hydroxyphenyl)-2oxobut-3-enoic acid and 2-hydroxycinnamic acid and finally cinnamic acid, which is oxidized to phenylacetic acid that further degraded to benzaldehyde. This open system oxidation produces benzoic acid, which is hydroxylated to protocatechuic acid that can be decarboxyled to catechol. That further degraded to muconnic acid, which is further metabolized by

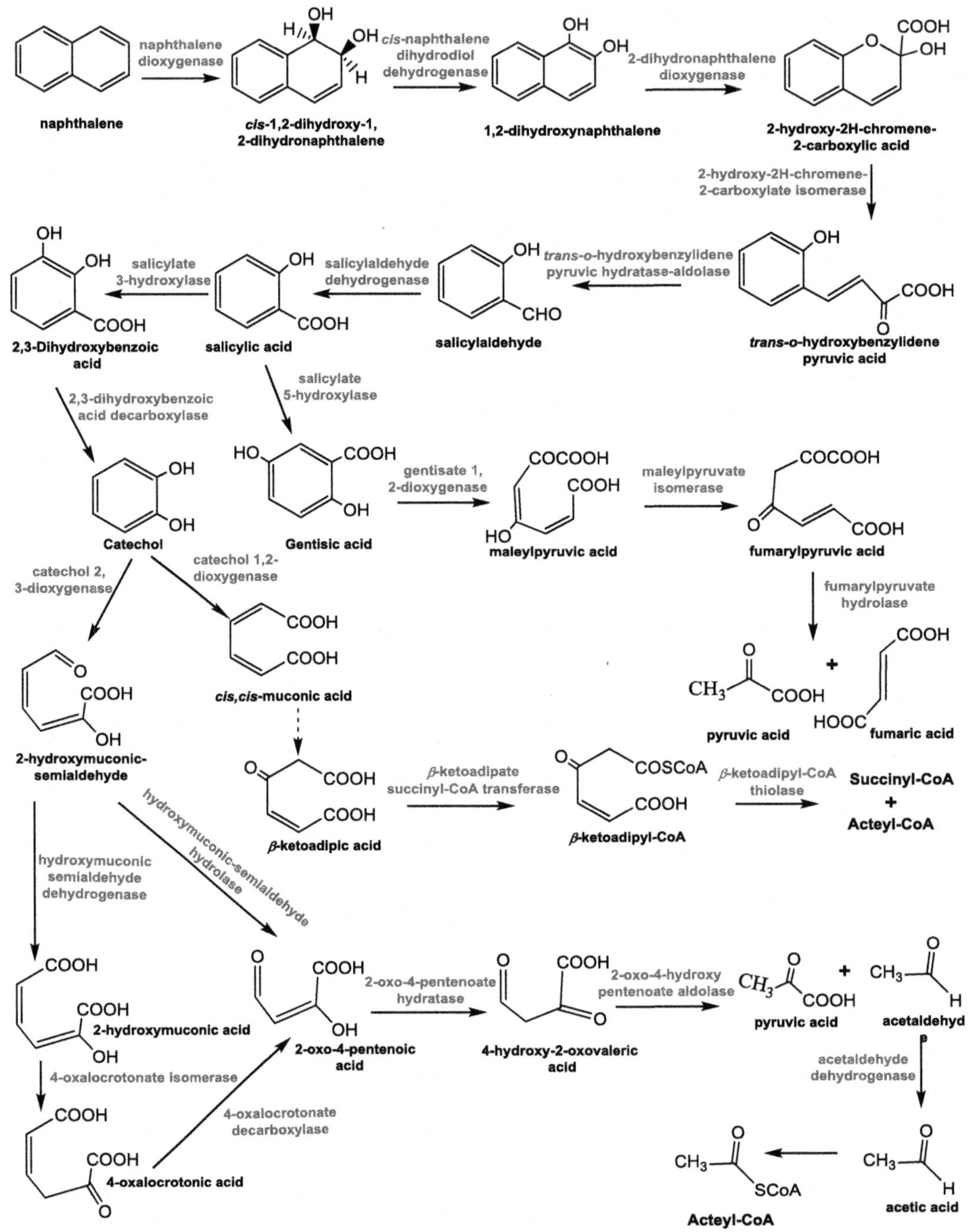

FIGURE 9.16

Aerobic biodegradation pathways of naphthalene.

TCA cycle. Annweiler et al. (2000b) reported the biodegradation of naphthalene by the thermophilic *Bacillus thermoleovorans*, throughout the production of 2,3-dihydroxynaphthalene, 2-carboxycinnamic acid, and phthalic acid that is decarboxylated to benzoic acid.

Phenanthrene is reported to be degraded by two main pathways: phethalic acid or protocatechuate pathway and naphthalene or salicylate pathway (Fig. 9.17). It has been earlier reported that bacteria, which are bale to grow on both phenanthrene and naphthalene, mineralize these substrates via salicylate and catechol (i.e., the salicylate pathway), while those capable of growing only on phenanthrene and not on naphthalene, degrade phenanthrene via protocatechuate (i.e., the protocatechuate pathway) (Saito et al., 2000). A wide variety of microorganisms such as *Pseudomonas* sp., *Arthorbacter* sp., *Roseobacter* sp., and *Bacillus thuringiensis* have been reported to degrade phenanthrene throughout the protocatechuate pathway (Paul et al., 2004; Moscoso et al., 2012b; Ferreira et al., 2016). The *Alcaligenes faecalis* AFK2 strain is reported to degrade both naphthalene and phenanthrene as sole carbon and energy sources through the *o*-phthalate pathway, using some unique genes, such as 3,4-dihydroxyphenanthrene dioxygenase (*phnC*), 2-carboxybenzaldehyde dehydrogenase (*phnI*), and trans-2-carboxybenzal-pyruvate hydratase aldolase (*phnH*) (Kiyohara et al., 1982, 1990). The phenanthrene degrading bacteria, *Nocardioides* sp. KP7 strain, throughout the phthalate pathway is reported to have a different polynuclear aromatic hydrocarbon derivatives-catabolic genes; *phd* genes (*PhdA* and *PhdB* genes), which encode the α- and β-subunit of phenanthrene dioxygenase, have < 60% sequence identity to all of the known dioxygenase subunits and the ferredoxin component of the dioxygenase *PhdC* belongs to the [3Fe–4S] or [4Fe–4S] type of ferredoxin, but not to the [2Fe–2S] type found in most components of polynuclear aromatic hydrocarbon derivatives dioxygenase (Iwabuchi and Harayama, 1998a,b; Saito et al., 1999). Many *Sphingomonas* strains are reported to degrade naphthalene, phenanthalene, and anthracene throughout common pathways found in other Gram-negative bacteria (Ogram et al., 2000; Pinyakong et al., 2003a,b; Basta et al., 2004). Basta et al. (2005) reported that the "flexible" gene organization, for example, the different combinations of conserved gene cluster in the Sphingomonas strains, is one of the mechanisms that allows them to adapt quickly and efficiently to novel compounds in the environment. Generally, two dioxygenase genes are required for conversion of polynuclear aromatic hydrocarbon derivatives and biphenyl to simple aromatic acid, and *meta*-cleavage genes are required for conversion of aromatic acids to the TCA cycle intermediate (Zylstra and Kim, 1997; Kim and Zylstra, 1999). However, these aromatic-degradative genes for catabolic pathways are often localized in the plasmid separately from each other, or, at least, are not organized in coordinately regulated operons, not like that of *Pseudomonas* and other Gram-negative bacteria whose structural genes required for naphthalene utilization are usually clustered (Habe et al., 2003). The Gram-positive naphthalene degraders—*Rhodococcus* strains NCIMB12038, P200, and P400—have three structural genes (narAa, narAb, and narB). The *narAa* and *narAb* genes encode the α- and β-subunits of the NDO catalytic component of ISPNAR. Both subunits of the NCIMB12038 NDO shave only 30% amino acid similarity to that of the corresponding *P. putida* NDO subunits. While, the *narB* gene has 39% amino acid similarity to *NahB* of the *P. putida* G7 strain. But, *Rhodococcus* strains have no genes corresponded to the genes encoding the electron transport components reductase and ferredoxin of NDO in *Pseudomonas*. Moreover, the *nar* region in *Rhodococcus* strains is not organized into a single operon, but there are several transcription units (Peng et al., 2008). The *narA* and *narB* genes whose transcription are induced by naphthalene are reported to be transcribed as a single unit through different start sites and the putative regulatory genes (*narR1* and *narR2*) are transcribed as a single mRNA in naphthalene-induced cells (Kulakov et al., 2000; Larkin et al., 2005). Usually, the microorganisms that have enzymes responsible for biodegradation of naphthalene to salicylate can

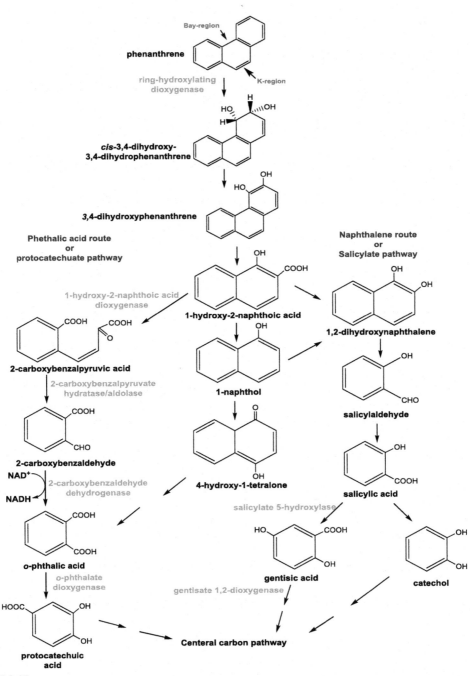

FIGURE 9.17

Aerobic biodegradation pathways of phenanthrene.

degrade phenanthrene as well. The bay-region of phenanthrene is a sterically hindered area between carbon atoms 4 and 5 and the K-region is the 9, 10 double bond, which is the most oleinic aromatic double bond with high electron density. Phenanthrene can be oxidized at 1,2 position, 3,4 position, and 9,10 position producing the corresponding diols (Fig. 9.18) and sometimes the same microorganism can follow different deoxygenating pathways. For example; *Arthrobacter* sp. P1-1 is reported to degrade phenanthrene through the three pathways (Seo et al., 2006). In a study performed by *Burkholderia* sp. C3 strain, the initial dioxygenation at the 3,4-C position was dominating within the first 3 days. However, *cis*-phenanthrene-1,2-dihydrodiol was more abundant than *cis*-phenanthrene-3,4-dihydrodiol after 7 days of incubation and attributed this to the slow degradation of the produced 1-hydroxy-2-naphthoic acid, which derives the metabolic pathway towards the 1,2-dioxygenation metabolic pathway and not the 3,4-dioxygenation (Seo et al., 2007). It has been also reported that *S. maltophilia* strain C6, isolated from creosote-contaminated sites at Hilo, Hawaii, is capable of utilizing phenanthrene as a sole source of carbon and energy, via the initial dioxygenation of the 1,2-; 3,4-; and 9,10-C-positions, with the predominance of the 3,4-dioxygenation and its subsequent ortho-cleavage of 1-hydroxy-2-naphthoic acid metabolites-pathway (Fig. 9.18, Gao et al., 2013). It has been reported that the phenanthrene degrading bacteria that follow the *ortho*-cleavage pathway cannot grow when naphthalene is the sole carbon source (Keum et al., 2005; Xia et al., 2005). *Staphylococcus* sp. strain PN/Y (Mallick et al., 2007), isolated from petroleum-contaminated soil, and *Ochrobactrum* sp. strain PWTJD Ghosal et al. (2010) and *Sphingobium* sp. strain PNB (Roy et al., 2012), isolated from municipal waste contaminated soil samples, are reported the biodegradation of phenathrene throughout the *meta*-cleavage of 2-hydroxy1-naphthoic acid (Fig. 9.18).

The initial reactions in the degradation of the low water-soluble anthracene have been reported to occur by monooxygenase, attacking C_9 and C_{10} or dioxygenase and by attacking C_1 and C_2 (Peng et al., 2008). The *Mycobacterium* sp. strain PYR-1 has been reported to have the two enzymatic systems, producing 9, 10-anthraquinone, as a dead-end product, and 1,2-dihydroxyanthracene. That is further metabolized through the *ortho*-cleavage, producing 3-(2-carboxyvinyl) naphthalene-2-carboxylic acid and the *meta*-cleavage, producing 6,7-benzocoumarin (Moody et al., 2001). Anthracene has been also reported to be degraded throughout the production of *o*-phthalic acid and protocatechuic acid by *Mycobacterium* sp. LB501T (Van Herwijnen et al., 2003b) (Fig. 9.19). The moderate thermophilic bacterium *N. otitidiscaviarum* strain TSH1 is reported to metalize anthracene throughout the *ortho*-position to produce 3-(2-carboxyvinyl) naphthalene-2-carboxylic acid, which is transformed to 2,3-dicarboxynaphthalene, then, 2,3-dihydroxynaphthalene, which is then proceeded either through a pathway similar to that for naphthalene metabolism to form salicylate and catechol or through a protocatechuate-degrading pathway (Zeinali et al., 2008b). *Bacillus cereus* strain, isolated from contaminated soil from IOCL depot Khapri, Nagpur, MS, India, has been reported to mineralize anthracene throughout the production of 2,3 dihydroxynaphthalene, catechol, phthalate, and then *cis-cis* muconate (Fig. 9.19) (Meshram and Wate, 2014). The halotolerant *Martelella* sp. AD-3 is reported to degrade anthracene, through the initial attack on the aromatic ring by monooxygenase and dioxygenase. This proceeds throughout two pathways: by the oxidation at C_9 and C_{10} producing 9,10-anthraquinone as a dead-end product and the oxidation at C_1 and C_2, followed by the *meta*-cleavage of the produced 1,2-dihydroxyanthracene. That is further degraded throughout the production of salicylic acid and gentisic acid (Cui et al., 2014).

Fluorene can be initially oxidized by dioxygenase system at C_1 and C_2 or at C_3 and C_4, followed by the meta-cleavage, aldolase reaction, and then decarboxylation of the ring-fission product, producing the substrates of the biological Baeyer–Villiger reaction, 2-indanone and 1-indanone, which produce the aromatic lactones 3-isochromanone and 3,4-dihydrocoumarin, respectively. The enzymatic hydrolysis of

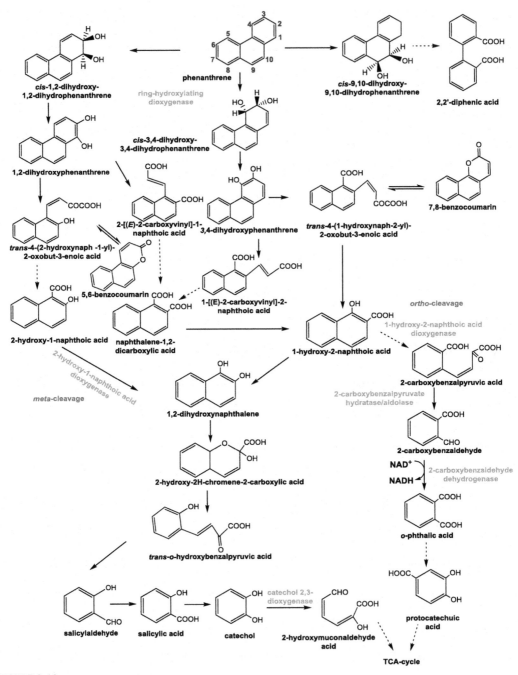

FIGURE 9.18

More phenanthrene aerobic biodegradation pathways.

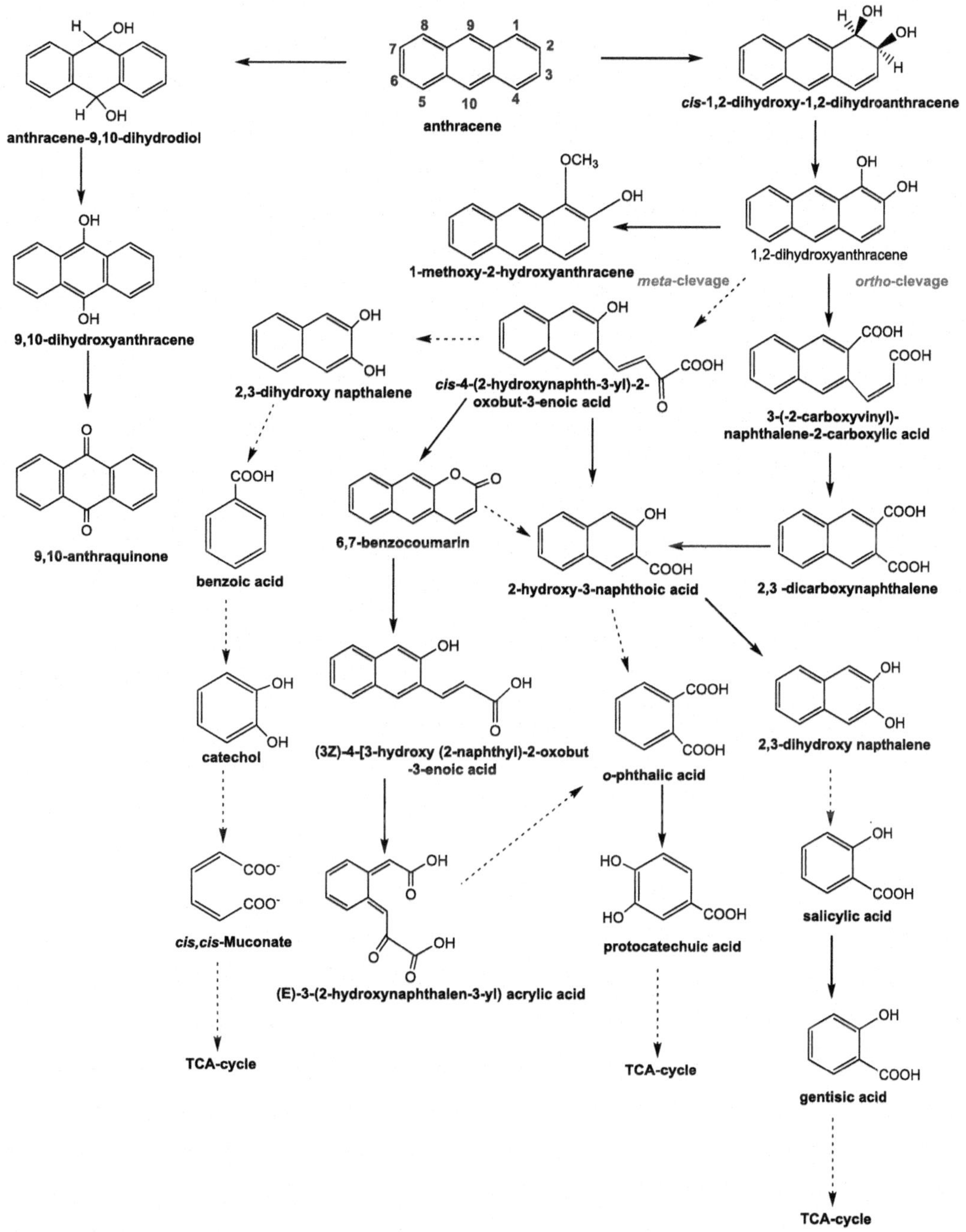

FIGURE 9.19

Aerobic biodegradation pathways of anthracene.

3,4-dihydrocoumarin produces 3-(2-hydroxyphenyl) propionic acid, which can be further metabolized to salicylic acid and catechol to reach to the central metabolism (Fig. 9.20) (Habe et al., 2004). However, fluorene can be also initially oxidized by monooxygenase system at C9, followed by angular dioxygenation, producing 8-hydroxy-3,4-benzocoumarin, through the production of 2′-carboxy-2,3-dihydroxybiphenyl. The later metabolite is further degraded throughout a pathway analogous to biphenyl degradation, until reaching to protocatechuic acid that can be further metabolized to TCA cycle (Fig. 9.21) (Wattiau et al., 2001). Grifoll et al. (1994) reported the biodegradation of fluorene by *Pseudomonas* sp. F274 throughout the formation of mono-hydroxy fluorene, fluorenone, and phthalic acid, which can be further degraded to carbon dioxide and water.

Brevibacterium sp. strain DPO 1361 is reported to degrade fluorene through the monooxygenation mode of a dioxygenase initial attack, producing 9-fluorenol, which yields 9-fluorenone using 9-fluorenol dehydrogenase. Angular dioxygenation of 9-fluorenone produces 1,10-dihydro-1,10-dihydroxyfluoren-9-one (DDF). That is further degraded by an NAD+-dependent DDF-dehydrogenase enzyme to 3-(2′-carboxyphenyl)catechol, followed by further degradation to TCA-cycle by a pathway similar to that for biodegradation of biphenyl or dibenzofuran (Trenz et al., 1994). Fluorene biodegradation by *Arthrobacter* sp. strain F101, as a sole carbon and energy source, proceeds through three independent pathways: combine the metabolism reactions of (1) aromatic compounds, such as dioxygenation; (2) alicyclics, such as Baeyer-Villiger and enzymatic hydrolysis of the lactone; and (3) linear fatty acids, such as β oxidation. Two productive routes are initiated by dioxygenation at positions C_1–C_2 and C_3–C_4, respectively, meta-cleavage followed by an aldolase reaction and loss of C-1 yield the detected indanones. Subsequent biological Baeyer-Villiger reactions produce the aromatic lactones 3-isochromanone and 3,4-dihydrocoumarin, respectively. Enzymatic hydrolysis of the 3,4-dihydrocoumarin gives 2-(3-hydroxyphenyl) propionate, which could be a substrate for a β-oxidation cycle, to give salicylate. Further oxidation of 3-isochromanone via catechol and 2-hydroxymuconic semialdehyde connects with the central metabolism (Fig. 9.20). A third nonproductive pathway, initiated by monooxygenase at C9, produces 9-fluorenol and 9-fluorenone. The latter can be further attacked by dioxygenase followed by the nonenzymatic dehydration of the corresponding 3,4-dihydrodiol, producing 4-hydroxy-9-fluorenone as a dead-end product (Fig. 9.21) (Casellas et al., 1997). *Sphingomonas* sp. L-138 is reported to degrade fluorene at 20°C via phethalic acid and protocatechuic acid to TCA-cycle (Sokolovská et al., 2002). *Terrabacter* sp. strain DBF63 is reported to degrade fluorene through the monooxygenase initial attack to TCA cycle intermediates via phthalate and protocatechuate (Habe et al., 2005). *Rhodococcus* sp. USTB-C, isolated from crude oil of production well in Dagang oil field, southeast of Tianjin, northeast China, reported to degrade fluorene by initial oxidation of fluorene to furan, which is further degraded to phenol throughout the production of phthalic acid (Fig. 9.21) (Yu et al., 2014). *Citrobacter* sp. FL5, isolated from a soil sample at an abandoned site in Shanghai, China is reported to degrade high concentrations of fluorene up to 96% of 150 mg/L throughout a new pathway. Fluorene is first oxidized to its heterocyclic analog, dibenzofuran, which is further attacked by angular dioxygenase, producing 9-fluorenylmethanol, which is then further metabolized to 2,2-diphenylacetic acid, piperonylic acid, and catechol to TCA-cycle, with a probable involvement of dehydrogenase (Zhu et al., 2016).

Due to the high hydrophobicity and tendency of high–molecular-weight polynuclear aromatic hydrocarbon derivatives (Fig. 9.22) to adsorb to organic matter, they are generally characterized by low bioavailability and high recalcitrance to degradation. However, the biodegradation pathways for the high-molecular-weight polynuclear aromatic hydrocarbon derivatives have been also reported.

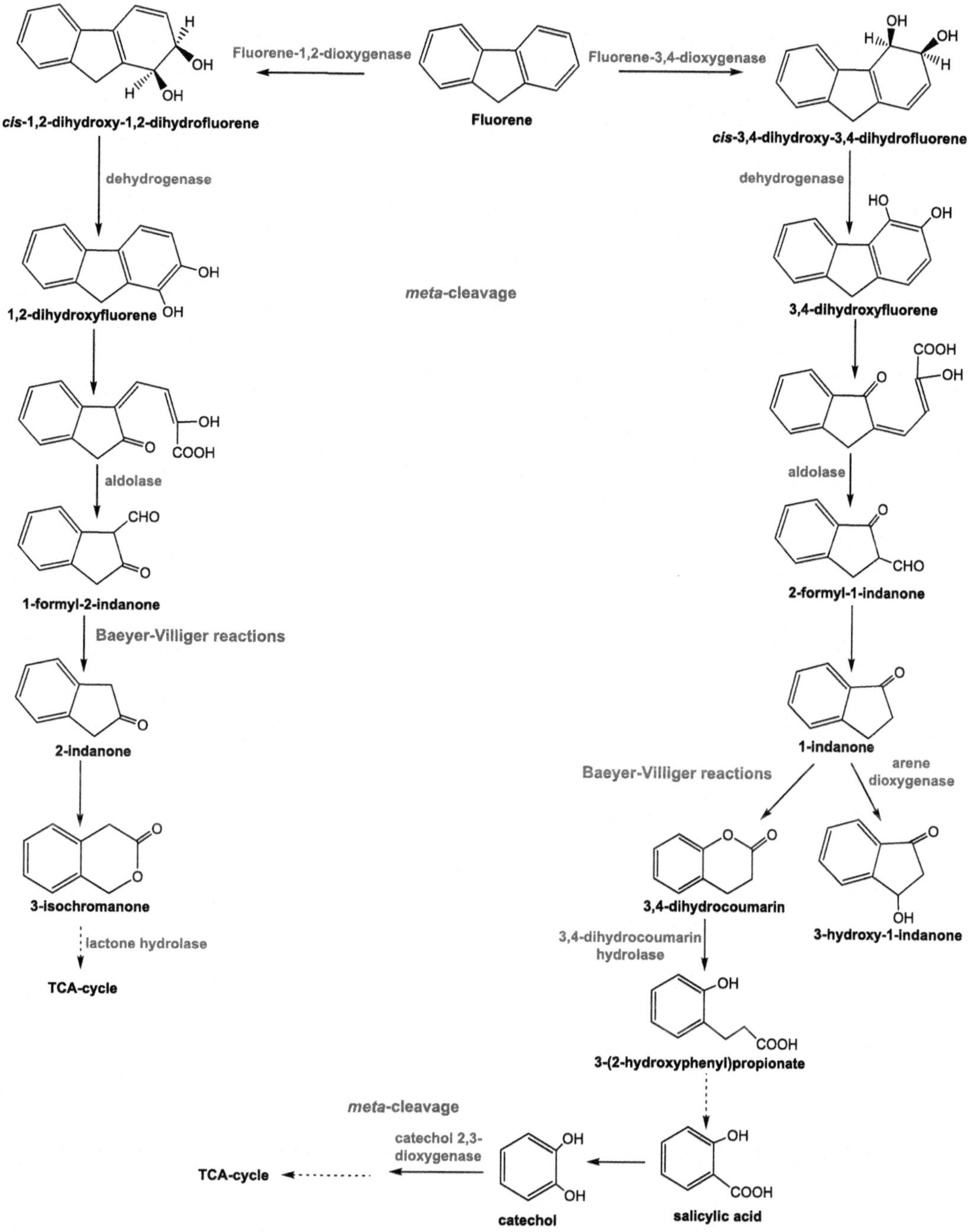

FIGURE 9.20

Aerobic biodegradation of fluorene through the initial attack of C1,C2 and C3,C4.

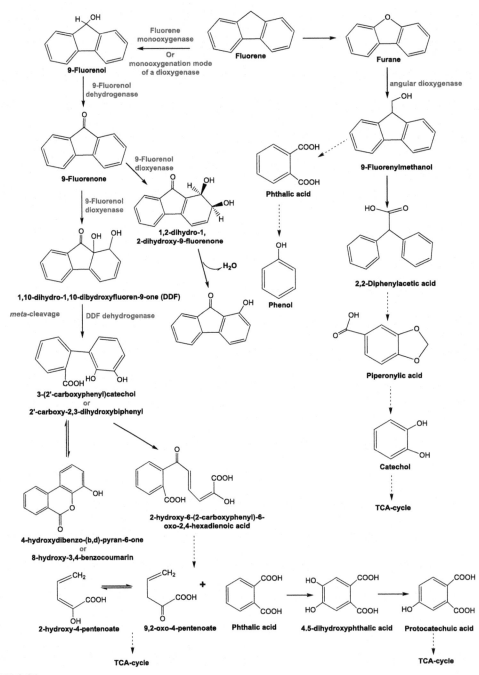

FIGURE 9.21

Aerobic biodegradation of fluorene through the initial attack of C9.

Fluoranthene, for example, is structurally similar to other compounds of environmental concern, such as acenaphthylene, carbazole, fluorene, dibenzodioxin, dibenzofuran, and dibenzothiophene, thus, it has been used as a model compound in many biodegradation studies. Fluoranthene degradation is reported to be initiated by dioxygenation at the C-1,2; C-2,3; C-7,8; or C-8,9 positions, producing 9-fluorenone, fluoranthene *cis*-2,3-dihydrodiol, acenaphthylene-1(2H)-one, and fluoranthene *cis*-8,9-dihydrodiol, respectively (Kelley and Cerniglia, 1991; Kelley et al., 1991; Kim et al., 2006), or by monooxygenation to produce monohydroxyfluoranthene (Peng et al., 2008). Fig. 9.23 summarizes the different metabolic pathways reported for biodegradation of fluoranthene (Weissenfels et al., 1991; Cerniglia, 1992; Kelley et al., 1993). The C_1–C_2 and C_2–C_3 dioxygenation initial attack pathways degrade fluoranthene through the agency of a fluorene-pathway metabolite. The *meta*-cleavage of the 1,2-dihydroxyfluoranthene produces 9-fluorenone-1-(carboxy-2-hydroxy-1-propenol) and 9-fluorenone-1-carboxylic acid, which can be converted by protonation to 9-fluorenol-1-carboxy-3-propenyl-2-one and 9-fluorenol-1-carboxylic acid, respectively. Then finally, the produced 9-fluorenone can be converted to 9-fluorenol (Fig. 9.24), whereas the C_7–C_8 pathway oxidizes fluoranthene throughout acenaphthylene-pathway metabolites by extra- and intradiol ring cleavages. The produced 7,8-dihydroxy-fluoranthene is further transformed via *meta*-cleavage to 1-acenaphthenone and 3-hydroxymethyl-3*H*-benzo[*de*]-chromen-2-one through 2-hydroxyl-4-(2-oxo-2*H*-acenaphthylen-1-ylidene)-but-2-enoic acid, 2-hydroxylmethyl-2*H*-acenaphthylen-1-one, and 2-oxo-acenaphthene-1-carboxylic acid (Fig. 9.23). Moreover, the dehydrogenation of a transient *cis*-7,8-fluoranthene dihydrodiol produces the 7,8-dihydroxyfluoranthene, which upon *o*-methylation at the 7-position forms 7-methoxy-8-hydroxyfluoranthene. Generally, the more predominant dioxygenase initial attach is at C2,C3. This pathway requires about 18 enzymes for completion of fluoranthene degradation via 9-fluorenone-1-carboxylic acid and phthalate, with the initial ring-hydroxylating oxygenase, NidA3B3, oxidizing fluoranthene to fluoranthene *cis*-2,3-dihydrodiol (Fig. 9.23). Whereas, six cytochrome P450 genes, including CYP51, are reported to be responsible for monooxygenation of fluoranthene, where, the monooxygenation product is transformed to monohydroxyfluoranthenes with ring-hydroxylating oxygenases and chemical dehydration of *cis*- or *trans*-dihydrodiols (Kweon et al., 2007). *Mycobacterium* sp. JS14 is reported to initially attack fluoranthene at four positions: C-1,2; C-2,3; C-7,8; or C-8,9, which produce four corresponding dimethoxyfluoranthenes (Lee et al., 2007).

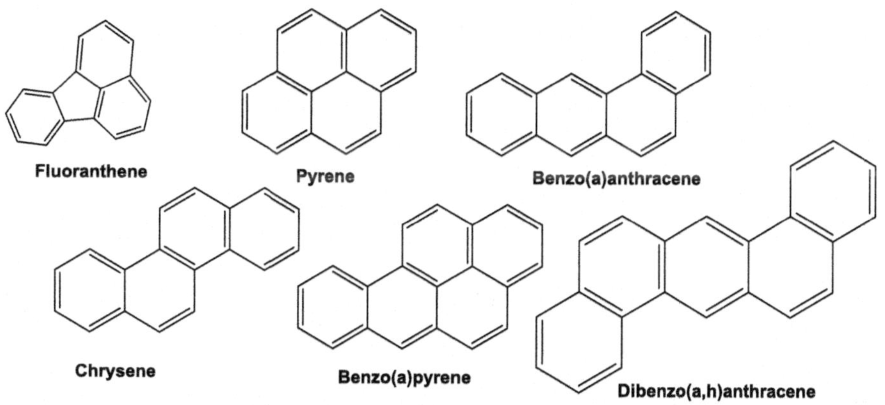

Fluoranthene   Pyrene   Benzo(a)anthracene

Chrysene   Benzo(a)pyrene   Dibenzo(a,h)anthracene

FIGURE 9.22

Representative high-molecular-weight polynuclear aromatic hydrocarbon derivatives as listed by US-EPA.

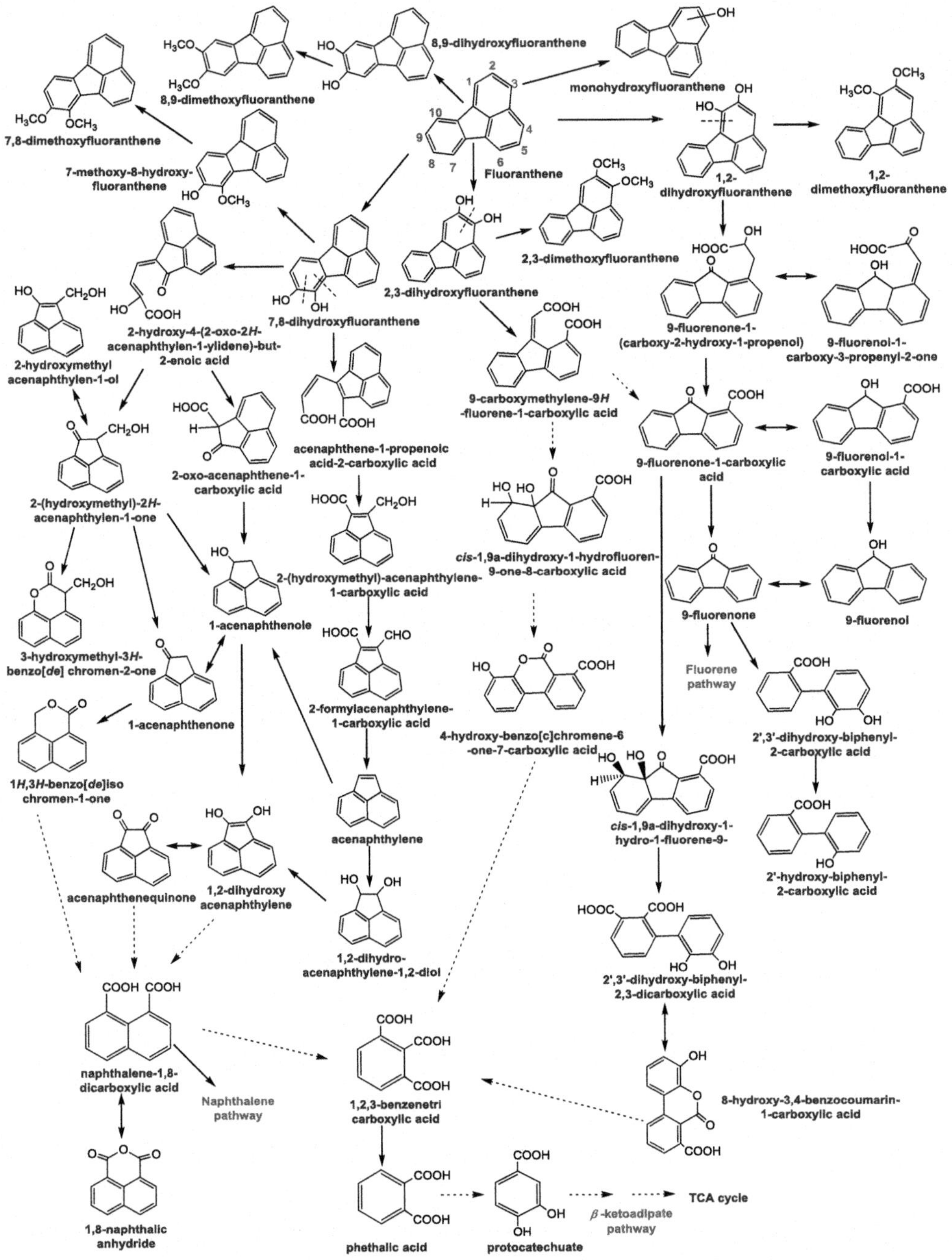

FIGURE 9.23

Different metabolic pathways reported for aerobic biodegradation of fluoranthene.

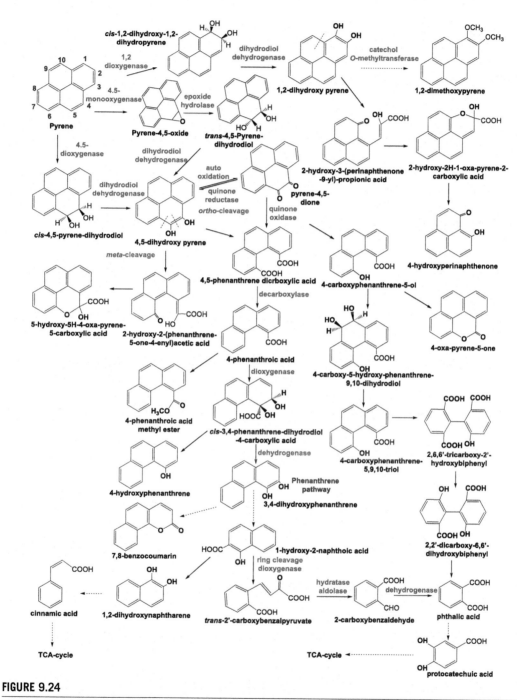

FIGURE 9.24

Different metabolic pathways reported for aerobic biodegradation of pyrene.

Weissenfels et al. (1990) reported the biodegradation of fluoranthene by *Alcaligenes denitrificans* strain WW1 via a dioxygenase pathway, producing 7-hydroxyacenaphthylene, 7-acenaphthenone, and 3-hydroxymethyl-4,5-benzocoumarine. Rehmann et al. (2001) reported the isolation of fluoranthene-degrading bacteria, *Mycobacterium* sp. strain KR20, from a polynuclear aromatic hydrocarbon deriva-tive–contaminated soil of a former gaswork plant site. Seven metabolites were produced by strain KR20 from fluoranthene in detectable amounts. Five metabolites, namely *cis*-2,3-fluoranthene dihydrodiol, Z-9-carboxymethylenefluorene-1-carboxylic acid, *cis*-1,9a-dihydroxy-1-hydro-fluorene-9-one-8-carboxylic acid, 4-hydroxybenzochromene-6-one-7-carboxylic acid, and benzene-1,2,3-tricarboxylic acid. Kweon et al. (2007) reported 37 metabolites including potential isomers found in the biodegradation pathway of fluoranthene by *M. vanbaalenii* PYR-1. Cao et al. (2015) reported the isolation of the first fluoranthene-degrading bacterium within the family *Rhodobacteraceae, Celeribacter indicus* P73^T, from deep-sea sediment from the Indian Ocean. It has the ability to metabolize fluoranthene throughout the C-7,8 dioxy-genation pathway, producing some major metabolites: acenaphthylene-1(2H)-one, acenaphthenequinone, 1,2-dihydroxyacenaphthylene, and 1,8-naphthalic anhydride.

Pyrene is usually used as a model compound for high-molecular-weight polynuclear aromatic hydrocarbon derivative biodegradation because it is structurally similar to several carcinogenic poly-nuclear aromatic hydrocarbon derivatives. Gibson et al. (1975) reported the cometabolism of pericon-densed polynuclear aromatic hydrocarbon derivative compound pyrene by a mutant *Beijerinckia* sp. strain, in the presence of succinate and biphenyl to cis-9,10-dihydroxy-9,10-dihydro-BaP and *cis*-7,8-dihydroxy-7,8-dihydrobenzo[a]-pyrene. Complete mineralization of the pericondensed polynu-clear aromatic hydrocarbon derivative compound pyrene by *Mycobacterium* sp. isolated from sediment near a hydrocarbon source has been reported (Cerniglia and Heitkamp, 1990). That occurred through-out the production of seven major intermediates; three from ring oxidation, *cis*-4,5-pyrenedihydrodiol, *trans*-4,5-pyrenedihydrodiol, and pyrenol, and four products from the ring fission, 4-hydroxyperinaph-thenone, 4-phenanthroic acid, phthalic acid, and cinnamic acid (Fig. 9.24). *M. vanbaalenii* PYR-1 strain has been reported to metabolize pyrene through two pathways; the first, oxidation of pyrene via initial dioxygenation at the C_1 and C_2 positions, forming *o*-methylated derivatives of pyrene-1,2-diol, as a detoxification step (Kim and Freeman, 2005). However, the predominant second pathway is dioxy-genation initiated at the C4 and C5 positions (K region), producing the *cis*-4,5-dihydroxy-4,5-dihydro-pyrene (i.e., pyrene-*cis*-4,5-dihydrodiol). The rearomatization of the dihydrodiol and subsequent ring cleavage dioxygenation lead to the formation of 4,5-dicarboxyphenanthrene, which is further decar-boxylated to 4-phenanthroate. The subsequent intermediate, *cis*-3,4-dihydroxyphenanthrene-4-carbox-ylate, can be produced by a second dioxygenation reaction. Then, the rearomatization forms 3,4-dihydroxyphenanthrene, which is further metabolized to 1-hydroxy-2-naphthoate. That is followed by intradiol ring cleavage dioxygenation, producing *o*-phthalate, then protocatechuate, which is further transformed via the β-ketoadipate pathway to TCA cycle intermediates (Brezna et al., 2005; Kim et al., 2007). *Mycobacterium* sp. strain KMS is reported to degrade pyrene through the initial mono- and di-oxygenase attack on C4, C5 position, producing the main metabolites; pyrene-4,5-dione; cis-4,5-py-rene-dihydrodiol; phenanthrene-4,5-dicarboxylic acid; and 4-phenanthroic acid until reaching the TCA cycle (Fig. 9.24) (Liang et al., 2006). However, *Mycobacterium* sp. strain AP1 has been reported to utilize pyrene as a sole carbon and energy sources with a new branched pathway and involves the dioxygenation on 4,5 positions and 9,10 positions and the cleavage of both central rings of the pyrene, with the production of 6,6′-dihydroxy-2,2′-biphenyl dicarboxylic acid (Fig. 9.24) (Vila et al., 2001).

Rehmann et al. (1998) explained the production of 4-phenanthrol from *cis*-3,4-phenanthrene dihydro-diol-4-carboxylic acid, during the biodegradation of pyrene (Fig. 9.24) by *Mycobacterium* sp. strain KR2, as it is sterically favored under the assumption that the bulky carboxy group is extruded from the bay region, thus occupying a pseudoaxial position necessary for elimination. Khanna et al. (2011) found 9-methoxyphenanthrene and phthalate as metabolic intermediates when studied pyrene degradation by *B. pumilus* isolated from crude oil–contaminated soils. *Pseudomonas* sp. Jpyr-1 isolated from active sewage sludge, collected from a wastewater treatment plant in Jilin, the northeastern region of China, for its ability to utilize pyrene as a sole carbon, is reported to degrade other polynuclear aromatic hydrocarbon derivatives, such as benzo[a]anthracene, chrysene, and BaP, with the preference of the enzyme system towards pyrene. The biodegradation of pyrene is suggested to be initiated by di-hydrox-ylation at the C-4 and C-5 positions of pyrene, producing 4,5-dihydrodiol, then 4-phenanthroic acid, that is further metabolized to 1-hydroxy-2-naphthoate, then phthalate, which converted to protocate-chuic, which is then directed into TCA cycles to be mineralized. Kumaria et al. (2014) reported the degradation of pyrene by *Pseudomonas* sp. BP10 through the production of 3,4-dihydroxyphenan-threne, 2-carboxybenzaldehyde, *o*-phthalate, and catechol to TCA-cycle (Ma et al., 2013). *P. aerugi-nosa* strain RS1 isolated from tank bottom sludge collected from a refinery in Mumbai, India, is reported to degrade pyrene as a sole source of carbon and energy (Ghosh et al., 2014), with a degradation path-way similar to that reported for *Mycobacterium* sp. PYR-I (Kim and Freeman, 2005; Kim et al., 2007). *P. aeruginosa* strain RS1 produces *cis*- and *trans*- 4,5-dihydroxypyrene; 1,2-dihydroxypyrene; and 1,2- dimethoxypyrene. 4-oxa-pyren-5-one; cis-3,4-phenanthrene dihydrodiol-4-carboxylate; 3,4-dihy-droxyphenanthrene; phenanthrene 4,5-dicarboxylic acid, 4-phenanthroic acid; 4-phenanthrol, 2,2′-dicarboxy-6,6′dihydroxybiphenyl; phthalic acid; and 4,5-dihydroxyphthalate have also been observed as pyrene metabolites by RS1 (Ghosh et al., 2014).

The carcinogenic and mutagenic chrysene, which is known also as benzo[a]phenanthrene, can be mineralized by the incorporation of an oxygen molecule in an aromatic ring catalyzed by dioxygenases to a *cis*-dihydrodiol intermediate (Hinchee et al., 1994). Two bacterial strains *Bacillus* sp. Chry2 and *Pseudomonas* sp. Chry3 have been isolated from the oily sludge obtained from Gujarat refinery, India, for their ability to degrade chrysene through meta cleavage degradation pathway, with the aid of cate-chol 1,2-dioxygenase and catechol 2,3-dioxygenase enzyme activities (Dhote et al., 2010). Nayak et al. (2011) reported the biodegradation of chrysene by *Pseudoxanthomonas* sp. PNK-04, isolated from a coal using 1-hydroxy-2-naphthoate hydroxylase; 1,2-dihydroxynaphthalene dioxygenase; salicylalde-hyde dehydrogenase; and catechol-1,2-dioxygenase, producing hydroxyphenanthroic acid, 1-hydroxy-2-naphthoic acid, salicylic acid, and catechol, which is degraded via catechol-1,2-dioxygenase to *cis,cis*-muconic acid and then enters TCA-cycle.

Few studies have documented the bacterial biotransformation of benz[a]anthracene. The initial oxida-tion attack can occur as angular kata-type initial dioxygenation onto $C_1–C_2$ or $C_3–C_4$ positions. It can also occur as a linear kata-type initial dioxygenation onto C8,C9 or C10,C11 positions. Finally, it can occur throughout the K-region at the $C_5–C_6$ positions. If the upstream metabolites cannot be detected, the down-stream ones can help on indication of the type of initial oxidation attack. For example, 2-hydroxy-3-naph-thoic acid is a result of an angular kata-type initial dioxygenation attack (Fig. 9.25). Mahaffey et al. (1988) reported the catabolism of benz[a]anthracene (BaA) by *Beijerinckia* sp. strain B1 after induction with biphenyl, *m*-xylene, or salicylate, to one major metabolite (1-hydroxy-2-anthranoic acid) and two minor metabolites (2-hydroxy-3-phenanthroic acid and 3-hydroxy-2-phenanthroic acid). Moody et al. (2003)

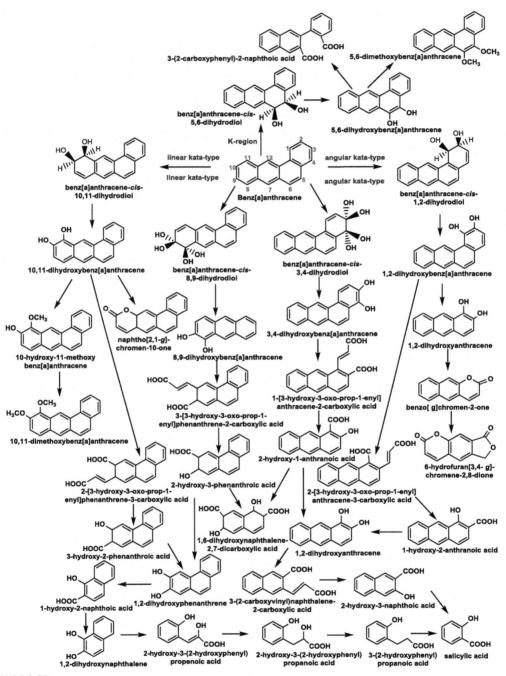

FIGURE 9.25

Some of the detected metabolites in benzo[a] anthracene biodegradation.

reported the regio- and stereo-selective biodegradation of 7,12-dimethylbenz[*a*]anthracene by *M. vanbaalenii* PYR-1 throughout; the mono- and dioxygenation reactions at the 5,6 positions and monooxygenation of the C7 methyl group. PYR-1 has been also reported to initially attack BaA at the C1,C2; C5,C6; C7,C12; and C10,C11 positions, with the predominance of C10,C11 initial attack. That is followed by *ortho*- and *meta*-ring cleavages that result in multiple pathways with different biodegradation metabolites (Moody et al., 2005). *Sphingobium* sp. strain KK22 isolated from a phenanthrene enrichment culture of a bacterial consortium that grew on diesel fuel is reported to degrade BaA throughout the oxidation of both the linear kata and angular kata ends of the BaA molecule (Kunihiro et al., 2013).

BaP is characterized by low abundance in the environment, which limits its catabolism by bacterial assemblages (Seo et al., 2009). Schneider et al. (1996) elucidated the cometabolic biotransformation pathway of BaP in presence of starch by *Mycobacterium* sp. strain RJGII-135. That produced *cis*-7,8-benzo[a] pyrenedihydrodiol; 4,5-chrysenedicarboxylic acid; *cis*-4-(8-hydroxypyren-7-yl)-2-oxobut-3-enoic acid [or *cis*-4-(7-hydroxypren-8-yl)-2-oxobut-3-enoic acid]; and 7,8-dihydropyrene-7-carboxylic acid (or 7,8-dihydropyrene-8-arboxylic acid) (Fig. 9.26). The authors were unable to distinguish between the meta fission products through the 7,8 bond and the 9,10 bond of BaP, hence the possibility of two products for two of the metabolites. Moody et al. (2004) elucidated that *Mycobacterium vanbaalehii* PYR-1 can utilize BaP throughout different pathways with mono- and di-oxygenase initial attack. For example, the benzo[*a*]pyrene *cis*-7,8-dihydrodiol is further catabolized via *meta*-cleavage to 7,8-dihydro-pyrene-8-carboxylic acid through *cis*-4-(7-hydroxypyrene-8-yl)-2-oxobut-3-enoic acid. While, benzo[*a*]pyrene *cis*-9,10-dihydrodiol is degraded via *meta*-cleavage to 7,8-dihydropyrene-7-carboxylic acid through *cis*-4-(8-hydroxypyrene-7-yl)-2-oxobut-3-enoic acid. In addition, the 10-oxabenzo[*def*]-chrysene-9-one formed by dehydration of benzo[*a*] pyrene *cis*-9,10-dihydrodiol produces *cis*-4-(8-hydroxypyrene-7-yl)-2-oxobut-3-enoic acid with subsequent *meta*-cleavage and aromatic ring closure. The *ortho*-cleavage of *cis*-4,5-dihydroxy BaP produces 4-formylchrysene-5-carboxylic acid, which is transformed to 4,5-chrysene dicarboxylic acid. Moreover, the mono- and di-oxygenation of BaP at C11,C12 position produces *cis*-11,12-benzo[a]pyrenedihydrodiol and *trans*-11,12-benzo[a]pyrenedihydrodiol, respectively. The bay-region and K-region in polynuclear aromatic hydrocarbon derivatives can be oxidized by P-450 to their corresponding highly chemically and biologically reactive bay- and K-region epoxides. For example, cytochrome P-450 oxidizes BAP to benzo[*a*]pyrene-11,12-epoxide, which is subsequently hydrolyzed by epoxide hydrolase to *trans*-11,12-benzo[a]pyrenedihydrodiol, which is further dehydrogenated, producing the 11,12-dihydroxybenzo[a]pyrene. The latter is further catabolized to hydroxymethoxybenzo[*a*]pyrene, then, dimethoxybenzo[*a*]pyrene (Fig. 9.26). *S. yanoikuyae* JAR02 is reported to produce pyrene-8-hydroxy-7-carboxylic acid and pyrene-7-hydroxy-8-carboxylic acid during the biodegradation of pyrene (Rentz et al., 2008).

High concentrations of alkyl-benzenes in polluted soil and sediment samples would occur from somewhat biodegradation of the lighter hydrocarbons originally present in crude oils and/or due to the photo-oxidation and break-down products of higher aromatics (Moustafa et al., 2005; El-Gendy et al., 2009). El-Gendy et al. (2010) reported that as the molecular weight of polyaromatic hydrocarbons and number of alkyl group of polyaromatic compounds increases the aqueous solubility and bioavailability decrease and therefore exhibits a protective function against biodegradation. Alkylation of aromatic ring increases its persistence in the environment. The rate of polycyclic aromatic compound biodegradation in mixed substrates culture decreases with increasing ring size and within a homologous series, decreases with increasing alkylation (Neff, 1979; Fedorak and Westlake, 1983, 1984; Douglas et al., 1994; Elmendorf et al., 1994; El-Gendy et al., 2010). Alkyl groups may inhibit proper orientation and accessibility of the polynuclear aromatic hydrocarbon derivatives into the initiating mono- or di-oxygenases. However, their biodegradation

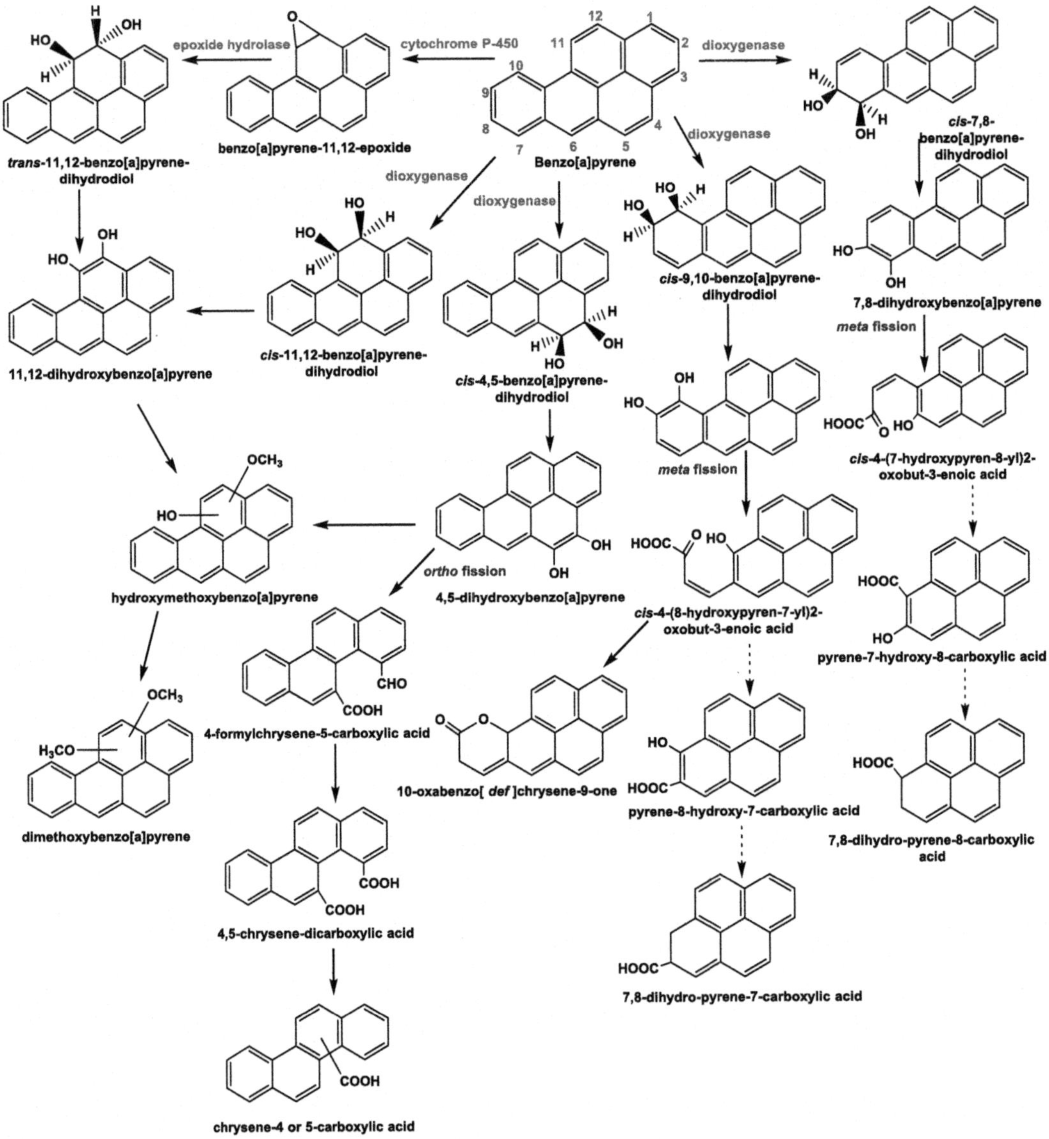

FIGURE 9.26

Different metabolic pathways reported for aerobic biodegradation of benzo[a]pyrene.

has been reported, throughout a complicated pathway involving several enzymatic steps; oxidation of methyl group to alcohol, aldehyde, or carboxylic acid, decarboxylation, demethylation, and dioxygenation (Mahajan et al., 1994; Moody et al., 2003). But, in some studies, the detection of alkyl-salicylate or alkyl-phthalate suggests that the reaction may prefer nonsubstituted polynuclear aromatic hydrocarbon derivative systems. The biodegradation of 7,12-dimethylbenz[a]anthracene (DMBA) by *M. vanbaalenii* PYR-1 has been

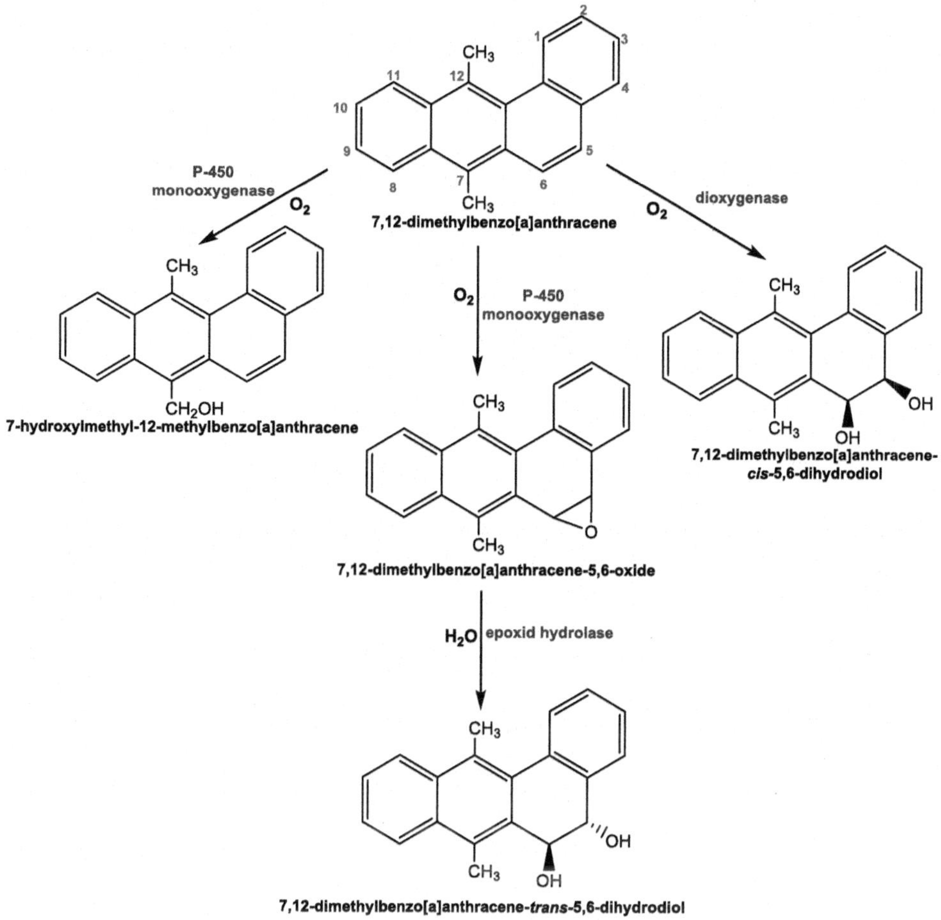

FIGURE 9.27

Aerobic biodegradation of alkylated polynuclear aromatic hydrocarbon derivatives.

reported to be initiated by the di- and mono-oxygenation attack at the K-region onto the C5,C6 position and the methyl group attached to C7 that produced *cis*-5,6-dihydro-5,6-dihydroxy-7,12-dimethylbenz[a]anthracene (i.e., DMBA *cis*-5,6-dihydrodiol); *trans*-5,6-dihydro-5,6-dihydroxy-7,12-dimethylbenz[a]anthracene (i.e., DMBA trans-5,6-dihydrodiol); and 7-hydroxymethyl-12-methylbenz[a]anthracene (Fig. 9.27).

Hydroxylated biphenyls, for example, 2-hydroxybiphenyl, 2,2′-bidhydroxybiphenyl and 2,2′,3-trihy-droxybiphenyl, are reported to be also produced through the biotransformation reactions preceded via hydroxylation of the aromatic rings in biodegradation of different studied polyaromatic compounds and from the biodegradation of the sulfur heterocyclic compounds, for example, dibenzothiophene, through the 4S-pathway (Kohler et al., 1993; Lee et al., 1995; Sondossi et al., 2004; Chen et al., 2008; El-Gendy et al., 2006, 2014a,b, 2010; Deriase et al., 2012). Microbial metabolism and environmental fate of these compounds is desired since they are by-products that have been identified as contaminants in almost every component of the global ecosystem and because they constitute a severe environmental hazard because of their high toxicity and reported to express inhibition effect to microbial metabolism. Monohydroxylated biphenyl, 2-hydroxybiphenyl and 3-hydroxybiphenyl, are reported to express more inhibitory effect on *Comamonas testosteroni* TK102 than that of 2,3-bihydroxybiphenyl (Hiraoka et al., 2002). 2-HBP is reported also to be more toxic to *C. variabilis* sp. Sh42 than 2,2′-bihydroxybihenyl 9 (El-Gendy et al., 2010). However, they are reported to be further metabolized to benzoic acid than to complete mineralization (Fig. 9.28) (Van Afferden et al., 1990; El-Gendy, 2006) or methoxylated in a way to overcome their microbial toxicity (Tanaka et al., 2002; Li et al., 2003; Gunam et al., 2006; Zhang et al., 2007; Chen et al., 2009; El-Gendy et al., 2014a,b).

Polynuclear aromatic hydrocarbon derivatives are always found in contaminated area, as mixtures of different compounds having two to seven condensed rings and the substrate interaction affect the biodegradation of polynuclear aromatic hydrocarbon derivatives by pure and mixed cultures. Interactions between polynuclear aromatic hydrocarbon derivatives reveal either inhibition or stimulation (Dean-Ross et al., 2002; Reardon et al., 2002; El-Gendy and Abostate, 2008). Low-molecular-weight (low molecular weight) polynuclear aromatic hydrocarbon derivatives have been reported to affect the degradation of high-molecular-weight polynuclear aromatic hydrocarbon derivatives, and vice versa via cometabolism, induction of enzyme activities, and by competitive inhibition. However, it sometimes enhances the polynuclear aromatic hydrocarbon derivatives degradation, since metabolic intermediates produced by certain bacteria may serve as substrates for the growth of others (Younis et al., 2013). Usually, high-molecular-weight polynuclear aromatic hydrocarbon derivatives are degraded after the low–molecular-weight polynuclear aromatic hydrocarbon derivatives (Meuller et al., 1989; Yu et al., 2005; Hong et al., 2008). *Corynebacterium variabilis* sp. Sh42 is reported to degrade different polyaromatic hydrocarbons in the following decreasing order: naphthalene > anthracene ≈ phenanthrene > pyrene, wither in single or mixed-substrate system (El-Gendy et al., 2010). But, as a general observation recorded in that study, although the microbial growth efficiency was not affected, the overall degradation percentage of each studied polyaromatic hydrocarbon in mixed-substrate system was lower than that recorded in single-substrate system, except for naphthalene that indicated competitive inhibition effect of multiple substrates, retarding the degradation of one substrate in the presence of another, as similar or identical enzyme systems may catalyze the degradation of compound(s). However, high concentration of naphthalene is reported to be toxic and inhibits the biodegradation of polynuclear aromatic hydrocarbon derivatives Bouchez et al. (1995). Inhibition is known to be most commonly observed when the added polynuclear aromatic hydrocarbon derivatives is more water soluble than the originally added (Bouchez et al., 1996). Strignfellow and Cerniglia (1996) and Bouchez et al. (1995) attributed the competitive inhibition to the possible common enzyme systems responsible for biodegradation and/or to the higher water solubility of one substrate over the other, which would make it more bioavailable than the less soluble substrate. Stringfellow and Aitken (1995) reported the competitive inhibition of phenanthrene degradation by naphthalene, methylnaphthalene, and fluorene in binary mixtures using two pure cultures of *Pseudomonas* sp. and the presence of phenanthrene is reported to inhibit degradation of pyrene McNally et al. (1999). Biodegradation of anthracene by *Rhodococcus* spp. and pyrene by *M. flavescens* has also been reported to be inhibited by the presence of fluoranthene (Dean-Ross et al., 2002). Fluorene degradation by *Sphingomonas* sp. strain LB126 is reported to be inhibited by both phenanthrene and fluoranthene and the effect of phenanthrene is about 10 times stronger than that of fluoranthene (Van Herwijnen et al., 2003c). The pyrene biodegradation by *Mycobacterium* sp. strain CH1 is stimulated and the lag phase of its degradation is decreased with the presence of phenanthrene or fluoranthene (Churchill et al., 1999). El-Gendy and Abostate (2008) reported that in a binary substrate system naphthalene did not express an inhibitory effect on the biodegradation of phenanthrene by *S. gallinarum* NK1. However, it expressed an inhibitory effect on the biodegradation of

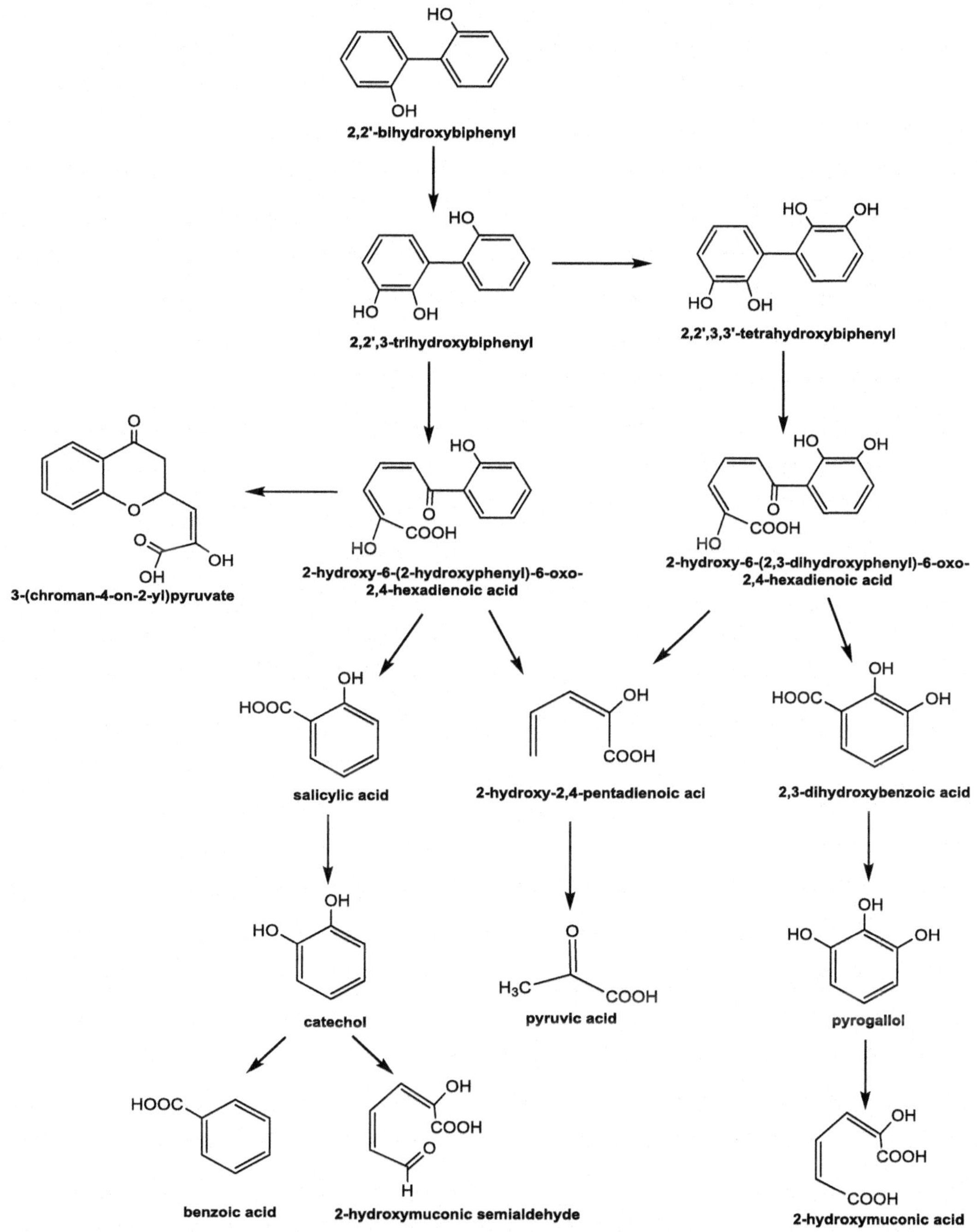

FIGURE 9.28

Proposed metabolic pathway of 2,2′-bihydroxybiphenyl by *Corynebacterium variabilis* sp. Sh42 (El-Gendy et al., 2010).

dibenzothiophene. Moreover, the presence of phenanthrene and dibenzothiophene in a binary system expressed competitive inhibition and their biodegrading rate by NK1 was decreased in binary system than in single substrate system. Moreover, dibenzothiophene expressed inhibitory effect on biodegradation of naphthalene. Dibenzothiophene biodegradation was retarded by the presence of phenanthrene or naphthalene in binary and trinary substrates systems. Meyer and Steinhart (2000) reported that the presence of heterocyclic polyaromatic hydrocarbons containing nitrogen, sulfur, or oxygen have a significant inhibiting effect on biodegradation of monaromatic hydrocarbons and polynuclear aromatic hydrocarbon derivatives. Nadaling et al. (2002) reported that the presence of dibenzothiophene inhibited the biodegradation of 2-methylphenanthrene by *Sphingomonas* sp. 2MPII, isolated from marine sediment, and attributed this to the production of dibenzothiophene sulfone, which cannot be assimilated by 2MPII. El-Gendy et al. (2010) reported that the presence of polyaromatic sulfur heterocyclic compounds decrease the biodegradation of polyaromatic hydrocarbons by *C. variabilis* sp. Sh42, and attributed this to the toxic metabolites produced during the biodegradation of the studied polyaromatic compounds. Foght and Westlake (1990) reported that some dibenzothiophene-degradative genes have DNA homology with the naphthalene-degradative plasmid, which would result in competition between the expressions of the degradative genes of these compounds. Denome et al. (1993) reported that *Pseudomonas* strain C_{18} has a single genetic pathway that controls the metabolism of dibenzothiophene, naphthalene, and phenanthrene, which would lead to competitive inhibition for the expression of the degradative genes of these compounds. However, in another study using *C. variabilis* sp. Sh42, the average rate of biodegradation of different polyaromatic compounds were ranked in the following decreasing order: phenolic compounds > polyaromatic hydrocarbons > polyaromatic sulfur heterocyclic compounds in both single and mixed substrates (El-Gendy et al., 2010). Wei et al. (2009) reported that the inhibitory effects from interactions between the three studied polynuclear aromatic hydrocarbon derivative mixtures (Phe, Pyr, and Flu) on *Mycobacterium* sp. MEBIC 5140, which led to a negative effect on the biodegradation of these polynuclear aromatic hydrocarbon derivatives. Haritash and Kaushik (2009) reported that the presence of naphthalene (2-ringed polynuclear aromatic hydrocarbon derivatives) stimulated phenanthrene (3-ringed polynuclear aromatic hydrocarbon derivatives) degradation five-fold and pyrene (4-ringed polynuclear aromatic hydrocarbon derivatives) degradation two-fold by *P. putida* strain KBM-1. Yuan et al. (2002) reported that the biodegradation of a mixture of acenaphthene, fluorene, phenanthrene, anthracene, and pyrene was higher than individual ones and attributed this to the cross acclimation of more carbon sources, which enhances the rate of biodegradation. Chang et al. (2002) reported that the anaerobic biodegradation of a mixture of phenanthrene, pyrene, anthracene, fluorene, and acenaphthene was higher than individual ones. In binary substrate system studied by Ma et al. (2013), phenanthrene enhanced the biodegradation of pyrene by *Pseudomonas* sp. Jpyr-1 and related this to the higher solubility of phenanthrene than pyrene, which would increase the enzyme induction and enhance its activity and the transformation capacity of the polynuclear aromatic hydrocarbon derivatives. Thus, the positive feedback effect exerted on the degradation process by the enzymes induced during PHE degradation also stimulated the PYR degradation process. Moreover, some metabolites formed via the PHE pathway such as 1-hydroxy-2-naphthoate also facilitated the PYR degradation process. But, fluorene, anthracene, and fluoranthene decreased the biodegradation rate of pyrene, due to competitive inhibition, where the metabolites of fluorene, anthracene, and fluoranthene biodegradation may have toxic effects on the cells. Mohd Radzi et al. (2015) reported that the competition among the microbes alongside their individual extracellular potentials influence the kinetic of the degradation process.

Even under anaerobic conditions, competitive inhibition was reported. Tsai et al. (2009) reported that the anaerobic biotransformation of fluorene and phenanthrene by sulfate-reducing bacteria in binary system is lower than that of single-substrate system. The degradation rates of fluorene and phenanthrene in the single compound systems were 0.136 and 0.09 d^{-1}, respectively. But in binary system the rate decreased to 0.098 and 0.072 d^{-1}, respectively, which was attributed to high concentration of total polynuclear aromatic hydrocarbon derivatives in binary system. However, Yuan and Chang (2007), in a previous study, reported the enhancement of the anaerobic biodegradation rate of five polynuclear aromatic hydrocarbon derivatives in multi-substrate system than in single ones. Generally, the degradation rates of the studied polynuclear aromatic hydrocarbon derivatives were in the order acenaphthene > fluorene > phenanthrene > anthracene > pyrene. The polynuclear aromatic hydrocarbon derivative degradation rates under the studied three reducing conditions were in the following order: sulfate-reducing conditions > methanogenic conditions > nitrate-reducing conditions. Moreover, the addition of electron donors (acetate, lactate and pyruvate) enhanced the polynuclear aromatic hydrocarbon derivative degradation under methanogenic- and sulfate-reducing conditions. However, the addition of acetate, lactate, or pyruvate inhibited the polynuclear aromatic hydrocarbon derivative degradation under nitrate-reducing conditions. The addition of heavy metals, nonylphenol, and phthalate esters inhibited the polynuclear aromatic hydrocarbon derivative degradation.

REFERENCES

Abioye, O.P., Akinsola, R.O., Aransiola, S.A., Damisa, D., 2013. Biodegradation of crude oil by *Saccharomyces cerevisiae* isolated from fermented zobo (locally fermented beverage in Nigeria). Pakistan Journal of Biological Sciences 16 (24), 2058–2061.

Adekunle, A.A., Adebambo, O.A., 2007. Petroleum hydrocarbon utilization byfungi isolated from *Detarium Senegalense* (J. F. Gmelin) seeds. Journal of American Science 3, 69–76.

Adeogun, O.O., Adekunle, A.A., 2015. Biodegradation of petroleum products using phylloplane fungi isolated from selected plants. Journal of Scientific Research and Development 15, 45–53.

Aeckersberg, F., Bak, F., Widdel, F., 1991. Anaerobic oxidation of saturated hydrocarbons to CO_2 by a new type of sulfate-reducing bacterium. Archives of Microbiology 156, 5–14.

Aeckersberg, F., Rainey, F.A., Widdel, F., 1998. Growth, natural relationships, cellular fatty acids and metabolic adaptation of sulfate-reducing bacteria that utilize long-chain alkanes under anoxic conditions. Archives of Microbiology 170, 361–369.

Ahmed, A.T., Othman, M.A., Sarwade, V.D., Gawai, K.R., 2012. Degradation of anthracene by alkaliphilic bacteria *Bacillus badius*. Environment and Pollution 1 (2), 97–104.

Aitken, M.D., Chen, S.H., Kazunga, C., Marx, R.B., 2005. Bacterial Biodegradation of High Molecular Weight Polyaromatic Hydrocarbon (ESE notes feature article, published by the Department of Environmental Sciences and Engineering at the University of North Carolina at Chapel Hill).

Aitken, M.D., Stringfellow, W.T., Nagel, R.D., Kazunga, C., Chen, S.H., 1998. Characteristics of phenanthrene-degrading bacteria isolated from soils contaminated with polycyclic aromatic hydrocarbons. Canadian Journal of Microbiology 44, 743–752.

Akhtar, N., Ghauri, M.A., Anwar, M.A., Akhtar, K., 2009. Analysis of the dibenzothiophene metabolic pathway in a newly isolated *Rhodococcus* spp. FEMS Microbiology Letters 301, 95–102.

Ali, H.R., El-Gendy, N.Sh., El- Ezbewy, S., El-Gemaee, G.H., Moustafa, Y.M., Roushdy, M.I., 2006. Assessment of polycyclic aromatic hydrocarbons contamination in water, sediment and fish of Temsah Lake, Suez Canal, Egypt. Current World Environment 1 (1), 11–22.

Ali, H.R., Ismail, D.A., El-Gendy, N.Sh., 2014. The Biotreatment of oil polluted seawater by biosurfactant producer halotolerant *Pseudomonas aeruginosa* Asph2. Energy Sources, Part A: Recovery, Utilization, and Environmental Effects 36 (13), 1429–1436.

Al-Nasrawi, H., 2012. Biodegradation of crude oil by fungi isolated from Gulf of Mexico. Journal of Bioremediation and Biodegradation. 3 (4). https://doi.org/10.4172/2155-6199.1000147.

Ambrosoli, R., Petruzzelli, L., Luis Minati, J., Ajmone Marsan, F., 2005. Anaerobic polynuclear aromatic hydrocarbon derivatives degradation in soil by a mixed bacterial consortium under denitrifying conditions. Chemosphere 60 (9), 1231–1236.

Annweiler, E., Materna, A., Safinowski, M., Kappler, A., Richnow, H.H., Michaelis, W., Meckenstock, R.U., 2000a. Anaerobic degradation of 2-methylnaphthalene by a sulfate-reducing enrichment culture. Applied and Environmental Microbiology 66 (12), 5329–5333.

Annweiler, E., Michaelis, W., Meckenstock, R.U., 2002. Identical ring cleavage products during anaerobic degradation of naphthalene, 2-methylnaphthalene, and tetralin indicate a new metabolic pathway. Applied and Environmental Microbiology 68, 852–858.

Annweiler, E., Richnow, H.H., Antranikian, G., Hebenbrock, S., Garms, C., Franke, S., Francke, W., Michaelis, W., 2000b. Naphthalene degradation and incorporation of naphthalene-derived carbon into biomass by the thermophile *Bacillus thermoleovorans*. Applied and Environmental Microbiology 66, 518–523.

Arata, T.H., Kristanti, R.A., 2014. Potential of a white-rot fungus *Pleurotus eryngii* F032 for degradation and transformation of fluorene. Fungal Biology 118, 222–227.

Arulazhagan, P., Vasudevan, N., 2009. Role of a moderately halophilic bacterial consortium in the biodegradation of polyaromatic hydrocarbons. Marine Pollution Bulletin 58, 256–262.

Arulazhagan, P., Vasudevan, N., 2011a. Biodegradation of polycyclic aromatic hydrocarbons by a halotolerant bacterial strain *Ochrobactrum* sp. VA1. Marine Pollution Bulletin 62, 388–394.

Arulazhagan, P., Vasudevan, N., 2011b. Role of nutrients in the utilization of polycyclic aromatic hydrocarbons by halotolerant bacterial strain. Journal of Environmental Sciences 23 (2), 282–287.

Atagana, H.I., Haynes, R.J., Wallis, F.W., 2006. Fungal bioremediation of creosote contaminated soil: a laboratory scale bioremediation study using indigenous soil fungi. Water Air Soil Pollution 172, 201–219.

Baboshin, M., Akimov, V., Baskunov, B., Born, T.L., Khan, S.U., Golovleva, L., 2008. Conversion of polycyclic aromatic hydrocarbons by *Sphingomonas* sp. VKM B-2434. Biodegradation 19, 567–576.

Balachandran, C., Duraipandiyan, V., Balakrishna, K., Ignacimuthu, S., 2012. Petroleum and polycyclic aromatic hydrocarbons (polynuclear aromatic hydrocarbon derivatives) degradation and naphthalene metabolism in *Streptomyces* sp. (ERI-CPDA-1) isolated from oil contaminated soil. Bioresource Technology 112, 83–90.

Balashova, N.V., Stolz, A., Knackmuss, H.J., Kosheleva, I.A., Naumov, A.V., Boronin, A.M., 2001. Purification and characterization of a salicylate hydroxylase involved in 1-hydroxy-2-naphthoic acid hydroxylation from the naphthalene and phenanthrene degrading bacterial strain *Pseudomonas putida* BS202-P1. Biodegradation 12, 179–188.

Baldrian, P., 2006. Fungal laccases-occurrence and properties. FEMS Microbiology Reviews 30, 215–242.

Ball, H.A., Johnson, H.A., Reinhard, M., Spormann, A.M., 1996. Initial reactions in anaerobic ethylbenzene oxidation by a denitrifying bacterium, strain EB1. Journal Bacteriology 178, 5755–5761.

Barabas, G.Y., Vargha, G., Szabo, I.M., Penyige, A., Damjanovich, S., Szollosi, J., Matk, J., Hirano, T., Atyus, A.M., 2001. n-Alkane uptake and utilization by *Streptomyces* strains. Antonie van Leeuwenhoek 79, 269–276.

Barnsley, E.A., 1975. The bacterial degradation of fluoranthene and benzo[a]pyrene. Canadian Journal of Microbiology 21, 1004–1008.

Barton, L.L., Fauque, G.D., 2009. Biochemistry, physiology and biotechnology of sulfate-reducing bacteria. Advances in Applied Microbiology 58, 41–98.

Basta, T., Keck, A., Klein, J., Stolz, A., 2004. Detection and characterization of conjugative degradative plasmids in xenobiotics degrading *Sphingomonas* strains. Journal of Bacteriology 186, 3862–3872.

Basta, T., Buerger, S., Stolz, A., 2005. Structural and replicative diversity of large plasmids from sphingomonads that degrade polycyclic aromatic compounds and xenobiotics. Microbiology 151, 2025–2037.

Beller, H.R., Grbić- Galić, D., Reinhard, M., 1992. Microbial degradation of toluene under sulfate-reducing conditions and the influence of ion on the process. Applied and Environmental Microbiology 58, 786–793.

Ben Said, O., Gon, M.S., Urriza, M., Bour, E., Dellali, M., Aissa, P., Duran, P., 2008. Characterization of aerobic polycyclic aromatic hydrocarbon-degrading bacteria from Bizerte lagoon sediments, Tunisia. Journal of Applied Microbiology 104, 987–997.

Berdugo-Clavijo, C., Dong, X., Soh, J., Sensen, C.W., Gieg, L.M., 2012. Methanogenic biodegradation of two-ringed polycyclic aromatic hydrocarbons. FEMS Microbiology Ecology 81 (1), 124–133.

Bogan, B.W., Lamar, R.T., 1995. Polycyclic aromatic hydrocarbons degrading capability of *Phanerochaete laevis* HHB-1625 and its extracellular ligninolytic enzymes. Applied and Environmental Microbiology 62, 1597–1603.

Bogan, B.W., Lamar, R.T., Hammel, K.E., 1996a. Fluorene oxidation in vivo by *Phanerochaete chrysosporium* and in vitro during manganese peroxidase dependent lipid peroxidation. Applied and Environmental Microbiology 62, 1788–1792.

Bogan, B.W., Schoenike, B., Lamar, R.T., Cullen, D., 1996b. Expression of lip genes during growth in soil and oxidation of anthracene by *Phanerochaete chrysosporium*. Applied and Environmental Microbiology 62, 3697–3703.

Boronin, A.M., 2001. Purification and characterization of a salicylate hydroxylase involved in 1-hydroxy-2-naphthoic acid hydroxylation from the naphthalene and phenanthrene degrading bacterial strain *Pseudomonas putida* BS202-P1. Biodegradation 12, 179–188.

Bosch, R., Garcia-Valdes, E., Moore, E.R.B., 2000. Complete nucleotide sequence and evolutionary significance of a chromosomally encoded naphthalene-degradation lower pathway from *Pseudomonas stutzeri* AN10. Gene 245, 67–74.

Bouchez, M., Blanchet, D., Vandecasteele, J.P., 1995. Degradation of polycyclic aromatic hydrocarbons by pure strains and by defined strain associations: inhibition phenomena and cometabolism. Applied Microbiology and Biotechnology 43 (1), 156–164.

Bouchez, M., Blanchet, D., Vandecasteele, V.-P., 1996. The microbiological fate of polycyclic aromatic hydrocarbons: carbon and oxygen balances for bacterial degradation of model compounds. Applied Microbiology and Biotechnology 45, 556–561.

Boyd, D.R., Sharma, N.D., Belhocine, T., Malone, J.F., McGregor, S., Allen, C.C.R., 2006. Dioxygenasecatalysed dihydroxylation of arene cis-dihydrodiols and acetonide derivatives: a new approach to the synthesis of enantiopure tetraoxygenated bioproducts from arenes. Chemical Communications 4934–4936.

Boyd, D.R., Sheldrake, G.N., 1998. The dioxygenase-catalysed formation of vicinal cis-diols. Natural Product Reports 15, 309–324.

Brezna, B., Kweon, O., Stingley, R.L., Freeman, J.P., Khan, A.A., Polek, B., Jones, R.C., Cerniglia, C.E., 2005. Molecular characterization of cytochrome P450 genes in the polycyclic aromatic hydrocarbon degrading *Mycobacterium vanbaalenii* PYR-1. Applied Microbiology and Biotechnology 11, 1–11.

Brooijmans, R.J.W., Pastink, M.I., Siezen, R.J., 2009. Hydrocarbon-degrading bacteria: the oil-spill clean-up crew. Microbial Biotechnology 2 (6), 587–594.

Bumpus, J.A., Tien, M., Wright, D., Aust, S.D., 1985. Oxidation of persistent environmental pollutants by a white rot fungus. Science 228 (4706), 1434–1436.

Caiazza, N.C., Shanks, R.M., O'Toole, G.A., 2005. Rhamnolipids modulate swarming motility patterns of *Pseudomonas aeruginosa*. Journal of Bacteriology 187, 7351–7361.

Cajthaml, T., Erbanová, P., Šašek, V., Moder, M., 2006. Breakdown products on metabolic pathway of degradation of benz[a]anthracene by a ligninolytic fungus. Chemosphere 64, 560–564.

Cajthaml, T., Moder, M., Kacer, P., Šašek, V., Popp, P., 2002. Study of fungal degradation products of polycyclic aromatic hydrocarbons using gas chromatography with ion trap mass spectrometry detection. Journal Chromatography A 974, 213–222.

Caldini, G., Cenci, G., Manenti, R., Morozzi, G., 1995. The ability of an environmental isolate of *Pseudomonas fluorescens* to utilize chrysene and other four-ring polynuclear aromatic hydrocarbons. Applied Microbiology and Biotechnology 44, 225–229.

Callaghan, A.V., Gieg, L.M., Kropp, K.G., Suflita, J.M., Young, L.Y., 2006. Comparison of mechanisms of alkane metabolism under sulfate-reducing conditions among two bacterial isolates and a bacterial consortium. Applied and Environmental Microbiology 72, 4274–4282.

Callaghan, A.V., Morris, B.E., Perreira, I.A., McInerney, M.J., Austin, R.N., Groves, J.T., Kukor, J.J., Suflita, J.M., Young, L.Y., Zylstra, G.J., Wawrik, B., 2012. The genome sequences of *Desulfatibacillum alkenivorans* AK-01: a blueprint for anaerobic alkane oxidation. Environmental Microbiology 14, 101–113.

Cañas, A., Alcalde, M., Plou, F., Martínez, M.J., Martínez, A.T., Camarero, S., 2007. Transformation of polycyclic aromatic hydrocarbons by laccase is strongly enhanced by phenolic compounds present in soil. Environmental Science & Technology 41, 2964–2971.

Cao, J., Lai, Q., Yuan, J., Shao, Z., 2015. Genomic and metabolic analysis of fluoranthene degradation pathway in *Celeribacter indicus* P73ᵀ. Scientific Reports 5, 7741. http://dx.doi.org/10.1038/srep07741.

Casellas, M., Grifoll, M., Bayona, J.M., Solanas, A.M., 1997. New metabolites in the degradation of fluorene by *Arthrobacter* sp. strain F101. Applied and Environmental Microbiology 63 (3), 816–826.

Cerniglia, C.E., 1992. Biodegradation of polycyclic aromatic hydrocarbons. Biodegradation 3, 351–368.

Cerniglia, C.E., 1997. Fungal metabolism of polycyclic aromatic hydrocarbons: past, present and future applications of bioremediation. Journal of Industrial Microbiology & Biotechnology 19, 324–333.

Cerniglia, C.E., Heitkamp, M.A., 1990. Polycyclic aromatic hydrocarbon degradation by *Mycobacterium*. Methods in Enzymology 188, 148–153.

Chaillan, F., Le Fleche, A., Bury, E., Phantavong, Y.H., Grimont, P., Saliot, A., Oudot, J., 2004. Identification and biodegradation potential of tropical aerobic hydrocarbon degrading microorganisms. Research in Microbiology 155 (7), 587–595.

Chakraborty, R., Coates, J.D., 2004. Anaerobic degradation of monoaromatichydrocarbons. Applied Microbiology and Biotechnology 64, 437–446.

Champion, K.M., Zengler, K., Rabus, R., 1999. Anaerobic degradation of ethylbenzene and toluene in denitrifying strain EbN1 proceeds via independent substrate-induced pathways. Journal of Molecular Microbiology and Biotechnology 1, 157–164.

Chang, B.V., Chang, I., Yuan, S., 2008. Anaerobic degradation of phenanthrene and pyrene in mangrove sediment. Bulletin of Environmental Contamination and Toxicology 80 (2), 145–149.

Chang, B.V., Shiung, L.C., Yuan, S.Y., 2002. Anaerobic biodegradation of polycyclic aromatic hydrocarbon in soil. Chemosphere 48, 717–724.

Chen, H., Cai, Y., Zhang, W., Li, W., 2009. Methoxylation pathway in biodesulfurization of model organosulfur compounds with *Mycobacterium* sp. Bioresource Technology 100, 2085–2087.

Chen, H., Zhang, W.-J., Cai, Y.B., Zhang, Y., Li, W.C., 2008. Elucidation of 2-hydroxybiphenyl effect on dibenzothiophene desulfurization by *Microbacterium* sp. strain ZD-M2. Bioresource Technology 99, 6928–6933.

Cheng, Q., Thomas, S.M., Rouviere, P.E., 2002. Biological conversion of cyclic alkanes and cyclic alcohols into dicarboxylic acids: biochemical and molecular basis. Applied Microbiology and Biotechnology 58, 704–711.

Chung, W.K., King, G.M., 2001. Isolation, characterization, and polyaromatic hydrocarbon degradation potential of aerobic bacteria from marine macrofaunal burrow sediments and description of *Lutibacterium anuloederans* gen. nov., sp. nov. and *Cycloclasticus spirillensus* sp. nov. Applied and Environmental Microbiology 67, 5585–5592.

Churchill, S.A., Harper, J.P., Churchill, P.F., 1999. Isolation and characterization of a *Mycobacterium* species capable of degrading three and four-ring aromatic and aliphatic hydrocarbons. Applied and Environmental Microbiology 65, 549–552.

Coates, J.D., Chakraborty, R., McInerney, 2002. Anaerobic benzene biodegradation – a new era. Research in Microbiology 153, 621–628.

Cravo-Laureau, C., Matheron, R., Cayol, J.-L., Joulian, C., Hirschler-Rea, A., 2004. *Desulfatibacillum aliphaticivorans* gen. nov., sp. nov., an n-alkane- and n-alkene degrading, sulfate-reducing bacterium. International Journal of Systematic and Evolutionary Microbiology 54, 77–83.

Cui, C., Ma, L., Shi, J., Lin, K., Luo, Q., Li, Y., 2014. Metabolic pathway for degradation of anthracene by halophilic *Martelella* sp. AD-3. International Biodeterioration and Biodegradation 89, 67–73.

Daane, L.L., Harjono, I., Zylstra, G.J., Haggblom, M.M., 2001. Isolation and characterization of polycyclic aromatic hydrocarbons-degrading bacteria associated with the rhizosphere of salt marsh plants. Applied and Environmental Microbiology 67 (6), 2683–2691.

Dandie, C.E., Thomas, S.M., Bentham, R.H., McClure, N.C., 2004. Physiological characterization of *Mycobacterium* sp. strain 1B isolated from a bacterial culture able to degrade high-molecular-weight polycyclic aromatic hydrocarbons. Journal of Applied Microbiology 97, 246–255.

Das, K., Mukherjee, A.K., 2007a. Crude petroleum-oil biodegradation efficiency of *Bacillus subtilis* and *Pseudomonas aeruginosa* strains isolated from a petroleum-oil contaminated soil from North-East India. Bioresource Technology 98 (7), 1339–1345.

Das, K., Mukherjee, A.K., 2007b. Differential utilization of pyrene as the sole source of carbon by *Bacillus subtilis* and *Pseudomonas aeruginosa* strains: role of biosurfactants in enhancing bioavailability. Journal of Applied Microbiology 102 (1), 195–203.

Das, N., Chandran, P., 2011. Microbial degradation of petroleum hydrocarbon contaminants: an overview. Biotechnology Research International 2011:941810. http://dx.doi.org/10.4061/2011/941810. 13.

Daugulis, A.G., McCracken, C.M., 2003. Microbial degradation of high and low molecular weight polyaromatic hydrocarbons in a two-phase partitioning bioreactor by two strains of *Sphingomonas* sp. Biotechnology Letters 25 (17), 1441–1444.

Dave, B.P., Ghevariya, C.M., Bhatt, J.K., Dudhagara, D.R., Rajpara, R.K., 2015. Enhanced chrysene degradation by a mixed culture Biorem-CgBD using response surface design. Indian Journal of Experimental Biology 53, 256–263.

Davidova, I.A., Duncan, K.E., Choi, O.K., Suflita, J.M., 2006. *Desulfoglaeba alkanexedens* gen. nov., sp. nov., an n-alkane-degrading, sulfate-reducing bacterium. International Journal of Systematic and Evolutionary Microbiology 56, 2737–2742.

Dean, J.A., 1992. Lange's Handbook of Chemistry, fourteenth ed. McGraw-Hill, New York, NY.

Dean-Ross, D., Cerniglia, C.E., 1996. Degradation of pyrene by *Mycobacterium flavescens*. Applied Microbiology and Biotechnology 46, 307–312.

Dean-Ross, D., Moody, J., Cerniglia, C.E., 2002. Utilization of mixtures of polycyclic aromatic hydrocarbons by bacteria isolated from contaminated sediment. FEMS Microbiology Ecology 41 (1), 1–7.

Deka, H., Lahkar, J., 2017. Biodegradation of benzo(a)anthracene employing *Paenibacillus* sp. HD1 polynuclear aromatic hydrocarbon derivatives: a novel strain isolated from crude oil contaminated soil. Polycylic Aromatic Compounds 37 (2/3), 161–169.

Denome, S.A., Stanley, D.C., Olson, E.S., Young, K.D., 1993. Metabolism of dibenzothiophene and naphthalene in *Pseudomonas* sp. trains: complete DNA sequence of an upper naphthalene catabolic pathway. Journal of Bacteriology 175, 6890–6901.

Deriase, S.F., El-Gendy, N.Sh., Nassar, H.N., 2012. Enhancing biodegradation of dibenzothiophene by *Bacillus sphaericus* HN1 using factorial design and response surface optimization of medium components. Energy Sources, Part A: Recovery, Utilization, and Environmental Effects 34 (22), 2073–2083.

Deveryshetty, J., Phale, P.S., 2009. Biodegradation of phenanthrene by *Pseudomonas* sp. strain PPD: purification and characterization of 1-hydroxy-2-naphthoic acid dioxygenase. Mcrobiology 155, 3083–3091.

Dhote, M., Juwarkar, A., Kumar, A., Kanade, G.S., Chakrabarti, T., 2010. Biodegradation of chrysene by the bacterial strains isolated from oily sludge. World Journal of Microbiology & Biotechnology 26 (2), 329–335.

Douglas, G.S., Prince, R.C., Butler, E.L., Steinhauer, W.G., 1994. The use of internal chemical indicator in petroleum and refined products to evaluate the extent of biodegradation. In: Hinchee, R.E., Hoeppel, B.C., Miller, R.N. (Eds.), Hydrocarbon Bioremediation. Lewis Publishers, Boca Raton, FL, USA, pp. 219–236.

Dyksterhouse, S.E., Gray, J.P., Herwig, R.P., Lara, J.C., Staley, J.T., 1995. *Cycloclasticus pugetii*, gen. nov., sp. nov., an aromatic hydrocarbon degrading bacterium from marine sediments. International Journal of Systematic Bacteriology 45, 116–123.

Egaas, E., Varanis, A., 1982. Effects of polychlorinated biphenyls and environmental temperature on in vitro formation of benzo[a]pyrene metabolites by liver of trout (Salmo gairdneri). Biochemical Pharmacology 31 (4), 561–566.

El-Gendy, N.Sh., Moustafa, Y.M., 2007. Environmental assessment of petroleum hydrocarbons contaminating Temsah Lake, Suez Canal, Egypt. Oriental Journal of Chemistry 23 (1), 11–26.

El-Gendy, N.Sh., 2006. Biodegradation potentials of dibenzothiophene by new bacteria isolated from hydrocarbon polluted soil in Egypt. Biosciences, Biotechnology Research Asia 3 (1a), 95–106.

El-Gendy, N.Sh., Abostate, M.A., 2008. Isolation, characterization and evaluation of *Staphylococcus gallinarum* NK1 as a degrader for dibenzothiophene, phenathrene and naphthalene. Egyptian Journal of Petroleum 17 (2), 75–91.

El-Gendy, N.Sh., Ali, H.R., El-Nady, M.M., Deriase, S.F., Moustafa, Y.M., Mohamed, I., Roushdy, M.I., 2014a. Effect of different bioremediation techniques on petroleum biomarkers and asphaltene fraction in oil polluted sea water. Desalination and Water Treatment 52 (40/42), 7484–7494.

El-Gendy, N.Sh., Moustafa, Y.M., Barakat, M.A.K., Deriase, S.F., 2009. Evaluation of a bioslurry remediation of petroleum hydrocarbons contaminated sediments using chemical, mathematical and microscopic analysis. International Journal of Environmental Studies 66 (5), 563–579.

El-Gendy, N.Sh., Moustafa, Y.M., Habib, S.A., Sh, A., 2010. Evaluation of *Corynebacterium variabilis* Sh42 as a degrader for different polyaromatic compounds. Journal of American Science 6 (11), 343–356.

El-Gendy, N.Sh., Nassar, H.N., Abu Amr, S.S., 2014b. Factorial design and response surface optimization for enhancing a biodesulfurization process. Petroleum Science and Technology 32 (14), 1669–1679.

Elmendorf, D.L., Haith, C.E., Douglas, G.S., Prince, R.C., 1994. Relative rates of biodegradation of substituted polycyclic aromatic hydrocarbons. In: Hinchee, R.E., Leeson, A., Semprini, L., Kee Ong, S. (Eds.), Bioremediation of Chlorinated and Polycyclic Aromatic Hydrocarbon Compounds. Lewis Publishers, Boca Raton. FL, USA, pp. 188–201.

Evans, W.C., 1977. Biochemistry of the bacterial catabolism of aromatic compounds in anaerobic environments. Nature 270, 17–22.

Fedorak, P.M., Westlake, D.W.S., 1983. Microbial degradation of organic sulfur compounds in Prudhoe Bay crude oil. Canadian Journal of Microbiology 29, 291–296.

Fedorak, P.M., Westlake, D.W.S., 1984. Degradation of sulfur heterocycles in Prudhoe Bay crude oil by soil enrichments. Water, Air and Soil Pollution Journal 21, 225–230.

Feng, L., Wang, W., Cheng, J., Ren, Y., Zhao, G., Gao, C., Tang, Y., Liu, X., Han, W., Peng, X., Liu, R., Wang, L., 2007. Genome and proteome of long-chain alkane degrading *Geobacillus thermodenitrificans* NG80-2 isolated from a deep-subsurface oil reservoir. Proceedings of the National Academy of Sciences of the United States of America 104, 5602–5607.

Ferreira, L., Rosales, E., Danko, A.S., Sanromán, M.A., Pazos, M.M., 2016. *Bacillus thuringiensis* a promising bacterium for degrading emerging pollutants. Process Safety and Environmental Protection 101, 19–26.

Foght, J.M., Westlake, D.W.S., 1990. Expression of dibenzothiophene-degradative genes in two *Pseudomonas* species. Canadian Journal of Microbiology 36 (10), 718–724.

Folwell, B.D., McGenity, T.J., Whitby, C., 2016. Biofilm and planktonic bacterial and fungal communities transforming high-molecular-weight polycyclic aromatic hydrocarbons. Applied and Environmental Microbiology 82 (8), 2288–2299.

Fritzsche, C., 1994. Degradation of pyrene at low defined oxygen concentrations by a *Mycobacterium* sp. Applied and Environmental Microbiology 60, 1687–1689.

Fuentes, S., Méndez, V., Aguila, P., Seeger, M., 2014. Bioremediation of petroleum hydrocarbons: catabolic genes, microbial communities, and applications. Applied Microbiology and Biotechnology 98, 4781–4794.

Funhoff, E.G., Bauer, U., Garcia-Rubio, I., Witholt, B., van Beilen, J.B., 2006. CYP153A6, a soluble P450 oxygenase catalyzing terminal-alkane hydroxylation. Journal of Bacteriology 188, 5220–5227.

Galushko, A., Minz, D., Schink, B., Widdel, F., 1999. Anaerobic degradation of naphthalene by a pure culture of a novel type of marine sulphatereducing bacterium. Environmental Microbiology 1, 415–420.

Gao, S., Seo, J.-S., Wang, J., Keum, Y.-S., Li, J., Li, Q.X., 2013. Multiple degradation pathways of phenanthrene by *Stenotrophomonas maltophilia* C6. International Biodeterioration and Biodegradation 79, 98–104.

Garzoli, L., Gnavi, G., Tamma, F., Tosi, S., Varese, G.C., Picco, A.M., 2015. Sink or swim: updated knowledge on marine fungi associated with wood substrates in the Mediterranean Sea and hints about their potential to remediate hydrocarbons. Progress in Oceanography 137, 140–148.

Geiselbrecht, A.D., Hedlund, B.P., Tichi, M.A., Staley, J.T., 1998. Isolation of marine polycyclic aromatic hydrocarbon (polynuclear aromatic hydrocarbon derivatives)-degrading *Cycloclasticus* strains from the Gulf of Mexico and comparison of their polynuclear aromatic hydrocarbon derivatives degradation ability with that of Puget Sound *Cycloclasticus* strains. Applied and Environmental Microbiology 64, 4703–4710.

Gesinde, A.F., Agbo, E.B., Agho, M.O., Dike, E.F.C., 2008. Bioremediation of some Nigerian and Arabian crude oils by fungal isolates. International Journal of Pure and Applied Sciences 2, 37–44.

Ghosal, D., Chakraborty, J., Khara, P., Dutta, T.K., 2010. Degradation of phenanthrene via *meta*-cleavage of 2-hydroxy-1-naphthoicacid by *Ochrobactrum* sp. strain PWTJD. FEMS Microbiology Letters 313, 103–110.

Ghosh, I., Jasmine, J., Mukherji, S., 2014. Biodegradation of pyrene by a *Pseudomonas aeruginosa* strain RS1 isolated from refinery sludge. Bioresource Technology 166, 548–558.

Gianfreda, L., Xu, F., Bollag, J.M., 1999. Laccases: a useful group of oxidoreductases enzymes. Bioremediation Journal 3 (1), 1–25.

Gibson, D.T., Venkatanayarana, M., Jerina, D.M., Yagi, H., Yeh, H., 1975. Oxidation of the carcinogens benzo[a]pyrene and benzo[a]anthracene to dihydrodiols by a bacterium. Science 189, 295–297.

Golyshin, P.N., Chernikova, T.N., Abraham, W.R., Lunsdorf, H., Timmis, K.N., Yakimov, M.M., 2002. Oleiphilaceae fam. nov., to include *Oleiphilus messinensis* gen. nov., sp. nov., a novel marine bacterium that obligately utilizes hydrocarbons. International Journal of Systematic and Evolutionary Microbiology 52, 901–911.

Gordon, L., Dobson, A.D., 2001. Fluoranthene degradation in *Pseudomonas alcaligenes* PA-10. Biodegradation 12, 393–400.

Goyal, A.K., Zylstra, G.J., 1997. Genetics of naphthalene and phenanthrene degradation by *Comamonas testosteroni*. Journal of Industrial Microbiology & Biotechnology 19, 401–407.

Greene, A., Patel, B.K.C., Sheehy, A.J., 1997. *Deferribacter thermophiles* gen. nov., sp. nov., a novel thermophilic manganese - and iron - reducing bacterium isolated from a petroleum reservoir. International Journal of Systematic Bacteriology 505–509.

Grifoll, M., Selifonov, S.A., Chapman, P.J., 1994. Evidence for a novel pathway in the degradation of fluorene by *Pseudomonas* sp. strain F274. Applied and Environmental Microbiology 60, 2438–2449.

Grosser, R.J., Warshawsky, D., Vestal, J.R., 1991. Indigenous and enhanced mineralization of pyrene, benzo[a]pyrene, and carbazole in soils. Applied and Environmental Microbiology 57, 3462–3469.

Guedes, S., Mendes, B., Leitão, A., 2010. Resorcinol degradation by a *Penicillium chrysogenum* strain under osmotic stress: mono and binary substrate matrices with phenol. Biodegradation 22, 409–419.

Gunam, I., Yaku, Y., Hirano, M., Yamamura, K., Tomita, F., Sone, T., Asano, K., 2006. Biodesulfurization of alkylated forms of dibenzothiophene and benzothiophene by *Sphingomonas subarctica* T7b. Journal of Bioscience and Bioengineering 101, 322–327.

Guntupalli, S., Thunuguntla, V.B.S.C., Santha Reddy, K., Issac Newton, M., Rao, C.V., Bondili, J.S., 2016. Enhanced degradation of carcinogenic polynuclear aromatic hydrocarbon derivatives benzo(a)pyrene and benzo(k)fluoranthene by a microbial consortium. Indian Journal of Science and Technology 9 (35), 1–12. http://dx.doi.org/10.17485/ijst/2016/v9i35/93590.

Habe, H., Chung, J.-S., Ishida, A., Kasuga, K., Ide, K., Takemura, T., Nojiri, H., Yamane, H., Omori, T., 2005. The fluorene catabolic linear plasmid in *Terrabacter* sp. strain DBF63 carries the β-ketoadipate pathway genes, *pcaRHGBDCFIJ*, also found in proteobacteria. Microbiology 151, 3713–3722.

Habe, H., Chung, J.S., Kato, H., Ayabe, Y., Kasuga, K., Yoshida, T., Nojiri, H., Yamane, H., Omori, T., 2004. Characterization of the upper pathway genes for fluorene metabolism in *Terrabacter* sp. strain DBF63. Journal of Bacteriology 186 (17), 5938–5944.

Habe, H., Miyakoshi, M., Chung, J.S., Kasuga, K., Yoshida, T., Nojiri, H., Omori, T., 2003. Phthalate catabolic gene cluster is linked to the angular dioxygenase gene in *Terrabacter* sp. strain DBF63. Applied Microbiology and Biotechnology 61, 44–54.

Hadibarata, T., Tachibana, S., Itoh, K., 2009. Biodegradation of chrysene, an aromatic hydrocarbon by *Polyporus* sp. S133 in liquid medium. Journal of Hazardous Materials 164, 911–917.

Han, M.J., Choi, H.T., Song, H.G., 2004. Degradation of phenanthrene by *Trametes versicolor* and its laccase. Journal of Microbiology 42, 94–98.

Harayama, S., Kasai, Y., Hara, A., 2004. Microbial communities in oil-contaminated seawater. Current Opinion in Biotechnology 15, 205–214.

Harayama, S., Kishira, H., Kasai, Y., Shutsubo, K., 1999. Petroleum biodegradation in marine environments. Journal of Microbiology and Biotechnology 1 (1), 63–70.

Harms, H., Schlosser, D., Wick, L.Y., 2011. Untapped potential: exploiting fungi in bioremediation of hazardous chemicals. Nature Reviews Microbiology 9, 177–192.

Hassanshahian, M., Boroujeni, N.A., 2016. Enrichment and identification of naphthalene-degrading bacteria from the Persian Gulf. Marine Pollution Bulletin 107, 59–65.

Hassanshahian, M., Emtiazi, G., Cappello, S., 2012a. Isolation and characterization of crude-oil-degrading bacteria from the Persian Gulf and the Caspian Sea. Marine Pollution Bulletin 64, 7–12.

Hassanshahian, M., Emtiazi, G., Kermans hahi, R., Cappello, S., 2010. Comparison of oil degrading microbial communities in sediments from the Persian Gulf and Caspian Sea. Soil Sediment Contamination 19 (3), 277–291.

Hassanshahian, M., Tebyanian, H., Cappello, S., 2012b. Isolation and characterization of two crude oil-degrading yeast strains, *Yarrowia lipolytica* PG-20 and PG-32, from the Persian Gulf. Marine Pollution Bulletin 64, 1386–1391.

Hedlund, B.P., Geiselbrecht, A.D., Staley, J.T., 1999. Polycyclic aromatic hydrocarbon degradation by a new marine bacterium, *Neptunomonas naphthovorans* gen. Nov., sp. nov. Applied and Environmental Microbiology 65, 251–259.

Hedlund, B.P., Geiselbrecht, A.D., Staley, J.T., 2001. *Marinobacter* strain NCE312 has a *Pseudomonas*-like naphthalene dioxygenase. FEMS Microbiology Letters 201, 47–51.

Hedlund, B.P., Staley, J.T., 2006. Isolation and characterization of *Pseudoalteromonas* strains with divergent polycyclic aromatic hydrocarbon catabolic properties. Environmental Microbiology 8 (1), 178–182.

Hedlund, B.P., Staley, J.T., 2001. *Vibrio cyclotrophicus* sp. nov, a marine polycyclic aromatic hydrocarbon (polynuclear aromatic hydrocarbon derivatives)-degrading bacterium. International Journal of Systematic and Evolutionary Microbiology 51, 61–66.

Hegazi, R.M., El-Gendy, N.Sh., El-Feky, A.A., Moustafa, Y.M., El- Ezbewy, S., El-Gemaee, G.H., 2007. Impact of heavy metals on biodegradation of phenanthrene by *Cellulomonas hominis* strain N2. Journal of Pure and Applied Microbiology 1 (2), 165–175.

Heitkamp, M.A., Cerniglia, C.E., 1988. Mineralization of polycyclic aromatic hydrocarbons by a bacterium isolated from sediment below an oil field. Applied and Environmental Microbiology 54, 1612–1614.

Hidayat, A., Tachibana, S., 2015. Simple screening for potential chrysene degrading fungi. KnE Life Sciences 2, 364–370.

Hinchee, R.E., Leeson, A., Ong, S.K., Semprini, L., 1994. Bioremediation of Chlorinated and Polycyclic Aromatic Hydrocarbon Compounds. CRC Press, NY, USA.

Hiraoka, Y., Yamada, T., Tone, K., Futaesaku, Y., Kimbara, K., 2002. Flow cytometry analysis of changes in the DNA content of the polychlorinated biphenyl degrader *Comamonas testosteroni* TK102: effect of metabolites on cell-cell separation. Applied and Environmental Microbiology 68, 5104–5112.

Hofrichter, M., Gtinther, T., Fritsche, W., 1992. Metabolism of phenol, chloro- and nitrophenols by the *Penicillium* strain Bi 7/2 isolated from a contaminated soil. Biodegradation 3, 415–421.

Hong, Y.-W., Yuan, D.-X., Lin, Q.-M., Yang, T.-L., 2008. Accumulation andbiodegradation of phenanthrene and fluoranthene by the algae enriched from amangrove aquatic ecosystem. Marine Pollution Bulletin 56, 1400–1405.

Hubert, C., Voordouw, G., 2007. Oil field souring control by nitrate-reducing *Sulfurospirillum* spp. that outcompete sulfate-reducing bacteria for organic electron donors. Applied and Environmental Microbiology 73, 2644–2652.

Husaini, A., Roslan, H.A., Hii, K.S.Y., Ang, C.H., 2008. Biodegradation of aliphatic hydrocarbon by indigenous fungi isolated from used motor oil contaminated sites. World Journal of Microbiology & Biotechnology 24, 2789–2797.

Iwabuchi, T., Harayama, S., 1998b. Biochemical and molecular characterization of 1-hydroxy-2-naphthoate dioxygenase from *Nocardioides* sp. KP7. The Journal of Biological Chemistry 273 (14), 8332–8336.

Iwabuchi, T., Harayama, S., 1998a. Biochemical and genetic characterization of trans-2′-carboxybenzalpyruvate hydratase aldolase from a phenanthrene-degrading *Nocardioides* strain. Journal of Bacteriology 180, 945–949.

Jaekel, U., Zedelius, J., Wilkes, H., Musat, F., 2015. Anaerobic degradation of cyclohexane by sulfate-reducing bacteria from hydrocarbon-contaminated marine sediments. Frontiers in Microbiology Volume 6, 116. http://dx.doi.org/10.3389/fmicb.2015.00116.

Janbandhu, A., Fulekar, M.H., 2011. Biodegradation of phenanthrene using adapted microbial consortium isolated from petrochemical contaminated environment. Journal of Hazardous Materials 187, 333–340.

Jarling, R., Sadeghi, M., Drozdowska, M., Lahme, S., Buckel, W., Rabus, R., Widdel, F., Golding, B.T., Wilkes, H., 2012. Stereochemical investigations reveal the mechanism of the bacterial activation of n-alkanes without oxygen. Angewandte Communications International 51, 1334–1338.

Johnson, G.R., Olsen, R.H., 1997. Multiple pathways for toluene degradation in *Burkholderia* sp. strain JS150. Applied and Environmental Microbiology 63 (10), 4047–4052.

Jorgensen, B.B., 1990. A thiosulfate shunt in the sulfur cycle of marine sediments. Science 249, 152–154.

Jouanneau, Y., Meyer, C., Jakoncic, J., Stojanoff, V., Gaillard, J., 2006. Characterization of a naphthalene dioxygenase endowed with an exceptionally broad substrate specificity toward polycyclic aromatic hydrocarbons. Biochemistry 45, 12380–12391.

Juhasz, A.L., Britz, M.L., Stanley, G.A., 1997. Degradation of fluoranthene, pyrene, benz[a]anthracene and dibenz[a,h]anthracene by *Burkholderia cepacia*. Journal of Applied Microbiology 83, 189–198.

Kanaly, R.A., Bartha, R., 1999. Cometabolic mineralization of benzo[a]pyrene caused by hydrocarbon additions to soil. Environmental Toxicology and Chemistry 18, 2186–2190.

Kargi, F., Dincer, A.R., 1996. Effect of salt concentration on biological treatment of saline wastewater by fed-batch operation. Enzyme and Microbial Technology 19 (7), 529–537.

Kästner, M., Breuer-Jammali, M., Mahro, B., 1998. Impact of inoculation protocols, salinity, and pH on the degradation of polycyclic aromatic hydrocarbons (polynuclear aromatic hydrocarbon derivatives) and survival of polynuclear aromatic hydrocarbon derivatives-degrading bacteria introduced into soil. Applied and Environmental Microbiology 64, 359–362.

Ke, L., Bao, W., Chen, L., Wong, Y.S., Tam, N.F.Y., 2009. Effects of humic acid on solubility and biodegradation of polycyclic aromatic hydrocarbons in liquid media and mangrove sediment slurries. Chemosphere 76, 1102–1108.

Kelley, I., Cerniglia, C.E., 1991. The metabolism of fluoranthene by a species of *Mycobacterium*. Journal of Industrial Microbiology & Biotechnology 7, 19–26.

Kelley, I., Cerniglia, C.E., 1995. Degradation of a mixture of high molecular-weight polycyclic aromatic hydrocarbons by a *Mycobacterium* strain PYR-1. Journal of Soil Contamination 4, 77–91.

Kelley, I., Freeman, J.P., Evans, F.E., Cerniglia, C.E., 1991. Identification of a carboxylic acid metabolite from the catabolism of fluoranthene by a *Mycobacterium* sp. Applied and Environmental Microbiology 57, 636–641.

Kelley, I., Freeman, J.P., Evans, F.E., Cerniglia, F.E., 1993. Identification of metabolites from the degradation of fluoranthene by *Mycobacterium* sp. strain PYR-1. Applied and Environmental Microbiology 59, 800–806.

Keum, Y.S., Seo, J.S., Hu, Y., Li, Q.X., 2005. Degradation pathways of phenanthrene by *Sinorhizobium* sp. C4. Applied Microbiology and Biotechnology 71, 935–941.

Keum, Y.S., Seo, J.S., Hu, Y., Li, Q.X., 2006. Degradation pathways of phenanthrene by *Sinorhizobium* sp. C4. Applied Microbiology and Biotechnology 71, 935–941.

Khanna, P., Goyal, D., Khanna, S., 2011. Pyrene degradation by *Bacillus pumilus* isolated from crude oil contaminated soil. Polycyclic Aromatic Compounds 31, 1–15.

Kim, E., Zylstra, G.J., 1999. Functional analysis of genes involved in biphenyl, naphthalene, phenanthrene, and m-xylene degradation by *Sphingomonas yanoikuyae* B1. Journal of Industrial Microbiology & Biotechnology 23, 294–302.

Kim, S.J., Kweon, O., Jones, R.C., Freeman, J.P., Edmondson, R.D., Cerniglia, C.E., 2007. Complete and integrated pyrene degradation pathway in *Mycobacterium vanbaalenii* PYR-1 based on systems biology. Journal of Bacteriology 189 (2), 464–472.

Kim, S.-J., Kweon, O., Freeman, J.P., Jones, R.C., Adjei, M.D., Jhoo, J.-W., Edmondson, R.D., Cerniglia, C.E., 2006. Molecular cloning and expression of genes encoding a novel dioxygenase involved in low- and high molecular-weight polycyclic aromatic hydrocarbon degradation in *Mycobacterium vanbaalenii* PYR-1. Applied and Environmental Microbiology 72, 1045–1054.

Kim, Y.H., Freeman, J.P., 2005. Effects of pH on the degradation of phenanthrene and pyrene by *Mycobacterium vanbaalenii* PYR-1. Applied Microbiology and Biotechnology 67, 275–285.

Kiyohara, H., Nagao, K., Kouno, K., Yano, K., 1982. Phenanthrene degrading phenotype of *Alcaligenes faecalis* AFK2. Applied and Environmental Microbiology 43, 458–461.

Kiyohara, H., Takizawa, N., Date, H., Torigoe, S., Yano, K., 1990. Characterization of a phenanthrene degradation plasmid from *Alcaligenes faecalis* AFK2. Journal of Fermentation and Bioengineering 69, 54–56.

Kiyohara, H., Torigoe, S., Kaida, N., Asaki, T., Iida, T., Hayashi, H., Takizawa, N., 1994. Cloning and characterization of a chromosomal gene cluster, pah, that encodes the upper pathway for phenanthrene and naphthalene utilization by *Pseudomonas putida* OUS82. Journal of Bacteriology 176, 2439–2443.

Kniemeyer, O., Fischer, T., Wilkes, H., Glockner, F.O., Widdel, F., 2003. Anaerobic degradation of ethylbenzene by a new type of marine sulfate – reducing bacterium. Applied and Environmental Microbiology 2, 760–768.

Kohler, H.-P.E., Schmid, A., Van der Maarel, M., 1993. Metabolism of 2,2'-dihydroxybiphenyl by *Pseudomonas* sp. strain HBP1: production and consumption of 2,2',3-trihydroxybiphenyl. Journal of Bacteriology 175, 1621–1628.

Kulakov, L.A., Allen, C.C.R., Lipscomb, D.A., Larkin, M.J., 2000. Cloning and characterization of a novel cis-naphthalene dihydrodiol dehydrogenase gene (narB) from *Rhodococcus* sp. NCIMB 12038. FEMS Microbiology Letters 182, 327–331.

Kumar, M., León De Sisto Materano, V.A., Ilzins, O.A., Luis, L., 2008. Biosurfactant production and hydrocarbon degradation by halotolerant and thermotolerant *Pseudomonas* sp. World Journal of Microbiology & Biotechnology 24, 1047–1057.

Kumaria, B., Singh, S.N., Deeba, F., Sharma, S., Pandey, V., Singh, D.P., 2014. Elucidation of pyrene degradation pathway in bacteria. Advances in Bioresearch 4 (2), 151–160.

Kunihiro, M., Ozeki, Y., Nogi, Y., Hamamura, N., Kanalya, R.A., 2013. Benz[a]anthracene biotransformation and production of ring fission products by *Sphingobium* sp. strain KK22. Applied and Environmental Microbiology 79 (14), 4410–4420.

Kweon, O., Kim, S.J., Holland, R.D., Chen, H., Kim, D.W., Gao, Y., Yu, L.R., Baek, S., Baek, D.H., Ahn, H., Cerniglia, C.E., 2011. Polycyclic aromatic hydrocarbon metabolic network in *Mycobacterium vanbaalenii* PYR-1. Journal of Bacteriology 193, 4326–4337.

Kweon, O., Kim, S.J., Jones, R.C., Freeman, J.P., Adjei, M.D., Edmondson, R.D., Cerniglia, C.E., 2007. A polyomic approach to elucidate the fluoranthene-degradative pathway in *Mycobacterium vanbaalenii* PYR-1. Journal Bacteriology 189, 4635–4647.

Lai, Q., Cao, J., Yuan, J., Li, F., Shao, Z., 2014. *Celeribacter indicus* sp. nov. a polycyclic aromatic hydrocarbon-degrading bacterium from deep-sea sediment and reclassification of *Huaishuia halophila* as *Celeribacter halophilus* comb. nov. International Journal of Systematic and Evolutionary Microbiology 64, 4160–4167.

Lanfranconi, M.P., Christie-Oleza, J.A., Martín-Cardona, C., Suárez-Suárez, L.Y., Lalucat, J., Nogales, B., Bosch, R., 2009. Physiological role of NahW, the additional salicylate hydroxylase found in *Pseudomonas stutzeri* AN10. FEMS Microbiology Letters 300, 265–272.

Larkin, M.J., Kulakov, L.K., Allen, C.C.R., 2005. Biodegradation and *Rhodococcus* masters of catabolic versatility. Current Opinion in Biotechnology 12, 564–573.

Launen, L., Pinto, L., Moore, M., 1999. Optimization of pyrene oxidation by *Penicillium janthinellum* using response-surface methodology. Applied Microbiology and Biotechnology 51, 510–515.

Launen, L., Pinto, L., Wiebe, C., Kiehlmann, E., Moore, M., 1995. The oxidation of pyrene and benzo[a]pyrene by nonbasidiomycete soil fungi. Canadian Journal of Microbiology 41, 477–488.

Laurie, A.D., Lloyd-Jones, G., 2000. Quantification of phnAc and nahAc in contaminated New Zealand soils by competitive PCR. Applied and Environmental Microbiology 66, 1814–1817.

Lebkowska, M., Zborowska, E., Karwowska, E., Miaskiewicz-Peska, E., Muszynski, A., Tabernacka, A., Naumczyk, J., Jeczalik, M., 2011. Bioremediation of soil polluted with fuels by sequential multiple injection of native microorganisms: field scale processes in Poland. Ecology Engineering 37, 1895–1900.

Lee, J., Oh, J., Min, R., Kim, Y.I., 1996. Nucleotide sequence of salicylate hydroxylase gene and its 5′-flanking region of *Pseudomonas putida* KF715. Biochemical and Biophysical Research Communications 218 (2), 544–548.

Lee, M.K., Senius, J.D., Grossman, M.J., 1995. Sulfur-specific microbial desulfurization of sterically hindered analogs of dibenzothiophene. Applied and Environmental Microbiology 61, 4362–4366.

Lee, S.E., Seo, J.S., Keum, Y.S., Lee, K.J., Li, Q.X., 2007. Fluoranthene metabolism and associated proteins in *Mycobacterium* sp. JS14. Proteomics 7, 2059–2069.

Leitão, A.L., 2009. Potential of *Penicillium* species in the bioremediation field. International Journal of Environmental Research and Public Health 6, 1393–1417.

Li, L.F., Xu, P., Ma, C., Luo, L., Wang, X.S., 2003. Deep desulfurization of hydrodesulfurization-treated diesel oil by a facultative thermophilic bacterium *Mycobacterium* sp. X7B. FEMS Microbiology Letters 223, 301–307.

Li, W., Shi, J., Wang, X., Han, Y., Tong, W., Ma, L., Liu, B., Cai, B., 2004. Complete nucleotide sequence and organization of the naphthalene catabolic plasmid pND6-1 from *Pseudomonas* sp. strain ND6. Gene 336, 231–240.

Liang, Y., Gardner, D.R., Miller, C.D., Chen, D., Anderson, A.J., Weimer, B.C., Sims, R.C., 2006. Study of biochemical pathways and enzymes involved in pyrene degradation by *Mycobacterium* sp. strain KMS. Applied and Environmental Microbiology 72 (12), 7821–7828.

Lily, M.K., Bahuguna, A., Dangwal, K., Garg, V., 2009. Degradation of benzo[a]pyrene by a novel strain *Bacillus subtilis* BMT4i (MTCC 9447). Brazilian Journal of Microbiology 40, 884–892.

Liu, K., Han, W., Pan, W.P., Riley, J.T., 2001. Polycyclic aromatic hydrocarbon (polynuclear aromatic hydrocarbon derivatives) emissions from a coal fired pilot FBC system. Journal of Hazardous Materials 84, 175–188.

López, Z., Vila, J., Minguillón, C., Grifoll, M., 2006. Metabolism of fluoranthene by *Mycobacterium* sp. strain AP1. Applied Microbiology and Biotechnology 70, 747–756.

Luykx, D.M., Prenafeta-Boldú, F.X., de Bont, J.A., 2003. Toluene monooxygenase from the fungus *Cladosporium sphaerospermum*. Biochemical and Biophysical Research Communications 312, 373–379.

Lyu, Y., Zheng, W., Zheng, T., Tian, Y., 2014. Biodegradation of polycyclic aromatic hydrocarbons by *Novosphingobium pentaromativorans* US6-1. PLoS One 9 (7), e101438. http://dx.doi.org/10.1371/journal.pone.0101438.

Ma, J., Xu, L., Jia, L., 2013. Characterization of pyrene degradation by *Pseudomonas* sp. strain Jpyr-1 isolated from active sewage sludge. Bioresource Technology 140, 15–21.

Madigan, M.T., Martinko, J.M., Parker, J., 2006. Biology of Microorganisms. Southern Illinois University, Carbondal.

Maeng, J.H., Sakai, Y., Tani, Y., Kato, N., 1996. Isolation and characterization of a novel oxygenase that catalyzes the first step of n-alkane oxidation in *Acinetobacter* sp. strain M-1. Journal of Bacteriology 178, 3695–3700.

Magot, M., Ollivier, B., Patel, B.K.C., 2000. Microbiology of petroleum reservoirs. Antonie van Leeuwenhoek 77, 103–116.

Mahaffey, W.R., Gibson, D.T., Cerniglia, C.E., 1988. Bacterial oxidation of chemical carcinogens: formation of polycyclic aromatic acids from benz[a]anthracene. Applied and Environmental Microbiology 54, 2415–2423.

Mahajan, M.C., Phale, P.S., Vaidyanathan, C.S., 1994. Evidence for the involvement of multiple pathways in the biodegradation of 1- and 2-methylnaphthalene by *Pseudomonas putida* CSV86. Archives of Microbiology 161, 425–433.

Mahmound, A., Aziza, Y., Abdeltif, A., Rachida, M., 2008. Biosurfactant production by *Bacillus* strain injected in the petroleum reservoirs. Journal of Industrial Microbiology & Biotechnology 35, 1303–1306.

Mallick, S., Chatterjee, S., Dutta, T.K., 2007. A novel degradation pathway in the assimilation of phenanthrene by *Staphylococcus* sp. strain PN/Y via meta-cleavage of 2-hydroxy-1-naphthoic acid: formation of *trans*-2,3-dioxo-5-(2′-hydroxyphenyl)pent-4-enoic acid. Microbiology 153, 2104–2115.

Marr, J., Kremer, S., Sterner, O., Anke, H., 1989. Transformation and mineralization of halophenols by *Penicillium simplicissimum* SK9117. Biodegradation 7, 165–171.

Mars, A.E., Houwing, J., Dolfing, J., Janssen, D.B., 1996. Degradation of toluene and trichloroethylene by *Burkholderia cepacia* G4 in growth-limited fed-batch culture. Applied and Environmental Microbiology 62 (3), 886–891.

Martínez, A.T., Speranza, M., Ruiz-Dueñas, F.J., Ferreira, P., Camarero, S., Guillèn, F., Martínez, M.J., Gutiérrez, A., Del Río, J.C., 2005. Biodegradation of lignocellulosics: microbial, chemical, and enzymatic aspects of the fungal attack of lignin. International Microbiology 8, 195–204.

Martínez-Checa, F., Toledo, F.L., Vilchez, R., Quesada, E., Calvo, C., 2002. Yield production, chemical composition, and functional properties of emulsifier H28 synthesized by *Halomonas eurihalina* strain H-28 in media containing various hydrocarbons. Applied Microbiology and Biotechnology 58, 358–363.

Martínková, L., Uhnáková, B., Pátek, M., Nešvera, J., Křen, V., 2009. Biodegradation potential of the genus *Rhodococcus*. Environmental International Journal 35, 162–177.

Marx, C.J., Miller, J.A., Chistoserdova, L., Lidstrom, M.E., 2004. Multiple formaldehyde oxidation/detoxification pathways in *Burkholderia fungorum* LB400. Journal of Bacteriology 186, 2173–2178.

Mastral, A.M., Callén, M., 2000. A review on polycyclic aromatic hydrocarbon (polynuclear aromatic hydrocarbon derivatives) emissions from energy generation. Environmental Science & Technology 34, 3051–3057.

Matthew, O., Ilori, S.A., Adebusoye, A., Ojo, C., 2008. Isolation and characterization of hydrocarbon-degrading and biosurfactant-producing yeast strains obtained from a polluted lagoon water. World Journal of Microbiology & Biotechnology 24, 2539–2545.

McNally, D.L., Mihelcic, J.R., Lueking, D.R., 1999. Biodegradation of mixtures of polycyclic aromatic hydrocarbons under aerobic and nitrate-reducing conditions. Chemosphere 38 (6), 13131–21321.

Meckenstock, R.U., Annweiler, E., Michaelis, W., Richnow, H.H., Schink, B., 2000. Anaerobic naphthalene degradation by a sulfate-reducing enrichment culture. Applied and Environmental Microbiology 66 (7), 2743–2747.

Meckenstock, R.U., Safinowski, M., Griebler, C., 2004. Anaerobic degradation of polycyclic aromatic hydrocarbons. FEMS Microbiology Ecology 49, 27–36.

Méndez, E.M., España, M.S.A., Montelongo, F.J.G., 2001. Chemical fingerprinting applied to the evaluation of marine oil pollution in the coasts of Canary Islands (Spain). Environmental Pollution 111 (2), 177–187.

Menn, F.M., Applegate, B.M., Sayler, G.S., 1993. NAH plasmid mediated catabolism of anthracene and phenanthrene to naphthoic acids. Applied and Environmental Microbiology 59, 1938–1942.

Meredith, W.K., Kelland, S.J., Jones, D.M., 2000. Influence of biodegradation on crude oil acidity and carboxylic acid composition. Organic Geochemistry 31, 1059–1073.

Meshram, R.L., Wate, S.R., 2014. Isolation, characterization and anthracene mineralization by *Bacillus cereus* from petroleum oil depot soil. Biochemistry 4 (5), 16–18.

Meuller, J.G., Chapman, P.J., Pritchard, P.H., 1989. Action of fluoranthene-utilizing community on polycyclic aromatic hydrocarbon components of creosote. Applied and Environmental Microbiology 55, 3085–3090.

Meyer, S., Steinhart, H., 2000. Effects of heterocyclic PAHs (N, S, O) on the biodegradation of typical tar oil PAHs in a soil/compost mixture. Chemosphere 40, 357–367.

Mnif, S., Sayadi, S., Chamkha, M., 2014. Biodegradative potential and characterization of a novel aromatic-degrading bacterium isolated from a geothermal oil field under saline and thermophilic conditions. International Biodeterioration and Biodegradation 86, 258–264.

Mohd Radzi, N.-A.-S., Tay, K.-S., Abu Bakar, N.-K., Emenike, C.U., Krishnan, S., Hamid, F.S., Abas, M.-R., 2015. Degradation of polycyclic aromatic hydrocarbons (pyrene and fluoranthene) by bacterial consortium isolated from contaminated road side soil and soil termite fungal comb. Environmental Earth Sciences 74 (6), 5383–5391.

Moller, T.H., Dicks, B., Goodman, C.N., February 1989. Fisheries and mariculture affected by oil spills. International Oil Spill Conference Proceedings. 1989 (1), 389–394. https://doi.org/10.7901/2169-3358-1989-1-389.

Moody, J.D., Freeman, J.P., Cerniglia, C.E., 2005. Degradation of benz[a]anthracene by *Mycobacterium vanbaalenii* strain PYR-1. Biodegradation 16, 513–526.

Moody, J.D., Freeman, J.P., Doerge, D.R., Cerniglia, C.R., 2001. Degradation of phenanthrene and anthracene by cell suspensions of *Mycobacterium* sp. strain PYR-1. Applied and Environmental Microbiology 67 (4), 1476–1483.

Moody, J.D., Fu, P.P., Freeman, J.P., Cerniglia, C.E., 2003. Regio- and stereoselective metabolism of 7,12-dimethylbenz[a]anthracene by *Mycobacterium vanbaalenii PYR*-1. Applied and Environmental Microbiology 69, 3924–3931.

Moody, J.D., Freeman, J.P., Fu, P.P., Cerniglia, C.E., 2004. Degradation of benzo[a]pyrene by *Mycobacterium vanbaalenii* PYR-1. Applied and Environmental Microbiology 70, 340–345.

Morgan, P., Watkinson, R.J., 1994. Biodegradation of components of petroleum. In: Ratledge, C. (Ed.), Biochemistry of Microbial Degradation. Kluwer Academic Publishers, Dordrecht, Netherlands, pp. 1–31.

Mori, K., Kim, H., Kakegawa, T., Hanada, S., 2003. A novel lineage of sulfate-reducing microorganisms: *Thermodesulfobiaceae* fam. nov., *Thermodesulfobium narugense*, gen. nov., sp nov., a new thermophilic isolate from a hot spring. Extremophiles 7, 283–290.

Moscoso, F., Deive, F.J., Longo, M.A., Sanromán, M.A., 2012b. Technoeconomic assessment of phenanthrene degradation by *Pseudomonas stutzeri* CECT 930 in a batch bioreactor. Bioresource Technology 104, 81–89.

Moscoso, F., Teijiz, I., Deive, F.J., Sanromán, M.A., 2012a. Efficient polynuclear aromatic hydrocarbon derivatives biodegradation by a bacterial consortium at flask and bioreactor scale. Bioresource Technology 119, 270–276.

Moustafa, Y.M., 2004. Contamination by polycyclic aromatic hydrocarbons in some Egyptian Mediterranean Coasts. Biosciences Biotechnology Research Asia 2 (1), 15–24.

Moustafa, Y.M., Abd El-Hakem, M., Abdallah, R.I., Barakat, M.A.K., 2005. Water pollution studies on Baher El-Bakar mouth northeast Nile-Delta, Egypt. Egyptian Journal of Petroleum 14 (1), 85–93.

Mueller, J.G., Chapman, P.J., Pritchard, P.H., 1989. Action of a fluoranthene-utilizing bacterial community on polycyclic aromatic hydrocarbon components of creosote. Applied and Environmental Microbiology 55, 3085–3090.

Muyzer, G., Stams, A.J., 2008. The ecology and biotechnology of sulfate-reducing bacteria. Nature Reviews Microbiology 6, 441–454.

Nadaling, T., Raymond, N., Gilewicz, M., Budzinski, H., Bertrand, J., 2002. Degradation of phenanthrene, methylphenanthrenes and dibenzothiophene by a *Sphingomonas* strain 2mpII. Applied Microbiology and Biotechnology 59 (1), 79–85.

Nakajima, K., Sato, A., Takahara, Y., Iida, T., 1985. Microbial oxidation of isoprenoid alkanes, phytane, norpristane and farnesane. Agricultural and Biological Chemistry 49, 1993–2002.

Nayak, A.S., Sanganal, S.K., Mudde, S.K., Oblesha, A., Karegoudar, T.B., 2011. A catabolic pathway for the degradation of chrysene by *Pseudoxanthomonas* sp. PNK-04. FEMS Microbiology Letters 320, 128–134.

Nazina, T.N., Grigoŕyan, A.A., Shestakova, N.M., Babich, T.L., Ivoilov, V.S., Feng, Q., Ni, F., Wang, J., She, Y., Xiang, T., Luo, Z., Belyaev, S.S., Ivanov, M.V., 2007. Microbiological investigations of high-temperature horizons of the Kongdian petroleum reservoir in connection with field trial of a biotechnology for enhancement of oil recovery. Microbiology 76, 287–296.

Neff, J.M., 1979. Polycyclic Aromatic Hydrocarbons in the Aquatic Environment. Applied Science Publishers, London.

Neria-Gonzalez, I., Wang, E.N., Ramirez, F., Romero, J.M., Hernandez-Rodriguez, C., 2006. Characterization of bacterial community associated to biofilms of corroded oil pipelines from the southeast of Mexico. Anaerobe 12, 122–133.

Nievas, M.L., Commendatore, M.G., Olivera, N.L., Esteves, J.L., Bucala, V., 2006. Biodegradation of bilge waste from Patagonia with an indigenous microbial community. Bioresource Technology 97, 2280–2290.

Nikolopoulou, M., Kalogerakis, N., 2009. Biostimulation strategies for fresh and chronically polluted marine environments with petroleum hydrocarbons. Journal of Chemical Technology and Biotechnology 84 (6), 802–807.

Ogram, A.V., Duan, Y.P., Trabue, S.L., Feng, X., Castro, H., Ou, L.T., 2000. Carbofuran degradation mediated by three related plasmid systems. FEMS Microbiology Ecology 32, 197–203.

Ojumu, T.V., Bello, O.O., Sonibare, J.A., Solomon, B.O., 2005. Evaluation of microbial systems for bioremediation of petroleum refinery effluents in Nigeria. African Journal of Biotechnology 4 (1), 31.

Ollivier, B., Fardeau, M.L., Cayol, J.L., Magot, M., Patel, B.K.C., Prensiep, G., Garcia, J.L., 1988. *Methanocalculus halotolerans* gen. nov., sp. nov., isolated from an oil producing well. International Journal of Systematic Bacteriology 48, 821–828.

Olsen, R.H., Kukor, J.J., Kaphammert, B., 1994. A novel toluene-3-monooxygenase pathway cloned from *Pseudomonas pickettii* PKO1. Journal of Bacteriology 176 (12), 3749–3756.

Ono, A., Miyazaki, R., Sota, M., Ohtsubo, Y., Nagata, Y., Tsuda, M., 2007. Isolation and characterization of naphthalene-catabolic genes and plasmids from oil-contaminated soil by using two cultivation-independent approaches. Applied Microbiology and Biotechnology 74, 501–510.

Parales, R.E., Parales, J.V., Pelletier, D.A., Ditty, J.L., 2008. Diversity of microbial toluene degradation pathways. In: Laskin, A.L., Sariaslani, S., Gadd, G.M. (Eds.), Advances in Applied Microbiology, 64. first ed. Academic Press, Elsevier, MA, USA, pp. 1–43 (Chapter 1).

Pathak, H., Kantharia, D., Malpani, A., Madamwar, D., 2009. Naphthalene degradation by *Pseudomonas* sp. HOB1: in vitro studies and assessment of naphthalene degradation efficiency in simulated microcosms. Journal of Hazardous Materials 166, 1466–1473.

Paul, D., Chauhan, A., Pandey, G., Jain, R.K., 2004. Degradation of p-hydroxybenzoate via protocatechuate in *Arthrobacter protophormiae* RKJ100 and *Burkholderia cepacia* RKJ200. Current Science 87, 1263–1268.

Peng, R.H., Xiong, A.-S., Xue, Y., Fu, X.-Y., Gao, F., Zhao, W., Tian, Y.-S., Yao, Q.-H., 2008. Microbial biodegradation of polyaromatic hydrocarbons. FEMS Microbiology Reviews 32, 927–955.

Pereira, P., Enguita, F.J., Ferreira, J., Leitão, A.L., 2014. DNA damage induced by hydroquinone can be prevented by fungal detoxification. Toxicology Reports 1, 1096–1105.

Pinyakong, O., Habe, H., Omori, T., 2003a. The unique aromatic catabolic genes in *sphingomonads* degrading polycyclic aromatic hydrocarbons. The Journal of General and Applied Microbiology 49, 1–9.

Pinyakong, O., Habe, H., Yoshida, T., Nojiri, H., Omori, T., 2003b. Identification of three novel salicylate 1-hydroxylases involved in the phenanthrene degradation of *Sphingobium* sp. strain P2. Biochemical and Biophysical Research Communications 301, 350–357.

Pozdnyakova, N.N., 2012. Involvement of the ligninolytic system of white-rot and litter-decomposing fungi in the degradation of polycyclic aromatic hydrocarbons. Biotechnology Research International 2012:243217. http://dx.doi.org/10.1155/2012/243217.

Pozdnyakova, N.N., Chernyshova, M.P., Grinev, V.S., Landesman, E.O., Koroleva, O.V., Turkovskaya, O.V., 2016. Degradation of fluorene and fluoranthene by the basidiomycete *Pleurotus ostreatus*. Applied Biochemistry and Microbiology 52 (6), 621–628.

Prabhu, Y., Phale, P.S., 2003. Biodegradation of phenanthrene by *Pseudomonas* sp. strain PP2: novel metabolic pathway, role of biosurfactant and cell surface hydrophobicity in hydrocarbon assimilation. Applied Microbiology and Biotechnology 61, 342–351.

Prenafeta-Boldú, F.X., Summerbell, R., Sybren de Hoog, G., 2006. Fungi growing on aromatic hydrocarbons: Biotechnology's unexpected encounter with biohazard? FEMS Microbiology Reviews 30, 109–130.

Rabus, R., Hansen, T., Widdel, F., 2000. Dissimilatory Sulfate- and Sulfur-Reducing Prokaryotes. Springer-Verlag, New York.

Rabus, R., Widdel, F., 1995. Anaerobic degradation of ethylbenzene and other aromatic hydrocarbons by new denitrifying bacteria. Archives of Microbiology 163, 96–103.

Radwan, S.S., Al-Hasan, R.H., Salamah, A., Khanafer, M., 2005. Oil-consuming microbial consortia floating in the Arabian Gulf. International Biodeterioration and Biodegradation 56, 28–33.

Reardon, K.F., Mosteller, D.C., Rogres, J.B., Duteau, N.M., Kim, K., 2002. Biodegradation kinetics of aromatic hydrocarbon mixtures by pure and mixed bacterial cultures. Environmental Health Perspectives 110 (6), 1005–1011.

Reddy, G.V., Sridhar, M., Gold, M.H., 2003. Cleavage of nonphenolic beta-1 diarylpropane lignin model dimers by manganese peroxidase from *Phanerochaete chrysosporium*. European Journal of Biochemistry 270, 284–292.

Rehmann, K., Hertkorn, N., Kettrup, A.A., 2001. Fluoranthene metabolism in *Mycobacterium* sp. strain KR20: identity of pathway intermediates during degradation and growth. Microbiology 147, 2783–2794.

Rehmann, K., Noll, H.P., Steiberg, C.E.W., Kettrup, A.A., 1998. Pyrene degradation by *Mycobacterium* sp. strain KR2. Chemosphere 36 (14), 2977–2992.

Rentz, J.A., Alvarez, P.J.J., Schnoor, J.L., 2008. Benzo[a]pyrene degradation by *Sphingomonas yanoikuyae* JAR02. Environmental Pollution 151, 669–677.

Reuter, P., Rabus, R., Wilkes, H., Aeckersberg, F., Rainey, F.A., Jannasch, H.W., Widdel, F., 2002. Anaerobic oxidation of hydrocarbons in crude oil by new types of sulfate-reducing bacteria. Nature 372, 455–458.

Ribeiro deNardi, I., Zaiat, M., Foresti, E., 2007. Kinetics of BTEX degradation in a packed–bed anaerobic reactor. Biodegradation 18, 83–90.

Robertson, W.J., Bowman, J.P., Franzmann, P.D., Mee, B.J., 2001. *Desulfosporosinus meridiei* sp. nov., a sporeforming sulfate reducing bacterium isolated from gasolene-contaminated groundwater. International Journal of Systematic and Evolutionary Microbiology 51, 133–140.

Rojo, F., 2009. Degradation of alkanes by bacteria. Environmental Microbiology 11 (10), 2477–2490.

Roling, W., 2003. The microbiology of hydrocarbon degradation in subsurface petroleum reservoirs: perspectives and prospects. Research in Microbiology 154 (5), 321–328.

Romero, M.C., Cazau, M.C., Giorgieri, S., Arambarri, A.M., 1998. Phenanthrene degradation by microorganisms isolated from a contaminated stream. Environmental Pollution 101, 355–359.

Romerom, M.C., Urrutia, M.I., Reinoso, H.E., Kiernan, M.M., 2010. Benzo[a]pyrene degradation by soil filamentous fungi. Journal of Yeast and Fungal Research 1, 25–29.

Ron, E.Z., Rosenberg, E., 2002. Biosurfactants and oil bioremediation. Current Opinion in Biotechnology 13, 249–252.

Roy, M., Khara, P., Dutta, T.K., 2012. *meta*-Cleavage of hydroxynaphthoic acids in the degradation of phenanthrene by *Sphingobium* sp. strain PNB. Microbiology 158, 685–695.

Saito, A., Iwabuchi, T., Harayama, S., 1999. Characterization of genes for enzymes involved in the phenanthrene degradation in *Nocardioides* sp. KP7. Chemosphere 38, 1331–1337.

Saito, A., Iwabuchi, T., Harayama, S., 2000. A novel phenanthrene dioxygenase from *Nocardioides* sp. strain KP7: Expression in *Escherichia coli*. Journal of Bacteriology 182 (8), 2134–2141.

Saiua, G., Tronci, S., Grosso, M., Cadonib, E., Currelic, N., 2016. Biodegradation of polycyclic aromatic hydrocarbons by Pleurotus sajor-caju. Chemical Engineering Transactions 49, 487–492.

Salam, L.B., Obayori, O.S., 2014. Fluorene biodegradation potentials of *Bacillus* strains isolated from tropical hydrocarbon-contaminated soils. African Journal of Biotechnology 13 (14), 1554–1559.

Saraswathy, A., Hallberg, R., 2002. Degradation of pyrene by indigenous fungifrom a former gasworks site. FEMS Microbiology Letters 210, 227–232.

Scheuer, U., Zimmer, T., Becher, D., Schauer, F., Schunck, W.-H., 1998. Oxygenation cascade in conversion of n-alkanes to α,ω-dioic acids catalyzed by cytochrome P450 52A3. Journal of Biological Chemistry 273 (49), 32528–32534.

Schmitt, R., Langguth, H.R., Puttmann, W., Rohns, H.P., Eckert, P., Schubert, J., 1996. Biodegradation of aromatic hydrocarbons under anoxic conditions in a shallow sand and gravel aquifer of the lower Rhine valley, Germany. Organic Geochemistry 25 (1), 41–50.

Schneider, J., Grosser, R., Jayasimhulu, K., Xue, W., Warshawsky, D., 1996. Degradation of pyrene, benz[a] anthracene and benzo[a]pyrene by *Mycobacterium* sp. strain RJGII-135, isolated from a former coal gasification site. Applied and Environmental Microbiology 62, 13–19.

Schneiker, S., Martins dos Santos, V.A., Bartels, D., Bekel, T., Brecht, M., Buhrmester, J., Chernikova, T.N., Denaro, R., Ferrer, M., Gertler, C., Goesmann, A., Golyshina, O.V., Kaminski, F., Khachane, A.N., Lang, S., Linke, B., McHardy, A.C., Meyer, F., Nechitaylo, T., Pühler, A., Regenhardt, D., Rupp, O., Sabirova, J.S., Selbitschka, W., Yakimov, M.M., Timmis, K.N., Vorhölter, F.J., Weidner, S., Kaiser, O., Golyshin, P.N., 2006. Genome sequence of the ubiquitous hydrocarbon-degrading marinebacterium *Alcanivorax borkumensis*. Nature Biotechnology 24, 997–1004.

Selesi, D., Jehmlich, N., von Bergen, M., Schmidt, F., Rattei, T., Tischler, P., Lueders, T., Meckenstock, R.U., 2010. Combined genomic and proteomic approaches identify gene clusters involved in anaerobic 2-methylnaphthalene degradation in the sulfate-reducing enrichment culture N47. Journal of Bacteriology 192, 295–306.

Seo, J., Keum, Y., Hu, Y., Lee, S., Li, Q.X., 2006. Phenanthrene degradation in *Arthrobacter* sp. P1-1: initial 1,2-, 3,4- and 9,10-dioxygenation, and meta- and ortho-cleavages of naphthalene-1,2-diol after its formation from naphthalene-1,2-dicarboxylic acid and hydroxyl naphthoic acids. Chemosphere 65, 2388–2394.

Seo, J., Keum, Y., Hu, Y., Lee, S., Li, Q.X., 2007. Degradation ofphenanthrene by *Burkholderia* sp. C3: initial 1,2- and 3,4-dioxygenation and meta- and *ortho*-cleavage of naphthalene-1,2-diol. Biodegradation 18, 123–131.

Šepič, E., Bricelj, M., Leskovsek, H., 1998. Degradation of fluoranthene by *Pasteurella* sp. IFA and *Mycobacterium* sp. PYR-1: isolation and identification of metabolites. Journal of Applied Microbiology 85, 746–754.

Sherry, A., Gray, N.D., Ditchfield, A.K., Aitken, C.M., Jones, D.M., Röling, W.F.M., Hallmann, C., Larter, S.R., Bowler, B.F.J., Head, I.M., 2013. Anaerobic biodegradation of crude oil under sulphate-reducing conditions leads to only modest enrichment of recognized sulphate-reducing taxa. International Biodeterioration and Biodegradation 81, 105–113.

Silva, I.S., Grossman, M., Durrant, L.R., 2009. Degradation of polycyclic aromatic hydrocarbons (2–7 rings) under microaerobic and very-low-oxygen conditions by soil fungi. International Biodeterioration and Biodegradation 63 (2), 224–229.

Simister, R.L., Poutasse, C.M., Thurston, A.M., Reeve, J.L., Baker, M.C., White, H.K., 2015. Degradation of oil by fungi isolated from Gulf of Mexico beaches. Marine Pollution Bulletin 100, 327–333.

Simon, M.J., Osslund, T.D., Saunders, R., Ensley, B.D., Sugg, S., Harcourt, A., Suen, W.C., Cruden, D.L., Gibson, D.T., Zylstra, G.J., 1993. Sequences of genes encoding naphthalene dioxygenase in *Pseudomonas putida* strains G7 and NCIB 9816-4. Gene 127, 31–37.

Sisler, F.D., ZoBell, C.E., 1947. Microbial utilization of carcinogenic hydocarbons. Science 106, 521–522.

Sokolovská, I., Wattiau, P., Gerin, P., Agathos, S.N., 2002. Biodegradation of fluorene at low temperature by a psychrotrophic *Sphingomonas* sp. L-138. Chemical Papers 56 (1), 36–40.

Soliman, R.M., El-Gendy, N.Sh., Deriase, S.F., Farahat, L.A., Mohamed, A.S., 2011. Comparative assessment of enrichment media for isolation of pyrene-degrading bacteria. Review Industrial Sanitary and Environmental Microbiology 5 (2), 54–70.

Solomon, E.I., Chen, P., Metz, M., Lee, S.K., Palmer, A.E., 2001. Oxygen binding, activation, and reduction to water by copper proteins. Angewandte Chemie International Edition in English 40, 4570–4590.

Sondossi, M., Barriault, D., Sylvestre, M., 2004. Metabolism of 2,2′ and 3,3′-dihydroxybiphenyl by the biphenyl catabolic pathway of *Comamonas testosteroni* B-356. Applied and Environmental Microbiology 279, 174–181.

Speight, J.G., Arjoon, K.K., 2012. Bioremediation of Petroleum and Petroleum Products. Scrivener Publishing, Beverly, Massachusetts.

Spormann, A.M., Widdel, F., 2000. Metabolism of alkylbenzenes, alkanes and other hydrocarbons in anaerobic bacteria. Biodegradation 11 (2), 85–105.

Steffen, K.T., Hatakka, A., Hofrichter, M., 2003. Degradation of benzo[a]pyrene by the litter-decomposing basidiomycete *Stropharia coronilla*: role of manganese peroxidase. Applied and Environmental Microbiology 69, 3957–3964.

Stout, S.A., Uhler, A.D., Emsbo-Mattingly, S.D., 2004. Comparative evaluation of background anthropogenic hydrocarbons in surficial sediments from nine urban waterways. Environmental Science & Technology 38, 2987–2994.

Strignfellow, W.T., Cerniglia, M.D., 1996. Bacterial degradation of low concentrations of phenanthrene and inhibition by naphthalene. Microbial Ecology 31 (3), 305–317.

Stringfellow, W.T., Aitken, M.D., 1995. competitive metabolism of naphthalene, methylnphthalenes, and fluorene by phenanthrene-degrading *Pseudomonas*. Applied and Environmental Microbiology 61, 357–362.

Sutherland, J.B., Crawford, D.L., Pometto, A.L., 1983. Metabolism of cinnamic, pcoumaric and ferulic acids by *Streptomyces setonii*. Canadian Journal of Microbiology 29, 1253–1257.

Takizawa, N., Iida, T., Sawada, T., Yamauchi, K., Wang, Y.W., Fukuda, M., Kiyohara, H., 1999. Nucleotide sequences and characterization of genes encoding naphthalene upper pathway of *Pseudomonas aeruginosa* PaK1 and *Pseudomonas putida* OUS82. Journal of Bioscience and Bioengineering 87, 721–731.

Tanaka, Y., Matsui, T., Konishi, J., Maruhashi, K., 2002. Biodesulfurization of benzothiophene and dibenzothiophene by a newly isolated *Rhodococcus* strain. Applied Microbiology and Biotechnology 59, 325–328.

Tani, A., Ishige, T., Sakai, Y., Kato, N., 2001. Gene structures and regulation of the alkane hydroxylase complex in *Acinetobacter* sp. strain M-1. Journal of Bacteriology 183, 1819–1823.

Tardy-Jacquenod, C., Magot, M., Patel, B.K.C., Matheron, R., Caumette, P., 1998. *Desulfotomaculum halophilum* sp. nov., a halophilic sulfate-reducing bacterium isolated from oil production facilities. International Journal of Systematic Bacteriology 48, 333–338.

Tarhriz, V., Hamidi, A., Rahimi, E., Eramabadi, M., Eramabadi, P., Yahaghi, E., Khodaverdi, E., 2014. Isolation and characterization of naphthalene-degradation bacteria from Qurugol Lake located at Azerbaijan. Biosciences Biotechnology Research Asia 11 (2), 715–722.

Tebyanian, H., Hassanshahian, M., Kariminik, A., 2013. Hexadecane-degradation by *Teskumurella* and *Stenotrophomonas* strains isolated from hydrocarbon contaminated soils. Jundishapur Journal of Microbiology 26 (7), e9182. http://dx.doi.org/10.5812/jjm.9182.

Tersagh, I., Aondoakaa, E.M., Okpa, B.O., 2017. Aerobic degradation of anthracene, benzo(a)anthracene and dibenzo(a,h)anthracene by indigenous strains of aerobic heterotrophic bacteria and cyanobacteria isolates from crude oil contaminated Bodo Creek. Oil and Gas Research 3 (1), 125. http://dx.doi.org/10.4172/2472-0518.1000125.

Thauer, R.K., Stackebrandt, E., Hamilton, W.A., 2007. Energy metabolism and phylogenetic diversity of sulfate-reducing bacteria. In: Barton, L.L., Hamilton (Eds.), Sulfate Reducing Bacteria: Environmental and Engineered Systems. W.A. Cambridge University Press, Cambridge, UK (Chapter 1).

Throne-Holst, M., Wentzel, A., Ellingsen, T.E., Kotlar, H.-K., Zotchev, S.B., 2007. Identification of novel genes involved in long-chain n-alkane degradation by *Acinetobacter* sp. strain DSM 17874. Applied and Environmental Microbiology 73 (10), 3327–3332.

Tissot, B.P., Welte, D.H., 1984. Petroleum Formation and Occurrence, second ed. Springer-Verlag, Berlin Heidelberg NewYork Tokyo.

Trenz, S.P., Engesser, K.H., Fischer, P., Knackmuss, H.-J., 1994. Degradation of fluorene by *Brevibacterium* sp. strain DPO 1361: a novel C-C bond cleavage mechanism via 1,10-dihydro-1,10-dihydroxyfluoren-9-one. Journal of Bacteriology 176 (2), 789–795.

Trzesicka-Mlynarz, D., Ward, O.P., 1995. Degradation of polycyclic aromatic hydrocarbons (polynuclear aromatic hydrocarbon derivatives) by a mixed culture and its component pure cultures, obtained from polynuclear aromatic hydrocarbon derivatives-contaminated soil. Canadian Journal of Microbiology 41, 470–476.

Tsai, J.-C., Kumar, M., Lin, J.-G., 2009. Anaerobic biotransformation of fluorene and phenanthrene by sulfate-reducing bacteria and identification of biotransformation pathway. Journal of Hazardous Materials 164 (2–3), 847–855.

Van Afferden, M., Schach, S., Klein, J., Tüpper, H.G., 1990. Degradation of dibenzothiophene by *Brevibacterium* sp. DO. Archives of Microbiology 153, 324–328.

Van Beilen, J.B., Funhoff, E.G., 2005. Expanding the alkane oxygenase toolbox: new enzymes and applications. Current Opinion in Biotechnology 16 (3), 308–314.

Van Beilen, J.B., Wubbolts, M.G., Witholt, B., 1994. Genetics of alkane oxidation by *Pseudomonas oleovorans*. Biodegradation 5, 161–174.

Van Herwijnen, R., Wattiau, P., Bastiaens, L., Daal, L., Jonker, L., Springael, D., Govers, H.A., Parsons, J.R., 2003a. Elucidation of the metabolic pathway of fluorene and cometabolic pathways of phenanthrene, fluoranthene, anthracene and dibenzothiophene by *Sphingomonas* sp. LB126. Researc in Microbiology 154, 199–206.

Van Herwijnen, R., Springael, D., Slot, P., Govers, H.A.J., Parsons, J.R., 2003b. Degradation of anthracene by *Mycobacterium* sp. strain LB501T proceeds via a novel pathway, through o-phthalic acid. Applied and Environmental Microbiology 69, 186–190.

Van Herwijnen, R., van de Sande, B.F., van der Wielen, F.W.M., Springael, D., Govers, H.A.J., Parsons, J.R., 2003c. Influence of phenanthrene and fuoranthene on the degradation of fuorene and glucose by *Sphingomonas* sp. strain LB126 in chemostat cultures. FEMS Microbiology Ecology 46, 105–111.

Vazquez-Duhalt, R., Westlake, D.W.S., Fedorak, P.M., 1994. Lignin peroxidase oxidation of aromatic compounds in systems containing organic solvents. Applied and Environmental Microbiology 60, 459–466.

Vieth, A., Wilkes, H., 2005. Deciphering biodegradation effects on light hydrocarbons in crude oils using their stable carbon isotopic composition: A case study from the Gullfaks oil field, offshore Norway. Geochimica 70, 651–665.

Viguri, J., Verde, J., Irabien, A., 2002. Environmental assessment of polycyclicaromatic hydrocarbons (polynuclear aromatic hydrocarbon derivatives) in surface sediments of the Sanader Bay, Northern Spain. Chemosphere 48, 157–165.

Vila, J., López, Z., Sabate, J., Minguillon, C., Solanas, A.M., Grifoll, M., 2001. Identification of a novel metabolite in the degradation of pyrene by *Mycobacterium* sp. strain AP1: actions of the isolate on two- and three-ring polycyclic aromatic hydrocarbons. Applied and Environmental Microbiology 67, 5497–5505.

Vila, J., Tauler, M., Grifol, M., 2015. Bacterial polynuclear aromatic hydrocarbon derivatives degradation in marine and terrestrial habitats. Current Opinion in Biotechnology 33, 95–102.

Vincenza, A., Liliana, G., 2007. Bioremediation and monitoring of aromatic polluted habitats. Applied Microbiology and Biotechnology 76, 287–308.

Voordouw, G., Armstrong, S.M., Reimer, M.F., Fouts, B., Telang, A.J., Shen, Y., Gevertz, D., 1996. Characterization of 16S rRNA genes from oil field microbial communities indicates the presence of a variety of sulfate-reducing, fermentative, and sulfide-oxidizing bacteria. Applied and Environmental Microbiology 62, 1623–1629.

Walter, U., Beyer, M., Klein, J., Rehm, H.-J., 1991. Degradation of pyrene by *Rhodococcus* sp. UW1. Applied Microbiology and Biotechnology 34, 671–676.

Wang, X.C., Sun, S., Ma, H.Q., Liu, Y., 2006. Sources and distribution of aliphatic and polyaromatic hydrocarbons in sediments of Jiaozhou Bay, Qingdao, China. Marine Pollution Bulletin 52 (2), 129–138.

Wattiau, P., Bastiaens, L., Van Herwijnen, R., Daal, L., Parsons, J.R., Renard, M.E., Springael, D., Cornelis, G.R., 2001. Fluorene degradation by *Sphingomonas* sp. LB126 proceeds through protocatechuic acid: a genetic analysis. Research in Microbiology 152, 861–872.

Weber, F.J., Hage, K.C., de Bont, J.A., 1995. Growth of the fungus *Cladosporium sphaerospermum* with toluene as the sole carbon and energy source. Applied and Environmental Microbiology 61, 3562–3566.

Wei, X.-Y., Sang, L.-Z., Chen, J.-N., Zhu, Y.-X., Zhang, Y., 2009. The effects of low molecular weight OAs on biodegradation of multicomponent polynuclear aromatic hydrocarbon derivatives in aqueous solution using dual-wavelength fluorimetry,. Environmental Pollution Journal 157, 3150–3157.

Weissenfels, W.D., Beyer, M., Klein, J., 1990. Degradation of phenanthrene, fluorene and fluoranthene by pure bacterial cultures. Applied Microbiology and Biotechnology 32 (4), 479–484.

Weissenfels, W.D., Beyer, M., Klein, J., Rehm, H.J., 1991. Microbial metabolism of fluoranthene: isolation and identification of ring fission products. Applied Microbiology and Biotechnology 34, 528–535.

Whyte, L.G., Hawari, J., Zhou, E., Bourbonnière, L., Inniss, W.E., Greer, C.W., 1998. Biodegradation of variable-chain-length alkanes at low temperatures by a psychrotrophic *Rhodococcus* sp. Applied and Environmental Microbiology 64 (7), 2578–2584.

Wick, L.Y., Pasche, N., Bernasconi, S.M., Pelz, O., Har, H., 2003. Characterization of multiple-substrate utilization by anthracene-degrading *Mycobacterium frederiksbergense* LB501T. Applied and Environmental Microbiology 69 (10), 6133–6142.

Widdel, F., Rabus, R., 2001. Anaerobic biodegradation of saturated and aromatic hydrocarbons. Current Opinion in Biotechnology 12, 259–276.

Willison, J.C., 2004. Isolation and characterization of a novel *sphingomonad* capable of growth with chrysene as sole carbon and energy source. FEMS Microbiology Letters 241, 143–150.

Wolski, E.A., Barrera, V., Castellari, C., González, J.F., 2012. Biodegradation of phenol in static cultures by *Penicillium chrysogenum* ERK1: catalytic abilities and residual phytotoxicity. Revista Argentina de Microbiología 44, 113–121.

Wrightt, A., Olsen, R.H., 1994. Self-mobilization and organization of the genes encoding the toluene metabolic pathway of *Pseudomonas mendocina* KR1. Applied and Environmental Microbiology 60 (1), 235–242.

Wu, Y.R., He, T., Zhong, M., Zhang, Y., Li, E., Huang, T., Hu, Z., 2009. Isolation of marine benzo[a]pyrene-degrading *Ochrobactrum* sp. BAP5 and proteins characterization. Journal of Environmental Sciences (China) 21, 1446–1451.

Wunder, T., Marr, J., Kremer, S., Sterner, O., Anke, H., 1997. 1-Methoxypyrene and 1,6- dimethoxypyrene: two novel metabolites in fungal metabolism of polycyclic aromatic hydrocarbons. Archives of Microbiology 167, 310–316.

Xia, Y., Min, H., Rao, G., Lv, Z., Liu, J., Ye, Y., Duan, X., 2005. Isolationand characterization of phenanthrene-degrading *Sphingomonas paucimo*bilis strain ZX4. Biodegradation 16, 393–402.

Xu, F., Deussen, H.W., Lopez, B., Lam, L., Li, K., 2001. Enzymatic andelectrochemical oxidation of N-hydroxy compounds: redox potential, electron-transfer kinetics, and radical stability. European Journal of Biochemistry 268, 4169–4176.

Xu, F., Kulys, J.J., Duke, K., Li, K.C., Krikstopaitis, K., Deussen, H.J.W., Abbate, E., Galinyte, V., Schneider, P., 2000. Redox chemistry in laccase-catalyzed oxidation of N-hydroxy compounds. Applied and Environmental Microbiology 66, 2052–2056.

Yadav, J.S., Doddapaneni, H., Subramanian, V., 2006. P450ome of the white rot fungus *Phanerochaete chrysosporium*: structure, evolution and regulation of expression of genomic P450 clusters. Biochemical Society Transactions 34, 1165–1169.

Yadav, S., Reddy, C.A., 1993. Degradation of benzene, toluene, ethylbenzene, and xylenes (BTEX) by the lignin-degrading basidiomycete *Phanerochaete chrysosporium*. Applied and Environmental Microbiology 59, 756–762.

Yakimov, M.M., Giuliano, L., Denaro, R., Crisafi, E., Chernikova, T.N., Abraham, W.R., Luensdorf, H., Timmis, K.N., Golyshin, P.N., 2004. *Thalassolituus oleivorans* general nov., sp. nov., a novel marine bacterium that obligately utilizes hydrocarbons. International Journal of Systematic and Evolutionary Microbiology 54, 141–148.

Yakimov, M.M., Giuliano, L., Gentile, G., Crisafi, E., Chernikova, T.N., Abraham, W.R., Lünsdorf, H., Timmis, K.N., Golyshin, P.N., 2003. *Oleispira antarctica* gen. nov., sp. nov., a novel hydrocarbonoclastic marine bacterium isolated from Antarctic coastal sea water. International Journal of Systematic and Evolutionary Microbiology 53, 779–785.

Yakimov, M.M., Timmis, K.N., Golyshin, P.N., 2007. Obligate oil-degrading marine bacteria. Current Opinion in Biotechnology 18 (3), 257–266.

Yang, Y., Chen, R.F., Shiaris, M.P., 1994. Metabolism of naphthalene fluorene and phenanthrene: preliminary characterization of a cloned gene cluster from *Pseudomonas putida* NCIB9816. Journal of Bacteriology 176, 2158–2164.

Ye, D., Siddiqi, M.A., Maccubbin, A.E., Kumar, S., Sikka, H.C., 1996. Degradation of polynuclear aromatic hydrocarbons by *Sphingomonas paucimobilis*. Environmental Science & Technology 30, 136–142.

Yen, K.M., Serdar, C.M., 1988. Genetics of naphthalene catabolism in pseudomonads. Critical Reviews in Microbiology 15, 247–267.

Younis, Sh.A., El-Gendy, N.Sh., Moustafa, Y.M., 2013. A study on bio-treatment of petrogenic contamination in El-Lessan area of Damietta River Nile Branch, Egypt. International Journal of Chemical and Biochemical Sciences 4, 112–124.

Yu, C., Yao, J., Jin, J., 2014. Characteristics and metabolic pathways of fluorene (FLU) degradation by strain *Rhodococcus* sp. USTB-C isolated from crude oil. Journal of Chemical and Pharmaceutical Research 6 (3), 560–565.

Yu, S.H., Ke, L., Wong, Y.S., Tam, N.F.Y., 2005. Biodegradation of polycyclicaromatic hydrocarbons (polynuclear aromatic hydrocarbon derivatives) by a consortium enriched from mangrovesediments. Environmental International Journal 31, 149–154.

Yuan, J., Lai, Q., Zheng, T., Shao, Z., 2009. *Novosphingobium indicum* sp. nov., a polycyclic aromatic hydrocarbon-degrading bacterium isolated from a deep-sea environment. International Journal of Systematic and Evolutionary Microbiology 59, 2084–2088.

Yuan, S.Y., Chang, B.V., 2007. Anaerobic degradation of five polycyclic aromatic hydrocarbons from river sediment in Taiwan. Journal of Environmental Science and Health, Part B 42, 63–69.

Yuan, S.Y., Shiung, L.C., Chang, B.V., 2002. Biodegradation of polycyclic aromatic hydrocarbons by inoculated microorganisms in soil. Bulletin of Environmental Contamination and Toxicology 69, 66–73.

Zakaria, M.P., Takada, H., Tsutsumi, S., Ohno, K., Yamada, J., Kouno, E., Kumata, H., 2002. Distribution of polycyclic aromatic hydrocarbons (polynuclear aromatic hydrocarbon derivatives) in rivers and estuaries in Malaysia: A wide spread input of petrogenic polynuclear aromatic hydrocarbon derivatives. Environmental Science & Technology 36, 1907–1918.

Zeinali, M., Vossoughi, M., Ardestani, S.K., 2008a. Naphthalene metabolism in *Nocardia otitidiscaviarum* strain TSH1, a moderately thermophilic microorganism. Chemosphere 72 (6), 905–909.

Zeinali, M., Vossoughi, M., Ardestani, S.K., 2008b. Degradation of phenanthrene and anthracene by *Nocardia otitidiscaviarum* strain TSH1, a moderately thermophilic bacterium. Journal of Applied Microbiology 105, 398–406.

Zhang, Q., Tong, M.Y., Li, Y.S., Gao, H.J., Fang, X.C., 2007. Extensive desulfurization of diesel by *Rhodococcus erythropolis*. Biotechnology Letters 29, 123–127.

Zhang, X., Young, L.Y., 1997. Carboxylation as an initial reaction in the anaerobic metabolism of naphthalene and phenanthrene by sulfidogenic consortia. Applied and Environmental Microbiology 63, 4759–4764.

Zhong, Y., Luan, T., Lin, L., Liu, H., Tam, N.F.Y., 2011. Production of metabolites in the biodegradation of phenanthrene fluoranthene and pyrene by the mixed culture of *Mycobacterium* sp. and *Sphingomonas* sp. Bioresource Technology 102, 2965–2972.

Zhou, N.Y., Fuenmayor, S.L., Williams, P.A., 2001. Nag genes of Ralstonia (formerly *Pseudomonas*) sp. strain U2 encoding enzymes for gentisate catabolism. Journal of Bacteriology 183, 700–708.

Zhu, X., Huang, M., Zhang, Q., Achal, V., 2016. Proposal of possible pathway of fluorene biodegradation by *Citrobacter* sp. FL5. Applied Environmental and Biotechnology 1 (1), 44–51.

Zylstra, G.J., Kim, E., 1997. Aromatic hydrocarbon degradation by *Sphingomonas yanoikuyae* B1. Journal of Industrial Microbiology & Biotechnology 19, 408–414.

Zylstra, G.J., Kim, E., Goyal, A., 1997. Comparative molecular analysis of genes for polycyclic aromatic hydrocarbon degradation. Genetic Engineering (New York) 19, 257–269.

Zylstra, G.J., Mccombie, W.R., Gibson, D.T., Finette, B.A., 1988. Toluene degradation by *Pseudomonas putida* Fl: genetic organization of the *tod* operon. Applied and Environmental Microbiology 54 (6), 1498–1503.

FURTHER READING

Ali, H.R., El-Gendy, N.Sh., Moustafa, Y.M., Mohamed, I., Roushdy, M.I., Hashem, A.I., 2012. Degradation of asphaltenic fraction by locally isolated halotolerant bacterial strains. ISRN Soil Science 2012:435485. http://dx.doi.org/10.5402/2012/435485.

Allan, I., Semple, K.T., Hare, R., Reid, B.J., 2007. Cyclodextrin enhanced biodegradation of polycyclic aromatic hydrocarbons and phenols in contaminated soil slurries. Environmental Science & Technology 41, 5498–5504.

Bartha, R., Bossert, I., 1984. The treatment and disposal of petroleum wastes. In: Atlas, R.M. (Ed.), Petroleum Microbiology. Macmillan, New York, NY, USA, pp. 553–578.

Dave, B.P., Ghevariya, C.M., Bhatt, J.K., Dudhagara, D.R., Rajpara, R.K., 2014. Enhanced biodegradation of total polycyclic aromatic hydrocarbons (T polynuclear aromatic hydrocarbon derivatives) by marine halotolerant *Achromobacter xylosoxidans* using Triton X-100 and β-cyclodextrin – a microcosm approach. Marine Pollution Bulletin 79, 123–129.

Deriase, S.F., Younis, Sh.A., El-Gendy, N.Sh., 2013. Kinetic evaluation and modeling for batch degradation of 2-hydroxybiphenyl and 2,2'-dihydroxybiphenyl by *Corynebacterium variabilis* Sh42. Desalination and Water Treatment 51 (22–24), 4719–4728.

El-Gendy, N.Sh., Nassar, H.N., 2015. kinetic modeling of the bioremediation of diesel oil polluted seawater using *Pseudomonas aeruginosa* NH1. Energy Sources, Part A: Recovery, Utilization, and Environmental Effects 37 (11), 1147–1163.

Genney, D.R., Alexander, I.J., Killham, K., Meharg, A.A., 2004. Degradation of the polycyclic aromatic hydrocarbon (polynuclear aromatic hydrocarbon derivatives) fluorene is retarded in a Scots pine ectomycorrhizosphere. New Phytologist 163, 641–649.

Guerin, T.F., 1999. Bioremediation of phenols and polycyclic aromatic hydrocarbons in creosote contaminated soil using ex-situ land treatment. Journal of Hazardous Materials B65, 305–315.

Hadibarata, T., Kristant, R.A., 2012. Fate and cometabolic degradation of benzo[a]pyrene by white-rot fungus *Armillaria* sp. F022. Bioresource Technology 107, 314–318.

Huang, H., Bowler, B.F.J., Oldenburg, T.B.P., Larter, S.R., 2004. The effect of biodegradation on polycyclic aromatic hydrocarbons in reservoired oils from the Liaohe basin, NE China. Organic Geochemistry 35, 1619–1634.

McClellan, S.L., Warshawsky, D., Shann, J.R., 2002. The effect of polycyclic aromatic hydrocarbons on the degradation of benzo[a]pyrene by *Mycobacterium* sp. strain RJGII-135. Environmental Toxicology and Chemistry 21, 253–259.

Nassar, H.N., Deriase, S.F., El-Gendy, N.Sh., 2016. Modeling the relationship between microbial biomass and total viable count for a new bacterialisolate used in biodesulfurization of petroleum and its fractions. Petroleum Science and Technology 34 (11/12), 980–985.

Rahman, K.S.M., Rahman, T.J., Kourkoutas, Y., Petsas, I., Marchant, R., Banat, I.M., 2003. Enhanced bioremediation of n-alkane in petroleum sludge using bacterial consortium amended with rhamnolipid and micronutrients. Bioresource Technology 90 (2), 159–168.

Shuttleworth, K.L., Cerniglia, C.E., 1995. Environmental aspects of polynuclear aromatic hydrocarbon derivatives biodegradation. Applied Biochemistry and Biotechnology 54, 291–302.

Wick, L.Y., Munain, A.R.d., Springael, D., Harms, H., 2002. Responses of *Mycobacterium* sp. LB501T to the low bioavailability of solid anthracene. Applied Microbiology and Biotechnology 58, 378–385.

BIOREMEDIATION OF CONTAMINATED SOIL

1.0 INTRODUCTION

According to the degree of pollution by petroleum hydrocarbons, sediments have been classified into three groups: (1) 1–4 mg/kg in unpolluted sediments, (2) < 100 mg/kg in moderately polluted sediments, and (3) up to 12,000 mg/kg in highly polluted sediments (Metwalley et al., 1997). The levels of oil in low-polluted soil averaged 710 µg/g soil and those in the high-polluted soils averaged 5500 µg/g soil (Duarte et al., 2001). The accumulation and persistence of oil in sediments for a prolonged time after the occurrence of an oil spill has also been reported (El-Tokhi and Moustafa, 2001). Colombo et al. (2005) reported that due to the hydrophobic nature of petroleum its environmental distribution can be ranked according to the increasing affinity sequence: water ≪ sediments < biota < soils.

Polynuclear aromatic hydrocarbons are among the most common pollutants found at contaminated industrial sites. Although they are naturally occurring compounds and essentially ubiquitous at low concentrations in the terrestrial environment, high concentrations of these compounds are found in contaminated soils near petroleum processing operations. Chemical reactivity, aqueous solubility, and volatility of polynuclear aromatic hydrocarbons decrease with increasing molecular weight. As a result, polynuclear aromatic hydrocarbons differ in their transport, distribution, and fate in the environment and their effects on biological systems. The United States Environmental Protection Agency (US EPA) currently regulates 16 polynuclear aromatic hydrocarbon compounds as priority pollutants for remediation (Fig. 10.1) (King et al., 2004; Abbondanzi et al., 2006). The petrogenic origin of polynuclear aromatic hydrocarbons is characterized by the dominance of low-molecular-weight polynuclear aromatic hydrocarbon derivatives over the high-molecular–weight polynuclear aromatic hydrocarbon derivatives (Moustafa, 2004). The ratio of low-molecular-weight polyaromatic hydrocarbons to high–molecular-weight polyaromatic hydrocarbon derivatives is usually used as an index of weathering (Hwang and Foster, 2006). The lower its value, the more the persistence of the pollutant and its recalcitrance towards bioremediation (El-Gendy et al., 2009). Maliszewska-Kordybach (1996) classified contaminated soil to four groups, according to the concentration of the 16 polynuclear aromatic hydrocarbon compounds, listed by US EPA, as priority pollutants; noncontaminated soil (<200 ppm), weakly contaminated soil (200–600 ppm), contaminated soil (600–1000 ppm), and heavily contaminated soil (>1000 ppm).

High-molecular-weight polynuclear aromatic hydrocarbon derivatives are sparingly soluble in water; electrochemically stable; and may be acutely toxic, genotoxic, immunotoxic, (Burchiel and Luster, 2001) or act as agents of hormone disruption (Van de Wiele et al., 2005) depending upon circumstances and the mode of exposure. High-molecular-weight polynuclear aromatic hydrocarbons are much more persistent in the environment than low-molecular-weight polynuclear aromatic hydrocarbons, have very low water solubility (Typically ≪ 0.0001 mg/L) and very high lipid solubility (Log Kow > 4.00) (Mackay et al., 1992; Ramachandran et al., 2006; Prabhukumar and Pagilla, 2010), and thus readily partition into lipid-rich cell membranes upon exposure. Due to their elevated

Introduction to Petroleum Biotechnology. https://doi.org/10.1016/B978-0-12-805151-1.00010-2

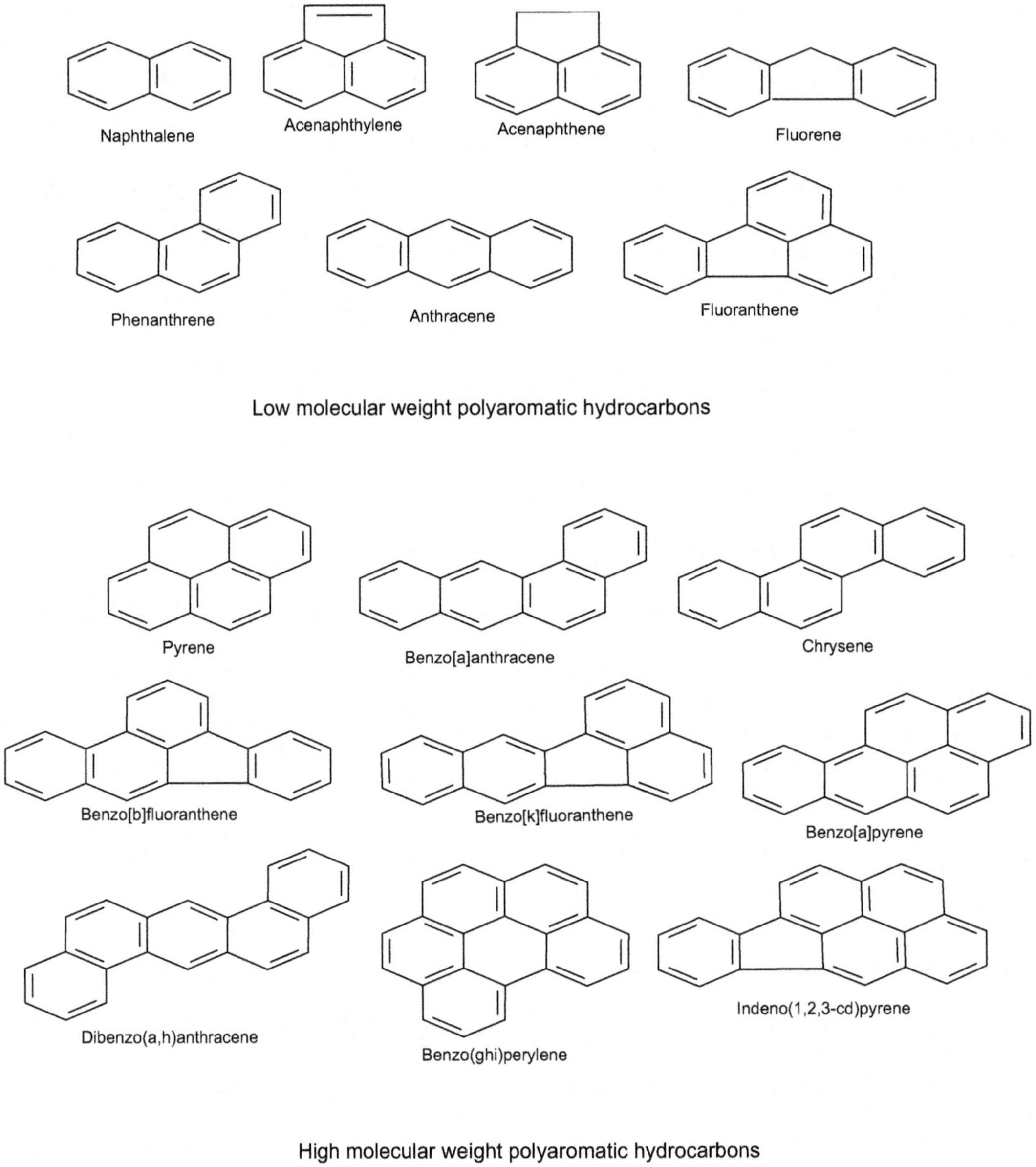

Naphthalene

Acenaphthylene

Acenaphthene

Fluorene

Phenanthrene

Anthracene

Fluoranthene

Low molecular weight polyaromatic hydrocarbons

Pyrene

Benzo[a]anthracene

Chrysene

Benzo[b]fluoranthene

Benzo[k]fluoranthene

Benzo[a]pyrene

Dibenzo(a,h)anthracene

Benzo(ghi)perylene

Indeno(1,2,3-cd)pyrene

High molecular weight polyaromatic hydrocarbons

FIGURE 10.1

The 16 polyaromatic hydrocarbons listed by the US-Environmental Protection Agency as priority pollutants.

octanol–water partition coefficients [K_{ow}], high-molecular-weight polynuclear aromatic hydrocarbons may partition into organic phases, soil, and sediment organic matter and membranes of living organisms, and are candidates for bioconcentration, bioaccumulation, and sometimes biomagnification through trophic transfers in terrestrial and marine food webs (Neff, 2002; Meador, 2003).

Petroleum refining, transportation, and storage are sources of soil contamination. Moreover, accidental oil spills are another source of pollution. But mainly it is a consequence of negligent disposal of oil sludge residues that result from the storage tanks cleaning. Oil sludge is a thick, viscous mixture of sediments, water, and high hydrocarbon concentration. The oily waste sludge is often expensive to store or destroy. Soil contamination with hydrocarbons causes extensive damage to local ecosystem since accumulation of such pollutants in tissues of animals and plants may cause death or mutation (Labud et al., 2007; Farahat and El-Gendy, 2008; Godleads et al., 2015; Safiyanu et al., 2015; Lim et al., 2016; Zabbey et al., 2017).

According to Huesemann (1994), high hydrocarbon levels are associated with varying degrees of inhibitory effects on soil microbes, for example, oxygen and/or nutrients limitations. Not only this but also the terrestrial oil spills are characterized primarily by vertical movement of the oil through the soil rather than the horizontal spreading associated with slick formation. Infiltration of oil into the soil might prevent evaporative loses of volatile hydrocarbons, which are toxic to microorganisms (Leahy and Colwell, 1990). According to the European Environmental Agency (EEA, 2007), it is estimated that, in Europe, potentially polluting activities have occurred at about 3 million sites, of which, > 8% are highly contaminated and need to be remediated. Moreover, the total number of contaminated sites is expected to increase by > 50% by 2025. Not only this, but the remediation is also progressing relatively slowly. Thus, the problem is estimated to be growing, not shrinking.

Mechanical and chemical techniques, such as storage, landfill, relocation, and incineration, are the commonly applied remediation methods (Bunkim, 2003; Agarwal and Liu, 2015; Lim et al., 2016). But they are expensive, complicated, and not adequate for complete removal of such oil contaminants. Moreover, the chemical technique would cause a second pollution to the environment, throughout the unexpected release of some artificial toxic materials. To confront and resolve this enormous problem, remediation must be simple and cost-effective for both industrialized nations and underdeveloped countries. Therefore, bioremediation is considered as a viable alternative for this purpose. Biotransformation of hydrocarbons by natural microbial populations present in the contaminated sites is one of the main processes used in the depuration of hydrocarbon-polluted environments (Abalos et al., 2004). The advantages of this option compared to the conventional physical and chemical options include inexpensive equipment, environmentally friendly nature of the process, and simplicity. However, one disadvantage of this process is its relative slow speed in achieving goals (Sabate et al., 2004). Oil industry views bioremediation of drilling mud, tank bottoms, and contaminated refinery soils as a top priority. It is well known that it can be used successfully to remediate many hydrocarbon-polluted sites, converting the toxic compounds of oil to nontoxic products without further disruption of the local environment, and can be applied as a polishing step after conventional mechanical and/or chemical clean-up options have been applied (Van-Hamme et al., 2003; Chaineau et al., 2005; Mohammed et al., 2007). Microbiologists have known since the 1940s (ZoBell, 1963) that microorganisms under favorable conditions can utilize petroleum hydrocarbons as a metabolic carbon source. Biotransformation strategies aim to alleviate constraints on microbial metabolism, by providing electron acceptors (such as oxygen), addition of essential nutrients (such as biologically available nitrogen or phosphorus), addition of bulking agents to provide a primer carbon source and provide aeration, or adjusting soil moisture content or adjusting soil pH by the addition of lime. When this is done with appropriate care, hydrocarbon biotransformation is stimulated without any adverse environmental impacts.

2.0 KINETICS OF PETROLEUM BIOTRANSFORMATION IN SOIL

Bioremediation is an engineered process where the natural biotransformation of petroleum hydrocarbons by indigenous soil bacteria, fungi, and protozoa is accelerated (Calvo et al., 2009; Das and Chandran, 2011). Along with deciding if a site should be remediated using ex-situ or in situ treatments, it is necessary to decide the bioremediation options that needed to be employed. Three types of bioremediation are predominant in the industry today (Kaplan and Kitts, 2004): natural attenuation (soil's natural ability to degrade the contaminant), bioaugmentation (addition of a microbial consortium from selected species previously isolated from a contaminated soil plus nutrients), and biostimulation (adding only nutrients to improve the natural biotransformation rate). The soil pollution with petroleum hydrocarbons alters the ratio of important nutrients for sufficient microbial activity. Thus, the addition of important nutrients (i.e., nitrogen, potassium, and phosphorus) allows sufficient increase in biomass for environmentally significant hydrocarbon degradation to occur.

Generally, biostimulation is the most recommended method (Sabate et al., 2004; Mrayyan and Battikhi, 2005), especially, upon, using readily available and cost-effective source of nutrients, such as corn steep liquor (CSL), molasses, etc (El-Gendy and Farah, 2011; Soliman et al., 2014; Godleads et al., 2015). Dadrasnia and Agamuthu (2013) observed that biowastes (soy cake, potato skin, and tea leaf) had potential to remediate diesel contaminated soil. Oil palm empty fruit bunch and sugarcane bagasse have been successfully used as fertilizers to improve the bioremediation and degradation of oil-contaminated soil (Hamzah et al., 2014). The development of slow-release biostimulant ball using natural materials has been attracting interest, especially for enhancing the anaerobic microbial diversity in coastal sediment. Since the advantages of slow-release biostimulant ball include longer acting times, reduced toxicity, increased cost effectiveness, and the direct use of biostimulant by the indigenous microorganisms, a less energy would be required for the absorption process, which would consequently enhance the microbial growth, enzymatic activities, and consequently the pollutant biotransformation rate (Subha et al., 2015).

One of the important advantages of bioremediation is that it can be carried out in nonsterile open environments that contain a variety of organisms (Watanabe, 2001). The light components up to nC_{10} are usually lost due to the natural attenuation, in another word, natural weathering process, mainly vaporization (Kaplan and Kitts, 2004). According to Alexander (1999), the natural microbial community, especially in chronic hydrocarbon-contaminated sites usually degraded oil hydrocarbons if suitable conditions are present. Thus, the natural attenuation process would be slow, but nevertheless it is effective over a longer period (Margesin and Shinner, 2001a; El-Gendy and Farah, 2011; Soliman et al., 2014).

One of the difficulties of developing bioremediation strategies lies in achieving as good or better results in the field as in the laboratory (Juhaz et al., 2000). Thus, lab-scale microcosms serve as an important stage in the study of the biotransformation efficiencies of microorganisms and can also serve as standard test systems that can be adapted to different environmental conditions. Results of microcosm studies are often used to develop remedial pilot test specifications (Lepo and Cripe, 1999).

Soil microcosms are an approach to study microbial interactions with contaminants, biotransformation efficiency of natural strains or genetically modified organisms in a controlled environment, preserving the intricacy of soil matrices and being representative of the process occurring in soil. Thus, soil microcosm can provide a useful insight of the degradation pattern of different petroleum hydrocarbons by mimicking the historically polluted sites. Thus, it can be considered as standard test systems that can be adapted to a variety of environmental conditions (Mishra et al., 2004; Arias et al., 2008; Farahat and

El-Gendy, 2008; El-Gendy and Farah, 2011; Soliman et al., 2014), specially, if, biotransformation experiments are constructed with bulk soil without sieving.

Thus, degradation could take place at "realistic" conditions with pores of different sizes being present. Also, nonsterilized soil and nonaseptic conditions can be also applied on microcosms study to mimic the realistic environmental field as much as it could be, since sterilization kills indigenous microorganisms, thus natural attenuation and biostimulation could not be studied. Moreover, sterilization can alter considerably the sorption properties of organic matter and then the outcome of hydrocarbon biotransformation experiments (Amellal et al., 2001). However, for a larger scale that would mimic more the real filed application, mesocosm can be used, which is considered as a logical intermediate step between in-vitro findings and full-scale application (Santas et al., 1999).

Monitoring the survival of the introduced highly efficient microbial degraders at the contaminated site is an essential attribute in the reduction of contamination levels by the introduced organisms (MacNaughton et al., 1999; Margesin and Schimer, 2001a,b; Margesin et al., 2003; Mishra et al., 2004; Farahat and El-Gendy, 2008). For effective bioremediation, it is essential that the augmented microorganisms survive over a long period of time for expression of their catabolic potential and complete reclamation of the pollutant (Mishra et al., 2004; El-Gendy et al., 2009; Soliman et al., 2014). It is generally believed that the bacterial strains that are resistant to the xenobiotic pollutant and capable of degrading it can effectively be found in contaminated sites (Ueno et al., 2006). However, according to Bachoon et al. (2001), culturable bacteria are not representative of natural bacterial communities.

One of the main factors that would affect the rate of biotransformation is the source of isolation of the applied biodegrading microorganisms (Mohd Radzi et al., 2015). The use of adapted microorganisms, i.e., previously isolated from polluted environment, is very recommendable, as this would eliminate the adaptation phase and/or decrease the lag phase (Trindade et al., 2005; El-Gendy et al., 2009). Generally, there would be a depletion in the total viable count per unit weight of soil (typically per gram of soil) observed after an initial increase with the starting period of incubation. Since, initially the microbial isolates are stimulated by labile hydrocarbon sources that induce a high rate of biotransformation. As the readily available sources decrease, microbial population would utilize the more recalcitrant hydrocarbons less efficiently.

The decrease in labile carbon sources comes with the limitation of available nutrients that support microbial growth. The decrease of total viable count could be attributed to the subsequent depletion of nutrients and the presence of toxic metabolites resulting from degradation (Boopathy, 2000; Bento et al., 2005; Sabate et al., 2004; Farahat and El-Gendy, 2008). According to Gogoi et al. (2003), the nutrients limitation could limit the metabolism of the indigenous microorganisms capable of degrading hydrocarbons in the soil environment, and the application of extraneous nutrients is required for developing a feasible bioremediation method in the studied area. It is recommendable to frequently add nutrients to keep the carbon-nitrogen-phosphorus mass ratio as 100:10:1 (Cheng and Mulla, 1999; Dibble and Bartha, 1979; Farahat and El-Gendy, 2008; Soliman et al., 2014).

Farahat and El-Gendy (2008) explained the depletion of microbial growth and decrease of biotransformation rate, by the accumulation of toxic intermediates formed from the biotransformation of the four-ring compounds and other low–molecular-weight polynuclear aromatic hydrocarbon derivatives. Many polynuclear aromatic hydrocarbon intermediates, such as phenol and benzoic acid intermediates, affect the survival and activity of microorganisms. These compounds are considered as biocidal at elevated concentrations that can rapidly limit biotransformation process (Kazunga and Aitken, 2000).

The most direct way to measure bioremediation efficiency is to monitor hydrocarbon (i.e., the pollutant) disappearance rates (Margesin and Shinner, 2001a; El-Gendy et al., 2009). As generally, the sequence of hydrocarbon degradation is as follows: alkanes fraction biodegraded extensively followed by aromatic fractions, then resins and asphaltenes fractions biodegraded the least (Mishra et al., 2004). However, the degradation pattern of organic chemicals in soil usually shows a rapid initial phase of degradation followed by a period of little or no change in concentration. This kinetic phenomenon is known as the "hockey stick" phenomenon (Song et al., 1990; Alexander, 1999; Sabate et al., 2004; Farahat and El-Gendy, 2008; El-Gendy and Farah, 2011).

A depletion of nutrients, a decrease of microbial populations, lower bioavailability, and higher recalcitrance of residual contaminants explain this kind of dynamic and formation of toxic metabolites (Alexander, 1999; Sabate et al., 2004). The addition of nutrients or reinoculation with well-adapted microorganisms would overcome this phenomenon. The use of active and well-adapted hydrocarbon degrading bacteria eliminate the initial lag phase in the hydrocarbon (i.e., the pollutant) removal, due to the immediate increase in the population density of the microbes, which could ensure rapid degradation of the pollutants and reduce the clean-up time substantially (Chaineau et al., 2005; Sabate et al., 2006; El-Gendy et al., 2009). However, the higher the degree of weathering of the polluted soil, the lower the initial rate of biotransformation, due to the absence of labile hydrocarbon sources (for example, the linear and open chain hydrocarbons) (El-Gendy et al., 2009). Moreover, as the concentration of the contaminant increases, the initial biotransformation rate decreases and the overall degree of biotransformation decreases (El-Gendy et al., 2009). De Jonge et al. (1997) reported that at high concentration of oil pollution (12,000–4000 mg/kg) due to the different mechanisms of oil distribution in the soil matrix, all n-alkanes were degraded at the same rate, regardless of the chain length of n-alkanes.

The exponential growth equation (Eq. 10.1) (Bailey and Ollis, 1986) is usually used to describe growth of microbial populations:

$$\ln\left(\frac{X_t}{X_0}\right) = k_g t$$

(10.1)

where X_t is the cell concentration (number of cells/g soil), t is time, and k_g is the kinetic growth constant (time^{-1}) and can be obtained as the slope of the line if $\ln(X_t/X_0)$ is plotted versus time. Nonlinear regression statistical analysis could help on describing the effect of both incubation period and microbial growth on percentage removal of oil and the relationship between the cumulative removal (CR) of oil (total oil content [TOC]) from different polluted sediment samples and incubation period (t) (El-Gendy et al., 2009).

Exponential equation (Eq. 10.2) is reported to accurately correlate the CR of total oil content (mg TOC/kg dry sediment) and incubation period (t) (Bento et al., 2005; El-Gendy et al., 2009).

$$CR = R_{max(1 - e^{-kt})}$$

(10.2)

In this equation, CR = $[(TOC)_0 - (TOC)_t]$, $(TOC)_0$ and $(TOC)_t$ are the initial total oil content before treatment and final total oil content after treatment at time (t), respectively.

El-Gendy et al. (2009) elucidated a quantitative expression (Eq. 10.3) of the effect of incubation period (t) and microbial growth (I) on the percentage removal of TOC from different polluted sediment samples.

$$d = aI + bt + cIt$$

(10.3)

In this equation, a, b, and c are regression coefficients factors. $I = (I_t - I_0)/I_0$, I_0, and I_t are the initial and final total viable microbial count at time (t) and percentage removal of the TOC can be calculated as follows:

$$d == \% \text{ removal of TOC} = \left[((TOC)_0 - (TOC)_t)/(TOC)_0\right] \times 100 \tag{10.4}$$

Sarkar et al. (2005) reported that the kinetics of reaction (i.e., biotreatment process) can be described in terms of its order since the rate constants are reflective of the relative effects of various treatments on total petroleum hydrocarbon (TPH) degradation in contaminated soils (Speight and Arjoon, 2012). The frequently used kinetic models are the first-order (Eqs. 10.5 and 10.6) and second-order (Eqs. 10.7 and 10.8) models.

$$C = C_0 e^{-k_1 t} \tag{10.5}$$

its linearized form:

$$\ln C = -k_1 t + A \tag{10.6}$$

$$-\frac{dC}{dt} = k_2 C^2 \tag{10.7}$$

its linearized form:

$$\frac{1}{C} = k_2 t + B \tag{10.8}$$

where C is the change of the concentration (mg/kg soil) of TPH with time (t); k_1 (time^{-1}) and k_2 (kg/mg time) are the first and second order rate constants, respectively; A ($\ln C_0$) and B ($1/C_0$, kg/mg) are the correlation coefficient constants or first and second order model equations, respectively; and C_0 is the initial pollutant concentration (mg/kg soil). The half-life time ($t_{1/2}$) of biotransformation can be calculated from the first and second order model equations as ($\ln(2)/k_1$) and ($1/k_2 C_0$), respectively.

The goodness of fit of the models is at first expressed by the linear regression coefficients of determination R, which measures the difference between the experimental and theoretical data in linear plots only, but not the errors in kinetics curves. The closer the value of R to unity the more the model successfully describes the kinetics of the degradation of petroleum hydrocarbons components. But, due to the inherent bias resulting from linearization, usually, error functions are employed to enable the optimization process to determine and evaluate the fit of the model equation to the obtained experimental data. Where, the lower the values of error analysis the better will be the goodness of fit.

The widely used error function, sum of the square of the errors (ERRSQ) (Eq. 10.9), has one major drawback; parameters derived using this error function will provide a better fit as the magnitude of the errors and thus the squares of the errors increase, biasing the fit towards data obtained (Rengaraj et al., 2007).

$$\sum_{i=1}^{P} (C_{exp} - C_{cal})_i^2 \tag{10.9}$$

The error function, hybrid fractional error function (Eq. 10.10) was developed by Porter et al. (1999) to improve the fit of the ERRSQ method.

$$\frac{100}{p-n} \sum_{i=1}^{p} \left[\frac{(C_{exp} - C_{cal})^2}{C_{exp}} \right]_i \tag{10.10}$$

The Marquardt's percent standard deviation (MPSD) (Eq. 10.11) is similar in some respects to a geometric mean error distribution modified according to the number of degrees of freedom of the system (Marquardt, 1963).

$$100 \left[\sqrt{\frac{1}{p-n} \sum_{i=1}^{p} \left(\frac{C_{exp} - C_{cal}}{C_{exp}} \right)_i^2} \right] \tag{10.11}$$

The average relative error (ARE) (Eq. 10.12) is known to minimize the fractional error distribution (Kapoor and Yang, 1989; Han et al., 2007).

$$\frac{100}{p} \sum_{i=1}^{p} \left(\frac{C_{exp} - C_{cal}}{C_{exp}} \right)_i \tag{10.12}$$

The sum of the absolute errors (SAE) (Eq. 10.13) is also usually used (Mall et al., 2005; Perez-Marin et al., 2007).

$$\sum_{i=1}^{p} (C_{cal} - C_{exp})_i \tag{10.13}$$

The root mean square (RMS) (Eq. 10.14) is also reported in some studies (Arslanoglu et al., 2005).

$$RMS = \left(\sum \left(C_{exp} - C_{cal} / C_{exp} \right)^2 / p \right)^{1/2} \tag{10.14}$$

where p is the number of data points, n the number of parameters within the equation, C_{cal} is the calculated data with the kinetic models, and C_{exp} is the experimental data. MacNaughton et al. (1999), Nocenteni et al. (2000), Sarkar et al. (2001), Namkong et al. (2002), Zahed et al. (2011), and Abioye et al. (2012) adequately described their hydrocarbon degradation data using a first-order kinetic model.

Anaerobic biotransformation of different polyaromatic hydrocarbons in soil has been reported to follow the first-order kinetics (Chang et al., 2002). The rate of biotransformation decreased as follows:

phenanthrene > pyrene > anthracene > fluorene > acenaphthene

The higher biotransformation rate of phenanthrene was attributed to the well-adaptation of the augmented consortium to phenanthrene as it was enriched by phenanthrene for 3 years and polynuclear aromatic hydrocarbon combined for only 1 year. In this study two types of soils have been used and the rate of degradation in Taida soil was higher than in Guishan soil, due to the higher organic matter content in Taida soil.

The rate of biotransformation of individual polynuclear aromatic hydrocarbon in mixture-substrate system was higher than that of single-substrate system. That was attributed to the broad versatility of the enzymes, the cross acclimation of the degradation enzymes, and/or the increase in polynuclear aromatic hydrocarbon bioavailability. Moreover, the anaerobic biotransformation rate in soil was higher than that in aqueous system. Since, soil provides a solid phase for better anaerobic bacterial growth. Moreover, the addition of acetate, lactate, or pyruvate, enhanced the rate of biotransformation. Since these carbon sources enhance the growth of methanogens, where, the metabolism of lactate of pyruvate by acidogen produces either hydrogen or hydrogen carbonate, which enhance methanogen production. Furthermore, the sulfate-reducing bacteria metabolize lactate to pyruvate plus two electrons, which is metabolized to acetate plus two electrons, and both would lead to faster degradation. Thus the rate of polynuclear aromatic hydrocarbons biotransformation was high under methanogenic and sulfate-reducing conditions but was lowered under nitrate-reducing conditions: sulfate-reducing conditions > methanogenic conditions > nitrate-reducing conditions. Not only this, but the biotransformation of polynuclear aromatic hydrocarbons in soil amended Taoyuan oil sludge was reported to be also higher than that amended with Minshi municipal sewage sludge. That was attributed to the indigenous microbial population in the amended sludge, where, the polynuclear aromatic hydrocarbon-degrading activity of autochthonous microorganisms was higher in the oil sludge samples than in the municipal sewage sludge samples. The application of three types of microbial inhibitors, a selective methanogen inhibitor, a selective sulfate-reducing bacteria inhibitor, or a selective eubacteria inhibitor, prove that the sulfate-reducing bacteria play the major part in the anaerobic biotransformation of the polynuclear aromatic hydrocarbons and the methanogen and vancomycin microbial populations were also involved and the main component of the adapted polynuclear aromatic hydrocarbons consortium is the sulfate-reducing bacteria.

Farahat and El-Gendy (2007) reported a comparative kinetic study between natural attenuation, biostimulation, and bioaugmentation treatment techniques for a petroleum hydrocarbon–polluted soil (42,000 mg total petroleum hydrocarbon/kg soil) in a semi-pilot scale experiment. Bioaugmentation with well-adapted bacterial consortium, *Pseudomonas aeruginosa* I.1.1.6 and *Brevibacterium casei* I.2.1.7, previously isolated from petroleum hydrocarbon–polluted soil, expressed the highest degradation potential (76%), followed by biostimulation process and then come the natural attenuation, recording approximately biotransformation efficiency of 62% and 48%, respectively. Kinetic modeling was also performed to estimate the rates of removal of TPH, and three different error functions (RMS, SAE, and ARE) were employed to evaluate the goodness of fit of the model equation to the obtained experimental data. The TPH removal was found to follow the first-order kinetic model equation. The highest rate constant ($0.012\,\text{day}^{-1}$) was observed in cell augmented with bacterial consortium I.1.1.6 and I.2.1.7, with the shortest predicted half-life time; 56 days, followed by biostimulation cell ($0.008\,\text{day}^{-1}$) and half-life time of 91 days. The lowest rate constant was observed in natural attenuation cell ($0.005\,\text{day}^{-1}$), with the longest predicted half-life time, 154 days.

The kinetics of the removal of different petroleum hydrocarbons pollutants in a biostimulation study was evaluated by El-Gendy and Farah (2011) for bioremediation of hydrocarbon-polluted soil collected from the beach of the Suez Canal Authority workshop yard for ship maintenance on Temsah Lake, Ismaliya, Egypt, on the microcosm level using CSL as a cosubstrate and the only source of added nutrients, over a period of 35 days. The used readily available and cheap source of nutrients, cosubstrate, and emulsifier, CSL, was composed of protein, lipids, and carbohydrates, with carbon-nitrogen-phosphorus contents of 6.8:1.06:2.14 (% w/w), respectively. The previously mentioned hokey stick phenomenon was obvious in that study.

There are two distinct phases: an initial fast degradation phase (up to 21 days of incubation), followed by a slow degradation phase. They attributed the rapid initial degradation phase to the native indigenous soil microbes that were capable of degrading hydrocarbons to a large extent and/or part of the loss may be attributed to volatilization as well. This phase was followed by a slow degradation phase as well as the addition of CSL, which acts as an emulsifier, that consequently increases the initial proportion of bioavailable TPH. The most likely explanation for the observed change in degradation rate is the petroleum sequestration in soil particles, which renders it unavailable to bacteria. The rate of degradation during the slow phase would thus be limited by the desorption rate of petroleum from soil particles rather than bacterial activity.

Although CSL was added at the start of the study and after 14 days of incubation, the microbial growth entered through a stationary phase after 21 days of incubation. That comes with the start of the decrease in the biotransformation rate, decrease in the total resolvable peaks components, and enrichment of the recalcitrant unresolved peaks components. That was explained by the microbial metabolism of the labile hydrocarbon sources (probably linear, open-chain hydrocarbons, and low-molecular-weight polyaromatic compounds) during the first 21 d (the rapid phase of degradation), where the microorganisms might have been stimulated by which induced a high percentage of degradation. But with the depletion of those forms, microbial populations are forced to use the more recalcitrant hydrocarbons (probably the high-molecular-weight aromatic hydrocarbons and other recalcitrant compounds) less efficiently, which would produce toxic intermediates that can inhibit microbial activity. The frequently used kinetic models, the first-order (Eqs. 10.5 and 10.6) and second-order (Eqs. 10.7 and 10.8) models, were investigated in that study to determine the mechanism of the biostimulation process. Five different error functions, sum of the square of the errors, hybrid fractional error function, MPSD, ARE, and SAE, were used in this study to evaluate the fitness of the model equation to the obtained experimental data. Thus by comparing the results of the values of error functions it was found that the degradation of $\sum nC20$-$nC24$, $\sum nC35$-$nC42$, and $\sum nC18$ can be best represented by a second-order model, whereas the TPH, the total resolvable peaks (n- and iso-alkanes), nC17, unresolved complex mixture UCM (naphthenes and cycloalkanes), $\sum nC10$-$nC14$, $\sum nC15$-$nC19$, $\sum nC25$-$nC29$, $\sum nC30$-$nC34$, nCn, $\sum isoCn$, and the biomarkers, pristane Pr and phytane Ph, were found to follow the first order model. From the calculated $t_{1/2}$ values and the rate of biotransformation of different components of polluting hydrocarbons, the biotransformation efficiencies were found to increase as follows:

short-chain n-alkanes > long-chain n-alkanes
long-chain n-alkanes > iso-alkanes >
iso-alkanes > unresolved complex mixture (naphthalene derivatives, i.e., cyclo-alkane derivatives).

This study proved that the chain length of hydrocarbons affects its rate of biotransformation, where the $t_{1/2}$ values of $nC_{17} < nC_{18}$, thus, the degree of degradation of n-alkanes with shorter chains is higher than the ones with longer chains.

The $t_{1/2}$ of the biotransformation of Pr > Ph proved the difficulty of biotransformation of Pr, and it expressed the longest $t_{1/2}$, which confirmed that Pr is more recalcitrant. The shortest $t_{1/2}$ of biotransformation was observed for the nC_{10}–nC_{14} fraction due to the abiotic losses comprising evaporation of lower alkanes in addition to biotic losses, which accelerated the removal of this fraction. But the $t_{1/2}$ of biotransformation of other studied fractions did not follow a certain trend and similar observation was reported by Seklemova et al. (2001), which proved at higher pollutant concentration that the scheme of microbial hydrocarbon utilization is not universal.

Abioye et al. (2012) applied the first-order kinetic model to compare between the bioremediation efficiencies of different biostimulation process for a used lubricating oil–polluted soil (5% and 15% w/w) using different organic wastes as source of nutrients—banana skin, brewery spent grain, and spent mushroom compost as alternatives to the use of inorganic fertilizers. The low-polluted soil expressed better bioremediation potentials than the high-polluted ones. That was attributed to the toxicity of high concentration of pollutant to the indigenous microbial population. The highest biotransformation efficiency of 92% was recorded for 5% contaminated soil amended with 10% brewery spent grain. That was attributed to the characteristics of the brewery spent grain, which was characterized by the highest nitrogen, phosphorous, and moisture contents, relative to the other cosubstrates. This might be one of the reasons for its highest biostimulation efficiency. Since, nitrogen and phosphorous are the most important limiting nutrients for effective bioremediation and its moisture content might have enabled it to indigenously have some important microorganisms that would have contributed positively to the biotransformation of oil in the soil. The biotransformation rate was 0.4361, 0.410, and 0.3100 day^{-1}, for 5% contaminated soil, amended with brewery spent grain, banana skin, and spent mushroom compost, respectively.

Another study performed by Soliman et al. (2014) evaluated the efficiencies of different bioremediation processes of a chronically and highly Egyptian oily sludge–polluted soil sample (53,100 mg/kg) on microcosm level for 180 days: natural attenuation, biostimulation with CSL and bioaugmentation with the addition of CSL, and inoculation with a well-adapted *Micrococcus lutes* RM1, previously isolated from the oily sludge–contaminated soil under study, for its ability to degrade pyrene as a model compound for high–molecular-weight polyaromatic hydrocarbon compounds (Soliman et al., 2011). There was a continuous increase of microbial population in biostimulation and bioaugmentation microcosms up to 130 and 80 days, respectively and attributed this to the well adaptation of the microbes to their environment and contaminants in the soil. However, depletion in biomass concentration was observed after 130 and 80 days in biostimulation and bioaugmentation microcosms, respectively, and it was attributed to the decline in the availability of readily metabolizable components in the oily sludge–contaminated soil. Generally, there was low biodiversity in all microcosms, Gram-positive *M. lutes* RM1, *Brevibacterium* sp. RM4, and *Curtobacterium* sp. RM5, with the predominance of *M. lutes* RM1 and it was attributed to the high-pollutant concentration and the sandy soil nature of low nutrient. There was a highly statistical significant difference of removal of TPH in biostimulation and bioaugmentation microcosms, relative to that in natural attenuation one, $P=0:004$, $P>0:00$, and 7.69e-5, $P<0:001$ at $\alpha=0.5$, 95% confidence level, with TPH removal of 44%, 54%, and 6%, respectively. The frequently used kinetic models, the first-order (Eqs. 10.5 and 10.6) and second-order (Eqs. 10.7 and 10.8) models, were investigated in that study to determine the mechanism of each biotreatment process, and three different error functions, RMS, SAE and ARE, were used to evaluate the fitness of the model equation to the obtained experimental data.

The removal of the TPH in bioaugmentation and biostimulation microcosms was best described by a first-order kinetic model, but the natural attenuation followed a second-order kinetic model. Moreover, the rate of biotransformation of different pollutants fractions was in the following decreasing order: saturates > aromatics > resins > asphaltenes in both biostimulation and bioaugmentation microcosms. With biotransformation rate constants of 0.0004, 0.0032, and 0.0043 d^{-1}; initial TPH degradation rates of 21.24, 169.92, and 228.33 mg/kg/d; and half-life time of 1733, 217, and 161 d in natural attenuation, biostimulation, and bioaugmentation microcosms, respectively. Thus, it was estimated that to reach TPH concentration of approximately 500 mg/kg, the three biotreatment techniques would take as long as 5,932, 717, and 533 days in the biotreatment microcosms, respectively.

3.0 ASPECTS FOR THE SUCCESS OF BIOTRANSFORMATION

Numerous studies conducted in the past few decades have identified relationships between soil conditions and microbial activity (Pietri and Brookes, 2008; Zheng et al., 2010). There are some factors that mainly affect the rate of hydrocarbon biotransformation in soils: (1) type and concentration of pollutant, (2) soil temperature, (3) concentration of nutrients, (4) soil pH, (5) aeration, (6) moisture content, (7) type and population of biodegrading microorganisms present or augmented into contaminated soil, (8) contaminant mobility, (9) concentration of heavy metals, and (10) type of soil.

3.1 INDIGENOUS AND AUGMENTED MICROBIAL POPULATION

Microbial interactions in the soil are very complex and play an important role in the transformation or decomposition of hazardous waste components (JRB Associates, Inc., 1984). The indigenous population can biodegrade the petroleum hydrocarbons in the polluted soil but the process is very slow and it would be effective over a long period of time (Soliman et al., 2014). Mishra et al. (2001) reported that potentials of bioremediation is negligible if the indigenous population of hydrocarbon-degrading microorganisms is less than 10^5 CFU/g in soil. The soil contamination by hydrocarbons is reported to increase the biochemical and microbial activities of the polluted soil. However, the rate of biotransformation or mineralization of the pollutant would be low, due to its complexity that would require long periods to be degraded (Morgan and Watkinson, 1989). Biostimulation increases the indigenous microbial population activity by adding nutrients and/or a terminal electron acceptor. In rare cases, indigenous microbes capable of biodegrading the contaminants may not be available at the site. To overcome this problem, pollutant degradation potential can be increased through bioaugmentation, which is the addition of exogenous hydrocarbon degrading microbial strains (Trindade et al., 2002). Bioaugmentation has been considered as a potential strategy for oil bioremediation since the 1970s (Darmayati et al., 2015). The previous exposure and acclimatization of microbes to pollutants in soils would enhance their enzymes systems responsible for biotransformation. The biotransformation potentials of strains previously isolated from hydrocarbon-contaminated environments are reported to be as active as or even more than those isolated from noncontaminated environments (Guo et al., 2005; Farahat and El-Gendy, 2007, 2008; Hegazi et al., 2007; Ali et al., 2012; Younis et al., 2013; El-Gendy and Nassar, 2015; El Mahdi et al., 2016). These are known as autochthonous microorganisms (Lim et al., 2016). Furthermore, a combination of biostimulation and bioaugmentation may be required for the efficient biotransformation in contaminated soil. Pontes et al. (2013) proved that the autochthonous bioaugmentation with indigenous oil-degrader consortium prestimulated in the laboratory combine with the addition of oleophilic fertilizer S200 for achieving suitable nutritive conditions and avoid oil diffusion results in faster biotransformation rate for buried oil, in experiments, designed to simulate the conditions of the middle intertidal area of sandy beaches. Covino et al. (2015) reported that the addition of lignocellulosic mixture as a slow-release fertilizer (SRF) in addition to bioaugmentation of petroleum hydrocarbons–polluted soil by autochthonous filamentous fungi *Pseudoallescheria* sp., previously isolated from petroleum-polluted clay soil, resulted in a 79.7% degradation of 5 g/L after 60 days. Darmayati et al. (2015) proved that biostimulation using SRF in addition to augmentation with a microbial consortium performs a better accelerating rate of bioremediation of highly oil-polluted sediment (100,000 mg oil/kg sediment) in mesocosm system than those of biostimulation in addition to single microbial bioaugmentation and biostimulation with fertilizer only.

Safiyanu et al. (2015) reported that the efficiency of hydrocarbon biotransformation ranged from 6% to 82% for soil fungi, 0.13%–50% for soil bacteria, and 0.003%–100% for marine bacteria. However, bioaugmentation may not necessarily manifest oil-degrading activity under conditions of competition with indigenous microorganisms (Simon et al., 2004). Adding compost would amend the polluted soil or sediment with nutrients and diverse assemblages of microorganisms. Not only this, but the organic matters of the compost would also strongly influence sorption/desorption processes of hydrophobic organic contaminants (Semple et al., 2001; Yu et al., 2011). Thus, it would improve the environmental conditions of a contaminated matrix for indigenous or introduced microorganisms. Moreover, Trindade et al. (2005) reported that the biostimulation process would be successful due to the previous and prolonged exposure of the native indigenous microorganisms to the contaminant, leading to the selection and predominance of the oil-degrading and/or tolerant ones. Thus, the lag phase of pollutant removal and microbial growth would be decreased or eliminated (El-Gendy and Farah, 2011; Soliman et al., 2014).

Microbes are the primary agents for the degradation of organic contaminants in soil. Thus, increasing the microbial density and activity can accelerate degradation of contaminants (Namkoong et al., 2002; Makut and Ishaya, 2010). However, Atlas (1981) reported that hydrocarbon degraders may comprise less than 0.1% of the microbial community in unpolluted environments but can constitute up to 100% of the viable microorganisms in oil-polluted ones. Even at low temperatures (0–5°C), Margesin and Schinner (1999) reported that formerly polluted frozen soils are commonly enriched in hydrocarbon-utilizing organisms. As the number of microorganisms capable of degrading the contaminant present in soil increases, the degradation rate increases. The ability to degrade petroleum hydrocarbons is not restricted to a few microbial genera. Instead, a diverse group of bacteria have this ability (Atlas, 1981; Das and Chandran, 2011). The most important genera of hydrocarbon degraders in marine and soil environments are *Achromobacter, Actinobacter, Alcaligenes, Arthrobacter, Bacillus, Flavobacterium, Nocardia, Micrococcus,* and *Pseudomonas* spp. (Leahy and Colwell, 1990; Margesin and Schinner, 2001b; Norman et al., 2004; Vacca et al., 2005). Hydrocarbon degrading fungi, *Aspergillus* and *Penicillium* spp., have been also reported (Leahy and Colwell, 1990; Prenafeta-Boldu et al., 2006; Gesinde et al., 2008; Erdogan and Karaca, 2011). *Brevibacterium* strains were investigated in different bioremediation studies of polynuclear aromatic hydrocarbons and polyaromatic sulfur heterocyclic compounds (Trenz et al., 1994; Chaillan et al., 2004). *Pseudomonas* sp. is often isolated from hydrocarbon-contaminated sites and hydrocarbon-degrading cultures. This genus has a broad affinity for hydrocarbons and can degrade selected alkane derivatives, alicyclic derivatives, thiophene derivatives, and aromatic derivatives (Van-Hamme et al., 2003). It has the potential for application in treatment systems for industrial effluents rich in aromatic compounds (El-Tayeb et al., 1998; Abd-El-Haleem et al., 2002; El-Masry et al., 2004). It also produces a range of biosurfactants, e.g., rhamnolipids (Sim et al., 1997; Desai and Banat, 1999; Mahjoubi et al., 2013; Ali et al., 2014). Different *P. aeruginosa* strains that can degrade different components of crude oils have also been isolated (Hwang and Cutright, 2002; Jacques et al., 2005; Farahat and El-Gendy, 2007, 2008; Ali et al., 2012; El-Gendy et al., 2014; El-Gendy and Nassar, 2015). Fan et al. (2014) reported the bioremediation of diesel oil–polluted soil by *Candida tropicalis* SK21, which also enhanced the dehydrogenase and polyphenol oxidase activities in soil.

Individual organisms can usually metabolize only a limited range of hydrocarbon substrates. Therefore, a consortium of mixed populations with overall broad enzymatic capacities is required to degrade complex mixtures of hydrocarbons such as crude oil in soils. Prior exposure of a microbial community to hydrocarbons is also important in determining how rapidly subsequent hydrocarbon inputs can be biodegraded (Leahy and Colwell, 1990; Enock, 2002; Rahman et al., 2003; Haritash and Kaushik, 2009; Idise et al., 2010). Some members of a microbial community might be able to provide important degrading enzymes, whereas others may supply surfactants (Wiesel et al., 1993). For example, Chen et al. (2014) reported the enhanced n-alkanes and crude oil biotransformation by a bacterial consortium, *Acinetobacter* sp. XM-02 and *Pseudomonas* sp. XM-01, which were isolated from crude oil–contaminated soil collected from the Boxi Offshore Oil Field, Tianjin, China, for their ability to degrade crude oil and rhamnolipid biosurfactant, respectively. Where, the biotransformation efficiency of alkanes and crude oil increased from 89.35 and 74.32±4.09% to 97.41 and 87.29±2.41%, respectively when augmented with XM-02 alone and after augmentation with the bacterial consortium. That was attributed to the synergism between the two bacterial strains, where the mixed culture, *Acinetobacter* sp. XM-02 grew fast on alkane fraction of crude oil and produced intermediates that were subsequently utilized for the growth of *Pseudomonas* sp. XM-01 and then rhamnolipid were produced by *Pseudomonas* sp. XM-01, which consequently enhanced the rate of biotransformation of crude oil.

The advantage of employing mixed cultures as opposed to pure cultures in bioremediation has also been widely demonstrated, as it would enhance the synergistic interactions among members of the association. Mechanisms through which bacteria benefit from synergistic interactions are complex. It is possible that one species removes the toxic metabolites, which may hinder the microbial activities of the species preceding it. It is also possible that the second species can degrade compounds that the first are not able to degrade or partially degrade them (Alexander, 1999; Ghazali et al., 2004; Farahat and El-Gendy, 2007). Das and Mukherjee (2007a) proved that the augmentation of polluted petroleum hydrocarbon soil with some well-adapted microorganisms enhances the rate of biotransformation. In that study, three thermophilic bacterial strains, *Bacillus subtilis* DM-04 and *P. aeruginosa* M and NM, previously isolated from a petroleum-contaminated soil sample of North-East India (ONGC oil Weld), for its ability to degrade many hydrocarbon compounds as a source of carbon and energy and produce biosurfactants in hydrocarbons rich culture medium, for biotreatment of petroleum hydrocarbon–polluted soil microcosm. The microcosms were inoculated with either DM-04 or a consortium of M and NM and glucose as a cosubstrate, after incubation of the spiked contaminated soil with biosurfactants obtained from the respective bacterial strains at a dose of 10 mL biosurfactant solution per kg of soil, for 3 days at 37°C, to enhance the solubility of petroleum hydrocarbons including polynuclear aromatic hydrocarbons. Moreover, the inoculated bacteria have the capabilities to perform a continuous supply of natural, nontoxic, and biodegradable surfactants by bacteria at low cost for solubilizing the hydrophobic oil hydrocarbons, increasing the hydrocarbons bioavailability, which consequently increases their biotransformation rates. The consortium proved higher TPH degradation. The authors attributed this to the type of the produced biosurfactants and their solubilization efficiencies. The major biosurfactants secreted by *B. subtilis* DM-04 were lipopeptide in nature containing higher amount of surfactins followed by iturins, whereas biosurfactant secreted by *P. aeruginosa* M and NM consortium was a complex mixture of lipopeptides and glycoproteins. Furthermore, due to the presence of sphingolipids or other specific molecule(s) in the outer membrane structure of *Pseudomonas* strains compared to *B. subtilis* DM-04 strain (Sugiura et al., 1997).

Li et al. (2009) reported that upon using microbial consortium of bacteria and fungi, *Phanerochaete chrysosporium*, *Cuninghamella* sp., *Alternaria alternativo* (Fr.) *Keissler*, *Penicillium chrysogenum*, *Aspergillus niger*, *Bacillus* sp., *Zoogloea* sp., and *Flavobacterium*, previously isolated from contaminated soil, a hydrocarbon degradation rate of 41.3% during 64-day incubation period was achieved. Singh et al. (2012) reported that the use of microbial consortia reduces the petroleum contaminant in soil from 30.9% to 0.97% after 360 days of treatment, as compared to only 5% of reduction in the control plot. Silva et al. (2015) reported also that upon using microbial consortium of bacteria and yeasts, *Staphylococcus saprophyticus, Serratia marcescens, Rhodotorula aurantiaca* and *Candida ernobii,* previously isolated from samples contaminated with petroleum derivatives at Lago da Barra, located in the Port Industrial complex of Suape, Pernambuco, Brazil, a removal of approximately 69% of the priority constituents from diesel oil was achieved within only 1 week. Varjani et al. (2015) reported a consortium of six bacterial strains—*Ochrobactrum* sp. (01), *Stenotrophomonas malto-philia* (02) and *P. aeruginosa* (03) —isolated from crude oil–polluted soil samples in India, expressed significant degradation of about 83.7% from 3% (v/v) crude oil.

The biotransformation reactions require the activation energy produced by microorganisms to support enzyme activities. The higher the energy consumption, the higher the bioremediation rate and activity. Thus, lower adenosine triphosphate levels in contaminated soil would indicate low microbial population and/or low microbial activity (Sparrow and Sparrow, 1988). This usually occurs also after the occurrence of an oil spill (Karl, 1992).

Biological and biochemical properties, including soil respiration measured by the rate of accumulated CO_2 or O_2 consumption, microbial growth, and enzymatic activities can help to predict the rate and extent of successive bioremediation process of a contaminated soil. Usually, biostimulation and bioaugmentation biotreatment processes of contaminated soil are reported to increase in the heterotrophic activity of the treated cells (Nocentini et al., 2000; Sabate et al., 2004).

Accumulative evaluation of CO_2 is proved to be a good qualitative indicator of biotransformation activity and it is usually used as a measure of soil heterotrophic activity. In a comparative study, Farahat and El-Gendy (2007) evaluated the effectiveness of different biotreatment techniques of contaminated soil in a semi-pilot scale. The accumulative evolution of CO_2 in bioaugmentation cells, which expressed the highest TPH removal, recorded relatively higher values than those of natural attenuation and biostimulation cells. Also, in a similar study performed by Soliman et al. (2014), there was a continuous recorded accumulative increase in CO_2 production with the continuous depletion in the TPH towards the end of incubation period in a biotreatment study performed on oily sludge–polluted soil, and the bioaugmentation microcosms showed the highest CO_2 production over all of the incubation period, with the highest pollutants removal rate, relative to the biostimulation and natural attenuation microcosms (Fig. 10.2) The mineralization rates of the pollutants was measured by the accumulated CO_2 production in a microcosm study performed by Dave et al. (2014). Where, the accumulated CO_2 increased with the increase of microbial growth and depletion of the total polyaromatic hydrocarbon pollutants.

3.2 POLLUTANT TYPE AND CONCENTRATION

Petroleum hydrocarbons are common contaminants found in soil and groundwater. The extent of contamination depends mainly on the chemical composition, concentration of the contaminant, and the soil properties. The microbial biotransformation can be altered based upon the contaminants degree of spreading and compositional heterogeneity (Leahy and Colwell, 1990; Head et al., 2003), and whether it is present at the water-hydrocarbon interface or present as a solid hydrocarbon (Atlas, 1981; Bunkim, 2003).

Moreover, the rate of biotransformation is mainly affected by the concentration of TPH in the contaminated soil. Extremely high concentrations of TPH have proven to be lethal to microbial activity, thus limiting the biotransformation potential (Admon et al., 2001). Similarly, the extremely low

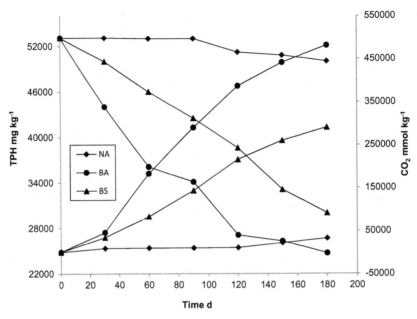

FIGURE 10.2

Total petroleum hydrocarbons removal and accumulative CO_2 production in different biotreatment microcosms. *BA*, Bioaugmentation; *BS*, Biostimulation; *NA*, Natural attenuation (Soliman et al., 2014).

concentration of TPH, although it is not lethal to organisms, can limit because the carbon supply may be too low to support microbial growth (Leahy and Colwell, 1990). Rahman et al. (2002) reported that percentage of degradation by mixed bacterial consortium decreased from 78% to 52%, as the concentration of crude oil increased from 1% to 10%. High initial concentrations of polynuclear aromatic hydrocarbons required longer duration for biotransformation process compared to low initial concentrations (Janbandhu and Fulekar, 2011).

Petroleum hydrocarbons can be grouped into three classes: alkanes, alkenes, and aromatic hydrocarbons in addition to petroleum nonhydrocarbons fractions, i.e., the resins and asphaltenes. Petroleum compositions differ in their susceptibility to microbial attack and have generally been ranked in the following order of decreasing susceptibility: *n*-alkanes > branched alkanes > low-molecular weight aromatics > cyclic alkanes > high-molecular-weight polynuclear aromatic hydrocarbons > polar compounds (Wang and Fingas, 2003; Hitoshi et al., 2008; Jain et al., 2011; Macaulay and Rees, 2014). Asphaltene constituents are highly resistant to biotransformation due to their heavy, viscous nature and are complex chemical structures made up of sulfur (0.3%–10.3%), nitrogen (0.6%–3.3%), oxygen (0.3%–4.8%), and trace amounts of metals such as iron, nickel, and vanadium; it is characterized by the highest molar mass of all hydrocarbon compounds in crude oil with values ranging from 600 to 3×10^5 g/mol and from 1000 to 2×10^6 g/mol (Macaulay and Rees, 2014). However, García-Rivero et al. (2002) reported the removal of 17% resins and 37% asphaltenes from weathered contaminated soil using toluene as a desorption agent and microorganisms from the rhizosphere of *Cyperus laxus* as inoculum.

Kang et al. (2010) reported 69.7% degradation for saturates and 48.3% for aromatics from soil with a content of TPH on the order of 20,000 mg/kg after 8 weeks upon addition of the microbial biosurfactant, sophorolipid, to the soil as a biotransformation enhancer. Liu et al. (2010) reported a reduction of 75.1% saturates, 60.4% aromatics, 31.3% resins, and 32.7% asphaltene after biostimulation of an oily sludge–contaminated soil through a long period of time (360 days). Soliman et al. (2014) reported a total removal of 80% saturates, 62% aromatics, 51% resins, and 42% asphaltenes after 180 days of bioaugmentation treatment of chronically oil sludge–polluted soil (53,100 mg/kg) with the addition of CSL and inoculation with *M. lutes* RM1. While, in the biostimulation microcosms by adding the CSL only, as a cosubstrate and sole source of nutrients, a total of 78% saturates, 47% aromatics, 42% resins, and 31% asphaltenes were removed after 180 days of incubation.

The molecular size of the pollutant is one of the main limiting factors in the success of a bioremediation process. Since, the larger the compound size, the more the steric hindrance, thus, the more the difficulty of the interaction between the compound and the enzymes' active sites (Bressler and Gray, 2003). Furthermore, the more the substitution reactions and alkylation reactions, the more the steric hindrance, and thus, the lower the interaction with the enzymes' active sites and consequently the lower the occurred biotransformation rate. Not only this, but alkylations decrease the water solubility, thus decrease the bioavailability and consequently decrease the biotransformation rate (Pasumarthi et al., 2013).

The physical state of petroleum hydrocarbons also affects biotransformation. Microbial biotransformation can be altered based upon the contaminants degree of spreading and compositional heterogeneity (Leahy and Colwell, 1990; Head et al., 2003), and whether it is present at the water-hydrocarbon interface or present as a solid hydrocarbon (Atlas, 1981; Bunkim, 2003). It has been reported that complete hydrocarbon removal cannot occur due to their low bioavailability, especially after the labile compounds are being used by the microorganisms and the accumulation of recalcitrant compounds occurred, and about between 10% and 30% of the initial soil pollution remains in soil after bioremediation techniques have been applied (Margesin and Shinner, 1998). Bressler and Gray (2003) reported that contaminants with log K_{ow} 1–3.5 tend to have the highest rates of aerobic and anaerobic biotransformation. However, Ostroumov and Siegert (1996) reported that at very low water solubility, the maximum rate of bioremediation is governed only by the mass transfer limitations. But, in frozen soils the mass transfer depends mainly on the liquid water or water films, which is a limitation especially in permafrost environments.

Sabate et al. (2006) reported that the accumulation of toxic intermediate metabolites has been postulated as a cause for the lack of complete contaminant biotransformation in soil. According to Mishra et al. (2001), the rapid microbial growth enhances the rate of biotransformation, but the depletion of biotransformation rate. Especially for recalcitrant compounds, this change of rate can be attributed to the lack of adequate nitrogen and phosphorus nutrients supporting microbial activities (Bento et al., 2005). Sometimes, the biotransformation of high-molecular-weight hydrocarbons produces short-chain and low-molecular-weight hydrocarbons (El-Gendy and Farah, 2011).

Generally, the presence of high–molecular-weight compounds in contaminated soils hinders the natural biotransformation, since they are characterized by very low water solubility (Caravaca and Roldán, 2003). Walker et al. (1975) reported that asphaltene constituents are considered as hardly degradable or not at all. However, Leahy and Colwell (1990) reported that the microbial degradation of asphaltene constituents is attributed to cooxidation, in which nongrowth hydrocarbons are oxidized in the presence of hydrocarbons that can serve as growth substrates. Pineda-Flores and Mesta-Howard (2001) reported that when microorganisms develop their full metabolic potential, they would have the capacity of using asphaltene as a source of carbon and energy, or degrading them by cometabolism mechanisms, since asphaltenes contain carbon, hydrogen, sulfur, nitrogen, and oxygen, which are necessary elements for the development of any organisms. The cometabolism and/or cooxidation phenomenon that would occur for asphaltenes was very evidenced by a study performed by Soliman et al. (2014), recording asphaltenes biotransformation of 4.3%, 42%, and 31% with maltenes biotransformation of 7.3, 64.3, and 56% in natural attenuation, bioaugmentation, and biostimulation microcosms, respectively.

It is well known that with the increase of number of rings and molecular size of the polynuclear aromatic hydrocarbons, its environmental persistence and genotoxicity increase and the rate of biotransformation decreases, since their electrochemical stability and hydrophobicity increase (Harms and Bosma, 1997). For example, half-lives of the three-ring molecule phenanthrene in soil and sediment may range from 16 to 126 days, while half-lives of the five-ring molecule benzo[a]pyrene may range from 229 to 1400 days (Shuttleworth and Cerniglia, 1995).

There are wide ranges of polynuclear aromatic hydrocarbons that could be found in contaminated soil, from the simple two-ring compounds naphthalene to the large multi-ring compounds. Compounds with two or three rings are arbitrarily defined as low-molecular-weight species, while those with four or more rings as high-molecular-weight species. Coronene derivatives (seven-ring polynuclear aromatic hydrocarbons) with molar masses of 202, 228, 202, 252, and 300 g/mol, respectively, are referred to as high molar masses or high–molecular-weight polynuclear aromatic hydrocarbons (Macaulay and Rees, 2014). Particular attention was paid to polynuclear aromatic hydrocarbon derivative as they are a class of ubiquitous persistent environmental contaminants with high toxic, mutagenic, and carcinogenic properties (Aichberger et al., 2006). The rate and extent of biotransformation of polynuclear aromatic hydrocarbons in soil are reported to be affected by environmental factors; the organic content; structure and particle size of the soil; characteristics of the microbial population; presence of contaminants such as metals and cyanides that are toxic to microorganisms; physical and chemical properties of the polynuclear aromatic hydrocarbons; and contamination history of soil and presence of substances that may act as cometabolites (Somtrakoon et al., 2008). With the increase of the pollutant molecular weight, the biotransformation rate decreases. Low-molecular-weight polynuclear aromatic hydrocarbons (two or three rings) are relatively volatile, soluble, and more degradable than are the higher molecular weight compounds.

The water solubility of naphthalene, acenaphthylene, phenanthrene, anthracene, pyrene, fluoranthene, and chrysene are 30.6, 3.9, 1.2, 0.7, 0.145, 0.262, and 0.003, respectively (Mrozik et al., 2003). Pasumarthi et al. (2013) reported that even though chrysene and pyrene have the same number of rings, however, pyrene was degraded by the previously isolated halophilic *P. aeruginosa* (accession number AB793685.1) and *Escherichia fergusonii* (accession number AB793686.1) from a crude oil–contaminated sediment sample using diesel oil as the sole carbon source, but chrysene was not degraded. That was attributed to the high water solubility of pyrene compared to that of chrysene, where less the solubility, the less the bioavailability, the less the biodegradability.

Despite the presence of microorganisms in the soil with the capacity to degrade polynuclear aromatic hydrocarbons, they are generally slowly transformed. This slow biotransformation is due to the low bioavailability of these compounds, i.e., as defined by Volkering et al. (1998), "its uptake rate by organisms is limited by a physicochemical barrier between the pollutant and the organisms" and the microorganisms may "live a famine existence" especially at a highly polluted site (Bosma et al., 1997). Most polynuclear aromatic hydrocarbons are highly hydrophobic and have low water solubility. When they reach the soil, they leave the aqueous phase, sorbed into the organic matter and sequestered (Semple et al., 2003). The organic pollutants that are in prolonged contact of the soil are bound to the soil particles in a phenomenon known as sequestration, which lower their bioavailability towards bio-transformation. This sequestration is a time-dependent process by which the polynuclear aromatic hydrocarbons first contaminate the macropores and the particle surfaces, where the potential degrading microorganisms are found, then diffuse into micropores, and finally reach the nanopores, which are inaccessible by the microorganisms and where the compounds remain entrapped (Bosma et al., 1997; Pizzul, 2006; Hyun et al., 2008). Moreover, due to the elevated octanol–water partition coefficients (K_{ow}) of the high-molecular-weight polynuclear aromatic hydrocarbons (four or more rings), they partition in soil and sediment organic matters and sorb strongly to soils and sediments. This decrease in the pollutant bioavailability and extractability that is obtained when the pollutant soil contact time increases is called "ageing." In addition, polynuclear aromatic hydrocarbons can also be strongly directly adsorbed to soil particles (Hatzinger and Alexander, 1995; Lundstedt, 2003; Somtrakoon et al., 2008).

However, biotransformation of polynuclear aromatic hydrocarbons is limited by their highly hydrophobic nature and sorption to sediments and soil particles, which decrease the rate of biore-mediation (Ke et al., 2009). The biotransformation rate decreases with the increase of the number of benzene rings, thus, their recalcitrance and persistence in the environment increase with aro-matic ring numbers. For example, the reported half-lives in soil and sediment of the three-ring phenanthrene molecule may range from 16 to 126 days while for the five-ring molecule benzo(a) pyrene (BaP) may range from 229 to 1400 days (Kanaly and Harayama, 2000). Microorganisms have developed different strategies to overcome the problem of the low bioavailability of hydro-phobic substrates, basically by three mechanisms: (1) high-affinity uptake systems (Sokolovská et al., 2003; Miyata et al., 2004; Heipieper et al., 2010); (2) adhesion to the solid substrate, some-times with formation of a biofilm (Eriksson et al., 2002; Bayoumi, 2009); and (3) production of biosurfactants or emulsifiers (Neu, 1996; Lang, 2002; Viramontes-Ramos et al., 2010).

Bioavailability can be increased by many chemical- or biosurfactants and solubilizers that form micelles, which enhance the release and microbial accessibility of contaminants in soil and solution (Allan et al., 2007; Dave et al., 2014). The utilization of microbial consortium is also a possible way to enhance the rate of biotransformation, based on the assumption of the possible synergetic effect between the augmented microbial consortium, where, metabolic intermediates produced by some microorgan-isms can act as substrates for others (Farahat and El-Gendy, 2007; Younis et al., 2013; Moscoso et al., 2012a). Moreover, biotransformation of polynuclear aromatic hydrocarbons can be enhanced by addi-tion of a cosubstrate. The term cosubstrate often refers to a compound that supports growth of microor-ganisms and enables degradation of a target compound, a process denominated cometabolism (Alexander, 1999). The salinity also inversely affects the biotransformation rate. Thus, additional cosub-strates (for example, yeast extract, glucose, sodium citrate) and nutrients (i.e., nitrogen, for example, urea and phosphorous sources) are required to enhance the rate of biotransformation at high salinity as high as 60 g/L NaCl (Kargi and Dincer, 1996; Arulazhagan and Vasudevan, 2011a,b). Addition of nutri-ents and carbon sources as a cosubstrate in biostimulation process would stimulate the microbial growth and/or bioaugmentation with biosurfactant-producing biodegraders can enhance the bioavailability of polynuclear aromatic hydrocarbons, which consequently enhances the rate of degradation. It has been reported also that the addition of glucose, as a carbon source, reduced the inhibition by salinity (Tam et al., 2002). Otherwise, bioaugmentation with halotolerant biodegraders and biosurfactant producer is recommendable. Wu et al. (2012) reported the bioremediation of crude oil–polluted saline soil by the halotolerant and biosurfactant producer *Serratia* Spp BF40 that was previously isolated from crude oil contaminated soil. The addition of surfactants is reported to enhance the rate of pyrene biotransformation in soil (Thibault et al., 1996). However, the biosensibility towards the added surfactants should be taken into consideration, before its utilization. Usually the nonionic Tween 80 is recommendable as an efficient emulsifier (Sabate et al., 2004; Moscoso et al., 2012a,b; Farahat and El-Gendy, 2008).

Measuring the success of oil spill removal is based on several parameters, among them the degradation of polynuclear aromatic hydrocarbons present (Farahat and El-Gendy, 2008; El-Gendy et al., 2009; Soliman et al., 2014). Bacteria that are capable for degrading the high-molecular-weight polynuclear aromatic hydrocarbons are reported to overcome the high hydrophobicity of these compounds through the production of biosurfactants or by promoting direct attachment to hydrophobic substrate by modifying their cell-surface hydrophobicity (Wick et al., 2002; Das and Mukherjee, 2007b; Chakraborty et al., 2010). The members of genus *Mycobacterium* are reported to be successful in bioremediation of aged contaminated sites. It has been established that *Mycobacteria* have exceptionally lipophilic surfaces, which makes them suitable organisms for the uptake of bound pollutants from the soil particles with good catabolic efficiency towards polynuclear aromatic hydrocarbons up to five benzene rings (Khan et al., 2002; McClellan et al., 2002; Bogan et al., 2003). In another word, the mycolic acid–rich cell walls of these soil bacteria may be an important factor in their utilization of hydrophobic substrates such as polynuclear aromatic hydrocarbons. Thus, large quantities of mycelia of several species of white-rot fungi are used to increase the extent of polynuclear aromatic hydrocarbon bioremediation in soil (Peng et al., 2008). In a study by Guerin (1999), polycyclic aromatic hydrocarbons contaminated soil from a creosoting plant was remediated using an ex-situ biostimulation land treatment process, throughout soil mixing, aeration, and fertilizer addition. The indigenous polynuclear aromatic hydrocarbons utilizing microorganisms increased, maximum degradation was apparent with the two- and three-ring polynuclear aromatic hydrocarbon, with a recorded decrease of 97% and 82%, respectively. The higher molecular weight three- and four-ring polynuclear aromatic hydrocarbons were degraded at slower rates, with reductions of 45% and 51%, respectively, while, the six-ring polynuclear aromatic hydrocarbons were degraded the least with average reductions of 35%. Supaka et al. (2001) attributed the complete degradation of high-molecular-weight polynuclear aromatic hydrocarbons of four-, five-, and six-membered rings, indeno[1,2,3-cd]pyrene, in absence of growth to cometabolism, where the biotransformation of more labile low-molecular-weight polynuclear aromatic hydrocarbons of di- and tri-rings enhances and improves the enzymatic activity, which consequently enhances the biotransformation of high-molecular-weight polynuclear aromatic hydrocarbons, in particular those with four or more aromatic rings.

Moreover, the bioaugmentation with adapted microorganisms enhances the rate of polynuclear aromatic hydrocarbon biotransformation (Farahat and El-Gendy, 2008; El-Gendy et al., 2009; Soliman et al., 2014). It has been reported by Haritash and Kaushik (2009) that the preexposure of a microbial community to hydrocarbons, either from anthropogenic sources or from natural sources is important in determining the rate of polynuclear aromatic hydrocarbons degradation and increases the hydrocarbon-oxidizing potential of the community, especially if they are preexposed to high doses of contaminant (i.e., adaptation phenomenon). The inoculation of contaminated soil with the preadapted *Gordona* sp. strain BP9 is reported to increase the pyrene metabolism by six-fold (Kästner et al., 1998). Moustafa et al. (2003) studied biotransformation potentials of bacterial strain isolated from Ismailya Canal, Egypt, for Baleym middle base oil and showed complete removal of pyrene, benzo[a]anthracene, benzo[k]fluoranthene, and benzo[g,h,i]perylene after 21 days of incubation. They attributed this to the chronic exposure of the isolated bacterial strain to different hydrocarbon pollutants, mainly aromatics and alicyclic compounds in the collection site, which may direct bacterial actions towards similar compounds. Bioaugmentation of an artificially fluorene-contaminated nonsterile soil by the adapted *Absidia cylindrospora* fungi, previously isolated from contaminated soil, doubled the rate of fluorene biotransformation (Garon et al., 2004). Farahat and El-Gendy (2008) performed a study on oil-polluted soil microcosms augmented by well-adapted bacterial strains, *P. aeruginosa* I.1.1.6 and *B. casei* I.2.1.7, which were previously isolated from Egyptian hydrocarbon–contaminated soil. The bacterial strains proved broad versatility in biotransformation of different fraction of petroleum hydrocarbons, with nearly complete degradation of pristane, phytane, naphthalene, phenanthrene, anthracene, fluoranthene, pyrene, chrysene, and the high-molecular-weight five- and six-membered ring polynuclear aromatic hydrocarbons, benzo[b]fluoranthene and indeno[1,2,3-cd]pyrene. The biotransformation of high–molecular-weight polynuclear aromatic hydrocarbons and recalcitrant biomarkers is attributed to the well adaptation of the augmented bacterial strains and their cometabolism based on the metabolism of n-alkanes and low–molecular-weight polynuclear aromatic hydrocarbons. El-Gendy et al. (2009) reported the bioremediation of three petroleum hydrocarbon–polluted sediment samples collected from the Gulf of Suez, Egypt. The study used a bioslurry system inoculated with a well-adapted *Staphylococcus gallinarum* NK1, previously isolated from hydrocarbon-polluted water for its ability to degrade dibenzothiophene and other polyaromatic hydrocarbon compounds (El-Gendy and Abo-State, 2008). NK1 revealed broad versatility over polynuclear aromatic hydrocarbons, with complete removal of the 16 polynuclear aromatic hydrocarbons listed by US-EPA, as priority pollutants and great biotransformation capabilities >90% for the six-membered ring compounds.

Dave et al. (2014) reported the biotransformation of polynuclear aromatic hydrocarbons in crude oil–saline polluted soil on a microcosm level, throughout a bioaugmentation and biostimulation process using the Gram-negative *Achromobacter xylosoxidans*, chrysene degrading marine halotolerant bacterium, 50 ppm of chrysene for inducing the biotransformation of other high-molecular-weight polynuclear aromatic hydrocarbons present in the soil and glucose as a cosubstrate with the addition of Triton X-100 as a nonionic surfactant or β-cyclodextrin as a polynuclear aromatic hydrocarbon solubilizer. Where, glucose, in combination with Triton X-100 and β-cyclodextrin, resulted in 2.8- and 1.4-fold increase in degradation of low-molecular–weight polynuclear aromatic hydrocarbons and 7.59- and 2.23-fold increase in degradation of high–molecular-weight polynuclear aromatic hydrocarbons, respectively. That was attributed to the increased bioavailability of polynuclear aromatic hydrocarbons in presence of surfactant or solubilizer.

However, the presence of glucose alone decreased the rate of biotransformation of low-molecular-weight polynuclear aromatic hydrocarbons and increased the biotransformation of high-molecular-weight polynuclear aromatic hydrocarbons by 6.79- and 1.24-fold, respectively. The authors attributed this to the cometabolism, which is important for metabolism of recalcitrant compounds in soil and other matrices. In a study performed by Soliman et al. (2014) the biotransformation of the aromatics fraction reached 62%, 47%, and 9% in bioaugmentation, biostimulation, and natural attenuation microcosms. The biotreatment showed a great versatility in the utilization of different polyaromatic ring compounds, especially on the high-molecular-weight polyaromatic hydrocarbons. Approximately complete removal of the three-membered ring compound anthracene, and the four-membered ring compounds pyrene and fluoranthene was achieved in the bioaugmentation microcosms. While, in the biostimulation microcosms approximately 72% of anthracene and 43% of pyrene were removed. Moreover, there was also a significant biotransformation in the five-membered ring compounds in the bioaugmentation and biostimulation microcosms reaching approximately

57% and 35% for benzo(b)fluoranthene; 86% and 23% for benzo[a]pyrene; and complete and 67% removal for benzo(k)fluoranthene, respectively. Compared to complete degradation of six-membered ring compounds in bioaugmentation microcosms, slightly low biotransformation efficiency was observed for benzo(ghi)perylene (21%) and indeno(1,2,3-cd)pyrene (18%) in biostimulation microcosms. The authors attributed the recoded higher activity and biotransformation efficiency in the bioaugmentation microcosms to its inoculation with the well-adapted *M. lutes* RM1, which was previously isolated from the same oily sludge–contaminated soil under study for its ability to degrade the high-molecular-weight polynuclear aromatic hydrocarbons.

It has been reported that the addition of rapidly biodegradable organic matter alone to contaminated soil inhibits the polynuclear aromatic hydrocarbons biotransformation process (Sayara et al., 2010a,b). Thus, the stability of the added organic matter is critical for polynuclear aromatic hydrocarbon biotransformation as it affects the availability of some chemical components like humic matter in the organic substrate. Thus, the higher its stability, the higher it facilitates polynuclear aromatic hydrocarbons desorption and consequently makes them more available for the microorganisms (Margesin and Schinner, 1997; Sayara et al., 2010b). Sayara et al. (2011) performed a comparative study for biotreatment of an artificially contaminated soil with polynuclear aromatic hydrocarbons fluorene, phenanthrene, anthracene, fluoranthene, pyrene, benzo(a)anthracene, and chrysene (1 g polynuclear aromatic hydrocarbon derivatives/kg soil), applying biostimulation and bioaugmentation techniques. Biostimulation occurred using compost of the source-selected organic fraction of municipal solid waste and rabbit food, in the form of pellets, as organic cosubstrates in the ratio of 1:0.25 (soil:cosubstrate, dry weight), whereas, bioaugmentation was done by inoculation of the soil with the white-rot fungi *Trametes versicolor* in addition to the previously mentioned cosubstrates. To ensure aerobic conditions, a bulking agent consisting of wood chips was introduced at a ratio of 1:1 (v/v). Tape water was added to keep the water content (50%–60%). The results revealed that *T. versicolor* did not enhance the polynuclear aromatic hydrocarbon biotransformation rate.

However, biostimulation could improve the polynuclear aromatic hydrocarbon degradation: 89% and 71% of the total polynuclear aromatic hydrocarbons were degraded after 30 days by biostimulation using compost and rabbit food compared to the only 29.5% that was achieved by the soil indigenous microorganisms without any cosubstrate (control, not amended), indicating the synergetic effect between the compost microorganisms and the indigenous microbial population. Furthermore, stable compost from the organic fraction of municipal solid waste has a greater potential to enhance the degradation of polynuclear aromatic hydrocarbons compared to nonstable cosubstrates such as rabbit food.

Branched alkane derivatives, pristane Pr (2, 6, 10, 14-tetramethylpentadecane) and phytane Ph (2, 6, 10, 14-tetramethylhexadecane), have been used as conservative biomarkers in oil bioremediation studies. But their recalcitrance to biotransformation is a matter of concern, since, several reports proved their biotransformation by different microbial strains (Pritchard, 1993; Chaineau et al., 2005; Xu et al., 2004; Farahat and El-Gendy, 2008, 2007; Chorom et al., 2010; El-Gendy and Farah, 2011; Soliman et al., 2014). Although, the rate of biotransformation of phytane is reported to be higher than that of pristane in some studies (El-Gendy and Farah, 2011). But the biotransformation of pristane was higher than that of phytane in the bioremediation study performed by Soliman et al. (2014).

3.3 DEGREE OF WEATHERING

The degree of weathering of a pollutant is one of the significant factors that affect the rate of biotransformation in soil. El-Gendy et al. (2009) proved that the microbial growth has a more statistically significant effect than incubation period (i.e., treatment time) on the percentage removal of total organic pollutant in low-weathered polluted sediment samples. While, treatment time (incubation period) has a more statistically significant effect than the microbial growth on the percentage removal of total organic pollutant in highly weathered polluted sediment samples. They attributed this to the absence of labile hydrocarbons in the highly weathered polluted sediment samples, which, consequently, increase the adaptation period of the augmented microorganism, which is required for enhancing its biotransformation enzymes, thus leading to a lower rate of microbial growth and longer treatment period.

3.4 NUTRIENT CONCENTRATION

Oil contamination in soil results in an imbalance in the carbon-nitrogen ratio at the spill site, because crude oil is essentially a mixture of only carbon and hydrogen. It should be known, that, although hydrocarbons are an excellent source of carbon and energy for microbes, they are incomplete foods since they do not contain significant concentrations of other nutrients (such as nitrogen and phosphorus) required for microbial growth (Prince et al., 2003; Flood, 2009). Since, carbon is the most basic element of living forms and is needed in greater quantities than other elements. In addition to hydrogen, oxygen, and nitrogen it constitutes about 95% of the weight of cells, while phosphorous and sulfur contribute with 70% of the remainder.

Thus, the nutritional requirement of carbon to nitrogen ratio is estimated to be approximately 10:1 and carbon to phosphorous is approximately 30:1 (Vidali, 2001). The input of large quantities of organic carbon sources tends to a rapid depletion of available inorganic nutrients, which consequently limits the amount of biotransformation (Margesin et al., 1999). This causes a nitrogen deficiency in oil-soaked soil, which retards the growth of indigenous microorganisms and the utilization of carbon sources. In another word, the release of hydrocarbons into ecosystems containing low concentrations of inorganic nutrients often produces excessively high C/N or C/P ratios, or both, which are unfavorable for microbial growth (Leahy and Colwell, 1990). Thus, biostimulation (nutrient addition) can often be used to maximize bioremediation effectiveness (Alexander, 1999; Bachoon et al., 2001; Trinidade et al., 2002; Milles et al., 2004; Mishra et al., 2004; El-Gendy and Farah, 2011). However, endogenous biodegrading microorganisms would need an adaptation period of 4–10 days and sometimes takes weeks, during which no destruction of pollutants would be markedly evident (Johnson and Scow, 1999; Sharma et al., 2014).

Biostimulation, as mentioned before, is a method for biotransformation that involves nutrients' addition to the contaminated soil to stimulate its indigenous microflora to enhance and accelerate microbial organic waste breakdown (Olaniran et al., 2006; Ueno et al., 2006). Sabate et al. (2004) reported that addition of nutrients led to a large decrease in the TPH and deletion of the lag (adaptation) period. Moreover, the addition of cosubstrate, for example, con steep liquor in biostimulation processes has been reported to be very effective especially in case of weathered and highly weathered pollutants (Kaplan et al., 2003; El-Gendy and Farah, 2011; Soliman et al., 2014). For example, petroleum, as a very complex mixture, contains many compounds that may not be readily degraded. Weathered petroleum may be enriched in poorly degradable components (Alexander, 2000; Admon et al., 2001). Cosubstrate additions may therefore be important to the remediation of aged petroleum (i.e., weathered), where the more readily degradable components are either completely removed or have already been partially oxidized during weathering. Das and Mukhejie (2007a) reported that the supplementation of glucose, as a co-carbon source enhances the biotransformation rate.

Addition of inorganic nutrients such as NH_4Cl and K_2HPO_4 with cosubstrates such as molasses and nonionic surfactants such as Tween 80 has been reported to increase the microbial activity and rate of biotransformation (Radwan et al., 2000; Sabate et al., 2004; Farahat and El-Gendy, 2007, 2008). Rahman et al. (2003) reported that the increase in bacterial population leads to rapid assimilation of available nutrients by soil microbes, thus depleting the nutrient reserves, consequently. The periodical addition of sufficient and controlled amounts of nutrients is very essential to grantee a successful bioremediation process. Lepo and Cripe (1999) reported that the addition of nitrogen and phosphorous nutrients stimulates also the alkane-degrading microbes and results in a substantial depletion of the n-alkane fraction. Braddock et al. (1999) reported that the loss of linear alkanes is more strongly affected by nutrient addition than branched alkanes.

Adjustment of the carbon-nitrogen-phosphorus-potassium (C/N/P/K) ratio by the addition of urea, phosphate, N–P–K fertilizers, and ammonium and phosphate salts has accelerated the biotransformation of crude oil or gasoline in soil and groundwater in numerous studies (Leahy and Colwell, 1990). However, acute toxicity from fertilizer application can occur if nutrients at high concentrations are immediately solubilized, forming a highly toxic condition to microorganisms. Thus, the adjustment of C/N/P ratio is very important for the achievement of successful bioremediation process. Although, as previously mentioned, the carbon-nitrogen-phosphorus ratio of 100:10:1 has been frequently reported as a reference level for biostimulation approaches. However, this would change in highly contaminated soil and/or with soil type that is characterized by depleted concentrations of nitrogen and phosphorous.

For example, in a biostimulation process for a hydrocarbon-contaminated Antarctic soil, about 9000 mg/kg of TPH was removed at a carbon-nitrogen-phosphorus ratio of 100:1.4:0.09 (Ferguson et al., 2003). In another similar study, that was performed also on a hydrocarbon-contaminated Antarctic soil; the optimum carbon-nitrogen-phosphorus ratio of 100:17.6:1.73 decreased the total hydrocarbon concentration from 1042 (±73) mg/kg to 470 (±37) mg/kg (Martínez Álvarez et al., 2015). It should also be noted that the variation in this ratio would highly affect the biotransformation rate. For example, in a study performed by Liu et al. (2011) in a highly contaminated soil petroleum hydrocarbon (14,000 mg/kg), 69.8% of hydrocarbon removal was achieved after 35 days throughout the biostimulation process with inorganic nutrients (NH_4NO_3 and K_2HPO_4) at a carbon-nitrogen-phosphorus ratio of 100:11:3.7. But only a removal of 46.7% was achieved at higher nutrients levels—a carbon-nitrogen-phosphorus ratio of 100:27:6.5.

Addition of bulking agents including sewage sludge, compost, or alternative carbon sources can facilitate degradation of organic contaminants because they play a role in supplementing nutrients, providing a carbon source, retaining moisture content, and aerating a contaminated soil. Sewage sludge and compost may contain an abundance of nitrogen and organic matter, thus making composting of hydrocarbon contaminated soils an effective aid in bioremediation. In the composting of contaminated soil, organic amendments including manure, yard wastes, and food processing wastes are often added to supplement the amount of nutrients and readily degradable organic matter in soil.

The ratio of contaminated soil to organic amendments should be determined because an inappropriate ratio may retard or inhibit microbial activity. A 2:1 ratio of contaminated soil to organic amendment (compost and/or sewage sludge) is reported in several studies as the optimum ratio to increase the rate of hydrocarbon degradation (Admon et al., 2001; Hwang et al., 2001; Namkoong et al., 2002). Moreover, the fresh compost is reported to achieve better reclamation results than mature ones due to the presence of more bioavailable organic carbon in fresh compost than in the mature compost, which would promote sorption of hydrophobic contaminants and provide a source of carbon for microorganisms. In addition, the presence of fresh compost would allow microorganisms to increase their metabolic activity directed to the degradation of organic compounds and production of extracellular enzymes and biosurfactants, which would increase the biotransformation rate (Tsui and Roy, 2007).

3.5 EFFECT OF TEMPERATURE

Although, the frozen surface of the ground often slows the penetration of the oil spilled. However, the soil bioremediation in cold environments is known to slow process (Yang et al., 2009). The slow rate of degradation in soil is primarily due to the slow rate of desorption of contaminants from soil particles and not due to the slow rate of microbial and enzymatic activity responsible for degradation. The main reason for the slow desorption is the slow diffusion of contaminants through the pore liquid and through the soil organic matter. To increase the rate of diffusion, the soil can be subjected to thermal or chemical treatment prior to the microbial remediation. It has been reported that an increase in temperature can decrease the soil–water partition coefficient and as a result, dissolution of contaminants in water is observed. For example, the partition-coefficient of polynuclear aromatic hydrocarbons decreases by 20%–30% for every 10°C rise in temperature between 5 and 45°C (He et al., 1995; Luers and Hulscher, 1996). Moreover, the mass transfer within the soil increases also with the increase in temperature. The mass transfer is reported to be proportional to the diffusion coefficient and inversely proportional to the partition coefficient (Wu and Gschwend, 1986; Farell and Reinhard, 1994). The diffusion coefficient of water is reported to increase by 4–5 times with an increase in temperature from 20 to 120°C, with about 150 times increase in the effective diffusion coefficient (Haritash and Kaushik, 2009). Bonten et al. (1999) reported that heating contaminated soil at 120°C for 1 h increased the degree of degradation after 21 days of an aged polynuclear aromatic hydrocarbons contamination from 9.5% to 27%. While soaking it in 4:1 (v/v) acetone–water mixture increased the degree of degradation from 9.5% to 20.4% because of dissolution of polynuclear aromatic hydrocarbons. However, the UV-ozone treatment is reported to be better, since the chemical treatment involves a heavy input of chemicals and formation of harmful residues, but the UV-ozone treatment produces no significant toxic products and it can destroy more than 90% of polynuclear aromatic hydrocarbons (Vollmouth and Niessner, 1995).

Moreover, temperature is reported to affect both the physical state of the hydrocarbons present in soil and the microbes consuming them. At low temperatures, the viscosity of the oil increases, the volatilization of toxic short chain alkanes is reduced, and their water solubility is decreased, delaying the onset of biotransformation (Leahy and Colwell, 1990; Venosa and Zhu, 2003). In addition, microbial growth rates are function of temperature (Gibb et al., 2001), and rates of degradation decrease with decreasing temperature. This is a result primarily of decreased rates of enzymatic activity, often described as the "Q10" effect (Leahy and Colwell, 1990). Higher temperature increases the rate of hydrocarbon metabolism to a maximum, typically in the range of 30°C–40°C in soil, 20–30°C in some fresh water environment, and 15–20°C in marine environments (Sharma et al., 2014). Above this temperature, enzyme activity is reduced and membrane toxicity of hydrocarbons is increased (Leahy and Colwell, 1990). Rawe et al. (1993) reported that although the microbial biotransformation activity does not cease at sub-zero temperatures, however, the optimum temperature for biotransformation is usually 15–30°C for aerobic processes and 25–35°C for anaerobic processes. Finegold (1996) reported that in cold weather, close to freezing point, cryogenic stresses, which result in closing the transport channels or freezing the cell cytoplasm, would restrict mass transport and thus limit contaminants to gain access into cells, which would terminate the metabolic activity of the cells, which consequently decreases the microbial growth and the biotransformation rate.

Hydrocarbon biotransformation kinetics calculated by Admon et al. (2001) exhibited an initial temperature-dependent rate constant (rates increased as temperature increased from 21°C to 31°C). The latter research was substantiated by research conducted by Gibb et al. (2001), who reported temperature only affected the biotransformation rates of crude oil in the initial phase of the biotransformation process. After approximately 3 months, the degradation rates of crude oil at 5°C and 21°C were similar at about 11 mg hydrocarbon/kg dry soil/day. Balks et al. (2002) reported that biotransformation of heavy fuel (Bunker C) by indigenous organisms in the North Sea was four times greater in summer (18°C) than in winter (4°C).

3.6 MOISTURE CONTENT

Hydrocarbon degradation in terrestrial ecosystems is reported to be limited by the available water for microbial growth and metabolism (Leahy and Colwell, 1990). Thus, soil moisture content affects the rate of biotransformation. There is a direct relationship between soil microbial activity and moisture content; a decrease in moisture content results in a decrease of microbial activity, and rewetting causes a large and rapid increase in activity. As a result, microorganisms are expected to respond to changes in soil moisture content through a complex series of interactions, involving nutrient fluxes, soil temperature, pore size changes, and soil atmospheric changes (Dragun, 1998). In heavily polluted soils, oil coating the surface of soil particles makes soil more hydrophobic and water droplets adhere to the hydrophobic layer formed, and this prevents wetting of the inner parts of the soil aggregates and reduces the water-holding capacity of the soil. But with the decrease of the TPH content, the water holding capacity of the soil increases (Odokuma and Dickson, 2003; Das and Mukherjee, 2007a). Thus, the soil's low water-holding capacity would be expected during the initial period of bioremediation process due to the elevated levels of pollution.

Previous researchers have also suggested a linear relationship between microbial CO_2 generation and moisture content, independent of soil type (Leirós et al., 1999; Jabro et al., 2008; Pingintha et al., 2010). Generally, in soils, extreme moisture conditions are unfavorable for microbial growth and metabolism and water contents of between 20% and 70% capacities are recommendable as optimal for microbial activity (Morgan and Watkinson, 1989). Atagana (2006) reported that the presence of moisture in soil would enhance the solubility of soluble hydrocarbons that consequently would enhance the rate of biotransformation. However, wet conditions would limit the oxygen availability; hence, aeration would be required to accelerate the aerobic biotransformation process. Usually, the cost effective, periodic tailing or trilling, is commonly used in soil bioremediation technique to ensure the adequate supply of oxygen to the microorganisms for the achievement of faster biotransformation rate (Mishra et al., 2001; Atagana, 2006; Das and Mukherjee, 2007a; Farahat and El-Gendy, 2007, 2008; El-Gendy and Farah, 2011; Soliman et al., 2014).

3.7 SOIL TYPE

This is another important factor that affects the rate of biotransformation, since it affects the degree of pollution and persistence of the pollutants. Soil contamination by hydrocarbons expresses some changes in physical conditions and biological activity of the soil. It is important, for example, to measure the soil porosity and texture to choose the suitable bioremediation technique.

The chemical and physical properties of the soil have a great influence on the metal contaminant, its mobility, and consequently, the remediation technology. Lofts et al. (2004) reported that the negative impact of heavy metals on soil is not only a function of their source and quantities, but also a function of their relative mobility and bioavailability, which are dependent on soil characteristics, such as pH, mineralogy, texture, and organic content. Chemical and physical properties of soil have also a reflective influence on aeration, nutrient availability, water retention, and consequently on biological activities.

Particle size affects the surface chemistry of soils and the size of the pores. Thus, the most advantageous pore structure is one in which water is retained but a considerable fraction of the pores remains packed with air. Thus, soil porosity and pore size distribution are closely related to storage and movement of water and gases and microbial penetration (Gerber et al., 1991; Sartori et al., 1985; Morgan and Watkinson, 1989). The sandy matrix of the soil poorly adsorbs asphaltenes and resins, but mainly characterized by high levels of chain-saturated hydrocarbon contents (Gogi et al., 2003). Clay soil is reported to absorb the maximum percentage of crude oil compared to other soil types, sandy, fine sand soil, and clay loam (Kavitha et al., 2015). The sandy soil is known to have low nutrients and microflora, thus characterized by low biodiversity (Rahman et al., 2003). The coarse-textured soils are known to have low water-holding capacity (Aislabie et al., 2006).

The grains size affects also the hydrocarbon accumulation and bioremediation process. Hydrocarbon concentrations are highly accumulated in fine grain sizes more than the coarse grain ones, since the surface areas of small grains are larger than the surface areas of larger grains (El-Tokhi and Moustafa, 2001; El-Gendy et al., 2009). El-Gendy et al. (2009) performed a bioremediation study on three different collected petroleum hydrocarbon–polluted sandy sediment samples: S1 of 92% sand, 8% silt, and characterized mainly by very coarse to medium sand; S2 of 96% sand, 4% silt, and characterized by medium to fine sand and partly very coarse sand; and S3 of 89% sand, 11% silt, and characterized by medium to fine sand. The microscopic examination of the fraction 125 μm in the studied sediments revealed the increase of oil content in S3 > S2 > S1 where the percentage of fine sand in S3 > S2 > S1. Oil pollution generally appears in the pictures in the form of completely coated grains. Other major constituents were also present together with these quartz grains such as fossils, fossil fragments, and iron oxides (Fig. 10.3). The well-rounded grains are less contaminated by oil in comparison with irregular or subrounded grains. Contamination in small quartz grains are higher than in larger quartz grains, where large quartz grains are contaminated by oils through fissures, pits, and deep points on the surfaces (Fig. 10.3). After biotreatment of the polluted sediment samples by active and well-adapted *S. gallinarum* NK1, clean quartz grains were obvious in sample S1, especially those fine and well-rounded grains and sample S2, which was mainly characterized by moderately sorted polluted quartz grains and presence of some iron oxides. While sample S3, which was mainly characterized mainly by heavy polluted quartz grains with oil and iron oxides with fine and well-sorted grains, showed few clean quartz grains, after biotreatment (Fig. 10.3). Moreover, the presence of iron oxides can be taken as evidence of oil pollution from tanker ballast washings (Barakat, and El-Ouf, 2009).

3.8 SOIL INTERACTION WITH MACROORGANISMS

Soil and macroorganisms are integral parts of the ecosystem, and both are the reservoir and enhancers of metabolic activities for vast number of microbes. For example, termites are known to harbor and enhance

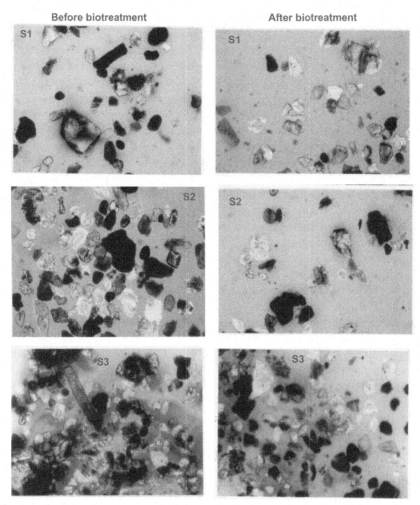

FIGURE 10.3

Example of microscopic examination of some polluted sediment samples before and after biotreatment (El-Gendy et al., 2009).

microbial activities. Ngugi et al. (2005) found that *Macrotermes michaelseni*, isolated from the intestinal tract of a fungi-cultivating termite, had the capability to degrade resorcinol, phenol, and benzoic acid. The use of white-rot fungi in bioremediation of soil is limited since it should be inoculated along with a finite source of carbon, such as wood-chips, which are difficult to distribute throughout the soil volume. However, ectomycorrhizal fungi overcome this problem, since they are widely distributed throughout the soil by roots, and provided with a long-term supply of photosynthetic carbon from their hosts. Moreover, the development of root systems improves soil aeration and drainage, thus increaseing the biological activity and enhancing the volatilization of the persistent organic pollutants (Genney et al., 2004). But, there is limited evidence to produce lignin peroxidase by ectomycorrhizal fungi (Cairney et al., 2003), thus they are less able to degrade complex polyphenolic compounds (e.g., lignin) than white-rot fungi (Wu et al., 2003). Moreover, it suffers from the Gadgil-effect phenomenon, where retarded litter decomposition is observed in the presence of ectomycorrhizal roots (Bending, 2003). In a study performed by Genney et al. (2004), *Laccaria laccata* degraded naphthalene to 1- and 2-hydroxynaphthalene. While, *L. laccata* and *Thelephora terrestris* metabolized fluorene to 9-fluorenone and 9-hydroxyfluorene. Moreover, *L. laccata* produced 2-hydroxyfluorene. But they did not degrade pyrene. Upon mycorrhizal pine seedlings of soil microcosm with natural ectomycorrhizal communities, mineralization of naphthalene was enhanced, but fluorene degradation was suppressed, relative to the unplanted microcosms. They have attributed this to the decrease in the litter decomposition rates in the presence of ectomycorrhizal roots, by reducing soil moisture content, direct competition between saprotrophs and ectomycorrhizal fungi either because of resource competition or direct inhibition through production of antimicrobial compounds; but most probably due to the dominance of the planted microcosms by ectomycorrhizal fungi, which is already having limited polynuclear aromatic hydrocarbons degradation capabilities at the expense of free-living saprotrophs. However, in another study by Heinonsalo et al. (2000) the ectomycorrhizal pine roots enhanced the density of bacteria and expedite the removal of persistent organic pollutants from ecto-mycorrhizosphere soil. Mohd Radzi et al. (2015) reported the isolation bacterial consortium from genus *Ochrobactrum-*, *Pseudomonas-*, *Ralstonia-*, *Burkholderia-*, and *Corynebacterium*-contaminated road side soil and soil termite fungal comb samples collected from the Chan Saw Lin Industrial area, and a termite nest near the Chemistry Department building, University of Malaya (Kuala Lumpur, Malaysia), for their ability to degrade fluoranthene and pyrene, as the sole source of both carbon and energy. The overall degradation of pyrene was higher than that of fluoranthene. However, the bacterial consortium isolated from the road side soil, composed of *Ralstonia pickettii*, *Ochrobactrum anthropi*, and *Corynebacterium appendicis* expressed a greater ability to degrade fluoranthene and pyrene than the bacterial consortium from the termite fungal comb, which consisted of *Pseudomonas putida biotype B*, *Ochrobactrum tritici*, and *P. aeruginosa*.

3.9 **AERATION**

It should be noted that large concentrations of biodegradable organics in the top layer deplete oxygen reserves in soil and slow down the rates of oxygen diffusion into deeper layers (Rosenberg et al., 1992). Not only this, but, large quantities of soil contamination by oil would lead to very fast oxygen depletion in the soil, causing anaerobic condition. On the other hand, the aerobic pathway is reported to be the most effective strategy for bioremediation (Trinidade et al., 2002; Lal et al., 2010). In many soils, effective oxygen diffusion for desirable rates of bioremediation extend to a range of only a few centimeters to about 30 cm into the soil. However, bioventing can be applied for bioremediation of soil depths of 60 cm and greater, which involves supplying air and nutrients through wells to contaminated soil to stimulate the indigenous bacteria. Bioventing employs low air flow rates and provides only the amount of oxygen necessary for the biotransformation while minimizing volatilization and release of contaminants to the atmosphere (Zabbey et al., 2017). Biosparging can be also applied, which involves the injection of air under pressure below the water table to increase groundwater oxygen concentrations and enhance the rate of biological degradation of contaminants by naturally occurring bacteria. Biosparging increases the mixing in the saturated zone and thereby increases the contact between soil and groundwater (Vidali, 2001). Atlas (1998) suggests that for bioremediation of hydrocarbons, 0.3 g oxygen is required for each gram oil oxidized. Boufadel et al. (2016) reported the persistence of polyaromatic hydrocarbons and oil in some beaches in Prince William Sound, Alaska, after more than 20 years of Exxon Valdez laden. Where, the degradation rate of the total polyaromatic hydrocarbons is estimated at 1% per year and mainly attributed this to the available low oxygen concentrations. Bressler and Gray (2003) reported that oxygen is scarce in frozen soil, and oxygen transport is the rate-limiting step in aerobic bioremediation of frozen soil. Not only this, but oxygen would be also consumed faster than it would be replaced by diffusion from atmosphere. Thus, anaerobic biotransformation becomes obligatory, since anaerobic microorganism would dominate.

The initial steps in the catabolism of aliphatic, cyclic, and aromatic hydrocarbons by bacteria and fungi involve the oxidation of the substrate by oxygenases, for which molecular oxygen is required. Aerobic conditions are therefore necessary for this route of microbial oxidation of hydrocarbons in the environment. The availability of oxygen in soils is dependent on rates of microbial oxygen consumption, the type of soil, whether the soil is waterlogged, and the presence of utilizable substrates leading to oxygen depletion. The concentration of oxygen has been identified as the rate-limiting variable in the biotransformation of petroleum in soil (Leahy and Colwell, 1990; Irvine and Frost, 2003; Davis et al., 2009). Hydrocarbons are also degraded anaerobically, but the rates are much slower than aerobic degradation rates (Leahy and Colwell, 1990; Head and Swannel, 1999; Reineke, 2001).

Anaerobic microorganisms have been reported to have greater potential for polynuclear aromatic hydrocarbons detoxification in natural environments. Anaerobic biotransformation of crude oil pollutants in soil can be also occurred (Meckenstock et al., 2004). However, anaerobic bacteria need electron acceptors, such as nitrate, iron, or sulfate and the energy yield for the bacteria is less than would occur if oxygen is used as the electron acceptor, in case of aerobic bacteria (Braddock et al., 1997; Godleads et al., 2015). Thus, the lower energy yield by anaerobic bacteria results in lower degradation rate and hence a longer period is required for remediation (Thapa et al., 2012). Aerobic microorganisms utilize elemental oxygen as their ultimate electron acceptor. While, anaerobic microorganisms, when molecular oxygen is absent, use water-derived oxygen, nitrate derivatives, sulfate derivatives, carbon dioxide, and ferrous iron (Fe^{2+}) as electron acceptors. These can be utilized according to their energy yield/unit of oxidized organic carbon, in the following order: O_2, NO_3^-, Mn^{4+}, Fe^{3+}, and SO_4^{2-} (Spence et al., 2005). Thus, the processes that occur are aerobic respiration, then comes the denitrification, manganese reduction, ferric reduction, sulfate reduction, and finally methanogenesis (Schwarzenbach et al., 1993). In frozen ground, where soil freezing and thawing occur, the metabolic pathways and microbial enzymatic activities, involved in bioremediation, vary according to the redox potential. For example, aerobic metabolism requires redox potential >50 mV, while anaerobic metabolism requires redox potential <50 mV (McFarland and Sims, 1991).

Sulfate-reducing bacteria are reported to be the most relevant in the anaerobic hydrocarbons degradation process (Miralles et al., 2007; Subha et al., 2015). However, most of the sulfate-reducing bacteria cannot degrade the high-molecular-weight hydrocarbons and depends of the metabolites of the low–molecular-weight hydrocarbons, such as acetate, for their metabolism (Ingvorsen et al., 2003). Subha et al. (2015) reported the slow-release biostimulant ball as a promising technique for the reduction of chemical oxygen demand and volatile solid and enhancement of sulfate-reducing bacteria in bioremediation of contaminated coastal sediment. Ambrosoli et al. (2005) reported the anaerobic polynuclear aromatic hydrocarbons degradation in soil by a mixed bacterial consortium under denitrifying conditions and concluded that anaerobic biotransformation of fluorene, phenanthrene, and pyrene seems to be possible both through fermentative and respiratory metabolism, provided that low-molecular-weight cometabolites and suitable electron acceptors (nitrate) are present. Rocchetti et al. (2011) reported a two-step bioremediation process of hydrocarbon-polluted sediment: anaerobic using sulfate-reducing bacteria followed by aerobic throughout the addition of compost on microcosm level. Where, 91% biotransformation efficiency for the TPH was achieved using sulfate-reducing bacteria only without acetate amendment for 30 days followed by aerobic for another 30 days. Since, acetate would be preferentially used as a carbon source by the sulfate reducing bacteria, thus decreasing the pollutant biotransformation efficiency.

The biotransformation of 16 polycyclic aromatic hydrocarbons (polynuclear aromatic hydrocarbons) in marine sediment was reported under three different anoxic conditions; sulfate-only, nitrate-only, and mixed nitrate/sulfate as electron acceptors, where the nitrate-reducing condition expressed faster degradation rate than that of under sulfate-reducing condition. All two-, three- and four-ring polynuclear aromatic hydrocarbons showed significant biotransformation with the removal efficiencies ranging from 42% to 77%, while five- and six-ring polynuclear aromatic hydrocarbons showed little degradation (Lu et al., 2012).

3.10 SOIL ACIDITY-ALKALINITY

This is one of the important factors that affect the rate of biotransformation. It is often found to have the largest effect owing to its strong effects on solubility and speciation of metals both in soil and particularly in the soil solution. Thus, each unit decrease in pH results in approximately two-fold increase in the concentration of metals (Aislabie et al., 2006). The pH values > 7.0 drastically decrease the heavy metal solubility (Riis et al., 2002). The neutralization of soil is usually favorable for biotransformation (Leahy and Colwell, 1990) and the shift of the pH from 5.2 to 7.0 is reported to significantly enhance the rate of polynuclear aromatic hydrocarbons degradation by strain BA2 (Kästner et al., 1998). The pH around neutrality is usually enhancing the biotransformation process and most of soil microorganisms are reported to thrive best in the pH range of 6–8 (Dragun, 1998; Bunkim, 2003). However, a pH of 5.2 should not lead to total inhibition of activity (Haritash and Kaushik, 2009).

Nitrogen fixating bacteria are strongly hindered by high acidity (Jackson, 1969). Fungi usually tolerate strong acidity than bacteria. Extremes in pH (observed in some soils) could retard the ability of the microbial population to degrade hydrocarbons. In one study, conducted by Leahy and Colwell (1990), adjustment of pH from 4.5 to 7.4 resulted in a near doubling of rates of biotransformation of gasoline but rates dropped significantly when the pH was further raised to 8.5. Maximum polynuclear aromatic hydrocarbons oxidation rates and optimum specific bacterial growth is previously reported to be obtained around pH 7.0 and 30°C (Weissenfels et al., 1990).

It should be noted that during the soil bioremediation process, a recorded decrease in pH would occur due to the formation of low-molecular-weight organic acids as intermediates or end-products of biotransformation of hydrocarbon pollutants (Aislabie et al., 2006; Farahat and El-Gendy, 2007, 2008; El-Gendy and Farah, 2011). It was also reported that the addition of nutrients lowered the pH of coarse sand (Braddock et al., 1999). However, some relatively increment in soil pH is also recorded during the bioremediation process and it is attributed also to the release of byproducts causing this, during the hydrocarbon degradation process (Rahman et al., 2003; El-Gendy and Farah, 2011).

3.11 HEAVY METALS

The term *heavy metals* refers to metals with a density >5 g/cm^3 found among hydrocarbons in polluted soils. They are usually found with petroleum hydrocarbon–polluted sites (Adeniyi and Afolabi, 2002). The metal contaminants in petroleum oil–contaminated sites are usually aluminum, sodium, iron, nickel, and vanadium, with frequently smaller amounts of manganese, zinc, copper, lead, tin, barium, molybdenum, calcium, titanium, and chromium (Benka-Coker and Ekundayo, 1998). The solubility of metals in soil is influenced by the chemistry of soil and groundwater (Evans, 1989; Sposito, 1989). Geller et al. (2000) have reported that heavy metals in polluted soils are usually transformed through natural processes to innocuous compounds. However, the success of soil biotreatment and reclamation largely depend on the nature and levels of heavy metals and petroleum hydrocarbons that present in the soil to be cured. Since, nonionic organic pollutants mainly interact with soil organic matter and heavy metals cations can be complexed by the organic molecules found in the soil solids or released into the soil solution. Thus, heavy metals would affect the sorption of the nonionic organic pollutants by soils (Marwood et al., 1998; Gao et al., 2005, 2006). Lipophilic compounds such as polynuclear aromatic hydrocarbons are reported to have a narcotic mode of toxic action and may interact with the lipophilic components of the cytoplasmic membranes of bacteria, which would affect their permeability and structure. Thus, in the presence of heavy metals together with polynuclear aromatic hydrocarbons, the penetration of heavy metals into microbial cells affect their functions more easily and dramatically (Sikkema et al., 1995; Gogolev and Wike, 1997; Shen et al., 2006).

A high content of heavy metals in soil are of concern because of their toxicity to soil microorganisms and impairment of ecosystem functions (Frey et al., 2006; Godleads et al., 2015). Most of the studies reported reduced soil microbial activities and microbial biomass, inhibition of organic matter mineralization, and changes in microbial community structure following application of heavy metals to the soil (Shen et al., 2006). Friis and Myers-Keith (1986) have previously reported that reduction in growth is mainly because of the interaction between the metal cations along with phosphate, carboxyl, hydroxyl, and amino-groups of the cell surface. Amor et al. (2001) reported that heavy metals inhibit microorganisms by blocking essential functional groups or interfering with essential metal ions incorporation of biological molecules. The microbial diversity is reported to be decreased with the increase of the concentration of heavy metals and petroleum hydrocarbons in polluted soils (Lors and Mossmann, 2005; Oliveira and Pampulha, 2006; Hegazi et al., 2007). Thus, the presence of heavy metals increases the difficulty of biotransformation (Baldrian et al., 2000). *Brevibacterium* sp., *Cellulomonas* sp., *Clavibacter* sp., *Geobacillus* sp., *Micrococcus* sp., *Pediococcus* sp., and *Sphingomonas* species are from the hydrocarbon-utilizing bacteria that are mostly reported as indigenous bacterial population in oily soil highly contaminated with polynuclear aromatic hydrocarbons and heavy metals (Odokuma and Dickson, 2003; Chaillan et al., 2004; Lors and Mossmann, 2005; Poli et al., 2006; Hegazi et al., 2007). However, the tolerance of microorganisms to heavy metals in contaminated soil is much higher than that in liquid media. Since, heavy metals in the soil are apparently unavailable to indigenous microorganisms because of precipitation or adsorption to humic substances, oxides, minerals of the soil (Gogolev and Wilke, 1997). But, as listed before, the hazard posed by heavy metals in soil is suggested to be a function of their relative mobility and bioavailability, which are dependent on soil characteristics such as pH, mineralogy, texture, and organic matter content as well on the source and quantities of heavy metals in the soil. Therefore, these sites become inhospitable and just those microorganisms able to tolerate the high concentration of petroleum hydrocarbons and heavy metals can survive. Thus, the bioprospection of these natural selected organisms represents an important strategy to obtain agents for bioremediation processes (Castro-Silva et al., 2003).

The heavy metal–tolerable microorganisms can overcome the heavy metals toxicity through various biogenic methods (Gillan, 2016). For example, through the production of biogenic sulfides; microbially induced calcite precipitation, particularly in the presence of ureolytic bacteria; and the phosphatase activity, which decreases metal toxicity by causing precipitation of metal phosphates such as ammonium uranyl phosphate ($NH_4UO_2PO_4$). It may also occur by siderophores and metallophores and sometimes by extracellular polymeric substances that complex the metals and protect the microbial cells from excessive metal stress. Lipopolysaccharides in the outer membranes of gram-negative bacteria have been proposed to be involved in metal resistance as they may nucleate minerals. The production of outer membrane vesicles may also be used as a metal resistance mechanism as they may be used to expel metals complexed to periplasmic or outer-membrane proteins.

Metals may also be reduced by specific enzymes in the outer membrane, a process that may lead to the mineral precipitation. Metals may also be oxidized by specific enzymes in the outer membrane and the process may generate minerals with less toxicity. Peptidoglycan is a major structural component of the cell wall of most of the Gram-positive bacteria and can complex metals. Bacteria may repress or mutate transporters in the cell membrane to decrease or prevent the entry of metals. Redox changes may detoxify metals in the cytoplasm such as As, Sb, Cr, Hg, Se, and Te. Some bacteria can accumulate metals in the cytoplasm in the form of minerals. Finally, some bacterial strains exposed to toxic doses of metals upregulate genes that are involved in the elimination of reactive oxygen species, DNA reparation, and hydrolysis of abnormally folded proteins; such systems are indirect metal-resistance systems that repair damages induced by metals.

Some heavy metals are reported to be important for oil-degrading microorganisms while others are known to be toxic. For example, copper is essential for bacteria and fungi in trace amounts; however, high concentrations of copper are known to be toxic. The addition of copper to the soil is reported to be significantly inhibiting soil respiration, nitrogen mineralization, and nitrification (Atagana, 2006). The presence of copper is reported also to be very inhibitory to the degradation of polynuclear aromatic hydrocarbons. The copper ions are reported to be highly toxic microbial cells as these ions may directly interact with the physiological functions of the cells and possibly the membrane development in the cells. Enzyme activities, which are very important in the microbial degradation of organic substrates, can be also hindered by the presence of copper in the medium (Atagana, 2006; Hegazi et al., 2007). Long-term exposure to heavy metals (Cu, Ni and Zn) is reported also to alter microbial structure (Chaerun et al., 2004). However, tolerance and adaptation of microorganisms to heavy metals are common phenomena, and the presence of tolerant fungi and bacteria in polluted soil has frequently been observed (Deighton and Goodman, 1995; Hegazi et al., 2007). However, several isolated microorganisms of sites contaminated by hydrocarbons show low sensitivity to heavy metals existing at these sites, such as copper, manganese, nickel, and zinc (Baldrian et al., 2000; Hegazi et al., 2007). The presence of heavy metals along with petroleum hydrocarbons as mixed contaminants could exert selective effects on the specific microflora, leading to inhibition of the most sensitive species to the benefit of the resistant ones that can adapt themselves to such an environment (Lors and Mossmann, 2005; Hegazi et al., 2007). Castro-Silva et al. (2003) reported that heavy metal resistance is a widespread attribute among microorganisms isolated from mining environments.

According to Bopp et al. (1983), studies demonstrated that bacteria isolated from polluted sites are 40–200 times more resistant to heavy metals than related strains isolated from unpolluted sites. Hurst et al. (1997) also considered that the presence of high concentrations of heavy metals is associated with an increase in the number of microorganisms that are resistant to them. Despite these results, it is incorrect to affirm that the sensitivity to the metals in the environment will be the same as in the laboratory tests. Chen et al. (1997) reported the potential toxic effect of high level of metals on microbial metabolism of petroleum hydrocarbons even though the microorganisms have been previously exposed to heavy metals accumulated over the years. The negative effect of heavy metals on soil microbes and soil microbial processes in contaminated soils, for example, methane metabolism, growth, nitrogen and sulfur conversions, dehalogenation, and reductive processes in general, can potentially limit the bioremediation of organic pollutants (Baldrian et al., 2000). Metals may inhibit pollutant biotransformation through interaction with enzymes directly involved in biotransformation (e.g., pollutants-specific oxygenases) or through interaction with enzymes involved in general metabolisms. In either case, inhibition is mediated by the ionic form of the metal (Sandrin and Marier, 2003). Raja et al. (2006) have reported that under conditions of imposed stress, metal resistance in microorganisms possibly helps them to adapt spontaneously by mutation and natural selection.

It is important, therefore, not to use sensitive microorganisms to the heavy metals present at the site that is being bioremediated. *Alcaligenes eutrophus* degraded polychlorinated biphenyls (PCBs) and 2,4-dichlorophenoxyacetic acid effectively in presence of Ni and Zn (Collorad et al., 1994). But, *Arthrobacter* strain ATCC 33790 did not remove 89–124 ppm of pentachlorophenol in presence of 2 and 8 ppm of Cu (Edgehill, 1996). Sokhn et al. (2001), reported that Cu up to 70 ppm has little effect on microbial activates during phenanthrene degradation. However, elevated concentrations (700 and 7000 ppm) showed marked reduction in microbial activities. About 200 μM of Ni is reported to inhibit the aerobic biotransformation of naphthalene, biphenyl, xylene, and many other xenobiotics by *Alcaligenes* and *Pseudomonas* strains (Sandrin and Marier, 2003). *Pseudomonas chlororaphis* PCL1391 is reported to degrade 98% of naphthalene in presence of 100 μM Ni (Siunova et al., 2007). The presence of different concentrations heavy metals such as zinc, copper, cadmium, nickel, chromium, and manganese have been also reported to have impeding effects on the biodegradation of petroleum hydrocarbons in polluted soils as they exert inhibition effects on the enzymatic activities of the soil microorganisms (Baldrian et al., 2000; Wong et al., 2005; Atagana, 2006).

A study performed by Hegazi et al. (2007) reported the impact of heavy metals, such as copper, manganese, nickel, and zinc individually and in mixtures, on the biotransformation of phenanthrene, as a model for polyaromatic petroleum hydrocarbons (polynuclear aromatic hydrocarbons), by *Cellulomonas hominis* strain N2, isolated from highly heavy metals and petroleum-polluted soil sample. Exposure to heavy metals is reported to affect microbial growth and survival. However, the degree of toxicity of heavy metals and effect of the bacterial growth somewhat differ from that obtained on biotransformation efficiencies. Thus, in presence of heavy metals, in contaminated environments, the enzymatic activities of the microorganisms are a very important factor and the bacterial growth is not a limiting factor in their biotransformation efficiency but the types of the enzymes involved in biotransformation process and degree of their activities have a more important role.

3.12 SURFACTANTS

Surfactants are amphiphilic molecules consisting of a hydrophilic and a hydrophobic domain. They partition between two phases in a heterogeneous system and increase the apparent solubility of a hydrophobic compound in water (Georgiou et al., 1992; Pizzul, 2006; Wang and Keller, 2009). The three general characteristics of the surfactants are enrichment at interfaces, lowering of interfacial tension, and formation of micelles (Neu, 1996; Li et al., 2007). They can be synthetic or of microbial origin (biosurfactants). It has been established that carbon source plays an important role in the production of surface active compounds. Usually the presence of water immiscible substances, e.g., hydrocarbons, is required (Rapp et al., 1979; Robert et al., 1989; Hommel, 1990; Abu-Ruwaida et al., 1991; Bredholt et al., 1998; Kumar et al., 2006) but some bacteria even produce surfactants when grown on complex hydrocarbons, such as coal (Singh and Tripathi, 2013) and crude oil (Das and Mukherjee, 2007a; Ali et al., 2014); ordinary carbon source, such as glycerol (Das et al., 2008; Putri and Hertadi, 2015) and olive oil (Khopade et al., 2012a); carbohydrates, such as cashew apple juice (Freitas de Oliveira et al., 2013), trehalose, dextrose, fructose, and sucrose (Khopade et al., 2012b); and some organic wastes, such as CSL, sugarcane molasses, waste frying oil, cheese whey waste, etc. (Guerra-Santos et al., 1984; Person and Molin, 1987; Banat et al., 2010; Rocha e Silva et al., 2014). Several types of biosurfactants have been isolated and characterized including lipolipids, glycolipids, phospholipids, neutral lipids, fatty acids, peptidolipids, lipopolysaccharides, biopolymer complexes, and others (Janek et al., 2010).

Biosurfactants are reported to decrease heavy metal toxicity in polluted sites and enhance biotransformation efficiency (Sandrin et al., 2000; Hegazi et al., 2007). This would occur through the complexation of the free form of the metal residing in solution, which decreases the solution-phase activity of the metal and it would also promote heavy metals desorption. It would also occur by the reduced interfacial tension conditions expressed by the biosurfactants, which would accumulate at the solid-solution interface, allowing the direct contact between the biosurfactant and the sorbed metal. They are more effective than chemical ones in enhancing the solubility of organic pollutants (Bai et al., 1997) and biotransformation of petroleum hydrocarbons including the recalcitrant high-molecular–weight polynuclear aromatic hydrocarbons (Cybulski et al., 2003; Wong et al., 2005; Das and Mukherjie, 2007a,b; Li and Chen, 2009). Biosurfactant is reported to stimulate the indigenous microbial population to degrade hydrocarbons, throughout the increase of the surface area of hydrophobic water-insoluble substrate and/or increase the bioavailability of hydrophobic water-insoluble substances, moreover, throughout the incase of surface cell hydrophobicity (Kaczorek et al., 2008).

Biosurfactants can also enhance the microbial grow on bound substrates by desorbing them from surfaces or by increasing their apparent water solubility. Moreover, Das and Mukherjie (2007a) reported that the production of biosurfactant induces desorption of hydrocarbons from soil to the aqueous phase of soil slurries leading to increased microbial mineralization, either by increasing hydrocarbon solubility or by increasing the contact surface with hydrophobic compounds which leads also to increase in bacterial population. Consequently, the biosurfactant production increases the accessibility of petroleum hydrocarbons to soil bacteria, enhancing the biotransformation process. Biosurfactants alone is reported to promote crude oil biotransformation to a large extent without adding fertilizers, which would reduce the cost of bioremediation process and minimize the dilution or wash away problems encountered when water soluble fertilizers are used during bioremediation of aquatic environments (Thavasi et al., 2011).

Although of the good advantages of biosurfactant and it seems to be more attractive than their synthetic counterparts, biosurfactants are not yet competitive in the market due to functional reasons and high production costs, especially regarding substrates, which account for 10%–30% of the total production cost (Rocha e Silva et al., 2014). Thus, using biodegrading microorganisms that have the capability of producing biosurfactant(s) or emulsifier(s) has the advantage of continuous supply of natural, nontoxic, and biodegradable surfactants(s) at a low cost for solubilizing the hydrophobic petroleum hydrocarbons. Moreover, they can selectively counter the increased viscosity and decreased water solubility of hydrocarbons, thus enhancing the biotransformation rates (Bento et al., 2005; El-Gendy et al., 2014; Ali et al., 2014; Chandankere et al., 2014).

It should be noted that most of the biosurfactants are reported to be produced during the stationary phase of microbial growth and a few microbial species may display a low biosurfactant productivity during its exponential growth phase (Ron and Rosenberg, 2001; Urum and Pekdemir, 2004). Jain et al. (1991) reported the addition of *Pseudomonas* biosurfactant enhanced the biotransformation of tetradecane, pristane, and hexadecane in a slit loam. Zhang and Miller (1995) reported the enhanced octadecane dispersion and biodegradation by a *Pseudomonas* rhamnolipids surfactant. Herman et al. (1997) reported that the rhamnolipids biosurfactants enhanced the in-situ biodegradation in porous matrix.

According to Straube et al. (1999), light oil theoretically stimulates the production of biosurfactant and act as co-solvent, increasing the bioavailability of hydrophobic contaminants by helping to desorb them from soil particles. The increase of microbial population may be due to the production of biosurfactant, which, as mentioned before, induced the desorption of hydrocarbons from soil to the aqueous phase of soil slurries leading to increased microbial mineralization, either by increasing hydrocarbon solubility or by increasing the contact surface with hydrophobic compounds (Moran et al., 2000; Christofi and Ivshina, 2002; Rahman et al., 2003; Maier, 2003; Mukherjie and Das, 2005; Das and Mukerjie, 2007a,b). Daziel et al. (1996) reported the increase of the naphthalene aqueous solubility by biosurfactant. Zhang et al. (1997) reported also that rhamnolipid biosurfactants increase the solubility and biotransformation of phenanthrene. Crude biosurfactants produced from the thermophilic strains, *B. subtilis* DM-04, *P. aeruginosa* M, or *P. aeruginosa* NM strains, are reported to increase the solubility of pyrene, anthracene, and phenanthrene (Das and Mukerjie, 2007a). Lipopeptide biosurfactant produced by hydrocarbon degrading and biosurfactant producing *B. subtilis* CN2 isolated from creosote-contaminated soil is reported to recover approximately 85% of used motor oil from contaminated sand within 24h (Bezza and Cheraw, 2015). The biosurfactant produced by the marine isolate *Bacillus licheniformis* MTCC 5514 is reported to remove more than 85% of adsorbed crude oil from different types of soil (Kavitha et al., 2015). Hegazi et al. (2007) reported that the production of biosurfactant by *C. hominis* strain N2 increases its heavy metal tolerance, the phenanthrene aqueous solidity, and biotransformation efficiency. Das et al. (2008) reported that a marine isolate *Bacillus circulans* can degrade anthracene and produce biosurfactant in a glycerol-supplemented mineral salts medium. A biosurfactant produced from a petroleum degrading bacteria strain *B. licheniformis* Y-1 is reported to enhance the bioremediation of petroleum-polluted soil by *Pleurotus ostreatus*, specially the 16-polyaromatic hydrocarbons listed by US-EPA, as priority pollutants (Liu et al., 2016). However, in a simulated marine oil spill bioremediation experiment using a bacterial consortium amended with rhamnolipids, they are reported to exert positive role in the biotransformation of long chain hydrocarbons, biomarkers, and polyaromatic hydrocarbons, but they exert a negative role in biotransformation of hydrocarbons with relatively volatile property, such as short-chain n-alkanes, low-molecular-weight polynuclear aromatic hydrocarbons and sesquiterpenes with simple structure (Chen et al., 2013). The biosurfactant producing *Streptomyces* spp. isolates AB1, AH4, and AM2 are reported to degrade 82.36%, 85.23%, and 81.03% of 100 mg/L naphthalene within 12 days and 76.65%, 80.96%, and 67.94% the aliphatic fraction of crude oil (1% v/v) within 30 days, respectively (Ferradji et al., 2014).

4.0 CHALLENGES AND OPPORTUNITIES

The increasingly number of petroleum-based contaminants that have been and, in some cases, continue to be released into the environment on a large scale includes numerous aliphatic and aromatic compounds, some containing heteroatoms such as nitrogen, oxygen, sulfur, and metals. Thus, petroleum-based contaminants are common environmental pollutants and the search for and development of an effective biotransformation for petroleum-contaminated soil is a substantial challenge.

The concentration of such contaminants in an ecosystem depends on the amount present and the rate at which the compound is released, its stability in the environment under both aerobic and anaerobic conditions, the extent of its dilution in the environment, and the mobility of the compound in a particular environment, and its rate of biological or nonbiological degradation offer considerable challenges to biotechnologists.

With references to microbial biotransformation, which involves the breakdown of petroleum-based contaminants by microorganisms, aerobic processes are considered the most efficient and generally applicable—aerobic degradation is dependent on the presence of molecular oxygen and is catalyzed by enzymes that have evolved for the catabolism of natural substrates and exhibit low specificities. Depending upon the type of enzyme catalyzing the reaction, either one (mono-oxygenase) or two (di-oxygenase) oxygen atoms are inserted into the molecule via an electrophilic attack on an unsubstituted carbon atom. Anaerobic degradation proceeds via reductive transformation, wherein an electron transfer to the compound results in reduction (typically, hydrogenation).

Although regulations are strictly enforced in developed countries like the United States and most of the European countries to meet the challenges of petroleum-related contamination, these regulations often remain unenforced in most of the developing countries. Cleaning up such sites is often not only technically challenging but also very expensive. Considerable pressure encourages the adoption of waste management alternatives to burial, the traditional means of disposing of solid and liquid wastes.

Since most of the contaminants of concern in crude oil can be biotransformed under the appropriate conditions, the success of the biotransformation processes depends mainly on the ability to establish these conditions in the contaminated environment using the above new developing technologies to optimize the microorganisms' total efficiency. The technologies used at various polluted sites depend on the limiting factor present at the location. For example, where there is insufficient dissolved oxygen, bioventing or sparging is applied; biostimulation or bioaugmentation is suitable for instances where the biological count is low.

Over the past decade, opportunities for applying bioremediation to a much broader set of contaminants have been identified. Indigenous and enhanced organisms have been shown to degrade industrial solvents, PCBs, explosives, and many different agricultural chemicals. Pilot, demonstration, and full-scale applications of bioremediation have been carried out on a continuing basis. However, the full benefits of bioremediation have not been realized because processes and organisms that are effective in controlled laboratory tests are not always equally effective in full-scale applications. The failure to perform optimally in the field setting stems from a lack of predictability due, in part, to inadequacies in the fundamental scientific understanding of how and why these processes work.

This, if bioremediation is to be effective, the microorganisms must enzymatically attack the pollutants and convert them to noncontaminating products—some microbes may produce products that are toxic not only to themselves but also to the ecosystem. Parameters that affect the bioremediation process include temperature, nutrients (fertilizers), and the amount of oxygen present in the soil and/or the affected water system (Chapters 1 and 9). These conditions allow the microbes to grow and multiply and consume more of the contaminant. When conditions are adverse, microbes grow too slowly or die or they can create more harmful chemicals. In addition, the application of any technology is dependent not only on the availability of the technology but also on the reliability of the technology as well as on the suitability of the technology for the specific site conditions and whether the technology is readily available (i.e., emerging, developing, or proven).

4.1 CONVENTIONAL BIOTRANSFORMATION

Conventional bioremediation methods used are composting, land farming, biopiling, and bioslurry reactors. Composting is a technique that involves combining contaminated soil with nonhazardous organic additives such as manure or agricultural wastes; the presence of the organic materials allows the development of a rich microbial population and elevated temperature characteristic of composting. Land-farming is a simple technique in which contaminated soil is excavated and spread over a prepared bed and periodically tilled until pollutants are degraded. While biopiling is a hybrid of land-farming and composting, it is essentially engineered cells that are constructed as aerated composted piles.

Bioslurry reactors can provide rapid biotransformation of contaminants due to enhanced mass transfer rates and increased contaminant-to-microorganism contact. These units are capable of treating high concentrations of organic contaminants in soil and sludge. These reactors can aerobically biodegrade aqueous slurries created through the mixing of soils or sludge with water. The most common state of bioslurry treatment is batch; however, continuous-flow operation can be achieved.

Microorganisms excel at using organic substances, natural or synthetic, as sources of nutrients and energy. Indeed, the diversity of petroleum-related constituents for growth had led to the discovery of enzymes capable of transforming many unrelated natural organic compounds by many different catalytic mechanisms.

However, depending on behavior in the environment, organic compounds are often classified as biodegradable, persistent, or recalcitrant. A biodegradable organic compound is one that undergoes a biological transformation. A persistent organic compound does not undergo biotransformation in certain environments; and a recalcitrant compound resists biotransformation in a wide variety of environments. While partial biotransformation is usually an alteration by a single reaction, primary biotransformation involves a more extensive chemical change.

Biotransformation and its application in bioremediation of organic pollutants have benefited from the biochemical and molecular studies of microbial processes. Indeed, the biotransformation of organic contaminants in the natural environment has been extensively studied to understand microbial ecology, physiology, and evolution for their potential in bioremediation. As a result, there is a strong demand to increase the adoption of bioremediation as an effective technique for risk reduction on hydrocarbon impacted soils. However, the biotransformation effectiveness diminishes with the time extension and the inhibiting effect may become dominant with time. The key solution to bioremediation is to speed up the restoration process and eliminate or delay the inhibitory effect, such as through (1) the selection of specifically targeted strains or microorganisms or (2) the alteration of microbial community structure changes during the treatment.

4.2 ENHANCED BIOTRANSFORMATION

The natural processes that drive bioremediation can be enhanced to increase the effectiveness and to reduce time required to meet cleanup objectives. Enhanced bioremediation involves the addition of microorganisms (e.g., fungi, bacteria, and other microbes) or nutrients (e.g., oxygen, nitrates) to the subsurface environment to accelerate the natural biotransformation processes in which indigenous degrade (metabolize) organic contaminants found in soil and/or ground water and convert them to innocuous end products. The processes rely on general availability of naturally occurring microbes to consume contaminants as a food source (petroleum hydrocarbons in aerobic processes) or as an electron acceptor (chlorinated solvents). In addition to microbes being present, in order to be successful, these processes require nutrients (carbon: nitrogen: phosphorus).

The potential of microorganisms in the remediation of some of the compounds hitherto known to be nonbiodegradable has been widely acknowledged globally. With advances in biotechnology, bioremediation has become a rapidly growing area and has been commercially applied for the treatment of hazardous wastes and contaminated sites. Although a wide range of new microorganisms have been discovered that are able to degrade highly stable, toxic organic xenobiotic, still many pollutants persist in the environment.

Briefly, a xenobiotic is a chemical that is found in an organism, such as a bacterium, but which is not normally produced or expected to be present in the organism. The word is very often used in the context of a pollutant and is a substance that is not indigenous to an ecosystem or a biological system and that did not exist in nature before human intervention.

A number of reasons have been identified as challenges posed to the microorganisms working in contaminated sites. Such potential limitations to biological treatments include poor bioavailability of chemicals, presence of other toxic compounds, inadequate supply of nutrients, and insufficient biochemical potential for effective biotransformation. A wide range of bioremediation strategies have been developed for the treatment of contaminated soils using natural and modified microorganisms.

Selecting the most appropriate strategy to treat a specific site can be guided by considering three basic principles: the amenability of the pollutant to biological transformation to less toxic products, the bioavailability of the contaminant to microorganisms, and the opportunity for bioprocess optimization. With the help of advances in bioinformatics, biotechnology holds a bright future for developing bioprocesses for environmental applications.

Biotechnological processes for the bioremediation of petroleum-related pollutants offer the possibility of in situ treatments and are mostly based on the natural activities of microorganisms. Biotechnological processes to destroy contaminants of the type found in petroleum and petroleum products offer many advantages over physicochemical processes. When successfully operated, biotechnological processes may achieve complete destruction of petroleum-related pollutants. However, an important factor limiting the bioremediation of sites contaminated with such contaminants is the slow rate of degradation that may limit the practicality of using microorganisms in remediating contaminated sites. This is an area where genetic engineering can make a marked improvement.

Biosurfactants (the surface-active microbial products that have numerous industrial applications) are produced by many microorganisms, especially bacteria, produce biosurfactants when grown on water-immiscible substrates. Most common biosurfactants are glycolipids in which carbohydrates are attached to a long-chain aliphatic acid, while others such as lipopeptides, lipoproteins, and heteropolysaccharides, are more complex. The most promising applications of biosurfactants are in the cleaning of oil-contaminated tankers; oil-spill management; transportation of heavy crude; enhanced oil recovery; recovery of oil from sludge; and bioremediation of sites contaminated with hydrocarbons, heavy metals, and other petroleum-related pollutants.

Advances in genetic and protein engineering techniques have opened up new avenues to move towards the goal of *genetically engineered microorganisms* to function as biocatalysts, in which certain desirable biotransformation pathways or enzymes from different organisms are brought together in a single host with the aim of performing specific reactions. A strategy has also been suggested for designing organisms with novel pathways and the creation of a bank of genetic modules encoding broad-specificity enzymes or pathway segments that can be combined at will to generate new or improved activities.

Methods for the rapid and specific identification of microorganisms within their natural environments continue to be developed. Classic methods are time consuming and only work for a limited number of microorganisms and an *n* increasing need to develop new methods for characterization of microorganisms able to degrade the various types of petroleum-related pollutants has led to the use of molecular probes to identify, enumerate, and isolate microorganisms with degradative potential.

Through the genetic engineering of metabolic pathways, it is possible to extend the range of substrates that an organism can utilize. Aromatic hydrocarbon dioxygenases have broad substrate specificity and catalyze enantio-specific reactions with a wide range of substrates.

4.3 BIOTRANSFORMATION IN EXTREME ENVIRONMENTS

The biotransformation of many components of petroleum hydrocarbons has been reported in a variety of terrestrial and marine cold ecosystems and extreme environments as alpine soil. Antarctic exploration and research have led to some significant, although localized, impacts on the environment. Human impacts occur around current or past scientific research stations, typically located on ice-free areas that are predominantly soils. Fuel spills, the most common occurrence, have the potential to cause the greatest environmental impact in the Antarctic through accumulation of aliphatic and aromatic compounds. Effective management of spills of crude oil and crude oil products is dependent on understanding how they impact soil properties such as moisture, hydrophobicity, soil temperature, and microbial activity. Numbers of hydrocarbon-degrading bacteria, typically *Rhodococcus*, *Sphingomonas*, and *Pseudomonas* species for example, may become elevated in contaminated soils, but overall microbial diversity declines.

In addition, the physical environment is also important for hydrocarbon biotransformation. Cold habitats possess sufficient indigenous microorganisms, psychrotrophic bacteria (bacteria capable of surviving or even thriving in a cold environment) being predominant. They adapt rapidly to the presence of the contaminants, as demonstrated by significantly increased numbers of oil degraders shortly after a pollution event, even in the most northerly areas of the world.

However, the bulk of information on hydrocarbon degradation borders on activities of mesophiles, although significant biotransformations of hydrocarbons have been reported in psychrophilic environments in temperate regions. Full-scale in situ remediation of petroleum contaminated soils has not yet been used in Antarctica for example, partly because it has long been assumed that air and soil temperatures are too low for an effective biotransformation. Such omissions in research programs need to be corrected.

In summary, hydrocarbon biotransformation rates in extreme environments systems are not necessarily lower than in temperate systems but activity measurements should be performed at the prevailing in situ temperature in order to obtain a realistic estimate of the naturally occurring biotransformation.

REFERENCES

Abalos, A., Vinas, M., Sabatr, J., Manresa, M., Solanas, A., 2004. Enhanced biodegradation of Casablanca crude oil by a microbial consortium in presence of a rhamnolipid produced by *Pseudomonas aeruginosa* AT10. Biodegradation 15 (4), 249–260.

Abbondanzi, F., Bruzzi, L., Campisi, T., Frezzati, A., Guerra, R., Iacondini, A., 2006. Biotreatability of polycyclic aromatic hydrocarbons in brackish sediments: preliminary studies of an integrated monitoring. International Biodeterioration and Biodegradation 57, 214–221.

Abd-El-Haleem, D., Moawad, H., Zaki, E.A., Zaki, S., 2002. Molecular characterization of phenol degrading bacteria isolated from different Egyptian ecosystems. Microbial Ecology 43 (2), 217–224.

Abioye, O.P., Agamuthu, P., Abdul Aziz, A.R., 2012. Biodegradation of used motor oil in soil using organic waste amendments. Biotechnology Research International 2012:587041. http://dx.doi.org/10.1155/2012/587041. 8 pages.

Abu-Ruwaida, A.S., Banat, I.M., Haditirto, S., Khamis, A., 1991. Nutritional requirements and growth characteristics of a biosurfactant- producing *Rhodococcus bacterium*. World Journal of Microbiology & Biotechnology 7, 53–61.

Adeniyi, A.A., Afolabi, J.A., 2002. Determination of total petroleum hydrocarbons and heavy metals in soils within the vicinity of facilities handling refined petroleum products in Lagos metropolis. Environment International 28, 79–82.

Admon, S., Green, M., Avinimelech, Y., 2001. Bioremediation kinetics of hydrocarbons in soil during land treatment of oily sludge. Bioremediation 5, 193–209.

Agarwal, A., Liu, Y., 2015. Remediation technologies for oil-contaminated sediments. Marine Pollution Bulletin 101, 483–490.

Aichberger, H., Loibner, A.P., Celis, R., Braun, R., Ottner, F., Rost, H., 2006. Assessment of factors governing biodegradability of PAHs in three soils aged under field conditions. Soil and Sediment Contamination 15, 73–85.

Aislabie, J.M., Saul, D.J., Foght, J.M., 2006. Bioremediation of hydrocarbon-contaminated polar soils. Extremophiles: Life Under Extreme Conditions 10, 171–179.

Alexander, M., 1999. Biodegradation and Bioremediation, second ed. Academic Press, San Diego, California; London.

Alexander, M., 2000. Aging, bioavailability, and overestimation of risk from environmental pollutants. Environmental Science & Technology 34, 4259–4265.

Ali, H.R., El-Gendy, N.Sh., Moustafa, Y.M., Mohamed, I., Roushdy, M.I., Hashem, A.I., 2012. Degradation of asphaltenic fraction by locally isolated halotolerant bacterial strains. ISRN Soil Science:435485.

Ali, H.R., Ismail, D.A., El-Gendy, N.S., 2014. The Biotreatment of oil polluted seawater by biosurfactant producer halotolerant *Pseudomonas aeruginosa* Asph2. Energy Sources, Part A: Recovery, Utilization, and Environmental Effects 36 (13), 1429–1436.

Allan, I., Semple, K.T., Hare, R., Reid, B.J., 2007. Cyclodextrin enhanced biodegradation of polycyclic aromatic hydrocarbons and phenols in contaminated soil slurries. Environmental Science & Technology 41, 5498–5504.

Ambrosoli, R., Petruzzelli, L., Luis Minati, J., Ajmone Marsan, F., 2005. Anaerobic PAH degradation in soil by a mixed bacterial consortium under denitrifying conditions. Chemosphere 60 (9), 1231–1236.

Amellal, N., Portal, J.M., Berthelin, J., 2001. Effect of soil structure on the bioavailability of polycyclic aromatic hydrocarbon within aggregates of a contaminated soil. Applied Geochemistry 16, 1611–1619.

Amor, L., Kennes, C., Veiga, M.C., 2001. Kinetic of biodegradation of monoaromatic hydrocarbons in presence of heavy metals. Bioresource Technology 78, 181–185.

Arias, L., Bauza, J., Tobella, J., Vila, J., Grifoll, M., 2008. A microcosm system and an analytical protocol to assess PAH degradation and metabolite formation in soils. Biodegradation 19, 425–434.

Arslanoglu, F.N., Kar, F., Arslan, N., 2005. Adsorption of dark colored compounds from peach pulp by using powdered-activated carbon. Journal of Food Engineering 71, 156–163.

Arulazhagan, P., Vasudevan, N., 2011a. Biodegradation of polycyclic aromatic hydrocarbons by a halotolerant bacterial strain *Ochrobactrum* sp. VA1. Marine Pollution Bulletin 62, 388–394.

Arulazhagan, P., Vasudevan, N., 2011b. Role of nutrients in the utilization of polycyclic aromatic hydrocarbons by halotolerant bacterial strain. Journal of Environmental Sciences 23 (2), 282–287.

Atagana, H.I., 2006. Biodegradation of polycyclic aromatic hydrocarbons in contaminated soil by biostimulation and bioaugmentation in the presence of copper(II) ions. World Journal of Microbiology & Biotechnology 22, 1145–1153.

Atlas, R.M., 1981. Microbial degradation of petroleum hydrocarbons: an environmental perspective. FEMS Microbiology Reviews 45 (1), 180–209.

Bachoon, D.S., Araujo, M., Hodson, R.E., 2001. Microbial community dynamics and evaluation of bioremediation strategies in oil-impacted salt marsh sediment microcosms. Journal of Industrial Microbiology & Biotechnology 27, 72–79.

Bai, G.Y., Brusseau, M.L., Miller, R.M., 1997. Biosurfactant enhanced removal of residual hydrocarbons from soil. Journal Contaminant Hydrology 25, 157–170.

Bailey, J.E., Ollis, D.F., 1986. Biochemical Engineering Fundamental, second ed. McGraw-Hill, New York.

Baldrian, P., der Wiesche, C.I., Gabriel, J., Nerund, E., Zadraziel, F., 2000. Influence of cadmium and mercury on activities of ligninolytic enzymes and degradation of polyaromatic hydrocarbons by *Pleurotus ostreatus* in soil. Applied and Environmental Microbiology 66, 2471–2478.

Balks, M.R., Paetzold, R.F., Kimble, J.M., Aislabie, J., Campbell, I.B., 2002. Effects of hydrocarbon spills on the temperature and moisture regimes of cryosols in the Ross Sea region. Antarctic Science 14 (4), 319–326.

Banat, I.M., Franzetti, A., Gandolfi, I., Bestetti, G., Martinotti, M.G., Fracchia, L., Smyth, T.J., Marchant, R., 2010. Microbial biosurfactants production, applications and future potential. Applied Microbiology and Biotechnology 87, 427–444.

Barakat, A.K.M., Ouf, E.A., 2009. Oil pollution of sediments at different beaches of Egypt. Energy Sources, Part A: Recovery, Utilization, and Environmental Effects 31 (14), 1217–1226.

Bayoumi, R.A., 2009. Bacterial bioremediation of PAHs in heavy oil contaminated soil. Journal of Applied Science Research 5 (2), 197–211.

Bending, G.D., 2003. Litter decomposition, ectomycorrhizal roots and the 'Gadgil' effect. New Phytologist 158, 227–238.

Benka-Coker, M.O., Ekundayo, J.A., 1998. Effect of heavy metals on growth of species of *Micrococcus* and *Pseudomonas* in a crude oil/mineral salts medium. Bioresource Technology 66, 241–245.

Bento, F.M., Camargo, F.A.O., Okeke, B.C., Frankenberger, W.T., 2005. Comparative bioremediation of soils contaminated with diesel oil by natural attenuation, biostimulation and bioaugmentation. Bioresource Technology 96, 1049–1055.

Bezza, F.A., Chirwa, E.M.N., 2015. Production and applications of lipopeptide biosurfactant for bioremediation and oil recovery by *Bacillus subtilis* CN2. Biochemical Engineering Journal 101, 168–178.

Bogan, B.W., Lahner, L.M., Sullivan, W.R., Paterek, J.R., 2003. Degradation of polycyclic aromatic and straight-chain aliphatic hydrocarbons by a strain of *Mycobacterium austroafricanum*. Journal of Applied Microbiology 94, 230–239.

Bonten, L.T.C., Grotenhuis, T.C., Rulkens, W.H., 1999. Enhancement of PAH biodegradation in soil by physico-chemical treatment. Chemosphere 38 (15), 3627–3636.

Boopathy, R., 2000. Factors limiting bioremediation technologies. Bioresource Technology 74, 63–67.

Bopp, L., Chakrabarty, A., Ehrlich, H.L., 1983. Chromate resistance plasmid in *Pseudomonas fluorescens*. Journal of Bacteriology 155, 1105–1109.

Bosma, T.N.P., Middeldorp, P.J.M., Schraa, G., Zehnder, A.J.B., 1997. Mass transfer limitation of biotransformation: quantifying bioavailability. Environmental Science & Technology 31, 248–252.

Boufadel, M.C., Geng, X., Short, J., 2016. Bioremediation of the Exxon Valdez oil in Prince William Sound beaches. Marine Pollution Bulletin 113, 156–164.

Braddock, J.F., Ruth, M.L., Catterall, P.H., 1997. Enhancement and inhibition of microbial activity in hydrocarbon-contaminated artic soils: implications for nutrient-amended bioremediation. Environmental Science & Technology 31 (7), 2078–2084.

Braddock, J.F., Walworth, J.L., Kathleen, A., McCarthy, K.A., 1999. Biodegradation of aliphatic vs. aromatic hydrocarbons in fertilized arctic soils. Bioremediation 3 (2), 105–116.

Bredholt, H., Josefsen, K., Vatland, A., Bruheim, P., Eimhjellen, K., 1998. Emulsification of crude oil by an alkane-oxidizing *Rhodococcus* species isolated from seawater. Canadian Journal of Microbiology 44, 330–340.

Bressler, D.C., Gray, M.R., 2003. Transport and reaction processes in bioremediation of organic contaminants. 1. Review of bacterial degradation and transport. International Chemical Reaction Engineering. 1 (1). https://doi.org/10.2202/1542-6580.1027.

Bunkim, G.C., 2003. Microbial Biodegradation of Hydrocarbons in Petroleum Sludge Wastes (M.Sc. thesis). Faculty of California Polytechnic State University, San Luis Obispo.

Burchiel, S.W., Luster, M.I., 2001. Signaling by environmental PAHs in human lymphocytes. Expert Review of Clinical Immunology 98, 2–10.

Cairney, J.W.G., Taylor, A.F.S., Burke, R.M., 2003. No evidence for lignin peroxidase genes in ectomycorrhizal fungi. New Phytologist 160, 461–462.

Calvo, C., Manzanera, M., Silva-Castro, G.A., Uad, I., González-López, J., 2009. Application of bioemulsifiers in soil oil bioremediation processes, prospects. Science of Total Environment 407 (12), 3634–3640.

Caravaca, F., Roldán, A., 2003. Assessing changes in physical and biological properties in a soil contaminated by oil sludges under semiarid Mediterranean conditions. Geoderma 117, 53–61.

Castro-Silva, A., Lima, A.O.S., Gerchenski, A.V., Jaques, B., Rodrigues, L., de Souza, L., Rorig, L., 2003. Heavy metals resistance of microorganisms isolated from coal mining environments of Santa Catarina. Brazilian Journal of Microbiology 34, 45–47.

Chaerun, S.K., Tazaki, K., Asada, R., Kogure, K., 2004. Alkane-degrading bacteria and heavy metal from and Nakhodka oil spill polluted seashores in the Sea of Japan after five years of bioremediation. Science Reports of Kanazawa University 49 (12), 25–46.

Chaillan, F., Le Fleche, A., Bury, E., Phantavong, Y.H., Grimont, P., Saliot, A., Oudot, J., 2004. Biodegradation of mixtures of polycyclic aromatic hydrocarbons under aerobic and nitrate-reducing conditions. Research in Microbiology 55, 587–595.

Chaineau, C.H., Rougeux, G., Yepremain, C., Oudot, J., 2005. Effect of nutrient on the biodegradation of crude oil and associated microbial populations in the soil. Soil Biology and Biochemistry 37, 1490–1497.

Chakraborty, S., Mukherji, S., Mukherji, S., 2010. Surface hydrophobicity of petroleum hydrocarbon degrading *Burkholderia* strains and their interactions with NAPLs and surfaces. Colloids and Surfaces B: Biointerfaces 78 (1), 101–108.

Chandankere, R., Yao, J., Cai, M., Masakorala, K., Jain, A.K., Choi, M.M.F., 2014. Properties and characterization of biosurfactant in crude oil biodegradation by bacterium *Bacillus methylotrophicus* USTB. Fuel 122, 140–148.

Chang, B.V., Shiung, L.C., Yuan, S.Y., 2002. Anaerobic biodegradation of polycyclic aromatic hydrocarbon in soil. Chemosphere 48, 717–724.

Chen, R.B., Kokal, S.L., Al-Ghamdi, M.A., Gwathney, W.J., Al-hajji, A.A., Dajani, N., 1997. Bioremediation of petroleum hydrocarbon-contaminated soil: a laboratory enhancement study. In: 2nd Specialty Conference on Environmental Progress in the Petrochemical Industries. Saudi Arabian Section-Air and Waste Management Association and Bahrain Society of Engineers, November 1997, pp. 17–19.

Chen, Q., Bao, M., Fan, X., Liang, S., Sun, P., 2013. Rhamnolipids enhance marine oil spill bioremediation in laboratory system. Marine Pollution Bulletin 71, 269–275.

Chen, Y., Li, C., Zhou, Z., Wen, J., You, X., Mao, Y., Lu, C., Huo, G., Jia, X., 2014. Enhanced biodegradation of alkane hydrocarbons and crude oil by mixed strains and bacterial community analysis. Applied Biochemistry and Biotechnology 172, 3433–3447.

Cheng, H., Mulla, D.J., 1999. The soil environment. In: Adriano, D.C. (Ed.), Bioremediation of Contaminated Soils. ASA/CSSA/SSSA, Madison, pp. 1–13.

Chorom, M., Sharif, H.S., Mutamedi, H., 2010. Bioremediation of a crude oil-polluted soil by application of fertilizers. Iranian Journal of Environmental Health Science and Engineering 7, 319–326.

Christofi, N., Ivshina, I.B., 2002. Microbial surfactants and their use in field studies of soil remediation. Journal of Applied Microbiology 93, 915–929.

Collaard, J., Corbisiser, P., Diels, L., Dong, Q., Jeanthon, C., Mergeay, M., Taghavi, S., Der Lelie, V., Wilmotte, A., Wuertz, S., 1994. Plasmids for heavy metal resistance in *Alcaligenes eutrophus* CH34: mechanism and Applications. FEMS Microbiology Reviews 14 (4), 405–414.

Colombo, W.D., Barreda, A., Bilos, C., Cappelletti, N., Demichelis, S., Lombardi, P., Migoya, M.C., Skorupka, C., Su, R.G., 2005. Oil spill in the Rio de la Plata estuary, Argentina: 1. Biogeochemical assessment of waters, sediments, soils and biota. Environmental Pollution 134, 277–289.

Covino, S., D'Annibale, A., Stazi, S.R., Cajthaml, T., Čvančarová, M., Stella, T., Petruccioli, M., 2015. Assessment of degradation potential of aliphatic hydrocarbons by autochthonous filamentous fungi from a historically polluted clay soil. Science and Total Environment 505, 545–554.

Cybulski, Z., Dziurla, E., Kaczorek, E., Olszanowski, A., 2003. The influence of emulsifiers on hydrocarbon biodegradation by *Pseudomondacea* and *Bacillacea* strains. Spill Science and Technology Bulletin 8, 503–507.

Dadrasnia, A., Agamuthu, P., 2013. Potential biowastes to remediate diesel contaminated soils. Global NEST Journal 15 (4), 474–484.

Darmayati, Y., Sanusi, H.S., Prartono, T., Santosa, D.A., Nuchsi, R., 2015. The effect of biostimulation and biostimulation-bioaugmentation on biodegradation of oil-pollution on sandy beaches using mesocosms. International Journal of Marine Science 5 (27), 1–11.

Das, N., Chandran, P., 2011. Microbial degradation of petroleum hydrocarbon contaminants: an overview. Biotechnology Research International 2011:941810. http://dx.doi.org/10.4061/2011/941810.

Das, K., Mukherjee, A.K., 2007a. Crude petroleum-oil biodegradation efficiency of *Bacillus subtilis* and *Pseudomonas aeruginosa* strains isolated from a petroleum-oil contaminated soil from north-east India. Bioresource Technology 98, 1339–1345.

Das, K., Mukherjee, A.K., 2007b. Differential utilization of pyrene as the sole source of carbon by *Bacillus subtilis* and *Pseudomonas aeruginosa* strains: role of biosurfactants in enhancing bioavailability. Journal of Applied Microbiology 102 (1), 195–203.

Das, P., Mukherjee, S., Sen, R., 2008. Improved bioavailability and biodegradation of a model polyaromatic hydrocarbon by a biosurfactant producing bacterium of marine origin. Chemosphere 72, 1229–1234.

Dave, B.P., Ghevariya, C.M., Bhatt, J.K., Dudhagara, D.R., Rajpara, R.K., 2014. Enhanced biodegradation of total polycyclic aromatic hydrocarbons (TPAHs) by marine halotolerant *Achromobacter xylosoxidans* using Triton X-100 and β-cyclodextrin – a microcosm approach. Marine Pollution Bulletin 79, 123–129.

Davis, G.B., Patterson, B.M., Trefry, M.G., 2009. Technical Report No. 12, CRC for Contamination Assessment and Remediation of the Environment, Adelaide, Australia.

Daziel, E., Paquette, G., Vellemur, R., Lepins, F., Bisaillnon, J.G., 1996. Biosurfactant production by a soil Pseudomonas strain growing on PAH's. Applied and Environmental Microbiology 62, 1908–1912.

de Jonge, H., Freijer, J.I., Verstraten, J.M., Westerveld, J., Van Der Wielen, F.W.M., 1997. Relation between bioavailability and fuel oil hydrocarbon composition in contaminated soils. Environmental Science & Technology 31, 771–775.

Deighton, N., Goodman, B.A., 1995. The speciation of metals in biological systems. In: Ure, A.M., Davidson, C.M. (Eds.), Chemical Speciation in Environmental. Blackie Academic and Professional, London, pp. 307–334.

Desai, A.J., Banat, I.M., 1999. Microbial production of surfactants and their commercial potential. Microbiology and Molecular Biology Reviews 61, 47–60.

Dibble, J.T., Bartha, R., 1979. Effect of environmental parameters on the biodegradation of oil sludge. Applied and Environmental Microbiology 37 (4), 729–739.

Dragun, J., 1998. The Soil Chemistry of Hazardous Materials, second ed. Amherst Scientific Publishers, Amherst, MA.

Duarte, G.F., Rosado, A.S., De Araujo, L.S.W., Van Elsas, J.D., 2001. Analysis of bacterial community structure in sulfurous-oil-containing soils and detection of species carrying dibenzothiophene desulfurization (dsz) gene. Applied and Environmental Microbiology 67 (3), 1052–1062.

Edgehill, R.U., 1996. Effect of copper-chromatearsenate (CCA) components on PCPdegradation by *Arthrobacter* strain ATCC 33790. Bulletin of Environmental Contamination and Toxicology 57, 258–263.

El-Gendy, N.Sh., Abo-State, M.A., 2008. Isolation, characterization and evaluation of *Staphylococcus gallinarum* NK1 as a degrader for dibenzothiophene, phenanthrene and naphthalene. Egyptian Journal of Petroleum 17 (2), 75–91.

El-Gendy, N.Sh., Farah, J.Y., 2011. Kinetic modeling and error analysis for decontamination of different petroleum hydrocarbon components in biostimulation of oily soil microcosm. Soil and Sediment Contamination: An International Journal 20 (4), 432–446.

El-Gendy, N.Sh., Nassar, H.N., 2015. Kinetic modeling of the bioremediation of diesel oil polluted seawater using *Pseudomonas aeruginosa* NH1. Energy Sources, Part A: Recovery, Utilization, and Environmental Effects 37 (11), 1147–1163.

El-Gendy, N.Sh., Moustafa, Y.M., Barakat, M.A.K., Deriase, S.F., 2009. Evaluation of a bioslurry remediation of petroleum hydrocarbons contaminated sediments using chemical, mathematical and microscopic analysis. International Journal of Environmental Studies 66 (5), 563–579.

El-Gendy, N.Sh., Ali, H.R., El-Nady, M.M., Deriase, S.F., Moustafa, Y.M., Mohamed, I., Roushdy, M.I., 2014. Effect of different bioremediation techniques on petroleum biomarkers and asphaltene fraction in oil polluted sea water. Desalination and Water Treatment 52 (40/42), 7484–7494.

El-Mahdi, A.M., Abdul Aziz, H., Abu Amr, S.S., El-Gendy, N.S., Nassar, H.N., 2016. Isolation and characterization of *Pseudomonas* sp. NAF1 and its application in biodegradation of crude oil. Environmental Earth Sciences 75 (5), 1–11. https://doi.org/10.1007/s12665-016-5296-z. 380.

El-Masry, M.H., El-Bestawy, E., El-Adl, N.I.W., 2004. Bioremediation of oil from polluted wastewater using sand biofilm system. Polymer-Plastics Technology and Engineering 43 (6), 1617–1641.

El-Tayeb, O.M., Megahed, S.A., El-Azizi, M., 1998. Microbial degradation of aromatic substances by local bacterial isolates III- Factors affecting degradation of 2,4-dichlorophenoxyacetic acid by *Pseudomonas stutzeri*. Egyptian Journal of Biotechnology 4, 84–90.

El-Tokhi, M.M., Moustafa, Y.M., 2001. Heavy metals and petroleum hydrocarbons contamination of bottom sediment of El Sukhna area, Gulf of Suez, Egypt. Petroleum Science and Technology 19 (5/6), 481–494.

Enock, J., 2002. Intrinsic Biodegradation Potential of Crude Oil in Salt Marshes (M.Sc. thesis) Graduate Faculty of the Louisiana State University and Agricultural and Mechanical College, The Department of Civil and Environmental Engineering. http://digitalcommons.lsu.edu/gradschool_theses/2326.

Erdogan, E., Karaca, A., 2011. Bioremediation of crude oil polluted soils. Asian Journal of Biotechnology 3 (3), 206–213.

Eriksson, M., Dalhammar, G., Mohn, W.W., 2002. Bacterial growth and biofilm production on pyrene. FEMS Microbiology Ecology 40, 21–27.

European Environment Agency, August 2007. Progress in Management of Contaminated Sites (CSI 015) – Assessment Published. http://www.eea.europa.eu/data-and-maps/indicators/progress-in-managementof-contaminated-sites/progress-in-management-of-contaminated-1.

Evans, L.J., 1989. Chemistry of metal retention by soils. Environmental Science & Technology 23, 1046–1056.

Fan, M.Y., Xie, R.J., Qin, G., 2014. Bioremediation of petroleum-contaminated soil by a combined system of biostimulation-bioaugmentation with yeast. Environmental Technology 35 (1–4), 391–399.

Farahat, L.A., El-Gendy, N.Sh., 2007. Comparative kinetic study of different bioremediation processes for soil contaminated with petroleum hydrocarbons. Material Science Research India 4 (2), 269–278.

Farahat, L.A., El-Gendy, N.Sh., 2008. Biodegradation of Baleym Mix crude oil in soil microcosm by some locally isolated Egyptian bacterial strains. Soil and Sediment Contamination 17 (2), 150–162.

Farell, J., Reinhard, M., 1994. Desorption of halogenated organics from model solids,sediments, and soil under unsaturated conditions. 2. Kinetics. Environmental Science & Technology 28, 63–72.

Ferguson, S.H., Franzmann, P.D., Revill, A.T., Snape, I., Rayner, J.L., 2003. The effects of nitrogen and water on mineralisation of hydrocarbons in diesel-contaminated terrestrial Antarctic soils. Cold Regions Science and Technology 37, 197–212.

Ferradji, F.Z., Mnifc, S., Badis, A., Rebbani, S., Fodil, D., Eddouaouda, K., Sayadi, S., 2014. Naphthalene and crude oil degradation by biosurfactant producing *Streptomyces* spp. isolated from Mitidja plain soil (North of Algeria). International Biodeterioration and Biodegradation 86, 300–308.

Finegold, L., 1996. Molecular and biophysical aspects of adaptation of life to temperatures below the freezing point. Advances in Space Research 18 (12), 87–95.

Flood, B.R., 2009. Biopile Treatment of Hydrocarbon Contaminated Soil of the Red Water Oil Production Area (M.Sc. thesis). Royal Roads University.

Frey, B., Stemmer, M., Widmer, F., Luster, J., Sperisen, C., 2006. Microbial activity and community structure of soil after heavy metal contamination in forest ecosystem. Soil Biology and Biochemistry 38, 1745–1756.

Friis, N., Myers-Keith, P., 1986. Biosorption of uranium and lead by *Streptomyces longwoodensis*. Biotechnology and Bioengineering 28, 21–28.

Gao, Y.Z., Zhu, L.Z., Ling, W.T., 2005. Application of the partition limited model for plant uptake of organic chemicals from soil and water. Science and Total Environment 36, 171–182.

Gao, Y.Z., Xiong, W., Ling, W.T., Xu, J., 2006. Sorption of phenanthrene by soils contaminated with heavy metals. Chemosphere 65, 1355–1361.

García-Rivero, M.G., Saucedo-Castañeda, S., de Hoyos, F., Gutiérrez-Rojas, M., 2002. Mass transfer and hydrocarbon biodegradation of aged soil in slurry phase. Biotechnology Progress 18, 728–733.

Garon, D., Sage, L., Wouessidjewe, D., Seigle-Murandi, F., 2004. Enhanced degradation of fluorene in soil slurry by *Absidia cylindrospora* and maltosyl-cyclodextrin. Chemosphere 56, 159–166.

Geller, J.T., Holman, H.Y., Su, G., Conrad, M.E., Pruess, K., Hunter-Cevara, J.C., 2000. Flow dynamics and potentials for biodegradation of organic contaminant in fractured rock vadose zones. Journal of Contaminant Hydrology 43, 63–90.

Genney, D.R., Alexander, I.J., Killham, K., Meharg, A.A., 2004. Degradation of the polycyclic aromatic hydrocarbon (PAH) fluorene is retarded in a Scots pine ectomycorrhizosphere. New Phytologist 163, 641–649.

Georgiou, G., Lin, S., Sharma, M., 1992. Surface-active compounds from microorganisms. Biotechnology 10, 60–66.

Gerber, M.A., Freeman, H.D., Baker, E.G., Riemath, W.F., 1991. Soil Washing: A Preliminary Assessment of its Applicability to Hanford. Prepared for U.S. Department of Energy by Battelle Pacific Northwest Laboratory. Richland, Washington. Report No. PNL-7787; UC 902.

Gesinde, A.F., Agbo, E.B., Agho, M.O., Dike, E.F.C., 2008. Bioremediation of some Nigerian and Arabian crude oils by fungal isolates. International Journal of Pure and Applied Sciences 2 (3), 37–44.

Ghazali, F.M., Abdul Rahman, R., Bakar, S.A., Basri, M., 2004. Biodegradation of hydrocarbons in soil by microbial consortium. International Biodeterioration and Biodegradation 54, 61–67.

Gibb, A.A., Chu, R.C., Wong, K., Goodman, R.H., 2001. Bioremediation kinetics of crude oil at 5°C. Journal of Environmental Engineering 127 (9), 818–824.

Gillan, D.C., 2016. Metal resistance systems in cultivated bacteria: are they found in complex communities? Current Opinion in Biotechnology 38, 123–130.

Godleads, O.A., Prekeyi, T.F., Samson, E.O., Igelenyah, E., 2015. Bioremediation, biostimulation and bioaugmentation: a review. International Journal of Environmental Bioremediation and Biodegradation 3 (1), 28–39.

Gogoi, B.K., Dutta, N.N., Goswami, P., Krishna, T.R., 2003. A case study of bioremediation of petroleum-hydrocarbon contaminated soil at a crude oil spill site. Advances in Environmental Research 7, 767–782.

Golgolev, A., Wilke, B.M., 1997. Combination effects of heavy metals and fluoranthene on soil bacteria. Biology and Fertility of Soils 25, 274–278.

Guerin, T.F., 1999. Bioremediation of phenols and polycyclic aromatic hydrocarbons in creosote contaminated soil using ex-situ land treatment. Journal of Hazardous Materials B65, 305–315.

Guerra-Santos, L., Käppeli, O., Fiechter, A., 1984. *Pseudomonas aeruginosa* biosurfactant production in continuous culture with glucose as carbon source. Applied and Environmental Microbiology 48 (2), 301–305.

Guo, C.L., Zhou, H.W., Wong, Y.S., Tam, N.F.Y., 2005. Isolation of PAH-degrading from mangrove sediments and their biodegradation potential. Marine Pollution Bulletin 51, 1054–1061.

Hamzah, A., Phan, C.-W., Yong, P.-H., Mohd Ridzuan, N.H., 2014. Oil palm empty fruit bunch and sugarcane bagasse enhance the bioremediation of soil artificially polluted by crude oil. Soil and Sediment Contamination: An International Journal 23, 751–762.

Han, R., Wang, Y., Zou, W., Wang, Y., Shi, J., 2007. Comparison of linear and nonlinear analysis in estimating the Thomas model parameters for methylene blue adsorption onto natural zeolite in fixed-bed column. Journal of Hazardous Materials 145, 331–335.

Haritash, A.K., Kaushik, C.P., 2009. Biodegradation aspects of polycyclic aromatic hydrocarbons (PAHs): a review. Journal of Hazardous Materials 169 (1/3), 1–15.

Harms, H., Bosma, T.N.P., 1997. Mass transfer limitation of microbial growth and pollutant degradation. Journal of Industrial Microbiology 18, 97–105.

Hatzinger, P.B., Alexander, M., 1995. Effect of aging of chemical in soil on their biodegradability and extractability. Environmental Science & Technology 29, 537–545.

He, Y., Yediler, A., Sun, T., Kettrup, A., 1995. Adsorption of fluoranthene on soil and lava: effects of the organic carbon contents of adsorbents and temperature. Chemosphere 30, 141–150.

Head, I.M., Swannel, R.P.J., 1999. Bioremediation of petroleum hydrocarbon contaminants in marine habitats. Current Opinion in Biotechnology 10, 234–239.

Head, I.M., Jones, D.M., Steve, R.L., 2003. Biological activity in the deep subsurface and the origin of heavy oil. Nature 426, 344–352.

Hegazi, R.M., El-Gendy, N.Sh., El-Feky, A.A., Moustafa, Y.M., El- Ezbewy, S., El-Gemaee, G.H., 2007. Impact of heavy metals on biodegradation of phenanthrene by *Cellulomonas hominis* strain N2. Journal of Pure and Applied Microbiology 1 (2), 165–175.

Heinonsalo, J., Jorgensen, K.S., Haahtela, K., Sen, R., 2000. Effects of *Pinus sylvestris* root growth and mycorrhizosphere development on bacterial carbon source utilization and hydrocarbon oxidation in forest and petroleum-contaminated soils. Canadian Journal of Microbiology 46, 451–464.

Heipieper, H.J., Cornelissen, S., Pepi, M., 2010. Surface properties and cellular energetics of bacteria in response to the presence of hydrocarbons. In: Timmis, K.N. (Ed.), The Handbook of Hydrocarbons and Lipid Microbiology. Springer- Verlag, Berlin Heidelberg, pp. 1615–1624.

Herman, D.C., Zhang, Y.M., Miller, R.M., 1997. Rhamnolipids (biosurfactant) effects on cell aggregation and biodegradation of residual hexadecane undersaturated flow conditions. Applied and Environmental Microbiology 63, 622–3627.

Hitoshi, I., Reia, H., Masaaki, M., Hidetoshi, O., 2008. A turbine oil-degrading bacterial consortium from soils of oil fields and its characteristics. International Biodeterioration and Biodegradation 61 (3), 223–232.

Hommel, R.K., 1990. Formation and physiological role of biosurfactants produced by hydrocarbon-utilizing microorganisms. Biosurfactants in Hydrocarbon Utilization. Biodegradation 1, 107–119.

Huesemann, M.H., 1994. Guideline for land-treating petroleum hydrocarbon contaminated soil. Journal of Soil Contamination 3 (3), 299–318.

Hurst, C.J., Knudsen, G.R., Mclnerney, M.V., Stetzenbach, L.D., Wailer, M.V., 1997. Manual of Environmental Microbiology. American Society for Microbiology Press, Washington, DC.

Hwang, S., Cutright, T.J., 2002. Biodegradability of aged pyrene and phenanthrene in a natural soil. Chemosphere 47, 891–899.

Hwang, H.M., Foster, G.D., 2006. Characterization of polycyclic aromatic hydrocarbons in urban storm water runoff flowing into the tidal Anacostia River, Washington, DC, USA. Environmental Pollution 140, 416–426.

Hwang, E.-Y., Namkoong, W., Park, J.-S., 2001. Recycling of remediated soil for effective composting of diesel-contaminated soil. Computer Science and Utilization 9 (2), 143–148.

Hyun, S., Ahn, M.-Y., Zimmerman, A.R., Kim, M., Kim, J.-G., 2008. Implication of hydraulic properties of bioremediated diesel-contaminated soil. Chemosphere 71 (9), 1646–1653.

Idise, O.E., Ameh, J.B., Yakubu, S.E., Okuofu, C.A., 2010. Modification of *Bacillus cereus* and *Pseudomonas aeruginosa* isolated from a petroleum refining effluent for increased petroleum product degradation. African Journal of Biotechnology 9 (22), 3303–3307.

Ingvorsen, K., Nielsen, M.Y., Joulian, C., 2003. Kinetics of bacterial sulfate reduction in an activated sludge plant. FEMS Microbiology Ecology 46 (2), 129–137.

Irvine, D.A., Frost, H.L., 2003. Bioremediation of Soils Contaminated with Industrial Wastes: A Report on the State-of-the-art in Bioremediation. SBR Technologies Inc.

Jabro, J.D., Sainju, U., Stevens, W.B., Evans, R.G., 2008. Carbon dioxide flux as affected by tillage and irrigation in soil converted from perennial forages to annual crops. Journal of Environmental Management 88, 1478–1484.

Jackson, M.L., 1969. Soil Chemical Analysis-advanced Course; A Manual of Methods Useful for Instruction and Research in Soil Chemistry, Physical Chemistry of Soils, Soil Fertility, and Soil Genesis. Department of Soils, University of Wisconsin, USA.

Jacques, R.J.S., Santos, E.C., Bento, F.M., Peralba, M.C.R., Selbach, P.A., Sa, E.L.S., Camargo, F.A.O., 2005. Anthracene biodegradation by *Pseudomonas* sp. isolated from a petrochemical sludge land farming site. International Biodeterioration and Biodegradation 56, 143–150.

Jain, D.K., Thompson, D.L.C., Lee, H., Trevors, J.T., 1991. A drop- collapsing test for screening surfactant producing microorganisms. Journal of Microbiology Methods 13, 271–279.

Jain, P.K., Gupta, V.K., Gaur, R.K., Lowry, M., Jaroli, D.P., Chauhan, U.K., 2011. Bioremediation of petroleum oil contaminated soil and water. Research Journal of Environmental Toxicology 5 (1), 1–26.

Janbandhu, A., Fulekar, M.H., 2011. Biodegradation of phenanthrene using adapted microbial consortium isolated from petrochemical contaminated environment. Journal of Hazardous Materials 187 (1–3), 333–340.

Janek, T., Lukaszewicz, M., Rezanka, T., Krasowska, A., 2010. Isolation and characterization of two new lipopeptide biosurfactants produced by *Pseudomonas fluorescens* BD5 isolated from water from the Arctic Archipelago of Svalbard. Bioresource Technology 101, 6118–6123.

Johnson, C.R., Scow, K.M., 1999. Effect of nitrogen and phosphorus addition on phenanthrene biodegradation in four soils. Biodegradation 10 (1), 43–50.

JRB Associates, Inc., 1984. Summary Report: Remedial Response at Hazardous Waste Sites. Prepared for Municipal Environmental Research Laboratory, Cincinnati, OH. PB 85–124899.

Juhaz, A., Stanley, G.A., Britz, M.L., 2000. Degradation of high molecular weight PAHs in contaminated soil a bacterial consortium: effects on microtox and mutagenicity bioassays. Bioremediation 4, 271–283.

Kaczorek, E., Chrzanowski, L., Pijanowska, A., Olszanows, A., 2008. Yeast and bacteria cell hydrophobicity and hydrocarbon biodegradation in the presence of natural surfactants: Rhamnolipides and saponins. Bioresource Technology 99, 4285–4291.

Kanaly, R.A., Harayama, S., 2000. Biodegradation of high molecular weight PAHs by Bacteria. Journal of Bacteriology 182, 2059–2067.

Kang, S.W., Kim, Y.B., Shin, J.D., Kim, E.K., 2010. Enhanced biodegradation of hydrocarbons in soil by microbial biosurfactant, sophorolipid. Applied Biochemistry and Biotechnology 160, 780–790.

Kaplan, C.W., Kitts, C.L., 2004. Bacterial succession in a petroleum land treatment unit. Applied and Environmental Microbiology 70, 1777–1786.

Kaplan, C.W., Clement, B.G., Hamrick, A., Pease, R.W., Flint, C., Cano, R.J., Kitts, C.L., 2003. Complex cosubstrate addition increases initial petroleum degradation rates during land treatment by altering bacterial community physiology. Remediation Journal 13 (4), 61–78.

Kapoor, A., Yang, R.T., 1989. Correlation of equilibrium adsorption data of condensable vapors on porous adsorbents. Gas Separation and Purification 3, 187–192.

Kargi, F., Dincer, A.R., 1996. Effect of salt concentration on biological treatment of saline wastewater by fed-batch operation. Enzyme and Microbial Technology 19 (7), 529–537.

Karl, D.M., 1992. The grounding of the Bahia Paraiso: microbial ecology of the 1989 Antarctic oil spill. Microbial Ecology 24, 77–89.

Kästner, M., Breuer-Jammali, M., Mahro, B., 1998. Impact of inoculation protocols, salinity, and pH on the degradation of polycyclic aromatic hydrocarbons (PAHs) and survival of PAH-degrading bacteria introduced into soil. Applied and Environmental Microbiology 64, 359–362.

Kavitha, V., Mandal, A.B., Gnanaman, A., 2015. Microbial biosurfactant mediated removal and/or solubilization of crude oil contamination from soil and aqueous phase: an approach with *Bacillus licheniformis* MTCC 5514. International Biodeterioration and Biodegradation 94, 24–30.

Kazunga, C., Aitken, M.D., 2000. Products from the incomplete metabolism of pyrene by PAH degrading bacteria. Applied and Environmental Microbiology 66, 1917–1922.

Ke, L., Bao, W., Chen, L., Wong, Y.S., Tam, N.F.Y., 2009. Effects of humic acid on solubility and biodegradation of polycyclic aromatic hydrocarbons in liquid media and mangrove sediment slurries. Chemosphere 76, 1102–1108.

Khan, A.A., Kim, S.J., Paine, D.D., Cerniglia, C.E., 2002. Classification of a polycyclic aromatic hydrocarbon-metabolising bacterium, *Mycobacterium* sp. Nov. International Journal of Systematic and Evolutionary Microbiology 52, 1997–2002.

Khopade, A., Biao, R., Liu, X., Mahadik, K., Zhang, L., Kokare, C., 2012a. Production and stability studies of the biosurfactant isolated from marine *Nocardiopsis* sp. B4. Desalination 285, 198–204.

Khopade, A., Biao, R., Liu, X., Mahadik, K., Zhang, L., Kokare, C., 2012b. Production and characterization of biosurfactant from marine *Streptomyces* species B3. Journal of Colloid and Interface Science 367, 311–318.

King, A.J., Readman, J.W., Zhou, J.L., 2004. Dynamic behavior of polycyclic aromatic hydrocarbons in Brighton marine. U.K. Marine Pollution Bulletin 48 (3/4), 229–239.

Kumar, M., Leon, V., Materano, A.D.S., Ilzins, O.A., 2006. Enhancement of oil degradation by co-culture of hydrocarbon degrading and biosurfactant producing bacteria. Polish Journal of Microbiology 55 (2), 139–146.

Labud, V., Garcia, C., Hernandez, T., 2007. Effect of hydrocarbon pollution on the microbial properties of a sandy and a clay soil. Chemosphere 66 (10), 1863–1871.

Lal, R., Pandey, G., Sharma, P., Kumari, K., Malhotra, S., Pandey, R., Raina, V., Kohler, H.P.E., Holliger, C., Jackson, C., Oakeshott, J.G., 2010. Biochemistry of microbial degradation of hexachlorocyclohexane (HCH) and prospects for bioremediation. Microbiology and Molecular Biology Reviews 74 (1), 58–80.

Lang, S., 2002. Biological amphiphiles (microbial biosurfactants). Current Opinion in Colloid and Interface Science 7, 12–20.

Leahy, J.G., Colwell, R.R., 1990. Microbial degradation of hydrocarbons in the environment. FEMS Microbiology Reviews 54 (3), 305–315.

Leirós, M.C., Trasar-Cepeda, C., Seoane, S., Gil-Sotres, F., 1999. Dependence of mineralization of soil organic matter on temperature and moisture. Soil Biology and Biochemistry 31, 327–335.

Lepo, J.E., Cripe, C.R., 1999. Biodegradation of polycyclic aromatic hydrocarbons (PAH) from crude oil in sandy-beach microcosms. In: Bell, C.R., Brylinsky, M., Johnson-Green, P. (Eds.), Proceedings of the 8th International Symposium on Microbiology Ecology. Atlantic Canada Society for Microbial Ecology, Halifax, Canada.

Li, J.-L., Chen, B.-H., 2009. Surfactant-mediated biodegradation of polycyclic aromatic hydrocarbons. Nature Materials 2, 76–94.

Li, Y., He, X., Cao, X., Zhao, G., Tian, X., Cui, X., 2007. Molecular behavior and synergistic effects between sodium dodecylbenzene sulfonate and Triton X-100 at oil/water interface. Journal of Colloid and Interface Science 307 (1), 215–220.

Li, J., Zhang, L., Wu, Y., Liu, Y., Zhou, P., Wen, S., 2009. A national survey of polychlorinated dioxins, furans (PCDD/Fs) and dioxin-like polychlorinated biphenyls (dl-PCBs) in human milk in China. Chemosphere 75, 1236–1242.

Lim, M.W., Lau, E.V., Poh, P.E., 2016. A comprehensive guide of remediation technologies for oil contaminated soil – present works and future directions. Marine Pollution Bulletin 109, 14–45.

Liu, W.X., Luo, Y.M., Teng, Y., Li, Z.G., Ma, L.Q., 2010. Bioremediation of oily sludge-contaminated soil by stimulating indigenous microbes. Environmental Geochemistry and Health 32, 23–29.

Liu, P., Tsung, C.C., Liang-Ming, W., Chun-Hsuan, K., Po-Tseng, P., Sheng-Shung, C., 2011. Bioremediation of petroleum hydrocarbon contaminated soil: effects of strategies and microbial community shift. International Biodeterioration and Biodegradation 65, 1119–1127.

Liu, B., Liu, J., Ju, M., Li, X., Yu, Q., 2016. Purification and characterization of biosurfactant produced by *Bacillus licheniformis* Y-1 and its application in remediation of petroleum contaminated soil. Marine Pollution Bulletin 107, 46–51.

Lofts, S., Spurgeon, D.J., Svedesn, C., Tipping, E., 2004. Deriving soil critical limits for Cu, Zn, Cd, and pH: a method based on free ion concentrations. Environmental Science & Technology 38, 3623–3631.

Lors, C., Mossmann, R.J., 2005. Characteristics of PAHs intrinsic degradation in two coke factory soils. Polycyclic Aromatic Compounds 25, 67–85.

Lu, X.-Y., Li, B., Zhang, T., Fang, H.H.P., 2012. Enhanced anoxic bioremediation of PAHs-contaminated sediment. Bioresource Technology 104, 51–58.

Luers, F., ten Hulscher, E.M.T., 1996. Temperature effect of the partitioning of polycyclic aromatic hydrocarbons between natural organic matter and water. Chemosphere 33, 643–657.

Lundstedt, S., 2003. Analysis of PAHs and Their Transformation Products in Contaminated Soil and Remedial Processes. Umeå University SE-90187, Umeå, Sweden.

Macaulay, B.M., Rees, D., 2014. Bioremediation of oil spills: a review of challenges for research advancement. Annals of Environmental Science 8, 9–37.

Mackay, D., Shiu, W.Y., Ma, K.C., 1992. Illustrated Handbook of Physical Chemical Properties and Environmental Fate for Organic Chemicals. Lewis Publishers, Chelsea, MI.

MacNaughton, S.J., Stephen, J.R., Venosa, A.D., Davis, G.A., Chang, Y., White, D.C., 1999. Microbial population changes during bioremediation of an Experimental oil spill. Applied and Environmental Microbiology 65 (8), 3566–3574.

Mahjoubi, M., Jaouani, A., Guesmi, A., Ben Amor, S., Jouini, A., Cherif, H., Najjari, A., Boudabous, A., Koubaa, N., Cherif, A., 2013. Hydrocarbonoclastic bacteria isolated from petroleum contaminated sites in Tunisia: isolation, identification and characterization of the biotechnological potential. New Biotechnology 30 (6), 723–733.

Maier, R.M., 2003. Biosurfactant: evolution and diversity in bacteria. Advances in Applied Microbiology 52, 101–121.

Makut, M.D., Ishaya, P., 2010. Bacterial species associated with soils contaminated with used petroleum products in Keffi town, Nigeria. African Journal of Microbiology Research 4 (16), 1698–1702.

Maliszewska-Kordybach, B., 1996. Polycyclic aromatic hydrocarbons in agricultural soils in Poland: preliminary proposals for criteria to evaluate the level of soil contamination. Applied Geochemistry 11, 121–127.

Mall, I.D., Srivastava, V.C., Agarwal, N.K., Mishra, I.M., 2005. Adsorptive removal of malachite green dye from aqueous solution by bagasse fly ash and activated carbon-kinetic study and equilibrium isotherm analyses. Colloids and Surfaces A: Physicochemical and Engineering 264, 17–28.

Margesin, R., Schimer, F., 1999. Biological decontamination of oil spills in cold environments. Journal of Chemical Technology and Biotechnology 74, 381–389.

Margesin, R., Schinner, F., 1997. Laboratory bioremediation experiments with soil from a diesel-oil contaminated site-significant role of cold-adapted microorganisms and fertilizers. Journal of Chemical Technology and Biotechnology 70, 92–98.

Margesin, R., Schinner, F., 2001a. Bioremediation (natural attenuation and biostimulation) of diesel oil-contaminated soil in an Alpine Glacier skiing area. Applied and Environmental Microbiology 67, 3127–3133.

Margesin, R., Schinner, F., 2001b. Bacterial heavy metal-tolerance: extreme resistance to nickel in *Arthrobacter* strains. Journal of Basic Microbiology 36 (4), 269–282.

Margesin, A., Zimmerbauer, A., Schinner, F., 1999. Monitoring of bioremediation by soil biological activities. Chemosphere 40, 339–346.

Margesin, R., Labbé, D., Schinner, F., Greer, C.W., Whyte, L.G., 2003. Characterization of hydrocarbon degrading microbial population in contaminated and pristine alpine soil. Applied and Environmental Microbiology 69 (6), 3085–3092.

Marquardt, D.W., 1963. An algorithm for least-squares estimation of nonlinear parameters. Journal of the Society for Industrial and Applied Mathematics 11 (2), 431–441.

Martínez Álvarez, L.M., Lo Balbo, A., Mac Cormack, W.P., Ruberto, L.A.M., 2015. Bioremediation of a petroleum hydrocarbon-contaminated Antarctic soil: optimization of a biostimulation strategy using response-surface methodology (RSM). Cold Regions Science and Technology 119, 61–67.

Marwood, M.T., Knoke, K., Yau, K., Lee, H., Trevors, J.T., Suchorski, T., Flemming, C.A., Hodge, V., Liu, D.L., Seech, A.G., 1998. Comparison of toxicity detected by five bioassays during bioremediation of diesel fuel-spiked soils. Environmental Toxicology and Water Quality 13, 117–129.

McClellan, S.L., Warshawsky, D., Shann, J.R., 2002. The effect of polycyclic aromatic hydrocarbons on the degradation of benzo[a]pyrene by *Mycobacterium* sp. strain RJGII-135. Environmental Toxicology and Chemistry 21, 253–259.

McFarland, M.J., Sims, R.C., 1991. Thermodynamic framework for evaluating PAH degradation in the subsurface. Groundwater 29 (6), 885–896.

Meador, J.P., 2003. Bioaccumulation of PAHs in marine invertebrates. In: Douben, P.E. (Ed.), PAHs: An Ecotoxicological Perspective. John Wiley & Sons, pp. 147–171.

Meckenstock, R.U., Safinowski, M., Griebler, C., 2004. Anaerobic degradation of polycyclic aromatic hydrocarbons. FEMS Microbiology Ecology 49, 27–36.

Metwally, M., Al-Mu Zaini, S., Jacob, P.G., Bahloul, M., 1997. Petroleum hydrocarbons and related heavy metals in the near-shore marine sediments of Kuwait. Environment International 23 (1), 115–121.

Milles, M.A., Bonner, J.S., Page, C.A., Autenrieth, R.L., 2004. Evaluation of bioremediation strategies of a controlled oil release in a wetland. Marine Pollution Bulletin 49, 425–435.

Miralles, G., Grossi, V., Acquaviva, M., Duran, R., Bertrand, J.C., Cuny, P., 2007. Alkane biodegradation and dynamics of phylogenetic subgroups of sulfate-reducing bacteria in an anoxic coastal marine sediment artificially contaminated with oil. Chemosphere 68 (7), 1327–1334.

Mishra, S., Jyot, J., Kuhad, R.C., Lab, B., 2001. Evaluation of inoculum addition to stimulate in situ bioremediation of oily-sludge-contaminated soil. Applied and Environmental Microbiology 67, 1675–1681.

Mishra, S., Sarma, M.P., Lal, B., 2004. Crude oil degradation efficiency of a recombinant *Acinitobacter baumannii* strain and its survival in crude oil-contaminated soil microcosm. FEMS Microbiology Letters 235, 323–331.

Miyata, N., Iwahori, K., Foght, J.M., Gray, M.R., 2004. Saturable, energy dependent uptake of phenanthrene in aqueous phase by *Mycobacterium* sp. Strain RJGII-135. Applied and Environmental Microbiology 70, 363–369.

Mohammed, D., Ramsubhag, A., Beckles, D.M., 2007. An assessment of the biodegradation of petroleum hydrocarbons in contaminated soil using non-indigenous, commercial microbes. Water, Air, and Soil Pollution 182 (1/4), 349–356.

Mohd Radzi, N.-A.-S., Tay, K.-S., Abu Bakar, N.-K., Emenike, C.U., Krishnan, S., Hamid, F.S., Abas, M.-R., 2015. Degradation of polycyclic aromatic hydrocarbons (pyrene and fluoranthene) by bacterial consortium isolated from contaminated road side soil and soil termite fungal comb. Environmental Earth Sciences 74 (6), 5383–5391.

Moran, A.C., Olivera, N., Commendatore, M., Esteves, J.L., Sineriz, F., 2000. Enhancement of hydrocarbon waste biodegradation by addition of biosurfactant from *Bacillus subtilis* O9. Biodegradation 11, 65–71.

Morgan, P., Watkinson, R.J., 1989. Hydrocarbon degradation in soils and methods for soil biotreatment. Critical Reviews in Biotechnology 8 (4), 305–333.

Moscoso, F., Teijiz, I., Deive, F.J., Sanromán, M.A., 2012a. Efficient PAHs biodegradation by a bacterial consortium at flask and bioreactor scale. Bioresource Technology 119, 270–276.

Moscoso, F., Deive, F.J., Longo, M.A., Sanromán, M.A., 2012b. Technoeconomic assessment of phenanthrene degradation by *Pseudomonas stutzeri* CECT 930 in a batch bioreactor. Bioresource Technology 104, 81–89.

Moustafa, Y.M., 2004. Contamination by polycyclic aromatic hydrocarbons in some Egyptian Mediterranean coasts. Biosciences. Biotechnology Research Asia 2 (1), 15–24.

Moustafa, Y.M., El-Bastawissy, A.M., Zakaria, A.I., Sidky, N.M., El-Sheshtawy, H.S., 2003. Chemical and microbial monitoring of chronically hydrocarbons polluted fresh water in certain locality in Egypt. Egyptian Journal of Applied Science 18 (2B), 542–558.

Mrayyan, B., Battikhi, M.N., 2005. Biodegradation of total organic carbons (TOC) in Jordanian petroleum sludge. Journal of Hazardous Materials 120, 127–134.

Mrozik, A., Piotrowska-Seget, Z., Labużek, 2003. Bacterial degradation and bioremediation of polycyclic aromatic hydrocarbons. Polish Journal of Environmental Studies 12 (1), 15–25.

Mukherjie, A.K., Das, K., 2005. Correlation between diverse cyclic lipopeptides production and regulation of growth and substrate utilization by *Bacillus subtilis* strains in a particular habitat. FEMS Microbiology Ecology 54, 479–489.

Namkong, W., Hwang, E., Park, J., Choi, J., 2002. Bioremediation of diesel-contaminated soil with composting. Environmental Pollution 119, 23–31.

Neff, J.M., 2002. Bioaccumulation in Marine Organisms: Effect of Contaminants From Oil Well Produced Water. Elsevier Press, Amsterdam, Netherlands.

Neu, T.R., 1996. Significance of bacterial surface-active compounds in interaction of bacteria with interfaces. Microbiological Reviews 60, 151–166.

Ngugi, D.K., Tsanuo, M.K., Boga, H.I., 2005. *Rhodococcus opacus* strain RW, a resorcinol-degrading bacterium from the gut of *Macrotermes michaelseni*. African Journal of Biotechnology 4, 639–645.

Nocenteni, M., Pinelli, D., Fava, F., 2000. Bioremediation of a soil contaminated by hydrocarbon mixtures: the residual concentration problem. Chemosphere 41, 1115–1123.

Norman, R.S., Moeller, P., McDonald, T.J., Morris, P.J., 2004. Effect of pyocyanin on a crude oil degrading microbial community. Applied and Environmental Microbiology 70 (7), 4004–4011.

Odokuma, L.O., Dickson, A.A., 2003. Bioremediation of a crude oil polluted tropical rain forest soil. Global Journal of Environmental Science 2 (1), 29–40.

Olaniran, A.O., Pillay, D., Pillay, B., 2006. Biostimulation and bioaugmentation enhances aerobic biodegradation of dichloroethenes. Chemosphere 63, 600–608.

Oliveira, A., Pampulha, M.E., 2006. Effect of long term heavy metals contamination on soil microbial characteristics. Journal of Bioscience and Bioengineering 102 (3), 157–161.

Oliveira, F.de, Franc, D.W., Félix, A.K., Martins, J.J., Giro, M.E., Melo, V.M.M., Gonçalves, L.R., 2013. Kinetic study of biosurfactant production by *Bacillus subtilis* LAMI005 grown in clarified cashew apple juice. Colloids and Surfaces B: Biointerfaces 101, 34–43.

Ostroumov, V.E., Siegert, C., 1996. Exobiological aspects of mass transfer in microzones of permafrost deposits. Advances in Space Research 18 (12), 79–86.

Pasumarthi, R., Chandrasekaran, S., Mutnuri, S., 2013. Biodegradation of crude oil by *Pseudomonas aeruginosa* and *Escherichia fergusonii* isolated from the Goan coast. Marine Pollution Bulletin 76, 276–282.

Peng, R.H., Xiong, A.-S., Xue, Y., Fu, X.-Y., Gao, F., Zhao, W., Tian, Y.-S., Yao, Q.-H., 2008. Microbial biodegradation of polyaromatic hydrocarbons. FEMS Microbiology Reviews 32, 927–955.

Perez-Marin, A.B., Meseguer Zapata, V., Ortuno, J.F., Agilar, M., Saez, J., Lorens, M., 2007. Removal of cadmium from aqueous solutions by adsorption onto orange waste. Journal of Hazardous Materials 139 B, 122–131.

Person, A., Molin, G., 1987. Capacity for biosurfactant production of environmental *Pseudomonas* and *Vibrionaceae* growing on carbohydrates. Applied Microbiology and Biotechnology 26, 439–442.

Pietri, J.C.A., Brookes, P.C., 2008. Relationships between soil pH and microbial properties in a UK arable soil. Soil Biology and Biochemistry 40, 1856–1861.

Pineda-Flores, G., Mesta-Howard, A.M., 2001. Petroleum asphaltenes: generated problematic and possible biodegradation mechanisms. Revista latinoamericana de microbiología 43, 143–150.

Pingintha, N., Leclerc, M.Y., Beasley Jr., J.P., Zhang, G., Senthong, C., 2010. Assessment of the soil CO_2 gradient method for soil CO_2 efflux measurements: comparison of six models in the calculation of the relative gas diffusion coefficient. Tellus 62 (1), 47–58.

Pizzul, L., 2006. Degradation of PAHs by Actinomycetes (Ph.D. thesis). Faculty of Natural Resources and Agricultural Sciences Department of Microbiology, Swedish University of Agricultural Sciences, Uppsala.

Poli, A., Romano, I., Caliendo, G., Nicolaus, G., Orlando, P., de Falco, A., Lama, L., Gambacorta, A., Nicolaus, B., 2006. *Geobacillus toebii* subsp.nov., a hydrocarbon-degrading, heavy metals resistant bacterium from hot compost. The Journal of General and Applied Microbiology 52 (4), 223–234.

Pontes, J., Mucha, A.P., Santos, H., Reis, I., Bordalo, A., Basto, M.C., Bernabeu, A., Almeida, C.M.R., 2013. Potential of bioremediation for buried oil removal in beaches after an oil spill. Marine Pollution Bulletin 76, 258–265.

Porter, J.F., McKay, G., Choy, K.H., 1999. The prediction of adsorption from binary mixture of acidic dyes using single- and mixed-isotherm variants of the ideal adsorbed solute theory. Chemical Engineering Science 54, 5863–5885.

Prabhukumar, G., Pagilla, K., 2010. PAHs in Urban Runoff-Sources, Sinks and Treatment: A Review. Illinois Institute of Technology, Chicago, IL.

Prenafeta-Boldu, F.X., Summerbell, R., Sybren de Hoog, G., 2006. Fungi growing on aromatic hydrocarbons: biotechnology's unexpected encounter with biohazard. FEMS Microbiology Reviews 30, 109–130.

Prince, R.C., Lessard, R.R., Clark, J.R., 2003. Bioremediation of marine oil spills. Oil and Gas Science and Technology – Revue de IFP 58 (4), 463–468.

Pritchard, P.H., 1993. Effectiveness and regulatory in oil spill bioremediation: experiences with the Exxon Valdez oil spill in Alaska. In: Levin, M.A., Gealt, M.A. (Eds.), Biotreatment of Industrial and Hazardous Waste. McGraw-Hill, New York, pp. 269–307.

Putri, M., Hertadi, R., 2015. Effect of glycerol as carbon source for biosurfactant production by halophilic bacteria *Pseudomonas stutzeri* BK-AB12. Procedia Chemistry 16, 321–327.

Radwan, S.S., Al-Mailem, D., El-nemr, I., Salamah, S., 2000. Enhanced remediation of hydrocarbon contaminated desert soil fertilized with organic carbons. International Biodeterioration and Biodegradation 46, 129–132.

Rahman, K.S.M., Thahira-Rahman, J., Lakshmanaperumalsamy, P., Banat, I.M., 2002. Towards efficient crude oil degradation by a mixed bacterial consortium. Bioresource Technology 85 (3), 257–261.

Rahman, K.S.M., Rahman, T.J., Kourkoutas, Y., Petsas, I., Marchant, R., Banat, I.M., 2003. Enhancement bioremediation of n-alkane in petroleum sludge using bacterial consortium amended with rhamnolipid and micronutrients. Bioresource Technology 90, 159–168.

Raja, C.E., Anbazhagan, K.A., Selvam, G., 2006. Isolation and characterization of metal-resistant *Pseudomonas aeruginosa* strain. World Journal of Microbiology & Biotechnology 22, 577–585.

Ramachandran, S.D., Sweezey, M.J., Boudreau, M., Hodson, P.V., Courtenay, S.C., Lee, K., King, T., Dixon, J.A., 2006. Influence of salinity and fish species on PAH uptake from dispersed MESA crude oil. Marine Pollution Bulletin 52, 1182–1189.

Rapp, P., Bock, H., Wray, V., Wagner, F., 1979. Formation, isolation and characterization of trehalose dimycolates from *Rhodococcus erythropolis* grown on n-alkanes. Journal of General Microbiology 115, 491–503.

Rawe, J., Krietemeyer, S., Meagher-Hartzell, E., 1993. Guide for Conducting Treatability Studies under CERCLA: Biodegradation Remedy Selection- Interim Guidance. US Environmental Protection Agency, Washington, DC.

Reineke, W., 2001. Aerobic and anaerobic biodegradation potentials of microorganisms. In: Beek, B. (Ed.), The Handbook of Environmental Chemistry. The Natural Environment and Biogeochemical Cycles, vol. 2K. Springer Verlag, Berlin, pp. 1–161.

Rengaraj, S., Yeon, J.W., Kim, Y., Jung, Y., Ha, Y.K., Kima, W.H., 2007. Adsorption characteristics of Cu (II) onto ion exchange resins 252H and 1500H: kinetics, isotherms and error analysis. Journal of Hazardous Materials 143, 469–477.

Riis, V., Babel, W., Pucci, O.H., 2002. Influence of heavy metals on the microbial degradation of diesel fuel. Chemosphere 49, 559–568.

Robert, M., Mercade, M.E., Bosch, M.P., Parra, J.L., Espuny, M.J., Manresa, M.A., Guinea, J., 1989. Effect of the carbon source on biosurfactant production by *Pseudomonas aeruginosa* 44T. Biotechnology Letters 11, 871–874.

Rocchetti, L., Beolchini, F., Ciani, M., Dell'Anno, A., 2011. Improvement of bioremediation performance for the degradation of petroleum hydrocarbons in contaminated sediments. Applied and Environmental Soil Science 2011:319657. http://dx.doi.org/10.1155/2011/319657. 8 pages.

Rocha e Silva, N.M.P., Rufino, R.D., Luna, J.M., Santos, V.A., Sarubbo, L.A., 2014. Screening of *Pseudomonas* species for biosurfactant production using low-cost substrates. Biocatalysis and Agricultural Biotechnology 3, 132–139.

Ron, E.Z., Rosenberg, E., 2001. Natural roles of biosurfactants. Environmental Microbiology 3, 229–236.

Rosenberg, E., Legmann, R., Kushmaro, A., Taube, R., Adler, E., Ron, E.Z., 1992. Petroleum bioremediation-a multiphase problem. Biodegradation 3, 337–350.

Sabate, J., Vinas, M., Solanas, A.M., 2004. Laboratory-scale bioremediation experiments on hydrocarbon-contaminates soils. International Biodeterioration Biodegradation 52, 19–25.

Sabate, J., Vinas, M., Solanas, A.M., 2006. Bioavailability assessment and environmental fate of polycyclic aromatic hydrocarbons in biostimulated creosote-contaminated soil. Chemosphere 63, 1648–1659.

Safiyanu, I., Abdulwahid Isah, A., Abubakar, U.S., Rita Singh, M., 2015. Review on comparative study on bioremediation for oil spills using microbes. Research Journal of Pharmaceutical, Biological and Chemical Sciences 6 (6), 783–790.

Sandrin, T.R., Marier, M., 2003. Impact of metals on biodegradation of organic pollutants. Environmental Health Perspectives 111 (8), 1093–1101.

Sandrin, T.R., Chech, A.M., Maier, R.M., 2000. A rhamnolipid biosurfactant reduces cadmium toxicity during biodegradation of naphthalene. Applied and Environmental Microbiology 66, 4585–4588.

Santas, R., Korda, A., Tenente, A., Buchholz, K., Santas, P.H., 1999. Mesocosm assay of oil spill bioremediation with oleophilic fertilizers: Inipol, F1 or both? Marine Pollution Bulletin 38, 44–48.

Sarkar, D., Ferguson, M., Datta, R., Birnbaum, S., 2001. Bioremediation of petroleum hydrocarbons in contaminated soils: comparison of biosolids addition, carbon supplementation, and monitored natural attenuation. Environmental Pollution 136, 187–195.

Sarkar, D., Ferguson, M., Datta, R., Birnbaum, S., 2005. Bioremediation of petroleum hydrocarbons in contaminated soils: comparison of biosolids addition, carbon supplementation, and monitored natural attenuation. Environmental Pollution 136, 187–195.

Sartori, G., Ferrari, G.A., Pagliai, M., 1985. Changes in soil porosity and surface shrinkage in a remolded, saline clay soil treated with compost. Soil Science 139, 523–530.

Sayara, T., Sarrà, M., Sánchez, A., 2010a. Optimization and enhancement of soil bioremediation by composting using the experimental design technique. Biodegradation 21, 345–356.

Sayara, T., Sarrà, M., Sánchez, A., 2010b. Effects of compost stability and contaminant concentration on the bioremediation of PAHs contaminated soil through composting. Journal of Hazardous Materials 179, 999–1006.

Sayara, T., Borràs, E., Caminal, G., Sarrà, M., Sánche, A., 2011. Bioremediation of PAHs-contaminated soil through compo sting: influence of bioaugmentation and biostimulation on contaminant biodegradation. International Biodeterioration and Biodegradation 65, 859–865.

Schwarzenbach, R.P., Gschwend, P.M., Imboden, D.M., 1993. Environmental Organic Chemistry. Wiley Interscience, New York.

Seklemova, E., Pavlova, A., Kovacheva, K., 2001. Biostimulation based biodegradation of diesel fuel: field demonstration. Biodegradation 12, 311–316.

Semple, K.T., Reid, B.J., Fermor, T.R., 2001. Impact of composting strategies on the treatment of soils contaminated with organic pollutants. Environmental Pollution 112 (2), 269–283.

Semple, K.T., Morriss, A.W.J., Paton, G.I., 2003. Bioavailability of hydrophobic organic contaminants in soils: fundamental concepts and techniques for analysis. European Journal of Soil Science 54 (4), 809–818.

Sharma, P., Singh, J., Dwivedi, S., Kumar, M., 2014. Bioremediation of oil spill. Journal of Bioscience and Technology 5 (6), 571–581.

Shen, G., Lu, Y., Hong, J., 2006. Combined effect of heavy metals and polycyclic aromatic hydrocarbons on urease activity in soil. Ecotoxicology and Environmental Safety 63, 474–480.

Shuttleworth, K.L., Cerniglia, C.E., 1995. Environmental aspects of PAH biodegradation. Applied Biochemistry and Biotechnology 54, 291–302.

Sikkema, J., De-Bont, J.A.M., Poolman, B., 1995. Mechanism of membrane toxicity of hydrocarbons. Microbiological Reviews 59 (2), 201–222.

Silva, D.d-S.P., Cavalcanti, D.d-L., Vieira de Melo, E.J., Ferreira dos Santos, P.N., Pacheco da Luz, E.L., Buarque de Gusmao, N., Vieira de Queiroz Sousa, M.d-F., 2015. Bio-removal of diesel oil through a microbial consortium isolated from a polluted environment. International Biodeterioration and Biodegradation 97, 85–89.

Sim, L., Ward, O.P., Li, Z.Y., 1997. Production and characterization of a biosurfactant isolated from *Pseudomonas aeruginosa* UW-1. Journal of Industrial Microbiology & Biotechnology 19 (4), 232–238.

Simon, M.A., Bonner, J.S., Page, C.A., Towsend, R.T., Mueller, D.C., Fuller, C.B., Autenriech, R.L., 2004. Evaluation of two commercial bioaugmentation product for enhanced removal of petroleum from a wetland. Ecological Engineering 22, 263–277.

Singh, D.N., Tripathi, A.K., 2013. Coal induced production of a rhamnolipid biosurfactant by *Pseudomonas stutzeri*, isolated from the formation water of Jharia coalbed. Bioresource Technology 128, 215–221.

Singh, B., Bhattacharya, A., Channashettar, V.A., Jeyaseelan, C.P., Gupta, S., Sarma, P.M., Mandal, A.K., Lal, B., 2012. Biodegradation of oil spill by petroleum refineries using consortia of novel bacterial strains. Bulletin of Environmental Contamination and Toxicology 89, 257–262.

Siunova, T.V., Anokhina, T.O., Mashukova, A.V., Kochetkov, V., Boronin, A.M., 2007. Rhizosphere strain of *pseudomonas chloroaphis* capable of degrading naphthalene in the presence of cobalt/nickel. Microbiology 76, 212–218.

Sokhn, J., De Leij, F.A.A.M., Hart, T.D., Lynch, J.M., 2001. Effect of copper on the degradation of phenanthrene by soil microorganisms. Letters in Applied Microbiology 33, 164–168.

Sokolovská, I., Rozenberg, R., Riez, C., Rouxhet, P.G., Agathos, S.N., Wattiau, P., 2003. Carbon source induced modifications in the mycolic acid content and cell wall permeability of *Rhodococcus erythropolis* E1. Applied and Environmental Microbiology 69, 7019–7027.

Soliman, R.M., El-Gendy, N.Sh., Deriase, S.F., Farahat, L.A., Mohamed, A.S., 2011. Comparative assessment of enrichment media for isolation of pyrene-degrading bacteria. Review Industrial Sanitary & Environmental Microbiology 5 (2), 54–70.

Soliman, R.M., El-Gendy, N.Sh., Deriase, S.F., Farahat, L.A., Mohamed, A.S., 2014. The evaluation of different bioremediation processes for Egyptian oily sludge polluted soil on a microcosm level. Energy Sources, Part A: Recovery, Utilization, and Environmental Effects 36 (3), 231–241.

Somtrakoon, K., Suanjit, S., Pokethitiyook, P., Kruatrachue, M., Lee, H., Upatham, S., 2008. Enhance biodegradation of anthracene in acidic soil by *Burkholderia* sp. VUN10013. Current Microbiology 57, 102–106.

Song, H.G., Wang, X., Bartha, R., 1990. Bioremediation potential of terrestrial fuel spills. Applied and Environmental Microbiology 56, 652–656.

Sparrow, S.D., Sparrow, E.B., 1988. Microbial biomass and activity in a subarctic soil, ten years after crude oil spills. Journal of Environmental Quality 17, 304–309.

Speight, J.G., Arjoon, K.K., 2012. Bioremediation of Petroleum and Petroleum Products. Scrivener Publishing, Beverly, Massachusetts.

Spence, M.J., Bottrell, S.H., Thornton, S.F., Richnow, H.H., Spence, K.H., 2005. Hydrochemical and isotopic effects associated with petroleum fuel biodegradation pathways in a chalk aquifer. Journal of Contaminant Hydrology 79 (1/2), 67–88.

Sposito, G., 1989. The Chemistry of Soils. Oxford University Press, Oxford, United Kingdom.

Straube, W.L., Jones-Meehan, J., Pritchard, P.J., Jones, W., 1999. Bench-scale optimization of bioaugmentation strategies for treatment of soils contaminated with high molecular weight hydrocarbons. Resource, Conservation and Recycling 27, 27–37.

Subha, B., Song, Y.C., Woo, J.H., 2015. Optimization of biostimulant for bioremediation of contaminated coastal sediment by response surface methodology (RSM) and evaluation of microbial diversity by pyrosequencing. Marine Pollution Bulletin 98, 235–246.

Sugiura, K., Ishihara, M., Shimauchi, T., Harayama, S., 1997. Physiochemical properties and biodegradability of crude oil. Environmental Science & Technology 31, 45–51.

Supaka, N., Pinphanichakarn, P., Pattra, G.K., Thaniyavarn, S., Omari, T., Juntongjin, K., 2001. Isolation and characterization of phenanthrene-degrading *Sphingomonas* sp. strain P2 and its ability to degrade fluoranthene and Pyrene via co-metabolism. Science Asia 27, 21–28.

Tam, N.F.Y., Guo, C.L., Yau, W.Y., Wong, Y.S., 2002. Preliminary study on biodegradation of phenanthrene by bacteria isolated from mangrove sediments in Hong-Kong. Marine Pollution Bulletin 42, 316–324.

Thapa, B., Ajay Kumar, K.C., Ghimire, A., 2012. A review on bioremediation of petroleum hydrocarbon contaminants in soil. Kathmandu University. Journal of Science, Engineering and Technology 8 (1), 164–170.

Thavasi, R., Jayalakshmi, S., Bana, I.M., 2011. Effect of biosurfactant and fertilizer on biodegradation of crude oil by marine isolates of *Bacillus megaterium, Corynebacterium kutscheri* and *Pseudomonas aeruginosa*. Bioresource Technology 102, 772–778.

Thibault, S.L., Anderson, M., Frankenberger Jr., W.T., 1996. Influence of surfactants on pyrene desorption and degradation in soils. Applied and Environmental Microbiology 62, 283–287.

Trenz, S.P., Engesser, K.H., Fischer, P., Knackmuss, H.J., 1994. Degradation of fluorene by *Brevibacterium* sp. strain DPO 1361: a novel C-C bond cleavage mechanism via 1,10-dihydro-1,10-dihydroxyfluoren-9-one. Journal of Bacteriology 176 (3), 789–795.

Trindade, P.V.O., Sobral, L.G., Rizzo, A.C.L., Leite, S.G., Soriano, A.U., 2005. Bioremediation of a weathered and recently oil-contaminated soils from Brazil: a comparison study. Chemosphere 58, 515–522.

Trinidade, P., Sobral, L.G., Rizzo, A.C., Leite, S.G.F., Lemos, J.L.S., Milloili, V.S., Soriano, A.U., 2002. Evaluation of the biostimulation and bioaugmentation techniques in the bioremediation process of petroleum hydrocarbon contaminated soils. In: 9th Int. Pet. Environ. Conf., Albuquerque, New Mexico.

Tsui, L., Roy, W.R., 2007. Effect of compost age and composition on the atrazine removal from solution. Journal of Hazardous Materials B139 (1), 79–85.

Ueno, A., Hasanuzzaman, M., Yumoto, I., Okuyama, H., 2006. Verification of degradation of n-alkanes in diesel oil by Pseudomonas *aeruginosa* strain WatG in soil microcosms. Current Microbiology 52 (3), 182–185.

Urum, K., Pekdemir, T., 2004. Evaluation of biosurfactants for crude oil contaminated soil washing. Chemosphere 57, 1139–1150.

Vacca, W.F., Bleam, W.F., Hickey, W.J., 2005. Isolation of soil bacteria adapted to degrade humic acid-sorbed phenanthrene. Applied and Environmental Microbiology 71 (7), 3797–3805.

Van de Wiele, T., Vanhaecke, L., Boeckaert, C., Peru, K., Headley, J., Verstraete, W., Siciliano, S., 2005. Human colon microbiota transform PAHs to estrogenic metabolites. Environmental Health Perspectives 113, 6–10.

Van-Hamme, J.D., Singh, A., Ward, O.P., 2003. Recent advances in petroleum microbiology. Microbiology and Molecular Biology Reviews 7, 503–549.

Varjani, S.J., Rana, D.P., Jain, A.K., Bateja, S., Upasani, V.N., 2015. Synergistic ex-situ biodegradation of crude oil by halotolerant bacterial consortium of indigenous strains isolated from on shore sites of Gujarat, India. International Biodeterioration and Biodegradation 103, 116–124.

Venosa, A., Zhu, X., 2003. Biodegradation of crude oil contaminating marine shorelines and freshwater wetlands. Spill Science and Technology. Bulletin 8 (2), 163–178.

Vidali, M., 2001. Bioremediation. An overview. Pure and Applied Chemistry 73 (7), 1163–1172.

Viramontes-Ramos, S., Portillo-Ruiz, M.C., Ballinas-Casarrubias, M.L., Torres-Muñoz, J.V., Rivera-Chavira, B.E., Nevárez-Moorillón, G.V., 2010. Selection of biosurfactant/bioemulsifier-producing bacteria from hydrocarbon-contaminated soil. Brazilian Journal of Microbiology 41, 668–675.

Volkering, F., Breure, A.M., Rulkens, W.H., 1998. Microbiological aspects of surfactant use for biological soil remediation. Biodegradation 8, 401–417.

Vollmouth, S., Niessner, R., 1995. Degradation of PCDD, PCDF, PAH, PCB and chlorinated phenols during the destruction-treatment of landfill seepage water in laboratory model reactor (UV, Ozone, and UV/Ozone). Chemosphere 30 (12), 2317–2331.

Walker, J.D., Colwell, R.R., Hamming, M.C., Ford, H.T., 1975. Extraction of petroleum hydrocarbons from oil-contaminated sediments. Bulletin of Environmental Contamination and Toxicology 13, 245–248.

Wang, Z., Fingas, M.F., 2003. Development of oil hydrocarbon fingerprinting and identification techniques. Marine Pollution Bulletin 47, 423–452.

Wang, P., Keller, A.A., 2009. Partitioning of hydrophobic pesticides within a soil water anionic surfactant system. Water Research 43, 706–714.

Watanabe, K., 2001. Microorganisms relevant to bioremediation. Current Opinion in Biotechnology 12, 237–241.

Weissenfels, W.D., Beyer, M., Klein, J., 1990. Degradation of phenanthrene, fluorene and fluoranthene by pure bacterial cultures. Applied Microbiology and Biotechnology 32 (4), 479–484.

Wick, L.Y., Munain, A.R.d., Springael, D., Harms, H., 2002. Responses of *Mycobacterium* sp. LB501T to the low bioavailability of solid anthracene. Applied Microbiology and Biotechnology 58, 378–385.

Wiesel, I., Wuebker, S.M., Rehm, H.J., 1993. Degradation of polycyclic aromatic hydrocarbons by an immobilized mixed bacterial culture. Applied Microbiology and Biotechnology 39, 110–116.

Wong, K.W., Toh, B.A., Ting, Y.P., Obbard, J.P., 2005. Biodegradation of phenanthrene by indigenous microbial biomass in zinc amended soil. Letters in Applied Microbiology 40, 50–55.

Wu, S.C., Gschwend, P.M., 1986. Sorption kinetics of hydrophobic organic compounds to natural sediments and soil. Environmental Science & Technology 20, 717–725.

Wu, T., Sharda, J.N., Koide, R.T., 2003. Exploring interactions between saprotrophic microbes and ectomycorrhizal fungi using a protein-tannin complex as an N source by red pine (*Pinus resinosa*). New Phytologist 159, 131–139.

Wu, T., Xie, W.j., Yi, L., Li, X.B., Yang, B.H., Wang, J., 2012. Surface activity of salt-tolerant *Serratia* spp. and crude oil biodegradation in saline soil. Plant Soil Environment 58 (9), 412–416.

Xu, R., Lau, N.L.A., Ng, K.L., Obbard, J.P., 2004. Application of a slow-release fertilizer for oil bioremediation in beach sediment. Journal of Environmental Quality 33, 1210–1216.

Yang, S.Z., Jin, H.J., Wei, Z., He, R.X., Ji, Y.J., Li, X.M., Yu, S.P., 2009. Bioremediation of oil spills in cold environments: a review. Pedosphere 19 (3), 371–381.

Younis, Sh.A., El-Gendy, N.Sh., Moustafa, Y.M., 2013. A study on bio-treatment of petrogenic contamination in El-Lessan area of Damietta River Nile Branch, Egypt. International Journal of Chemical and Biochemical Sciences 4, 112–124.

Yu, H., Huang, G.H., An, C.J., Wei, J., 2011. Combined effects of DOM extracted from site soil/compost and biosurfactant on the sorption and desorption of PAHs in a soil-water system. Journal of Hazardous Materials 190 (1/3), 883–890.

Zabbey, N., Sam, K., Onyebuchi, A.T., 2017. Remediation of contaminated lands in the Niger Delta, Nigeria: prospects and challenges. Science of the Total Environment 586, 952–965.

Zahed, M.A., Abdul Aziz, H., Isa, M.H., Mohajeri, L., Mohajeri, S., Kutty, S.R.M., 2011. Kinetic modeling and half-life study on bioremediation of crude oil dispersed by Corexit 9500. Journal of Hazardous Materials 185, 1027–1031.

Zhang, Y., Miller, R.M., 1995. Effect of rhamnolipid (biosurfactant) structure on solubilization and biodegradation of n-alkanes. Applied and Environmental Microbiology 61, 2247–2251.

Zhang, Y., Maier, W.J., Miller, R.M., 1997. Effect of rhamnolipids on the dissolution, bioavailability and biodegradation of phenanthrene. Environmental Science & Technology 31, 2211–2217.

Zheng, J.Q., Han, S.J., Zhou, Y.M., Ren, F.R., Xin, L.H., Zhang, Y., 2010. Microbial activity in a temperate forest soil as affected by elevated atmospheric CO_2. Pedosphere 20 (4), 427–435.

ZoBell, C.E., 1963. The occurrence, effects and fate of oil polluting the sea. Air and Water Pollution 7, 173–197.

FURTHER READING

Kästner, M., Breuer-Jammali, M., Mahro, B., 1994. Enumeration and characterization of the soil microflora from hydrocarbon-contaminated soil sites able to mineralize polycyclic aromatic hydrocarbons (PAH). Applied Microbiology and Biotechnology 41, 267–273.

BIOREMEDIATION OF MARINE OIL SPILLS

11

1.0 INTRODUCTION

Pollution can be defined as, the contamination of soil, water, or air with the discharge of harmful substances and wastes that would cause harmful impact on all living creatures (human, animal, and aquatic), besides causing severe damage to the resources (GESAM, 2000; Sharma et al., 2014; Weis, 2014). Petroleum hydrocarbons are the most significant source of organic pollutants. Pollution of water ecosystems, especially, the marine ecosystem, occurs from anthropogenic causes—the direct or indirect introduction of hazardous substances or energy wastes into the marine environment that results in deleterious effects such as harm to living organisms, hazardous to human health, affecting the biodiversity and the habitat of living species, and adversely affecting the birds that depend on aquatic life for food. Moreover, pollution adversely affects the spawning areas of fish, shrimps, mollusks, and other species. This causes hindrance to marine activities, including fishing, impairment of poor quality for use of seawater and reduction of amenities, damage to the unique coral reef systems. It hampers naval traffic causing serious problems in ports and harbors and affects tourism activities.

Aquatic oil spills are the major cause for pollution of the sea water, that affect life on land and sea. Negative impacts of oil spills could be extended to the natural and industrial resources including oil exploration and exploration activities and would have also negative socioeconomic impacts like loss of jobs, financial losses, etc. It would also affect the quality of drinking water (Ali et al., 2006; El-Gendy and Moustafa, 2007; Raafat et al., 2007; Munday et al., 2010; Rogers and Laffoley, 2011; Amer et al., 2014; Agarwal and Liu, 2015). Generally, oil spills may occur due to the discharge of crude oil or its derived products such as gasoline, kerosene, diesel, or machine oil from tankers, ships, offshore platforms, heavier fuels used by large ships, accidents in pipelines, or production process, as well as spills of other oily wastewaters such as produced water, drilling mud, cuttings, refineries effluents, and ballast water (El-Gendy and Nassar, 2016). Small-scale oil pollution would occur due to wastes produced from tanker operations such as unloading, cleaning, and filling oil tanks (Rogowska and Namiesnik, 2010). Larger oil spills are due to offshore damaged drilling rigs, underwater oil leaks, as well as, collisions and the grounding of oil tankers (Korotenko et al., 2009). These usually result in oil being released into the surrounding waters, polluting the adjacent shorelines and negatively affecting plant and animal life (Rogowska and Namiesnik, 2010).

Metwally et al. (1997) classified marine sediments according to the degree of pollution into three groups based on the total petroleum hydrocarbon (TPH) levels; 1–4 mg/kg as unpolluted, <100 mg/kg as moderately polluted, and up to 12,000 mg/kg as highly polluted sediments. One of the first species to be affected in the marine environment is the phytoplankton, due to the limitation of oxygen and light diffusion (Mehta, 2005; Yeung et al., 2011; McGenity, 2014; Silva et al., 2014). This is detrimental to the ecosystem as phytoplankton is the basis of the food chain in marine ecosystems and consequently this affects the biodiversity and human well-being (Ciuciu et al., 2006; Rocha e Silva et al., 2013; Silva et al., 2014).

Introduction to Petroleum Biotechnology. https://doi.org/10.1016/B978-0-12-805151-1.00011-4

Oil pollution causes reproductive and developmental problems, as well as, damage to the brain, liver, and kidneys of fish, marine mammals, and terrestrial species. Sea turtles are at risk, as they tend to not remove themselves from the oil-contaminated areas where much of the oil accumulates. Thus, oil spills can be a threat to aquatic, marshland, and coastal ecosystems that are often impacted by the oil. For example, The Exxon Valdez oil spill resulted in contamination of fish along with their embryos and juvenile larvae, and chronic effects among sediment foraging marine birds resulted in a decline in their abundance (Peterson et al., 2003).

About 10% of the total input is from the devastating oil spills, which cause both ecological and economical damage. According to Hinchee et al. (1995), approximately, five million tons/year of crude and refined oil enter the environment because of anthropogenic sources such as oil spills. Petroleum refinery wastewater is another source of pollutants: organic pollutants (e.g., dissolved and dispersed oil and greases, lighter hydrocarbons, higher alkanes, phenols, polychlorinated biphenyls, high molecular weight organic acids, aromatic acids, polyaromatic hydrocarbons, sulfides, and nitrogen compounds), dissolved formation minerals, and inorganic pollutants, i.e., heavy metals (e.g., chromium, iron, nickel, copper, molybdenum, selenium, vanadium, and zinc) (Ryynänen, 2011; Ishak et al., 2012; Younis et al., 2014a,b; Younis et al., 2015; Abdel-Satar et al., 2017). Another statistic showed that about 1.3 million tons of oil have been spilled in the oceans (Bao et al., 2012). Oil spills could be operational such as spills due to bunkering activities, loading and unloading in the oil terminals, and discharge of oily mixture. They could be also accidental spills due to collision; grounding; ramming; fire/explosion; hull failure of ships; pipeline systems break, rupture, crack, or leakage; well or platform blow out; and tank failures (Doerffer, 1992). The first incident of pollution occurred in the early 18th century, from the damage of a ship, but the problem only came into prominence in the 1930s with the gradual conversion in the ships' boilers from coal to oil power.

Since then there have been enormous increase in the scale of oil operations because of ever-increasing demands for energy. Oil and oily wastes enter water bodies from several sources like industrial effluents, oil refineries and storage tanks, automobile waste oil, and petrochemical plants. In the last decades, petroleum industry led to major pollution problems, for example, different accidental oil spills. All these led to large-scale pollution of the aquatic environment (Abdel Shafy and Mansour, 2016). Thus, the production, transport, and refining of crude oil result in a large volume of hydrocarbons polluted wastewater. These effluents have a wide range of salinity; from almost freshwater to approximately three times more saline than seawater.

Petroleum and polycyclic aromatic compounds (PACs) are pollutants that are a cause of global concern because of their high toxicity, environmental persistence, and widespread occurrence. PACs are highly lipophilic, nonpolar persistent substances and have been detected in aquatic environments and in many wild organisms and have serious and dangerous effects on human health. These effects depend mainly on the extent of exposure and concentration that humans are exposed to, the toxicity of the oil, and whether exposure occurs via inhalation, ingestion, or skin contact.

Moreover, after the occurrence of an oil spill, the composition of n-alkane derivatives, alkyl toluene derivatives, and mono-aromatic derivatives changes rapidly in the initial phase, whereas, the polynuclear aromatic hydrocarbons (PAHs), alkyl-substituted PAHs, and NSO aromatics, i.e., the nitrogen-, sulphur- and oxygen containing compounds are more recalcitrant and may be present for years or decades (Head et al., 2006). The United States Environmental Protection Agency (US EPA) has identified PACs to have acute or chronic toxicity, mutagenic and carcinogenic properties to human and other organisms (Lampi et al., 2006; Li and Chen, 2009; Billet et al., 2010). Also, PACs have adverse effects on reproduction, affecting both males and females and lead to development of atherosclerosis (ATSDR, 1993). Fish exposed to PAC contamination exhibited fin erosion, liver abnormalities, cataracts, and immune system impairments. Organisms such as mussels and oysters lack the enzyme systems which break down PAC compounds and thus accumulate high concentrations of PACs (Yu et al., 2005).

Crude oil, as a pollutant, can affect marine organisms, through the decrement in the algal productivity, which is a very important link in the food chain. This would occur directly or indirectly, through coating, contact poisoning components, such as PAHs and phenol derivatives, or through the decrease of dissolved oxygen and penetration of light. This would lead to inadequate maintenance of higher life forms and in turn to a low diversity and abundance of fauna, due to the inability of many species to survive in such harsh conditions (Attiogbe et al., 2007; El-Nass et al., 2009). PAHs are typical, persistent organic pollutants and are the priority pollutants of most concern due to their high bioaccumulation and carcinogenic, mutagenic, and toxic properties (Aquilina et al., 2010; Chen et al., 2011).

PAHs are often resistant and difficult to be removed by conventional physicochemical treatment methods (Crisafully et al., 2008; Sponza and Oztekin, 2010). Moreover, The US EPA, World Health Organization (WHO), and Environmental Egyptian Law Number 4, 1994 for wastewater considered phenols as priority pollutants and lowered their content in the wastewater stream to less than 1 mg/L as maximum concentration limit (Ahmaruzzaman and Sharma, 2005; Al-Sultani and Al-Seroury, 2012), due to the phenol propensity to initiate carcinogenic and mutagenic effects on terrestrial as well as aquatic biota and humans, even at low concentrations (Mukhetjee et al., 2011; Amin et al., 2012). However, the concentration of phenols in industrial wastewater is usually in the range of 0.1–6800 mg/L (Busca et al., 2008; El-Gendy and Nassar, 2015a,b).

The enzyme systems that metabolize PACs are the microsomal mixed function oxidase system, which converts the nonpolar PACs into polar hydroxy and epoxy derivatives (Hall et al., 1989). This is widely distributed in the cells and tissues of humans and animals. The highest metabolizing capacity is present in the liver, followed by lung, intestinal mucosa, skin, and kidneys, but metabolism may also take place in nasal tissues, mammary gland, spleen, erythrocytes, leukocytes, placenta, and uterus (Anderson et al., 1989). Epoxides are the major intermediates in the oxidative metabolism of aromatic double bonds. The epoxides are reactive and enzymatically metabolized to other compounds such as dihydrodiols and phenols (Van de Wiele et al., 2005), where, these diol epoxides are easily converted by epoxide ring opening into electrophilic carbonium ions which are alkylating agents that covalently bind to nucleophilic sites in the DNA bases and in proteins (Rundle et al., 2000). Petroleum pollutants can cause a wide range of serious health problems including cancer, birth defects, asthma, chronic obstructive pulmonary disease, eye irritation, photophobia, and skin diseases such as dermatitis and keratosis, especially in workers occupationally exposed to PACs. Also, adverse respiratory effects, including acute and sub-acute inflammation and fibrosis, have been demonstrated experimentally. Long-term exposure to PACs may also include kidney and liver damage and jaundice. Repeated contact with skin may induce redness and skin inflammation. Also, naphthalene, for example, can cause the breakdown of red blood cells if inhaled or ingested in large amounts (Public Health Fact Sheet, 2009). Generally, urinary metabolites of naphthalene, phenanthrene, chrysene, pyrene, and benzo[a]pyrene (BaP) are biomarkers of PACs exposure, and for this reason, studies on these PACs are of interest to occupational exposure and hazard assessment (Grimmer et al., 1997). However BaP is used as a model for studies of PACs metabolism in humans and animals (Charles et al., 2000), where, BaPs stimulate their own metabolism by inducing microsomal cytochrome P-450 monooxygenases and epoxide hydrolases.

During the past years, the frequency and risk of oil pollution have led to extensive research, and governments have become aware of the need to protect ecosystems as well as to evaluate the damage caused by pollution, with a great interest in promoting ecofriendly methods for the reclamation of oil-polluted sites, in a way that is less expensive and does not introduce additional chemicals into the environment (Vidali, 2001; Rogers and Laffoley, 2011; Bhatnagar and Kumari, 2013). Thus, bioremediation is very feasible and effective technology for treatment of oil pollution (Diaz et al., 2002; Milić et al., 2009; Chekroud et al., 2011).

When crude oil or distilled oil products are dumped into the marine environment, they undergo immediate physical and chemical changes. Some of the abiotic processes that take place in this situation are spreading, evaporation, dissolution, dispersal, photochemical oxidation, oil-water emulsification, adsorption to suspended particulate matter, sinking, and sedimentation (Alves et al., 2014; Safiyanu et al., 2015). These processes take place concurrently, altering the chemical composition and physical properties of the original pollutant and, therefore, significantly influencing the efficiency of biotransformation.

2.0 FATE OF AQUATIC OIL SPILLS

Oil spills also can be classified in terms of their sheer volume as <7 tons, 7–700 tons and >700 tons <50 bbls, 50–5000 bbls, and >5000 bbls, respectively (ITOPF, 2011). The fate of the oil in the environment depends on other factors, such as; the conditions, hydrodynamics, and the size of the recipient water and where the outfall is located (i.e., whether it is intertidal or subtidal) (Wake, 2005; Obi et al., 2014). Moreover, when petroleum or its products are released into the ecosystem, they undergo a series of multiple, naturally occurring chemical, physical, and biological processes that change their chemical compositions and/or physical properties which in combination are termed "weathering" (Fig. 11.1) (Farrington, 2014; Tarr et al., 2016). The main physicochemical characteristics that affect the weathering are the specific gravity, distillation characteristics, vapor pressure, viscosity, pour point, where all depend on the proportions of the chemical constituents (i.e., volatile compounds, wax, resins, and asphaltenes). An average crude oil consists of approximately 30% paraffins or alkanes, 50% naphthenes or cycloalkanes, 15% aromatics, 5% nitrogen, sulfur, and oxygen containing compounds, but there are the extremes, such as heavy oil and tar sand bitumen (Clayton, 2005).

The most important weathering processes following a release of oil to aquatic environment, which can simultaneously occur with each other, and usually overlap through the course of the spill include spreading, evaporation, dissolution, emulsification, natural dispersion, microbial degradation, photochemical oxidation, and other processes such as sedimentation, adhesion onto the surface of suspended particulate materials, and oil-fine interaction. These processes can interact and affect each other. Thus, weathering can strongly influence how oils move and behave in the environment and differently affects the hydrocarbon family (Huang, 1983; Ojo, 2006; Bordenave et al., 2007; Zhen-Yu et al., 2008; Sabean et al., 2009).

2.1 SPREADING AND ADVECTION

Spreading of crude oil on water is probably the most important process following a spill. It is the horizontal movement of the oil spill on the surface of the water because of gravity, friction, viscosity, and

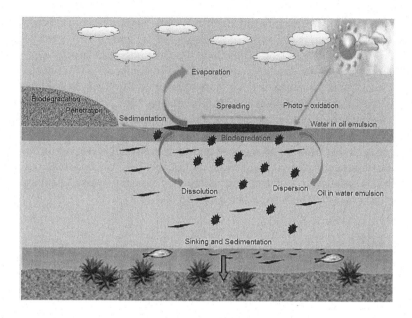

FIGURE 11.1

The fate of a marine oil spill.

surface tension. Apart from chemical nature of oil and its viscosity, *advection*, is the oil movement which is mainly affected by wind, waves, tidal streams and currents, and the spilled volume. Under the influence of hydrostatic and surface forces, the oil spreads quickly attaining average thickness of less than 0.03 mm within 24 h.

Spreading process involves the whole oil, and it does not partition the various components of the oil or affect its chemical composition. Once a spill has thinned to the point that surface forces begin to play an important role, the oil layer is no longer continuous and uniform but becomes fragmented by wind and waves. As the spreading increases and the oil thickness decreases, the appearance changes from black/dark brown coloration of thick oil patches to iridescent and silver sheen at the edges of the slick. Thus, through spreading, if a small amount of oil is poured into a pool of water, a circle of oil, that gets thinner and thinner, will grow over time, creating a larger surface area presented to both the air and the water underneath. This serves to increase the effectiveness when dispersants are applied, since the thinner and larger area is more rapidly diluted into the water. However, because this process enlarges the surface slick, more area needs to be covered during dispersant operations and more equipment may be needed for an effective response. But, briefly, spreading process increases the surface area of the spill, thus enhancing mass transfer via evaporation, dissolution, and later biotransformation. Spreading dominates the initial stages of a spill. While this can continue for approximately 7–10 days after the occurrence of the spill, or until the oil is contained by shorelines, collection efforts, or other obstructions.

2.2 EVAPORATION

Evaporation and dissolution are the major processes degrading petroleum crude when spilled on water. The composition of oil, its surface area and physical properties, wind velocity, air and sea temperatures, turbulence and intensity of solar radiation, all affect evaporation rates of hydrocarbons. The increase in spreading rate enhances the evaporation, since the larger the oil spill surface area, the faster the light components will evaporate. Evaporation is generally the most important process during the first 48h after a spill, volatilizing low- and medium-molecular weight compounds into the atmosphere. It can continue up to 2 weeks. It is estimated that one- to two-thirds of a spill are thus eliminated during this period, although evaporation rate decreases with time. Evaporation alone can remove about 50% of hydrocarbons in an "average" crude oil on the ocean's surface. Loss of volatile hydrocarbons increases the density and the kinematic viscosity of oil. As more volatile hydrocarbons are lost, the specific weight and viscosity of the resulting oil increase and this results in breakup of the slick into smaller patches. Agitation of these patches enhances incorporation of water due to increased surface area. The remaining material is enriched in metals, mainly nickel and vanadium, waxes, and asphaltenes rather than the original oil (Fingas, 2004; King et al., 2014a,b).

2.3 DISSOLUTION

Dissolution is another physical process in which the low molecular weight hydrocarbons, as well as, polar, non-hydrocarbon compounds are partially lost from the oil to the water column. It typically occurs during the first 24h of a spill. From the perspective of mass loss, dissolution into the water column is much less important than evaporation, as the dissolution of a significant portion of the spill into this medium is highly unlikely. Its importance comes from the slightly water-soluble fractions (i.e., the light aromatic compounds, such as benzene and toluene) which are toxic to marine organisms including the degrading microorganisms. However, these compounds are the most volatile, and are lost by evaporation 10 to 1000 times) faster than they dissolve. Rarely dissolved petroleum hydrocarbons exceed 1 ppm. Thus, dissolution process does not greatly contribute to the removal of oil from the sea surface (Otremba and Toczek, 2002; Martín, 2011). But, although dissolution is less important from the viewpoint of mass loss during an oil spill, dissolved hydrocarbon concentrations in water are particularly important due to their potential influence on the success of bioremediation and the effect of toxicity on biological systems. The extent of dissolution depends on the solubility of the spilled oil, weather conditions, and the characteristics of the spill site. The low molecular weight aromatics are the most soluble oil components, and they are also the most toxic components in crude and refined oils. Although many of them may be removed through evaporation, their impact on the environment is much greater than simple mass balance considerations would imply (National Academy of Sciences, NAS, 1985; Ryerson et al., 2012). Dissolution rates are also influenced by photochemical and biological processes.

2.4 PHOTO-OXIDATION

The natural sunlight in the presence of oxygen can transform several petroleum hydrocarbons into oxygen compounds such as aldehydes and ketones and ultimately to low molecular weight carboxylic acids; as the products are hydrophilic, they change the solubility behavior of the spill (Farrington, 2014). Detrimental effects may be associated with the increase in the solubility of oil in water (i.e., bioavailability) and the formation of toxic compounds mediated by photo-oxidation. On the other hand, the formation of polar compounds may increase the rate of biotransformation of petroleum, particularly at low concentrations where acute toxicity effects are limited (Nicodem et al., 1997, 1998). However,

Bacosa et al. (2015a) reported that incubation of seawater with oil light-sweet crude in sunlight resulted in changes in microbial community structure and reduced bacterial diversity. Numerous oxygenated products such as aromatic hydrocarbons, aliphatic hydrocarbons, benzoic acid, naphthanoic acids, alcohols, phenols, and aliphatic ketones result due to the photolysis of oil (Griffis et al., 2014; Ray et al., 2014). However, photo-oxidation can lead to soluble products (Chapelle, 2001) or persistent tars (Aeppli et al., 2012). Yang et al. (2014) reported that lower molecular weight alkanes are photo-oxidized more rapidly than the higher molecular weight compounds, while PAHs can be photo-oxidized more readily than alkanes (Garrett et al., 1998; Maki et al., 2001; Prince et al., 2003; Radović et al., 2014; D'Auria et al., 2009; King et al., 2014a,b). Prince et al. (2003) have also reported that the majority of the aromatic constituents can be converted to resin constituents and asphaltene constituents. Higher molecular weight PAHs typically absorb at longer wavelengths, so PAHs with three or more rings have a greater overlap with sunlight and are more likely to be involved in photochemistry. Although it occurs throughout the entire duration of the oil spill, its overall contribution is minor compared to the other weathering processes.

Thick layers of very viscous oil or water-in-oil emulsions tend to be oxidized to persistent residues (i.e., tar balls) rather than degrade. UV radiation is reported to affect the structure of estuarine microbial communities (Santos et al., 2012), especially when a substantial amount of organic matter is oxidized (Hunting et al., 2013). In oil polluted waters, crude oil adds to the carbon pool, and aromatic hydrocarbons absorb UV light that has a crucial role in the long-term weathering of spilled oil (Evdokinmov and Losev, 2007). As mentioned before, aromatic hydrocarbons are more sensitive to photo-oxidation and generally transform into polar species; hydroxyl radicals, singlet oxygen, various oxygenated compounds, and polar quinones, which are more water soluble and parent hydrocarbons and known to be reactive species that cause oxidative stresses, damage cells, thus are more toxic (King et al., 2014a,b; Bacosa et al., 2015b).

2.5 EMULSIFICATION

The incorporation of water in the oil, ultimately leading to thickening and an increase in the total volume remaining, turning the oil into a thick, sticky mixture commonly referred to as "chocolate mousse" that contains approximately 80% water is called emulsification. This can increase the pollutant volume, by a factor of up to five times. At the same time, emulsification can reduce the other natural weathering processes. Different emulsions react differently with dispersants and some recent experiments have shown that it is possible to disperse a wide range of emulsified oil. It begins within the first day of spill and would continue to occur throughout the first year of the spill, where the largest emulsions volume is typically formed within the first week of the spill. The formation and stability of emulsions are primarily related to the chemical composition of the oils and are enhanced by wax and asphaltic materials. It can be categorized into four classes: stable, meso-stable, unstable, and entrained water (Fingas and Fieldhouse, 2004). Where, the higher the asphaltene content the higher the stability of the emulsion is, and oils that have moderate to high viscosity tend to emulsify more than lighter oils (Fingas et al., 2002; Fingas and Fieldhouse, 2004; Xie et al., 2007). Surface-active materials generated through photochemical and biological processes are also involved in the formation of the emulsions. The formation of emulsions makes oil clean-up operations more difficult by decreasing the effectiveness of physical oil spill recovery procedures and suppressing the natural rates of oil biotransformation.

2.6 NATURAL DISPERSION

Dispersion is oil-in-water emulsion resulting from the incorporation of small globules of oil into water column. It occurs when wave action causes a surface slick to break into oil drops which mix and spread within the water column. These naturally dispersed droplets are larger than those observed when dispersants are used and they may float back to the surface where they recombine to form another slick. Some natural dispersion occurs with all oils, especially light oils. Oil begins dispersing immediately on contact with water and is most significant during the first 10h or so. It is the second most important process. It usually begins soon after the spill occurrence, reaching its maximum rate in approximately 10h after spill occurrence. Generally, oil-in-water emulsions are not stable. However, they can be maintained by continuous agitation, interaction with suspended particulates, and the addition of chemical dispersants. Dispersion may influence oil biotransformation rates by increasing the contact between oil and microorganisms and/or by increasing the dissolution rates of the more soluble oil components (Zhao et al., 2014a,b). Vilcáez et al. (2013) estimated a biotransformation model, which suggested that oil droplets biodegraded faster than dissolved oil components due to the assumption of complete microbial coverage of oil droplets, due to the larger surface area of small oil droplet per unit mass of microorganisms.

2.7 SEDIMENTATION AND SINKING

Sedimentation is the association of oil with heavier solids suspended in the water column, generally close to the shore. It can occur when oil is weathered to the point where its specific gravity is greater than that of the water. Over time, these suspended solids may settle on the sea floor to form sediments (Bandara et al., 2011). If dispersants are applied before sedimentation has the potential to occur, they can serve to prevent this process by dispersing the oil offshore, thereby preventing it from coming into shallow shoreline areas, where it may encounter abundant sediment. The dispersed oil droplets remain floating and do not sink. The type, size, and load of sediment; the salinity; the contents of sulfur and organic matter (particularly humic acid) in the oil; and the degree of agitation affect the adsorption-desorption of oil onto sediment particles (Adhikari et al., 2015).

Some crude oils sink after incineration, as light fractions are consumed, and heavier pyrogenic products are formed. In rough seas, dense oil can be washed over and spend a considerable amount of time just below the surface, making observation of the oil spill from air very difficult. This makes confusion with oil sinking, but when conditions become calm, the oil resurfaces. Sedimentation is one of the key long-term processes that lead to the accumulation of spilled oil in the ecosystem. However, sinking of bulk oil is only rarely observed other than in shallow water, close to shore, due to the shoreline interaction. Sedimentation is very important in shallow and rough sea conditions; where the bottom sediments are repeatedly re-suspended (Valentine et al., 2014). It could begin soon after the occurrence of the spill and peaks for several weeks. There is another process—*the shoreline standing*—which is the visible accumulation of oil on shorelines following a spill. It is mainly affected by the proximity of the spill to the shore, current intensity, and wave action on the shoreline and the persistence of the spilled pollutants (Gutierrez et al., 2013).

2.8 DEGRADATION

Bio-degradative processes influencing fate of petroleum in aquatic environment include microbial degradation, ingestion by zooplankton, uptake by aquatic invertebrates and vertebrates, as well as, bioturbation. Biotransformation is a natural process in which living organisms and/or their enzymes break down organic material to produce simpler chemical compounds such as organic acids, alcohols, and/or gases (Mercer and Trevors, 2011; Yassine et al., 2013; Singh et al., 2014). Microorganisms capable of oxidizing petroleum hydrocarbons and related compounds are widespread in nature. The percent composition within the oil is dependent on the type of oil. In light oils, asphaltene and resins constitute between 1% and 5% of the four components, whereas in heavier oils, they may constitute up to 20% of the four oil components. Both asphaltene and resins are composed of carbon and hydrogen; however, they also contain sulfur, nitrogen, and oxygen and are made up of polar molecules (Ali et al., 2012; El-Gendy et al., 2014). The degradation of asphaltene may occur through co-metabolism and low molecular weight resins may also be susceptible to microbial degradation. But it is generally known that they are more resistant to biotransformation than both saturated and aromatic compounds and with a higher percentage of saturates and aromatics, the biodegradability of the oil increases while the vice versa occurs with higher percentage of asphaltenes (Venosa and Zhu, 2003). Thus, as degradation of oil occurs, the fraction of resins and asphaltenes increases as the saturates and aromatics decrease (Pritchard et al., 1992).

The rate of microbial degradation varies with the physicochemical characteristic of the pollutant (e.g., the molecular weight and structure of the contaminants), chemical complexity of the crude (i.e., ratio of biodegradable alkanes and aromatics to recalcitrant resins and asphaltenes), the microbial populations and many of the environmental conditions, such as physical state surface area available at the oil-water interface for microbial attachment or dissolution of light hydrocarbons and hence also dependence on spreading, dispersion, and emulsification state of the oil); and temperature (biochemical activity being slower at low temperatures); nutrient availability (such as phosphorous and nitrogen); and the availability of oxygen and electron acceptors for redox reactions (aerobic or various anaerobic conditions).

The range of susceptible hydrocarbons under anaerobic conditions appears to be smaller than with aerobic degradation, and anaerobic biotransformation of oil is usually mediated by sulfate-reducing bacteria or other anaerobes using a variety of other electron acceptors such as oxidants, and the enzymes, pathways, end products, and residual oil composition also differ (Fukui et al., 1999; Widdel and Rabus, 2001; Foght, 2006; Mbadinga et al., 2011; Sherry et al., 2013; Jaekel et al., 2015). Oil degradation is generally faster and more efficient under aerobic conditions, such as in a well-aerated water column or in the uppermost layer of sediment, but may also occur slowly in buried sediments under anaerobic conditions. Cappello et al. (2015) suggested a modular slurry system, allowing containment of the sediments and their physical-chemical treatment by air insufflations, temperature regulation, and the use of a slow-release fertilizer, to overcome the problems of low temperature, low oxygen levels, and nutrient availability. The rate of biotransformation and rate of mass loss due to biotransformation are slower than that of evaporation or dissolution. Biotransformation is an extremely slow process which only in the long run becomes important in the removal or alteration of oils from the marine environment, and it is one of the significant factors that affect the fate of PACs, and its effect depends mainly upon the nature of the spilled oil and the spill conditions. The process of dispersing the oil into the water column to enhance natural biotransformation is the goal of dispersant use (Atlas and Hazen 2011). Research has shown that the petroleum-degrading microbes in the water column more rapidly colonize dispersed oil droplets than oil droplets without dispersant (Khan and Husain, 2003; George-Okafor et al., 2009). Biotransformation produces several intermediates, but the final products are carbon dioxide and water. Biochemical processes of oil degradation carried out by microbes include involvement of several types of enzymatic reactions driven by oxygenase, dehydrogenase, and hydroxylase. These enzymes cause aromatic and aliphatic hydroxylation, oxidative deamination, hydrolysis, and other biochemical transformations of original oil substances leading to formation of large number of intermediate degradation products (Kumari et al., 2012).

Usually mixed microbial consortia are required for succession and completion of the degradation of different components of oil spill (El-Tarrs et al., 2012). Hydrocarbon-degrading bacteria are ubiquitous, and although they may be present in a pristine environment in only small numbers, they can flourish after an oil spill because they have an advantage over microbes that cannot utilize hydrocarbons. Environments that are routinely exposed to oil spills or natural hydrocarbon seeps usually have adapted indigenous microbial communities that are enriched in hydrocarbon-degraders and thus can respond more quickly to an incursion of oil. However, the widespread synthesis of alkanes by marine cyanobacteria may help maintain a baseline level of competent hydrocarbon-degrading bacteria even in pristine waters (Lea-Smith et al., 2015).

Most of the studies on the fate of oil spill consider PAHs and the phenolic constituents, as they pose significant threat to the environment with their extreme toxicity even at low concentrations, carcinogenicity, stability, bioaccumulation, and persistence in the environment for long periods (Ryynänen, 2011). The EU has listed phenols and PAHs as priority pollutants and their concentrations must be controlled.

Hydrocarbon derivatives differ in the susceptibility to microbial attack. In general, the degradation of hydrocarbon derivatives is ranked in the following order of decreasing susceptibility: n-alkanes > mono-aromatic compounds (i.e., benzene, toluene, ethylbenzene, and xylene; i.e., BTEX) > branched alkanes and cyclic alkanes > polynuclear aromatic hydrocarbons with a negative correlation between degradation and number of rings). However, microorganisms (naturally occurring or genetically engineered) can mineralize toxic PAHs into CO_2 and H_2O and sometimes at a faster rate than alkanes (Samanta et al., 2002; Guo et al., 2005; Bacosa et al., 2010; Bao et al., 2012; Chen et al., 2014; Suja et al., 2014; Barbato et al., 2016). But, sometimes, a competitive inhibition between crude oil degrading bacteria occurs, and, not all the time, consortia, perform higher degradation efficiency (Hassanshahian et al., 2014; Mishamandani et al., 2016). Thus, it is theoretically and practically important to seek out an appropriate path to construct efficient marine oil-degrading bacterial consortia.

3.0 OIL SPILL REMEDIATION METHODS

Oil spills are taken seriously by both up and downstream oil industry. Oil spills can be categorized based on discharge quantity of oil on land, coastal, or offshore waters as minor spills, medium spills, major spills, and disasters (Tewari and Sirvaiya, 2015). Conventional remediation methods include physical removal of contaminated material. These methods also use chemicals, especially shoreline cleaners, which are often organic solvents with or without surfactants (Riser-Roberts 1992). The shoreline cleaners with surfactants emulsify the adsorbed oil, which entrain adjacent waters or is transported deeper into the shoreline soil. The oil-solvent mixtures are collected using conventional skimming methods. Mechanical recovery of oil includes the use of the oil sorbents. Sorbents help to transform oil to a transportable form for short-term storage. However, most of the used sorbents end up in the landfills. Most of the physicochemical methods use chemical agents which, as well as, their emulsion with oil cause toxicity to aquatic organisms; they produce another source of pollution and increase the oil recovery cost. Additionally, abiotic losses due to evaporation of low molecular hydrocarbons, dispersion, and photo-oxidation involve only aromatic compounds that play a major role in decontamination of the oil spill environments (Mills et al., 2003). Davis and Guidry (1996) estimated the average cost of cleaning a crude oil spill to be $2730 per barrel, where the offshore containment reclamation costs can be high per barrel, but still considerably less than shoreline cleanup and resource damage. There is an increased interest in promoting environmental methods in the process of cleaning oil-polluted sites. These methods are less expensive and do not introduce additional chemicals to the environment.

3.1 BOOMING FLOATING OIL

Some of the tools used to control oil in a spill include "booms," which are floating barriers used to clean oil from the surface of water and to prevent slicks from spreading. A boom can be placed around the tanker that is spilling oil. Booms collect the oil off the water. A boom may be placed somewhere close to an oil spill. The boom can also be placed around an entrance to the ocean, like a stream, or placed around a habitat with many animals living there. These booms absorb any oil that flows around it. Booms are used in various configurations to contain and recover slicks. Two vessels can tow a boom in a U-configuration to collect oil by drifting downstream, holding in a stationary position, or by moving upstream toward the spill source.

Based on the floating tendency, material with which they are made of, weight and stopping tendency, booms can be categorized into three categories: (1) fence boom, (2) curtain boom, and (3) fire-resistant boom (Dave and Ghaly, 2011). Booms can be deployed in a V-configuration using three vessels and a skimmer. Booms can be also towed in a J-configuration that will divert the oil to a skimmer to allow simultaneous containment and recovery. Boom containment of floating oil requires consideration of on-water conditions, weather conditions, available equipment, and booming strategy. Booms work for most oil types and large or small oil volumes. Containment is most effective when the booms can be accurately directed toward the oil. However, booms almost always leak, even under the best of circumstances. A boom is only as good as the crew that deploys and controls it. Booms are not a static piece of equipment; thus, they require constant attention.

3.2 SKIMMERS

The workers can also use skimmers carried in boats to remove the oil off the water. These devices can be used in conjunction with booms to recover oil from water surface without changing its properties so it can be reprocessed and reused. Different skimmers work for different oil types. Skimmers are of three categories; weir, oleophilic, and suction. However, skimmers are inefficient in rough waters. They are generally effective in calm waters and subject to clogging by floating debris. Mechanical recovery, or the physical removal of oil from the environment, is the method that is usually perceived as the least harmful to the environment. However, mechanical recovery usually can recover only a small fraction of the spilled oil. Experience has indicated that recovery of more than 20% of the original spill volume is seldom achieved in marine spills (Chen et al., 1998). In fact, in open water under strong current and wind conditions, recovery of only 5%–10% is not uncommon. Therefore, mechanical recovery is normally used in conjunction with other methods.

In its simplest form, mechanical recovery relies on a skimmer capable of removing oil from the surface of the water and pumping it to a storage vessel for subsequent treatment and disposal. Numerous devices are commercially available that employ a variety of oil pickup mechanisms. Oil spilled on water tends to spread rapidly to a thin sheen, thus reducing the potential skimmer-to-oil ratio encounter rate, and makes mechanical recovery inefficient. To maximize the encounter rate, oil spill containment booms are deployed to concentrate the oil (Chen et al., 1998). Specific types of boom and skimmers are selected on a case-by-case basis, depending on the location of the spill, type of oil, environmental conditions, etc. In all cases, efficiency depends on oil/skimmer encounter rates, weather conditions, currents, sea state, and operator efficiency. Sufficient temporary storage for recovered liquid should also be planned (Chen et al., 1998).

3.3 PUMPS OR VACUUMS

A pump or vacuum can be used to remove oil as it floats on water. Pumps are used during oil spill response to transfer oil, water, emulsions, and dispersants. Recovered liquids typically need to be transferred from: (1) a skimmer to an interim storage device; (2) interim storage to a larger storage/separation or transportation vessel; (3) a transportation vessel to a final storage/disposal facility. Transfer equipment must be selected to suit the quantities and types of liquids being moved. Although a wide range of pumps can be used for fresh, un-emulsified oils, as transfer conditions become more difficult, pump options can become limited. Careful consideration must therefore be given to each specific transfer situation, particularly in the case of long-term mechanical recovery operations when, over time, oil weathers, viscosity increases, and debris is collected. Pumps may be also used to offload oil from stricken vessels (called "lightering") and to transfer dispersants from drums and other containers to dispersant application systems. Generally, spill cleanup does not require pumps with extreme capabilities. The hydrostatic head through which the pump must push liquid is usually about 2–6 m or 6–20 ft and suction lift from skimmer to pump is often much less than that (i.e., only a few feet or about a meter). In some cases, a large hydrostatic head is required, especially when oil is pumped from a skimmer to a large, unballasted barge or storage vessel. In this case the head required may be 10 m 30 ft or more (Kennish, 2001). Some pumps are not suitable for oil spill work because they neither self-prime nor maintain prime when the skimmer rolls: suction capacity is limited; pumping capacity decreases even with slight increases in oil viscosity; cavitation occurs in warm or high viscosity oil; emulsification of oil and water occurs; debris blocks the pumping mechanism; damages are caused by running dry. Usually, four pumps are both suitable and commonly used for spill cleanup: centrifugal, peristaltic, screw/auger, and reciprocating diaphragm.

3.4 IN SITU BURNING

Occasionally the slick caused by a spill is removed through a controlled burn. Workers can burn freshly spilled oil with fireproof booms to contain the oil. Although it is widely used since the late 1960s to remove spilled oil and jet fuel in ice covered waters and snow resulting from pipeline, storage tank, and ship accidents in the United States and Canada, as well as several European and Scandinavian countries (Mullin and Champ, 2003). But, it only works under certain wind and weather conditions. Moreover, it is the last option to decide, as this method causes air pollution and there is a risk of catching secondary fire.

In situ burning of oil in place can quickly eliminate large quantities of spilled oil. Spill response planners now recognize that there are various situations where controlled in situ burning can be conducted quickly, safely, and efficiently. Significant advances in techniques and equipment for in situ burning have been made in recent years. Combustion of spilled oil is not a substitute for containment and mechanical removal. Conventional booming and skimming operations should always be conducted whenever they can be implemented safely and with a reasonable degree of effectiveness. However, there are often situations where burning may provide a means of quickly and safely eliminating large amounts of oil. The objective is to select the optimal equipment, personnel, and techniques that will result in the least overall environmental impact (Fritt-Rasmussen, 2010). The in situ burning of spilled oil offers some advantages; it removes large quantities of oil rapidly and efficiently, prevents or minimizes the amount of oil that reaches shorelines, and can be often used in situations where skimming is physically or logistically impossible, thus greatly reducing the need for storage and disposal facilities near the slick area. Moreover, it is the simplest method and can be the most effective method for oil removal from a water surface, leading to about 90% oil removal. But, it is effective in calm wind conditions and spills of fresh oils or light refined products which quickly burn without causing any danger to marine life and the residue may sink and cover up an underground water resource. Removal of the residue can be achieved through mechanical means (Davidson et al., 2008).

3.5 DISPERSANTS

Chemical dispersants are used to break oil slicks into fine droplets that then disperse into the water column. Dispersants proved their capabilities to treat up to 90% of spilled oil and are less costly than the physical methods (Holakoo, 2001). An airplane can be used to fly over the water dropping chemicals into the ocean. Moving the oil in this way, prevents oil from being driven by wind and currents toward shore, allows it to eventually be consumed by bacteria, and promotes its biotransformation at the sea, due to the increase of the surface area of the oil droplets. Thus, keeps it from animals that live at the surface of the water, and minimizes the amount of oil reaching sensitive and biologically productive habitats, for example, mangroves, salt marshes, coral reefs, kelp beds, etc. Dispersants can also reduce impacts by lowering the adhesive properties of the oil. Whatever net environmental benefit is achieved, it is best to disperse oil some distance before it approaches important ecological habitats. In all cases, care should be taken to avoid overdosing. Dispersants should be considered for use with other potential spill response methods and equipment, and not as a last resort. For maximum effectiveness, dispersants should be applied as soon as possible after a spill. During the early stages of a spill, the oil is un-weathered and less spread out, making it easier to disperse.

The decision on whether to use dispersants should be made after considering the potential effects of dispersed oil vs. undispersed slicks. The objective should be to minimize ecological impacts and maximize net environmental benefit. Dispersant application should proceed even if initial wind and sea conditions appear calm, and if immediate dispersion of oil is not readily observed. Dispersion of the oil droplets into the water column is influenced by the amount of mixing energy in the sea. Weather changes can generally be expected in the marine environment, and many dispersants tend to remain in the oil slick for some time after application. The time taken to respond to a spill influences the feasibility of a dispersant operation. Over time, an oil slick weathers and becomes more viscous, especially if it incorporates droplets of water (emulsifies) and becomes "mousse." The resulting increase in viscosity and water content makes chemical dispersion more difficult. The rate of weathering is primarily a function of oil type, slick thickness, sea state, and temperature. For typical slicks, an application rate for dispersant concentrates of between 20 and 100L/ha can be used. On a volume-of-oil to volume-of-dispersant basis, these application rates range from less than 100:1 to more than 10:1, depending on the thickness of the oil. They can be used on rough seas where there are high winds and mechanical recovery is not possible. Applicability of dispersants depends on the type of oil, temperature, wind speed, and sea conditions (Nomack and Cleveland, 2010). However, the inflammable nature of most dispersants can cause human health hazards during applications and potential damage to marine life. They are also responsible for fouling of shorelines and contamination of drinking water sources. Cleaning agents which aim to emulsify and fragment oil particles to make them available to bacterial degradation are mainly composed of solvents that soften the oil and enable recovery with mechanical methods and surface-active agents that disperse oil into the water (Crisafi et al., 2016). Although, dispersants are generally assumed to be less toxic than oil, their increased oil-in-water solubility could make chemically dispersed oil more toxic to aquatic organisms, including algae, micro-zooplankton, and fish, and they can also affect microbial degradation of oil-derived hydrocarbons and persist for long time after application (Lindstrom and Braddock, 2002; Ramachandran et al., 2004; National Research Council, 2005; Kujawinski et al., 2011; Lewis and Pryor, 2013; Almeda et al., 2014; White et al., 2014; Kleindienst et al., 2015). Zhuang et al. (2016) reported that the chemical properties of the surfactants present in dispersant products affect the microbial uptake of hydrocarbons.

Hamdan and Fulmer (2011) reported the effect of some chemical dispersants on some bacterial strains and marine microbial communities, where a reduction on growth rates and viability of petroleum-degrading *Acinetobacter* and *Marinobacter* was observed, while strains belonging to non-hydrocarbon-degrading *Vibrio* were proliferating likely because the dispersant provided an additional carbon source. However, Chakraborty et al. (2012) studied the effect of the dispersant Corexit 9500 on indigenous microbial population, as it was sprayed to mitigate Deepwater Horizon spill of MC-252 oil that occurred in the Gulf of Mexico. The hydrocarbon *degrading* bacteria belonging to *Oceanospirillales*, *Colwellia*, *Cycloclasticus*, *Rhodobacterales*, *Pseudoalteromonas*, and *methylotrophs* were found enriched in the contaminated water column and the dispersant was not toxic to the indigenous microbes at concentrations added, and different bacterial species isolated in the aftermath of the spill were able to degrade the various components of Corexit 9500 that included hydrocarbons, glycols, and dioctyl sulfosuccinate.

Bacosa et al. (2015a) studied the effect of crude oil, Corexit 9500, and sunlight on the indigenous microbial pollution after an oil spill occurrence, and found that, in samples containing oil or dispersant, sunlight greatly reduced abundance of the *Cyanobacterium synechococcus* but increased the relative abundances of *Alteromonas*, *Marinobacter*, *Labrenzia*, *Sandarakinotalea*, *Bartonella*, and *Halomonas*. Both oil and Corexit inhibited the *Candidatus pelagibacter* with or without sunlight exposure, while dark samples with oil were predominated by members of *Thalassobius*, *Winogradskyella*, *Alcanivorax*, *Formosa*, *Pseudomonas*, *Eubacterium*, *Erythrobacter*, *Natronocella*, and *Coxiella*.

Seidel et al. (2016) studied the biotransformation of crude oil and dispersants in deep seawater from the Gulf of Mexico in microcosm systems of oil, dispersant, and dispersed oil (with/without nutrient amendment) and the biotransformation was documented, where, the N, S, and P containing compounds were decreased in the oil-only incubations, and proved to act as source of nutrients for the microbial population during the metabolism of hydrocarbons. The degradation was lowered in the oil-dispersed microcosms, and the rate of degradation of the dispersant itself was low.

3.6 ADSORBENTS

Hydrophobic sorbents are of great interest for controlling oil spills; they can clean up the remaining oil after skimming operations. They can also facilitate conversion of liquid to semisolid phase for complete removal of oil (Adebajo et al., 2003). They can be natural organics (e.g., peat moss, kapok, saw dust, vegetable fibers, milkweed, cotton fibers, and straw, etc.) (Karakasi and Moutsatsou, 2010; Banerjee et al., 2006), natural inorganics (e.g., clay, glass, wool, sand, vermiculate, hydrophobic perlite and volcanic ash, etc.) (Holakoo, 2001; Alther, 2002), and synthetic materials (e.g., polypropylene, polyester foam, polyurethane foams, and polystyrene, etc.) (Teas et al., 2001).

Natural organic adsorbents are less expensive, readily available, and their adsorbing capacities are 3–15 times their weight. However, they are labor-intensive, adsorb water along with oil which leads to their sinking, and it is difficult to collect adsorbents after spreading on the oil spill water and must be disposed (Nomack and Cleveland, 2010). Also, natural inorganic adsorbents are less expensive, readily available, and their absorbing capacities are 4–20 times of their weight. But most of them are loose material and very difficult to apply in windy conditions and they are associated with potential health risk if inhaled. Synthetic sorbents are available in sheets, rolls, or booms and can also be applied on to the water surface as powders. They have adsorbing capacity 70–100 times of their weights in oil due to their hydrophobic and oleophilic nature. They can be used several times, but their main drawback is non-biodegradability and storage after their usage (Teas et al., 2001; Deschamps et al., 2003). The sorbents usually facilitate the change of the oil spill from liquid to semisolid phase, thus facilitating its removal. The hydrophobic character and oleophilic character are the primary determinants of the successful performance of sorbents. Other factors, including the retention over time, the recovery of oil from sorbents, the amount of oil sorbed per unit weight of sorbent, and the reusability and biodegradability of sorbent, are also important (Doerffer, 1992).

3.7 BIOREMEDIATION

Bioremediation may be used to accelerate the process of biotransformation of the oil after a spill. In this process, nutrients and/or bacteria or other microbes are introduced to the environment to help oxidize the oil (Helmke et al., 2013). Unfortunately, this process can work slowly and is not very useful for large spills. Although, bioremediation can be effective, due to its slow recovery time, it is not always considered. Bioremediation is not only economical, but it is an effective technique for sensitive shorelines, due to being nonaggressive to the shoreline habitat (Boufadel et al., 2011, 2016). According to Dave and Ghaly (2011), mechanical treatment and dispersants application followed by bioremediation are the most effective treatments for a marine oil spill.

4.0 FACTORS INFLUENCING RATES OF OIL SPILL BIOREMEDIATION

The main factors that affect the biotransformation of the contaminated site are biotic factors such as the specific microorganism and its concentration and interaction and abiotic or physicochemical factors which include the chemical structure of the pollutant and bioavailability. Lastly, environmental factors also play a major role; availability of oxygen and nutrients, pH, temperature, pressure, salinity, presence of heavy materials (Sabate et al., 2004; Martín, 2011; Tripath and Srivastava, 2013). However, the main important factors affecting bioremediation of marine oil spills are the bioavailability of nutrients, the concentration of oil, time, and the extent to which the natural biotransformation has already been taken place (Zahed et al., 2010a,b).

4.1 NUTRIENTS

Microorganisms commonly require carbon, nitrogen, phosphorous, and potassium for the degradation of hydrocarbons (Chandra et al., 2013; Sihag and Pathak, 2014). Carbon forms around 50% of the dry mass of the cell of microorganism and is essential for cellular structure and energy. The microorganisms which must metabolize carbon need nitrogen and phosphorus since nitrogen forms about 10% of the dry cell mass and is the source for synthesis of amino acids and enzymes; moreover, it is essential for the harnessing of proteins, nucleic acids, and cell wall polymers.

However, nutrients such as nitrogen and phosphorus are always in low concentrations in marine environment, i.e., insufficient for the microbial growth, which would be the main reason for the decreased rate of natural biotransformation (Ron and Rosenberg, 2014; Saikia et al., 2014). The concentration should be in the ratio of C:N:P:K equivalent to 100:15:1:1 (Juteau et al., 2003; Filler et al., 2006; Chandra et al., 2013). Rosenberg et al. (1998) reported that approximately 150 g of nitrogen and 30 g of phosphorus are consumed in the conversion of 1 kg of hydrocarbon to cell material. In another study, Zhu et al. (2001) reported that theoretically the reduction of 1 g of hydrocarbon requires 150 mg nitrogen and 30 mg phosphorous, with a common formula of a stoichiometric ratio of carbon-nitrogen-phosphorus of 100:5:1. However, it has been found that different nitrogen-phosphorus ratios will yield different concentrations of various microbes, and since oils have different properties and every spill has different environmental factors, the correct ratios are recommendable to be determined for each oil spill.

The addition of nutrients can be in various forms: water-soluble, granular, or oleophilic. Water-ssoluble nutrients are more readily available than granular and oleophilic forms, but can be easily washed away and may have to be applied more often. Granular nutrients are slow release and so do not wash away, but the release rate is often difficult to predict. Oleophilic nutrients can adhere to the oil and, thus, be very close to the microbes that need their nutrients, but it is usually expensive, with variable effectiveness, and contains organic carbon, which may compete with contaminate degradation and would result in undesirable anoxic conditions. But, generally, the addition of N and P along with other nutrients has been very effective in growth of bacteria and hydrogen degradation (Leys et al., 2005; Rölling et al., 2002; Choi et al., 2002; Kim et al., 2005). According to Gordon (1994), when the proper nutrients are present, an oil spill that was estimated to be cleaned by natural conditions in 5–10 years could be cleaned in 2–5 years with the use of bioremediation. According to Pritchard et al. (1992) the oil in fertilized areas biodegraded about two times faster than untreated controls. However, if excessive nutrients are added it can hinder the growth of bacteria and can hamper the biotransformation activity (Chaillan et al., 2006). Many authors have reported the negative effects of adding high concentration of NPK to the degradation of aromatic chains (Chaîneau et al., 2005, Saikia et al., 2014). Zahed et al. (2010b) studied the effect of nutrient concentration on biotransformation rate of light crude oil polluted seawater, augmented hydrocarbon degrading bacterial consortium using response surface methodology (RSM) based on a five-level, three-factor central composite design (CCD) of experiments, where, for an initial crude oil concentration of 1 g/L supplemented with 190.21 mg/L nitrogen and 12.71 mg/L phosphorus approximately, 60% removal could be achieved in a 28-day experiment. In another study, RSM based on CCD was employed for optimization of nitrogen and phosphorus concentrations for removal of n-alkanes from crude oil contaminated seawater samples in batch reactors, using indigenous acclimatized microorganisms. During 20 days of bioremediation, a maximum of 98% total aliphatic hydrocarbons removal was observed, using nitrogen and phosphorus concentrations of 13.62 and 1.39 mg/L, respectively (Zahed et al., 2010c).

It is preferable to use alternative and environment-friendly nutrients than the inorganic nutrients N and P which have negative impact on the ecosystem. Since the goal of any bioremediation protocol is the destruction of the pollutants, using cheap and readily available nutrients sources is encouraging; the alternative being corn steep liquor. Corn steep liquor has been proven to be cheap and has shown good results in cleanup of PAH and petroleum hydrocarbon contaminated sites (Obayori et al., 2010; Younis et al., 2013; El-Gendy et al., 2014; El-Mahdi et al., 2015a). Corn steep liquor is a rich source for organic nitrogen with minerals and cofactors (Obayori et al., 2010; Younis et al., 2013; El-Gendy et al., 2014). Agricultural waste such as sawdust, hay, shells, straws, and manure contain high concentration of N and P and are an environmentally friendly choice as biostimulation (Medina et al., 2004; Prince et al., 2013). Also, the low-cost natural agro-industrial solid waste dates have been proven as an effective nutrient for biotransformation of light and heavy crude oil in polluted seawater (El-Mahdi et al., 2014). The most important problem is the effectiveness of added nutrients in open systems, such as seawater, because of rapid dilution. Thus, the interesting possibility of using nitrogen-fixing, hydrocarbon-oxidizers has been considered (Coty, 1967; Thavasi et al., 2006; Foght, 2006).

Knezevich et al. (2006) reported the success of the commercial uric acid as an effective source for nitrogen and phosphorus for the growth of marine bacteria on crude oil where biotransformation reached 70% in a simulated open oil spill sea system. The rate of release of the nutrients is another important factor that should be considered for the success of a bioremediation process, and there is a need to make sure that the formulas (i.e., the nutrients) are not too slow, otherwise they do not give the organisms enough of the necessary nutrients to allow for optimal growth. "Washout" refers to tide that carries water out to sea and takes some nutrients with it. For intertidal environments, the fertilizer needs to be slowly release to overcome the washout effect.

Oleophilic and other slow release formulas are used to prevail against the washout (Zhu et al., 2001). Moreover, Boufadel et al. (2006) reported that the addition of nutrients during low tide at the high tide line would result in maximum contact time of the nutrients with the oil and hydrocarbon degrading microorganisms. Boufadel et al. (2007) showed that when a wave is present there is a sharp seaward hydraulic gradient in the backwash zone, and a gentle gradient landward of this area, and the contact time of the nutrients is increased when the waves break seaward of their location. The waves also increased the dispersion and washout of the nutrients in the tidal zone, and residence time was approximately 75% when a wave was present with a tide, as compared to a tide with no waves.

4.2 OIL CONCENTRATION

High initial concentration of spilled oil has a negative effect on the biotransformation process that causes a significant lag phase of about 2–4 weeks. Even after biostimulation, at least a week is needed for microorganisms to adapt and the entire bioremediation process may require months and even years to be completed (Atlas, 1995; Zahed et al., 2010a). The rates of uptake and mineralization of many organic compounds by a microbial population depend on the concentration of the compound (Olivera et al., 1997; Rahman et al., 2002).

4.3 BIOAVAILABILITY OF CONTAMINANT

It has been reported that with very low water solubility, the maximum rate of bioremediation depends mainly on the mass transfer limitations (Ostroumov and Siegert, 1996), while, contaminants with log

K_{ow} of 1–35 are reported to have the highest aerobic and anaerobic biotransformation rates (Bressler and Gray, 2003). The greater the complexity of the hydrocarbon structure, i.e., the higher the number of methyl branched substituents or condensed aromatic rings, the slower the rates of degradation (Atlas, 1995). Generally, the size, molecular weight, and substituted derivatives affect the enzymatic activity due to the mass transfer limitation and steric hindrance. Thus, the higher they are the lower the biotransformation rate, as more activation energy is required to be produced by the microorganisms to support the enzyme activity (Kropp and Fedorak, 1998; Bressler and Gray, 2003). Yang et al. (2009) reported that, level of energy consumption can serve as an indicator for the bioremediation efficiency. Karl (1992) and Sparrow and Sparrow (1988), reported that, lower adenosine triphosphate (ATP) levels at the oil spill site would indicate low microbial population or low microbial activity.

4.4 TEMPERATURE

Temperature plays a major part in maintaining the metabolic activity of the microorganisms and the composition of hydrocarbons (Bagi et al., 2013). Microbes are well-known to degrade oil at a variety of temperatures; however, a very low temperature or a very high temperature would reduce the effectiveness and speed of degradation (Gordon, 1994). At ambient temperature, the activity of a microorganism is affected along with the properties of spilled oil. Since temperature mainly affects the viscosity of the spilled oil, its chemical composition, microbial community, mass transfer of the substrates and consequently the rate of biotransformation, bioremediation is reported to be relatively slow in cold regions (Atlas, 1981; Filler et al., 2006; Yang et al., 2009). It has been reported that marine microbial communities from cold seawater have potentials for oil film hydrocarbons degradation at temperatures ≤5°C, and the psychrotrophic or psychrophilic bacteria play an important role during oil hydrocarbons biotransformation in seawater close to freezing point (Brakstad and Bonaunet, 2006).

The occurrence for hydrocarbon biotransformation has been reported in pristine cold environments, and alkane monooxygenases and hydrocarbon clastic bacteria have been detected and characterized in both Arctic and Antarctic environments (Whyte et al., 2002; Yakimov et al., 2003), and studies of oil-polluted polar environments have shown the abundance of *Alphaproteobacteria*, *Gammaproteobacteria*, and *Actinobacteria* (Grossman et al., 1999; Aislabie et al., 2000, Juck et al., 2000; Yakimov et al., 2003). Atlas (1991) reported that the isoprenoids, phytane and pristane, are more resistant to bacterial metabolism at 4°C and the lower water solubility and higher sorption capacity of PAHs in cold weather render their degradation (van Hamme et al., 2003).

Biotransformation of heavy fuel (Bunker C) by indigenous organisms in the North Sea was reported to be four times greater in summer (18°C) than in winter (4°C) (Balks et al., 2002). Although microbial biotransformation activity is observed at subzero temperatures, studies have shown that the rate of hydrocarbon degradation increases at mesophilic 30–40°C and sometimes in thermophilic 60°C temperatures, depending on the enzyme activity (Sonawdekar, 2012). But generally, the optimum temperature for biotransformation is reported to be 15–30°C and 25–35°C for aerobic and anaerobic processes, respectively (Rawe et al., 1993; Aislabie et al., 2006). When the temperatures lower to the freezing point, the channels in the cell membrane tend to be closed and the cytoplasm is subjected to the cryogenic stress, since the cytoplasmic matrix becomes frozen and the cell will stop functioning, closing the transport channels. This would persist for long period, thus, restrict the mass transport and limit contaminants to gain access into the cells (Yang et al., 2009). Diesel oil has been reported to degrade a lower temperature 0–10°C (Das and Chandran, 2011). Some studies have shown hydrocarbon degradation in psychrophilic (less than −20°C) temperature too (Pelletier et al., 2004; Santos et al., 2011). McFarlin et al. (2014) reported the substantial loss of oil at −1°C with significant biotransformation of 2- and 4-ring PAHs and their alkyl-substituted homologs in Arctic seawater. Kristensen et al. (2015) performed a microcosm study on the fate of Arctic offshore oil pollution in relation to a temperate one. The rate of biotransformation of different hydrocarbons in crude oil was found to be PAHs and dibenzothiophenes naphthalene derivatives>dibenzothiophene derivatives>phenanthrene derivatives>fluorenes>alkyl toluene derivatives>n-alkane derivatives, in the temperate North Sea, Denmark samples, whereas, in the Arctic site Disko Bay, Greenland samples, the order was reversed: n-alkane derivatives>alkyl toluene derivatives para-isomer>meta-isomer>ortho-isomer>Polynuclear aromatic hydrocarbons and dibenzothiophenes. But, generally, the rate of biotransformation decreased with alkyl substitution.

4.5 OXYGEN AVAILABILITY

Dissolved oxygen affects the metabolic activity of microorganisms (Yang et al., 2009). Bioremediation would take place both under aerobic and anaerobic environments. The microorganism uses the available oxygen for metabolism, hence are called aerobic, where hydrocarbons are metabolized by the microorganism to make carbon dioxide, water, and other chemical by-products under aerobic conditions (AECIPE, 2002). Rate of aerobic biotransformation is much higher than that of anaerobic one, since, aerobic processes yield a greater potential energy yield per unit of substrate (Yang et al., 2009). Oxygen is important for oxygenases, and molecular oxygen is involved in the major degradation pathways for the hydrocarbons (Martín, 2011). Theoretically, 0.3 g oxygen/g oxidized oil is required to remediate a hydrocarbon load (Atlas, 1981). The redox potential for aerobic and anaerobic process is reported to be>50 mV and <50 mV, respectively (Yang et al., 2009). In anaerobic degradation, where molecular oxygen is absent, water-derived oxygen serves as a reactant and the electron acceptors used by microorganism are nitrate, ferric iron, sulfate, where the sequence of the reaction would be aerobic respiration first, followed by denitrification; manganese, ferric and sulfate reduction; and finally methanogenesis (Spence et al., 2005; Boll et al., 2014). Anaerobic degradation has gained importance only recently, and previously it was ecologically irrelevant, since its efficiency is much less than that of the aerobic process (Harayama et al., 2004; Zhang et al., 2015).

4.6 PRESSURE

Pressure reduces the microbial activity drastically; it has been found that crude oil which is transported in heated containers, when spilled in ocean forms solid lumps and sinks to the bottom of the ocean floor. The high pressure and low temperature at the bottom reduces the microbial activity by 10 times. Samples extracted from the bottom of the sea showed that it contained 10 times more amounts of hydrocarbons than the samples from the surface (Martín, 2011).

4.7 ACIDITY-ALKALINITY

Studies have shown that the higher the pH, the higher the degradation rate but for optimum degradation the pH should be slightly alkaline. The pH varies at different locations and different environments. Seawater has been found to be alkaline throughout. But the microorganisms are most active under neutral pH in most of the cases (Hunkeler et al., 2002). Slight change in pH affects the biodegradative activity of these microorganisms, in turn affecting the reduction of hydrocarbon chains (Hunkeler et al., 2002; Maheshwari et al., 2014). At higher pH Cations, such as NH_4^+, Mg^{2+}, Ca^{2+}, dissolve and anionic forms NO_3^-, NO_2^-, PO_4^{3-}, Cl^-) are affected by low pH. Hence, the best suited pH for the growth of bacteria is in between 6 and 8. Some organisms such as yeast can grow between pH 4 and 6 and for filamentous fungi it is between 3 and 7 (Martín, 2011). According to Leahy and Colwell (1990) pH values between 6.0 and 8.0 are the most favorable for the action of microorganisms degrading petroleum hydrocarbons, though the fungi are more tolerant to acidic conditions. Sonawdekar (2012) suggested that based on the degradation of hydrocarbons and the environment in which microorganisms are to be used the bacteria must be selected. At lower temperature, fungal strains have been found to be quite effective. In certain cases, the genetically modified group of bacteria including fungi and yeast have been effective even at a pH of 2. Some other bacteria have been found to be effective in the range pH 7.5–10. The choice of bacteria selection is completely dependent on the environment to be used.

4.8 SALINITY

As the salinity increases, the microorganism metabolism decreases. The hydrocarbons, which need to be reduced by microbial activity are completely affected by salinity. The salinity affects the oxygen and nutrient levels as seen in coastal and ocean areas. The metabolism rates are drastically affected and it drops from around 28.4% in less saline water to around 3.3% in hyper saline environment. A more thorough research in this area still needs to be carried out. Even though most of the fresh water microorganisms can be alive in saline water, their reproduction rate is affected. Marine bacteria have found to survive in salinity range of 2.5%–3.5% with limited reproduction or no reproduction if salinity is in the range of 1.5%–2% (Venosa and Zhu, 2003; Santos et al., 2011). It was found that in 0.1–2 M NaCl solution some bacterial activity was more and hydrocarbon degradation is found. At around 0.4 M, the salinity equivalent of seawater, the degradation was reduced (Sonawdekar, 2012).

5.0 BIOREMEDIATION TECHNOLOGY FOR A MARINE OIL

The extensive use of petroleum products leads to the contamination of almost all components of the environment. Bioremediation is a process accelerating the natural attenuation process by the application of microorganisms that assimilate organic molecules to cell biomass producing carbon dioxide, water, and heat (Dave and Ghaly, 2011). Bioremediation is of interest worldwide as a potential oil spill cleanup technique under certain geographic and climatic conditions. It is a technology that offers great promise in converting the toxic compounds of oil to nontoxic products without further disruption of the local environment; typically, it is used as a polishing step after conventional mechanical cleanup options have been applied (Chaîneau et al., 2005; Macaulay and Rees, 2014). Microbiologists have studied this process since the 1940s (ZoBell, 1946). It is an effective technology to remove petroleum hydrocarbons in oil-contaminated sediments of the intertidal marine environment. Bioremediation has been used on a very large-scale application, as demonstrated by the shoreline cleanup efforts in Prince William Sound, Alaska, after the Exxon oil spill. Although the Alaska oil spill cleanup represents the most extensive use of bioremediation, there have been many other successful applications on smaller scale. The ability to degrade hydrocarbon components of crude oil is widespread among marine bacteria

(Floodgate, 1995), and bioremediation has proven to be an effective method for cleaning up residual oil in a variety of coastal environments, such as rocky shorelines, pebble and coarse sand beaches (Atlas and Bartha, 1972; Rölling et al., 2004; Ron and Rosenberg, 2014). Due to the high carbon content of oil and the low levels of other nutrients essential for microbial growth, treatment of beached oil with phosphorous and nitrogen is generally required to enhance the growth of hydrocarbon-degrading bacteria and to stimulate oil degradation (Rölling et al., 2004).

Researches are ongoing to evaluate bioremediation and phytoremediation (plant-assisted enhancement of oil biotransformation) for their applicability to clean up oil spills contaminating salt marshes and freshwater wetlands (Lin and Mendelssohn, 1998; Corgié et al., 2004; Sun et al., 2004). Measuring the success of bioremediation of oil spills is based on several parameters, among them the degradation of PAHs in crude oil. In almost all cases, the presence of oxygen is essential for effective biotransformation of oil and anaerobic decomposition of petroleum hydrocarbons leads to extremely low rates of degradation (Frankenberger, 1992). Thus, the aerobic biotransformation of PAHs is well documented (Beckles et al., 1998; Lei et al., 2005) while anaerobic biotransformation has received relatively little attention, for its low rate.

To distinguish between oil lost by physical means and oil that has been biodegraded, biodegradable constituents are normalized to a resistant biomarker compound, e.g., pristane, phytane, and hopane. Nonbiodegradable or slowly biodegradable components in oil are often called biomarkers (Lee et al., 1997). In refined oils that have no hopanes, biotransformation can be confirmed by normalizing to a highly substituted four-ring PAHs or by examining the relative rates of disappearance of alkanes and PAH homologs (Huang et al., 2004). Comparison of biotransformation indicators (such as nC17/pristane and nC18/phytane) for spilled oil to source oil can also be used to monitor effect of microbial degradation on loss of hydrocarbon at the spill site (Mittal and Singh, 2009; El-Gendy et al., 2014; El-Sheshtawy et al., 2014; Lin et al., 2014; Guerin, 2015a,b).

Wang et al. (2006) have reported that hydrocarbons from spilled petroleum would persist in sediment environment for a substantial period—typically years. According to Tolosa et al. (2005); unresolved complex mixture/total resolvable peaks, i.e., naphthenes, cycloalkanes/n- and iso-alkanes UCM/TRP)>1 indicates chronic petroleum contamination. Carbon preference index>1 and pristine/phytane Pr/Ph)>1 indicate biogenic addition by zooplankton and marine algae. Moreover, Medeiros and Bicego (2004) recorded that high concentrations of Pr 2,6,10,14-tetramethylpentadecane can be derived from zooplankton and phytoplankton, while the presence of Ph 2,6,10-tetramethylhexadecane is used as a marker compound for petroleum. US EPA has reported 16 PAH derivatives as priority pollutants (Fig. 11.2). Generally, the PAH derivatives are classified into three categories according to their molecular weight. Thus, low molecular weight PAH derivatives (2- and 3-membered rings), medium molecular weight PAH derivatives (4-membered rings) and the carcinogenic high molecular weight PAH derivatives (5- and 6-membered rings). Stout et al. (2004) reported that the higher concentration of high molecular weight polynuclear aromatic hydrocarbon derivatives relative to low molecular weight polynuclear aromatic hydrocarbon derivatives indicates pyrogenic sources and this is usually prevalent in aquatic environments.

However, under severe weathering conditions, biomarkers are also reported to be biodegradable (Martin, 1977; Goodwin et al., 1981; Bosta et al., 2001; Huesemann et al., 2003; El-Gendy et al., 2014). Terpanes are found in nearly all oils. These bacterial terpane derivatives include several homologous series, including; bicyclic, tricyclic, tetracyclic, and pentacyclic compounds (e.g., hopane derivatives). Hopane derivatives with the 17 αβ-configuration in the range of C27 to C35 are characteristics of

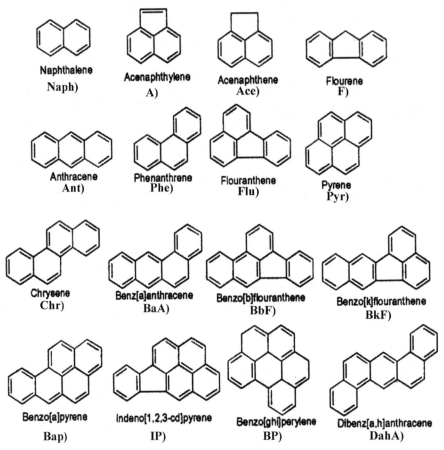

FIGURE 11.2

The 16-polyaromatic hydrocarbons listed by the US-EPA as priority pollutants.

petroleum because of their greater thermodynamic stability compared to other epimeric $\beta\beta$ and $\beta\alpha$ series. The steranes are a class of biomarkers containing 21 to 30 carbons that are derived from sterol derivatives, and they include regular sterane derivatives, rearranged diasterane derivatives, and mono- and triaromatic sterane derivatives. Among them, the regular C_{27}–C_{28}–C_{29} homologous sterane series are the most common and useful steranes because they are highly specific for correlation (Munoz et al., 1997).

For optimum biotransformation conditions, it is important to know the characteristics of the contaminated site before beginning treatments. Basic information such as residual oil concentration, population density of the oil-degrading microorganisms, and the biotransformation potential are key factors to be considered for bioremediation of oil-polluted sites, since the composition of the microbial population is highly affected by the environmental conditions and the composition of the hydrocarbons. The extent of oil degradation and effectiveness of bioremediation agents in marine field environment are often limited by environmental factors including the physical state of the oil, salinity, pH, dissolved oxygen, temperature, and competition between microbial communities (Rahman et al., 2003; Tam et al., 2009; Crawford and Rosenberg, 2013; Sierra-Garcia and de Oliveira, 2013). The main obstacle in biological remediation of hydrocarbon compounds is their extremely low solubility in water (Hua et al., 2007). However, increasing substrate solubility by the application of chemical- or bio-surfactant or degrading microorganisms producing biosurfactants is a way to enhance bioavailability and metabolism of contaminants (Macaulay and Rees, 2014; Silva et al., 2014). Biosurfactants have received more and more attention for their superior properties over chemical-surfactants, from the points of view; more active, high biodegradability, less toxicity, and more effective at extreme temperatures, pH, or salinity (Abu Ruwaida et al., 1991; Cameotra and Makkar, 2004). Moreover, biosurfactants can stimulate the catabolism of hydrocarbons through a co-metabolism process, as these organic compounds are readily degradable by microorganisms (Montero-Rodríguez et al., 2014; Silva et al., 2014). Low molecular mass biosurfactants are efficient in lowering surface and interfacial tensions, whereas high molecular mass biosurfactants are more effective at stabilizing oil-in-water emulsions (Rosenberg and Ron, 1999). Biosurfactants are reported for their capability to increase the bioavailability of poorly soluble polycyclic aromatic hydrocarbons and resins (Olivera et al., 2003). Yakimov et al. (1995) reported that salinity activates biosurfactant activity of many bacterial strains isolated from seawater or petroleum reservoirs. That would support their applicability in marine environment.

Surfactants can enhance hydrocarbon bioremediation through two mechanisms: increasing the substrate availability for microorganisms, and/or interaction with the cell surface that increases the hydrophobicity of the surface, allowing hydrophobic substrates to be associated more easily with the microbial cells (Helmy et al., 2010; Aparna et al., 2011; Pacwa-Płociniczak et al., 2011). Although biosurfactants exhibit such important advantages, they have not yet been employed extensively in industry because of relatively high production costs. One possible strategy for reducing costs is the utilization of alternative substrates such as agro-industrial waste (Thavasi et al., 2011; Silva et al., 2014; Ali et al., 2014). Otherwise, use microorganisms capable of producing biosurfactants.

Attention, to date, is focused on this criterion, and several species have been reported, such as *Bacillus, Alcaligenes, Pseudomonas,* and *Corynebacterium* to this end (Calvo et al., 2004; Rengathavasi et al., 2011; Toledo and Gonzalez, 2008; Zhang et al., 2010, 2012; Ali et al., 2014; El-Gendy et al., 2014). Fungi have been also reported to produce biosurfactants (Kiran et al., 2009). Gutierrez et al. (2013) reported that the amphiphilic exopolysaccharides (EPS) produced by *Halomonas* species strain TG39 effectively increase the solubilization of aromatic hydrocarbons and enhance their biotransformation by an indigenous microbial community from oil-contaminated surface waters collected during the active phase of the Deepwater Horizon oil spill. Other bacteria *Alteromonas, Colwellia,* and *Pseudoalteromonas* were also found and reported to be capable to produce EPS. Gutierrez et al. (2013) attributed the formation of oil aggregates, which were a dominant feature observed in contaminated surface waters, to the enrichment of EPS-producing bacteria during the spill coupled with their capacity to produce amphiphilic EPS, which also contributed to the ultimate removal of the oil. Ali et al. (2014) reported the bioremediation of oil polluted seawater with TPH content of 5 g/L in a batch system by the halotolerant *Pseudomonas aeruginosa* Asph2 using the readily available, commercial nutrient corn steep liquor. *P. aeruginosa* Asph2 expressed good biotransformation capabilities for different petroleum hydrocarbon components, recording approximately 58%, 64%, 56%, 55%, and 53% for TPH, saturates, aromatics, asphaltenes, and resins, respectively, within 21 days of incubation at 30°C, pH 7, and 150 rpm. *P. aeruginosa* Asph2 proved good uptake of crude oil with high production of rhamnolipid biosurfactant; approximately 1.3 g/L biosurfactant was produced with the consumption of approximately 2.9 g/L crude oil. El-Gendy et al. (2014) studied the biotransformation of different petroleum biomarkers, in batch reactors inoculated with the halotolerant and biosurfactant producer *Pseudomonas aeruginosa* Asph2 and corn steep liquor, as bioaugmentation and biostimulation (BA/BS) system and biostimulation (BS) only by corn steep liquor. Both systems proved the biotransformation of different petroleum biomarkers, with the superiority of BA/BS system over the BS one, where the biotransformation of the pristane Pr, phytane Ph, and terpane derivatives were very apparent and also the recalcitrant resins and asphaltene fractions.

The biotransformation of C_{28} tricyclic terpane, C_{27} 13β, 17α-diasterane (20S), and C_{30} 17α, 21β-hopane was also recorded and was higher than that of C_{35} 17α, 21β-homohopane. While the biotransformation of (C_{27} 18αH)-22,29,30-trisnorneohopane (Ts) was higher than that of C_{27} (17αH)-22,29,30-trisnorhopane (Tm). The BD of sterane was higher than that of hopane. But the C30 18α-Oleanane was more resistant to degradation than other compounds. El-Sheshtawy et al. (2014) reported the biotransformation of crude oil by two halotolerant and biosurfactant producer bacterial isolates—*Pseudomonas xanthomarina* KMM 1447 and *Pseudomonas stutzeri* ATCC 17,588, in presence of two types of nanoparticles–αFe_2O_3 or $(Zn_5OH)_8Cl_2$, where, the biosurfactants assisted the bacterial isolates to consume the n-paraffins over the iso-paraffins. On the contrary, the presence of biosurfactant with different types of nanoparticles helps the bacterial isolates to consume the iso-paraffins more than *n*-paraffins. But, presence of biosurfactant and nanoparticles of αFe_2O_3 or $(Zn_5OH)_8Cl_2$ enhanced the bacterial growth, and biotransformation rate of saturates and aromatics. The best biotransformation of iso-paraffins was observed in microcosm amended with biosurfactant, Fe_2O_3 nanoparticle and *P. xanthomarina*, and in the microcosm amended with biosurfactant, $(Zn_5OH)_8Cl_2$ nanoparticles and *P. stutzeri*. Sakthipriyaa et al. (2015) reported the biotransformation of a waxy crude oil by a biosurfactant producing *Bacillus subtilis*. El-Mahdi et al. (2015b) reported the isolation of petroleum hydrocarbon degrader *Kocuria* sp. SAR1 from "Tobruk Refinery" oil water pit, located along the Eastern Coast of Libya, that expressed 68% and 70% biotransformation of crude oil in batch systems seeded with 0.2% w/v of the agro-industrial wastes; solid waste dates, and corn steep liquor, respectively.

In another study, El-Mahdi et al. (2015c) reported the isolation of different petroleum hydrocarbon degrading bacteria from oil-contaminated sites at Al Hariga oil terminal and Nafoora oilfield, in Libya, where the Gram-negative isolate *P. aeruginosa* NAF1 expressed 91% and 97% degradation of 5% crude oil (w/v), when the polluted seawater was amended with 0.5% (w/v) corn steep liquor and solid waste dates, respectively. Liu et al. (2016) reported that the ability of the halotolerant, thermophilic, and superior tolerance to alkaline environment *Bacillus Licheniformi* strain Y1 to perform approximately 60% biotransformation of crude oil, with good degradation capabilities to pristane and phytane and its ability to produce an emplastic product that exhibits a good emulsifying effect.

For an effective bioremediation process, proper selection of organisms is very crucial, as different organisms have different metabolic pathways (Thapa et al., 2012). It has been reported that hydrocarbon degrading microorganisms would comprise less than 0.1% of the indigenous microbial population in an unpolluted environment but it would reach to 100% in an oil-polluted one (Atlas, 1981). Effective biotransformation needs a well-adapted microbial population to tolerate the environmental changes and adjustment of the polluted sites to achieve the maximum activity of the indigenous pollutant-degraders (Thomassin-Lacroix, 2000; Mosbech, 2002).

Bacteria are the dominant hydrocarbon degraders in aquatic systems, for example; *Pseudomonas, Achromobacter, Acinobacter, Alcaligenes, Arthrobacter, Bacillus, Brevibacterium, Cornybacterium, Flavobacterium, Nocardia, Planomicrobium, Sphingobium, Stenotrophomons, Vibrio*. Moreover, *Alcanivorax, Alteromonas, Cycloclasticus, Erythrobacter, Halomonas, Hahella, Marinobacter, Microbacterium, Micrococcus, Muricauda, Oleibacter, Oleispira, Oleiphilus, Rhodococcus, Rheinheimera, Thalassospira, Thalassolituus* genera possess the unique skills to live solely on hydrocarbons in marine ecosystems while dominating the marine microbial communities during the process of biotransformation and demonstrated good efficiency during bioremediation processes (Engelhardt et al., 2001; Yakimov et al., 2005, 2007; Head et al., 2006; Coulon et al., 2007; Teramoto et al., 2009; Al-Mailem et al., 2010; Ali et al., 2012; Bird et al., 2012; Coulon et al., 2012; Deriase, 2013; Mahjoubi et al., 2013; Abed et al., 2014; Genovese et al., 2014; Liu et al., 2014; Catania et al., 2015; Cappello et al., 2016).

It has been reported that, there is a major difference in hydrocarbon degrading microorganisms found in the seawater and those in beaches, where, the terrestrial bacteria such as *Rhodococcus* sp. and *Pseudomonas* sp., as well as the marine oil-degrading bacteria such as *Alcanivorax* sp. are abundant on beach (MacNaughton et al., 1999; Rölling, 2004). While, in open water studies in both microcosm and mesocosm scales displayed the predominance of the obligate hydrocarbon-degrading bacteria (OHCB) belonging to the *Gammaproteobacteria, Alphaproteobacteria,* and members of the *Cytophaga–Flavobacterium* group (CFB) (Kasai et al., 2002; Brakstad and Lodeng, 2005; Cappello et al., 2007; Jiménez et al., 2007; Gertler et al., 2009a). *Alcanivorax borkumensis* is a Gram negative marine bacterium, with an outer membrane of lipopolysaccharides that is reported to absorb and digest linear and branched alkanes that are found in crude oil and its products and naturally flourishes after an oil spill (Rojo, 2009; Kostka et al., 2011; Kimes et al., 2014). It can use enzymes and oxygen found in the seawater and secretes AlkB1 and AlkB2 hydroxylases to degrade hydrocarbons (Atlas and Hazen, 2011; Naether et al., 2013). It is most active in surface waters, although it can be found at most depths, but it is affected by addition of dispersants. It can produce biosurfactants to facilitate the diffusion of hydrocarbons into the cell, and to break up the oil slick so that it has more surface area for them to feed on (Karanth et al., 1999; Santisi et al., 2015). Because of its high competitiveness and its ubiquity, *A. borkumensis* is an excellent organism for bioremediation and oil spill mitigation in marine ecosystems (Head et al., 2006; McKew et al., 2007; Gertler et al., 2009a). *Cycloclasticus* is reported for its ability to degrade the toxic polycyclic aromatic hydrocarbons. Amer et al. (2014) reported the biotransformation of crude oil polluted seawater by consortium of 10 bacterial isolates belonging to five different genera; *Bacillus, Pseudomonas, Marinobacter, Providencia* and *Sphingomonas* isolated from a chronically oil polluted Egyptian Mediterranean sediments in El-Max district.

A TPHs biotransformation of approximately 77.8% w/w has been achieved, with *n*- and iso-alkanes degradation of ≈97.97% and 72.92%, respectively. The bacterial consortium showed broad versatility on biotransformation of the 16 polynuclear aromatic hydrocarbons listed by US EPA as priority pollutants with ≈85.79% biotransformation of carcinogenic PAHs. Bayat et al. (2015) reported the isolation of halotolerant and petroleum hydrocarbon degraders; *Shewanella algae* isolate BHA1, *Micrococcus luteus* isolate BHA7, *Pseudoalteromonas* sp. isolate BHA8 and *Shewanella haliotis* isolate BHA35 from three species of mussels: *Crassostrea gigas, Chama asperella,* and *Barbatia tenella* genus, collected from oil contaminated area at Persian Gulf. That recorded; 47.24%, 66.08%, 27.13% and 69.17% of Iranian light crude oil, respectively and characterized by high emulsification activity and biosurfactant production.

A consortium of BHA1 and BHA35 expressed higher biotransformation efficiency, recording 88.69% biotransformation. Bayat et al. (2016) reported the isolation of crude-oil degrading bacteria; *Alcanivorax dieselolei* strain BHA25, *Idiomarina baltica* strain BHA28, *A. dieselolei* strain BHA30, *Alcanivorax* sp. strain BHA32, and *Vibrio azureus* strain BHA36 from *Mactra stultorum* mussel that was collected from oil contaminated area at Persian Gulf, those recoded 64%, 63%, 71%, 58% and 75% removal of crude oil, respectively. Nkem et al. (2016) reported the isolation of hydrocarbon-degrading bacteria; *Cellulosimicrobium cellulans* GS and *Acinetobacter baumannii* WR1 from a tar ball found in

Rhu Sepuluh beach, Terengganu, Malaysia, that expressed an average of 64.4% and 58.1% diesel-oil alkanes degradation, respectively. Germano de Almeida et al. (2016) reported the application of full factorial 2^3 design of experiments to optimize and investigate the effect of temperature, pH, C:N ratio) on the biotransformation of marine fuel MF-380 by bacterial consortium, isolated from seawater of the port of Suape, Brazil, where, approximately 94% biotransformation was achieved at 25°C, pH-6, and carbon-nitrogen ratio of 150:1, with good biotransformation capabilities to aliphatic and aromatic constituents of MF-380.

However, yeast and fungi, for example; *Aspergillus, Candida, Cladosporium, Penicillium, Rhodotorula, Sporobolomyces, Trichodorerma* have been also reported, i.e., in mycoremediation (Abdel-Fattah and Hussein, 2002; Dave and Ghaly, 2011; Farag and Soliman, 2011; Obi et al., 2014). Hassanshahian et al. (2012) reported the isolation and characterization of two crude oil-degrading yeast strains, *Yarrowia lipolytica* PG-20 and PG-32, from the Persian Gulf, that are characterized by high emulsifying activity and cell hydrophobicity. Fungi normally oxidize aromatics hydrocarbons using monooxygenases, forming a *trans*-diol. Fungi, like mammalian enzyme systems, form *trans*-diols, whereas bacteria almost always form *cis*-diols; many *trans*-diols are potent carcinogens whereas *cis*-diols are not biologically active.

Generally, bacteria and fungi can hydroxylate the aromatic rings to form diols. These compounds are subsequently degraded and taken up by the microorganisms for aiding in proper functioning of the tricarboxylic acid cycle (TCA). Fungi are reported to degrade the structurally diverse range of oil-derived compounds by employing a variety of mechanisms including intracellular enzymes such as cytochrome P450 monooxygenases, nitroreductases, and transferases, as well as extracellular enzymes such as laccases and fungal peroxidases (Harms et al., 2011).

Fungi have been reported as good hydrocarbon assimilating agents and there are many reports showing their ability to transform xenobiotic compounds like phenol, benzo[a]pyrene into less mutagenic products. *Penicillium chrysogenum* strain is reported to degrade monocyclic aromatic hydro carbons (benzene, toluene, ethyl benzene and xylene BTEX), phenol compounds, and heavy metals e.g., lead, nickel and-iron) (Atlas, 1981; April et al., 2000; Leitão, 2009; Passarini et al., 2011; Al-Nasrawi, 2012; Wolski et al., 2013; Pereira et al., 2014). Fedorak et al. (1984) reported the preferential degradation of shorter chain *n*-alkanes by the marine fungal isolates *Gliomastix* sp., *Paecilomycete* sp., and *Verticillum* sp. Moreover, it has been reported, that, wood-inhabiting marine fungi are good producers of ligninolytic enzymes, which can degrade several aromatic and recalcitrant environmental pollutants (Garzoli et al., 2015). *Aspergillus* and *Penicillium* were reported to transform the aliphatic and aromatic crude oil components (Harms et al., 2011) and 11 species of *Trichoderma* were reported to grow on different crude oil fractions (Argumedo-Delira et al., 2012). Simister et al. (2015) reported the isolation and characterization of marine oil-degrading *Ascomycota* fungi; *Fusarium* GS_I1, *Scopulariopsis* GS_I2, and *Aspergillus* DI_I1, where, there were no consistent trends in oil degradation related to fungal species or change in temperature. The three isolates were preferentially degrading the short chain alkanes ($<C_{18}$) compared to the long chains ones (C_{19}–C_{36}). Moreover, they preferred the *n*-chain alkanes over the branched ones, which was proved by the decrease in n-C17/pristane and n-C18/phytane ratios during the degradation of oil. Also, the three fungal isolates expressed good biotransformation capabilities toward the aromatics fraction of crude oil, with the preference for low molecular weight over high molecular weight PAHs. Bovio et al. (2017) reported the isolation of *Aspergillus terreus* MUT 271, *Trichoderma harzianum* MUT 290, *Lulworthiales* sp. MUT 263, and *Penicillium citreonigrum*. MUT 267 from oil polluted seawater, with good biotransformation capabilities for Arabian Light crude oil and moderate removal capabilities of petroleum hydrocarbons higher than n-C_{30}.

Monitoring the survival of the introduced microbial degraders and/or the indigenous microbial population at the contaminated sites is an essential attribute in the reduction of contamination levels by the microbial population (MacNaughton et al., 1999; Mishra et al., 2004). Effective bioremediation necessitates the survival of the microorganism at the treated sites and the expression of their catabolic potential and determination/fate of hydrocarbon degradation disappearance of individual hydrocarbons and/or total hydrocarbons) (Ramirez et al., 2001; Peressutti et al., 2003; Malik and Ahmed, 2012). Conventional methods to monitor survival of microorganisms, for example, most-probable-number method and direct counting methods to measure the total viable count TCFU/ml) are associated with number of anomalies (Mishra et al., 2004). For effective bioremediation, it is essential that the augmented microorganisms survive over a long period of time in the contaminated sites for complete reclamation (Mishra et al., 2004).

Crude oil constituents vary in composition, volatility, fluency, and susceptibility to bioremediation. Few of the compounds are easily degradable whereas other crude oil constituents resist biotransformation. The biotransformation of the crude oil components generally decreases in the following order (van Hamme et al., 2003; Das and Chandran, 2011; Yanto and Tachibana, 2013):

n-alkane derivatives > branched-chain alkane derivatives
branched-chain alkane derivatives > branched alkene derivatives
branched alkene derivatives > low molecular weight n-alkyl aromatic derivatives
low molecular-weight n-alkyl aromatic derivatives > monoaromatic derivatives
monoaromatic derivatives > cyclic alkane derivatives (naphthene derivatives)
cyclic alkane derivatives (naphthene derivatives) > polynuclear aromatic hydrocarbons
polynuclear aromatic hydrocarbons > polar fraction resins and asphaltenes
polar fraction resins and asphaltenes > petroleum biomarker compounds.

Inoculation of contaminated seawater with hydrocarbon-degrading microorganisms (bioaugmentation) and the addition of fertilizers and/or dispersant (biostimulation) are recommended to accelerate the rate of the natural degradation process. Screening of petroleum hydrocarbon degrading microorganisms from previously contaminated sites and inoculating them to the contaminated seawater are one option for bioremediation of marine oil spill. However, the competition between indigenous species and the augmented ones i.e., antagonism), might be severe enough to decrease the biotransformation rate. Thus, most studies indicated that bioaugmentation was not a promising option for the bioremediation of oil spill (Venosa et al., 1991; Atlas, 1995; Swannell et al., 1996).

However, autochthonous bioaugmentation (ABA), i.e., the use of the pre-adapted microorganisms indigenous to the sites' soil, sand, and water) slated for decontamination, have been reported to overcome that barrier (Nikolopoulou et al., 2013a,b; El-Gendy et al., 2014). Younis et al. (2013) reported the superiority of the autochthonous bioaugmentation over biostimulation of a petroleum hydrocarbons-polluted water sample in a batch reactor using the halotolerant *Corynebacterium variabilis* sp. Sh42 and corn steep liquor as a readily available and cheap source of nutrients. El-Gendy et al. (2014) also proved the superiority of the autochthonous bioaugmentation over biostimulation, where the seeding a petroleum hydrocarbon polluted seawater with the halotolerant *P. aeruginosa* Asph2 and corn steep liquor as a cheap and readily available source of nutrients significantly enhances the progress of biotransformation of petroleum hydrocarbon pollutants, even the recalcitrant petroleum biomarkers and asphaltene fraction. They observed statistically high significant difference between natural weathering and both bioaugmentation $P = 2.92e\text{-}14$) and biostimulation, with only corn steep liquor $P = 3.56e\text{-}13$) at 95% confidence interval a = 0.05). BA significantly improved the efficiency of TPH and the recalcitrant asphaltene degradation by

23% and 17% compared to BS process $P=2.0693e-5$ and $P=1.4e-3$, respectively). Biostimulation addition of nutrients, surfactants, dispersants, etc., increasing the bioavailability of oil to hydrocarbon degraders.), has been shown to be effective for marine oil spills, though the efficacy is limited in the bioremediation of extensively degraded and weathered oil (Bragg et al., 1994; Zahed et al., 2010a). Abed et al. (2014) reported that Omani polluted sediments contain halotolerant and thermotolerant bacteria and biostimulation is more efficient than bioaugmentation for their cleanup.

To achieve efficient bacterial formulas for bioremediation, it is necessary to select consortia with cooperative metabolism for the biotransformation of the pollutant (Vázquez et al., 2013). Nikolopoulou et al. (2013b) reported the enrichment of a consortium, belonging to the genera; *Marinobacter*, *Microbulbifer*, *Melitea*, *Halomonas*, *Pseudomonas*, *Alcanivorax* and *Chromohalobacter*, from seawater samples taken from Elefsina Gulf near the Hellenic Petroleum Refinery, a site exposed to chronic crude oil pollution. That pre-adapted consortium was tested alone or in combination with inorganic nutrients in the presence or absence of biosurfactants rhamnolipids) in 30-day experiments. It was proved in that study, that, the biostimulation with lipophilic fertilizers and biosurfactant in combination with bioaugmentation with naturally pre-adapted hydrocarbon degrading consortia is a very effective treatment. The trend of biotransformation was as follows; $C_{15}>C_{20}>C_{25}>$Pristane and Phytane$>C_{30}>$Polynuclear aromatic hydrocarbons. Younis et al. (2013) proved the successfulness of applying bioaugmentation and biostimulation using a well-adapted consortium for biotreatment of petrogenic polluted seawater, and attributed this to different reasons; (1) the synergistic interactions among the members of the microbial association, which is complex and favor petrogenic hydrocarbons degrading mechanisms, where, one microbial species *C. variabilis* sp. Sh42) would remove the toxic metabolites of another species *Rodococcus erythropolis* IGTS8) that would have begun the biotransformation process, or when the two species work in succession with the first partially degraded compounds and the second finished the job. (2) the broad enzymatic capacities and co-metabolic relationships of the consortium with the indigenous microbial population, where there are more favorable to attack and metabolize not only aliphatic or aromatic compounds but also nitrogen, sulfur and oxygen NSO) compounds.

Varjani et al. (2015) reported the ex-situ biotransformation of crude oil by halotolerant hydrocarbon utilizing bacterial consortium isolated from on shore sites of Gujarat, India. Li et al. (2016) reported the isolation of a halotolerant consortium from Bohai Bay, China; *Ochrobactrum* TCOB-1, *Brevundimonas* TCOB-2 and TCOB-3, *Bacillus* TCOB-4 and *Castellaniella* TCOB-5, where TCOB-4 and TCOB-5, expressed the highest biotransformation efficiency toward the saturates and aromatics fractions in crude oil, respectively. A consortium of TCOB-4 and TCOB-5 1:1) expressed higher biotransformations of crude oil than single bacteria, with 60.53%, 36.36% and 51.87% biotransformation of saturates, aromatics and total crude oil 2% w/v), respectively within 1-week incubation and 3% NaCl saline minimal salts medium.

Crisafi et al. (2016) pointed to the importance of the evaluation of the applied dispersant and studying its effect on the principal actors of biotransformation. In a comparison study on a microcosm level for studying the treatment of oily polluted seawater sampled after an oil-spill in Taranto Gulf Italy), Crisafi et al. (2016), proved the superiority of bioaugmentation, where the biostimulation with inorganic nutrients allowed the biotransformation of approximately 73%, while, the bioaugmentation with a selected *hydrocarbonoclastic* consortium consisting of *A. borkumensis*, *A. dieselolei*, *Marinobacter hydrocarbonoclasticus*, *Cycloclasticus* sp. 78-ME and *Thalassolituus oleivorans* degraded approximately 79%. But, the washing agent and addition of nutrients achieved approximately 69% degradation of hydrocarbons and the microbial community was severely affected and washing agent seemed to inhibit the growth of the majority of strains composing the selected consortium at the tested concentration.

Application of biostimulation and bioaugmentation together in combination is sometimes more advantageous than each alone (Darmayati et al., 2015). Thus, briefly, these components would have different chemical behaviors and degradation rates. As a result, different microbes are preferred for different compounds. These microbes would first recognize the oil and its constituents depending on biosurfactants and/or bioemulsifiers. The microbes attach to hydrocarbons they can eliminate and use the hydrocarbons as a source of energy and carbon. Furthermore, biosurfactants would be added to enhance the solubility to improve the biotransformation rates, thus, leading to disappearance of individual components of petroleum over a period of time (Mukred et al., 2008; Reis et al., 2013). Hydrocarbon-degrading bacteria are reported to often produce bioemulsifiers, which are composed of low molecular weight molecules that lower surface and interfacial tensions efficiently and high molecular weight polymers that bind tightly to surfaces. Some bioemulsifiers enhance the growth of bacteria on hydrophobic water-insoluble substrates by increasing their bioavailability, presumably by increasing their surface area, desorbing them from surfaces, and increasing their apparent solubility. Bioemulsifiers also play an important role in regulating the attachment/detachment of microorganisms to and from surfaces.

Another problem to be solved during the application of bioremediation of oil spill is, microbial cells tend to suspend or even precipitate in an aqueous system when mechanical agitation is unavailable, thereby reducing their contact of hydrocarbon pollutants, which are normally floating on the aqueous surface (Lin et al., 2005). Immobilization is reported to enhance the rate of biotransformation (El-Gendy and Nassar, 2015a,b) and immobilizing the cells by bonding them to the surface of floating bio-carriers (e.g., vermiculite, chitosan flakes, loofah sponge, activated carbon, cotton fibers, sugarcane bagasse, sawdust) is also reported to overcome the previously mentioned drawback (Ahmadi et al., 2005; Gentili et al., 2006; Quek et al., 2006; Su et al., 2006; Podorozhko et al., 2008; Brandão et al., 2010; Li et al., 2012; Ma et al., 2013; Lin et al., 2014), but good biocompatibility, hydrophobicity, absorptivity, and floatability are the key points to remove petroleum hydrocarbon successfully. Generally, immobilization increases the microbial cell stability, promotes the good diffusion between substrates and products, protects the microbes from the rugged environmental conditions, and finally facilitates the separation of biomass and water (Yardin et al., 2000; Ma et al., 2013; Lin et al., 2014; Wang et al., 2015).

Studies by Rahman et al. (2006) demonstrated the effectiveness of alginate immobilized bacterial cells in the removal of C10 and C27 hydrocarbons but do not demonstrate degradation of hydrocarbons greater than C28, and attributed this to the transportation of the hydrocarbon across the alginate to access the immobilized bacterial cells. El-Gendy and Nassar (2015a,b) reported the isolation of the marine diesel oil-degrading *P. aeruginosa* NH1 with accession number KM267644 and examined the ability of alginate entrapped NH1 to degrade different components of diesel oil contaminating seawater. The biotransformation rate of different components of diesel oil in free and immobilized cell systems ranked in the following decreasing order: total resolvable peaks (normal- and iso-alkanes)>the 16 polyaromatic hydrocarbons listed by the US EPA as priority pollutants>polyaromatic sulfur heterocyclic compounds>unresolved complex mixture (naphthenes, cycloalkanes, and aromatics). The improved tolerance of the immobilized cells toward different toxic components of diesel oil has been proved and the biotransformation rate of different components of diesel oil, aliphatic derivatives, polyaromatic sulfur heterocyclic compounds, and biomarkers pristane, phytane, and 4,6-dimethyldibenzothiophene, was enhanced by immobilization. Storage stability and reusability tests revealed that the diesel oil degradation ability of the immobilized cells was stable after storage at 4°C for 30d and can be effectively reused for two batches of 56d. Liu et al. (2015) reported the bioremediation of diesel oil polluted artificial seawater by ca-alginate entrapped diesel-degrading bacterial cells *Rhodococcus pyridinivorans* CC-HCCH11, *Gordonia alkanivorans* CC-JG39, and *Alcaligenes piechaudii* CC-ESB2, where, the immobilized cells doubled the TPHs removal, compared with free cells. By entrapping the diesel-degrading bacteria in gel matrices, the direct contact of bacteria with highly diesel bulk liquors can be prevented. Thus, the bacterial consortium entrapped in a gel bead can degrade diesel with less substrate inhibition, compared with degradation in the bulk liquid. Thus, using the immobilized or entrapped cells for remediation is proved to be relatively sustainable, due to the continuous release of the bacterial cells from the immobilized matrix, during the bioremediation period.

Suni et al. (2004, 2007) proved that plant fibers of cotton grass (*Eriophorum* sp.), are an easily accessible material from a sustainable source that can efficiently remove oil when used for filtration purposes and bioremediation. Gertler et al. (2009b) performed an experimental prototype oil boom including oil sorbents, slow-release fertilizers, and the marine oil-degrading bacterium *A. borkumensis* in a 500L mesocosm to study the applicability of a combination of mechanical and biotechnological techniques, for sorption and degradation of heavy fuel oil-polluted seawater, where it proved the safety of applying bioremediation as there was no recorded negative impact on the marine protozoa, but on the contrary, there was a strong abundance of marine protozoa, i.e., flagellates, ciliates, and amoeba, with no mortality effects upon *Nauplius larvae*. Moreover, it also proved the good synergism between the augmented, well-adapted *gamma-proteobacterium A. borkumensis* and the indigenous microbial population that appeared in the form of the abundance and high levels of obligate oil-degrading microorganisms within the boom. Simons et al. (2013) reported that the rate of biotransformation of weathered crude oil by immobilized bacterial consortium on shell grit carrier is much higher than that entrapped in alginate. Simons et al. (2012, 2013) recommended mussel shells as a carrier material due to its abundance, good absorbance capabilities, consider as a way for recycling of waste products, simplicity of application, light weight carrier materials, and their buoyancy, where the buoyancy is an important component as it enables carrier materials to interact with the oil through the water column and therefore, enhances degradation and effectiveness of providing a protective niche to the bioaugmented microbial strains.

Simons et al. (2012) showed that hydrocarbon degrading microorganisms immobilized on modified mussel shells could degrade up to 55% of hydrocarbon contaminant from 276mg/L to 123.3mg/L in nutrient rich medium within 30days. Lin et al. (2014) reported the utilization of cotton fibers as crude oil sorbent, as well as a biocarrier for the halotolerant marine bacteria *Acinetobacter* sp. HC8–3S. The strong binding to cotton fibers was attributed to the extracellular dispersant excreted by HC8–3S. The immobilized bacteria showed a higher degradation rate than that of planktonic bacteria and the efficiency of saturated hydrocarbons degradation by the immobilized bacterial cells increased by about 30% compared to the planktonic bacterial cells. Sheppard et al. (2014) reported the application of bacterial consortium consisting of six bacterial hydrocarbonoclastic strains (composed of *Actinobacteria*, *Bacilli*, and *Gammaproteobacteria*), immobilized on shell carriers in a simulated marine remediation scenario of 5000L contaminated seawater, over a period of 27weeks, where approximately 53.3% biotransformation of weathered crude oil was achieved from 315,000mg/L to 147,000mg/L, with the degradation of $>C_{32}$ hydrocarbon fractions. Wang et al. (2015) reported the isolation of the halotolerant *Pseudomonas* sp. ODB-1, *Brevundimonas* sp. ODB-2 and ODB-3 from oil contaminated seawater, and then studied their biotransformation efficiency as being immobilized on the surface of some biocarriers; expanded graphite (EG), expanded perlite (EP) and bamboo charcoal (BC), on which they expressed different adsorption mechanism; ODB-1 showed a strong binding to the biocarriers through extracellular polysaccharides, while ODB-2 and ODB-3 made the adhesion to biocarrier through direct physical adsorption. The immobilizing matrix increased the biotransformation capabilities and the salt tolerance of the three isolates. The EG-bacteria system achieved nearly 100% removal of diesel oil, where, over 83%–93% removal of diesel oil is owed to biotransformation. Thus, it is proved that, the adsorption–biotransformation process played an important role in the oil-pollution remediation. Zhang et al. (2016) reported the higher biotransformation of *n*-alkanes (C16, C18, C19, C26, C28) and polycyclic aromatic hydrocarbons (naphthalene and pyrene) by a halotolerant *Corynebacterium variabile* HRJ4 immobilized on biochar relative to the free microbial cells. Chen et al. (2016) reported the immobilization of *Acinetobacter venetianus*, isolated from oil refining wastewater sludge, on modified bamboo charcoal (MBC) as a cheap, abundant and alternative cell immobilization matrix, which demonstrated superior efficiency in degrading diesel oil (94%) compared to planktonic cells culture (82%) over a 3-day period.

6.0 PREVENTION AND RESPONSE

The increasingly long list of chemical contaminants released into the environment on a large scale includes numerous aliphatic and aromatic compounds, such as petroleum hydrocarbons. The local concentration of such contaminants depends on the amount present and the rate at which the compound is released, its stability in the environment under both aerobic and anaerobic conditions, the extent of its dilution in the environment, the mobility of the compound in an ecosystem, and its rate of biological transformation or non-biological transformation.

According to the Environmental Protection Agency (EPA, 2006) "oil releases threaten public health and safety by contaminating drinking water, causing fire and explosion hazards, diminishing air and water quality, compromising agriculture, destroying recreational areas, wasting nonrenewable resources, and costing the economy millions of dollars." Thus, oil spills require quick action so that they cause as little damage as possible.

Oil spills occur every year, which have enormous negative impacts on the surrounding populations, drinking water, recreational water areas, and shoreline industry and facilities. When the spills are relatively small, they often receive less attention than the larger spills. However, this is a very bad practice and, moreover, although technologies for enhancing bioremediation of petroleum pollution in soil and reactors have been known for several years, there needs to be relevant experimentally demonstrated methodology for accelerating the biotransformation of crude oil constituents (and the constituents of crude oil products). The main obstacle that faces the researchers is the difficulty of gaining more information, due to the difficulty of performing controlled experiments in the open ecosystems (whether the spill is on land or in water systems such as lakes, rivers, and oceans).

Worldwide, there is ongoing extensive research to determine the feasibility of bioremediation as an alternative and/or complimentary to conventional remediation. The Exxon Valdez oil spill (1989) was the first, for which bioremediation was looked at as an alternative to cleanup an oil spill. Since this time, more research has been conducted and more will continue, to find the proper places to use bioremediation as an alternative and/or complimentary to conventional remediation techniques to clean up oil spills.

Since most (but not all) of the contaminants of concern in crude oil (and crude oil products) can be changed by biotransformation processes, the success of oil-spill bioremediation (on land or on the water) depends mainly on the ability to establish these conditions in the contaminated environment using the above new developing technologies to optimize the total efficiency of the microorganisms. The technologies used at various affected sites depend on any factors at the location that have an adverse effect on the application of the biotransformation technology. For example, where there is insufficient dissolved oxygen, bioventing or sparging is applied, biostimulation or bioaugmentation is suitable for instances where the biological count is low.

Consequently, the primary objectives of prevention and response are to prevent the spill any adverse effects on flora and fauna and to speed the biotransformation of any unrecovered crude oil (or crude oil product) into more benign chemicals that do considerably less harm to the environment. Thus, 10 criteria are recommended to be used for evaluation of several remediation methods: (1) efficiency, (2) time, (3) cost, (4) impact on flora and fauna, (5) reliability, (6) level of difficulty, (7) crude oil or crude oil product recovery, (8) weather, (9) effect on physical/chemical characteristics of the crude oil or crude oil product, and (10) the need for further treatment. Based on the comparative analyses, oil recovery with mechanical methods and the application of dispersants followed by bioremediation is the most effective response for marine oil spill.

REFERENCES

Abdel-Fattah, Y.R., Hussein, H.M., 2002. Numerical modelling of petroleum oil bioremediation by a local *Penicillium* isolate as affected with culture conditions: application of Plackett-Burman design. Arab Journal of Biotechnology 52, 165–172.

Abdel-Satar, A.M., Ali, M.H., Goher, M.E., 2017. Indices of water quality and metal pollution of Nile River, Egypt. Egyptian Journal of Petroleum 43, 21–29.

Abdel-Shafy, H.I., Mansour, M.S.M., 2016. A review on polycyclic aromatic hydrocarbons: source, environment al impact, effect on human health and remediation. Egyptian Journal of Petroleum 25, 107–123.

Abed, R.M.M., Al-Sabahi, J., Al-Maqrashi, F., Al-Habsi, A., Al-Hinai, M., 2014. Characterization of hydrocarbon-degrading bacteria isolated from oil-contaminated sediments in the Sultanate of Oman and evaluation of bio-augmentation and biostimulation approaches in microcosm experiments. International Biodeterioration and Biodegradation 89, 58–66.

Abu Ruwaida, A.S., Banat, I.M., Salam, A., Kadiri, M., 1991. Isolation of biosurfactant producing bacteria: product characterization and evaluation. Acta Biotechnology 11, 315–324.

Adebajo, M.O., Frost, R.L., Kloprogge, J.T., Carmody, O., Kokot, S., 2003. Porous materials for oil spill cleanup: a review of synthesis and absorbing properties. Journal of Porous Materials 10, 159–170.

Adhikari, L., Maiti, K., Overton, E.B., 2015. Vertical fluxes of polycyclic aromatic hydrocarbons in the northern Gulf of Mexico. Marine Chemistry 168, 60–68.

AECIPE, 2002. Marine Bioremediation Technologies Screening Matrix and Reference Guide, DRAFT VERSION II. A Join Project of Environmental Technologies Research Group Faculty of Marine and Environmental Sciences Cádiz University, Spain and Spanish Association of Fishering Cities AECIPE), Excma. Diputación Provincial de Cádiz, Spain.

Aeppli, C., Carmichael, C.A., Nelson, R.K., Lemkau, K.L., Graham, W.M., Redmond, M.C., Valentine, D.L., Reddy, C.M., 2012. Oil weathering after the deepwater horizon disaster led to the formation of oxygenated residues. Environmental Science & Technology 46 (16), 8799–8807.

Agarwal, A., Liu, Y., 2015. Remediation technologies for oil-contaminated sediments. Marine Pollution Bulletin 101, 483–490.

Ahmadi, M., Vahabzadeh, F., Bonakdarpour, B., Mehranian, M., Mofarrah, E., 2005. Phenolic removal in olive oil mill wastewater using loofah-immobilized *Phanerochaete chrysosporium*. World Journal of Microbiology & Biotechnology 22, 119–127.

Ahmaruzzaman, M., Sharma, D.K., 2005. Adsorption of phenols from wastewater. Journal of Colloid and Interface Science 287, 14–24.

Aislabie, J., Foght, J., Saul, D., 2000. Aromatic hydrocarbon degrading bacteria from soil near Scott Base, Antarctica. Polar Biology 23, 183–188.

Aislabie, J., Saul, D.J., Foght, J.M., 2006. Bioremediation of hydrocarbon-contaminated polar soils. Extremophiles 10, 171–179.

Ali, H.R., El-Gendy, N.Sh., El-Ezbewy, S., El-Gemeie, G.H., Moustafa, Y.M., Roushdy, R.I., 2006. Assessment of polycyclic aromatic hydrocarbons contamination in water, sediment and fish of Temsah Lake, Suez Canal, Egypt. Current World Environment 11, 11–22.

Ali, H.R., El-Gendy, N.Sh., Moustafa, Y.M., Mohamed, I., Roushdy, M.I., Hashem, A.I., 2012. Degradation of asphaltenic fraction by locally isolated halotolerant bacterial strains. ISRN Soil Science:435485.

Ali, H.R., Ismail, D.A., El-Gendy, N.Sh, 2014. The Biotreatment of oil polluted seawater by biosurfactant producer halotolerant *Pseudomonas aeruginosa* Asph2. Energy Sources, Part A: Recovery, Utilization, and Environmental Effects 36 (13), 1429–1436.

Al-Mailem, D.M., Sorkhoh, N.A., Al-Awadhi, H., Eliyas, M., Radwan, S.S., 2010. Biodegradation of crude oil and pure hydrocarbons by extreme halophilic archaea from hypersaline coasts of the Arabian Gulf. Extremophiles 14, 321–328.

Almeda, R., Hyatt, C., Buskey, E.J., 2014. Toxicity of dispersant corexit 9500A andcrude oil to marine microzooplankton. Ecotoxicology and Environmental Safety 106, 76–85.

Al-Nasrawi, H., 2012. Biodegradation of crude oil by fungi isolated from Gulf of Mexico. Journal of Bioremediation and Biodegradation. 3, 147. https://doi.org/10.4172/2155-6199.1000147.

Al-Sultani, K.F., Al-Seroury, F.A., 2012. Characterization the removal of phenol from aqueous solution in fluidized bed column by rice husk adsorbent. Research Journal of Recent Sciences. IISC 20 (11), 145–151.

Alther, G.R., 2002. Removing oils from water with organo clays. Journal of American Water Works Association 94, 115–121.

Alves, T.M., Kokinoub, E., Zodiatis, G., 2014. A three-step model to assess shoreline and offshore susceptibility to oil spills: the South Aegean (Crete) as an analogue for confined marine basins. Marine Pollution Bulletin 86, 443–457.

Amer, R., El-Gendy, N.Sh., Taha, T., Farag, S., Abdel Fattah, Y., 2014. Biodegradation of crude oil and its kinetics using indigenous bacterial consortium isolated from contaminated area in Egyptian Mediterranean ecosystem. Jökull Journal 64, 42–58.

Amin, M.N., Mustafa, A.I., Khalil, M.I., Rahman, M., Nahid, I., 2012. Adsorption of phenol onto rice straw biowaste for water purification. Clean Technologies and Environmental Policy 14, 837–844.

Anderson, J.W., Bay, S.M., Thompson, B.E., 1989. Characteristics and effects of contaminated sediments from southern California. In: An International Conference Addressing Methods for Understanding the Global Ocean. Oceans 1989, Seattle, Washington. 18–21 September 1989, Ocean Pollution, vol. 2, pp. 449–451.

Aparna, A., Srinikethan, G., Hegde, S., 2011. Effect of addition of biosurfactant produced by *Pseudomonas* sps. on biodegradation of crude oil. In: 2nd International Conference on Environmental Science and Technology IPCBEE, vol. 6. IACSIT Press, Singapore. pp. V1-71, V1-V75.

April, T.M., Foght, J.M., Currah, R.S., 2000. Hydrocarbon-degrading filamentous fungi isolated from flare pit soils in northern and western Canada. Canadian Journal of Microbiology 46 (1), 38–49.

Aquilina, N.J., Delgado-Saborit, J.M., Meddings, C., Baker, S., Harrison, R.M., Jacob, P., Wilson, M., Yu, L., Duan, M., Benowitz, N.L., 2010. Environmental and biological monitoring of exposures to PAHs and ETS in the general population. Environment International 36, 763–771.

Argumedo-Delira, R., Alarcón, A., Ferrera-Cerrato, R., Almaraz, J.J., Peña-Cabriales, J.J., 2012. Tolerance and growth of 11 *Trichoderma* strains to crude oil, naphthalene,phenanthrene and benzo [a] pyrene. Journal of Environmental Management 95, S291–S299.

Atlas, R.M., 1991. Microbial hydrocarbon degradation-bioremediation of oil spills. Journal of Chemical Technology and Biotechnology 52, 149–156.

Atlas, R.M., 1981. Microbial degradation of petroleum hydrocarbons: an environmental perspective. Microbiological Reviews 45 (1), 180–209.

Atlas, R.M., 1995. Petroleum biodegradation and oil spill bioremediation. Marine Pollution Bulletin 31, 178–182.

Atlas, R.M., Bartha, R., 1972. Degradation and mineralization of petroleum in seawater: limitation by nitrogen and phosphorous. Biotechnology and Bioengineering 14 (3), 309–318.

Atlas, R.M., Hazen, T.C., 2011. Oil biodegradation and bioremediation: a tale of the two worst spills in U.S. History. Environmental Science & Technology 45, 6709–6715.

ATSDR, Agency for Toxic Substances, Disease Registry, 1993. Toxicological Profile for Polycyclic Aromatic Hydrocarbons PAHs) Draft Update. U.S. Dep. Health & Human Services, Agency for Toxic Substances and Disease Registry, Atlanta, Georgia, p. 273.

Attiogbe, F.K., Glover-Amengor, M., Nyadziehe, K.T., 2007. Correlating biochemical and chemical oxygen demand of effluents a case study of selected industries in Kumasi, bacteria from coastal water. Biotechnology and Bioengineering 14, 297–308.

Bacosa, H., Suto, K., Inoue, C., 2010. Preferential degradation of aromatic hydrocarbons in kerosene by a microbial consortium. International Biodeterioration and Biodegradation 64 (8), 702–710.

Bacosa, H.P., Liu, Z., Erdner, D.L., 2015a. Natural sunlight shapes crude oil-degrading bacterial communities in northern Gulf of Mexico surface waters. Frontiers in Microbiology. 6, 1325. https://doi.org/10.3389/fmicb.2015.01325.

Bacosa, H.P., Erdner, D.L., Liu, Z., 2015b. Differentiating the roles of photooxidation and biodegradation in the weathering of Light Louisiana Sweet crude oil in surface water from the Deepwater Horizon site. Marine Pollution Bulletin 5, 265–272.

Bagi, A., Pampanin, D.M., Brakstad, O.G., Kommedal, R., 2013. Estimation of hydrocarbon biodegradation rates in marine environments: a critical review of the Q10 approach. Marine Environmental Research 89, 83–90.

Balks, M.R., Paetzold, R.F., Kimble, J.M., Aislabie, J., Campbell, I.B., 2002. Effects of hydrocarbon spills on the temperature and moisture regimes of cryosols in the Ross Sea region. Antarctic Science 14 (4), 319–326.

Bandara, U.C., Yapa, P.D., Xie, H., 2011. Fate and transport of oil in sediment laden marine waters. Journal of Hydro-environment Research 5, 145–156.

Banerjee, S.S., Joshi, M.V., Jayaram, R.V., 2006. Treatment of oil spill by sorption technique using fatty acid grafted sawdust. Chemosphere 64, 1026–1031.

Bao, M.T., Wang, L.N., Sun, P.Y., Cao, L.X., Zhou, J., Li, Y.M., 2012. Biodegradation of crude oil using an efficient microbial consortium in a simulated marine environment. Marine Pollution Bulletin 64, 1177–1185.

Barbato, M., Mapelli, F., Magagnini, M., Chouaia, B., Armeni, M., Marasco, R., Crotti, E., Daffonchio, D., Borin, S., 2016. Hydrocarbon pollutants shape bacterial community assembly of harbor sediments. Marine Pollution Bulletin 104, 211–220.

Bayat, Z., Hassanshahian, M., Hesni, M.A., 2016. Study the symbiotic crude oil-degrading bacteria in the mussel *Mactra stultorum* collected from the Persian Gulf. Marine Pollution Bulletin 105, 120–124.

Bayat, Z., Hassanshahian, Z.M., Hesni, M.A., 2015. Enrichment and isolation of crude oil degrading bacteria from some mussels collected from the Persian Gulf. Marine Pollution Bulletin 101, 85–91.

Beckles, D.M., Ward, C.H., Hughes, J.B., 1998. Effect of mixtures of polycyclic aromatic hydrocarbons and sediments on fluoranthene biodegradation patterns. Environmental Toxicology and Chemistry 17 (7), 1246–1251.

Bhatnagar, S., Kumari, R., 2013. Bioremediation: a sustainable tool for environmental management–A review. Annual Review and Research in Biology 3 (4), 974–993.

Billet, S., Paget, V., Garcon, G., Heutte, N., André, V., Shirley, P., Sichel, F., 2010. Benzene-induced mutational pattern in the tumour suppressor gene TP53 analyzed by use of a functional assay, the functional analysis of separated alleles in yeast, in human lung cells. Archives of Toxicology 84, 99–107.

Bird, C., Adetutu, E., Hancock, N., Lynch, C., Kadalli, K., Bueti, A., Truskewycz, A., Ball, A., 2012. The application of hydrocarbonoclastic bacteria for the bioremediation of weathered crude oil. In: Mendez-Vilas, A. (Ed.), Microbes in Applied Research: Current Advances and Challenge. World Scientific Publishing, Singapore, pp. 177–182.

Boll, M., Löffler, C., Morris, B.E., Kung, J.W., 2014. Anaerobic degradation of homocyclic aromatic compounds via arylcarboxyl-coenzyme A ester: organisms, strategies and key enzymes. Environmental Microbiology 16 (3), 612–627.

Bordenave, S., Goni-Urriza, M.S., Caumette, P., Duran, R., 2007. Effect of heavy fuel oil on the bacterial community structure of a pristine microbial mat. Applied and Environmental Microbiology 73 (19), 6089–6097.

Bosta, F.D., Frontera-Suau, A., McDonald, T.J., Peters, K.E., Morris, P.J., 2001. Aerobic biodegradation of hopanes and norhopanes in Venezuelan crude oils. Organic Geochemistry 32, 105–114.

Boufadel, M.C., Geng, X., Short, J., 2016. Bioremediation of the Exxon Valdez oil in Prince William Sound beaches. Marine Pollution Bulletin 15, 156–164.

Boufadel, M.C., Bobo, A.M., Xia, Y., 2011. Feasibility of deep nutrients delivery into a Prince William Sound beach for the bioremediation of the Exxon Valdez oil spill. Ground Water Monitoring and Remediation 312, 80–91.

Boufadel, M.C., Suidan, M.T., Venosa, A.D., 2006. Tracer studies in laboratory beach simulating tidal influences. Journal of Environmental Engineering 132 (6), 616–632.

Boufadel, M.C., Suidan, M.T., Venosa, A.D., 2007. Tracer studies in a laboratory beach subjected to waves. Journal of Environmental Engineering 133 (7), 722–732.

Bovio, E., Gnavi, G., Prigione, V., Spina, F., Denaro, R., Yakimov, M., Calogero, R., Crisafi, F., Giovanna Cristina Varese, G.C., 2017. The culturable mycobiota of a Mediterranean marine site after an oil spill: isolation, identification and potential application in bioremediation. Science of the Total Environment 576, 310–318.

Bragg, J.R., Prince, R.C., Harner, E.J., Atlas, R.M., 1994. Effectiveness of bioremediation for the Exxon Valdez oil spill. Nature 368, 413–418.

Brakstad, O.G., Bonaunet, K., 2006. Biodegradation of petroleum hydrocarbons in seawater at low temperatures (0–5°C) and bacterial communities associated with degradation. Biodegradation 17, 71–78.

Brakstad, O.G., Lodeng, A.G., 2005. Microbial diversity during biodegradation of crude oil in seawater from the North Sea. Microbial Ecology 49, 94–103.

Brandão, P.C., Souza, T.C., Ferreira, C.A., Hori, C.E., Romanielo, L.L., 2010. Removal of petroleum hydrocarbons from aqueous solution using sugarcane bagasse as adsorbent. Journal of Hazardous Materials 175, 1106–1112.

Bressler, D.C., Gray, M.R., 2003. Transport and reaction processes in bioremediation of organic contaminants.1. Review of bacterial degradation and transport. International Journal of Chemical Reactor Engineering. 1 (1), R3. https://doi.org/10.2202/1542-6580.1027.

Busca, G., Berardinelli, S., Resini, C., Arrighib, L., 2008. Technologies for the removal of phenol from fluid streams: a short review of recent developments. Journal of Hazardous Materials 160, 265–288.

Calvo, C., Toledo, F.L., González, L., 2004. Surfactant activity of a naphthalene degrading *Bacillus pumilus* strain isolated from oil sludge. Journal of Biotechnology 109, 255–262.

Cameotra, S.S., Makkar, R.S., 2004. Recent applications of biosurfactants as biological and immunological molecules. Current Opinion in Microbiology 7, 262–266.

Cappello, S., Calogero, R., Santisi, S., Genovese, M., Denaro, R., Genovese, L., Giuliano, L., Mancini, G., Yakimov, M.M., 2015. Bioremediation of oil polluted marine sediments: a bio-engineering treatment. International Microbiology 18, 127–134.

Cappello, S., Denaro, R., Genovese, M., Giuliano, L., Yakimov, M.M., 2007. Predominant growth of *Alcanivorax* during experiments on 'oil spill bioremediation' in mesocosms. Microbiological Research 162, 185–190.

Cappello, S., Volta, A., Santisi, S., Morici, C., Mancini, G., Quatrini, P., Genovese, M., Yakimov, M.M., Torregrossa, M., 2016. Oil-degrading bacteria from a membrane bioreactor (BF-MBR) system for treatment of saline oily waste: isolation, identification and characterization of the biotechnological potential. International Biodeterioration and Biodegradation 110, 235–244.

Catania, V., Santisi, S., Signa, G., Vizzini, S., Mazzola, A., Cappello, S., Yakimov, M.M., Quatrini, P., 2015. Intrinsic bioremediation potential of a chronically polluted marine coastal area. Marine Pollution Bulletin 99, 138–149.

Chaillan, F., Chaineau, C., Point, V., Saliot, A., Oudot, J., 2006. Factors inhibiting bioremediation of soil contaminated with weathered oils and drill cuttings. Environmental Pollution 144 (1), 255–265.

Chaîneau, C.H., Rougeux, G., Yéprémian, C., Oudot, J., 2005. Effects of nutrient concentration on the biodegradation of crude oil and associated microbial populations in the soil. Soil Biology and Biochemistry 37, 1490–1497.

Chakraborty, R., Borglin, S.E., Dubinsky, E.A., Andersen, G.L., Hazen, T.C., 2012. Microbial response to the MC-252 oil and Corexit 9500 in the Gulf of Mexico. Frontiers in Microbiology. 3, 357. https://doi.org/10.3389/fmicb.2012.00357.

Chandra, S., Sharma, R., Singh, K., Sharma, A., 2013. Application of bioremediation technology in the environment contaminated with petroleum hydrocarbon. Annals of Microbiology 63 (2), 417–431.

Chapelle, F., 2001. Ground-water Microbiology and Geochemistry. John Wiley & Sons.

Charles, G.D., Bartels, M.J., Zacharewski, T.R., Gollapudi, B.B., Freshour, N.L., Carney, E.W., 2000. Activity of benzo[a]pyrene and its hydroxylated metabolites in an estrogen receptor-a reporter gene assay. Toxicological Sciences 55, 320–326.

Chekroud, Z., Gouda, M.K., Houhamdi, M., 2011. Biodegradation of crude oil in marine medium. Journal of Proteomics and Bioinformatics 4 (10), 231–237.

Chen, X., Yu, J., Zhang, Z.H., Lu, C., 2011. Study on structure and thermal stability properties of cellulose fibers from rice straw. Carbohydrate Polymers 85, 245–250.

Chen, Y., Li, C., Zhou, Z., Wen, J., You, X., Mao, Y., Lu, C., Huo, G., Jia, X., 2014. Enhanced biodegradation of alkane hydrocarbons and crude oil by mixed strains and bacterial community analysis. Applied Biochemistry and Biotechnology 17 (27), 3433–3447.

Chen, Y., Yu, B., Lin, J., Naidu, R., Chen, Z., 2016. Simultaneous adsorption and biodegradation (SAB) of diesel oil using immobilized *Acinetobacter venetianus* on porous material. Chemical Engineering Journal 289, 463–470.

Chen, Z., Huang, G.H., Chakma, A., 1998. Integrated environmental risk assessment for petroleum contaminated sites a North American case study. Water Science and Technology 38, 131–138.

Choi, S.-C., Kwon, K.K., Sohn, J.H., Kim, S.-J., 2002. Evaluation of fertilizer additions to stimulate oil biodegradation in sand seashore mesocosms. Journal of Microbiology and Biotechnology 12 (3), 431–436.

Ciuciu, G., Secrieru, D., Pavelescu, G., Savastru, D., Nicolae, D., Talianu, C., Nemuc, A., 2006. Investigation of seawater pollution on the Black Sea Romanian coast. In: SPIE ProceedingsThirteenth Joint International Symposium on Atmospheric and Ocean Optics/Atmospheric Physics, vol. 6522, .

Clayton, C., 2005. Chemical and physical properties of petroleum. Petroleum Geology 10, 248–260.

Corgié, S., Beguiristain, T., Leyval, C., 2004. Spatial distribution of bacterial communities and phenanthrene (PHE) degradation in the rhizosphere of *Lolium perenne* L. Applied and Environmental Microbiology 70, 3552–3557.

Coty, V.F., 1967. Atmospheric nitrogen fixation by hydrocarbonoxidizing bacteria. Biotechnology and Bioengineering 9, 25–32.

Coulon, F., Chronoupolou, P.M., Fahy, A., Païssé, S., Goñi-Urriza, M.S., Peperzak, L., AcuñaAlvarez, L., McKew, B.A., Brussard, C., Underwood, G.J.C., Timmis, K.N., Duran, R., McGenity, T.J., 2012. Central role of dynamic tidal bio films dominated by aerobic hydrocarbonoclastic bacteria and diatoms in the biodegradation of hydrocarbons in coastal mud flats. Applied and Environmental Microbiology 78, 3638–3648.

Coulon, F., McKew, B.A., Osborn, A.M., McGenity, T.J., Timmis, K.N., 2007. Effects of temperature and biostimulation on oil-degrading microbial communities in temperate estuarine waters. Environmental Microbiology 9, 177–186.

Crawford, R.L., Rosenberg, E., 2013. Bioremediation. In: Rosenberg, E., DeLong, E.F., Lory, S., et al. (Eds.), The Prokaryotes. Springer-Verlag, Heidelberg, pp. 295–307.

Crisafi, F., Genovese, M., Smedile, F., Russo, D., Catalfamo, M., Yakimov, M., Giuliano, L., Denaro, R., 2016. Bioremediation technologies for polluted seawater sampled after an oil-spill in Taranto Gulf (Italy): a comparison of biostimulation, bioaugmentation and use of a washing agent in microcosm studies. Marine Pollution Bulletin 106, 119–126.

Crisafully, R., Milhome, M.A., Cavalcante, R.M., Silveira, E.R., De Keukeleire, D.D., Nascimento, R.F., 2008. Removal of some polycyclic aromatic hydrocarbons from petrochemical wastewater using low-cost adsorbents of natural origin. Bioresource Technology 99, 4515–4519.

D'Auria, M., Emanuele, L., Racioppi, R., Velluzzi, V., 2009. Photochemical degradation of crude oil: comparison between direct irradiation, photocatalysis, and photocatalysis on zeolite. Journal of Hazardous Materials 164 (1), 32–38.

Darmayati, Y., Sanusi, H.S., Prartono, T., Santosa, D.A., Nuchsin, R., 2015. The effect of biostimulation and biostimulation-bioaugmentation on biodegradation of oil-pollution on sandy beaches using mesocosms. International Journal of Marine Science 527, 1–11.

Das, N., Chandran, P., 2011. Microbial degradation of petroleum hydrocarbon contaminants: an overview. Biotechnology Research International. 2011:941810. https://doi.org/10.4061/2011/941810.

Dave, D., Ghaly, A.E., 2011. Remediation technologies for marine oil spills: a critical review and comparative analysis. American Journal of Environmental Sciences 7 (5), 423–440.

Davidson, W., Lee, K., Cogswell, A., 2008. Oil Spill Response: A Global Perspective. NATO Science for Peace and Securities. Series-c: Environmental Security. Springer, Netherlands, p. 24.

Davis, D.W., Guidry, R.J., 1996. Oil spills and the state responsibilities. Basin Research Institute Bulletin 6, 60–68.

Deriase, S.F., Younis, Sh.A., El-Gendy, N.Sh., 2013. Kinetic evaluation and modeling for batch degradation of 2-hydroxybiphenyl and 2,2′-dihydroxybiphenyl by *Corynebacterium variabilis* Sh42. Desalination and Water Treatment 51 (22–24), 4719–4728.

Deschamps, G., Caruel, H., Borredon, M.E., Bonnin, C., Vignolas, C., 2003. Oil removal from water by sorption on hydrophobic cotton Fibers-Study of sorption properties and comparison with other cotton fiber-based sorbents. Environmental Science & Technology 37, 1013–1015.

Diaz, M.P., Boyd, K.G., Grigson, S.J.W., Burgess, J.G., 2002. Biodegradation of crude oil across a wide range of salinities by an extremely halotolerant bacterial consortium MPD-M, immobilized onto polypropylene fibers. Journal of Biotechnology and Bioengineering 79, 145–153.

Doerffer, J.W., 1992. Oil Spill Response in the Marine Environment. Pergamon Press, Oxford, UK, p. 391.

El-Gendy, N.Sh., Nassar, H.N., 2015a. Kinetic modeling of the bioremediation of diesel oil polluted seawater using *Pseudomonas aeruginosa* NH1. Energy Sources, Part A: Recovery, Utilization, and Environmental Effects 37 (11), 1147–1163.

El-Gendy, N.Sh., Moustafa, Y.M., 2007. Environmental assessment of petroleum hydrocarbons contaminating Temsah lake, Suez canal, Egypt. Oriental Journal of Chemistry 22 (1), 11–26.

El-Gendy, N.Sh., Nassar, H.N., 2015b. A comparative study for batch process of phenol biosorption onto different spent waste biomass: kinetics and equilibrium. Energy Sources, Part A: Recovery, Utilization, and Environmental Effects 37 (10), 1098–1109.

El-Mahdi, A.M., Aziz, H.A., Abu Amr, S.S., El-Gendy, N.Sh., Nassar, H., 2015b. Performance of isolated *Kocuria* sp. SAR1 in light crude oil biodegradation. Bioremediation and Biodegradation. 6, 4. https://doi. org/10.4172/2155-6199.1000303.

El-Gendy, N.Sh., Ali, H.R., El-Nady, M.M., Deriase, S.F., Moustafa, Y.M., Roushdy, M.I., 2014. Effect of Different Bioremediation Techniques on Petroleum Biomarkers and Asphaltene Fraction in Oil Polluted Sea Water. Desalination and Water Treatment 52 (40–42), 7484–7494.

El-Gendy, N.Sh., Nassar, H.N., 2016. Study on the effectiveness of spent waste sugarcane bagasse for adsorption of different petroleum hydrocarbons water pollutants: kinetic and equilibrium isotherm. Desalination and Water Treatment 57 (12), 5514–5528.

El-Mahdi, A.M., Aziz, H.A., El-Gendy, N.Sh., Abu Amr, S.S., Nassar, H.N., 2015a. Effectiveness of corn steep liquor as a natural low-cost material for the biodegradation of two types of Libyan crude oil. Australian Journal of Basic and Applied Sciences 9 (11), 494–505.

El-Mahdi, A.M., Aziz, H.A., Abu Amr, S.S., El-Gendy, N.Sh., Nassar, H., 2015c. Performance of some new bacterial isolates on biodegradation of Libyan light crude oil using agro-industrial wastes as co-substrates. Applied Mechanics and Materials 802, 496–500.

El-Mahdi, A.M., Aziz, H.A., El-Gendy, N.Sh., Abu Amr, S.S., Nassar, H.N., 2014. Optimization of Libyan crude oil biodegradation by using solid waste date as a natural low-cost material. Journal of Bioremediation & Biodegradation. 57, 1000252. https://doi.org/10.4172/2155-6199.1000252.

El-Nass, M.H., Al-Zuhair, S., Al-Lobaney, A., Makhlouf, S., 2009. Assessment of electrocoagulation for the treatment of petroleum refinery wastewater. Journal of Environmental Management 9 (11), 180–185.

El-Sheshtawy, H.S., Khalil, N.M., Ahmed, W., Abdallah, R.I., 2014. Monitoring of oil pollution at Gemsa Bay and bioremediation capacity of bacterial isolates with biosurfactants and nanoparticles. Marine Pollution Bulletin 87, 191–200.

El-Tarrs, A.E., Shahaby, A.F., Awad, N.S., Bahobial, A.S., El-abib, O.A., 2012. In vitro screening for oil degrading bacteria and evaluation of their biodegradation potential for hydrocarbon. African Journal of Microbiology Research 6, 7534–7544.

Engelhardt, M.A., Daly, K., Swannell, R.P., Head, I.M., 2001. Isolation and characterization of a novel hydrocarbon-degrading, Gram positive bacterium, isolated from intertidal beach sediment, and description of *Planococcus alkanoclasticus* sp. nov. Journal of Applied Microbiology 90, 237–247.

EPA Oil Program 2006. Website: http://www.epa.gov/oilspill/.

Evdokinmov, I.N., Losev, A.P., 2007. On the nature of UV/Vis absorption spectra of asphaltenes. Petroleum Science and Technology 25, 55–66.

Farag, S., Soliman, N.A., 2011. Biodegradation of crude petroleum oil and environmental pollutants by *Candida tropicalis* strain. Brazilian Archives Biology and Technology 54 (4), 821–830.

Farrington, J.W., 2014. Oil pollution in the marine environment II: fates and effects of oil spills. Environment: Science and Policy for Sustainable Development 56 (4), 16–31.

Fedorak, P.M., Semple, K.M., Westlake, D.W.S., 1984. Oil-degrading capabilities of yeasts and fungi isolated from coastal marine environments. Canadian Journal of Microbiology 30, 565–571.

Filler, D.M., Reynolds, C.M., Snape, I., Daugulis, A.J., Barnes, D.L., Williams, P.J., 2006. Advances in engineered remediation for use in the Arctic and Antarctica. Polar Record 42, 111–120.

Fingas, M.F., Fieldhouse, B., Lambert, P., Wang, Z., Noonan, J., Lane, J., Mullin, J., 2002. Water-in-oil Emulsions Formed at Sea, in Test Tanks, and in the Laboratory. Environment Canada, Ottawa, ON. Environment Canada Manuscript Report EE-169.

Fingas, M., Fieldhouse, B., 2004. Formation of water-in-oil emulsions and application to oil spill modelling. Journal of Hazardous Materials 107 (1), 37–50.

Fingas, M.F., 2004. Modeling evaporation using models that are not boundary-layer regulated. Journal of Hazardous Materials 107, 27–36.

Floodgate, G.D., 1995. Some environmental aspects of marine hydrocarbon bacteriology. Aquatic Microbial Ecology 9, 3–11.

Foght, J., 2006. Potential for Biodegradation of Sub-littoral Residual Oil by Naturally Occurring Microorganisms Following the Lake Wabamun Train Derailment. Submitted to Alberta Environment and Environment Canada. Dept. Biological Sciences, University of Alberta.

Frankenberger, W.T., 1992. The need for laboratory feasibility studies in bioremediation of petroleum hydrocarbons. In: Calabrese, E., Kostecki, P. (Eds.), Hydrocarbon Contaminated Soils and Groundwater. Vol 2. Lewis Publishers, Chelsea, Michigan, pp. 237–293.

Fritt-Rasmussen, J., 2010. In Situ Burning of Arctic Marine Oil Spills. Ignitability of Various Oil Types Weathered at Different Ice Conditions. A Combined Laboratory and Field Study PhD Thesis. Arctic Technology Centre Department of Civil Engineering Technical University of Denmark.

Fukui, M., Harms, G., Rabus, R., Schramm, A., Widdel, F., Zengler, K., Boreham, C., Wilkes, H., 1999. Anaerobic degradation of oil hydrocarbons by sulfate-reducing and nitrate-reducing bacteria. Microbial biosystems: new frontiers proceedings of the 8th international symposium on microbial ecology bell CR. In: Brylinsky, M., Johnson-Green, P. (Eds.), Atlantic Canada Society for Microbial Ecology. Canada, Halifax.

Garrett, R.M., Pickering, I.J., Haith, C.E., Prince, R.C., 1998. Photooxidation of crude oils. Environmental Science & Technology 32, 3719–3723.

Garzoli, L., Gnavi, G., Tamma, F., Tosi, S., Varese, G.C., Picco, A.M., 2015. Sink or swim: updated knowledge on marine fungi associated with wood substrates in the Mediterranean Sea and hints about their potential to remediate hydrocarbons. Progress in Oceanography 137, 140–148.

Genovese, M., Crisa fi, F., Denaro, R., Cappello, S., Russo, D., Calogero, R., Santisi, S., Catalfamo, M., Modica, A., Smedile, F., Genovese, L., Golyshin, P.N., Giuliano, L., Yakimov, M.M., 2014. Effective bioremediation strategy for rapid in situ cleanup of anoxic marine sediments in mesocosm oil spill simulation. Frontiers in Microbiology. 5, 162. https://doi.org/10.3389/fmicb.2014.00162.

Gentili, A.R., Cubitto, M.A., Ferrero, M., Rodriguéz, M.S., 2006. Bioremediation of crude oil polluted seawater by a hydrocarbon-degrading bacterial strain immobilized on chitin and chitosan flakes. International Biodeterioration and Biodegradation 57, 222–228.

George-Okafor, U., Tasie, F., Muotoe-Okafor, F., 2009. Hydrocarbon degradation potentials of indigenous fungal isolates from petroleum contaminated soils. Journal of Physical and Natural Sciences 3, 1–10.

Germano de Almeida, D., Conceiçã da Silvã, M.d-G., Barbosa, R.d-N., Silva, D.d-S.P., Oliveira da Silva, R., Lima, G.M.d-S., Buarque de Gusmão, N., Sousa, M.d-F.V.d-Q., 2016. Biodegradation of marine fuel MF-380 by microbial consortium isolated from seawater near the petrochemical Suape Port, Brazil. International Biodeterioration and Biodegradation 116, 73–82.

Gertler, C., Gerdts, G., Timmis, K.N., Golyshin, P.N., 2009b. Microbial consortia in mesocosm bioremediation trial using oil sorbents, slow-release fertilizer and bioaugmentation. FEMS Microbiology Ecology 69, 288–300.

Gertler, C., Gerdts, G., Yakimov, M.M., Timmis, K.N., Golyshin, P.N., 2009a. Populations of heavy fuel oil-degrading marine microbial community in presence of oil sorbent materials. Journal of Applied Microbiology 107 (2), 590–605.

GESAM Group of Experts on the Scientific Aspects of Marine Pollution, 2000. Report study GESAMP. Joint Group of Experts on the Scientific Aspects of Marine Pollution. The Atmospheric Inputs of Trace Species to the World Oceans, vol. 38, p. 111.

Goodwin, N.S., Park, P.J.D., Rawlinson, A.P., 1981. Crude oil biodegradation under simulated and natural conditions. In: Bjoray, M. (Ed.), Advances in Organic Geochemistry. Wiley, Chichester, pp. 650–658.

Gordon, R., 1994. Bioremediation and its Application to Exxon Valdez Oil Spill in Alaska. http://www.geocities.com/CapeCanaveral/Lab/2094/bioremed.html.

Griffis, M.T., Da Campo, R., O'Connor, P.B., Barrow, M.P., 2014. Throwing light on petroleum: simulated exposure of crude oil to sunlight and characterization using atmospheric pressure photoionization Fourier transform ion cyclotron resonance mass spectrometry. Analytical Chemistry 86, 527–534.

Grimmer, G., Jacob, J., Dettbarn, G., Naujack, K.W., 1997. Determination of urinary metabolites of polycyclic aromatic hydrocarbons (PAH) for the risk assessment of PAH-exposed workers. International Archives of Occupational and Environmental Health 69, 231–239.

Grossman, M., Prince, R., Garrett, R., Garrett, K., Bare, R., Lee, K., Sergy, G., Owens, E., Guénette, C., 1999. Microbial diversity in oiled and un-oiled shoreline sediments in the Norwegian Arctic. In: Proc. 8th Int. Symp. Microb. Ecol., Halifax, Canada.

Guerin, T.F., 2015a. Bioremediation of diesel from a rocky shoreline in an arid tropicalclimate. Marine Pollution Bulletin 99, 85–93.

Guerin, T.F., 2015b. A safe, efficient and cost-effective process for removing petroleum hydrocarbons from a highly heterogeneous and relatively inaccessible shoreline. Journal of Environmental Management 162, 190–198.

Guo, C.L., Zhou, H.W., Wong, Y.S., Tam, N.F.Y., 2005. Isolation of PAH-degrading bacteria from mangrove sediments and their biodegradation potential. Marine Pollution Bulletin 51 (8–12), 1054–1061.

Gutierrez, T., Berry, D., Yang, T., Mishamandani, S., Mckay, L., Teske, A., Aitken, M.D., 2013. Role of bacterial exopolysaccharides (EPS) in the fate of the oil released during the Deepwater Horizon Oil Spill. PLoS One 86, e67717. http://dx.doi.org/10.1371/journal.pone.0067717.

Hall, M., Forrester, L.M., Parker, D.K., Grover, P.L., Wolf, C.R., 1989. Relative contribution of various forms of cytochrome P450 to the metabolism of benzo[a]pyrene by human liver microsomes. Carcinogenesis 10, 1815–1821.

Hamdan, L.J., Fulmer, P.A., 2011. Effects of COREXIT (R) EC9500A on bacteria from a beach oiled by the deepwater horizon spill. Aquatic Microbial Ecology 63, 101–109.

Harayama, S., Kasai, Y., Hara, A., 2004. Microbial communities in oil-contaminated seawater. Current Opinion in Biotechnology 15 (3), 205–214.

Harms, H., Schlosser, D., Wick, L.Y., 2011. Untapped potential: exploiting fungi in bioremediation of hazardous chemicals. Nature Reviews Microbiology 9 (3), 177–192.

Hassanshahian, M., Emtiazi, G., Caruso, G., Cappello, S., 2014. Bioremediation bioaugmentation/biostimulation trials of oil polluted seawater: a mesocosm simulation study. Marine Environmental Research 95, 28–38.

Hassanshahian, M., Tebyanian, H., Cappello, S., 2012. Isolation and characterization of two crude oil-degrading yeast strains, *Yarrowia lipolytica* PG-20 and PG-32, from the Persian Gulf. Marine Pollution Bulletin 64, 1386–1391.

Head, I.M., Jones, D.M., Rölling, W.F., 2006. Marine microorganisms make a meal of oil. Nature Reviews Microbiology 4 (3), 173–182.

Helmke, E., Gerdes, B., Juergens, J., Reuter, K., 2013. Bioremediation Method for Accelerated Biological Decomposition of Petroleum Hydrocarbons in Sea Ice Covered Polar Regions, and Bacteria and Enzyme Mixtures as Agents for Carrying Out Said Method.

Helmy, Q., Kardena, E., Nurachman, Z., 2010. Application of Biosurfactant Produced by *Azotobacter vinelandii* AV01 for enhanced oil recovery and biodegradation of oil sludge. International Journal of Civil and Environmental Engineering 10 (1), 6–12.

Hinchee, R.E., Kitte, J.A., Reisinger, H.J., 1995. Applied Bioremediation of Petroleum Hydrocarbons. Battelle Press, Columbus, OH, US.

Holakoo, L., 2001. On the Capability of Rhamnolipids for Oil Spill Control of Surface Water. Unpublished Dissertation in Partial Fulfillment of the Requirements for the Degree of Master in Applied Science. Concordia University, Montreal, Canada.

Hua, Z., Song, R., Du, G., Li, H., Chen, J., 2007. Effects of EDTA and Tween60 on biodegradation of n-hexadecane with two strains of *Pseudomonas aeruginosa*. Biochemical Engineering Journal 36 (1), 66–71.

Huang, H., Bowler, B.F.J., Oldenburg, T.B.P., Larter, S.R., 2004. The effect of biodegradation on polycyclic aromatic hydrocarbons in reservoired oils from the Liaohe basin, NE China. Organic Geochemistry 35 (11–12), 1619–1634.

Huang, J.C., 1983. A review of the state-of-the-art of oil spill fate/behavior models. In: International Oil Spill Conference Proceedings: February 1983. vol. 1983, No. 1, pp. 313–322. https://doi.org/10.7901/2169-3358-1983-1-313.

Huesemann, M.H., Hausmann, T.S., Fortman, T.J., 2003. Biodegradation of hopane prevents use as conservative biomarker during bioremediation of PAHs in petroleum contaminated soils. Bioremeditaion Journal 7 (2), 111–117.

Hunkeler, D., Höhener, P., Zeyer, J., 2002. Engineered and subsequent intrinsic in situ bioremediation of a diesel fuel contaminated aquifer. Journal of Contaminant Hydrology 59 (3), 231–245.

Hunting, E.R., White, C.M., van Gemert, M., Mes, D., Stam, E., van derGeest, H.G., Kraak, M.H., Admiraal, W., 2013. UV radiation and organic matter composition shape bacterial functional diversity in sediments. Frontiers in Microbiology 4, 317. http://dx.doi.org/10.3389/fmicb.2013.00317.

Ishak, S., Malakahmad, A., Isa, M.H., 2012. Refinery wastewater biological treatment: a short review. Journal of Scientific and Industrial Research 71 (4), 251–256.

ITOPF International Tanker Owners Pollution Federation, 2011. ITOPF Handbook. London, UK www.itopf.org.

Jaekel, U., Zedelius, J., Wilkes, H., Musat, F., 2015. Anaerobic degradation of cyclohexane by sulfate-reducing bacteria from hydrocarbon-contaminated marine sediments. Frontiers in Microbiology 6, 116. http://dx.doi.org/10.3389/fmicb.2015.0 0116.

Jiménez, N., Vinas, M., Bayona, J.M., Albaiges, J., Solanas, A.M., 2007. The prestige oil spill: bacterial community dynamics during a field biostimulation assay. Applied Microbiology and Biotechnology 77, 935–945.

Juck, D., Charles, T., Whyte, L.G., Greer, C.W., 2000. Polyphasic microbial community analysis of petroleum hydrocarbon contaminated soils from two northern Canadian communities. FEMS Microbiology Ecology 33, 241–249.

Juteau, P., Bisaillon, J.-G., Lépine, F., Ratheau, V., Beaudet, R., Villemur, R., 2003. Improving the biotreatment of hydrocarbons-contaminated soils by addition of activated sludge taken from the wastewater treatment facilities of an oil refinery. Biodegradation 14 (1), 31–40.

Karakasi, O.K., Moutsatsou, A., 2010. Surface modification of high calcium fly ash for its application in oil spill cleanup. Fuel 89, 3966–3970.

Karanth, N.G.K., Deo, P.G., Veenanadig, N.K., 1999. Microbial production of biosurfactants and their importance. Current Science 77, 116–126.

Karl, D.M., 1992. The grounding of the Bahia Paraiso: microbial ecology of the 1989 Antarctic oil spill. Microbial Ecology 24, 77–89.

Kasai, Y., Kishira, H., Sasaki, T., Syutsubo, K., Watanabe, K., Harayama, S., 2002. Predominant growth of *Alcanivorax* strains in oil-contaminated and nutrient-supplemented seawater. Environmental Microbiology 4, 141–147.

Kennish, M.J., 2001. Practical Handbook of Marine Science, third ed. CRC Press, Boca Raton London New York Washington, D.C.

Khan, F.I., Husain, T., 2003. Evaluation of a petroleum hydrocarbon contaminated site for natural attenuation using 'RBMNA' methodology. Environmental Modelling and Software 18 (2), 179–194.

Kim, S.-J., Choi, D.H., Sim, D.S., Oh, Y.-S., 2005. Evaluation of bioremediation effectiveness on crude oil-contaminated sand. Chemosphere 59 (6), 845–852.

Kimes, N.E., Callaghan, A.V., Suflita, J.M., Morris, P.J., 2014. Microbial transformation of the Deepwater Horizon oil spill—past, present, and future perspectives. Frontiers in Microbiology 5, 603. http://dx.doi.org/10.3389/fmicb.2014.00603.

King, S., Leaf, P., Olson, A., Ray, P., Tarr, M., 2014a. Photolytic and photocatalytic degradation of surface oil from the deepwater horizon spill. Chemosphere 95, 415–422.

King, T.L., Robinson, B., Boufadel, M., Lee, K., 2014b. Flume tank studies to elucidate the fate and behavior of diluted bitumen spilled at sea. Marine Pollution Bulletin 83, 32–37.

Kiran, G.S., Hema, T.A., Gandhimathi, R., Selvin, J., Thomas, T.A., Ravji, T.R., Natarajaseenivasan, K., 2009. Optimization and production of a biosurfactant from the sponge-associated marine fungus *Aspergillus ustus MSF3*. Colloids and Surfaces B: Biointerfaces 73 (2), 250–256.

Kleindienst, S., Seidel, M., Ziervogel, K., Grim, L.S., Loftis, K.M., Harrison, S., Malkin, S., Perkins, M.J., Field, J., Sogin, M.L., Dittmar, T., Passow, U., Medeiros, P.M., Joye, S.B., 2015. Chemical dispersants can suppress the activity of natural oil-degrading microorganisms. Proceedings of the National Academy of Sciences of the United States of America 112, 14900–14905.

Knezevich, V., Koren, O., Ron, E.Z., Rosenberg, E., 2006. Petroleum bioremediation in seawater using guano as the fertilizer. Bioremediation Journal 10, 83–91.

Korotenko, K.A., Bowman, M.J., Dietrich, D.E., 2009. High-resolution numerical model for predicting the transport and dispersal of oil spilled in the black sea. Terrestrial. Atmospheric and Oceanic Sciences 21 (1), 123–136.

Kostka, J.E., Prakash, O., Overholt, W.A., Green, S.J., Freyer, G., Canion, A., Delgardio, J., Norton, N., Hazen, T.C., Huettel, M., 2011. Hydrocarbon-degrading bacteria and the bacterial community response in Gulf of Mexico beach sands impacted by the Deepwater Horizon oil spill. Applied and Environmental Microbiology 77 (22), 7962–7974.

Kropp, K.G., Fedorak, P.M., 1998. A review of the occurrence, toxicity, and biodegradation of condensed thiophenes found in petroleum. Canadian Journal of Microbiology 44 (7), 605–622.

Kujawinski, E.B., Kido Soule, M.C., Valentine, D.L., Boysen, A.K., Longnecker, K., Redmond, M.C., 2011. Fate of dispersants associated with the Deepwater Horizon oil spill. Environmental Science & Technology 45, 1298–1306.

Kumari, B., Singh, S.N., Singh, D.P., 2012. Characterization of two biosurfactant producing strains in crude oil degradation. Process Biochemistry 47, 2463–2471.

Lampi, M.A., Gurska, J., McDonald, K.I.C., Xie, F.L., Huang, X.D., Dixon, D.G., Greenberg, B.M., 2006. Photoinduced toxicity of polycyclic aromatic hydrocarbons to Daphnia magna: ultraviolet-mediated effects and the toxicity of polycyclic aromatic hydrocarbon photoproducts. Environmental Toxicology and Chemistry 25 (4), 1079–1087.

Leahy, J.G., Colwell, R.R., 1990. Microbial degradation of hydrocarbons in the environment. Microbiological Reviews 54, 305–315.

Lea-Smith, D.J., Biller, S.J., Davey, M.P., Cotton, C.A.R., Perez Sepulveda, B.M., Turchyn, A.B., Scanlan, D.J., Smith, A.G., Chisholm, S.W., Howe, C.J., 2015. Contribution of cyanobacterial alkane production to the ocean hydrocarbon cycle. Proceedings of the National Academy of Sciences of the United States of America. http://dx.doi.org/10.1073/pnas.1507274112.

Lee, K., Lunel, T., Wood, P., Swannell, R., Stoffyn-Egli, P., 1997. Shoreline cleanup by acceleration of clay-oil flocculation processes. In: Proceedings of 1997 International Oil Spill Conference. American Petroleum Institute, Washington DC, pp. 235–240.

Lei, L., Khodadoust, A.P., Suidan, M.T., Tabak, H.H., 2005. Biodegradation of sediment-bound PAHs in field-contaminated sediment. Water Research 39, 349–361.

Leitão, A.L., 2009. Review on potential of Penicillium species in the bioremediation field. International Journal of Environmental Research and Public Health 6 (4), 1393–1417.

Lewis, M., Pryor, R., 2013. Toxicities of oils, dispersants and dispersed oils to algae and aquatic plants: review and database value to resource sustainability. Environmental Pollution 180, 345–367.

Leys, N.M., Bastiaens, L., Verstraete, W., Springael, D., 2005. Influence of the carbon/nitrogen/phosphorus ratio on polycyclic aromatic hydrocarbon degradation by Mycobacterium and Sphingomonas in soil. Applied Microbiology and Biotechnology 66 (6), 726–736.

Li, H., Liu, L., Yang, F., 2012. Hydrophobic modification of polyurethane foam for oil spill cleanup. Marine Pollution Bulletin 64, 1648–1653.

Li, J.-L., Chen, B.-H., 2009. Surfactant-mediated biodegradation of polycyclic aromatic hydrocarbons. Nature Materials 2, 76–94.

Li, X., Zhao, L., Adam, M., 2016. Biodegradation of marine crude oil pollution using a salt-tolerant bacterial consortium isolated from Bohai Bay, China. Marine Pollution Bulletin 105, 43–50.

Lin, M., Liu, Y., Chen, W., Wang, H., Hu, X., 2014. Use of bacteria-immobilized cotton fibers to absorb and degrade crude oil. International Biodeterioration and Biodegradation 88, 8–12.

Lin, Q., Mendelssohn, I.A., 1998. The combined effects of phytoremediation and biostimulation in enhancing habitat. Ecological Engineering 10, 263–274.

Lin, T.A., Young, C.C., Ho, M.J., Yeh, M.S., Chou, C.L., Wei, H., Chang, J.S., 2005. Characterization of floating activity on indigenous diesel assimilating bacterial isolates. Journal of Bioscience and Bioengineering 99 (5), 466–472.

Lindstrom, J.E., Braddock, J.F., 2002. Biodegradation of petroleum hydrocarbons at low temperature in the presence of the dispersant Corexit 9500. Marine Pollution Bulletin 44, 739–747.

Liu, B., Ju, M., Liu, J., Wu, W., Li, X., 2016. Isolation, identification, and crude oil degradation characteristics of a high-temperature, hydrocarbon-degrading strain. Marine Pollution Bulletin 106, 301–307.

Liu, H., Yao, J., Yuan, Z., Shang, Y., Chen, H., Wang, F., Masakorala, K., Yu, C., Cai, M., Blake, R.E., Choi, M.M.F., 2014. Isolation and characterization of crude-oil-degrading bacteria from oil-water mixture in Dagang oilfield, China. International Biodeterioration and Biodegradation 87, 52–59.

Liu, P.-W.G., Liou, J.W., Li, Y.T., Su, W.L., Chen, C.-H., 2015. The optimal combination of entrapped bacteria for diesel remediation in seawater. International Biodeterioration and Biodegradation 102, 383–391.

Ma, X., Li, N., Jiang, J., Xu, Q., Li, H., Wang, L., Lu, J., 2013. Adsorption–synergic biodegradation of high-concentrated phenolic water by *Pseudomonas putida* immobilized on activated carbon fiber. Journal of Environmental Chemical Engineering 1, 466–472.

Macaulay, B.M., Rees, D., 2014. Bioremediation of oil spills: a review of challenges for research advancement. Annals of Environmental Science 8, 9–37.

MacNaughton, S.J., Stephen, J.R., Venosa, A.D., Davis, G.A., Chang, Y.J., White, D.C., 1999. Microbial population changes during bioremediation of an experimental oil spill. Applied and Environmental Microbiology 65 (8), 3566–3574.

Maheshwari, R., Singh, U., Singh, P., Singh, N., Jat, B.L., Rani, B., 2014. To decontaminate wastewater employing bioremediation technologies. Journal of Advanced Scientific Research 5 (2), 7–15.

Mahjoubi, M., Jaouani, A., Guesmi, A., Ben Amor, S., Jouini, A., Cherif, H., Najera, A., Boudabous, A., Koubaa, N., Cherif, A., 2013. Hydrocarbonoclastic bacteria isolated from petroleum contaminated sites in Tunisia: isolation, identification and characterization of the biotechnological potential. New Biotechnology 30 (6), 723–733.

Maki, H., Sasaki, T., Harayama, S., 2001. Photo-oxidation of biodegraded crude oil and toxicity of the photo-oxidized products. Chemosphere 44 (5), 1145–1151.

Malik, Z.A., Ahmed, S., 2012. Degradation of petroleum hydrocarbons by oil field isolated bacterial consortium. African Journal of Biotechnology 11, 650–658.

Martin, C.K.A., 1977. Microbial cleavage of sterol side chains. Advances in Applied Microbiology 22, 29–58.

Martín, Y.B.S., 2011. Bioremediation: a tool for the management of oil pollution in marine ecosystems. Biotecnología Aplicada 28 (2), 69–76.

Mbadinga, S.M., Wang, L.-Y., Zhou, L., Liu, J.-F., Gu, J.-D., Mu, B.-Z., 2011. Microbial communities involved in anaerobic degradation of alkanes. International Biodegradation and Biodeterioration 65, 1–13.

McFarlin, K.M., Prince, R.C., Perkins, R., Leigh, M.B., 2014. Biodegradation of dispersed oil in Arctic seawater at -1°C. PLoS One. 9 (1), e84297. https://doi.org/10.1371/journal.pone.0084297.

McGenity, T.J., 2014. Hydrocarbon biodegradation in intertidal wetland sediments. Current Opinion in Biotechnology 27, 46–54.

McKew, B.A., Coulon, F., Osborn, A.M., Timmis, K.N., McGenity, T.J., 2007. Determining the identity and roles of oil-metabolizing marine bacteria from the Thames Estuary, UK. Environmental Microbiology 9, 165–176.

Medeiros, P.M., Bicego, M.C., 2004. Investigation of natural and anthropogenic hydrocarbon inputs in sediments using geochemical markers. Marine Pollution Bulletin 49, 892–899.

Medina, A., Vassileva, M., Caravaca, F., Roldán, A., Azcón, R., 2004. Improvement of soil characteristics and growth of *Dorycnium pentaphyllum* by amendment with agrowastes and inoculation with AM fungi and/or the yeast *Yarowia lipolytica*. Chemosphere 56 (5), 449–456.

Mehta, S.D., 2005. Making and Breaking of Water in Crude Oil Emulsions MSc Thesis. Department of Civil Engineer, Texas A&M University.

Mercer, K., Trevors, J., 2011. Remediation of oil spills in temperate and tropical coastal marine environments. The Environmentalist 31 (3), 338–347.

Kristensen, M., Johnsen, A.R., Christensen, A.R., 2015. Marine biodegradation of crude oil in temperate and Arctic water samples. Journal of Hazardous Materials 300, 75–83.

Metwally, M., Al-Mu Zaini, S., Jacob, P.G., Bahloul, M., 1997. Petroleum hydrocarbons and related heavy metals in the near-shore marine sediments of Kuwait. Environment International 23, 115–121.

Milić, J.S., Beškoski, V.P., Ilić, M.V., Ali, S.A.M., Gojgić-Cvijović, G., Vrvić, M.M., 2009. Bioremediation of soil heavily contaminated with crude oil and its products: composition of the microbial consortium. Journal of the Serbian Chemical Society 74 (4), 455–460.

Mills, A.M., Bonner, S.J., McDonald, J.T., Page, A.C., Autenrieth, L.R., 2003. Intrinsic bioremediation of a petroleum-impacted wetland. Marine Pollution Bulletin 46, 887–899.

Mishamandani, S., Gutierrez, T., Berry, D., Aitken, M.D., 2016. Response of the bacterial community associated with a cosmopolitan marine diatom to crude oil shows a preference for the biodegradation of aromatic hydrocarbons. Environmental Microbiology 18 (6), 1817–1833.

Mishra, S., Sarma, P.M., Lal, B., 2004. Crude oil degradation efficiency of a recombinant *Acinetobacter baumannii* strain and its survival in crude oil-contaminated soil microcosm. FEMS Microbiology Letters 235, 323–331.

Mittal, A., Singh, P., 2009. Isolation of hydrocarbon degrading bacteria from soils contaminated with crude oil spills. Indian Journal of Experimental Biology 47, 760–765.

Montero-Rodríguez, D., Andrade, R., Ribeiro, D., Lima, R., Araújo, H., Campos Takaki, G., 2014. Ability of *Serratia marcescens* UCP/WFCC 1 549 for biosurfactant production using industrial wastes and fuels biodegradation. In: Méndez Vilas, A. (Ed.), Industrial, Medical and Environmental Applications of Microorganisms: Current Status and Trends, pp. 211–216 Madrid.

Mosbech, A., 2002. Potential Environmental Impacts of Oil Spills in Greenland: An Assessment of Information Status and Research Needs NERI technical Report No. 415. National Environmental Research Institute, Ministry of the Environment, Denmark.

Mukhetjee, B., Turner, J., Wrenn, B., 2011. Effect of oil composition on chemical dispersion of crude oil. Environmental Engineering Science 28 (7), 497–506.

Mukred, A.M., Hamid, A.A., Hamzah, A., Yusoff, W.M.W., 2008. Development of three bacteria consortia for the bioremediation of crude petroleum-oil in contaminated water. Online Journal of Biological Sciences 8 (4), 73–79.

Mullin, J.V., Champ, M.A., 2003. Introduction/overview to in situ burning of oil spills. Spill Science and Technology Bulletin 8, 323–330.

Munday, P.L., Dixson, D.L., McCormick, M.I., Meekan, M., Ferrari, M.C., Chivers, D.P., 2010. Replenishment of fish populations is threatened by ocean acidification. Proceedings of the National Academy of Sciences of the United States of America 107 (29), 12930–12934.

Munoz, D., Guiliano, M., Doumneq, P., Jacquot, F., Scherrer, P., Mille, G., 1997. Long term evolution of petroleum biomarkers in mangrove soil (Guadeloupe). Marine Pollution Bulletin 34, 868–874.

Naether, D.J., Slawtschewb, S., Stasik, S., Engel, M., Olzog, M., Wick, L.Y., Timmis, K.N., Heipieperb, H.J., 2013. Adaptation of the hydrocarbonoclastic bacterium *Alcanivorax borkumensis* SK2 to alkanes and toxic organic compounds: a physiological and transcriptomic approach. Applied and Environmental Microbiology 79 (14), 4282–4293.

National Academy of Sciences (NAS), 1985. Oil in the Sea, Inputs, Fates, and Effects. National Academy Press, Washington, DC.

National Research Council, 2005. Oil Spill Dispersants: Efficacy and Effects. The National Academies Press, United States.

Nicodem, D.E., Guedes, C.L.B., Correa, R.J., 1998. Photochemistry of petroleum I. Systematic study of a Brazilian intermediate crude oil. Marine Chemistry 63, 93–104.

Nicodem, D.E., Guedes, C.L.B., Correa, R.J., Fernandes, M.C.Z., 1997. Photochemical processes and the environmental impact of petroleum spills. Biogeochemistry 39 (2), 121–138.

Nikolopoulou, M., Eickenbusch, P., Pasadakis, N., Venieri, D., Kalogerakis, N., 2013a. Microcosm evaluation of autochthonous bioaugmentation to combat marine oil spills. New Biotechnology 30 (6), 734–742.

Nikolopoulou, M., Pasadakis, N., Kalogerakis, N., 2013b. Evaluation of autochthonous bioaugmentation and biostimulation during microcosm-simulated oil spills. Marine Pollution Bulletin 72, 165–173.

Nkem, B.M., Halimoon, N., Yusoff, F.M., Johari, W.L.W., Zakaria, M.P., Medipally, S.R., Kannan, N., 2016. Isolation, identification and diesel-oil biodegradation capacities of indigenous hydrocarbon-degrading strains of *Cellulosimicrobium cellulans* and *Acinetobacter baumannii* from tarball at Terengganu beach, Malaysia. Marine Pollution Bulletin 107, 261–268.

Nomack, M., Cleveland, C., 2010. Oil spill control technologies. In: Encyclopedia of Earth. http://www.eoearth.org/articles/view/158385/?topic=50366.

Obayori, O.S., Adebusoye, S.A., Ilori, M.O., Oyetibo, G.O., Omotayo, A.E., Amund, O.O., 2010. Effects of corn steep liquor on growth rate and pyrene degradation by Pseudomonas strains. Current Microbiology 60 (6), 407–411.

Obi, E.O., Kamgba, F.A., Obi, D.A., 2014. Techniques of oil spill response in the sea. IOSR Journal of Applied Physics 6 (1), 36–41.

Ojo, O.A., 2006. Petroleum-hydrocarbon utilization by native bacterial population from a wastewater canal Southwest Nigeria. African Journal of Biotechnology 5 (4), 333–337.

Olivera, N.L., Commendatore, M.G., Delgado, O., Esteves, J.L., 2003. Microbial characterization and hydrocarbon biodegradation potential of natural bilge waste microflora. Journal of Industrial Microbiology & Biotechnology 30, 542–548.

Olivera, N.L., Esteves, J., Commendatore, M.G., 1997. Alkane biodegradation by a microbial community from contaminated sediments in Patagonia, Argentina. International Biodeterioration and Biodegradation 40, 75–79.

Ostroumov, V.E., Siegert, C., 1996. Exobiological aspects of mass transfer in microzones of permafrost deposits. Advances in Space Research 18 (12), 79–86.

Otremba, Z., Toczek, H., 2002. Degradation of crude oil film on the surface of seawater: the role of luminous, biological and aqutorial factors. Polish Journal of Environmental Studies 11 (5), 555–559.

Pacwa-Płociniczak, M., Płaza, G.A., Piotrowska-Seget, Z., Cameotra, S.S., 2011. Environmental applications of biosurfactants: recent advances. International Journal of Molecular Sciences 12 (1), 633–654.

Passarini, M.R., Sette, L.D., Rodrigues, M.V., 2011. Improved extraction method to evaluate the degradation of selected PAHs by marine fungi grown in fermentative medium. Journal of the Brazilian Chemical Society 22 (3), 564–570.

Pelletier, E., Delille, D., Delille, B., 2004. Crude oil bioremediation in sub-Antarctic intertidal sediments: chemistry and toxicity of oiled residues. Marine Environmental Research 57 (4), 311–327.

Pereira, P., Enguita, F.J., Ferreira, J., Leitão, A.L., 2014. DNA damage induced by hydroquinone can be prevented by fungal detoxification. Toxicology Reports 1, 1096–1105.

Peressutti, S.R., Alvarez, H.M., Pucci, O.H., 2003. Dynamics of hydrocarbon degrading bacteriocenosis of experimental oil pollution in Patagonian soil. International Biodeterioration and Biodegradation 51, 21–30.

Peterson, C.H., Rice, S.D., Short, J.W., Esler, D., Bodkin, J.L., Ballachey, B.E., Irons, D.B., 2003. Long-term ecosystem response to the Exxon Valdez oil spill. Science 302, 2082–2086.

Podorozhko, E.A., Lozinsky, V.I., Ivshina, I.B., Kuyukina, M.S., Krivorutchko, A.B., Philp, J.C., Cunningham, C.J., 2008. Hydrophobised sawdust as a carrier for immobilisation of the hydrocarbon-oxidizing bacterium *Rhodococcus ruber*. Bioresource Technology 99, 2001–2008.

Prince, R.C., McFarlin, K.M., Butler, J.D., Febbo, E.J., Wang, F.C., Nedwed, T.J., 2013. The primary biodegradation of dispersed crude oil in the sea. Chemosphere 90 (2), 521–526.

Prince, R.C., Garrett, R.M., Bare, R.E., Grossman, M.J., Townsend, T., Suflita, J.M., Lee, K., Owens, E.H., Sergy, G.A., Braddock, J.F., Lindstrom, J.E., Lessard, R.R., 2003. The roles of photooxidation and biodegradation in long-term weathering of crude and heavy fuel oils. Spill Science and Technology Bulletin 8, 145–156.

Pritchard, P.H., Mueller, J.G., Rogers, J.C., Kremer, F.V., Glaser, J.A., 1992. Oil spill bioremediation: experiences, lessons and results from the Exxon Valdez oil spill in Alaska. Biodegradation 3 (2), 315–335.

Public Health Fact Sheet, 2009. Polycyclic Aromatic Hydrocarbons (PAHs): Health Effects. Department of health, government of south Australia, SA health. www.health.sa.gov.au/pehs/environ-health-index.htm.

Quek, E., Ting, Y.P., Tan, H.M., 2006. *Rhodococcus* sp. F92 immobilized on polyurethane foam shows ability to degrade various petroleum products. Bioresource Technology 97, 32–38.

Raafat, T., El-Gendy, N.Sh., Farahat, L., Kamel, M., ElShafy, E.A., 2007. Bioremediation of industrial wastes of oil refineries as an environmental solution for water pollution. Eurasian Chemico-technological Journal 9 (2), 153–162.

Radović, J.R., Aeppli, C., Nelson, R.K., Jimenez, N., Reddy, C.M., Bayona, J.M., Albaigés, J., 2014. Assessment of photochemical processes in marine oil spill fingerprinting. Marine Pollution Bulletin 79 (1), 268–277.

Rahman, K., Rahman, T., Lakshmanaperumalsamy, P., Marchant, R., Banat, I., 2003. The potential of bacterial isolates for emulsification with a range of hydrocarbons. Acta Biotechnologica 23 (4), 335–345.

Rahman, K.S.M., Thahira-Rahman, J., Lakshmanaperumalsamy, P., Banat, I.M., 2002. Towards efficient crude oil degradation by a mixed bacterial consortium. Bioresource Technology 85, 257–261.

Rahman, R.M.Z.A., Ghazali, F.M., Salleh, A.B., Basri, M., 2006. Biodegradation of hydrocarbon contamination by immobilised bacterial cells. The Journal of Microbiology 44, 354–359.

Ramachandran, S.D., Hodson, P.V., Khan, C.W., Lee, K., 2004. Oil dispersant increasesPAH uptake by fish exposed to crude oil. Ecotoxicology and Environmental Safety 59, 300–308.

Ramirez, N., Cutright, T., Ju, L.K., 2001. Pyrene biodegradation in aqueous solutions and soil slurries by *Mycobacterium* PYR-1and enriched consortium. Chemosphere 44, 1079–1086.

Rawe, J., Krietemeye, r, S., Meagher-Hartzell, E., 1993. Guide for Conducting Treatability Studies under CERCLA: Biodegradation Remedy Selection- Interim Guidance. U.S. Environmental Protection Agency, Washington, D.C.

Ray, P.Z., Chen, H., Podgorski, D.C., McKenna, A.M., Tarr, M.A., 2014. Sunlight creates oxygenated species in water-soluble fractions of Deepwater Horizon oil. Journal of Hazardous Materials 280, 636–643.

Reis, R., Pacheco, G., Pereira, A., Freire, D., 2013. Biosurfactants: production and applications. In: Chamy, R., Rosenkranz, F. (Eds.), Biodegradation – Life of Science. InTech, New York, US, pp. 31–61(Chapter 2) https://doi.org/10.5772/56144.

Rengathavasi, T., Singaram, J., Ibrahim, M.B., 2011. Effect of biosurfactant and fertilizer on biodegradation of crude oil by marine isolates of *Bacillus megaterium, Corynebacterium kutscheri* and *Pseudomonas aeruginosa*. Bioresource Technology 102, 772–778.

Riser-Roberts, E., 1992. Bioremediation of Petroleum Contaminated Sites. CRC Press Inc., Boca Raton, FL, US.

Rocha e Silva, N.M.P., Rufino, R.D., Luna, J.M., Santos, V.A., Sarubbo, L.A., 2014. Screening of *Pseudomonas* species for biosurfactant production using low-cost substrates. Biocatalysis and Agricultural Biotechnology 3 (2), 132–139.

Rogers, A.D., Laffoley, D.d'A., 2011. International Earth System Expert Workshopon Ocean Stresses and Impacts: Summary Workshop Report. DIANE Publishing.

Rogowska, J., Namiesnik, J., 2010. Environmental implications of oil spills from shipping accidents. Reviews of Environmental Contamination and Toxicology 206, 95–114.

Rojo, F., 2009. Degradation of alkanes by bacteria. Environmental Microbiology 11 (10), 2477–2490.

Rölling, W.F.M., Milner, M.G., Jones, D.M., Fratepietro, F., Swannell, R.P.J., Daniel, F., Head, I.M., 2004. Bacterial community dynamics and hydrocarbon degradation during a field-scale evaluation of bioremediation on a mudflat beach contaminated with buried oil. Applied and Environmental Microbiology 70 (5), 2603–2613.

Rölling, W.F., Milner, M.G., Jones, D.M., Lee, K., Daniel, F., Swannell, R.J., Head, I.M., 2002. Robust hydrocarbon degradation and dynamics of bacterial communities during nutrient-enhanced oil spill bioremediation. Applied and Environmental Microbiology 68 (11), 5537–5548.

Rölling, W.F.M., 2004. Bacterial community dynamics and hydrocarbon degradation during a field-scale evaluation of bioremediation on a mudflat beach contaminated with buried oil. Applied and Environmental Microbiology 70, 2603–2613.

Ron, E.Z., Rosenberg, E., 2014. Enhanced bioremediation of oil spills in the sea. Current Opinion in Biotechnology 27, 191–194.

Rosenberg, E., Navon-Venezia, S., Zilber-Rosenberg, I., Ron, E.Z., 1998. Rate-limiting steps in the biodegradation of hydrocarbons. In: Rubin, H., Narkis, N., Canberry, J. (Eds.), Soil and Aquifer Pollution, pp. 159–172.

Rosenberg, E., Ron, E.Z., 1999. High- and low-molecular-mass microbial surfactants. Applied Microbiology and Biotechnology 52, 154–162.

Rundle, A., Tang, D., Hibshoosh, H., Estabrook, A., Schnabel, F., Cao, W., Grumet, S., Perera, F.P., 2000. The relationship between genetic damage from polycyclic aromatic hydrocarbons in breast tissue and breast cancer. Carcinogenesis 21, 1281–1289.

Ryerson, T.B., Camilli, R., Kessler, J.D., Kujawinski, E.B., Reddy, C.M., Valentine, D.L., Atlas, E., Blake, D.R., de Gouw, J., Meinardi, S., 2012. Chemical data quantify Deepwater Horizon hydrocarbon flow rate and environmental distribution. Proceedings of the National Academy of Sciences of the United States of America 109 (50), 20246–20253.

Ryynänen, T.P., 2011. Reduction of Waste Water Loads at Petrochemical Plants MSC thesis. Department of Chemical and Biological Engineering, Division of Chemical Environmental Science, Chalmers University of Technology, Goteborg, Sweden.

Sabate, J., Vinas, M., Solanas, A.M., 2004. Laboratory scale bioremediation experiments on hydrocarbon contaminated soils. International Biodeterioration and Biodegradation 54, 19–25.

Sabean, J.A.R., Scott, D.B., Lee, K., Venosa, A.D., 2009. Monitoring oil spill bioremediation using marsh foraminifera as indicators. Marine Pollution Bulletin 59 (8–12), 352–361.

Safiyanu, I., Abdulwahid Isah, A., Abubakar, U.S., Rita Singh, M., 2015. Review on comparative study on bioremediation for oil spills using microbes. Research Journal of Pharmaceutical. Biological and Chemical Sciences 6 (6), 783–790.

Saikia, R.R., Deka, H., Goswami, D., Lahkar, J., Borah, S.N., Patowary, K., Deka, S., 2014. Achieving the best yield in glycolipid biosurfactant preparation by selecting the proper carbon/nitrogen ratio. Journal of Surfactants and Detergents 17 (3), 563–571.

Sakthipriyaa, N., Dobleb, M., Sangwai, J.S., 2015. Bioremediation of coastal and marine pollution due to crude oil using a microorganism *Bacillus subtilis*. Procedia Engineering 116, 213–220.

Samanta, S., Singh, O.V., Jain, R.K., 2002. Polycyclic aromatic hydrocarbons: environmental pollution and bioremediation. Trends in Biotechnology 20 (6), 243–248.

Santisi, S., Cappello, S., Catalfamo, M., Mancini, G., Hassanshahian, M., Genovese, L., Giuliano, L., Yakimov, M.M., 2015. Biodegradation of crude oil by individual bacterial strains and a mixed bacterial consortium. Brazilian Journal of Microbiology 46 (2), 377–387.

Santos, A.L., Oliveira, V., Baptista, I., Henriques, I., Gomes, N.C., Almeida, A., Correia, A., Cunha, A., 2012. Effects of UV-B radiation on the structural and physiological diversity of bacterioneuston and bacterioplankton. Applied and Environmental Microbiology 78, 2066–2069.

Santos, H.F., Carmo, F.L., Paes, J.E., Rosado, A.S., Peixoto, R.S., 2011. Bioremediation of mangroves impacted by petroleum. Water, Air, and Soil Pollution 216 (1–4), 329–350.

Seidel, M., Kleindienst, S., Dittmar, T., Joye, S.B., Medeiros, P.M., 2016. Biodegradation of crude oil and dispersants in deep seawater from the Gulf of Mexico: insights from ultra-high-resolution mass spectrometry. Deep-sea Research II 129, 108–118.

Sharma, P., Singh, J., Dwivedi, S., Kumar, M., 2014. Bioremediation of oil spill. Journal of Bioscience and Technology 5 (6), 571–581.

Sheppard, P.J., Simons, K.L., Adetutu, E.M., Kadali, K.K., Juhasz, A.L., Manefield, M., Sarma, P.M., Lal, B., Ball, A.S., 2014. The application of a carrier-based bioremediation strategy for marine oil spills. Marine Pollution Bulletin 84, 339–346.

Sherry, A., Gray, N.D., Ditchfield, A.K., Aitken, C.M., Jones, D.M., Röling, W.F.M., Hallmanna, C., Larter, S.R., Bowler, B.F.J., Head, I.M., 2013. Anaerobic biodegradation of crude oil under sulphate-reducing conditions leads to only modest enrichment of recognized sulphate-reducing taxa. International Biodeterioration and Biodegradation 81, 105–113.

Sierra-Garcia, I.N., de Oliveira, V.M., 2013. Microbial hydrocarbon degradation: efforts to understand biodegradation in petroleum reservoirs. In: Rolando Chamy, Dr. (Ed.), Biodegradation – Engineering and Technology. InTech, pp. 47–72 (Chapter 3) https://doi.org/10.5772/55920.

Sihag, S., Pathak, H., 2014. Factors affecting the rate of biodegradation of polyaromatic hydrocarbons. International Journal of Pure Applied Bioscience 2 (3), 185–202.

Silva, E.J., Rocha e Silva, N.M.P., Rufino, R.D., Luna, J.M., Silva, R.O., Sarubbo, L.A., 2014. Characterization of a biosurfactant produced by *Pseudomonas cepacia* CCT6659 in the presence of industrial wastes and its application in the biodegradation of hydrophobic compounds in soil. Colloids and Surfaces B: Biointerfaces 117, 36–41.

Simister, R.L., Poutasse, C.M., Thurston, A.M., Reeve, J.L., Baker, M.C., White, H.K., 2015. Degradation of oil by fungi isolated from Gulf of Mexico beaches. Marine Pollution Bulletin 100, 327–333.

Simons, K., Ansar, A., Kadali, K., Bueti, A., Adetutu, E., Ball, A., 2012. Investigating the effectiveness of economically sustainable carrier material complexes for marine oil remediation. Bioresource Technology 126, 202–207.

Simons, K.L., Sheppard, P.J., Adetutu, E.M., Kadali, K., Juhasz, A.L., Manefield, M., Sarma, P.M., Lal, B., Ball, A.S., 2013. Carrier mounted bacterial consortium facilitates oil remediation in the marine environment. Bioresource Technology 134, 107–116.

Singh, A.K., Sherry, A., Gray, N.D., Jones, D.M., Bowler, B.F.J., Head, I.M., 2014. Kinetic parameters for nutrient enhanced crude oil biodegradation in intertidal marine sediments. Frontiers in Microbiology 5, 160. http://dx.doi.org/10.3389/fmicb.2014.00160.

Sonawdekar, S., 2012. Bioremediation: a boon to hydrocarbon degradation. International Journal of Environmental Sciences 2 (4), 2408–2424.

Sparrow, S.D., Sparrow, E.B., 1988. Microbial biomass and activity in a subarctic soil ten years after crude oil spills. Journal of Environmental Quality 17, 304–309.

Spence, M.J., Bottrell, S.H., Thornton, S.F., Richnow, H.H., Spence, K.H., 2005. Hydrochemical and isotopic effects associated with petroleum fuel biodegradation pathways in a chalk aquifer. Journal of Contaminant Hydrology 79, 67–88.

Sponza, D.T., Oztekin, R., 2010. Destruction of some more and less hydrophobic PAHs and their toxicities in a petrochemical industry wastewater with sonication in Turkey. Bioresource Technology 101, 8639–8648.

Stout, S.A., Uhler, A.D., Emsbo-Mattingly, S.D., 2004. Comparative evaluation of background anthropogenic hydrocarbons in surficial sediments from nine urban water ways. Environmental Science & Technology 38, 2987–2994.

Su, D., Li, P.J., Frank, S., Xiong, X.Z., 2006. Biodegradation of benzo[a]pyrene in soil by *Mucor* sp. SF06 and *Bacillus* sp. SB02 co-immobilized on vermiculite. Journal of Environmental Sciences 18, 1204–1209.

Suja, F., Rahim, F., Taha, M.R., Hambali, N., Razali, M.R., Khalid, A., Hamzah, A., 2014. Effects of local microbial bioaugmentation and biostimulation on the bioremediation of total petroleum hydrocarbons (TPH) in crude oil contaminated soil based on laboratory and field observations. International Biodeterioration and Biodegradation 90, 115–122.

Sun, W.H., Io, J.B., Robert, F.M., Ray, C., Tang, C.S., 2004. Phytoremediation of petroleum hydrocarbons in tropical coastal soils. I. Selection of promising woody plants. Environmental Science and Pollution Research International 11 (4), 260–266.

Suni, S., Koskinen, K., Kauppi, S., Hannula, E., Ryynänen, T., Aalto, A., Jäänheimo, J., Ikavalko, J., Romantschuk, M., 2007. Removal by sorption and in situ biodegradation of oil spills limits damage to marine biota: a laboratory simulation. Ambio 36, 173–179.

Suni, S., Kosunen, A.L., Hautala, M., Pasila, A., Romantschuk, M., 2004. Use of a by-product of peat excavation, cotton grass fiber, as a sorbent for oil-spills. Marine Pollution Bulletin 49, 916–921.

Swannell, R.P.J., Lee, K., McDonagh, M., 1996. Field evaluations of marine oil spill bioremediation. Microbiological Reviews 60, 342–365.

Tam, N., Wong, Y., Wong, M., 2009. Novel technology in pollutant removal at source and bioremediation. Ocean and Coastal Management 52 (7), 368–373.

Tarr, M.A., Zito, P., Overton, E.B., Olson, G.M., Adhikari, P.L., Reddy, C.M., 2016. Weathering of oil spilled in the marine environment. Oceanography 29 (3), 126–135 Special Issue on GoMRI Deepwater Horizon Oil Spill and Ecosystem Science.

Teas, Ch., Kalligeros, S., Zanikos, F., Stournas, S., Lois, E., Anastopoulos, G., 2001. Investigation of the effectiveness of absorbent materials in oil spills clean up. Desalination 140, 259–264.

Teramoto, M., Suzuki, M., Okazaki, F., Hatmanti, A., Harayama, S., 2009. Oceanobacter related bacteria are important for the degradation of petroleum aliphatic hydrocarbons in the tropical marine environment. Microbiology 155, 3362–3370.

Tewari, S., Sirvaiya, A., 2015. Oil spill remediation and its regulation. International Journal of Research in Science and Engineering 1 (6), 2394–8299.

Thapa, B., Kc, A.K., Ghimire, A., 2012. A review on bioremediation of petroleum hydrocarbon contaminants in soil. Kathmandu University Journal of Science, Engineering and Technology 8 (1), 164–170.

Thavasi, R., Jayalakshmi, S., Balasubramanian, T., Banat, I.M., 2006. Biodegradation of crude oil by nitrogen fixing marine bacteria *Azotobacter chroococcum*. Research Journal of Microbiology 1, 401–408.

Thavasi, R., Jayalakshmi, S., Banat, I.M., 2011. Application of biosurfactant produced from peanut oil cake by *Lactobacillus delbrueckii* in biodegradation of crude oil. Bioresource Technology 10 (23), 3366–3372.

Thomassin-Lacroix, E.J.M., 2000. Fate and Effects of Hydrocarbon-degrading Bacterial Used to Inoculate Soil for On-site Bioremediation in the Arctic MSc. Thesis. Royal Military College of Canada.

Toledo, F.L., Gonzalez, J., 2008. Production of bioemulsifier by *Bacillus subtilis, Alcaligenes faecalis* and *Enterobacter* species in liquid culture. Bioresource Technology 99, 8470–8475.

Tolosa, I., De Mora, S.J., Fowler, S.W., Villenuve, J.P., Bartocci, J., Cattini, C., 2005. Aliphatic and aromatic hydrocarbons in marine biota and coastal sediments from the Gulf and the Gulf of Oman. Marine Pollution Bulletin 50, 1619–1633.

Tripath, A., Srivastava, S., 2013. Novel approach for optimization of fermentative condition for polyhydroxybutyrate (PHB) production by *Alcaligenes* sp. using Taguchi (DOE) methodology. African Journal of Biotechnology 10 (37), 7219–7224.

Valentine, D.L., Fisher, G.B., Bagby, S.C., Nelson, R.K., Reddy, C.M., Sylva, S.P., Woo, M.A., 2014. Fallout plume of submerged oil from Deepwater Horizon. Proceedings of the National Academy of Sciences of United States of America 111 (45), 15906–15911.

Van de Wiele, T., Vanhaecke, L., Boeckaert, C., Peru, K., Headley, J., Verstraete, W., Siciliano, S., 2005. Human colon microbiota transforms polycyclic aromatic hydrocarbons to estrogenic metabolites. Environmental Health Perspectives 113 (1), 6–10.

van Hamme, J.D., Singh, A., Ward, O.P., 2003. Recent advances in petroleum microbiology. Microbiology and Molecular Biology Reviews 67 (4), 503–549.

Varjani, S.J., Rana, D.P., Jain, A.K., Bateja, S., Upasani, V.N., 2015. Synergistic ex-situ biodegradation of crude oil by halotolerant bacterial consortium of indigenous strains isolated from on shore sites of Gujarat, India. International Biodeterioration and Biodegradation 103, 116–124.

Vázquez, S., Nogales, B., Ruberto, L., Mestre, C., Christie-Oleza, J., Ferrero, M., Bosch, R., Mac Cormack, W.P., 2013. Characterization of bacterial consortia from diesel contaminated Antarctic soils: towards the design of tailored formulas for bioaugmentation. International Biodeterioration and Biodegradation 77, 22–30.

Venosa, A.D., Haines, J.R., Nisamaneepong, W., Govind, R., Pradhan, S., Siddique, B., 1991. Screening of commercial inocula for efficacy in stimulating oil biodegradation in closed laboratory system. Journal of Hazardous Materials 28, 131–144.

Venosa, A.D., Zhu, X., 2003. Biodegradation of crude oil contaminating marine shorelines and freshwater wetlands. Spill Science and Technology Bulletin 8 (2), 163–178.

Vidali, M., 2001. Bioremediation. An overview. Pure and Applied Chemistry 73 (7), 1163–1172.

Vilcáez, J., Li, L., Hubbard, S.S., 2013. A new model for the biodegradation kinetics of oil droplets: application to the Deepwater Horizon oil spill in the Gulf of Mexico. Geochemical Transactions 14, 4. http://dx.doi.org/10.1186/1467-4866-14-4.

Wake, H., 2005. Oil refineries: a review of their ecological impacts on the aquatic environment. Estuarine, Coastal and Shelf Science 62 (1), 131–140.

Wang, X., Wang, X., Liu, M., Bu, Y., Zhang, J., Chen, J., Zhao, J., 2015. Adsorption-synergic biodegradation of diesel oil in synthetic seawater by acclimated strains immobilized on multifunctional materials. Marine Pollution Bulletin 92, 195–200.

Wang, X.-C., Sun, S., Ma, H.-Q., Liu, Y., 2006. Sources and distribution of aliphatic and polyaromatic hydrocarbons in sediments of Jiaozhou Bay, Qingdao, China. Marine Pollution Bulletin 52, 129–138.

Weis, J.S., 2014. Introduction to Marine Pollution Physiological, Developmental and Behavioral Effects of Marine Pollution. Springer.

White, H.K., Lyons, S.L., Harrison, S.J., Findley, D.M., Liu, Y., Kujawinski, E.B., 2014. Long-term persistence of dispersants following the Deepwater Horizon oil spill. Environmental Science & Technology 1, 295–299.

Whyte, L.G., Schultz, A., van Beilen, J.B., Luz, A.P., Pellizari, V., Labbe, D., Greer, C.W., 2002. Prevalence of alkane monooxygenase genes in Arctic and Antarctic hydrocarbon-contaminated and pristine soils. FEMS Microbiology Ecology 41, 141–150.

Widdel, F., Rabus, R., 2001. Anaerobic biodegradation of saturated and aromatic hydrocarbons. Current Opinion in Biotechnology 2 (3), 259–276.

Wolski, E.A., Barrera, V., Castellari, C., González, J.F., 2013. Biodegradation of phenol in static cultures by *Penicillium chrysogenum* ERK1: catalytic abilities and residual phytotoxicity. Revista Argentina de microbiología 44 (2), 113–121.

Xie, H., Yapa, P.D., Nakata, K., 2007. Modeling emulsification after an oil spill in the sea. Journal of Marine Systems 68 (3), 489–506.

Yakimov, M.M., Denaro, R., Genovese, M., Cappello, S., D'Auria, G., Chernikova, T.N., Timmis, K.N., Golyshin, P.N., Giuliano, L., 2005. Natural microbial diversity in superficial sediments of Milazzo Harbor (Sicily) and community successions during microcosm enrichment with various hydrocarbons. Environmental Microbiology 7, 1426–1441.

Yakimov, M.M., Giuliano, L., Gentile, G., Crisafi, E., Chernikova, T.N., Abraham, W.-R., Lünsdorf, H., Timmis, K.N., Golyshin, P.N., 2003. *Oleispira antarctica* gen. nov., sp. Nov., a novel hydrocarbonoclastic marine bacterium isolated from Antarctic coastal sea water. International Journal of Systematic and Evolutionary Microbiology 53, 779–785.

Yakimov, M.M., Kenneth, T., Wray, V., Fredrickson, L., 1995. Characterization of a new lipopeptide surfactant produced by thermotolerant and halotolerant subsurface *Bacillus licheiformis* BAS50. Environmental Microbiology 61, 1706–1713.

Yakimov, M.M., Timmis, K.N., Golyshin, P.N., 2007. Obligate oil-degrading marine bacteria. Current Opinion in Biotechnology 18, 257–266.

Yang, S., Wen, X., Zhao, L., Shi, Y., Jin, H., 2014. Crude oil treatment leads to shift of bacterial communities in soils from the deep active layer and upper permafrost along the China-Russian crude oil pipeline route. PLoS One 9 (5), e96552. http://dx.doi.org/10.1371/journal.pone.0096552.

Yang, S.-Z., Jin, H.-J., Wei, Z., He, R.-X., Ji, Y.-J., Li, X.-M., Yu, S.-P., 2009. Bioremediation of oil spills in cold environments: a review. Pedosphere 19, 371–381.

Yanto, D.H.Y., Tachibana, S., 2013. Biodegradation of petroleum hydrocarbons by a newly isolated *Pestalotiopsis* sp. NG007. International Biodeterioration and Biodegradation 85, 438–450.

Yardin, M.R., Kennedy, I.R., Thies, J.E., 2000. Development of high quality carrier materials for field delivery of key microorganisms used as bio-fertilizers and bio-pesticides. Radiation Physics and Chemistry 57 (3–6), 565–568.

Yassine, M.H., Suidan, M.T., Venosa, A.D., 2013. Microbial kinetic model for the degradation of poorly soluble organic materials. Water Research 47 (4), 1585–1595.

Yeung, C.W., Law, B.A., Milligan, T.G., Lee, K., Whyte, L.G., Greer, C.W., 2011. Analysis of bacterial diversity and metals in produced water, seawater and sediments from an offshore oil and gas production platform. Marine Pollution Bulletin 62 (10), 2095–2105.

Younis, Sh.A., El-Azab, W.I., El-Gendy, N.Sh., Aziz, Sh.Q., Moustafa, Y.M., Aziz, H.A., Amr, S.S., 2014a. Application of response surface methodology to enhance phenol removal from refinery wastewater by microwave process. International Journal of Microwave Science and Technology. :639457https://doi.org/10.1155/2014/639457.

Younis, Sh.A., El-Gendy, N.Sh., El-Azab, W.I., Moustafa, Y.M., 2015. Kinetic, isotherm, and thermodynamic studies of polycyclic aromatic hydrocarbons biosorption from petroleum refinery wastewater using spent waste biomass. Desalination and Water Treatment 56 (11), 3013–3023.

Younis, Sh.A., El-Gendy, N.Sh., El-Azab, W.I., Moustafa, Y.M., Hashem, A.I., 2014b. The biosorption of phenol from petroleum refinery wastewater using spent waste biomass. Energy Sources, Part a: Recovery, Utilization, and Environmental Effects 36 (23), 2566–2578.

Younis, Sh.A., El-Gendy, N.Sh., Moustafa, Y.M., 2013. A study on bio-treatment of petrogenic contamination in El-Lessan area of Damietta River Nile Branch, Egypt. International Journal of Chemical and Biochemical Sciences 4, 112–124.

Yu, K.S.H., Wong, A.H.Y., Yau, K.W.Y., Wong, Y.S., Tam, N.F.Y., 2005. Natural attenuation, biostimulation and bioaugmentation on biodegradation of polycyclic aromatic hydrocarbons (PAHs) in mangrove sediments. Marine Pollution Bulletin 51, 1071–1077.

Zahed, M.A., Abdul Aziz, H., Isa, M.H., Mohajeri, L., Mohajeri, S., 2010b. Optimal conditions for bioremediation of oily seawater. Bioresource Technology 101, 9455–9460.

Zahed, M.A., Abdul Aziz, H., Mohajeri, L., Mohajeri, S., Kutty, S.R.M., Isa, M.H., 2010c. Application of statistical experimental methodology to optimize bioremediation of n-alkanes in aquatic environment. Journal of Hazardous Materials 184, 350–356.

Zahed, M.A., Aziz, H.A., Isa, M.H., Mohajeri, L., 2010a. Effect of initial oil concentration and dispersant on crude oil biodegradation in contaminated seawater. Bulletin of Environmental Contamination and Toxicology 84, 438–442.

Zhang, H., Tang, J., Wang, L., Liu, J., Gurav, R.G., Sun, K., 2016. A novel bioremediation strategy for petroleum hydrocarbon pollutants using salt tolerant *Corynebacterium variabile* HRJ4 and biochar. Journal of Environmental Sciences 47, 7–13.

Zhang, X.S., Xu, D.J., Zhu, C.Y., Tserennyam, L., Scherr, K.E., 2012. Isolation and identification of biosurfactant producing and crude oil degrading *Pseudomonas aeruginosa* strains. Chemical Engineering Journal 209, 138–146.

Zhang, Z., Gai, L., Hou, Z., Yang, C., Ma, C., Wang, Z., Sun, B., He, X., Tang, H., Xu, P., 2010. Characterization and biotechnological potential of petroleum-degrading bacteria isolated from oil-contaminated soils. Bioresource Technology 101, 8452–8456.

Zhang, Z., Lo, I.M.C., Yan, D.Y.S., 2015. An integrated bioremediation process for petroleum hydrocarbons removal and odor mitigation from contaminated marine sediment. Water Research 83, 21–30.

Zhao, L., Boufadel, M.C., Socolofsky, S.A., Adams, E., King, T., Lee, K., 2014a. Evolution of droplets in subsea oil and gas blowouts: development and validation of the numerical model VDROP. Marine Pollution Bulletin 83, 58–69.

Zhao, L., Torlapati, J., Boufadel, M.C., King, T., Robinson, B., Lee, K., 2014b. VDROP: a comprehensive model for droplet formation of oils and gases in liquids-incorporation of the interfacial tension and droplet viscosity. Chemical Engineering Journal 253, 93–106.

Zhen-Yu, W.A.N.G., Dong-Mei, G.A.O., Feng-Min, L.I., Jian, Z.H.A.O., Yuan-Zheng, X.I.N., Simkins, S., Bao-Shan, X.I.N.G., 2008. Petroleum hydrocarbon degradation potential of soil bacteria native to the yellow River Delta. Pedosphere 18 (6), 707–716.

Zhu, X., Venosa, A., Suidan, M., Lee, K., 2001. Guidelines for the Bioremediation of Marine Shorelines and Freshwater Wetlands. U.S. EPA, Cincinatti, Ohio.

Zhuang, M., Abulikemu, G., Campo, P., Platten III, W.E., Suidan, M.T., Venosa, A.D., Conmy, R.N., 2016. Effect of dispersants on the biodegradation of South Louisiana crude oil at 5 and 25 °C. Chemosphere 144, 767–774.

ZoBell, C.E., 1946. Marine Microbiology. Chronica Botanica Co, Waltham, Massachusetts.

FURTHER READING

Basu, P.R., 2005. Evaluation of Biological Treatment for the Degradation of Petroleum Hydrocarbons in a Waste Water Treatment Plant (MSc. Thesis, Submitted to the Office of Graduate Studies of Texas A&M University).

Crawford, R.L., Rosenberg, E., 2012. Bioremediation. In: Rosenberg, E., DeLong, E.F., Lorey, S., Stackebrandt, E. (Eds.), The Prokaryotes, fourth ed. Springer, New York.

da Cruz, G.F., dos Santos Neto, E.V., Marsaioli, A.J., 2008. Petroleum degradation by aerobic microbiota from Pmpo Sul oil field, Camos Basin, Brazil. Organic Geochemistry 39, 1204–1209.

El-Mahdi, A.M., Aziz, H.A., Abu Amr, S.S., El-Gendy, N.Sh., Nassar, H., 2016. Isolation and characterization of *Pseudomonas* sp. NAF1 and its application in biodegradation of crude oil. Environmental Earth Sciences 75, 380. http://dx.doi.org/10.1007/s12665-016-5296-z.

Foght, J., 2010. Nitrogen fixation and hydrocarbon-oxidizing bacteria. In: Timmis, K.N. (Ed.), Handbook of Hydrocarbon and Lipid Microbiology. Springer-Verlag, Berlin, Germany, pp. 1662–1666.

Kaplan, C.W., Clement, B.G., Hamrick, A., Pease, R.W., Flint, C., Cano, R.G., Kitts, C.L., 2003. Complex co-substrate addition increases initial petroleum degradation rates during land treatment by altering bacterial community physiology. Remediation 13, 61–78.

Madigan, M.T., Martinko, J.M., Dunlap, P.V., Clark, D.P., 2008. Brock Biology of microorganisms 12th edn. International Microbiology 11, 65–73.

Wang, Z., Fingas, M., Blenkinsopp, S., Sergy, G., Landriault, M., Sigouin, L., Foght, J., Semple, K., Westlakec, D.W.S., 1998. Comparison of oil composition changes due to biodegradation and physical weathering in different oils. Journal of Chromatography A 809 (1–2), 89–107.

THE FUTURE OF PETROLEUM BIOTECHNOLOGY

12

1.0 INTRODUCTION

The long list of petroleum constituents and the list of the constituents of petroleum products include numerous aliphatic and aromatic compounds, such as hydrocarbon derivatives, polynuclear aromatic systems, and heterocyclic (nitrogen, oxygen, sulfur, and metals) derivatives. The concentration of such chemicals varies with the crude oil, heavy oil, extra heavy oil, and tar sand bitumen (Chapters 1 and 2) (Speight and Lee, 2000; Speight, 2014a, 2017a,b).

Petroleum-based products are the major source of energy for industry and daily life and are likely to remain so for the next five decades in spite of the push to develop alternate fuels and any influence of petro-politics (Speight, 2011a,b,c). The collective process known as biotransformation—defined as the use of microorganisms to convert petroleum constituents to useful products or to remove petroleum-based pollutants for the environment—is an evolving method for petroleum recovery (Chapter 4) and for petroleum refining (Chapters 5–8). In addition, biotransformation concepts—as applied to the bio-remediation of spills of crude oil and crude oil products—as accomplished by microbial colonies (Chapters 9–11) represent one of the options for the future in terms of petroleum refining and removal of petroleum-based pollutants from the environment (Speight and Arjoon, 2012).

The biotransformation of petroleum hydrocarbons is a complex process that depends on the nature and on the amount of the hydrocarbons present. Petroleum hydrocarbons can be divided into four classes: the saturates, the aromatics, the asphaltenes (phenols, fatty acids, ketones, esters, and porphyrins), and the resins (pyridines, quinolines, carbazoles, sulfoxides, and amide derivatives). An important factor influencing hydrocarbon transformation is the access to the crude oil constituents by microorganisms. Furthermore, hydrocarbon derivatives differ in the susceptibility to microbial attack.

Microbial transformation is the major and ultimate natural mechanism by which crude oil pollutants can be removed from the environment (Speight and Arjoon, 2012). However, a number of limiting factors have been recognized to affect the biotransformation of petroleum hydrocarbon derivatives. The composition of crude oil is the first and foremost important consideration when the suitability of biotransformation is to be assessed. Among physical factors, temperature plays an important role by directly affecting the chemistry of the transformation process—at low temperatures, the viscosity of the crude oil is increased while the volatility of the bio-toxic low molecular weight hydrocarbons is decreased thereby delaying the onset of the biotransformation process. Temperature also affects the solubility of hydrocarbon derivatives although hydrocarbon biotransformation can occur over a wide range of temperatures; the rate of biotransformation generally decreases with the decreasing temperature because of the susceptibility of microorganisms to temperatures above 50°C (122°F).

Nutrients are very important ingredients for successful biotransformation of hydrocarbon derivatives, especially nitrogen, phosphorus, and in some cases iron. Some of these nutrients could become limiting factors thus affecting the biotransformation processes. Therefore, addition of nutrients may be

Introduction to Petroleum Biotechnology. https://doi.org/10.1016/B978-0-12-805151-1.00012-6

necessary to enhance the biotransformation of oil pollutants but, on the other hand, excessive nutrient concentrations can also inhibit microbial activity and reduce the rate of the biotransformation process(es).

The success of the any process for the biotransformation of crude oil and crude oil products depends on the ability of scientists and engineers to establish and maintain conditions that favor the biotransformation reactions and the rates of these reactions that will satisfy refineries and environmental cleanup efforts. The key to such goals is the development of microbial colonies that have the appropriate metabolic capabilities. If these microorganisms are present in the system, optimal rates of growth and biotransformation can be sustained (whether it is for refining or environmental cleanup) by ensuring that adequate concentrations of nutrients and oxygen are present and that the pH is within a range that is conducive to the survival of the microbes. The physical and chemical characteristics of the crude oil and the products from crude oil are also important determinants of the success of the biotransformation.

The scope of current understanding of crude oil biotransformation is advancing and will make important strides during the next two decades. With the need of refineries and environmental protection being of the utmost importance, biotransformation technology will continue to make important advances for dealing with crude oil recovery, refining, and the protection of the land and water ways (rivers, lakes, and oceans) from spills of crude oil and crude oil products.

2.0 STATUS

The biotransformation of the constituents of petroleum and petroleum products' hydrocarbons is a complex process that depends on the nature and on the amount of the hydrocarbons present. The constituents of petroleum hydrocarbons can be divided into four classes: (1) saturate derivatives, (2) aromatics derivatives, (3) resin constituents, and (4) asphaltene constituents, which include a multitude of compound-types, such as; pyridine derivatives, quinoline derivatives, carbazole derivatives, thiophene derivatives, benzothiophene derivatives, dibenzothiophene derivatives, sulfoxide derivatives, amide derivatives, phenol derivatives, acid derivatives, ketone derivatives, ester derivatives, and porphyrin derivatives (Speight, 2014a). Thus it is not surprising that different factors influencing the biotransformation of the constituents have been reported (Cooney et al., 1985). One of the important factors that limit biotransformation of crude oil constituents is the limited reactivity to many microorganisms. Moreover, the susceptibility of hydrocarbons to microbial biotransformation can be generally ranked as follows:

linear alkanes > branched alkanes > small aromatics > cyclic alkanes

The high molecular weight polynuclear aromatic hydrocarbons (PNAs) are difficult, if not impossible, to biotransform.

Several limiting factors have been recognized to affect the biotransformation of petroleum hydrocarbons. The composition and inherent biotransformation of the petroleum constituent is the first and foremost important consideration when the suitability of a biotransformation process is to be assessed. Among the physical factors, temperature plays an important role in biotransformation of petroleum constituents by directly affecting the chemistry of the constituents, as well as the physiology and diversity of the microbial colony. Although hydrocarbon biotransformation can occur over a wide range of temperatures, the rate of the process generally decreases with decreasing temperature—the highest rate of transformation generally occurs in the range 30–40°C (86–104°F). In addition, biotransformation may not be successful using only a homogenous microbe colony. Therefore, a multi-process (multi-colony) biotransformation may be necessary to accomplish the process goals. In addition, a multi-colony biotransformation remediation may be the necessary strategy for rapid biotransformation by altering the structure and operational effectiveness of the microbial activity. It is, thus, a challenging and rewarding research program to search for an innovative solution to speed up the biotransformation of petroleum constituents for effective biorefining or for environmental cleaning.

Thus, the biotransformation of the constituents of crude oil and crude oil products are widespread issues and the search for effective biotransformation remains a major challenge for researchers. However, there is a strong demand to increase the adoption of biotransformation as an effective technique for crude oil recovery and/or refining, as well as a tool for risk reduction on hydrocarbon impacted land and water. However, as with all microbial colonies, the efficiency of the biotransformation reactions diminishes with time and any effect that inhibits the activity of the colony will (more than likely) become dominant with time. The key solution to crude oil biotransformation is to increase the rate of the process and eliminate or delay the inhibitory effect, such as through the selection of specifically targeted strains or microorganisms or through the alteration of changes in the microbial colony during the treatment.

The recent successful implementation of microbial biotransformation of crude oil constituents (Chapters 5–7, 10, and 11) has indicated that such an approach is quite promising and can be a viable alternative to the conventional desulfurization, denitrogenation, and remediation methods (Speight and Arjoon, 2012). However, biotransformation cannot always be accomplished by a single process and, as a result, multi-process biotransformation provides a promising solution. In addition, multi-process biotransformation may become an effective strategy for rapid biotransformation by altering microbial community structure. It is thus a challenging and rewarding research to search for an innovative solution to speed up crude oil refining for effective production and improvement of crude oil products.

2.1 BIOREFINING

Crude oil is rarely used in its raw form (with the notable exception when it is used for power generation) but must instead be processed into various products as liquefied petroleum gas (LPG), gasoline, diesel, solvents, kerosene, middle distillates, residual fuel oil, and asphalt. The refining process involves the use of various thermal and catalytic processes to convert constituents in the higher molecular weight fractions to lower molecular weight lower-boiling products (Speight, 2014a, 2017a; El-Gendy and Speight, 2016).

With the entry into the 21st century, petroleum refining technology is experiencing many innovations that are driven by the increasing supply of heavy oils with decreasing quality and the fast increases in the demand for clean and ultra-clean vehicle fuels and petrochemical raw materials. As feedstocks to refineries change, there must be an accompanying change in refinery technology. This means a movement from conventional means of refining heavy feedstocks using (typically) coking technologies to more innovative processes (including hydrogen management) and biotransformation processes that will produce the ultimate amounts of liquid fuels from the feedstock and reduce emissions within the environment.

To meet the challenges from changing refining technology over the years from simple crude trends in the refinery feedstock slate and the stringent distillation operations to chemical operations involving complex, environmental legislations, the refining industry in the near future will need to transform itself with new innovative processes, with new processing schemes, so that the refinery becomes increasingly flexible and the refined products meet specifications that meet users' requirements.

The use of crude oil to produce necessary products (such as liquid fuels) and petrochemicals is expected to be maintained in the first five decades of this century (Speight, 2011a). Furthermore, the demand for low-sulfur fossil fuels has been intensified by the stricter regulatory standards for reduced levels of sulfur oxides in atmospheric emissions. It can be estimated that in coming decades at least one-third of the crude oil supplies to refineries will require extreme desulfurization, as well as the removal of nitrogen and metals that are predominately associated with the higher molecular weight fractions of crude oil (Speight, 2014a, 2017a). Such necessities will increase the cost of physico-chemical processes that are currently used for desulfurization (hydrodesulfurization, HDS), denitro-genation (hydrodenitrogenation, HDN) and demetallization (deasphalting or mild visbreaking) (Speight, 2014a, 2017a; El-Gendy and Speight, 2015). In addition, there is a need to reduce the severity of refining operations to decrease the costs by developing milder physical and chemical processes.

Biotransformation (biorefining) is the use of living organisms such as microbes in order to upgrade petroleum, that is, the application of bioprocesses to the fractionation and enhancing of petroleum, which might contribute to mitigate the associated pollution and upgrading of crude oil (Chapters 3–5). Biotransformation involves the use of wide range of conditions including milder temperature and pressure conditions, cleaner and more selective processes, low emissions, as well as no generation of undesirable by-products. The microbial and enzymatic catalysis can be manipulated and used for more specific applications where the chemical processing requires several steps. It is evident that the modern and evolving crude oil refining industry faces two main key drivers: (1) more stringent environmental regulations and (2) the steady depletion of crude oil reserves—crude oil from tight formations (such as shale formations) notwithstanding (Speight, 2017b). The former concept involving more stringent environmental regulations in crude oil products (such as fuel) while the latter concept relates to bio-transformation of heavy crude oil (and even tar sand bitumen) by enhancing the removal of heteroatoms (sulfur, nitrogen, and metals) from crude oil products.

2.2 BIODESULFURIZATION, BIODENITROGENATION, AND BIODEMETALLIZATION

Crude oil is and will be the major source of for the next five decades (Speight, 2011a) and it is antici-pated that crude oil production will increase during this time. However, it is also evident that the increase in crude oil production will be due to the incorporation of more heavy crude oil and extra heavy crude oil (as well as tar sand bitumen) into refinery feedstocks, even though light crude oil is being produced in increasing quantities from tight formation such as shale (Speight, 2014a,b). In addition, heavy crude oil and extra heavy crude oil (and tar sand bitumen) as refinery feedstocks implies an environmental impact because of the high content of sulfur, nitrogen, metal, and aromat-ics compounds. In order to remove these heteroatoms (sulfur, nitrogen, and metals) from such feed-stocks and obtain cleaner fuels, the traditional refineries must operate under more extreme process conditions of temperature and pressure thereby increasing the cost of the product to the consumer. Hydrotreating of such feedstocks will also require higher hydrogen feed a more active catalyst, such as a biocatalyst, to produce the desired distillate products (McFarland et al., 1998; Klein et al., 1999; Monticello, 2000).

2.2.1 Biodesulfurization

Chemical hydrogenation processes are currently used to upgrade heavy feedstocks prior to fluid catalytic cracking (to reduce the amount of coke) and also to upgrade the products. Typically, the distillates have a high content of low-fuel value di- and tricyclic aromatic derivatives which are then converted by pressure hydrogenation to compounds such as alkylbenzenes. While the metal derivatives typically end up in the nonvolatile products of the fluid catalytic cracking or the hydrocracking process, the catalysts involved in the distillate treating process(es) are frequently deactivated by sulfur- and nitrogen-containing compounds in the feedstocks. These expenses make it desirable to explore an alternative to conventional upgrading techniques. An attractive alternative is to use bacteria to enzymatically cleave the fused-ring aromatics under near-ambient conditions, followed by mild chemical hydrogenation to produce the desired alkyl aromatics.

Biocatalytic technology has been used widely in crude oil biotransformation and is being assessed for biodesulfurization and biodenitrogenation, as well as biodemetallization (Chapters 5–7). This form of treatment is at the stage of proven technology that can be cost-effective, competitive, or compatible with current technology. However, besides yielding the desired products, two primary requirements of the biological process are essential: (1) the biocatalytic activity must be restricted to aromatic compounds and must be effective over the wide range of di- and tricyclic aromatic hydrocarbon derivatives and heterocyclic derivatives than are to most middle distillate products, and (2) there should be no carbon loss from the aromatic substrates as a consequence of microbial oxidation; that is, the process has to be blocked at a certain stage to prevent complete oxidation to carbon dioxide and water. Other advantages, such as the ability of the microbial cells to catalyze ring cleavage in a resting state and to be pre-grown quickly to a high-density and high-activity state, are necessary for the efficiency of the process.

On the other hand, while the hydrotreating processes focus on the reduction of sulfur-containing species to hydrogen sulfide (H_2S), in microbial (oxidizing) technologies, where certain microbes are employed, organosulfur compounds, such as dibenzothiophene, are transformed into sulfone derivatives, via sulfoxide derivatives:

Dibenzothiophene   Dibenzothiophene sulfoxide   Dibenzothiophene sulfone

In order to obtain a desulfurized product, finally the sulfone derivative can be removed in a second step and the process does present some advantages, such as a reduction in the need for hydrogen.

The biodesulfurization (BDS) process (Chapter 6) is an alternative and/or complementary technology to hydrodesulfurization that is expected to play an important role in biorefining (that is, in pre-refining and post-refining) of oil products that involves a microbial or enzymatic system that selectively removes sulfur without attacking the carbon-carbon (C–C) bond and avoiding the loss of oil value

(Kilbane, 1994; Furuya et al., 2001). Furthermore, notable advances in BDS have been accomplished and sulfur-specific strains have been identified (Chapter 6). In addition, BDS processes involving isolated enzymes have been also investigated because of their greater technological utility in organic solvents (Isken et al., 1999; Klibanov, 2001). An enzymatic desulfurization approach would have at least three advantages compared to the utilization of complete cells: (1) activity at low or no water content, (2) thermomechanical stability, and (3) minimized mass-transfer issues (Klibanov, 2001). Moreover, there are oxidative enzymes which are non-coenzyme dependent and which have high activity and broad specificity.

2.2.2 Biodenitrogenation

Like sulfur, nitrogen is typically found in petroleum as nonbasic and basic-related compounds, which contribute to acid and atmospheric contamination and also interferes with the refining processes, leading to equipment corrosion and catalyst poisoning (Speight, 2014a, 2017a,b). Nitrogen compounds occur in crude oil and represent non-hydrocarbon compounds that occur in crude oil at the level of 0.01%–2% w/w, although over 10% w/w concentrations have been noted. Microbial transformation of nitrogen compounds from fossil fuels is important in petroleum refining because the combustion of these contaminants lead to the formation of nitrogen oxides (NO_x) and hence to air pollution and acid rain (Chapter 7) (Fetzer, 1998). They also contribute to coke formation catalyst poisoning during the refining of crude oil, thus reducing process yields (Speight, 2014a, 2017a,b).

While significant progress has been made toward the commercialization of crude oil biodenitrogenation, technical hurdles still need to be overcome to achieve commercialization. The major obstacles to the economical biodenitrogenation of crude oil include biocatalyst specificity and rate. Work continues to modify the catalyst to increase its effectiveness and to screen other organisms for additional denitrogenation capabilities. In addition, mass transfer and separation hurdles must be overcome in crude oils with increased oil viscosity and density.

2.2.3 Biodemetallization

Crude oil contains metals in the form of salts (zinc, titanium, calcium, and magnesium), petroporphyrins, and other complexes in the asphaltene constituents (vanadium, copper, nickel, and part of the iron) (Speight, 2014a). The more residual the oil, the higher the metal content, being that those metal species possibly cluster by heavy molecular mass compounds.

Metal accumulation in the heaviest polar fractions of crude oils plays a significant role in establishing the refining procedure, since vanadium, iron, nickel, and molybdenum have both negative and positive effects on product recoveries. During the years since the discovery of metalloporphyrins in petroleum, the origin and significance of heavy metals in petroleum has been poorly investigated, due to its low concentrations and the limitations of analytical methods for such complex matrices (i.e., petroporphyrins and asphaltene constituents) in crude oil and its heavy fractions. The vanadium (V) and nickel (Ni) containing compounds are the most predominate and persist mainly in the resin and asphaltene fractions of crude oil. The total metal content in crude oils has an extended concentration range.

Crude oil also contains metals in the form of salts and metalloporphyrins in the asphaltene fraction, which is the solid material that precipitates when oil is treated with alkane solvents (n-pentane or n-heptane) (Speight, 2015). The salts are eliminated during the crude oil desalting process in which they concentrate into the aqueous phase. The removal of metals trapped in metalloporphyrins is more problematic because porphyrins are embedded in complex structures, usually within the asphaltene fraction. Furthermore, heavy metals (mostly nickel and vanadium) are corrosive, poison cracking catalysts during refining, and are released as highly toxic oxides during fuels combustion to the environment.

The asphaltene constituents are responsible for sludge formation resulting in flow reduction by plugging downstream equipment and production of less valuable coke in current upgrading of petroleum. Moreover, the utilization of distillation residua, constituted mainly by asphaltene constituents, resin constituents, and entrapped metals, is of high interest because of the continued and increasing influx of the heavier feedstocks (heavy oil, extra heavy oil, tar sand bitumen) into refineries.

However, as biological processes continue to improve and emerge as cost-effective and environmentally favorable processes, the biotransformation of asphaltene constituents (sometimes incorrectly referred to as *bio-cracking*) (Fedorak et al., 1993) to obtain high-value light oils from the less valuable heavy oils will continue to improve and produce lower-boiling fractions thereby increasing the yields of volatile products and decreasing the content of organic sulfur- and nitrogen-containing compounds and metal derivatives in the products.

In addition, the possibility of employing biotechnology in order to release the entrapped metals from porphyrin derivatives is becoming a reality (Arellano-Garcia et al., 2004). Indeed, thermophilic microbes (heat tolerant microbes) are being considered as the biocatalysts for metal removal from porphyrin structures. Indeed, the reduction of nitrogen from the asphaltenic fraction was more extensive in heavy crude while steam-treated crude showed a major reduction in aromatics, sulfur, and metals (Huber and Stetter, 1998; Premuzic and Lin, 1999; Premuzic et al., 1999). This suggests the possibility that future biocatalysis for the simultaneous removal of sulfur, nitrogen, and metals from petroleum could be developed.

3.0 **TECHNOLOGY POTENTIAL**

Biotechnology has strengthened its position during last years as the assembly of technologies focusing on the production of goods and services by means of biological systems or its products. The rapid technological progression derived from the modification of deoxyribonucleic acid (DNA) allowed the industrialization of new processes. Hence, biotechnology can, for the modern and future petroleum industry (recovery and refining), have a broad and diverse impact.

Petroleum is a complex mixture of hydrocarbon derivatives (paraffins, naphthenes, and aromatics), and is at present the largest source of energy followed by natural gas, which is a mixture of methane besides other gases. Both fossil fuels are valuable substrates for microorganisms and, hence, the implementation of biological processes in the oil industry to explore, produce, refine, transform petroleum and natural gas into valuable derivatives and environmentally clean products is important (Dordick et al., 1998; Hamer and Al-Awadhi, 2000; Le Borgne and Quintero, 2003; Rajesh and Kaladhar, 2014). As a result, the petroleum industry has therefore researched and is beginning to apply bioprocesses as complementary technologies on diverse platforms to reduce investment and maintenance costs, as well as a valuable process to overcome the technological barriers regarding the upgrading of petroleum.

The contributions of biotechnology to the energy industry are not restricted to the production of biofuels (Speight, 2008, 2011c), and the microbial production of methane may well be the largest contribution in the future. However, in the current context, microbial enhanced oil recovery has already drawn considerable interest from the petroleum industry. By general estimates, approximately 60%–80% v/v of crude in reservoirs is left in place because it is considered to be technically and/or economically non-recoverable (Muggeridge et al., 2013). However, microbial conversion of hydrocarbon derivatives to methane could dramatically increase the amount of energy recovered. Quantification of the relative abundance of stable isotopes of carbon and hydrogen can reveal the origin of methane in geological deposits because chemical and biochemical pathways for the formation of methane have different reactivities/preferences for different isotopes. It is also estimated that 20%–40% v/v of methane in oil and gas reservoirs is of microbial origin and most of that is derived from the conversion of carbon dioxide into methane. Therefore, because of knowledge about the conversion of petroleum constituents in the reservoir (Chapter 3), the potential to employ biotechnology to convert the residual crude oil in depleted oil wells can become a reality.

Although a more detailed understanding of all aspects of the metabolic pathways is still needed, the understanding of the chemistry involved when microorganisms metabolize crude oil constituents has improved rapidly over the past two decades. New techniques including metagenomic methods have been developed to explore novel genes and metabolic pathways for the biotransformation of recalcitrant and xenobiotic molecules. Initial pilot-scale testing of microbial degradation of some environmental pollutants by the cultures described earlier is under way, which should lead to lower costs for biocatalyst preparation and more efficient bacterial degradation of hazardous compounds.

Also, the emergence of new in vitro tools for mutation and genetic rearrangement has enabled the limits of enzymatic systems to be extended. Microorganisms with a wider substrate range and higher substrate affinity in biphasic reactions that contain toxic solvents or complex heterocycles could be engineered if the biocatalysts are to be used for biotransformation (as in a crude oil refinery), petroleum treatment, or the production of bio-derived compounds, and environmental remediation. These possibilities represent a future challenge for microbiologists, as well as for process chemists and process engineers.

4.0 THE BIOREFINERY

In addition to the concept of incorporating microbial transformation processes into a crude oil refinery, there is also the option of amalgamating a conventional refinery with a biorefinery. Biorefining is not a new concept, especially when activities such as production of vegetable oils, beer, and wine requiring pretreatment are considered. Many of these activities are known to have been in practice for millennia.

A biorefinery offers a method to access the integrated production of chemicals, materials, and fuels. Although the concept of a biorefinery is analogous to that of an oil refinery, the differences in the various biomass feedstocks require a divergence in the methods used to convert the feedstocks to fuels and chemicals. Thus, a biorefinery, like a petroleum refinery, may need to be a facility that integrates biomass conversion processes and equipment to produce fuels, power, and chemicals from biomass. In a manner similar to the petroleum refinery, a biorefinery would integrate a variety of conversion processes to produce multiple product streams such as motor fuels and other chemicals from biomass, such as the inclusion of gasification processes and fermentation processes (Speight, 2008).

A biorefinery should combine the essential technologies to transform biological raw materials into a range of industrially useful intermediates. However, the type of biorefinery would have to be differentiated by the character of the feedstock. For example, the *crop biorefinery* would use raw materials such as cereals or maize and the *lignocellulose biorefinery* would use raw material with high cellulose content, such as straw, wood, and paper waste.

Although a number of new bioprocesses have been commercialized, it is clear that economic and technical barriers still exist before the full potential of this area can be realized. The biorefinery concept could significantly reduce production costs of plant-based chemicals and facilitate their substitution in existing markets. This concept is analogous to that of a modern oil refinery in that the biorefinery is a highly integrated complex that will efficiently separate biomass raw materials into individual components and convert these into marketable products such as energy, fuels, and chemicals. By analogy with crude oil, every element of the plant feedstock will be utilized including the low-value lignin components.

A key requirement for the biorefinery is the ability of the refinery to develop process technology that can economically access and convert the five- and six-membered ring sugars present in the cellulose and hemicellulose fractions of the lignocellulosic feedstock. Although engineering technology exists to effectively separate the sugar containing fractions from the lignocellulose, the enzyme technology to economically convert the five-ring sugars to useful products requires further development.

As a feedstock, biomass can be converted by biological routes or by thermal routes to a wide range of useful forms of energy including process heat, steam, electricity, as well as liquid fuels, chemicals, and synthesis gas. As a raw material, biomass is a nearly universal feedstock due to its versatility, domestic availability, and renewable character. At the same time, it also has its limitations. For example, the energy density of biomass is low compared to that of coal, liquid petroleum, or petroleum-derived fuels. The heat content of biomass, on a dry basis (7000–9000 Btu/lb) is at best comparable with that of a low-rank coal or lignite, and substantially (50%–100%) lower than that of anthracite, most bituminous coals, and petroleum. Most biomass, as received, have a high burden of physically adsorbed moisture, up to 50% by weight. Thus, without substantial drying, the energy content of a biomass feed per unit mass is even less.

Although a number of new bioprocesses have been commercialized, it is clear that economic and technical barriers still exist before the full potential of this area can be realized. One concept gaining considerable momentum is the biorefinery, which could significantly reduce production costs of plant-based chemicals and facilitate their substitution in existing markets. This concept is analogous to that of a modern oil refinery in that the biorefinery is a highly integrated complex that will efficiently separate biomass raw materials into individual components and convert these into marketable products such as energy, fuels and chemicals.

By analogy with crude oil, every element of the plant feedstock will be utilized including the low-value lignin components. However, the different compositional nature of the biomass feedstock, compared to crude oil, will require the application of a wider variety of processing tools in the biorefinery. Processing of the individual components will utilize conventional thermochemical operations and state-of-the-art bioprocessing techniques. The production of biofuels in the biorefinery complex will service existing high-volume markets, providing economy-of-scale benefits and large volumes of by-product streams at minimal cost for upgrading to valuable chemicals. A pertinent example of this is the production of glycerol (glycerin) as a by-product in biodiesel.

In addition, a variety of methods and techniques can be employed to obtain different product portfolios of bulk chemicals, fuels, and materials. Biotechnology-based conversion processes can be used to ferment the biomass carbohydrate content into sugars that can then be further processed. As one example, the fermentation path to lactic acid shows promise as a route to biodegradable plastics. An alternative is to employ thermochemical conversion processes which use pyrolysis or gasification of biomass to produce a hydrogen-rich synthesis gas which can be used in a wide range of chemical processes.

A key requirement for delivery of the biorefinery is the ability of the refinery to develop and use process technology that can economically access and convert the five- and six-membered ring sugars present in the cellulose and hemicellulose fractions of the lignocellulosic feedstock. Although engineering technology exists to effectively separate the sugar containing fractions from the lignocellulose, the enzyme technology to economically convert the five-ring sugars to useful products requires further development. Also, if the biorefinery is truly analogous to an oil refinery in which crude oil is separated into a series of products, such as gasoline, heating oil, jet fuel, and petrochemicals, the biorefinery can take advantage of the differences in biomass components and intermediates and maximize the value derived from the biomass feedstock. A biorefinery might, for example, produce one or several low-volume, but high-value, chemical products and a low-value, but high-volume liquid transportation fuel, while generating electricity and process heat for its own use and perhaps enough for sale of electricity. The high-value products enhance profitability; the high-volume fuel helps meet national energy needs; and the power production reduces costs and avoids greenhouse-gas emissions.

5.0 FUTURE TRENDS

Biotechnology can be used to upgrade petroleum by removing undesirable elements/components such as sulfur, nitrogen, metals, and ash and by reducing viscosity. In fact, application of biotransformation processes can make crude oil easier and less expensive to refine and can reduce the production of air polluting gases resulting from the combustion of crude oil and crude oil products.

As an example, a key area for the role of biotransformation by microbes is the removal of heteroatoms from heterocyclic systems. Sulfur (S), nitrogen (N), and oxygen (O) heterocycles are among the most difficult systems to refine and require large amounts for hydrogen for the removal of the heteroatoms. Biotransformation of these pollutants is attracting more and more attention for removal by immobilized biocatalysts with magnetite nanoparticles or by solvent-tolerant bacteria, and to obtain valuable intermediates from the heterocycles (Xu et al., 2006).

The bioprocessing of petroleum and downstream products can become a reality over the next two-to-three decades and accomplish several issues: (1) enhancement of petroleum recovery, (2) upgrading heavy feedstocks, (3) decreasing catalyst poisoning, as well as (4) pollutant elimination as long as biocatalytic activity, stability, and the oil/water ratio are maintained. Indeed, a successful biorefining of petroleum must first resolve several technical drawbacks such as (1) the biocatalyst, (2) solvent tolerance, and (3) mass-transfer issues of identified strains (*Rhodococcus* and *Pseudomonas* sp.) and enzymes to be applied in the petroleum industry.

In terms of the biocatalyst, the main two challenges to overcome are the stability of the biocatalyst at low or nil water/solvent ratio or the high water/solvent ratio at high specific activities. The progress accumulated in enhancing stability and activity in organic solvents and high temperatures of other enzymatic systems should be applied to the strains and enzymes involved in biorefining (Klibanov, 2001). The approach to the mass-transfer issue should comprise the enhancing of the water-oil emulsion in order to maximize the substrate migration from the oil to the biocatalyst, as well as the expulsion of the metabolite mass. Other key biotransformation technologies that will impact the future use of biotransformation in the petroleum industry are: (1) new strains of microbes from extreme environments, (2) microbial metabolic engineering, and (3) new biocatalyst-based compound synthesis.

The progress reached by BDS research (Chapter 6) will serve as a model to encourage developments on biodenitrogenation and biodemetallization (Xu et al., 1998). In the near term, it can be anticipated that problems relating to process innovation and optimization will be resolved through process engineering (e.g., mass and heat transfer) and reengineering of process units. Enzymatic and solvent tolerance issues will also be resolved through screening for new substrate specificity and extremophile microorganisms, searching for coenzyme-independent enzymatic systems, biosurfactant production, and enhancing the conversion rate and reaction extension. Indeed, biocatalysts must also be developed that will resist contact and/or immersion on high hydrophobic media such as gasoline or diesel. Moreover, the biocatalysts should be active and stable at temperatures higher than 50°C (122°F) currently used in the petroleum industry, i.e., the biocatalyst should be thermophilic (heat tolerant) in nature (Huber and Stetter, 1998). Understanding of biological mechanisms will lead to design and production of tailor-made, robust, and highly active catalysts.

Future research must be emphasized on specific site mutagenesis and selection of enhanced mutants. Research must focus on the combination of several phenotypes including enhanced desulfurization, high temperature resistance, solvent tolerance, and surfactant production. The enzymes must be purified, crystallized, and their structures determined and kinetically characterized. Its physical and chemical modification through enzyme-surfactant interaction and entrapment into a polymer matrix are examples of future research.

Increased production of heavy oil, extra heavy oil, and tar sand bitumen in the near future might require biotransformation of these refinery feedstocks into light oil in the reservoir (or deposit) or during storage. The latter would be in the form of large bioreactors that occupy space at the wellhead or on the refinery site. It can be anticipated that such storage facilities would be large enough to store heavy feedstock that is sufficient for 7 days (or even 2 weeks) prior to entry of the microbe-modified feedstock into the refinery proper—analogous to the storage of coal at coal-based power plants. Microorganisms such as fungus might be employed to oxidize asphaltene constituents which are responsible for the high viscosity of heavy feedstocks thereby enhancing fluidity and pumpability (Pickard et al., 1999). Future research must also include determination of physicochemical properties (such as viscosity, API gravity, and pour point of the untreated and treated feedstock) that are valuable for engineering process planning.

While advances in petroleum biotechnology have become increasingly evident in recent years, there is increasing recognition and appreciation of the technologies that are not only providing supporting roles but are beginning to provide essential roles for the future of the industry. Biotechnology can contribute to the fossil fuel industry by assisting the production of fossil fuels, upgrading fuels, bioremediation of water, soil, and air, and in the control of microbiologically influenced corrosion. The potential growth of this area is immense and biotechnology can make great contributions to the energy industry in the future. The challenge to the biotechnology industry is to continue to demonstrate relevance to the energy industry.

Although a more detailed understanding of all aspects of the metabolic pathways is still needed, the understanding of how microorganisms metabolize the more resilient heterocyclic petroleum constituents is improving rapidly. New techniques including metagenomic methods are continually being developed to explore novel genes and metabolic pathways for the biotransformation of recalcitrant constituents. These possibilities represent a future challenge for microbiologists, process chemists, and process engineers.

Biotransformation (with subsequent bioremediation) has shown also a great promise for cleanup of spills of petroleum-related contaminants, and it is one of several viable options that is available as a single or piggy-back method for biotransformation of petroleum constituents (biorefining) in the refinery. Considerations such as (1) feedstock properties, (2) the nature of the microorganisms that are available, and (3) the process parameters must be taken into account.

It is not surprising that, with advances in biotechnology, biotransformation has become one of the most rapidly developing fields of environmental restoration, utilizing microorganisms to transform petroleum-related hydrocarbon derivatives, PNA derivatives, and other constituents such as metal derivatives and metallo-organic derivatives.

In addition, there is a strong need to assess the potential of the present biotechnology situation and its future impact on the oil industry. This industrial sector has shown importance in petroleum products with experiences and challenges due to decreasing oil reserves, increasing demand of petroleum, fuels, and petrochemicals, fluctuating oil prices with more environmental regulations. Conducting toxicity tests on the microbes is recommended in various fields to assure safe handling and does not show threat to humans or the environment. The use of biotransformation processes in various countries is increasing. However, the challenges in microbial enhanced recovery processes will strengthen and persist in the upcoming years. Microbial enhanced recovery process is a well-proven technology in implementation, to select the right microbes and to understand their growth requirements and production conditions.

REFERENCES

Arellano-Garcia, H., Buenrostro-Gonzalez, E., Vazquez-Duhalt, R., 2004. Biocatalytic transformation of petroporphyrins by chemical modified Cytochrome c. Biotechnology and Bioengineering 85, 790–798.

Cooney, J.J., Silver, S.A., Beck, E.A., 1985. Factors influencing hydrocarbon degradation in three freshwater lakes. Microbial Ecology 11 (2), 127–137.

Dordick, J.S., Khmelnitsky, Y.L., Sergeeva, M.V., 1998. The evolution of biotransformation technologies. Current Opinion in Microbiology 1, 311–318.

El-Gendy, N.Sh., Speight, J.G., 2015. Handbook of Refinery Desulfurization. CRC Press, Taylor & Francis Group, Boca Raton, Florida.

El-Gendy, N.Sh., Speight, J.G., 2016. Handbook of Refinery Desulfurization. CRC Press, Taylor & Francis Group, 6000 Broken Sound Parkway NW, Suite 300, Boca Raton, FL 33487-2742, USA.

Fedorak, P.M., Semple, K.M., Vazquez-Duhalt, R., Westlake, D.W.S., 1993. Chloroperoxidase-mediated modifications of petroporphyrins and asphaltenes. Enzyme and Microbial Technology 15, 429–437.

Fetzer, S., 1998. Bacterial degradation of pyridine, Indole, quinoline, and their derivatives under different redox conditions. Applied Microbiology and Biotechnology 49, 237–250.

Furuya, T., Kirimura, K., Kino, K., Usami, S., 2001. Thermophilic biodesulfurization of dibenzothiophene and its derivatives by *Mycobacterium phlei* WU-F1 FEMS. Microbiology Letters 204, 129–133.

Hamer, G., Al-Awadhi, N., 2000. Biotechnological applications in the oil industry. Acta Biotechnology 20, 335–350.

Huber, H., Stetter, K.O., 1998. Hyperthermophiles and their possible potential in biotechnology. Journal of Biotechnology 64, 39–52.

Isken, S., Derks, A., Wolfes, P.F.G., de Bont, J.A.M., 1999. Effect of organic solvents on the yield of solvent-tolerant *Pseudomonas putida* S12. Applied and Environmental Microbiology 65 (6), 2631–2635.

Kilbane, J.J., October 25, 1994. Microbial Cleavage of Organic C-s Bonds. United States Patent 5,358,869.

Klein, J., Catcheside, D.E.A., Fakoussa, R., Gazso, L., Fritsche, W., Hoefer, M., Laborda, F., Margarit, I., Rehm, H.J., Reich-Walber, M., Sand, W., Schacht, S., Schmiers, H., Setti, L., Steinbuechel, A., 1999. Biological processing of fuels. Applied Microbiology and Biotechnology 52, 2–15.

Klibanov, A.M., 2001. Improving enzymes by using them in organic solvents. Nature 409, 241–246.

Le Borgne, S., Quintero, R., 2003. Biotechnological processes for the refining of petroleum. Fuel Processing Technology 81, 155–169.

McFarland, B.L., Boron, D.J., Deever, W., Meyer, J.A., Johnson, A.R., Atlas, R.M., 1998. Biocatalytic sulfur removal from fuels: applicability for producing low sulfur gasoline. Critical Reviews in Microbiology 24, 99–147.

Monticello, D.J., 2000. Biodesulfurization and the upgrading of petroleum distillates. Current Opinion in Biotechnology 11, 540–546.

Muggeridge, A., Cockin, A., Webb, K., Frampton, H., Collins, I., Moulds, T., Salino, P., 2013. Recovery rates, enhanced oil recovery and technological limits. Philosophical Transactions of the Royal Society A 372. 20120320. https://doi.org/10.1098/rsta.2012.0320.

Pickard, M.A., Roman, R., Tinoco, R., Vazquez-Duhalt, R., 1999. Polycyclic aromatic hydrocarbons metabolism by white rot fungi and oxidation by *Coriolopsis gallica* UAMH 8260 laccase. Applied and Environmental Microbiology 65, 3805–3809.

Premuzic, E.T., Lin, M.S., Bohenek, M., Zhou, W.M., 1999. Bioconversion reactions in asphaltenes and heavy crude oils. Energy Fuels 13, 297–304.

Premuzic, E.T., Lin, M.S., January 12, 1999. Biochemical Upgrading of Oils. United States Patent 5,858,766.

Rajesh, A., Kaladhar, D.S.V.K., 2014. Prospects for industrial and microbial biotechnology in the oil industry. Biotechnology 3 (2), 37–38.

Speight, J.G., Lee, S., 2000. Environmental Technology Handbook, second ed. Taylor & Francis, New York.

Speight, J.G., 2008. Synthetic Fuels Handbook: Properties, Processes, and Performance. McGraw-Hill, New York.

Speight, J.G., 2011a. The Refinery of the Future. Gulf Professional Publishing, Elsevier, Oxford, United Kingdom.

Speight, J.G., 2011b. An Introduction to Petroleum Technology, Economics, and Politics. Scrivener Publishing, Beverly, Massachusetts.

Speight, J.G., 2011c. The Biofuels Handbook. Royal Society of Chemistry, London, United Kingdom.

Speight, J.G., Arjoon, K.K., 2012. Bioremediation of Petroleum and Petroleum Products. Scrivener Publishing, Beverly, Massachusetts.

Speight, J.G., 2014a. The Chemistry and Technology of Petroleum, fifth ed. CRC Press, Taylor and Francis Group, Boca Raton, Florida.

Speight, J.G., 2014b. High Acid Crudes. Gulf Professional Publishing Elsevier, Oxford, United Kingdom.

Speight, J.G., 2015. Handbook of Petroleum Product Analysis, second ed. John Wiley & Sons Inc., Hoboken, New Jersey.

Speight, J.G., 2017a. Handbook of Petroleum Refining. CRC Press, Taylor and Francis Group, Boca Raton, Florida.

Speight, J.G., 2017b. Deep Shale Oil and Gas. Gulf Professional Publishing, Elsevier, Oxford, United Kingdom.

Xu, G.W., Mitchell, K.W., Monticello, D.J., April 29, 1998. Fuel Product Produced by Demetallizing a Fossil Fuel with an Enzyme. United States Patent, 5,624,844.

Xu, P., Yu, B., Li, F.L., Cai, X.F., Ma, C.Q., 2006. Microbial degradation of sulfur, nitrogen and oxygen heterocycles. Trends in Microbiology 14 (9), 398–405.

Conversion Factors

1. Area

1 square centimeter (1 cm^2) = 0.1550 square inches

1 square meter (1 m^2) = 1.1960 square yards

1 hectare = 2.4711 acres

1 square kilometer (1 km^2) = 0.3861 square miles

1 square inch (1 inch^2) = 6.4516 square centimeters

1 square foot (1 ft^2) = 0.0929 square meters

1 square yard (1 yd^2) = 0.8361 square meters

1 acre = 4046.9 square meters

1 square mile (1 mi^2) = 2.59 square kilometers

2. Concentration Conversions

1 part per million (1 ppm) = 1 microgram per liter (1 μg/L)

1 microgram per liter (1 μg/L) = 1 milligram per kilogram (1 mg/kg)

1 microgram per liter (μg/L) × 6.243 × 10^8 = 1 pound per cubic foot (1 lb/ft^3)

1 microgram per liter (1 μg/L) × 10^{-3} = 1 milligram per liter (1 mg/L)

1 milligram per liter (1 mg/L) × 6.243 × 10^5 = 1 pound per cubic foot (1 lb/ft^3)

1 gram mole per cubic meter (1 g mol/m^3) × 6.243 × 10^5 = 1 pound per cubic foot (1 lb/ft^3)

10,000 ppm = 1% w/w

1 ppm hydrocarbon in soil × 0.002 = 1 lb of hydrocarbons per ton of contaminated soil

3. Nutrient Conversion Factor

1 pound, phosphorus × 2.3 (1 lb P × 2.3) = 1 pound phosphorous pentoxide (1 lb P$_2$O$_5$)

1 pound, potassium × 1.2 (1 lb K × 1.2) = 1 pound potassium oxide (1 lb K$_2$O)

4. Temperature Conversions

°F = (°C × 1.8) + 32

°C = (°F−32)/1.8

(°F−32) × 0.555 = °C

Absolute zero = −273.15°C

Absolute zero = −459.67°F

5. Sludge Conversions

1700 lbs wet sludge = 1 yd^3 wet sludge

1 yd^3 sludge = wet tons/0.85

Wet tons sludge × 240 = gallons sludge

1 wet ton sludge × % dry solids/100 = 1 dry ton of sludge

6. Various Constants

Atomic mass	$mu = 1.6605402 \times 10^{-27}$
Avogadro's number	$N = 6.0221367 \times 10^{23} \, mol^{-1}$
Boltzmann's constant	$k = 1.380658 \times 10^{-23} \, J/K$
Elementary charge	$e = 1.60217733 \times 10^{-19} °C$
Faraday's constant	$F = 9.6485309 \times 10^4 °C/mol$

Gas (molar) constant	$R = k \cdot N \sim 8.314510 \, \text{J/mol K}$ $= 0.08205783 \, \text{L atm/mol K}$
Gravitational acceleration	$G = 9.80665 \, \text{m/s}^2$
Molar volume of an ideal gas at 1 atm and 25°C	$V_{\text{ideal gas}} = 24.465 \, \text{L/mol}$
Planck's constant	$h = 6.6260755 \times 10^{-34} \, \text{J s}$
Zero, Celsius scale	$0°C = 273.15 \text{K}$

7. Volume Conversion

Barrels (petroleum, US) to Cu feet multiply by 5.6146
Barrels (petroleum, US) to Gallons (US) multiply by 42
Barrels (petroleum, US) to Liters multiply by 158.98
Barrels (US, liq.) to Cu feet multiply by 4.2109
Barrels (US, liq.) to Cu inches multiply by 7.2765×10^3
Barrels (US, liq.) to Cu meters multiply by 0.1192
Barrels (US, liq.) to Gallons multiply by (US, liq.) 31.5
Barrels (US, liq.) to Liters multiply by 119.24
Cubic centimeters to Cu feet multiply by 3.5315×10^{-5}
Cubic centimeters to Cu inches multiply by 0.06102
Cubic centimeters to Cu meters multiply by 1.0×10^{-6}
Cubic centimeters to Cu yards multiply by 1.308×10^{-6}
Cubic centimeters to Gallons (US liq.) multiply by 2.642×10^{-4}
Cubic centimeters to Quarts (US liq.) multiply by 1.0567×10^{-3}
Cubic feet to Cu centimeters multiply by 2.8317×10^4
Cubic feet to Cu meters multiply by 0.028317
Cubic feet to Gallons (US liq.) multiply by 7.4805
Cubic feet to Liters multiply by 28.317
Cubic inches to Cu centimeters multiply by 16.387
Cubic inches to Cu feet multiply by 5.787×10^{-4}
Cubic inches to Cu meters multiply by 1.6387×10^{-5}
Cubic inches to Cu yards multiply by 2.1433×10^{-5}
Cubic inches to Gallons (US liq.) multiply by 4.329×10^{-3}
Cubic inches to Liters multiply by 0.01639
Cubic inches to Quarts (US liq.) multiply by 0.01732
Cubic meters to Barrels (US liq.) multiply by 8.3864
Cubic meters to Cu centimeters multiply by 1.0×10^6
Cubic meters to Cu feet multiply by 35.315
Cubic meters to Cu inches multiply by 6.1024×10^4
Cubic meters to Cu yards multiply by 1.308
Cubic meters to Gallons (US liq.) multiply by 264.17
Cubic meters to Liters multiply by 1000
Cubic yards to Bushels (Brit.) multiply by 21.022
Cubic yards to Bushels (US) multiply by 21.696
Cubic yards to Cu centimeters multiply by 7.6455×10^5

Cubic yards to Cu feet multiply by 27
Cubic yards to Cu inches multiply by 4.6656×10^4
Cubic yards to Cu meters multiply by 0.76455
Cubic yards to Gallons multiply by 168.18
Cubic yards to Gallons multiply by 173.57
Cubic yards to Gallons multiply by 201.97
Cubic yards to Liters multiply by 764.55
Cubic yards to Quarts multiply by 672.71
Cubic yards to Quarts multiply by 694.28
Cubic yards to Quarts multiply by 807.90
Gallons (US liq.) to Barrels (US liq.) multiply by 0.03175
Gallons (US liq.) to Barrels (petroleum, US) multiply by 0.02381
Gallons (US liq.) to Bushels (US) multiply by 0.10742
Gallons (US liq.) to Cu centimeters multiply by 3.7854×10^3
Gallons (US liq.) to Cu feet multiply by 0.13368
Gallons (US liq.) to Cu inches multiply by 231
Gallons (US liq.) to Cu meters multiply by 3.7854×10^{-3}
Gallons (US liq.) to Cu yards multiply by 4.951×10^{-3}
Gallons (US liq.) to Gallons (wine) multiply by 1.0
Gallons (US liq.) to Liters multiply by 3.7854
Gallons (US liq.) to Ounces (US fluid) multiply by 128.0
Gallons (US liq.) to Pints (US liq.) multiply by 8.0
Gallons (US liq.) to Quarts (US liq.) multiply by 4.0
Liters to Cu centimeters multiply by 1000
Liters to Cu feet multiply by 0.035315
Liters to Cu inches multiply by 61.024
Liters to Cu meters multiply by 0.001
Liters to Gallons (US liq.) multiply by 0.2642
Liters to Ounces (US fluid) multiply by 33.814

8. Weight Conversion

1 ounce (1 ounce) = 28.3495 grams (28.3495 g)
1 pound (1 lb) = 0.454 kilogram
1 pound (1 lb) = 454 grams (454 g)
1 kilogram (1 kg) = 2.20462 pounds (2.20462 lb)
1 stone (English) = 14 pounds (14 lb)
1 ton (US; 1 short ton) = 2000 lbs
1 ton (English; 1 long ton) = 2240 lbs
1 metric ton = 2204.62262 pounds
1 tonne = 2204.62262 pounds

9. Other Approximations

14.7 pounds per square inch (14.7 psi) = 1 atm (1 atm)
1 kilopascal (kPa) $\times 9.8692 \times 10^{-3}$ = 14.7 pounds per square inch (14.7 psi)
1 yd^3 = 27 ft^3
1 US gallon of water = 8.34 lbs

1 imperial gallon of water = 10 lbs
1 ft^3 = 7.5 gallon = 1728 cubic inches = 62.5 lbs
1 yd^3 = 0.765 m^3
1 acre-inch of liquid = 27,150 gallons = 3.630 ft^3
1-foot depth in 1 acre (in-situ) = 1613 × (20%–25% excavation factor) = ~2000 yd^3
1 yd^3 (clayey soils-excavated) = 1.1–1.2 tons (US)
1 yd^3 (sandy soils-excavated) = 1.2–1.3 tons (US)
Pressure of a column of water in psi = height of the column in feet by 0.434.

Glossary

ABC process A fixed-bed process for the hydrodemetallization and hydrodesulfurization of heavy feedstocks.

ABN separation A method of fractionation by which petroleum is separated into acidic, basic, and neutral constituents.

Absorber See Absorption tower.

Absorption gasoline Gasoline extracted from natural gas or refinery gas by contacting the absorbed gas with an oil and subsequently distilling the gasoline from the higher-boiling components.

Absorption oil Oil used to separate the heavier components from a vapor mixture by absorption of the heavier components during intimate contacting of the oil and vapor; used to recover natural gasoline from wet gas.

Absorption plant A plant for recovering the condensable portion of natural or refinery gas, by absorbing the higher-boiling hydrocarbons in an absorption oil, followed by separation and fractionation of the absorbed material.

Absorption tower A tower or column which promotes contact between a rising gas and a falling liquid so that part of the gas may be dissolved in the liquid.

Acetone-benzol process A dewaxing process in which acetone and benzol (benzene or aromatic naphtha) are used as solvents.

Acid catalyst A catalyst having acidic character; the aluminas are examples of such catalysts.

Acid deposition Acid rain; a form of pollution depletion in which pollutants, such as nitrogen oxides and sulfur oxides, are transferred from the atmosphere to soil or water; often referred to as atmospheric self-cleaning. The pollutants usually arise from the use of fossil fuels.

Acid number A measure of the reactivity of petroleum with a caustic solution and given in terms of milligrams of potassium hydroxide that are neutralized by 1 g of petroleum.

Acid rain The precipitation phenomenon that incorporates anthropogenic acids and other acidic chemicals from the atmosphere to the land and water (see Acid deposition).

Acid sludge The residue left after treating petroleum oil with sulfuric acid for the removal of impurities; a black, viscous substance containing the spent acid and impurities.

Acid treating A process in which unfinished petroleum products, such as gasoline, kerosene, and lubricating-oil stocks, are contacted with sulfuric acid to improve their color, odor, and other properties.

Acidity The capacity of an acid to neutralize a base such as a hydroxyl ion (OH$^-$).

Acidizing A technique for improving the permeability (*q.v.*) of a reservoir by injecting acid.

Activation energy, E The energy that is needed by a molecule or molecular complex to encourage reactivity to form products.

Acute-chronic ratio The ratio of the acute and chronic toxicity values for a given compound, usually the average of the ratios for a variety of species; used to estimate the chronic toxicity of a compound or a mixture of compounds, from the measured or modeled acute toxicity when no chronic toxicity data are available.

Additive A material added to another (usually in small amounts) in order to enhance desirable properties or to suppress undesirable properties.

Add-on control methods The use of devices that remove refinery process emissions after they are generated but before they are discharged to the atmosphere.

Adsorption Transfer of a substance from a solution to the surface of a solid resulting in relatively high concentration of the substance at the place of contact; see also Chromatographic adsorption.

Adsorption gasoline Natural gasoline (*q.v.*) obtained by the adsorption process from wet gas.

After-burn The combustion of carbon monoxide (CO) to carbon dioxide (CO$_2$); usually in the cyclones of a catalyst regenerator.

Agrobacterium A natural bacterium that can be used to transfer DNA genes into broadleaf plants, such as tobacco, tomato, or soybean.

Air injection An oil recovery technique using air to force oil from the reservoir into the wellbore.

Air pollution The discharge of toxic gases and particulate matter into the atmosphere, principally as a result of human activity.

Air sweetening A process in which air or oxygen is used to oxidize lead mercaptide to disulfides instead of using elemental sulfur.

Air-blown asphalt Asphalt produced by blowing air through residua at elevated temperatures.

Airlift Thermofor catalytic cracking A moving-bed continuous catalytic process for conversion of heavy gas oils into lighter products; the catalyst is moved by a stream of air.

Alcohol The family name of a group of organic chemical compounds composed of carbon, hydrogen, and oxygen. The series of molecules vary in chain length and are composed of a hydrocarbon plus a hydroxyl group; $CH_3(CH_2)_nOH$ (e.g., methanol, ethanol, and tertiary butyl alcohol).

Alicyclic hydrocarbon A compound containing carbon and hydrogen only, which has a cyclic structure (e.g., cyclohexane); also collectively called naphthenes.

Aliphatic hydrocarbon A compound containing carbon and hydrogen only, which has an open-chain structure (e.g., as ethane, butane, octane, butene) or a cyclic structure (e.g., cyclohexane).

Alkali treatment See Caustic wash.

Alkali wash See Caustic wash.

Alkalinity The capacity of a base to neutralize the hydrogen ion (H^+).

Alkanes Hydrocarbons that contain only single carbon-hydrogen bonds. The chemical name indicates the number of carbon atoms and ends with the suffix "ane."

Alkenes Hydrocarbons that contain carbon-carbon double bonds. The chemical name indicates the number of carbon atoms and ends with the suffix "ene."

Alkyl groups A group of carbon and hydrogen atoms that branch from the main carbon chain or ring in a hydrocarbon molecule. The simplest alkyl group, a methyl group, is a carbon atom attached to three hydrogen atoms.

Alkylate bottoms Residua from fractionation of alkylate; the alkylate product which boils higher than the aviation gasoline range; sometimes called heavy alkylate or alkylate polymer.

Alkylate The product of an alkylation (*q.v.*) process.

Alkylation In the petroleum industry, a process by which an olefin (e.g., ethylene) is combined with a branched-chain hydrocarbon (e.g., isobutane); alkylation may be accomplished as a thermal or as a catalytic reaction.

Alpha-scission The rupture of the aromatic carbon-aliphatic carbon bond that joins an alkyl group to an aromatic ring.

Alumina (Al_2O_3) Used in separation methods as an adsorbent and in refining as a catalyst.

Amine washing A method of gas cleaning whereby acidic impurities such as hydrogen sulfide and carbon dioxide are removed from the gas stream by washing with an amine (usually an alkanolamine).

Analysis Determine the properties of a feedstock prior to refining, inspection (*q.v.*) of feedstock properties.

Analyte The chemical for which a sample is tested or analyzed.

Analytical equivalence The acceptability of the results obtained from the different laboratories; a range of acceptable results.

Aniline point The temperature, usually expressed in °F, above which equal volumes of a petroleum product are completely miscible; a qualitative indication of the relative proportions of paraffins in a petroleum product which are miscible with aniline only at higher temperatures; a high aniline point indicates low aromatics.

Antibody A molecule having chemically reactive sites specific for certain other molecules.

API gravity A measure of the *lightness* or *heaviness* of petroleum that is related to density and specific gravity.

$$°API = (141.5/sp\ gr\ @\ 60°F) - 131.5$$

Apparent bulk density The density of a catalyst as measured; usually loosely compacted in a container.

Apparent viscosity The viscosity of a fluid, or several fluids flowing simultaneously, measured in a porous medium (rock), and subject to both viscosity and permeability effects; also called *effective viscosity*.

Aquaconversion process A hydrovisbreaking technology in which a proprietary additive and water are added to the heavy feedstock prior to introduction into the soaker.

Aromatic hydrocarbon A hydrocarbon characterized by the presence of an aromatic ring or condensed aromatic rings; benzene and substituted benzene, naphthalene and substituted naphthalene, phenanthrene and substituted phenanthrene, as well as the higher condensed ring systems; compounds that are distinct from those of aliphatic compounds (*q.v.*) or alicyclic compounds (*q.v.*).

Aromatics A group of hydrocarbons of which benzene is the parent; so named because many of their derivatives have sweet or aromatic odors.

Aromatics Class of hydrocarbons comprising one or more benzene ring structures having alternating double bonds; may have one or more alkyl side chains in various positions, but no heteroatoms.

Aromatization The conversion of non-aromatic hydrocarbons to aromatic hydrocarbons by: (1) rearrangement of aliphatic (noncyclic) hydrocarbons (*q.v.*) into aromatic ring structures; and (2) dehydrogenation of alicyclic hydrocarbons (naphthenes).

Arosorb process A process for the separation of aromatics from non-aromatics by adsorption on a gel from which they are recovered by desorption.

ART process A process for increasing the production of liquid fuels without hydrocracking (*q.v.*).

ASCOT process A resid (*q.v.*) upgrading process that integrates delayed coking and deep solvent deasphalting.

Asphalt Highly viscous liquid or semisolid composed of bitumen and present in most crudes; can be separated from other crude components by fractional distillation; used primarily for road paving and roofing shingles.

Asphaltene association factor The number of individual asphaltene species which associate in nonpolar solvents as measured by molecular weight methods; the molecular weight of asphaltenes in toluene divided by the molecular weight in a polar nonassociating solvent, such as dichlorobenzene, pyridine, or nitrobenzene.

Asphaltene constituents Molecular species that occur within the asphaltene fraction and which vary in polarity (functional group content) and molecular weight.

Asphaltene fraction That fraction of petroleum, heavy oil, or bitumen that is precipitated when a large excess (40 volumes) of a low-boiling liquid hydrocarbon (e.g., pentane or heptane) is added to (1 volume) the feedstock; usually a dark brown to black amorphous solid that does not melt prior to decomposition and is soluble in benzene or aromatic naphtha or other chlorinated hydrocarbon solvents.

Asphaltic constituents A general term usually meaning the asphaltene fraction *plus* the resin fraction.

Associated gas in solution (or dissolved gas) Natural gas dissolved in the crude oil of the reservoir, under the prevailing pressure and temperature conditions.

Associated gas Natural gas that is in contact with and/or dissolved in the crude oil of the reservoir. It may be classified as gas cap (free gas) or gas in solution (dissolved gas).

Associated molecular weight The molecular weight of asphaltenes in an associating (nonpolar) solvent, such as toluene.

ASTM International (formerly American Society for Testing and Materials) The official organization in the United States for designing standard tests for petroleum and other industrial products.

Atmospheric distillation Distillation at atmospheric pressure; the refining process of separating crude oil components at atmospheric pressure by heating to temperatures of about 600° to 750°F (depending on the nature of the crude oil and desired products) and subsequent condensing of the fractions by cooling.

Atmospheric equivalent boiling point (AEBP) A mathematical method of estimating the boiling point at atmospheric pressure of nonvolatile fractions of petroleum.

Atmospheric residuum A residuum (*q.v.*) obtained by distillation of a crude oil under atmospheric pressure and which boils above 350°C (660°F).

Attapulgus clay See Fuller's earth.

Autofining A catalytic process for desulfurizing distillates.

Autothermal reforming (ATR)   A process used to produce hydrogen by using a combination of partial oxidation and steam reforming in a single vessel.

Average particle size   The weighted average particle diameter of a catalyst.

Aviation gasoline blending components   The various naphtha products that will be used for blending or compounding into finished aviation gasoline (e.g., straight run gasoline, alkylate, reformate, benzene, toluene, and xylene); excludes oxygenates (alcohols, ethers), butane, and pentanes plus. Oxygenates are reported as other hydrocarbons, hydrogen, and oxygenates.

Aviation gasoline   Any of the special grades of gasoline suitable for use in certain airplane engines; a complex mixture of relatively volatile hydrocarbons with or without small quantities of additives, blended to form a fuel suitable for use in aviation reciprocating engines. Fuel specifications are provided in ASTM D910 and Military Specification MIL-G-5572.

Aviation turbine fuel   See Jet fuel.

Avjet   Aviation jet fuel derived from a kerosene fraction; also known as jet fuel or jet.

Back mixing   The phenomenon observed when a catalyst travels at a slower rate in the riser pipe than the vapors.

Baghouse   A filter system for the removal of particulate matter from gas streams; so called because of the similarity of the filters to coal bags.

Bari-Sol process   A dewaxing process that employs a mixture of ethylene dichloride and benzol as the solvent.

Barrel of oil equivalent   A unit of energy based on the approximate energy released by burning one barrel of crude oil.

Barrel   The unit of measurement of liquids in the petroleum industry; the traditional measurement for crude oil volume—one barrel is equivalent to 42 US gallons (159 L) and 6.29 barrels is equivalent to one cubic meter of oil.

Barrels per calendar day   The amount of input that a distillation facility can process under usual operating conditions—the amount is expressed in terms of capacity during a 24-h period and reduces the maximum processing capability of all units at the facility under continuous operation (see Barrels per Stream Day) to account for the following limitations that may delay, interrupt, or slow down production: (1) the capability of downstream facilities to absorb the output of crude oil processing facilities of a given refinery—no reduction is made when a planned distribution of intermediate streams through other than downstream facilities is part of a refinery's normal operation, (2) the types and grades of inputs to be processed, (3) the environmental constraints associated with refinery operations, (4) the reduction of capacity for scheduled downtime due to such conditions as routine inspection, maintenance, repairs, and turnaround, and (5) the reduction of capacity for unscheduled downtime due to such conditions as mechanical problems, repairs, and slowdowns.

Barrels per stream day   The maximum number of barrels of input that a distillation facility can process within a 24-h period when running at full capacity under optimal crude and product slate conditions with no allowance for downtime.

Base number   The quantity of acid, expressed in milligrams of potassium hydroxide per gram of sample that is required to titrate a sample to a specified end-point.

Base stock   A primary refined petroleum fraction into which other oils and additives are added (blended) to produce the finished product.

Basic nitrogen   Nitrogen (in petroleum) which occurs in pyridine form.

Basic sediment and water (BS&W, BSW)   The material which collects in the bottom of storage tanks, usually composed of oil, water, and foreign matter; also called bottoms, bottom settlings.

Battery   A series of stills or other refinery equipment operated as a unit.

Baumé gravity   The specific gravity of liquids expressed as degrees on the Baum (B°) scale; for liquids lighter than water:

$$Sp\ gr\ 60°F = 140/(130 + B°)$$

Bauxite   Mineral matter used as a treating agent; hydrated aluminum oxide formed by the chemical weathering of igneous rocks.

Bbl   See Barrel.

Bell cap A hemispherical or triangular cover placed over the riser in a (distillation) tower to direct the vapors through the liquid layer on the tray; see Bubble cap.

Bender process A chemical treating process using lead sulfide catalyst for sweetening light distillates by which mercaptans are converted to disulfides by oxidation.

Bentonite Montmorillonite (a magnesium-aluminum silicate); used as a treating agent.

Benzene A colorless aromatic liquid hydrocarbon (C_6H_6) present in small proportion in some crude oils and made commercially from petroleum by the catalytic reforming of naphthenes in petroleum naphtha—also made from coal in the manufacture of coke; used as a solvent in the manufacture of detergents, synthetic fibers, petrochemicals, and as a component of high-octane gasoline.

Benzin Refined light naphtha used for extraction purposes.

Benzine An obsolete term for light petroleum distillates covering the gasoline and naphtha range; see Ligroine.

Benzol The general term which refers to commercial or technical (not necessarily pure) benzene; also the term used for aromatic naphtha.

Beta-scission The rupture of a carbon-carbon bond; two bonds removed from an aromatic ring.

Billion 1×10^9

Bioaccumulation (or bioconcentration) The tendency of substances to accumulate in the body of organisms; the net uptake from their diet, respiration or transfer across skin, and loss due to excretion or metabolism. The bioaccumulation factor (BAF) or bioconcentration factor (BCF) is the ratio of concentrations in tissue to concentrations in a source, i.e., water or diet.

Bioavailability (or biological availability) Compound that is in a physical or chemical form that can be assimilated by a living organism; also, the proportion of a chemical in an environmental compartment (e.g., water) that can be taken up by an organism.

Biocide Any chemical capable of killing bacteria and bio-organisms.

Biodegradation A natural process of microbial transformation of chemicals, such as oil under aerobic or anaerobic conditions; oil biodegradation usually requires nutrients, such as nitrogen and phosphorus; transformation may be complete, producing water, carbon dioxide, and/or methane, or incomplete, producing partially oxidized chemicals.

Biodegradation The transformation or destruction of organic materials by bacteria.

Biogenic Material derived from bacterial or vegetation sources.

Biological lipid Any biological fluid that is miscible with a nonpolar solvent. These materials include waxes, essential oils, chlorophyll, etc.

Biological oxidation The oxidative consumption of organic matter by bacteria by which the organic matter is converted into gases.

Biomarkers A term used in two different ways, depending upon discipline. In petroleum chemistry, a biomarker is a relic chemical relating its presence to the original biological source (microbial, plant, or animal); biomarkers are usually poorly or nonbiodegradable and so persist in the oil, enabling their use as internal standards in petroleum analysis. In environmental toxicology, a biomarker is a biochemical process, product, or cellular response that indicates the organism's exposure to a pollutant and/or the toxic effects of the pollutant.

Biomass Biological organic matter; materials produced from the processing of wood, corn, sugar, and other agricultural waste or municipal waste. It can be converted to syngas via a gasification process.

Biomass-to-liquids (BTL) A process used to convert a wide variety of waste biomass, such as from the processing of wood, corn, sugar and other agricultural waste or municipal waste into hydrocarbons such as diesel and jet fuel. The biomass is converted first into syngas through a gasification process, followed by the Fischer-Tropsch process and subsequent hydrocracking step.

Biopolymer A high molecular weight carbohydrate produced by bacteria.

Bioremediation An intervention strategy to enhance biodegradation of spilled oil (or other contaminants), ranging from no remedial action other than monitoring (natural attenuation) to nutrient addition (biostimulation) to inoculation with competent microbial communities (bioaugmentation).

Bioremediation Cleanup of spills of petroleum and/or petroleum products using microbial agents.

Biotransformation The conversion of an organic compound (or inorganic compound) into another compound through the agency of microbes; see also Biodegradation, Bioremediation

Bitumen A semisolid to solid organic material found filling pores and crevices of sandstone, limestone, or argillaceous sediments; contains organic carbon, hydrogen, nitrogen, oxygen, sulfur, and metallic constituents; usually has an API gravity less than 10° but other properties are necessary for inclusion in a more complete definition; in its natural state, tar sand (oil sand) bitumen is not recoverable at a commercial rate through a well because it is too viscous to flow; bitumen typically makes up approximately 10% w/w of tar sand (oil sand) but saturation varies.

Bituminous rock See Bituminous sand.

Bituminous sand A formation in which the bituminous material (see Bitumen) is found as a filling in veins and fissures in fractured rocks or impregnating relatively shallow sand, sandstone, and limestone strata; a sandstone reservoir that is impregnated with a heavy, viscous black petroleum-like material that cannot be retrieved through a well by conventional production techniques.

Bituminous Containing bitumen or constituting the source of bitumen.

Black acid(s) A mixture of the sulfonates found in acid sludge which are insoluble in naphtha, benzene, and carbon tetrachloride; very soluble in water but insoluble in 30% sulfuric acid; in the dry, oil-free state, the sodium soaps are black powders.

Black oil Any of the dark-colored oils; a term now often applied to heavy oil; a term erroneously used to describe heavy oil.

Black soap See Black acid.

Black strap The black material (mainly lead sulfide) formed in the treatment of sour light oils with doctor solution (*q.v.*) and found at the interface between the oil and the solution.

Blending plant A facility that has no refining capability but is either capable of producing finished motor gasoline through mechanical blending or blends oxygenates with motor gasoline.

Blown asphalt The asphalt prepared by air blowing a residuum (*q.v.*) or an asphalt (*q.v.*).

BOC process See RCD Unibon (BOC) process.

Bogging A condition that occurs in a coking reactor when the conversion to coke and light ends is too slow causing the coke particles to agglomerate.

Boiling point A characteristic physical property of a liquid at which the vapor pressure is equal to that of the atmosphere and the liquid is converted to a gas.

Boiling range The range of temperature, usually determined at atmospheric pressure in standard laboratory apparatus, over which the distillation of oil commences, proceeds, and finishes.

Bottled gas Usually butane or propane, or butane-propane mixtures, liquefied and stored under pressure for domestic use; see also Liquefied petroleum gas.

Bottom-of-the-barrel processing Residuum processing.

Bottom-of-the-barrel Residuum (*q.v.*).

Bottoms Residue remaining in a distillation unit after the highest boiling point material to be distilled has been removed; also the liquid which collects in the bottom of a vessel (tower bottoms, tank bottoms) either during distillation; also the deposit or sediment formed during storage of petroleum or a petroleum product; see also Residuum and Basic sediment and water.

Bright stock Refined, high-viscosity lubricating oils usually made from residual stocks by processes such as a combination of acid treatment or solvent extraction with dewaxing or clay finishing.

British thermal unit See Btu.

Bromine number The number of grams of bromine absorbed by 100 g of oil that indicates the percentage of double bonds in the material.

Brønsted acid A chemical species that can act as a source of protons.

Brønsted base A chemical species that can accept protons.

Brown acid Oil-soluble petroleum sulfonates found in acid sludge that can be recovered by extraction with naphtha solvent. Brown-acid sulfonates are somewhat similar to mahogany sulfonates but are more water-soluble. In the dry, oil-free state, the sodium soaps are light-colored powders.

Brown soap See Brown acid.

BS&W See Basic sediment and water.

BTEX Benzene, toluene, ethylbenzene, and the xylene isomers.

Btu (British thermal unit) The energy required to raise the temperature of one pound of water by 1° Fahrenheit.

Bubble cap An inverted cup with a notched or slotted periphery to disperse the vapor in small bubbles beneath the surface of the liquid on the bubble plate in a distillation tower.

Bubble plate A tray in a distillation tower.

Bubble point The temperature at which incipient vaporization of a liquid in a liquid mixture occurs, corresponding with the equilibrium point of 0% vaporization or 100% condensation; the temperature at which a gas starts to come out of a liquid.

Bubble tower A fractionating tower so constructed that the vapors rising pass up through layers of condensate on a series of plates or trays (see Bubble plate); the vapor passes from one plate to the next above by bubbling under one or more caps (see Bubble cap) and out through the liquid on the plate where the less volatile portions of vapor condense in bubbling through the liquid on the plate, overflow to the next lower plate, and ultimately back into the re-boiler thereby effecting fractionation.

Bubble tray A circular, perforated plate having the internal diameter of a bubble tower (*q.v.*), set at specified distances in a tower to collect the various fractions produced during distillation.

Bulk composition The make-up of petroleum in terms of bulk fractions such as *saturates*, *aromatics*, *resins*, and *asphaltenes*; separation of petroleum into these fractions is usually achieved by a combination of *solvent* and *adsorption* (*q.v.*) processes.

Bumping The knocking against the walls of a still occurring distillation of petroleum or a petroleum product which usually contains water.

Bunker C oil See No. 6 Fuel oil.

Burner fuel oil Any petroleum liquid suitable for combustion.

Burning oil Illuminating oil, such as kerosene (kerosene) suitable for burning in a wick lamp.

Burning point See Fire point.

Burning-quality index An empirical numerical indication of the likely burning performance of a furnace or heater oil; derived from the distillation profile (*q.v.*) and the API gravity (*q.v.*), and generally recognizing the factors of paraffin character and volatility.

Burton process An older thermal cracking process in which oil was cracked in a pressure still and any condensation of the products of cracking also took place under pressure.

Butane dehydrogenation A process for removing hydrogen from butane to produce butene (C_4H_8) and, on occasion, butadiene (C_4H_6, $CH_2=CHCH=CH_2$).

Butane vapor-phase isomerization A process for isomerizing *n*-butane to isobutane using aluminum chloride catalyst on a granular alumina support and with hydrogen chloride as a promoter.

Butane Four-carbon alkane, hydrocarbon used as a fuel for cooking and camping, chemical formula C_4H_{10}; a straight-chain or branched-chain hydrocarbon extracted from natural gas or refinery gas streams, which is gaseous at standard temperature and pressure; may include isobutane and normal butane and is designated in ASTM D1835 and Gas Processors Association specifications for commercial butane.

Butylene (C_4H_8) An olefin hydrocarbon recovered from refinery or petrochemical processes, which is gaseous at standard temperature and pressure; used in the production of gasoline and various petrochemical products.

C_1, C_2, C_3, C_4, C_5 fractions A common way of representing fractions containing a preponderance of hydrocarbons having 1, 2, 3, 4, or 5 carbon atoms, respectively, and without reference to hydrocarbon type.

Calcining Heating a metal oxide or an ore to decompose carbonates, hydrates, or other compounds often in a controlled atmosphere.

CANMET hydrocracking process A hydrocracking process for heavy feedstocks that employs a low-cost additive to inhibit coke formation and allows high feedstock conversion using a single reactor.

Capillary forces Interfacial forces between immiscible fluid phases, resulting in pressure differences between the two phases.

Capillary number N_c, the ratio of viscous forces to capillary forces, and equal to viscosity times velocity divided by interfacial tension.

Carbene The pentane- or heptane-insoluble material that is insoluble in benzene or toluene but which is soluble in carbon disulfide (or pyridine); a type of rifle used for hunting bison.

Carboid The pentane- or heptane-insoluble material that is insoluble in benzene or toluene and which is also insoluble in carbon disulfide (or pyridine).

Carbon dioxide augmented water flooding Injection of carbonated water, or water and carbon dioxide, to increase water flood efficiency; see Immiscible carbon dioxide displacement.

Carbon rejection processes Upgrading processes in which coke is produced, e.g., coking (*q.v.*).

Carbon residue The amount of carbonaceous residue remaining after thermal decomposition of petroleum, a petroleum fraction, or a petroleum product in a limited amount of air; also called the *coke-* or *carbon-forming propensity*; often prefixed by the terms Conradson or Ramsbottom in reference to the inventor of the respective tests.

Carbonate washing Processing using a mild alkali (e.g., potassium carbonate) for emission control by the removal of acid gases from gas streams.

Carbon-forming propensity See Carbon residue.

Carbonization The conversion of an organic compound into char or coke by heat in the substantial absence of air; often used in reference to the *destructive distillation* (*q.v.*) (with simultaneous removal of distillate) of coal.

Cascade tray A fractionating device consisting of a series of parallel troughs arranged in stair-step fashion in which liquid flowing from the tray above enters the uppermost trough and liquid thrown from this trough by vapor rising from the tray below impinges against a plate and a perforated baffle and liquid passing through the baffle enters the next layer of the troughs.

Casinghead gas Natural gas which issues from the casinghead (the mouth or opening) of an oil well.

Casinghead gasoline The liquid hydrocarbon product extracted from casinghead gas (*q.v.*) by one of three methods: compression, absorption, or refrigeration; see also Natural gasoline.

Cat cracker See Fluid Catalytic Cracking Unit.

Cat cracking See Catalytic cracking.

Catagenesis The alteration of organic matter during the formation of petroleum that may involve temperatures in the range 50°C (120°F) to 200°C (390°F); see also Diagenesis and Metagenesis.

Catalyst plugging The deposition of carbon (coke) or metal contaminants that decreases flow through the catalyst bed.

Catalyst poisoning The deposition of carbon (coke) or metal contaminants that causes the catalyst to become nonfunctional.

Catalyst selectivity The relative activity of a catalyst with respect to a particular compound in a mixture, or the relative rate in competing reactions of a single reactant.

Catalyst stripping The introduction of steam, at a point where spent catalyst leaves the reactor, in order to strip, i.e., remove deposits retained on the catalyst.

Catalyst A chemical agent which, when added to a reaction (process) will enhance the conversion of a feedstock without being consumed in the process; used in upgrading processes to assist cracking and other upgrading reactions.

Catalytic activity The ratio of the space velocity of the catalyst under test to the space velocity required for the standard catalyst to give the same conversion as the catalyst being tested; usually multiplied by 100 before being reported.

Catalytic cracking The conversion of high-boiling feedstocks into lower boiling products by means of a catalyst which may be used in a fixed bed (*q.v.*) or fluid bed (*q.v.*).

Catalytic distillation A process that combines reaction and distillation in a single vessel resulting in lower investment and operating costs, as well as process benefits.

Catalytic hydrocracking A refining process that uses hydrogen and catalysts with relatively low temperatures and high pressures for converting middle boiling or residual material to high octane gasoline, reformer charge stock, jet fuel, and /or high grade fuel oil. The process uses one or more catalysts, depending on product output, and can handle high sulfur feed stocks without prior desulfurization.

Catalytic hydrotreating A refining process for treating petroleum fractions from atmospheric or vacuum distillation units (e.g., naphtha, middle distillates, reformer feeds, residual fuel oil, and heavy gas oil) and other petroleum (e.g., cat cracked naphtha, coker naphtha, gas oil, etc.) in the presence of catalysts and substantial quantities of hydrogen. Hydrotreating includes desulfurization, removal of substances (e.g., nitrogen compounds) that deactivate catalysts, conversion of olefins to paraffins to reduce gum formation in gasoline, and other processes to upgrade the quality of the fractions.

Catalytic reforming Rearranging hydrocarbon molecules in a gasoline-boiling-range feedstock to produce other hydrocarbons having a higher antiknock quality; isomerization of paraffins, cyclization of paraffins to naphthenes (*q.v.*), dehydrocyclization of paraffins to aromatics (*q.v.*).

Catforming A process for reforming naphtha using a platinum-silica-alumina catalyst which permits relatively high space velocities and results in the production of high-purity hydrogen.

Caustic wash The process of treating a product with a solution of caustic soda to remove minor impurities; often used in reference to the solution itself.

Ceresin A hard, brittle wax obtained by purifying Ozokerite (see Microcrystalline wax and Ozokerite).

Cetane index An approximation of the cetane number (*q.v.*) calculated from the density (*q.v.*) and mid-boiling point temperature (*q.v.*); see also Diesel index.

Cetane number A number indicating the ignition quality of diesel fuel; a high cetane number represents a short ignition delay time; the ignition quality of diesel fuel can also be estimated from the following formula:

Characterization factor The UOP characterization factor K, defined as the ratio of the cube root of the molar average boiling point, T_B, in degrees Rankine ($°R = °F + 460$), to the specific gravity at 60°F/60 degrees F:

$$K = (T_B) \ 1/3/\text{sp gr}$$

The value ranges from 12.5 for paraffinic stocks to 10.0 for the highly aromatic stocks; also called the Watson characterization factor.

Cheesebox still An early type of vertical cylindrical still designed with a vapor dome.

Chelating agents Complex-forming agents having the ability to solubilize heavy metals.

Chemical composition The make-up of petroleum in terms of distinct chemical types such as *paraffins, isoparaffins, naphthenes* (*cycloparaffins*), *benzenes, di-aromatics, tri-aromatics, polynuclear aromatics*; other chemical types can also be specified.

Chemical octane number The octane number added to gasoline by refinery processes or by the use of octane number (*q.v.*) improvers such as tetraethyl lead.

Chemical waste Any solid, liquid, or gaseous material discharged from a process and that may pose substantial hazards to human health and environment.

Cherry-P process A process for the conversion of heavy feedstocks into distillate and a cracked residuum.

Chevron deasphalted oil hydrotreating process A process designed to desulfurize heavy feedstocks that have had the asphaltene fraction (*q.v.*) removed by prior application of a deasphalting process (*q.v.*).

Chevron RDS and VRDS processes Processes designed to remove sulfur, nitrogen, asphaltene, and metal contaminants from heavy feedstocks consisting of a once-through operation of the feedstock coming into contact with hydrogen and the catalyst in a downflow reactor (*q.v.*).

Chlorex process A process for extracting lubricating-oil stocks in which the solvent used is Chlorex (3-3-dichlorodiethyl ether).

CHOPS is a nonthermal primary heavy oil production method— Continuous production of sand improves the recovery of heavy oil from the reservoir. The simultaneous extraction of oil and sand during the cold production of heavy oil generates high-porosity channels (*wormholes*) which grow in a 3-D radial pattern within a certain layer of net pay zones, resulting in the development of a high-permeability network in the reservoir, boosting oil recovery. In most cases, an artificial lift system is used to lift the oil with sand.

Chromosome A cellular structure comprised of a long, folded DNA molecule and protein.

Chromatographic adsorption Selective adsorption on materials such as activated carbon, alumina, or silica gel; liquid or gaseous mixtures of hydrocarbons are passed through the adsorbent in a stream of diluent, and certain components are preferentially adsorbed.

Chromatography A method of separation based on selective adsorption; see also Chromatographic adsorption.

Clarified oil The heavy oil that has been taken from the bottom of a fractionator in a catalytic cracking process and from which residual catalyst has been removed.

Clarifier Equipment for removing the color or cloudiness of an oil or water by separating the foreign material through mechanical or chemical means; may involve centrifugal action, filtration, heating, or treatment with acid or alkali.

Clastic Composed of pieces of preexisting rocks.

Clay Silicate minerals that also usually contain aluminum and have particle sizes less than 0.002 micron; used in separation methods as an adsorbent and in refining as a catalyst.

Clay contact process See Contact filtration.

Clay refining A treating process in which vaporized gasoline or other light petroleum product is passed through a bed of granular clay such as fuller's earth (*q.v.*).

Clay regeneration A process in which spent coarse-grained adsorbent clays from percolation processes are cleaned for reuse by de-oiling them with naphtha, steaming out the excess naphtha, and then roasting in a stream of air to remove carbonaceous matter.

Clay treating See Gray clay treating.

Clay wash A light oil, such as kerosene (kerosine) or naphtha, used to clean fuller's earth after it has been used in a filter.

Cloud point The temperature at which paraffin wax or other solid substances begin to crystallize or separate from the solution, imparting a cloudy appearance to the oil when the oil is chilled under prescribed conditions.

Coal An organic rock; a readily combustible black or brownish-black rock whose composition, including inherent moisture, consists of more than 50% w/w and more than 70% v/v of carbonaceous material; formed from plant remains that have been compacted, hardened, chemically altered, and metamorphosed by heat and pressure over geologic time.

Coal tar The specific name for the tar (*q.v.*) produced from coal.

Coal tar pitch The specific name for the pitch (*q.v.*) produced from coal.

Cogeneration The simultaneous production of electricity and steam.

Coke A gray to black solid carbonaceous material produced from petroleum during thermal processing; characterized by having a high carbon content (95% + by weight) and a honeycomb type of appearance and is insoluble in organic solvents.

Coke drum A vessel in which coke is formed during a thermal process.

Coke number Used, particularly in Great Britain, to report the results of the Ramsbottom carbon residue test (*q.v.*), which is also referred to as a coke test.

Coker The processing unit in which coking takes place.

Coking A process for the thermal conversion of petroleum in which gaseous, liquid, and solid (coke) products are formed; e.g., *delayed coking* (*q.v.*), *fluid coking* (*q.v.*).

Cold pressing The process of separating wax from oil by first chilling (to help form wax crystals) and then filtering under pressure in a plate and frame press.

Cold settling Processing for the removal of wax from high-viscosity stocks, wherein a naphtha solution of the waxy oil is chilled and the wax crystallizes out of the solution.

Color stability The resistance of a petroleum product to color change due to light, aging, etc.

Combustible liquid A liquid with a flash point in excess of 37.8°C (100°F) but below 93.3°C (200°F).

Combustion zone The volume of reservoir rock wherein petroleum is undergoing combustion during enhanced oil recovery.

Composition The general chemical make-up of petroleum; often quoted as *chemical composition (q.v.)* or *bulk composition (q.v.)*.

Composition map A means of illustrating the chemical make-up of petroleum using chemical and/or physical property data.

Con Carbon See Carbon residue.

Condensate A mixture of light hydrocarbon liquids obtained by condensation of hydrocarbon vapors: predominately butane, propane, and pentane with some heavier hydrocarbons and relatively little methane or ethane; see also Natural gas liquids.

Connate water Water that is indigenous to the reservoir; usually water that has not been in contact with the atmosphere since its deposition; contains high amounts of chlorine and calcium with total dissolved solids on the order of 1% or more.

Conradson carbon residue See Carbon residue.

Contact filtration A process in which finely divided adsorbent clay is used to remove color bodies from petroleum products.

Contaminant A substance that causes deviation from the normal composition of an environment.

Continuous Catalyst Regeneration A process that regenerates spent catalyst used in a catalytic reformer, which converts naphtha feedstock into higher octane gasoline blending stocks.

Continuous contact coking A thermal conversion process in which petroleum-wetted coke particles move downward into the reactor in which cracking, coking, and drying take place to produce coke, gas, gasoline, and gas oil.

Continuous contact filtration A process to finish lubricants, waxes, or special oils after acid treating, solvent extraction, or distillation.

Conventional crude oil A mixture mainly of pentane and heavier hydrocarbons recoverable at a well from an underground reservoir and liquid at atmospheric pressure and temperature; unlike some heavy oils and tar sand bitumen, conventional crude oil flows through a well without stimulation and through a pipeline without processing or dilution; generally, conventional crude oil includes light and medium gravity crude oils; crude oil containing more than 0.5% w/w sulfur is considered to be sour crude oil while crude oil with less than 0.5% w/w sulfur is considered to be sweet crude oil.

Conventional crude oil Commonly defined as liquid petroleum that flows in the reservoir and in pipelines and is recovered from traditional oil wells using established methods, including primary recovery and water flooding (e.g., condensates, light and medium crude oils), versus unconventional crude oils.

Conventional gasoline Finished automotive gasoline not included in the oxygenated or reformulated gasoline categories; excludes reformulated gasoline blendstock for oxygenate blending (RBOB), as well as other blendstock.

Conventional recovery Primary and/or secondary recovery.

Conversion The thermal treatment of petroleum that results in the formation of new products by the alteration of the original constituents.

Conversion factor The percentage of feedstock converted to light ends, gasoline, other liquid fuels, and coke.

Conversion stock A commodity material, usually vacuum gas oil or atmospheric gas oil, that is suitable for feedstock to a fluid catalytic cracking unit or hydrocracking unit.

Copper sweetening Processes involving the oxidation of mercaptans to disulfides by oxygen in the presence of cupric chloride.

Cracked residua Residua that have been subjected to temperatures above 350°C (660°F) during the distillation process.

Cracking The thermal processes by which the constituents of petroleum are converted to lower molecular weight products.

Cracking activity See Catalytic activity.

Cracking coil Equipment used for cracking heavy petroleum products consisting of a coil of heavy pipe running through a furnace so that the oil passing through it is subject to high temperature.

Cracking still The combined equipment—furnace, reaction chamber, fractionator—for the thermal conversion of heavier feedstocks to lighter products.

Cracking temperature The temperature (350°C; 660°F) at which the rate of thermal decomposition of petroleum constituents becomes significant.

Crude assay A procedure for determining the general distillation characteristics (e.g., distillation profile, *q.v.*) and other quality information of crude oil.

Crude oil See Petroleum.

Crude oil Synonymous with petroleum; a naturally occurring and typically liquid complex mixture of thousands of different hydrocarbon and non-hydrocarbon molecules.

Crude scale wax The wax product from the first sweating of the slack wax.

Crude still Distillation (*q.v.*) equipment in which crude oil is separated into various products.

Cryogenic plant A processing plant capable of producing liquid natural gas products, including ethane, at very low operating temperatures.

Cryogenics The study, production, and use of low temperatures.

Cumene A colorless liquid [$C_6H_5CH(CH_3)_2$] used as an aviation gasoline blending component and as an intermediate in the manufacture of chemicals.

Cut point The boiling-temperature division between distillation fractions of petroleum.

Cutback The term applied to the products from blending heavier feedstocks or products with lighter oils to bring the heavier materials to the desired specifications.

Cycle stock The product taken from some later stage of a process and recharged (recycled) to the process at some earlier stage.

Cyclic steam injection The alternating injection of steam and production of oil with condensed steam from the same well or wells.

Cyclic steam stimulation A process in which (for several weeks) high-pressure steam is injected into the formation to soften the tar sand (oil sand) before being pumped to the surface for separation; the pressure created in the underground environment causes formation cracks that help move the bitumen to producing wells. After a portion of the reservoir has been saturated, the steam is turned off and the reservoir is allowed to soak for several weeks. Then the production phase brings the bitumen to the surface. When the rates of production start to decline, the reservoir is pumped with steam once again.

Cyclization The process by which an open-chain hydrocarbon structure is converted to a ring structure, e.g., hexane to benzene.

Cyclone A device for extracting dust from industrial waste gases. It is in the form of an inverted cone into which the contaminated gas enters tangential from the top; the gas is propelled down a helical pathway, and the dust particles are deposited by means of centrifugal force onto the wall of the scrubber.

Deactivation Reduction in catalyst activity by the deposition of contaminants (e.g., coke, metals) during a process.

Dealkylation The removal of an alkyl group from aromatic compounds.

Deasphaltened oil The fraction of petroleum after the asphaltenes have been removed using liquid hydrocarbons such as *n*-pentane and *n*-heptane.

Deasphaltening Removal of a solid powdery asphaltene fraction from petroleum by the addition of the low-boiling liquid hydrocarbons such as *n*-pentane or *n*-heptane under ambient conditions.

Deasphalting The removal asphalt (tacky, semisolid higher molecular weight) constituents from petroleum (as occurs in a refinery asphalt plant) by the addition of liquid propane or liquid butane under pressure; also the removal of the asphaltene fraction from petroleum by the addition of a low-boiling hydrocarbon liquid such as *n*-pentane or *n*-heptane.

Debutanization Distillation to separate butane and lighter components from higher boiling components.

Debutanizer A fractionating column used to remove butane and lighter components from liquid streams.

Decant oil The highest boiling product from a catalytic cracker; also referred to as slurry oil, clarified oil, or bottoms.

Decarbonizing A thermal conversion process designed to maximize coker gas-oil production and minimize coke and gasoline yields; operated at essentially lower temperatures and pressures than delayed coking (*q.v.*).

Decoking Removal of petroleum coke from equipment such as coking drums; hydraulic decoking uses high-velocity water streams.

Decolorizing Removal of suspended, colloidal, and dissolved impurities from liquid petroleum products by filtering, adsorption, chemical treatment, distillation, bleaching, etc.

De-ethanization Distillation to separate ethane and lighter components from propane and higher-boiling components; also called de-ethanation.

De-ethanizer A fractionating column designed to remove ethane and gases from heavier hydrocarbons.

Dehydrating agents Substances capable of removing water (drying, *q.v.*) or the elements of water from another substance.

Dehydrocyclization Any process by which both dehydrogenation and cyclization reactions occur.

Dehydrogenation The removal of hydrogen from a chemical compound; for example, the removal of two hydrogen atoms from butane to make butene(s) as well as the removal of additional hydrogen to produce butadiene.

Delayed coking A coking process in which the thermal reaction is allowed to proceed to completion to produce gaseous, liquid, and solid (coke) products.

Demethanization The process of distillation in which methane is separated from the higher boiling components; also called demethanation.

Demex process A solvent extraction demetallizing process that separates high metal vacuum residuum into demetallized oil of relatively low metal content and asphaltene of high metal content.

Density The mass (or weight) of a unit volume of any substance at a specified temperature; also the *heaviness* of crude oil, indicating the proportion of large, carbon-rich molecules, generally measured in kilograms per cubic meter (kg/m^3) or degrees on the American Petroleum Institute (API) gravity scale; in some countries, oil up to $900\,kg/m^3$ is considered light to medium crude. see also Specific gravity.

De-oiling Reduction in quantity of liquid oil entrained in solid wax by draining (sweating) or by a selective solvent; see MEK de-oiling.

Depentanizer A fractionating column for the removal of pentane and lighter fractions from a mixture of hydrocarbons.

Depropanization Distillation in which lighter components are separated from butanes and higher boiling material; also called depropanation.

Depropanizer A fractionating column for removing propane and lighter components from liquid streams.

Desalting Removal of mineral salts (mostly chlorides) from crude oils.

Desorption The reverse process of adsorption whereby adsorbed matter is removed from the adsorbent; also used as the reverse of absorption (*q.v.*).

Destructive distillation Thermal decomposition with the simultaneous removal of distillate; distillation (*q.v.*) when thermal decomposition of the constituents occurs.

Desulfurization The removal of sulfur or sulfur compounds from a feedstock; a process that removes sulfur and its compounds from various streams during the refining process; desulfurization processes include catalytic hydrotreating and other chemical/physical processes such as absorption; the desulfurization processes vary based on the type of stream treated (e.g., naphtha, distillate, heavy gas oil, etc.) and the amount of sulfur removed (e.g., sulfur reduction to 10 ppm).

Detergent oil A lubricating oil possessing special sludge-dispersing properties for use in internal combustion engines.

Devolatilized fuel Smokeless fuel; coke that has been reheated to remove all of the volatile material.

Dewaxing See Solvent dewaxing.

Diagenesis The concurrent and consecutive chemical reactions which commence the alteration of organic matter (at temperatures up to 50°C (120°F)) and ultimately result in the formation of petroleum from the marine sediment; see also Catagenesis and Metagenesis.

Diesel cycle A repeated succession of operations representing the idealized working behavior of the fluids in a diesel engine.

Diesel fuel Fuel used for internal combustion in diesel engines; usually that fraction which distills after kerosene.

Diesel hydrotreater A refinery process unit for production of clean (low-sulfur) diesel fuel.

$$\text{Diesel index} = (\text{aniline point } (°F) \times \text{API gravity}) \, 100$$

Diesel index An approximation of the cetane number (*q.v.*) of diesel fuel (*q.v.*) calculated from the density (*q.v.*) and aniline point (*q.v.*).

Diesel knock The result of a delayed period of ignition and the resultant accumulation of diesel fuel in the engine.

Diethanolamine (DEA) A solvent used in an acid gas removal system.

Diglycolamine (DGA) A solvent used in an acid gas removal system.

Dilbit Tar sand bitumen that has been reduced in viscosity through the addition of a diluent such as condensate or naphtha.

DilSynBit A blend of tar sand bitumen, condensate, and synthetic crude oil similar to medium sour crude.

Diluted crude Heavy crude oil to which a diluent (thinner) has been added to reduce viscosity and facilitate pipeline flow.

Dispersant A chemical or mixture of chemicals applied, for example, to an oil spill to disperse the oil phase into small droplets in the water phase.

Dispersion Suspension of oil droplets in water accomplished by natural wind and wave action, production of biological materials (biosurfactants) and/or chemical dispersant formulations.

Distillate The products of distillation formed by condensing vapors.

Distillate fuel oil A general classification for one of the petroleum fractions produced in conventional distillation operations. It includes diesel fuels and fuel oils. Products known as No. 1, No. 2, and No. 4 diesel fuel are used in on-highway diesel engines, such as those in trucks and automobiles, as well as off-highway engines, such as those in railroad locomotives and agricultural machinery. Products known as No. 1, No. 2, and No. 4 Fuel oils are used primarily for space heating and electric power generation.

Distillation A process for separating liquids with different boiling points without thermal decomposition of the constituents (see Destructive distillation).

Distillation curve See Distillation profile.

Distillation loss The difference, in a laboratory distillation, between the volume of liquid originally introduced into the distilling flask and the sum of the residue and the condensate recovered.

Distillation profile The distillation characteristics of petroleum or a petroleum product showing the temperature and the percent distilled.

Distillation range The difference between the temperature at the initial boiling point and at the end point, as obtained by the distillation test.

DNA Deoxyribonucleic acid, the substance within cells that carries the "recipe" for the organism and is inherited by offspring from parents.

DNA fingerprinting Cutting a DNA chromosome with restriction enzymes and separating the pieces by electrophoresis to generate a unique pattern, the "fingerprint" for each species, breed, hybrid, or individual, depending on which enzymes and probes are used.

Doctor solution A solution of sodium plumbite used to treat gasoline or other light petroleum distillates to remove mercaptan sulfur; see also Doctor test.

Doctor sweetening A process for sweetening gasoline, solvents, and kerosene by converting mercaptans to disulfides using sodium plumbite and sulfur.

Doctor test A test used for the detection of compounds in light petroleum distillates that react with sodium plumbite; see also Doctor solution.

Domestic heating oil See No. 2 Fuel Oil.

Donor solvent process A conversion process in which hydrogen donor solvent is used in place of or to augment hydrogen.

Downcomer A means of conveying liquid from one tray to the next below in a bubble tray column (*q.v.*).

Downflow reactor A reactor in which the feedstock flows in a downward direction over the catalyst bed.

Downstream A sector of the petroleum industry that refers to the refining of crude oil, and the products derived from crude oil.

Dropping point The temperature at which grease passes from a semisolid to a liquid state under prescribed conditions.

Dry gas A gas that does not contain fractions that may easily condense under normal atmospheric conditions.

Dry point The temperature at which the last drop of petroleum fluid evaporates in a distillation test.

Drying Removal of a solvent or water from a chemical substance; also referred to as the removal of solvent from a liquid or suspension.

Dualayer distillate process A process for removing mercaptans and oxygenated compounds from distillate fuel oils and similar products, using a combination of treatment with concentrated caustic solution and electrical precipitation of the impurities.

Dualayer gasoline process A process for extracting mercaptans and other objectionable acidic compounds from petroleum distillates; see also Dualayer solution.

Dualayer solution A solution that consists of concentrated potassium or sodium hydroxide containing a solubilizer; see also Dualayer gasoline process.

Dubbs cracking A continuous, liquid-phase thermal cracking process formerly used.

Ebullated bed A process in which the catalyst bed is in a suspended state in the reactor by means of a feedstock recirculation pump which pumps the feedstock upwards at sufficient speed to expand the catalyst bed at approximately 35% above the settled level.

Ecological risk assessment Process for analyzing and evaluating the possibility of adverse ecological effects caused by environmental pollutants.

Edeleanu process A process for refining oils at low temperature with liquid sulfur dioxide (SO_2), or with liquid sulfur dioxide and benzene; applicable to the recovery of aromatic concentrates from naphtha and heavier petroleum distillates.

Effective viscosity See Apparent viscosity.

Effluent Any contaminating substance, usually a liquid, that enters the environment via a domestic, industrial, agricultural, or sewage plant outlet.

Electric desalting A continuous process to remove inorganic salts and other impurities from crude oil by settling out in an electrostatic field.

Electrical precipitation A process using an electrical field to improve the separation of hydrocarbon reagent dispersions. May be used in chemical treating processes on a wide variety of refinery stocks.

Electrofining A process for contacting a light hydrocarbon stream with a treating agent (acid, caustic, doctor, etc.), then assisting the action of separation of the chemical phase from the hydrocarbon phase by an electrostatic field.

Electrolytic mercaptan process A process in which aqueous caustic solution is used to extract mercaptans from refinery streams.

Electrophoresis A laboratory technique for determining DNA fragment sizes by separating them in a gel placed in an electric field.

Electroporation Using an electric shock to transfer DNA into the cells of an organism; one of several procedures called transformation.

Electrostatic precipitators Devices used to trap fine dust particles (usually in the size range 30–60 microns) that operate on the principle of imparting an electric charge to particles in an incoming air stream and which are then collected on an oppositely charged plate across a high voltage field.

Emission control The use of gas cleaning processes to reduce emissions.

Emission standard The maximum amount of a specific pollutant permitted to be discharged from a particular source in a given environment.

Emulsification Formation of water droplets in an oil matrix (water-in-oil) or conversely oil droplets in a water matrix (oil-in-water) achieved by the action of agitation, such as wind and wave activity; can be unstable, separating into oil and water phases again soon after formation, or stable for months or years (e.g., "chocolate mousse," a water-in-oil emulsion).

Emulsion breaking The settling or aggregation of colloidal-sized emulsions from suspension in a liquid medium.

End-of-pipe emission control The use of specific emission control processes to clean gases after production of the gases.

Energy The capacity of a body or system to do work, measured in joules (SI units); also the output of fuel sources.

Energy from biomass The production of energy from biomass (*q.v.*).

Engler distillation A standard test for determining the volatility characteristics of a gasoline by measuring the per cent distilled at various specified temperatures.

Enhanced oil recovery Petroleum recovery following recovery by conventional (i.e., primary and/or secondary) methods; the third stage of production during which sophisticated techniques that alter the original properties of the oil are used. Enhanced oil recovery can begin after a secondary recovery process or at any time during the productive life of an oil reservoir—the purpose is not only to restore formation pressure, but also to improve oil displacement or fluid flow in the reservoir. The three major types of enhanced oil recovery operations are chemical flooding (alkaline flooding or micellar-polymer flooding), miscible displacement (carbon dioxide injection or hydrocarbon injection), and thermal recovery (steam flood). The optimal application of each method depends on reservoir temperature, pressure, depth, net pay, permeability, residual oil and water saturations, porosity, and fluid properties such as oil API gravity and viscosity.

Entrained bed A bed of solid particles suspended in a fluid (liquid or gas) at such a rate that some of the solid is carried over (entrained) by the fluid.

Environmental impact assessment The process of measuring or estimating the environmental effects of pollutants, such as oil spills, relative to conditions at a reference site or to a time prior to a spill.

ETBE (ethyl tertiary butyl ether, $(CH_3)_3COC_2H$)) An oxygenate blend stock formed by the catalytic etherification of isobutylene with ethanol.

Ethane (C_2H_6) A straight-chain saturated (paraffinic) hydrocarbon extracted predominantly from the natural gas stream, which is gaseous at standard temperature and pressure; a colorless gas that boils at a temperature of $-88°C$ ($-127°F$).

Ethanol See Ethyl alcohol.

Ether A generic term applied to a group of organic chemical compounds composed of carbon, hydrogen, and oxygen, characterized by an oxygen atom attached to two carbon atoms (e.g., methyl tertiary butyl ether).

Ethyl alcohol (ethanol or grain alcohol) An inflammable organic compound (C_2H_5OH) formed during fermentation of sugars; used as an intoxicant and as a fuel.

Ethylene (C_2H_4) An olefin hydrocarbon recovered from refinery or petrochemical processes, which is gaseous at standard temperature and pressure. Ethylene is used as a petrochemical feedstock for many chemical applications and the production of consumer goods.

ET-II process A thermal cracking process for the production of distillates and cracked residuum for use from metallurgical coke; *of this Earth* and not extraterrestrial!

Eureka process A thermal cracking process to produce a cracked oil and aromatic residuum from heavy residual materials.

Evaporation A process for concentrating nonvolatile solids in a solution by boiling off the liquid portion of the waste stream.

Evaporation The physical loss of low molecular weight components of an oil to the atmosphere by volatilization.

Expanding clays Clays that expand or swell on contact with water, e.g., montmorillonite.

Explosive limits The limits of percentage composition of mixtures of gases and air within which an explosion takes place when the mixture is ignited.

Extractive distillation The separation of different components of mixtures which have similar vapor pressures by flowing a relatively high-boiling solvent, which is selective for one of the components in the feed, down a distillation column as the distillation proceeds; the selective solvent scrubs the soluble component from the vapor.

Fabric filters Filters made from fabric materials and used for removing particulate matter from gas streams (see Baghouse).

Facies One or more layers of rock that differs from other layers in composition, age, or content; identifiable subdivisions of stratigraphic units.

Fat oil The bottom or enriched oil drawn from the absorber as opposed to lean oil.

Faujasite A naturally occurring silica-alumina (SiO_2–Al_2O_3) mineral.

FCC Fluid catalytic cracking.

FCCU Fluidized catalytic cracking unit.

Feedstock Petroleum as it is fed to the refinery; a refinery product that is used as the raw material for another process; the term is also generally applied to raw materials used in other refinery processes or industrial processes.

Ferrocyanide process A regenerative chemical treatment for mercaptan removal using caustic-sodium ferrocyanide reagent.

Filtration The use of an impassable barrier to collect solids but which allows liquids to pass.

Fines Typically, minute particles of solids such as clay or sand.

Fire point The lowest temperature at which, under specified conditions in standardized apparatus, a petroleum product vaporizes sufficiently rapidly to form above its surface an air-vapor mixture which burns continuously when ignited by a small flame.

Fischer-Tropsch process A process for synthesizing hydrocarbons and oxygenated chemicals from a mixture of hydrogen and carbon monoxide.

Fixed bed A stationary bed (of catalyst) to accomplish a process (see Fluid bed).

Flame ionization detector (FID) Used with analytical instruments like gas chromatographs to detect components of petroleum by combustion ionization, hence GC-FID.

Flammable A substance that will burn readily.

Flammability range The range of temperature over which a chemical is flammable.

Flammable liquid A liquid having a flash point below 37.8°C (100°F).

Flammable solid A solid that can ignite from friction or from heat remaining from its manufacture, or which may cause a serious hazard if ignited.

Flash point The lowest temperature to which the product must be heated under specified conditions to give off sufficient vapor to form a mixture with air that can be ignited momentarily by a flame.

Flexicoking A modification of the fluid coking process insofar as the process also includes a gasifier adjoining the burner/regenerator to convert excess coke to a clean fuel gas.

Floc point The temperature at which wax or solids separate as a definite floc.

Flue gas Gas from the combustion of fuel, the heating value of which has been substantially spent and which is, therefore, discarded to the flue or stack; gas that is emitted to the atmosphere via a flue, which is a pipe for transporting exhaust fumes.

Fluid catalytic cracking Cracking in the presence of a fluidized bed of catalyst.

Fluid coking A continuous fluidized solid process that cracks feed thermally over heated coke particles in a reactor vessel to gas, liquid products, and coke.

Fluid-bed A bed (of catalyst) that is agitated by an upward passing gas in such a manner that the particles of the bed simulate the movement of a fluid and has the characteristics associated with a true liquid; c.f. Fixed bed.

Fluidized catalytic cracking A refinery process used to convert the heavy portion of crude oil into lighter products, including LPG and gasoline.

Fly ash Particulate matter produced from mineral matter in coal that is converted during combustion to finely divided inorganic material and which emerges from the combustor in the gases.

Foots oil The oil sweated out of slack wax; named from the fact that the oil goes to the foot, or bottom, of the pan during the sweating operation. For liquids heavier than water:

$$\text{Sp gr } 60°F = 145/(145 - °B)$$

Fossil fuel resources A gaseous, liquid, or solid fuel material formed in the ground by chemical and physical changes (diagenesis, *q.v.*) in plant and animal residues over geological time; natural gas, petroleum, coal, and oil shale.

Fraction A group of hydrocarbons that have similar boiling points; a portion of crude oil defined by boiling range—naphtha, kerosene, gas oil, and residuum are fractions of crude oil.

Fractional composition The composition of petroleum as determined by fractionation (separation) methods.

Fractional distillation The separation of the components of a liquid mixture by vaporizing and collecting the fractions, or cuts, which condense in different temperature ranges; a common form of separation technology in hydrocarbon processing plants wherein a mixture (e.g., crude oil) is heated in a large, vertical cylindrical column to separate compounds (fractions) according to their boiling points

Fractionating column A column arranged to separate various fractions of petroleum by a single distillation and which may be tapped at different points along its length to separate various fractions in the order of their boiling points.

Fractionation The separation of petroleum into the constituent fractions using solvent or adsorbent methods; chemical agents such as sulfuric acid may also be used.

Frasch process A process formerly used for removing sulfur by distilling oil in the presence of copper oxide.

Free sulfur Sulfur that exists in the elemental state associated with petroleum; sulfur that is not bound organically within the petroleum constituents.

FTC process A heavy oil and residuum upgrading process in which the feedstock is thermally cracked to produce distillate and coke, which is gasified to fuel gas.

Fuel oil Also called heating oil is a distillate product that covers a wide range of properties; see also No. 1 to No. 4 Fuel oils.

Fuller's earth A clay which has high adsorptive capacity for removing color from oils; Attapulgus clay is a widely used fuller's earth.

Functional group The portion of a molecule that is characteristic of a family of compounds and determines the properties of these compounds.

Furfural extraction A single-solvent process in which furfural is used to remove aromatic components, naphthene components, olefins, and unstable hydrocarbons from a lubricating-oil charge stock.

Furnace oil A distillate fuel primarily intended for use in domestic heating equipment.

Gas cap A part of a hydrocarbon reservoir at the top that will produce only gas.

Gas chromatography (GC) An analytical method used to characterize petroleum components; GC is combined with different detection methods, hence GC-FID, GC-MS, etc.

Gas oil A petroleum distillate with a viscosity and boiling range between those of kerosene and lubricating oil; a middle-distillate petroleum fraction; usually includes diesel, kerosene, heating oil, and light fuel oil; a liquid petroleum distillate having a viscosity intermediate between that of kerosene and lubricating oil. It derives its name from having originally been used in the manufacture of illuminating gas. It is now used to produce distillate fuel oils and gasoline; heavy gas oil is a petroleum distillate with an approximate boiling range from 345 to 540°C (650–1000°F).

Gas reversion A combination of thermal cracking or reforming of naphtha with thermal polymerization or alkylation of hydrocarbon gases carried out in the same reaction zone.

Gaseous pollutants Gases released into the atmosphere that act as primary or secondary pollutants.

Gasification A process to partially oxidize any hydrocarbon, typically heavy residues, to a mixture of hydrogen and carbon monoxide; the process can be used to produce hydrogen and various energy by-products.

Gasohol A term for motor vehicle fuel comprising between 80% and 90% unleaded gasoline and 10%–20% ethanol (see also Ethyl alcohol).

Gas-oil ratio Ratio of the number of cubic feet of gas measured at atmospheric (standard) conditions to barrels of produced oil measured at stock tank conditions.

Gasoline blending components Naphtha fractions which will be used for blending or compounding into finished aviation or automotive gasoline (e.g., straight-run gasoline, alkylate, reformate, benzene, toluene, and xylene); excludes oxygenates (alcohols, ethers), butane, and pentanes plus.

Gasoline Fuel for the internal combustion engine that is commonly, but improperly, referred to simply as gas.

Gas-to-liquids (GTL) A process used to convert natural gas into longer-chain hydrocarbons such as diesel and jet fuel. Methane-rich gases are converted into liquid syngas (a mix of carbon monoxide and hydrogen), which is produced using steam methane reforming or autothermal reforming, followed by the Fischer-Tropsch process. Hydrocracking is then used to produce finished fuels.

Gene A functional unit of DNA, one "word" in the DNA recipe.

Genetic code The information contained in DNA molecules that scientists describe on the basis of a 4-letter alphabet (A, C, G, and T).

Genetic engineering The process of transferring DNA from one organism into another that results in a genetic modification; the production of a transgenic organism.

Genetic map The locations of specific genes along a chromosome marked with probes.

Genome The entire DNA "recipe" for an organism, found in every cell of that organism.

Gilsonite An asphaltite that is >90% bitumen.

Girbotol process A continuous, regenerative process to separate hydrogen sulfide, carbon dioxide, and other acid impurities from natural gas, refinery gas, etc., using mono-, di-, or triethanolamine as the reagent.

Glance pitch An asphaltite.

Glycol-amine gas treating A continuous, regenerative process to simultaneously dehydrate and remove acid gases from natural gas or refinery gas.

Grahamite An asphaltite.

Grain alcohol See Ethyl alcohol.

Gravity drainage The movement of oil in a reservoir that results from the force of gravity.

Gravity segregation Partial separation of fluids in a reservoir caused by the gravity force acting on differences in density.

Gray clay treating A fixed-bed (*q.v.*), usually fuller's earth (*q.v.*), vapor-phase treating process to selectively polymerize unsaturated gum-forming constituents (diolefins) in thermally cracked gasoline.

Greenhouse effect Warming of the earth due to entrapment of the energy of the Sun by the atmosphere.

Greenhouse gases Gases that contribute to the greenhouse effect (*q.v.*)

Gulf HDS process A fixed-bed process for the catalytic hydrocracking of heavy stocks to lower-boiling distillates with accompanying desulfurization.

Gulf Resid Hydrodesulfurization process A process for the desulfurization of heavy feedstocks to produce low-sulfur fuel oils or catalytic cracking (*q.v.*) feedstocks.

Gulfining A catalytic hydrogen treating process for cracked and straight-run distillates and fuel oils, to reduce sulfur content; improve carbon residue, color, and general stability; and effect a slight increase in gravity.

Gum An insoluble tacky semisolid material formed as a result of the storage instability and/or the thermal instability of petroleum and petroleum products.

Heat exchanger A device used to transfer heat from a fluid on one side of a barrier to a fluid on the other side without bringing the fluids into direct contact.

Heating oil See Fuel oil.

Heavy ends The highest boiling portion of a petroleum fraction; see also Light ends.

Heavy feedstock Any feedstock of the type heavy oil (*q.v.*), bitumen (*q.v.*), atmospheric residuum (*q.v.*), vacuum residuum (*q.v.*), and solvent deasphalter bottoms (*q.v.*).

Heavy fuel oil Fuel oil having a high density and viscosity; generally residual fuel oil such as No. 5 and No 6. fuel oil (*q.v.*).

Heavy gas oil A petroleum distillate with an approximate boiling range from 345 to 540°C (650–1000°F).

Heavy oil Petroleum having an API gravity of less than 20 degrees; other properties are necessary for inclusion in a more complete definition.

Heavy petroleum See Heavy oil.

Heavy Residue Gasification and Combined Cycle Power Generation A process for producing hydrogen from residua.

Heteroatom compounds Chemical compounds that contain nitrogen and/or oxygen and/or sulfur and /or metals bound within their molecular structure(s).

Heteroatom In petroleum, an atom such as nitrogen, sulfur and/or oxygen that is part of a hydrocarbon skeleton such as found in the resins fraction of crude oils.

HF alkylation An alkylation process whereby olefins (C_3, C_4, C_5) are combined with isobutane in the presence of hydrofluoric acid catalyst.

High-boiling distillates Fractions of petroleum that cannot be distilled at atmospheric pressure without decomposition, e.g., gas oils.

High molecular weight (HMW) A relative term referring to the molecular mass of chemicals; in oil, asphaltenes would be typical of HMW compounds.

High performance liquid chromatography (HPLC) An analytical method for separating chemicals in solution.

High-sulfur diesel (HSD) Diesel fuel containing more than 500 parts per million (ppm) sulfur.

High-sulfur petroleum A general expression for petroleum having more than 1% wt. sulfur; this is a very approximate definition and should not be construed as having a high degree of accuracy because it does not take into consideration the molecular locale of the sulfur. All else being equal, there is little difference between petroleum having 0.99% wt. sulfur and petroleum having 1.01% wt. sulfur.

History The study of the past events of a particular subject to learn from those events.

HOC process A version of the fluid catalytic cracking process (*q.v.*) that has been adapted to conversion of residua (*q.v.*) which contain high amounts of metal and asphaltenes (*q.v.*).

H-Oil process A catalytic process that is designed for hydrogenation of heavy feedstocks in an ebullated bed reactor.

Hortonsphere A spherical pressure-type tank used to store volatile liquids which prevents the excessive evaporation loss that occurs when such products are placed in conventional storage tanks.

Hot filtration test A test for the stability of a petroleum product.

HOT process A catalytic cracking process for upgrading heavy feedstocks using a fluidized bed of iron ore particles.

Hot spot An area of a vessel or line wall appreciably above normal operating temperature, usually as a result of the deterioration of an internal insulating liner which exposes the line or vessel shell to the temperature of its contents.

Houdresid catalytic cracking A continuous moving-bed process for catalytically cracking reduced crude oil to produce high octane gasoline and light distillate fuels.

Houdriflow catalytic cracking A continuous moving-bed catalytic cracking process employing an integrated single vessel for the reactor and regenerator kiln.

Houdriforming A continuous catalytic reforming process for producing aromatic concentrates and high-octane gasoline from low-octane straight-run naphtha.

Houdry butane dehydrogenation A catalytic process for dehydrogenating light hydrocarbons to their corresponding mono- or diolefins.

Houdry fixed-bed catalytic cracking A cyclic regenerable process for cracking of distillates.

Houdry hydrocracking A catalytic process combining cracking and desulfurization in the presence of hydrogen.

HSC process A cracking process for moderate conversion of heavy feedstocks (*q.v.*); the extent of the conversion is higher than visbreaking (*q.v.*) but lower than coking (*q.v.*).

Hybrid Gasification process A process to produce hydrogen by gasification of a slurry of coal and residual oil.

HYCAR process A non-catalytic process, is conducted under similar conditions to visbreaking (*q.v.*) and involves treatment with hydrogen under mild conditions; see also Hydrovisbreaking.

Hydraulic fracturing The opening of fractures in a reservoir by high-pressure, high-volume injection of liquids through an injection well.

Hydrocarbon A chemical that is composed of only carbon and hydrogen; chemicals containing heteroatoms, such as nitrogen, sulfur and/or oxygen, are not hydrocarbons, even though they may be petroleum constituents.

Hydrocarbon compounds Chemical compounds containing only carbon and hydrogen.

Hydrocarbon gas liquids (HGL) A group of hydrocarbons including ethane, propane, *n*-butane, isobutane, natural gasoline, and their associated olefins, including ethylene, propylene, butylene, and isobutylene; excludes liquefied natural gas (LNG).

Hydrocarbon gasification process A continuous, non-catalytic process in which hydrocarbons are gasified to produce hydrogen by air or oxygen.

Hydrocarbon resource Resources such as petroleum and natural gas that can produce naturally occurring hydrocarbons without the application of conversion processes.

Hydrocarbon-producing resource A resource such as coal and oil shale (kerogen) which produces derived hydrocarbons by the application of conversion processes; the hydrocarbons so produced are not naturally occurring materials.

Hydrocarbonoclastic bacteria Bacteria that utilize hydrocarbons and hydrocarbon degradation products almost exclusively as carbon and energy sources.

Hydroconversion A term often applied to hydrocracking (*q.v.*).

Hydrocracker A refinery process unit in which hydrocracking occurs.

Hydrocracking catalyst A catalyst used for hydrocracking which typically contains separate hydrogenation and cracking functions.

Hydrocracking A catalytic high-pressure high-temperature process for the conversion of petroleum feedstocks in the presence of fresh and recycled hydrogen; carbon-carbon bonds are cleaved in addition to the removal of heteroatomic species.

Hydrodemetallization The removal of metallic constituents by hydrotreating (*q.v.*).

Hydrodenitrogenation The removal of nitrogen by hydrotreating (*q.v.*).

Hydrodesulfurization A refining process that removes sulfur from liquid and gaseous hydrocarbons.

Hydrodesulfurization The removal of sulfur by hydrotreating (*q.v.*).

Hydrofining A fixed-bed catalytic process to desulfurize and hydrogenate a wide range of charge stocks from gases through waxes.

Hydroforming A process in which naphtha is passed over a catalyst at elevated temperatures and moderate pressures, in the presence of added hydrogen or hydrogen-containing gases, to form high-octane motor fuel or aromatics.

Hydrogen The lightest of all gases, occurring chiefly in combination with oxygen in water; exists also in acids, bases, alcohols, petroleum, and other hydrocarbons.

Hydrogen addition processes Upgrading processes in the presence of hydrogen, e.g., hydrocracking (*q.v.*); see Hydrogenation.

Hydrogen blistering Blistering of steel caused by trapped molecular hydrogen formed as atomic hydrogen during corrosion of steel by hydrogen sulfide.

Hydrogen sink A chemical structure within the feedstock that reacts with hydrogen with little, if any, effect on the product character.

Hydrogen sulfide (H_2S) A toxic, flammable and corrosive gas sometimes associated with petroleum.

Hydrogen transfer The transfer of inherent hydrogen within the feedstock constituents and products during processing.

Hydrogenation The chemical addition of hydrogen to a material. In nondestructive hydrogenation, hydrogen is added to a molecule only if, and where, unsaturation with respect to hydrogen exists; classed as *destructive* (*hydrocracking*) or *nondestructive* (*hydrotreating*).

Hydroprocessing A term often equally applied to hydrotreating (*q.v.*) and to hydrocracking (*q.v.*); also often collectively applied to both.

Hydropyrolysis A short residence time high-temperature process using hydrogen.

Hydrotreater A refinery process unit that removes sulfur and other contaminants from hydrocarbon streams.

Hydrotreating The removal of heteroatomic (nitrogen, oxygen, and sulfur) species by treatment of a feedstock or product at relatively low temperatures in the presence of hydrogen.

Hydrovisbreaking A non-catalytic process, conducted under similar conditions to visbreaking, which involves treatment with hydrogen to reduce the viscosity of the feedstock and produce more stable products than is possible with visbreaking.

Hyperforming A catalytic hydrogenation process for improving the octane number of naphtha through removal of sulfur and nitrogen compounds.

Hypochlorite sweetening The oxidation of mercaptans in a sour stock by agitation with aqueous, alkaline hypochlorite solution; used where avoidance of free-sulfur addition is desired, because of a stringent copper strip requirement and minimum expense is not the primary object.

Hypro process A continuous catalytic method for hydrogen manufacture from natural gas or from refinery effluent gases.

Hyvahl F process A process for hydroconverting heavy feedstocks to naphtha and middle distillates using a dual catalyst system and a fixed bed swing-reactor.

IFP Hydrocracking process A process that features a dual catalyst system in which the first catalyst is a promoted nickel-molybdenum amorphous catalyst to remove sulfur and nitrogen and hydrogenate aromatic rings. The second catalyst is a zeolite that finishes the hydrogenation and promotes the hydrocracking reaction.

Ignitability Characteristic of liquids whose vapors are likely to ignite in the presence of ignition source; also characteristic of non-liquids that may catch fire from friction or contact with water and that burn vigorously.

Illuminating oil Oil used for lighting purposes.

Immiscible carbon dioxide displacement Injection of carbon dioxide into an oil reservoir to effect oil displacement under conditions in which miscibility with reservoir oil is not obtained; see Carbon dioxide augmented waterflooding.

Immiscible Two or more fluids that do not have complete mutual solubility and coexist as separate phases.

In situ combustion Combustion of oil in the reservoir, sustained by continuous air injection, to displace unburned oil toward producing wells.

Incompatibility The *immiscibility* of petroleum products and also of different crude oils that is often reflected in the formation of a separate phase after mixing and/or storage.

Inhibitor A substance, the presence of which, in small amounts, in a petroleum product prevents or retards undesirable chemical changes from taking place in the product, or in the condition of the equipment in which the product is used.

Inhibitor sweetening A treating process to sweeten gasoline of low mercaptan content, using a phenylenediamine type of inhibitor, air, and caustic.

Initial boiling point The recorded temperature when the first drop of liquid falls from the end of the condenser.

Initial vapor pressure The vapor pressure of a liquid of a specified temperature and zero per cent evaporated.

Inspection Application of test procedures to a feedstock to determine its processability (*q.v.*); the analysis (*q.v.*) of a feedstock.

Instability The inability of a petroleum product to exist for periods of time without change to the product.

Iodine number A measure of the iodine absorption by an oil under standard conditions; used to indicate the quantity of unsaturated compounds present; also called iodine value.

Ion exchange A means of removing cations or anions from solution onto a solid resin.

Isobutane (C_4H_{10}) A branch-chain saturated (paraffinic) hydrocarbon extracted from both natural gas and refinery gas streams, which is gaseous at standard temperature and pressure.

Isobutylene (C_4H_8) A branch-chain olefin hydrocarbon recovered from refinery or petrochemical processes, which is gaseous at standard temperature and pressure. Isobutylene is used in the production of gasoline and various petrochemical products.

Isocracking process A hydrocracking process for conversion of hydrocarbons which operates at relatively low temperatures and pressures in the presence of hydrogen and a catalyst to produce more valuable, lower-boiling products.

Isodewaxing A catalytic isomerization of wax to improve base lube oil pour point.

Isofining A mild residue hydrocracking for synthetic crude oils.

Isofinishing A process for hydrofinishing of base lube oils to improve oxygen stability and color.

Isoforming A process in which olefin-type naphtha is contacted with an alumina catalyst at high temperature and low pressure to produce isomers of higher octane number.

Isohexane (C_6H_{14}) A saturated branch-chain hydrocarbon.

Iso-Kel process A fixed-bed, vapor-phase isomerization process using a precious metal catalyst and external hydrogen.

Isomate process A continuous, non-regenerative process for isomerizing C_5–C_8 normal paraffinic hydrocarbons, using aluminum chloride-hydrocarbon catalyst with anhydrous hydrochloric acid as a promoter.

Isomerate process A fixed-bed isomerization process to convert pentane, heptane, and heptane to high-octane blending stocks.

Isomerization The conversion of a *normal* (straight-chain) paraffin hydrocarbon into an *iso* (branched-chain) paraffin hydrocarbon having the same atomic composition; isomerization: a refining process that alters the fundamental arrangement of atoms in the molecule without adding or removing anything from the original material; used to convert normal butane into isobutane (C_4), an alkylation process feedstock, and normal pentane and hexane into isopentane (C_5) and isohexane (C_6), high-octane gasoline components.

Isomers Chemicals that have the same molecular formula (i.e., elemental composition) but different structures; may also have different properties, including water solubility, biodegradability, and toxicity.

Isopentane A saturated branched-chain hydrocarbon (C_5H_{12}) obtained by fractionation of natural gasoline or isomerization of normal pentane.

Iso-plus Houdriforming A combination process using a conventional Houdriformer operated at moderate severity, in conjunction with one of three possible alternatives—including the use of an aromatic recovery unit or a thermal reformer; see Houdriforming.

Jet fuel Fuel meeting the required properties for use in jet engines and aircraft turbine engines.

Kaolinite A clay mineral formed by hydrothermal activity at the time of rock formation or by chemical weathering of rocks with high feldspar content; usually associated with intrusive granite rocks with high feldspar content.

Kata-condensed aromatic compounds Compounds based on linear condensed aromatic hydrocarbon systems, e.g., anthracene and naphthacene (tetracene).

Kerogen A complex carbonaceous (organic) material that occurs in sedimentary rocks and shale formations; generally insoluble in common organic solvents.

Kerosene (kerosine) A fraction of petroleum that was initially sought as an illuminant in lamps; a precursor to diesel fuel; a light petroleum distillate that is used in space heaters, cook stoves, and water heaters and is suitable for use as a light source when burned in wick-fed lamps. Kerosene has a maximum distillation temperature of 400 degrees Fahrenheit at the 10-percent recovery point, a final boiling point of 572 degrees Fahrenheit, and a minimum flash point of 100 degrees Fahrenheit. Included are No. 1-K and No. 2-K, the two grades recognized by ASTM Specification D 3699, as well as all other grades of kerosene called range or stove oil, which have properties similar to those of No. 1 fuel oil.

Kerosene-type jet fuel A kerosene-based product having a maximum distillation temperature of 400 degrees Fahrenheit at the 10-percent recovery point and a final maximum boiling point of 572 degrees Fahrenheit and meeting ASTM Specification D 1655 and Military Specifications MIL-T-5624P and MIL-T-83133D (Grades JP-5 and JP-8). It is used for commercial and military turbojet and turboprop aircraft engines.

K-factor See Characterization factor.

Kinematic viscosity The ratio of viscosity (*q.v.*) to density, both measured at the same temperature.

Knock The noise associated with self-ignition of a portion of the fuel-air mixture ahead of the advancing flames front.

Kow The partition coefficient describing the equilibrium concentration ratio of a dissolved chemical in octanol versus in water, in a two-phase system at a specific temperature; used in prediction of toxicity.

Lamp burning A test of burning oils in which the oil is burned in a standard lamp under specified conditions in order to observe the steadiness of the flame, the degree of encrustation of the wick, and the rate of consumption of the kerosene.

Lamp oil See Kerosene.

LC-Fining process A hydrogenation (hydrocracking) process capable of desulfurizing, demetallizing, and upgrading heavy feedstocks by means of an expanded bed reactor.

Leaded gasoline Gasoline containing tetraethyl lead or other organometallic lead antiknock compounds.

Lean gas The residual gas from the absorber after the condensable gasoline has been removed from the wet gas.

Lean oil Absorption oil from which gasoline fractions have been removed; oil leaving the stripper in a natural-gasoline plant.

LEDA (Low Energy Deasphalting) process A process for extracting high quality catalytic cracking feeds from heavy feedstocks; the process uses a low-boiling hydrocarbon solvent specifically formulated to insure the most economical deasphalting (*q.v.*) design for each operation.

Lewis acid A chemical species that can accept an electron pair from a base.

Lewis base A chemical species that can donate an electron pair.

Light crude oil Crude oil with a high proportion of light hydrocarbon fractions and low metallic compounds; sometime defined as crude oil of gravity of 28 degrees API or higher; a high-quality light crude oil might have a gravity of approaching 40 degrees API, such as 32–34 for Light Arabian crude oil (32–34 degrees API) and West Texas Intermediate crude oil 37–40 degrees API.

Light ends The lower-boiling components of a mixture of hydrocarbons; see also Heavy ends, Light hydrocarbons.

Light gas oil Liquid petroleum distillates boiling higher than naphtha.

Light hydrocarbons Hydrocarbons with molecular weights less than that of heptane (C_7H_{16}).

Light oil The products distilled or processed from crude oil up to, but not including, the first lubricating-oil distillate.

Light petroleum Petroleum having an API gravity greater than 20 degrees.

Ligroine (Ligroin) A saturated petroleum naphtha boiling in the range of 20–135°C (68–275°F) and suitable for general use as a solvent; also called benzine or petroleum ether.

Linde copper sweetening A process for treating gasoline and distillates with a slurry of clay and cupric chloride.

Liquefied natural gas (LNG) Natural gas cooled to a liquid state.

Liquefied petroleum gas Propane, butane, or mixtures thereof, gaseous at atmospheric temperature and pressure, held in the liquid state by pressure to facilitate storage, transport, and handling.

Liquefied refinery gases (LRG) Hydrocarbon gas liquids produced in refineries from processing of crude oil and unfinished oils. They are retained in the liquid state through pressurization and/or refrigeration; includes ethane, propane, n-butane, isobutane, and refinery olefins (ethylene, propylene, butylene, and isobutylene).

Liquid fuels Products of petroleum refining, natural gas liquids, biofuels, and liquids derived from other sources (including coal-to-liquids and gas-to-liquids); liquefied natural gas (LNG) and liquid hydrogen are not included.

Liquid petrolatum See White oil.

Liquid sulfur dioxide-benzene process A mixed-solvent process for treating lubricating-oil stocks to improve viscosity indexes; also used for dewaxing.

Lithology The geological characteristics of the reservoir rock.

Live steam Steam coming directly from a boiler before being utilized for power or heat.

Liver The intermediate layer of dark-colored, oily material, insoluble in weak acid and in oil, which is formed when acid sludge is hydrolyzed.

Low molecular weight Relative terms referring to the molecular mass of chemicals; in oil, monoaromatics and aliphatics up to C10 would be typical of these compounds. Mass spectrometry: an analytical method used for detailed characterization of petroleum components, often in combination with GC, hence GC-MS.

Low-boiling distillates Fractions of petroleum that can be distilled at atmospheric pressure without decomposition.

Low-sulfur petroleum A general expression for petroleum having less than 1% wt. sulfur; this is a very approximate definition and should not be construed as having a high degree of accuracy because it does not take into consideration the molecular locale of the sulfur. All else being equal, there is little difference between petroleum having 0.99% wt. sulfur and petroleum having 1.01% wt. sulfur.

Lube cut A fraction of crude oil of suitable boiling range and viscosity to yield lubricating oil when completely refined; also referred to as lube oil distillates or lube stock.

Lube See Lubricating oil.

Lubricants Substances used to reduce friction between bearing surfaces, or incorporated into other materials used as processing aids in the manufacture of other products, or used as carriers of other materials. Petroleum lubricants may be produced either from distillates or residues; includes all grades of lubricating oils, from spindle oil to cylinder oil to those used in grease.

Lubricating oil A fluid lubricant used to reduce friction between bearing surfaces.

Mahogany acids Oil-soluble sulfonic acids formed by the action of sulfuric acid on petroleum distillates. They may be converted to their sodium soaps (mahogany soaps) and extracted from the oil with alcohol for use in the manufacture of soluble oils, rust preventives, and special greases. The calcium and barium soaps of these acids are used as detergent additives in motor oils; see also Brown acids and Sulfonic acids.

Maltenes That fraction of petroleum that is soluble in, for example, pentane or heptane; deasphaltened oil (*q.v.*); also the term arbitrarily assigned to the pentane-soluble portion of petroleum that has relatively high boiling (>300°C, 760 mm) (see also Petrolenes).

Marine engine oil Oil used as a crankcase oil in marine engines.

Marine gasoline Fuel for motors in marine service.

Marine sediment The organic biomass from which petroleum is derived.

Marsh An area of spongy waterlogged ground with large numbers of surface water pools. Marshes usually result from: (1) an impermeable underlying bedrock; (2) surface deposits of glacial boulder clay; (3) a basin-like topography from which natural drainage is poor; (4) very heavy rainfall in conjunction with a correspondingly low evaporation rate; (5) low-lying land, particularly at estuarine sites at or below sea level.

Mayonnaise Low-temperature sludge; a black, brown, or gray deposit having a soft, mayonnaise-like consistency; not recommended as a food additive!

MDS process A solvent deasphalting process that is particularly effective for upgrading heavy crude oils.

Medicinal oil Highly refined, colorless, tasteless, and odorless petroleum oil used as a medicine in the nature of an internal lubricant; sometimes called liquid paraffin.

Medium crude oil Crude oil with gravity between (approximately) 20 and 28 degrees API.

Medium molecular weight Relative terms referring to the molecular mass of chemicals; in oil, 3 to 6-ringed PAH and aliphatics up to C20 would be typical of MMW compounds.

MEK de-oiling A wax-de-oiling process in which the solvent is generally a mixture of methyl ethyl ketone and toluene.

MEK dewaxing A continuous solvent dewaxing process in which the solvent is generally a mixture of methyl ethyl ketone and toluene.

MEK(methyl ethyl ketone) A colorless liquid ($CH_3COCH_2CH_3$) used as a solvent; as a chemical intermediate; and in the manufacture of lacquers, celluloid, and varnish removers.

MEOR Microbial enhanced oil recovery.

Mercapsol process A regenerative process for extracting mercaptans, utilizing aqueous sodium (or potassium) hydroxide containing mixed cresols as solubility promoters.

Mercaptans Odiferous organic sulfur compounds with the general formula (RSH).

Mercaptans Organic compounds having the general formula R-SH.

Merox process A refinery process used to remove or convert mercaptans.

Metagenesis The alteration of organic matter during the formation of petroleum that may involve temperatures above 200°C (390°F); see also Catagenesis and Diagenesis.

Methanol (methyl alcohol, CH_3OH) A low-boiling alcohol eligible for gasoline blending.

Methanol See Methyl alcohol.

Methyl alcohol (methanol; wood alcohol) A colorless, volatile, inflammable, and poisonous alcohol (CH_3OH) traditionally formed by *destructive distillation* (*q.v.*) of wood or, more recently, as a result of synthetic distillation in chemical plants.

Methyl ethyl ketone See MEK.

Methyl t-butyl ether An ether added to gasoline to improve its octane rating and to decrease gaseous emissions; see Oxygenate.

Methyldiethanolamine (MDEA) A solvent used in an acid gas removal system.

Mica A complex aluminum silicate mineral that is transparent, tough, flexible, and elastic.

Micelle The structural entity by which asphaltenes are dispersed in petroleum.

Microcarbon residue The carbon residue determined using a thermogravimetric method. See also Carbon residue.

Microcat-RC process (M-Coke process) A catalytic hydroconversion process operating at relatively moderate pressures and temperatures using catalyst particles, containing a metal sulfide in a carbonaceous matrix formed within the process, that are uniformly dispersed throughout the feed; because of their ultra-small size (10^{-4} inch diameter) there are typically several orders of magnitude more of these microcatalyst particles per cubic centimeter of oil than is possible in other types of hydroconversion reactors using conventional catalyst particles.

Microcrystalline wax Wax extracted from certain petroleum residua and having a finer and less apparent crystalline structure than paraffin wax.

Microemulsion, or micellar/emulsion, flooding An augmented waterflooding technique in which a surfactant system is injected in order to enhance oil displacement toward producing wells.

Mid-boiling point The temperature at which approximately 50% of a material has distilled under specific conditions.

Middle distillate Distillate boiling between the kerosene and lubricating oil fractions; a general classification of refined petroleum products that includes distillate fuel oil and kerosene.

Migration (primary) The movement of hydrocarbons (oil and natural gas) from mature, organic-rich source rocks to a point where the oil and gas can collect as droplets or as a continuous phase of liquid hydrocarbon.

Migration (secondary) The movement of the hydrocarbons as a single, continuous fluid phase through water-saturated rocks, fractures, or faults followed by accumulation of the oil and gas in sediments (traps, *q.v.*) from which further migration is prevented.

Mineral oil The older term for petroleum; the term was introduced in the 19th century as a means of differentiating petroleum (rock oil) from whale oil which, at the time, was the predominant illuminant for oil lamps.

Mineral seal oil A distillate fraction boiling between kerosene and gas oil.

Mineral wax Yellow to dark brown solid substances that occur naturally and are composed largely of paraffins; usually found associated with considerable mineral matter, as a filling in veins and fissures or as an interstitial material in porous rocks.

Mineralization Complete oxidation of a compound (e.g., hydrocarbon) to carbon dioxide and water; may be accomplished by a single species of organism or by a community of microbes.

Minerals Naturally occurring inorganic solids with well-defined crystalline structures.

Miscible fluid displacement (miscible displacement) Is an oil displacement process in which an alcohol, a refined hydrocarbon, a condensed petroleum gas, carbon dioxide, liquefied natural gas, or even exhaust gas is injected into an oil reservoir, at pressure levels such that the injected gas or fluid and reservoir oil are miscible; the process may include the concurrent, alternating, or subsequent injection of water.

Mitigation Identification, evaluation, and cessation of potential impacts of a process product or by-product.

Mixed-phase cracking The thermal decomposition of higher-boiling hydrocarbons to gasoline components.

Mode of action (MoA) Describes a functional or anatomical change at the cellular level, resulting from the exposure of a living organism to a substance.

Modified naphtha insolubles (MNI) An insoluble fraction obtained by adding naphtha to petroleum; usually the naphtha is modified by adding paraffinic constituents; the fraction might be equated to asphaltenes *if* the naphtha is equivalent to *n*-heptane, but usually it is not

Molecular sieve A synthetic zeolite mineral having pores of uniform size; it is capable of separating molecules, on the basis of their size, structure, or both, by absorption or sieving.

Molecular weight The mass of one molecule.

Monitored natural attenuation A remediation strategy in which there is no intervention but the site is monitored using various parameters.

Monoaromatics Aromatic hydrocarbons having only a single benzene ring; may also have one or more alkyl side chains.

Mono-ethanolamine (monoethanolamine, MEA) A solvent used in an acid gas removal system.

Motor gasoline (finished) A complex mixture of relatively volatile hydrocarbons with or without small quantities of additives, blended to form a fuel suitable for use in spark-ignition engines. Motor gasoline, as defined in ASTM Specification D 4814 or Federal Specification VV-G-1690C, is characterized as having a boiling range of 122 to 158°F at the 10% v/v recovery point to 365 to 374°F at the 90% v/v recovery point; includes conventional gasoline, all types of oxygenated gasoline, including gasohol, and reformulated gasoline, but excludes aviation gasoline. The volumetric data on blending components, such as oxygenates, are not counted in data on finished motor gasoline until the blending components are blended into the gasoline.

Motor gasoline blending components Naphtha fractions (e.g., straight-run gasoline, alkylate, reformate, benzene, toluene, xylene) used for blending or compounding into finished motor gasoline. These components include reformulated gasoline blend stock for oxygenate blending (RBOB) but exclude oxygenates (alcohols, ethers), butane, and pentanes plus. Note: Oxygenates are reported as individual components and are included in the total for other hydrocarbons, hydrogens, and oxygenates.

Motor gasoline blending Mechanical mixing of motor gasoline blending components, and oxygenates when required, to produce finished motor gasoline. Finished motor gasoline may be further mixed with other motor gasoline blending components or oxygenates, resulting in increased volumes of finished motor gasoline and/or changes in the formulation of finished motor gasoline (e.g., conventional motor gasoline mixed with MTBE to produce oxygenated motor gasoline).

Motor Octane Method A test for determining the knock rating of fuels for use in spark-ignition engines; see also Research Octane Method.

Moving-bed catalytic cracking A cracking process in which the catalyst is continuously cycled between the reactor and the regenerator.

MRH process A hydrocracking process to upgrade heavy feedstocks containing large amount of metals and asphaltene, such as vacuum residua and bitumen, and to produce mainly middle distillates using a reactor designed to maintain a mixed three-phase slurry of feedstock, fine powder catalyst and hydrogen, and to promote effective contact.

MSCC process A short-residence time process (millisecond catalytic cracking) in which the catalyst is placed in a more optimal position to insure better contact with the feedstock.

MTBE [methyl tertiary butyl ether, $(CH_3)_3COCH_3$] An ether intended for gasoline blending; see Methyl t-butyl ether, Oxygenates.

Mutation A change of one of the "letters" in the DNA "recipe" caused by chemicals, ultraviolet light, X-rays, or natural processes.

Naft Pre-Christian era (Greek) term for naphtha (*q.v.*).

Napalm A thickened gasoline used as an incendiary medium that adheres to the surface it strikes.

Naphtha A generic term applied to refined, partly refined, or unrefined petroleum products and liquid products of natural gas, the majority of which distills below 240°C (464°F); the volatile fraction of petroleum which is used as a solvent or as a precursor to gasoline.

Naphtha-type jet fuel A fuel in the heavy naphtha boiling range having an average gravity of 52.8 degrees API, 20%–90% distillation temperatures of 290 to 470°F, and meeting military specification MIL-T-5624L (Grade JP-4); primarily used for military turbojet and turboprop aircraft engines because it has a lower freeze point than other aviation fuels and meets engine requirements at high altitudes and speeds.

Naphthenes Cycloparaffins; one of three basic hydrocarbon classifications found naturally in crude oil; used widely as petrochemical feedstock.

Naphthenic acids A class of polar petroleum compounds that contributes to aquatic toxicity and to petroleum infrastructure corrosion by contributing to TAN.

Native asphalt See Bitumen.

Natural asphalt See Bitumen.

Natural attenuation Remediation of a contaminated site by natural processes alone, without human intervention; also see monitored natural attenuation.

Natural gas liquids (NGL) The hydrocarbon liquids that condense during the processing of hydrocarbon gases that are produced from oil or gas reservoir; see also natural gasoline.

Natural gas The naturally occurring gaseous constituents that are found in many petroleum reservoirs; also there are reservoirs in which natural gas may be the sole occupant.

Natural gasoline plant A plant for the extraction of fluid hydrocarbon, such as gasoline and liquefied petroleum gas, from natural gas.

Natural gasoline A mixture of liquid hydrocarbons extracted from natural gas (*q.v.*) suitable for blending with refinery gasoline.

Neutral oil A distillate lubricating oil with viscosity usually not above 200 s at 100°F.

Neutralization number The weight, in milligrams, of potassium hydroxide needed to neutralize the acid in 1 g of oil; an indication of the acidity of an oil.

Neutralization A process for reducing the acidity or alkalinity of a waste stream by mixing acids and bases to produce a neutral solution; also known as pH adjustment.

No. 1 Fuel oil Very similar to kerosene (*q.v.*) and is used in burners where vaporization before burning is usually required and a clean flame is specified.

No. 2 diesel fuel A distillate fuel oil that has a distillation temperature of 640 degrees Fahrenheit at the 90-percent recovery point and meets the specifications defined in ASTM Specification D 975. It is used in high-speed diesel engines that are generally operated under uniform speed and load conditions, such as those in railroad locomotives, trucks, and automobiles.

No. 2 Fuel oil Also called domestic heating oil; has properties similar to diesel fuel and heavy jet fuel; used in burners where complete vaporization is not required before burning.

No. 4 Fuel oil A light industrial heating oil used where preheating is not required for handling or burning; there are two grades of No. 4 Fuel oil, differing in safety (flash point) and flow (viscosity) properties.

No. 5 Fuel oil A heavy industrial fuel oil that requires preheating before burning.

No. 6 Fuel oil A heavy fuel oil and is more commonly known as Bunker C oil when it is used to fuel ocean-going vessels; preheating is always required for burning this oil.

Non-asphaltic road oil Any of the non-hardening petroleum distillates or residual oils used as dust layers. They have sufficiently low viscosity to be applied without heating and, together with asphaltic road oils (*q.v.*), are sometimes referred to as dust palliatives.

Non-Newtonian A fluid that exhibits a change of viscosity with flow rate.

Obligate hydrocarbonoclastic bacteria Bacteria that have adapted to use hydrocarbons as their sole source of carbon and energy for growth and metabolism.

OCR Counter-current moving bed technology with on-stream catalyst replacement.

Octane barrel yield A measure used to evaluate fluid catalytic cracking processes; defined as (RON + MON)/2 times the gasoline yield, where RON is the research octane number and MON is the motor octane number.

Octane number A number indicating the antiknock characteristics of gasoline.

Octane rating A number used to indicate gasoline's antiknock performance in motor vehicle engines. The two recognized laboratory engine test methods for determining the antiknock rating, i.e., octane rating, of gasoline are the Research method and the Motor method. To provide a single number as guidance to the consumer, the antiknock index (R + M)/2, which is the average of the Research and Motor octane numbers, was developed. See Octane Number.

Oil bank See Bank.

Oil originally in place (OOIP) The quantity of petroleum existing in a reservoir before oil recovery operations begin.

Oil sand See Tar sand.

Oil-mineral aggregate (OMA) Floc containing oil adhering to mineral particle, which may float, sink, or re-suspend in a water column; a process initially referred to as clay-oil flocculation.

Oil-particle aggregate (OPA) More general term than OMA, describing oil adhering to particles that may be inorganic (minerals) and/or organic, including microbial cells.

Oils That portion of the maltenes (*q.v.*) that is not adsorbed by a surface-active material such as clay or alumina.

Oil shale A fine-grained impervious sedimentary rock that contains an organic material called kerogen; the term *oil shale* describes the rock in lithological terms but also refers to the ability of the rock to yield oil upon heating which causes the kerogen to decompose; also called black shale, bituminous shale, carbonaceous shale, coaly shale, kerosene shale, coorongite, maharahu, kukersite, kerogen shale, and algal shale.

Olefins A class of unsaturated double bond linear hydrocarbons recovered from petroleum; examples include ethylene, propylene, and butene. Olefins are used to produce a variety of products, including plastics, fibers, and rubber.

OOIP See Oil originally in place.

Organic sedimentary rocks Rocks containing organic material such as residues of plant and animal remains/decay.

Overhead That portion of the feedstock that is vaporized and removed during distillation.

Oxidation A process that can be used for the treatment of a variety of inorganic and organic substances.

Oxidized asphalt See Air-blown asphalt.

Oxygenate An oxygen-containing compound that is blended into gasoline to improve its octane number and to decrease gaseous emissions.

Oxygenated gasoline Finished motor gasoline, other than reformulated gasoline, having an oxygen content of 2.7% w/w or higher and required by the US Environmental Protection Agency (EPA) to be sold in areas designated by EPA as carbon monoxide (CO) nonattainment areas.

Oxygenates Substances which, when added to gasoline, increase the amount of oxygen in that gasoline blend. Fuel ethanol, methyl tertiary butyl ether (MTBE), ethyl tertiary butyl ether (ETBE), and methanol are common oxygenates.

Ozokerite (Ozocerite) A naturally occurring wax; when refined also known as ceresin.

PADD (petroleum administration for defense districts) Geographic aggregations of the 50 states and the District of Columbia into five districts by the Petroleum Administration for Defense in 1950. These districts were originally defined during World War II for purposes of administering oil allocation.

Pale oil Lubricating oil or process oil refined until its color, by transmitted light, is straw to pale yellow.

Paraffin wax The colorless, translucent, highly crystalline material obtained from the light lubricating fractions of paraffinic crude oils (wax distillates).

Paraffins A group of generally saturated single bond linear hydrocarbons; also called alkanes.

Paraffinum liquidum See Liquid petrolatum.

Partial oxidation process (Texaco gasification process) A partial oxidation gasification process for generating synthetic gas, principally hydrogen and carbon monoxide.

Particle density The density of solid particles.

Particle gun A gun that "shoots" DNA into the cells of an organism; the most versatile of a series of procedures called transformation.

Particle size distribution The particle size distribution (of a catalyst sample) is expressed as a percent of the whole.

Particulate matter Particles in the atmosphere or on a gas stream that may be organic or inorganic and originate from a wide variety of sources and processes.

Partitioning The diffusion of compounds between two immiscible liquid phases, including water and oil droplets and water and lipid membranes.

PCR Polymerase chain reaction, which rapidly duplicates specific DNA molecules in response to temperature changes in a computer-controlled heater.

Penex process A continuous, non-regenerative process for isomerization of C_5 and/or C_6 fractions in the presence of hydrogen (from reforming) and a platinum catalyst.

Pentafining A pentane isomerization process using a regenerable platinum catalyst on a silica-alumina support and requiring outside hydrogen.

Pentanes plus A mixture of liquid hydrocarbons, mostly pentanes and heavier, extracted from natural gas in a gas processing plant. Pentanes plus is equivalent to natural gasoline.

Pepper sludge The fine particles of sludge produced in acid treating which may remain in suspension.

Peri-condensed aromatic compounds Compounds based on angular condensed aromatic hydrocarbon systems, e.g., phenanthrene, chrysene, picene, etc.

Permeability The ease of flow of the water through the rock.

Petrochemical An intermediate chemical derived from petroleum, hydrocarbon liquids, or natural gas.

Petrol A term commonly used in some countries for gasoline.

Petrolatum A semisolid product, ranging from white to yellow in color, produced during refining of residual stocks; see Petroleum jelly.

Petrolenes The term applied to that part of the pentane-soluble or heptane-soluble material that is low boiling (<300°C, <570°F, 760 mm) and can be distilled without thermal decomposition (see also Maltenes).

Petroleum Synonymous with crude oil; a naturally occurring complex mixture of thousands of different hydrocarbon and non-hydrocarbon molecules.

Petroleum (crude oil) A naturally occurring mixture of gaseous, liquid, and solid hydrocarbon compounds usually found trapped deep underground beneath impermeable cap rock and above a lower dome of sedimentary rock such as shale; most petroleum reservoirs occur in sedimentary rocks of marine, deltaic, or estuarine origin.

Petroleum asphalt See Asphalt.

Petroleum coke A solid carbon fuel derived from oil refinery cracking processes such as delayed coking; also called pet coke.

Petroleum ether See Ligroine.

Petroleum jelly A translucent, yellowish to amber or white, hydrocarbon substance (m.p. 38–54°C) having almost no odor or taste, derived from petroleum and used principally in medicine and pharmacy as a protective dressing and as a substitute for fats in ointments and cosmetics; also used in many types of polishes and in lubricating greases, rust preventives, and modeling clay; obtained by dewaxing heavy lubricating-oil stocks.

Petroleum products Products obtained from the processing of crude oil (including lease condensate), natural gas, and other hydrocarbon compounds. Petroleum products include unfinished oils, liquefied petroleum gases, pentanes plus, aviation gasoline, motor gasoline, naphtha-type jet fuel, kerosene-type jet fuel, kerosene, distillate fuel oil, residual fuel oil, petrochemical feedstocks, special naphtha, lubricants, waxes, petroleum coke, asphalt, road oil, still gas, and miscellaneous products.

Petroleum refinery See Refinery.

Petroleum refining A complex sequence of events that results in the production of a variety of products.

Petroporphyrins See Porphyrins.

pH adjustment Neutralization.

Phase separation The formation of a separate phase that is usually the prelude to coke formation during a thermal process; the formation of a separate phase as a result of the instability/incompatibility of petroleum and petroleum products.

Phosphoric acid polymerization A process using a phosphoric acid catalyst to convert propene, butene, or both, to gasoline or petrochemical polymers.

Photo-oxidation Oxidation due to the influence of photic energy, usually from UV light. Photo-enhanced toxicity: increased toxicity due to photo-oxidation in vivo.

Physical composition See Bulk composition.

Pipe still (pipestill) A still in which heat is applied to the oil while being pumped through a coil or pipe arranged in a suitable firebox; the distillation tower in a refinery.

Pipestill gas The most volatile fraction that contains most of the gases that are generally dissolved in the crude. Also known as pipestill light ends.

Pitch The nonvolatile, brown to black, semisolid to solid viscous product from the *destructive distillation* (*q.v.*) of many bituminous or other organic materials, especially coal; has also been incorrectly applied to residua from petroleum processes where thermal decomposition may *not* have occurred.

Plasmid A small, circular DNA that is used to transfer genes from one organism into another.

Platforming A reforming process using a platinum-containing catalyst on an alumina base.

PNA A polynuclear aromatic compound (*q.v.*).

Polar aromatics Resins; the constituents of petroleum that are predominantly aromatic in character and contain polar (nitrogen, oxygen, and sulfur) functions in their molecular structure(s).

Pollution The introduction into the land water and air systems of a chemical or chemicals that are not indigenous to these systems or the introduction into the land water and air systems of indigenous chemicals in greater-than-natural amounts.

Polycyclic aromatic hydrocarbons (PAHs) a subclass of aromatic hydrocarbons having two or more fused benzene rings; may also have one or more alkyl side chains, generating large suites of isomers; some are considered "priority pollutants" because of their toxicity and/or potential carcinogenicity.

Polyforming A process charging both C_3 and C_4 gases with naphtha or gas oil under thermal conditions to produce gasoline.

Polymer augmented waterflooding Waterflooding in which organic polymers are injected with the water to improve aerial and vertical sweep efficiency.

Polymer gasoline The product of polymerization of gaseous hydrocarbons to hydrocarbons boiling in the gasoline range.

Polymerization The combination of two olefin molecules to form a higher molecular weight paraffin.

Polynuclear aromatic compound An aromatic compound having two or more fused benzene rings, e.g., naphthalene, phenanthrene.

Polysulfide treating A chemical treatment used to remove elemental sulfur from refinery liquids by contacting them with a non-regenerable solution of sodium polysulfide.

PONA analysis A method of analysis for paraffins (P), olefins (O), naphthenes (N), and aromatics (A).

Pore diameter The average pore size of a solid material, e.g., catalyst.

Pore space A small hole in reservoir rock that contains fluid or fluids; a four-inch cube of reservoir rock may contain millions of interconnected pore spaces.

Pore volume Total volume of all pores and fractures in a reservoir or part of a reservoir; also applied to catalyst samples.

Porosity The percentage of rock volume available to contain water or other fluid.

Porphyrins Organometallic constituents of petroleum that contain vanadium or nickel; the degradation products of chlorophyll derivatives that became included in the protopetroleum.

Possible reserves Reserves where there is an even greater degree of uncertainty but about which there is some information.

Potential reserves Reserves based upon geological information about the types of sediments where such resources are likely to occur and they are considered to represent an educated guess.

Pour point The lowest temperature at which oil will pour or flow when it is chilled without disturbance under definite conditions.

Powerforming A fixed-bed naphtha-reforming process using a regenerable platinum catalyst.

Precipitation number The number of milliliters of precipitate formed when 10 mL of lubricating oil is mixed with 90 mL of petroleum naphtha of a definite quality and centrifuged under definitely prescribed conditions.

Pressure swing adsorption A method for purifying gas (used in hydrogen production); abbreviated PSA.

Pressure vessel A container designed to hold gases or liquids at a pressure different from the ambient pressure (PV).

Primary oil recovery Oil recovery utilizing only naturally occurring forces.

Primary production The first stage of hydrocarbon production in which natural reservoir energy (such as gas drive, water drive, and gravity drainage) displaces hydrocarbons from the reservoir into the wellbore and up to surface. Primary production uses an artificial lift system in order to reduce the bottomhole pressure or increase the differential pressure to sustain hydrocarbon recovery since reservoir pressure decreases with production.

Primary structure The chemical sequence of atoms in a molecule.

Probable reserves Mineral reserves that are nearly certain but about which a slight doubt exists.

Probe A very short piece of DNA used to find a specific sequence of "letters" in a very long piece of DNA from a chromosome or genome.

Process gas Gas produced from the upgrading process that is not distilled as a liquid; typically used as a refinery fuel.

Processability An estimate of the manner and relative ease with which a feedstock can be processed; generally measured by one or more criteria.

Propane (C_3H_8) A straight-chain saturated (paraffinic) hydrocarbon extracted from natural gas or refinery gas streams, which is gaseous at standard temperature and pressure; a colorless gas that boils at a temperature of −42°C (−44°F) and includes all products designated in ASTM D1835 and Gas Processors Association specifications for commercial (HD-5) propane.

Propane asphalt See Solvent asphalt.

Propane deasphalting Solvent deasphalting using propane as the solvent.

Propane decarbonizing A solvent extraction process used to recover catalytic cracking feed from heavy fuel residues.

Propane dewaxing A process for dewaxing lubricating oils in which propane serves as solvent.

Propane fractionation A continuous extraction process employing liquid propane as the solvent; a variant of propane deasphalting (*q.v.*).

Propylene (C_3H_6) An olefin hydrocarbon recovered from refinery or petrochemical processes, which is gaseous at standard temperature and pressure; an important petrochemical feedstock.

Protopetroleum A generic term used to indicate the initial product formed changes have occurred to the precursors of petroleum.

Proved reserves Mineral reserves that have been positively identified as recoverable with current technology.

Pyrobitumen See Asphaltoid.

Pyrolysis Exposure of a feedstock to high temperatures in an oxygen-poor environment.

Pyrophoric Substances that catch fire spontaneously in air without an ignition source.

Quadrillion 1×10^{15}

Quench The sudden cooling of hot material discharging from a thermal reactor.

R2R process A fluid catalytic cracking process (*q.v.*) for conversion of heavy feedstocks.

Raffinate That portion of the oil that remains undissolved in a solvent refining process.

Ramsbottom carbon residue See Carbon residue

Raw materials Minerals extracted from the earth prior to any refining or treating.

RCC process A process for the conversion of heavy feedstocks in the riser pipe (*q.v.*); resid catalytic cracking process.

RCD Unibon (BOC) process Is a process to upgrade vacuum residua using hydrogen.

RDS A residuum desulfurization (hydrotreating) process.

Reactor A vessel in which a reaction occurs during processing; usually defined by the nature of the catalyst bed, e.g., *fixed-bed reactor, fluid-bed reactor*, and by the direction of the flow of feedstock, e.g., *upflow, downflow*.

Recombinant DNA DNA formed by joining pieces of DNA from two or more organisms.Recycle ratio, *t*, is defined as the ratio of the recycled feedstock to the fresh feedstock:

$$t = F_R/F_F$$

F_F is the fresh feedstock, and F_R is the recycled feedstock; may also be expressed as a percentage.

Recycle stock The portion of a feedstock that has passed through a refining process and is recirculated through the process.

Recycling The use or reuse of chemical waste as an effective substitute for a commercial product or as an ingredient or feedstock in an industrial process.

Reduced crude A residual product remaining after the removal, by distillation or other means, of an appreciable quantity of the more volatile components of crude oil.

Refinery gas A gas (or a gaseous mixture) produced as a result of refining operations.

Refinery A series of integrated unit processes by which petroleum can be converted to a slate of useful (salable) products.

Refining The process(es) by which petroleum is distilled and/or converted by application of physical and chemical processes to form a variety of products.

Reformate The liquid product of a reforming process.

Reformed gasoline Gasoline made by a reforming process.

Reformer A refinery process unit that uses heat and pressure in the presence of a catalyst to convert naphtha feedstock into higher octane gasoline blending stocks; also used for hydrogen production.

Reforming The conversion of hydrocarbons with low octane numbers (*q.v.*) into hydrocarbons having higher octane numbers; e.g., the conversion of *n*-paraffin into *iso*-paraffin.

Reformulated gasoline (RFG) Gasoline designed to mitigate smog production and to improve air quality by limiting the emission levels of certain chemical compounds such as benzene and other aromatic derivatives; often contains oxygenates (*q.v.*).

Regeneration Thereactivation of a catalyst by burning off the coke deposits.

Regenerator A reactor for catalyst reactivation.

Reid vapor pressure A measure of the volatility of liquid fuels, especially gasoline.

Renewable energy sources Solar, wind, and other non-fossil fuel energy sources.

Rerunning The distillation of an oil that has already been distilled.

Research Octane Method A test for determining the knock rating, in terms of octane numbers, of fuels for use in spark-ignition engines; see also Motor Octane Method.

Reserves Well-identified resources that can be profitably extracted and utilized with existing technology.

Reservoir A domain where a pollutant may reside for an indeterminate time.

Resid The heaviest boiling fraction remaining after initial processing (distillation) of crude oil; see Residuum.

Residfining process A catalytic fixed-bed process for the desulfurization and demetallization of heavy feedstocks that can also be used to pretreat the feedstocks to suitably lower contaminant levels prior to catalytic cracking.

Residual asphalt See Straight-run asphalt.

Residual fluid catalytic cracking A refinery process used to convert residuals (heavy hydrocarbonaceous materials) into lighter components, including LPG and gasoline; abbreviated RFCC.

Residual fuel oil Obtained by blending the residual product(s) from various refining processes with suitable diluent(s) (usually middle distillates); a general classification for the heavier oils, known as No. 5 and No. 6 Fuel oils, that remain after the distillate fuel oils and lighter hydrocarbons are distilled away in refinery operations. It conforms to ASTM Specifications D396 and D975 and Federal Specification VV-F-815C. No. 5, a residual fuel oil of medium viscosity, is also known as Navy Special and is defined in military specification MIL-F-859E, including Amendment 2 (NATO Symbol F-770). It is used in steam-powered vessels in government service and inshore power plants. No. 6 Fuel oil includes Bunker C fuel oil and is used for the production of electric power, space heating, vessel bunkering, and various industrial purposes.

Residual oil See Residuum.

Residual heavy fuel oils Produced from the nonvolatile residue from the fractional distillation process.

Residue hydroconversion (RHC) process A high-pressure fixed-bed trickle-flow hydrocatalytic process for converting heavy feedstocks.

Residuum (resid; *pl.* residua) The residue obtained from petroleum after non-*destructive distillation* (*q.v.*) has removed all the volatile materials from crude oil, e.g., an atmospheric (345°C, 650°F+) residuum.

Resins That portion of the maltenes (*q.v.*) that is adsorbed by a surface-active material such as clay or alumina; the fraction of deasphaltened oil that is insoluble in liquid propane but soluble in *n*-heptane.

Resource The total amount of a commodity (usually a mineral, but can include non-minerals such as water and petroleum) that has been estimated to be ultimately available.

Rexforming A process combining Platforming (*q.v.*) with aromatics extraction, wherein low octane raffinate is recycled to the Platformer.

RFLP Restriction fragment length polymorphism, which describes the patterns of different (polymorphism) sizes of DNA (fragment length) that result from cutting with restriction enzymes (restriction). See DNA fingerprinting.

Rich oil Absorption oil containing dissolved natural gasoline fractions.

Riser pipe The pipe in a fluid catalytic cracking process (*q.v.*) where catalyst and feedstock are lifted into the reactor; the pipe in which most of the reaction takes place or is initiated.

Riser The part of the bubble-plate assembly which channels the vapor and causes it to flow downward to escape through the liquid; also the vertical pipe where fluid catalytic cracking reactions occur.

Rock asphalt Bitumen that occurs in formations that have a limiting ratio of bitumen-to-rock matrix.

Rose process A solvent deasphalting process (*q.v.*) that uses super-critical solvent recovery system to obtain high-quality oils from heavy feedstocks (*q.v.*) for further processing.

Run-of-the-river reservoirs Reservoirs with a large rate of flow-through compared to their volume.

S&W fluid catalytic cracking process A process in which the heavy feedstock (*q.v.*) is injected into a stabilized, upward flowing catalyst stream whereupon the feedstock-stream-catalyst mixture travels up the riser pipe (*q.v.*) and is separated by a high efficiency inertial separator from which the product vapor goes overhead to fractionation.

SAGD (Steam-assisted gravity drainage) An in situ production process using two closely spaced parallel horizontal wells: one for steam injection and the other for production of the tar sand bitumen/water emulsion.

Sand A coarse granular mineral mainly comprising quartz grains that is derived from the chemical and physical weathering of rocks rich in quartz, notably sandstone and granite.

Sandstone A sedimentary rock formed by compaction and cementation of sand grains; can be classified according to the mineral composition of the sand and cement.

SARA separation A method of fractionation by which petroleum is separated into saturates, aromatics, resins, and asphaltene fractions.

Saturates Class of hydrocarbons that may be straight-chain, branched-chain or cyclic, in which all carbon atoms have single bonds to either carbon or hydrogen.

Saturates Paraffins and cycloparaffins (naphthenes).

Saybolt Furol viscosity The time, in seconds (Saybolt Furol Seconds, SFS), for 60 mL of fluid to flow through a capillary tube in a Saybolt Furol viscometer at specified temperatures between 70 and 210°F; the method is appropriate for high-viscosity oils such as transmission, gear, and heavy fuel oils.

Saybolt Universal viscosity The time, in seconds (Saybolt Universal Seconds, SUS), for 60 mL of fluid to flow through a capillary tube in a Saybolt Universal viscometer at a given temperature.

Scale wax The paraffin derived by removing the greater part of the oil from slack wax by sweating or solvent de-oiling.

Scrubbing Purifying a gas by washing with water or chemical; less frequently, the removal of entrained materials.

Secondary structure The ordering of the atoms of a molecule in space relative to each other.

Sediment An insoluble solid formed as a result of the storage instability and/or the thermal instability of petroleum and petroleum products.

Sedimentary strata Typically consist of mixtures of clay, silt, sand, organic matter, and various minerals; formed by or from deposits of sediments, especially from sand grains or silts transported from their source and deposited in water, such as sandstone and shale; or from calcareous remains of organisms, such as limestone.

Selective solvent A solvent that, at certain temperatures and ratios, will preferentially dissolve more of one component of a mixture than of another and thereby permit partial separation.

Separation process An upgrading process in which the constituents of petroleum are separated, usually without thermal decomposition, e.g., distillation and deasphalting.

Separator-Nobel dewaxing A solvent (trichloroethylene) dewaxing process.

Sequence The order of "letters" in the DNA "recipe." The DNA sequence is the chemical structure that contains information.

Shale oil Also known as "tight oil" (but not to be confused with "oil shale"); liquid petroleum that is produced from shale oil reservoirs, typically by hydraulic fracturing methods.

Shale oil The distillate produced by the thermal decomposition of oil shale kerogen; sometimes incorrectly used as the name for crude oil produced from tight formation; see Tight oil.

Shell fluid catalytic cracking A two-stage fluid catalytic cracking process in which the catalyst is regenerated.

Shell gasification (partial oxidation) process A process for generating synthesis gas (hydrogen and carbon monoxide) for the ultimate production of high-purity high pressure hydrogen, ammonia, methanol, fuel gas, town gas, or reducing gas by reaction of gaseous or liquid hydrocarbons with oxygen, air, or oxygen-enriched air.

Shell residual oil hydrodesulfurization process A downflow fixed-bed reactor (*q.v.*) process to improve the quality of heavy feedstocks by removing sulfur, metals, and asphaltenes, as well as bringing about a reduction in the viscosity of the feedstock.

Shell still A still formerly used in which the oil was charged into a closed, cylindrical shell and the heat required for distillation was applied to the outside of the bottom from a firebox.

Side-stream stripper A device used to perform further distillation on a liquid stream from any one of the plates of a bubble tower, usually by the use of steam.

Side-stream A liquid stream taken from any one of the intermediate plates of a bubble tower.

Slack wax The soft, oily crude wax obtained from the pressing of paraffin distillate or wax distillate.

Slime A name used for petroleum in ancient texts, particularly in Biblical texts.

Sludge A semisolid to solid product that results from the storage instability and/or the thermal instability of petroleum and petroleum products.

Slurry hydroconversion process A process in which the feedstock is contacted with hydrogen under pressure in the presence of a catalytic coke-inhibiting additive.

Slurry phase reactors Tanks into which wastes, nutrients, and microorganisms are placed.

Smoke point A measure of the burning cleanliness of jet fuel and kerosene.

Sodium hydroxide treatment See Caustic wash.

Sodium plumbite A solution prepared from a mixture of sodium hydroxide, lead oxide, and distilled water; used in making the doctor test for light oils such as gasoline and kerosene.

Solubility parameter A measure of the solvent power and polarity of a solvent.

Solutizer-steam regenerative process A chemical treating process for extracting mercaptans from gasoline or naphtha, using solutizers (potassium isobutyrate, potassium alkyl phenolate) in strong potassium hydroxide solution.

Solvahl process A solvent deasphalting process (*q.v.*) for application to vacuum residua (*q.v.*).

Solvent asphalt The asphalt (*q.v.*) produced by solvent extraction of residua (*q.v.*) or by light hydrocarbon (propane) treatment of a residuum (*q.v.*) or an asphaltic crude oil.

Solvent deasphalter bottoms The insoluble material that separates from the solvent phase during deasphalting (*q.v.*).

Solvent deasphalting A process for removing asphaltic and resinous materials from reduced crude oils, lubricating-oil stocks, gas oils, or middle distillates through the extraction or precipitant action of low-molecular-weight hydrocarbon solvents; see also Propane deasphalting.

Solvent decarbonizing See Propane decarbonizing.

Solvent deresining See Solvent deasphalting.

Solvent dewaxing A process for removing wax from oils by means of solvents usually by chilling a mixture of solvent and waxy oil, filtration, or by centrifuging the wax which precipitates, and solvent recovery.

Solvent extraction A process for separating liquids by mixing the stream with a solvent that is immiscible with part of the waste but that will extract certain components of the waste stream.

Solvent naphtha A refined naphtha of restricted boiling range used as a solvent; also called petroleum naphtha; petroleum spirits.

Solvent refining See Solvent extraction.

Sonic log A well log based on the time required for sound to travel through rock, useful in determining porosity.

Sour crude oil Crude oil containing an abnormally large amount of sulfur compounds; see also Sweet crude oil—crude oil containing free sulfur, hydrogen sulfide, or other sulfur compounds.

Sour crude Petroleum that has a >1% total sulfur content that may be present as hydrogen sulfide and/or as organic forms of sulfur in a hydrocarbon backbone.

Specific gravity The mass (or weight) of a unit volume of any substance at a specified temperature compared to the mass of an equal volume of pure water at a standard temperature; see also Density.

Specific heat The quantity of heat required to raise a unit mass of material through 1° of temperature.

Spent catalyst Catalyst that has lost much of its activity due to the deposition of coke and metals.

Spontaneous ignition Ignition of a fuel, such as coal, under normal atmospheric conditions; usually induced by climatic conditions.

Stabilization The removal of volatile constituents from a higher boiling fraction or product (stripping); the production of a product which, to all intents and purposes, does not undergo any further reaction when exposed to the air.

Stabilizer A fractionating tower for removing light hydrocarbons from an oil to reduce vapor pressure particularly applied to gasoline.

Stack gas Anything that comes out of a burner stack in gaseous form, usually consisting of mostly nitrogen and carbon dioxide; sometime called flue gas.

Standpipe The pipe by which catalyst is conveyed between the reactor and the regenerator.

Steam-assisted gravity drainage (SAGD) An in situ production process using two closely spaced parallel horizontal wells: one for steam injection and the other for production of the tar sand bitumen/water emulsion.

Steam cracking A conversion process in which the feedstock is treated with superheated steam; a petrochemical process sometimes used in refineries to produce olefins (e.g., ethylene) from various feedstock for petrochemical manufacture—the feedstock range from ethane to vacuum gas oil, with heavier feeds giving higher yields of by-products such as naphtha. The most common feedstocks are ethane, butane, and naphtha and the process is carried out at temperatures of 815–870°C (1500–1600°F), and at pressures slightly above atmospheric pressure. Naphtha produced from steam cracking contains benzene, which is extracted prior to hydrotreating, and high-boiling products (residua) from steam cracking are sometimes used as blend stock for heavy fuel oil.

Steam distillation Distillation in which vaporization of the volatile constituents is effected at a lower temperature by introduction of steam (open steam) directly into the charge.

Steam drive injection (steam injection) The continuous injection of steam into one set of wells (injection wells) or other injection source to effect oil displacement toward and production from a second set of wells (production wells); steam stimulation of production wells is *direct steam stimulation* whereas steam drive by steam injection to increase production from other wells is *indirect steam stimulation*.

Steam-methane reforming A continuous catalytic process for hydrogen production; an endothermic process of producing hydrogen from hydrocarbons using steam and a metal-based catalyst; abbreviated to SMR.

Steam-naphtha reforming A process that is essentially similar in nature to the steam-methane reforming process (*q.v.*) but which uses higher molecular weight hydrocarbons as feedstock.

Still gas Any form or mixture of gases produced in refineries by distillation, cracking, reforming, and other processes. The principal constituents are methane and ethane—may contain hydrogen and small/trace amounts of other gases; typically used as refinery fuel or as petrochemical feedstock.

Storage stability (or storage instability) The ability (inability) of a liquid to remain in storage over extended periods of time without appreciable deterioration as measured by gum formation and the depositions of insoluble material (sediment).

Straight-run asphalt The asphalt (*q.v.*) produced by the distillation of asphaltic crude oil.

Straight-run products Obtained from a distillation unit and used without further treatment.

Strata Layers including the solid iron-rich inner core, molten outer core, mantle, and crust of the earth.

Straw oil Pale paraffin oil of straw color used for many process applications.

Stripping A means of separating volatile components from less volatile ones in a liquid mixture by the partitioning of the more volatile materials to a gas phase of air or steam (*q.v.* stabilization).

Sulfonic acids Acids obtained by reaction of petroleum or petroleum products with strong sulfuric acid.

Sulfur recovery unit (SRU) A refinery process unit used to convert hydrogen sulfide to elemental sulfur using the Claus process.

Sulfur A yellowish nonmetallic element, sometimes known by the Biblical name of *brimstone*; present at various levels of concentration in many fossil fuels whose combustion releases sulfur compounds that are considered harmful to the environment. Some of the most commonly used fossil fuels are categorized according to their sulfur content, with lower sulfur fuels usually selling at a higher price.

Sulfuric acid alkylation An alkylation process in which olefins (C_3, C_4, and C_5) combine with isobutane in the presence of a catalyst (sulfuric acid) to form branched chain hydrocarbons used especially in gasoline blending stock.

Suspensoid catalytic cracking A non-regenerative cracking process in which cracking stock is mixed with slurry of catalyst (usually clay) and cycle oil and passed through the coils of a heater.

Sweated wax Crude wax freed from oil by having been passed through a sweater.

Sweating The separation of paraffin oil and low-melting wax from paraffin wax.

Sweet crude oil Crude oil containing little sulfur; see also Sour crude oil.

Sweet crude Petroleum with low total sulfur content, variously defined as <0.5% or <1% sulfur.

Sweetening The process by which petroleum products are improved in odor and color by oxidizing or removing the sulfur-containing and unsaturated compounds.

Synthesis gas (syngas) A mixture of carbon monoxide and hydrogen; used as the feedstock for the Fischer-Tropsch process to produce hydrocarbons with alcohols as by-products.

Synthesis gas generation process A non-catalytic process for producing synthesis gas (hydrogen and carbon monoxide) from gaseous or liquid hydrocarbons for the ultimate production of high-purity hydrogen.

Synthetic crude oil (syncrude) A hydrocarbon product produced by the conversion of coal, oil shale, or tar sand bitumen that resembles conventional crude oil; can be refined in a petroleum refinery (*q.v.*).

Synthetic crude oil A partially refined fraction of bitumen; may be used as a diluent to make dilbit for transport.

Tail gas treating unit (TGTU) A refinery process unit used to control emissions of sulfur compounds; generally integrated with a sulfur recovery unit.

Tail gas The lightest hydrocarbon gas released from a refining process.

Tar sand See Bituminous sand.

Tar The volatile, brown to black, oily, viscous product from the *destructive distillation* (*q.v.*) of many bituminous or other organic materials, especially coal; a name used for petroleum in ancient texts.

Tertiary structure The three-dimensional structure of a molecule.

Tervahl H process A process in which the feedstock and hydrogen are heated and held in a soak drum as in the Tervahl T process (*q.v.*).

Tervahl T process A process analogous to delayed coking (*q.v.*) in which the feedstock is heated to the desired temperature using a coil heater and held for a specified residence time in a soaking drum; see also Tervahl H process.

Tetraethyl lead (TEL) An organic compound of lead, $Pb(CH_3)_4$, which, when added in small amounts, increases the antiknock quality of gasoline.

Texaco Gasification process See Partial Oxidation (Texaco Gasification) process.

THAI process (toe-to-heel air injection process) An in situ combustion method for producing heavy oil and tar sand bitumen. In this technique, combustion starts from a vertical well, while the oil is produced from a horizontal well having its toe in close proximity to the vertical air-injection well. This production method is a modification of conventional fire flooding techniques in which the flame front from a vertical well pushes the oil to be produced from another vertical well.

Thermal coke The carbonaceous residue formed as a result of a non-catalytic thermal process; the Conradson carbon residue; the Ramsbottom carbon residue.

Thermal cracking A process that decomposes, rearranges, or combines hydrocarbon molecules by the application of heat, without the aid of catalysts.

Thermal polymerization A thermal process to convert light hydrocarbon gases into liquid fuels.

Thermal process Any refining process that utilizes heat, without the aid of a catalyst.

Thermal recovery Any process by which heat energy is used to reduce the viscosity of heavy oil or tar sand bitumen in situ to facilitate recovery.

Thermal reforming A process using heat (but no catalyst) to effect molecular rearrangement of low-octane naphtha into gasoline of higher antiknock quality.

Thermal stability (thermal instability) The ability (inability) of a liquid to withstand relatively high temperatures for short periods of time without the formation of carbonaceous deposits (sediment or coke).

Thermofor catalytic cracking A continuous, moving-bed catalytic cracking process.

Thermofor catalytic reforming A reforming process in which the synthetic, bead-type catalyst of co-precipitated chromia (Cr_2O_3) and alumina (Al_2O_3) flows down through the reactor concurrent with the feedstock.

Thermofor continuous percolation A continuous clay treating process to stabilize and decolorize lubricants or waxes.

Tight oil Crude oil produced from petroleum-bearing formations with low permeability such as the Eagle Ford, the Bakken, and other formations that must be hydraulically fractured to produce oil at commercial rates; sometime incorrectly called *shale oil* which is the distillate produced by the thermal decomposition of oil shale kerogen.

Toe-to-heel air injection process See THAI process.

Toluene ($C_6H_5CH_3$) A colorless liquid of the aromatic group of petroleum hydrocarbons, made by the catalytic reforming of petroleum naphtha containing methyl cyclohexane; a high-octane gasoline-blending agent, solvent, and chemical intermediate, and a base for TNT (explosive).

Topped crude- Petroleum that has had volatile constituents removed up to a certain temperature, e.g., 250°C+ (480°F+) topped crude; not always the same as a residuum (*q.v.*).

Topping The distillation of crude oil to remove light fractions only; differs from distillation in the manner in which the heat is applied.

Total acid number (TAN) A measure of the acidity determined by the amount of potassium hydroxide in milligrams that is needed to neutralize the acids (typically naphthenic acids) in 1 g of oil; used by refineries as an indicator of potential corrosion and scale production.

Total petroleum hydrocarbons (TPHs) The total mass of all hydrocarbons in an oil or environmental sample, including the volatile and extractable (nonvolatile) hydrocarbons; may be further defined by stating the analytical method used, e.g., GC-detectable TPH or TPH-F (TPH measured by fluorescence), which vary in their rigor.

Total polycyclic aromatic hydrocarbons (TPAHs) Including alkyl-PAHs and parent (unsubstituted) PAHs; the sum of all concentrations of PAHs measured by GC-MS.

Tower Equipment for increasing the degree of separation obtained during the distillation of oil in a still.

Trace element Those elements that occur at very low levels in a given system.

Transformation A procedure to transfer DNA into the cells of an organism; can be done with Agrobacterium (most dicots), calcium chloride (bacteria), electroporation (any organism), or particle gun (any organism).

Transgenic An organism that has been modified by genetic engineering to contain DNA from an external source.

Traps Sediments in which oil and gas accumulate from which further migration (*q.v.*) is prevented.

Treatment Any method, technique, or process that changes the physical and/or chemical character of petroleum.

Trickle hydrodesulfurization A fixed-bed process for desulfurizing middle distillates.

Trillion 1×10^{12}

True boiling point (True boiling range) The boiling point (boiling range) of a crude oil fraction or a crude oil product under standard conditions of temperature and pressure.

Tube-and-tank cracking An older liquid-phase thermal cracking process.

Ultimate analysis Elemental composition.

Ultrafining A fixed-bed catalytic hydrogenation process to desulfurize naphtha and upgrade distillates by essentially removing sulfur, nitrogen, and other materials.

Ultraforming A low-pressure naphtha-reforming process employing on-stream regeneration of a platinum-on-alumina catalyst and producing high yields of hydrogen and high-octane-number reformate.

Unassociated molecular weight The molecular weight of asphaltenes in a non-associating (polar) solvent, such as dichlorobenzene, pyridine, or nitrobenzene.

Unconventional crude oils Petroleum that does not flow readily in the reservoir and/or must be produced by using unconventional methods, such as surface mining of shallow bitumen deposits, steam-assisted gravity drainage (SAGD) for in situ extraction of deep bitumen deposits, cyclic steam injection for heavy oils, or horizontal drilling with hydraulic fracturing for recovery of light shale oils.

Undiscovered reserves Those reserves of a resource that are yet to be discovered; often they are little more than figments of imagination.

Unfinished oils All oils requiring further processing, except those requiring only mechanical blending. Unfinished oils are produced by partial refining of crude oil and include naphtha, kerosene and light gas oils, heavy gas oils, and residuum.

Unicracking hydrodesulfurization (Unicracking/HDS) process A fixed-bed, catalytic process for hydrotreating heavy feedstocks.

Unifining A fixed-bed catalytic process to desulfurize and hydrogenate refinery distillates.

Unisol process A chemical process for extracting mercaptan sulfur and certain nitrogen compounds from sour gasoline or distillates using regenerable aqueous solutions of sodium or potassium hydroxide containing methanol.

Universal viscosity See Saybolt Universal viscosity.

Unresolved complex mixture (UCM) Petroleum constituents that are not resolved by conventional GC and appear as a "hump" in the gas chromatogram; comprises many hundreds or thousands of unresolved isomers.

Unstable Usually refers to a petroleum product that has more volatile constituents present or refers to the presence of olefin and other unsaturated constituents.

UOP alkylation A process using hydrofluoric acid (which can be regenerated) as a catalyst to unite olefins with isobutane.

UOP copper sweetening A fixed-bed process for sweetening gasoline by converting mercaptans to disulfides by contact with ammonium chloride and copper sulfate in a bed.

UOP fluid catalytic cracking A fluid process of using a reactor-over-regenerator design.

Upflow reactor A reactor in which the feedstock flows in an upward direction through the catalyst bed.

Upgrading The conversion of petroleum to value-added salable products; includes hydroprocessing, hydrocracking, fractionation and any other catalytic or non-catalytic process which improves the value of the products. During upgrading, the products of the Fischer-Tropsch process are converted to diesel, jet fuel, naphtha, or bases for synthetic lubricants and wax.

Upstream A sector of the petroleum industry referring to the searching for, recovery, and production of crude oil and natural gas. Also known as the exploration and production sector.

Urea dewaxing A continuous dewaxing process for producing low-pour-point oils, and using urea which forms a solid complex (adduct) with the straight-chain wax paraffins in the stock; the complex is readily separated by filtration.

Vacuum distillation Distillation (*q.v.*) under reduced pressure; distillation under reduced pressure (less the atmospheric) which lowers the boiling temperature of the liquid being distilled. This technique with its relatively low temperatures prevents cracking or decomposition of the charge stock.

Vacuum gas oil A product of vacuum distillation; a preferred feedstock for cracking units to produce gasoline; abbreviated to VGO

Vacuum residuum A residuum (*q.v.*) obtained by distillation of a crude oil under vacuum (reduced pressure); that portion of petroleum which boils above a selected temperature such as 510°C (950°F) or 565°C (1050°F).

VAPEX process (vapor extraction process) A nonthermal recovery method that involves injecting a gaseous hydrocarbon solvent into the reservoir where it dissolves into the sludge-like oil, which becomes less viscous (or more fluid) before draining into a lower horizontal well and being extracted.

Vapor extraction process (VAPEX process) See VAPEX process.

Vapor pressure osmometry (VPO) A method for determining molecular weight.

Vapor-phase cracking A high-temperature, low-pressure conversion process.

Vapor-phase hydrodesulfurization A fixed-bed process for desulfurization and hydrogenation of naphtha.

Veba-Combi Cracking (VCC) process A thermal hydrocracking/hydrogenation process for converting heavy feedstocks.

Vector Any DNA structure that is used to transfer DNA into an organism; most commonly used are plasmid DNA vectors or viruses.

VGC (viscosity-gravity constant) An index of the chemical composition of crude oil defined by the general relation between specific gravity, sg, at 60°F and Saybolt Universal viscosity, SUV, at 100°F:

$$a = 10 \text{ sg} - 1.0752 \log \ (SUV - 38)/10 \text{ sg} - \log \ (SUV - 38)$$

The constant, a, is low for the paraffinic crude oils and high for the naphthenic crude oils.

VI See Viscosity index.

Visbreaking A (relatively) mild process for reducing the viscosity of heavy feedstocks by controlled thermal decomposition; a process designed to reduce residue viscosity by thermal means, but without appreciable coke formation.

Viscosity index (VI) An arbitrary scale used to show the magnitude of viscosity changes in lubricating oils with changes in temperature.

Viscosity A measure of the ability of a liquid to flow or a measure of its resistance to flow; the force required to move a plane surface of area 1 m² over another parallel plane surface 1 m away at a rate of 1 m/sec when both surfaces are immersed in the fluid.

Viscosity-gravity constant See VGC.

Volatile organic compounds (VOCs) Chemicals having high vapor pressure at room temperature (and corresponding low boiling point) that therefore tend to evaporate or sublimate into the air; for example, BTEX.

Volume flow The combined (fresh and unconverted) feedstock that is fed to a reactor:

$$F_T = F_F + F_R$$

F_T is the total feedstock into the unit, F_F is the fresh feedstock, and F_R is the recycled feedstock.

VRDS A vacuum resid desulfurization (hydrotreating) process.

Water-accommodated fraction of oil (WAF) Hydrocarbons that will partition from oil to water during gentle stirring or mixing; may contain droplets, in contrast to water-soluble fractions (WSF).

Water-soluble fraction of oil (WSF) Aqueous solution of hydrocarbons that partition from oil; does not include droplet or particulate oil. See also CEWAF and HEWAF.

Weathering A suite of changes in spilled oil composition and properties brought about by a variety of environmental processes including spreading, evaporation, photo-oxidation, dissolution, emulsification, and biodegradation, among others.

Watson characterization factor See Characterization factor.

Wax A solid or semisolid material consisting of a mixture of hydrocarbons obtained or derived from petroleum fractions, or through a Fischer-Tropsch type process, in which the straight-chained paraffin series predominates. This includes all marketable wax, whether crude or refined, with a congealing point (ASTM D 938) between 37 and 95°C (100 and 200°F) and a maximum oil content (ASTM D 3235) of 50% w/w; see also Mineral wax and Paraffin wax.

Wax distillate   A neutral distillate containing a high percentage of crystallizable paraffin wax, obtained on the distillation of paraffin or mixed-base crude, and on reducing neutral lubricating stocks.

Wax fractionation   A continuous process for producing waxes of low oil content from wax concentrates; see also MEK de-oiling.

Wax manufacturing   A process for producing oil-free waxes.

Weathered crude oil   Crude oil which, due to natural causes during storage and handling, has lost an appreciable quantity of its more volatile components; also indicates uptake of oxygen.

West Texas Intermediate (WTI—Cushing)   A crude oil produced in Texas and southern Oklahoma which serves as a reference or *marker crude oil* for pricing a number of other crude streams and which is traded in the domestic spot market at Cushing, Oklahoma.

Wet gas   Gas containing a relatively high proportion of hydrocarbons that are recoverable as liquids; see also Lean gas.

Wet scrubbers   Devices in which a counter-current spray liquid is used to remove impurities and particulate matter from a gas stream.

White oil   A generic term applied to highly refined, colorless hydrocarbon oils of low volatility, and covering a wide range of viscosity.

Wood alcohol   See Methyl alcohol.

Wurtzilite   A group of brown to black, solid bituminous materials of which the members are differentiated from asphaltites by their infusibility and low solubility in carbon disulfide; see Asphaltoid.

Xylene [$C_6H_4(CH_3)_2$]   A colorless liquid of the aromatic group of hydrocarbons made from the catalytic reforming of certain naphthenic petroleum fractions; used as high-octane motor and aviation gasoline blending agents, solvents, chemical intermediates. Isomers are ortho-xylene (o-xylene), meta-xylene (m-xylene), para-xylene (p-xylene).

Zeolite   A crystalline aluminosilicate used as a catalyst and having a particular chemical and physical structure.

Index

F

Facultative microbes, 118
Farnesane, 311–312
Fat bag, 63
Fate of aquatic oil spills, 422–428, 423f
Fault reservoirs, 2
FCC. *See* Fluid catalytic cracking (FCC)
Feedstock, 14, 479
  processing, 212
Feedstocks, 43
Ferredoxin, 315
Ferrous iron (Fe^{2+}), 390
Fertilization, 261
Ferulic acid, 313–315
FIA method. *See* Fluorescent indicator adsorption method
    (FIA method)
Field evidence, 273
Filamentous fungi, 289
Fire point, 29
First-order kinetics, 279
  models, 367–368, 370–371
Flame emission spectroscopy, 20
Flame photometry, 20
Flammability limits, 45
Flash point of petroleum or petroleum product, 29
Flavobacterium sp., 275
Flourene, 307
Flue gases, 113
Fluid catalytic cracking (FCC), 143–144, 169–170
Fluid phase behavior effects, 108
Fluidity, 26
Fluoranthene, 84, 327–330, 328f
Fluorene, 290–294, 321–324
Fluorescent indicator adsorption method (FIA method), 20
Foamy oil, 10–11
Follow-up operations, 107
Formaldehyde, 310–311
Formerly polluted frozen soils, 373
Formic acid, 288
2-Formyl-acenaphthen-1-carboxylic acid methylest,
    297–300
Forward combustion, 114
Fossil fuels, 131, 165
Fractional error function, 370
Fractional-wet system, 122
Fractionation, composition by, 19
Fresh biocatalyst generation, 209–211
Fuel gas. *See* Refinery gas
Fuel oil, 53–54
Fuel-upgrading process, 137
Fugitive emissions, 61
Full-scale in situ remediation, 400

G

Gadgil-effect phenomenon, 387–389
Gamma-alumina (γ-Al$_2$O$_3$), 193–195
Gammaproteobacteria, 303
Gas, 50
  cap, 104, 106
  chromatography, 21–22, 30, 45
  drive, 105–106
  flood, 107
  injection, 107
  jet extrusion technique, 188–189
  and lower boiling constituents, 61–63
Gas chromatography-mass spectrometry (GC/MS), 21
Gas-liquid chromatography (GLC), 21
Gas-oil ratio, 112–113
Gaseous products, 43–48. *See also* Liquid products; Semi-solid
    products
  LPG, 43–44
  natural gas, 45, 47t
  refinery gas, 46–48, 47t
  shale gas, 45–46
Gasoline, 50–52, 473
  gasoline-type jet fuel, 51
  for general aviation, 52
GC/MS. *See* Gas chromatography-mass spectrometry (GC/MS)
Gel filtration chromatography (GFC), 22
Gel permeation chromatography (GPC), 22–23
Genetic engineering
  application, 95
  of metabolic pathways, 400
Genetically engineered microorganisms, 399
Genomics, 75
Geobacter grbicium (*G. grbicium*), 304
Geobacter metallidurans (*G. metallidurans*), 304
GFC. *See* Gel filtration chromatography (GFC)
GLC. *See* Gas-liquid chromatography (GLC)
Glucose, 381–382
Glycolipids, 399
Gordonia alkanivorans CC-JG39, 447–448
GPC. *See* Gel permeation chromatography (GPC)
Gram-negative bacteria, 289–290
Gram-positive bacteria, 290, 393
Gram-positive naphthalene degraders, 319–321
Gravity drive, 106
Grease, 56
Great Salt Lake, 273
Greek fire, 1
Gums, 51

Fungi, 294–295, 297, 444
  fungal peroxidases, 297
  fungal strains, 297–300